INTERNATIONAL CONFERENCE ON PLASMA PHYSICS
ICPP 1994

AIP CONFERENCE PROCEEDINGS 345

INTERNATIONAL CONFERENCE ON PLASMA PHYSICS ICPP 1994

FOZ DO IGUAÇU, BRAZIL
OCTOBER-NOVEMBER 1994

EDITORS: **PAULO H. SAKANAKA**
UNIVERSIDADE ESTADUAL DE CAMPINAS

MICHAEL TENDLER
ROYAL INSTITUTE OF TECHNOLOGY

American Institute of Physics   Woodbury, New York

Authorization to photocopy items for internal or personal use, beyond the free copying permitted under the 1978 U.S. Copyright Law (see statement below), is granted by the American Institute of Physics for users registered with the Copyright Clearance Center (CCC) Transactional Reporting Service, provided that the base fee of $6.00 per copy is paid directly to CCC, 222 Rosewood Drive, Danvers, MA 01923. For those organizations that have been granted a photocopy license by CCC, a separate system of payment has been arranged. The fee code for users of the Transactional Reporting Service is: 1-56396-496-1/ 95/ $6.00.

© 1995 American Institute of Physics.

Individual readers of this volume and nonprofit libraries, acting for them, are permitted to make fair use of the material in it, such as copying an article for use in teaching or research. Permission is granted to quote from this volume in scientific work with the customary acknowledgment of the source. To reprint a figure, table, or other excerpt requires the consent of one of the original authors and notification to AIP. Republication or systematic or multiple reproduction of any material in this volume is permitted only under license from AIP. Address inquiries to Office of Rights and Permissions, 500 Sunnyside Boulevard, Woodbury, NY 11797-2999; phone 516-576-2268; fax: 516-576-2499; e-mail: rights@aip.org.

L.C. Catalog Card No. 95-78438
ISBN 1-56396-496-1
DOE CONF-9410249

Printed in the United States of America.

CONTENTS

List of Invited Papers

Preface xi

History xiii

Summary Talks

Fusion Plasma

F. Engelmann
Max-Planck Institut für Plasmaphysik, Garching, Germany
Title: *Fusion Plasmas* 3

Basic Plasma Physics and Applications

F. F. Chen
University of California, Los Angeles, USA
Title: *Summary of Basic Plasma Physics and Applications* 9

Space and Astrophysical Plasmas

R. Opher
University of São Paulo, São Paulo, Brazil
Title: *Space and Astrophysical Plasmas* 19

Opening Colloquium

W. Livingston
National Solar Observatory, Tucson, Arizona, USA
Title: *Solar Eclipses* 27

Fusion Plasmas

ITER

S. Putvinski
ITER San Diego Joint Work Site, USA
Title: *The Fundamentals of the ITER Physics* 37

Large Tokamak Plasmas

M. Keilhacker
JET Joint Undertaking, Abingdon, UK
Title: *JET Physics in Support of ITER: Results and Future Work* 45

N. L. Bretz
Princeton Plasma Physics Laboratory, Princeton, USA
Title: *Deuterium-Tritium Experiments on TFTR* 53

M. F. F. Nave
JET Joint Undertaking, Abingdon, UK
Title: *Observations of MHD Activity in JET High Performance Plasmas* 58

K. Ushigusa
Japan Atomic Energy Research Institute, Naka, Japan
Title: *Non-Inductive Current Drive and Profile Control in JT-60U* 66

R. J. Groebner
General Atomics, San Diego, USA
Title: *Evidence for Modified Transport Due to Sheared ExB Flows in High-Temperature Plasmas* 74

A. Bécoulet
Centre D'Edutes Nucleaire de Cadarache, France
Title: *Recent Results on Current Profile Shaping on TORE SUPRA* 83

Small Tokamak Plasmas

P. K. Kaw
Institute for Plasma Research, Bhat Gandhinagar, India
Title: *Physics Research with Small Tokamaks* 92

A. Sykes
Culham Laboratory, Oxfordshire, England
Title: *Tight Aspect Ratio Tokamak Experiments and Prospects for the Future* 110

Z. Sedlaček
Institute of Plasma Physics, Prague, Chek Republic (*)
Title: *Continuum Damping in Plasma Physics* 119

A. Hirose
University of Saskatchewan, Saskatoon, Canada
Title: *Long Wavelength Modes and Anomalous Transport in Tokamaks* 127

N. J. Lopes Cardozo
FOM-Nieuwegein, The Netherlands
Title: *Filamentation in Tokamaks* 135

V. E. Golant
A. F. Ioffe Physico Technical Institute of Plasma Physics, Saint Petersburg, Russia
Title: *Experimental Results from the Tuman-3 Tokamak* 143

S.-J. Qian
Southwestern Institute of Physics, Chengdu, China
Title: *Results of HL-1 Tokamak* 150

F. Crisanti
Centro Ricerche Energia Frascati, Italy
Title: *High Z Material Operation in FTU* 158

P. E. Vandenplas
Laboratoire de Physique des Plasmas, Ecole Royale Militaire, Brussels, Belgium
Title: *Recent Results on Heating and Confinement in TEXTOR* 166

K. H. Finken
Institut für Plasmaphysik, Jülich, Germany
Title: *Characteristics of Pellet Injected Discharges in TEXTOR* 172

K. Ida
National Institute for Fusion Science, Toki, Japan
Title: *Rotation and Ion Temperature Measurements in Tokamaks and Stellarators* 177

M. Tendler and J.-L. Lachambre
    Alfvén Laboratory, Stokholm, Sweden
    Centre Canadien de Fusion Magnétique, Varennes, Canada
    Title: *Theoretical and Experimental Studies of the Scrape-Off Layer Biasing in a Divertor Tokamak* 185

Non-Tokamak Plasmas

T. Tamano
    Tsukuba University, Tsukuba, Japan
    Title: *Review of GAMMA-10 Experiment* 192

R. N. Sudan
    Cornell University, Ithaca, USA
    Title: *Field Reversed Ion Rings* 200

A. Iiyoshi
    National Institute for Fusion Science, Toki, Japan
    Title: *Recent Studies of the Large Helical Device* 206

M. Wakatani
    Kyoto University, Uji, Japan
    Title: *Confinement Physics of Non-Axisymmetric Torus* 214

S. Ortolani
    ITER, San Diego, USA
    Title: *Reversed Field Pinch Physics* 222

Inertial Confinement Fusion Plasmas

K. Mima
    Institute of Laser Engineering, Osaka, Japan
    Title: *Review on Laser ICF Physics* 230

T. J. M. Boyd
    University of Essex, UK
    Title: *Generation of Megagauss Magnetic Fields in Laser-Target Plasmas* 238

Plasma Based Neutron Sources

E. P. Kruglyakov
    Institute of Nuclear Physics, Novosibirsk, Russia
    Title: *Plasma Based 14 MeV Neutron Sources* 247

M. G. Haines
    Imperial College of Science, London, UK
    Title: *Review of the Z-Pinch* 254

P. Choi
    Imperial College of Science, London, UK
    Title: *Dynamics of a Fibre Z-Pinch under Different Initial Conditions* 263

Basic Plasma Physics

M. von Hellermann
    JET Joint Undertaking, Abingdon, UK
    Title: *Progress in Spectroscopic Plasma Diagnostics and Atomic Data* 269

G. Laval
    Ecole Polytechnique, Palaiseau, France
    Title: *Induced Transparency and Parametric Instabilities in the Relativistic Regime of the Laser Plasma Interaction* 275

J. B. Taylor
    Culham Laboratory, Abingdon, Oxon, UK
    Title: *Structure of Short-Wavelength Modes in a Toroidal Plasma* 282

N. L. Tsintsadze
    Georgia Academy of Sciences, Georgia
    Title: *Strong Langmuir Turbulence* 290

R. L. Merlino
    University of Iowa, Iowa City, USA
    Title: *Laboratory Experiments in Dusty Plasmas* 295

A. Sen
    Institute for Plasma Research, Bhat Gandhinagar, India
    Title: *Short Pulse Laser Solitons* 303

A. G. Zagorodny
    Institute for Theoretical Physics, Kiev, Ukraine
    Title: *Fluctuation Phenomena in Dusty Plasmas* 311

S. Kuhn
    University of Innsbruck, Innsbruck, Austria
    Title: *Kinetic Theory of Bounded Plasmas* 319

L. Friedland
    Hebrew University of Jerusalem, Jerusalem, Israel
    Title: *Autoresonant Wave Interactions in Nonuniform Plasmas* 327

T. Sato
    National Institute for Fusion Studies, Nagoya, Japan
    Title: *Scenario of Self-Organization* 335

R. Pozzoli
    CNR-Istituto di Fisica del Plasma, Milan, Italy
    Title: *Open Field Lines Instabilities* 351

J. Juul Rasmussen
    Risoe National Laboratory, Roskilde, Denmark
    Title: *Coherent Structures in Plasmas and Fluids* 359

A. M. El-Nadi
    Cairo University, Cairo, Egypt
    Title: *Microinstabilities in Plasmas* 367

Space and Astrophysical Plasmas

Space Plasmas

P. K. Schukla
    Ruhr Universität Bochum, Bochum, Germany
    Title: *Theory of Auroral Arcs* 377

C. Uberoi
    Indian Institute of Sciences, Bangalese, India
    Title: *Alfvén Waves in Space Plasmas* 383

W. D. Gonzales
    Instituto Nacional de Pesquisas Espacianis, São José dos Campos, Brazil
    Title: *Review on Interplanetary-Magnetosphere Interactions* 391

H. R. Lewis
    Dartmouth College, Hanover, USA
    Title: *Characteristics of the Ion Pressure Tensor in the Earth's Magnetosheath: AMPTE/IRM Observations*. 398

Astrophysical Plasmas

R. Bingham
    Rutherford Appleton Laboratory, Chilton, England
    Title: *Anomalous Scattering and Absorption of Neutrinos in Dense Plasmas* 406

G. B. Field
    Harvard-Smithsonian Center for Astrophysics, Cambridge, USA
    Title: *Origin of Astrophysical Magnetic Fields* 416

E. M. de G. Dal Pino
    University of São Paulo, São Paulo, Brazil
    Title: *Protostellar Jets: The Best Laboratories to Investigate Astrophysical Jets* 427

C. Ferro-Fontán
    University of Buenos Aires, Buenos Aires, Argentina
    Title: *MHD Wind Solutions for Rotating Stars* 439

M. V. Goldman
    University of Colorado, Boulder, USA
    Title: *Langmuir Turbulence in Space Plasmas* 449

A. Ferrari
    Osservatorio Astronomico di Torino, Pino Torinese, Italy
    Title: *Numerical Simulations of Supersonic Directed Flows in Astrophysical Plasmas* 457

J. P. Goedbloed
    FOM-Instituut voor Plasmafysica, Netherlands
    Title: *Magnetohydrodynamic Waves in Fusion and Astrophysical Plasmas* 465

Plasma Applications

Y. Kawai
    Kyushu University, Fukuoka, Japan
    Title: *Production of Plasma with Large Area for Plasma Application* 479

R. N. Szente
    Instituto de Pesquisas Tecnológicas, São Paulo, Brazil
    Title: *Industrial Applications of Thermal Plasmas* 487

E. S. Wyndham
    Pontifícia Universidad Católica de Chile, Santiago, Chile
    Title: *Observations of Dense Plasma Formations and Beams in the Vacuum Spark*....... 495

1996 ICPP First Announcement 503

Author Index. 507

PREFACE

This Proceedings contains the invited papers delivered at the 1994 International Conference on Plasma Physics, held in Foz do Iguaçu, Brazil, from October 31st to November 4th 1994. This is the joint conference of the 10th Kiev International Conference on Plasma Theory and the 10th International Congress on Waves and Instabilities in Plasmas.

The Conference was organized by the National Institute of Space Research-INPE, the State University of Campinas-UNICAMP, and the Brazilian Vacuum Society, under the auspices of the International Union of Pure and Applied Physics. The Conference provided an environment of exciting scientific ideas and discoveries, formulas and theories. The volume aims to ponder the spirit of the event. The Plasma Physics almost eclipsed the "Solar Eclipse" (the photo to be found on the next page), which took place at the same time and place as the Conference.

The discipline of Plasma Physics, aimed at reflecting the objective reality of plasmas found in space, in astrophysical objects, in laboratories, and in fusion devices, has a collective cooperative character that is required in order to safeguard the objectivity. The scope of the Conference was broad and covered all aspects of plasma physics. The Program included 9 plenary lectures, 50 topical lectures, and over 400 contributed papers presented in poster sessions. The accumulation of experimental data, the painful assessment of difficulties, the injection of new ideas and new devices have all been manifested at the Conference. Resulting debates between different ideas and interpretations are very encouraging for progress in plasma physics. Physics has always progressed in this way. We hope that this Volume will survive the most difficult endeavor in science—the passage of time.

P. H. Sakanaka
M. Tendler

Campinas, April 22, 1995

HISTORY

1971 Kiev (USSR)
    Kiev International Conference on Plasma Theory

1973 Innsbruck (Austria)
    International Congress on Waves and Instabilities in Plasmas

1974 Kiev (USSR)
    2nd Kiev International Conference on Plasma Theory

1975 Innsbruck (Austria)
    2nd International Congress on Waves and Instabilities in Plasmas

1977 Trieste (Italy)
    3rd Kiev International Conference on Plasma Theory

1977 Palaiseau (France)
    3rd International Congress on Waves and Instabilities in Plasmas

1980 Nagoya (Japan)
    International Conference on Plasma Physics
    Joint Conference of the
    5th Kiev International Conference on Plasma Theory and the
    5th International Congress on Waves and Instabilities in Plasmas

1982 Göteborg (Sweden)
    Joint Conference of the
    5th Kiev International Conference on Plasma Theory and the
    5th International Congress on Waves and Instabilities in Plasmas

1984 Lausanne (Switzerland)
    International Conference on Plasma Physics
    Joint Conference of the
    6th Kiev International Conference on Plasma Theory and the
    6th International Congress on Waves and Instabilities in Plasmas

1887 Kiev (USSR)
    International Conference on Plasma Physics
    Joint Conference of the
    7th Kiev International Conference on Plasma Theory and the
    7th International Congress on Waves and Instabilities in Plasmas

1989 New Delhi (India)
    International Conference on Plasma Physics

1992 Innsbruck (Austria)
    International Conference on Plasma Physics
    Joint Conference of the
    9th Kiev International Conference on Plasma Theory and the
    9th International Congress on Waves and Instabilities in Plasmas
    Combined with the 19th EPS Conference on Controlled Fusion and Plasma Physics

1994 Foz do Iguaçu (Brazil)
    International Conference on Plasma Physics
    Joint Conference of the
    10th Kiev International Conference on Plasma Theory and the
    10th International Congress on Waves and Instabilities in Plasmas
    Combined with the VI Latin American Conference on Plasma Physics

SUMMARY TALKS

FUSION PLASMAS

F. Engelmann
The NET Team, c/o Max-Planck-Institut für Plasmaphysik, D-85748 Garching

While the borderline between fusion research and basic plasma physics is quite diffuse, about 130 of the papers presented at this Conference can be taken as belonging to the field of fusion plasma physics. This is about 35% of the total number of papers. However, among the invited papers the share of fusion research was much larger, viz. about 60%. This is a consequence of the fact that most of the large research projects have been reviewed in invited papers whereas the contributed papers mainly covered results from smaller devices and from theory.

All areas of fusion plasma physics were represented: there were experimental results on magnetic confinement (tokamaks; stellarators; mirrors; reversed field pinches; field reversed configurations; Z-pinches, with emphasis on the dense Z-pinch; plasma focus, etc.) and on inertial confinement; related modelling and diagnostics development; theory, as well as some technological activities (power generators; RF sources, etc.) and component (e.g. antennae) development for smaller fusion devices. In particular, fusion-related research in Latin America was exhaustively covered. In addition, large future projects in fusion research were summarized.

In the following, a synthetic review of the information reported at the Conference will be given. No attempt is made to summarize specific contributions; rather the material contributed will be looked at from a few different angles.

1. LARGE FUTURE DEVICES AND PROJECTS

Large devices are being constructed or designed on three lines of approach to fusion power. Most of these were presented at the Conference.

On the tokamak line, the engineering design of the International Thermonuclear Experimental Reactor (ITER) is underway as a collaborative effort of the European Union, Japan, the Russian Federation and the United States. Its first aim is to demonstrate controlled ignited burn (i.e. a fusion multiplication factor $Q \gtrsim 50$) of a DT plasma for at least 1000 s, a pulse length which allows to approach closely the operating conditions of a reactor. To reach ignition a device with a plasma current around 25 MA is required. A second aim of ITER is to demonstrate the technologies (power and particle exhaust, tritium breeding and handling, large-size superconducting magnets) needed for a fusion reactor.

To complement ITER, in Japan the device JT-60 SU [and on a smaller scale in the United States the device TPX] is being designed with the aim of demonstrating operation conditions (high plasma beta and self-generated "bootstrap" current) in a deuterium plasma as appropriate for a steady-state tokamak reactor.

A large stellarator, the Large Helical Device (LHD), is under construction at Toki in Japan. LHD is an optimized heliotron allowing by its very nature for steady-state operation; it aims at reaching plasma conditions corresponding to a fusion multiplication factor $Q \approx 0.1$. A modular coil concept attractive for a heliotron reactor was also presented. [Another large stellarator, W7-X, is being designed in the European Union; it is based on the optimized helias concept, already uses modular coils and has objectives similar to LHD].

Inertial confinement has also reached the point that planning of ignition devices could begin. In the United States the engineering design of the National Ignition Facility (NIF) is being started, and similar plans are being pursued in

© 1995 American Institute of Physics

Japan, with the facility KONGO, and in France. For NIF, both direct and indirect drive of pellet implosion is being considered. The aim is to obtain a fusion multiplication factor $Q \approx 10$. [The work on inertial confinement, in particular on indirect drive and on optimizing the target configuration, is expected to take an advantage of the recent declassification of results of the United States research programme.]

In parallel to operating pre-reactor devices, the development of fusion power requires a 14 MeV neutron source for material testing to the appropriate neutron fluences. While neutron sources in which a high energy ion beam interacts with a solid target are technically within reach, the large testing volume ($\gtrsim 30\,l$) eventually required, a neutron spectrum as present in a fusion reactor and the desired neutron beam uniformity can only be obtained in a plasma-based neutron source. Candidates for this are mirror confined plasmas and tokamaks. For the former approach, in Novosibirsk a preparatory experiment using the Gas Dynamic Trap (GDT) concept, the Hydrogen Prototype Neutron Source (HPNS), is underway and the neutron source projects such as GDT-2 and GDT-3 are being developed in collaboration with the Tsukuba group which also studies alternative configurations (FEF-II). On the tokamak line, a low-aspect-ratio device appears most attractive, but so far only preliminary experiments of small size, such as START at Culham, are being done; a next step on this line, the 2 MA device MAST, is being planned and a neutron source, the Component Test Facility (CTF), is being studied at Culham.

2. TRENDS

An obvious trend in the recent fusion plasma work as presented at the Conference is that plasma profile effects are emphasized in the experimental research. While the importance of profile effects has been clear from theoretical analyses for a long time, only the advent of active profile control methods (such as, e.g., noninductive current drive in tokamaks and pellet injection in various toroidal confinement devices) and improved diagnostics (e.g., methods using active beams and more sophisticated approaches to the interpretation of spectroscopical measurements as reported from JET at Abingdon, combined with better atomic data) has allowed profile effects to be addressed experimentally. This has lead both to a better understanding of plasma physics and to an improvement in plasma performance.

Three typical examples are:

(i) The role of magnetic shear in tokamak plasmas was studied controlling the current profile both directly by using non-inductive current drive by RF waves and neutral beams and indirectly by modifying the plasma resistivity changing the density and consequently the temperature profile by pellet injection. Results were reported from HL-1 at Chengdu, JT-60 U at Naka and Tore Supra at Cadarache. It emerged that confinement is improved if the magnetic shear is enhanced in the range between 0.4 and 0.8 of the minor radius while the shear in the core plasma may be low (Tore Supra) and that long-pulse discharges at high plasma pressure and with good confinement (in the so-called H-mode) can be sustained for appropriate current (and pressure) profiles (JT-60 U and JET).

(ii) Localized transport barriers can be generated in tokamak and stellarator plasmas by a sufficiently high shear in the poloidal rotation velocity caused by a negative shear in the radial electric field. Such fields can be produced and controlled by various means such as electric biasing of the edge plasma (DIII-D at San Diego; HL-1; JFT-2M at Tokai; STOR-M at Saskatoon; TdeV at Varennes; ...); momentum injection (DIII-D; JT-60 U; ...); gas puffing, pellet injection and magnetic compression (TUMAN-3 at St. Petersburg); ion Bernstein wave injection (PBX-M at Princeton); the creation of current pulses (STOR-M); and the generation of a net current in a stellarator (CHS at Nagoya). These experiments have provided a wide body of data on transport barrier formation both adjacent to the plasma edge (where such a barrier induces the H-mode or, if it penetrates the plasma more deeply, the VH-mode which has still better confinement properties than the H-mode) and deeper inside the discharge. A careful comparative study of rotation and temperature profiles in the tokamaks JFT-2M as well as JIPP T-IIU at Nagoya and in the stellarators CHS, Heliotron - E at Kyoto and W7-AS at Garching has confirmed earlier results and the findings from other devices on the role of the radial electric field in the mechanisms leading to the generation of the shear in plasma rotation and consequently of the transport barrier.

The research on transport barriers may provide a starting point for devising ways for active control of confinement in tokamaks and stellarators. In particular, it may be possible to lower the heating power threshold that exists for access into H-mode operation which for large plasmas such as in ITER is predicted to be at a sizeable fraction of the self-heating power generated during plasma burn. In this context, it should also be noted that the generation of fusion α-particles in a burning plasma by itself implies a transport of electrical charge and hence generates a radial electric field in plasma interior, while fast α-particle losses due to the magnetic field ripple contribute to generating a negative electric field at the plasma edge.

(iii) A better understanding of the magnetic field dynamics in a reversed field pinch (RFP) was achieved by measuring the radial profile of the MHD dynamo electric field adjacent to the plasma edge in MST at Madison, and of the spatial distribution of magnetic field fluctuations and its correlation with the profile of the current parallel to the magnetic field in EXTRAP T1 at Stockholm. The physics of the relaxation process being well understood by now, active control of profiles may allow optimization of confinement in a reversed field pinch configuration.

A second trend transparent in the contributions is the increased interest in compact (low aspect ratio) tokamak configurations. This configuration is attractive for its intrinsic capability to sustain high plasma pressure and to provide improved stability with respect to vertical displacement of elongated plasmas as well as a natural divertor; moreover, in the small START device it was found that in low aspect ratio tokamak plasmas energy confinement is comparatively good, better than the neo - ALCATOR scaling for ohmic discharges would predict, and most importantly, that the operational limits are marked by internal magnetic reconnection events (IREs) rather than by plasma disruption. Results on compact tokamak plasmas are by now available also from CDX-U at Princeton, HIT at Seattle, TS-3 at Tokyo and TUMAN-3. However, a better appreciation of the physics of compact tokamak plasmas, in particular information on the role of plasma and current profiles, is needed to be able to judge the potential of this configuration. This is expected to become available from the larger devices presently planned such as MAST; other compact tokamaks under consideration or construction are ETE at São José dos Campos, GLOBUS at St. Petersburg, as well as MRX and PSTX at Princeton.

3. ISSUES

The main issues of fusion research remain power and particle exhaust, plasma stability, and the plasma transport properties. These questions were addressed in many contributions to the Conference. The status of these subjects can be summarized as follows:

(i) Power and particle exhaust is the most demanding problem for reactor-grade magnetically confined plasmas. For tokamaks and stellarators, divertor configurations will be used, but only for tokamaks a broad programme for investigating plasma edge and exhaust issues is now in place. In ITER, like in a reactor, efficient helium exhaust must be provided while excessive localized heat fluxes to the divertor walls must be avoided, the limit being at ≈ 5 MW/m^2. This necessitates operation in a regime in which only cold plasma with a temperature in the eV range, or even better a hot gas, is in contact with the walls. Hence, the physics of partially ionized plasmas and especially the physics of atomic processes involving hydrogen, helium and various impurity species, has an important field of application here. To obtain such divertor plasma conditions, a large fraction, typically 70%, of the power transferred to the plasma by fusion α-particles must be exhausted by radiation. This does not appear unreasonable as results from TEXTOR at Jülich [and other devices] show that under special conditions more than 90% of the heating power can be radiated from the plasma, mainly from the edge region, without excessive impurity contamination of the core plasma. Operation with the plasma detached from the walls of the divertor chamber was demonstrated, e.g., in JET, but extrapolation to ITER conditions is not yet definitely possible. In particular, the need for edge localized modes (ELMs) in steady-state H-mode operation (which is the operation regime anticipated for ITER), leading to large fluctuations in the power flux to the divertor, is a critical problem. Also the optimum choice of the divertor wall material (low-Z materials like Be and/or carbon fibre composites versus tungsten) is not yet clear. Studies on high-Z materials and in particular on tungsten as a plasma-facing material are being intensified and were reported from FTU at Frascati and TEXTOR. Systematic studies of the edge and divertor physics are underway in the tokamaks ASDEX Upgrade at Garching, ALCATOR C-Mod at MIT, DIII-D, JET and JT-60 U and should produce the information required for ITER over the next four years.

(ii) The most important stability problem of tokamak plasmas is the avoidance of plasma disruption, a phenomenon that appears to have similarities with the relaxation process in the RFP. It seems that a coupling of m = 1 and m = 2 MHD modes, as analysed in detail in TEXT-U at Austin, leads to a collapse of the discharge; this is probably accompanied by a transient formation of a transport barrier at the plasma edge and followed by the loss, in less than 1 ms, of the plasma thermal energy to the wall and consequently quench of the plasma current on a resistive time scale. Mode locking, which can be caused by quite small external error fields in large devices in which plasma rotation is small, is often observed as a precursor to disruption. Disruption avoidance is an active field of investigation, as is also the phenomenology of disruptions which has not yet been sufficiently elucidated to be able to predict the consequences of plasma disruption in ITER with satisfactory accuracy.

Sawtooth oscillations and ELMs of large amplitude must also be avoided in a reactor-grade tokamak as they might trigger a disruption in a burning plasma. Results were reported from JET and JT-60 U [but are also available from other devices]. For avoiding large-amplitude sawtooth collapses (the physics of which again has similarities with RFP relaxation phenomena) fast ion effects may be helpful in a burning plasma, but some control of the current profile may also have to be exercised. As far as giant ELMs are concerned, these are limited to specific operation regimes which will have to be avoided in ITER and a reactor.

Plasma stability also remains an important question for efficient plasma heating during pellet compression in inertial confinement fusion. Nonuniformities in driving the compression lead to Rayleigh - Taylor instabilities and consequently mixing of the cold shell and the hot core of the pellet, strongly impairing the fusion output. This, in particular, is a problem for direct drive schemes. Simulation results from the ILE group at Osaka were presented at the Conference. Also, new concept to remedy the problem, recently suggested by the Livermore group, was described; this concept relies on ultra-intense pulses which heat the core directly via nonlinear interaction effects involving relativistic electrons. Overall, the relative merits of direct and indirect drive, viz. more efficient energy coupling versus better uniformity of compression, must be further traded off on the basis of future experiments. Therefore, for NIF presently both schemes are being considered. A vigorous research programme both on direct and indirect drive is being pursued in various count-ries (e.g., Gekko XII at ILE, NOVA at Livermore, OMEGA at Rochester etc.).

(iii) Plasma energy and particle transport is a subject of intense investigation in all magnetic confinement systems. There are two causes for enhanced transport in the plasma interior above that induced by particle collisions, namely plasma turbulence and imperfections in the structure of the confining magnetic field. A broad range of knowledge is by now available for tokamaks and stellarators. Experimentally, tokamak and stellarator plasmas behave similarly which could be taken as an indication that short wavelength (high-n) electrostatic modes are the main cause of the transport observed. However, it is also known that the formation of magnetic islands and the generation of regions where the magnetic field is stochastic leads to increased transport. From RTP at Nieuwegein observations of a fine structure in the electron temperature profile were reported which can be interpreted as being due to plasma filamentation; this suggests that the plasma transport coefficients may be strongly varying in space (much beyond the presence of a few specific flux tubes having enhanced confinement like in the "snake" phenomenon first observed in JET). It appears that transport effectively is determined by the interplay of various causes. To disentangle them, a more systematic and more detailed comparison of theoretical models and experiments in a wide parameter range will be needed and has recently been started, e.g. by the JET Team. Also the theoretical models need further development. A revision of the properties of high-n electrostatic modes by Taylor and Wilson, presented at the Conference, is a step in this direction. A specific unsolved problem is why transport is reduced with increasing mass of the hydrogenic isotope present in the plasma, as now also observed for a DT plasma during the recent experiments on TFTR at Princeton.

Other confinement issues of great practical importance for ITER and a reactor are: The physics of variants of the "normal" tokamak confinement regime, the "L-mode", which have enhanced confinement needs to be elucidated to be able to judge whether they could be appropriate for an ignition device. An example of such a regime is the I-mode reported from TEXTOR. The transport of (thermal) helium needs to be quantitatively understood since a helium contamination of 10% leads to a significant reduction of the quality of energy confinement and at a contamination larger than 20% ignition is no longer possible. It is for this reason that, on TEXTOR, experiments on RF assisted helium transport have been started, but it is anticipated that this scheme would require excessive RF power in ITER and a reactor. Finally, the effect of a substantial population of fast ions needs to be assessed to devise operation scenarios for ITER and a reactor. Apart from the sawtooth dynamics, the excitation of fishbone and toroidal Alfvén eigenmodes (in particular at high n) as well as the transport induced by these modes is an issue. As reported from TFTR DT experiments, no problem was encountered there, but extrapolation to the size of ITER requires further investigation. As also recalled at the Conference, a critical point in ITER is that enhanced transport of fast α-particles in the core plasma together with the α-particle losses caused by the magnetic field ripple at the plasma edge may lead to excessive localized heat loads on the walls. In order to limit the accumulation of both the thermal and fast helium population, operating ITER at comparatively low temperature (T $\approx$ 10 keV) and high density is to be preferred (also satisfactory divertor operation is more easily attainable in this case); therefore, this operation regime must be retained for ITER although it may imply, depending on the plasma transport properties, that the operation point is thermally unstable.

4. FUSION PERFORMANCE

As far as fusion performance is concerned, the most significant contribution to the Conference came from TFTR which by now has been operating routinely for about a year with a DT plasma containing up to 50% of tritium. Up to about 10 MW of fusion power were generated with a fusion multiplication coefficient Q up to 0.28. The best fusion performance earlier obtained in deuterium plasmas in JET and JT-60 U is equivalent to a Q-value of about 1.

From the tandem mirror Gamma 10 at Tsukuba, for the first time in such a device, the generation of thermonuclear fusion neutrons was reported; the plasma, in this case, was a deuterium/hydrogen mixture having a peak density of $2 \cdot 10^{18}$ m^{-3} and a peak ion temperature of 10 keV.

5. SMALL DEVICES

The Conference has provided a unique opportunity for displaying the wealth of results contributed by small confinement devices. These results can only be appreciated here in view of their overall value for fusion research. For small and medium sized tokamaks, an exhaustive review of the work and its results was given by P. K. Kaw. Also referring to his evaluation, small devices continue to have an important role in supporting performance oriented research in large devices; in testing new ideas; in doing more detailed studies necessary for the full understanding of the physics of fusion plasmas; in counteracting the formation of just one world-wide school of thinking, a risk which could arise from a close global collaboration as established for ITER; and in educating young plasma physicists on experiments which are transparent and which more easily allow integral insight into their plasma physics and its interplay with the operation of the device.

A very detailed account of the fusion-related experimental work in Latin America was given at the Conference. This work covers a broad larger device TCA/BR, taken over from spectrum of devices and research problems.

In Brazil, tokamak work is presently being done on TBR-1 which will be soon replaced by the larger device TCA/BR taken over from Lausanne. In addition, the compact tokamak ETE is being prepared. Other systems being studied include the Z-pinch, the reversed field configuration and the CECI device. In Argentina investigations of the Z-pinch and plasma focus are underway while in Chile Z-pinch studies are being performed. In Mexico a small tokamak, Novillo, is operated and plasma focus research is performed. All this experimental work is accompanied by intense theoretical research activities.

6. FINAL REMARK

The development of fusion power remains a challenge for plasma physics research. The steady progress in the field has led us to the point that projects for experimental fusion reactors like ITER and NIF could be started. This implies another change in the quality of the research on fusion plasmas as the size of the devices has to be upscaled again and the technology required becomes considerably more complex and demanding. While taking this step is necessary to make fusion power a reality, it must not jeopardize the smaller scale research needed as a scientific basis for the performance oriented work. Proceeding vigorously in this way is a duty at a moment when energy resources are still abundant, since it is only a question of time until this is no longer the case. Mankind is already critically dependent on fusion power as an energy source, as the participants were reminded by the solar eclipse which they could observe during the Conference, and this will become still more true in future when presently available energy sources on Earth become scarce.

SUMMARY OF BASIC PLASMA PHYSICS AND APPLICATIONS

FRANCIS F. CHEN
University of California, Los Angeles, California 90024-1594, USA

1. CONFERENCE STATISTICS

Is basic plasma physics still alive, and if so, where is it being pursued? Some answers to these questions can be found in the statistics of the papers presented at this conference. The contributed papers were divided into the 13 subject areas shown in Table I.

Table I: Subject Areas

TS:	Tokamaks and Stellarators
AS:	Alternate confinement Systems
HC:	Heating and Current drive
PE:	Plasma Edge physics
IC:	Inertial Confinement fusion
SP:	Space Plasmas
AP:	Astrophysical Plasmas
GT:	General plasma Theory
GD:	General plasma Diagnostics
PI:	Partially Ionized plasmas
EP:	Elementary Processes
BC:	Basic Collisionless plasmas
PA:	Plasma Applications

The last six of these areas (GT, GD, PI, EP, BC, and PA) concerned basic plasma physics and applications, especially the last four, with a few contributions from areas AS, PE, and IC. For the six areas concerned, Table II shows the number of abstracts submitted, the number of written papers delivered, and the number of papers actually presented. In addition, of the 56 plenary and invited papers, 10 were on basic plasma physics and applications, and two were on related subjects. Though fusion dominated the invited talks, these were mainly reviews of the major installations for the benefit of the participants.

Table II: Statistics of Poster Papers

Poster Session	Abstracts	Papers	Actually given
GT 1,2,3,4	79	48	23+ *
GD 1	24	16	15[†]
PI 1,2	31	19	16
EP 1	18	12	12[†]
BC 1,2	49	34	25
PA 1,2	33	27	25

*Count of General Theory papers is incomplete, since many were not to be included in this summary.
[†] Includes postdeadline papers.

Plenary and Invited Papers

On basic plasma physics: 10/56
On related subjects: 2/56

From Table II we can draw several conclusions. Only about half of those intending to come to ICPP'94 actually made it, probably because of financial difficulties and the remoteness of the site. It is encouraging, however, that a comparatively large number of papers on basic plasmas and applications, particularly applications, were actually delivered. This may show a shift of interest toward practical uses of plasma physics.

Further insight can be gained from the geographical distribution of the origin of the papers enumerated in Table II; this is shown in Table III. These numbers indicate not only the existence of basis research, but also the availability of adequate funding for the presentation of results.

Table III: Geographic Distribution of Basic Plasma Papers

	Poster	Invited
Japan	26½*	2
South & Central America	38½	2
Brazil	25	1
Argentina	6	
Mexico	3	
Venezuela	3	
Colombia and Chile	1½	1
Western Europe	20	2
Austria	6	1
Germany	5½	
Italy	3	
Denmark, Sweden, Norway	4½	1
Portugal	1	
North America (USA)	4	1
England	2½	
India	5	1
Former Soviet Union	5	2
Australia	2	
South Africa	2	
Egypt	2	1
Czech, Romania	1	1
Israel		1

* Fractions indicate multiple authorship.

We see that basic research is still actively pursued in Japan, and that many have found the means to travel from Japan to Brazil. It is encouraging that Japan is still able to support basic research in spite of its heavy commitment to fusion. Considerable activity in basic research is to be found in South and Central America, though the numbers here are biased in favor of Brazil, not only because of proximity, but also because of special government support for travel to this conference. The number of papers from Western Europe and the UK is lower than usual, and that from the U.S. has dropped off precipitously, owing to the budgetary difficulties there. As always, there are papers from countries all over the world. Plasma science is widely studied. Indeed, the ICPP is one of the few conferences in which plasma physicists from Third World countries can present their work to an international audience.

2. SELF-ORGANIZATION

The diversity of topics presented here makes it difficult to name one single new result that stands out as a highlight of the conference. If forced to choose, I would cite the advances in our understanding of the self-organization of plasmas. It has been 20 years since J.B. Taylor proposed his minimum energy state based on the conservation of helicity, a theory that lies behind work on reversed-field pinches. Tests of this theory can now by made by computer simulation and in experiment, with the result that the idea is basically confirmed, but with modifications in the details.

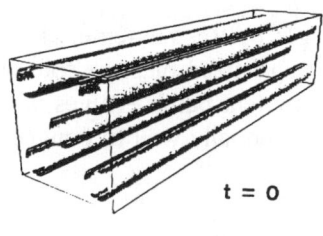

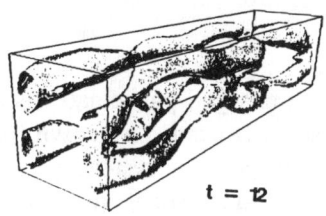

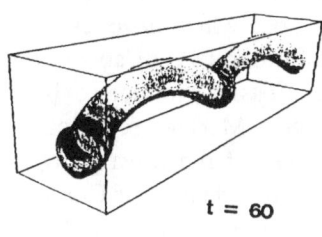

Fig. 1

An invited paper by T. Sato of the NIFS in Nagoya, Japan summarizes a large body of 3D computations clarifying the mechanisms of self-organization. If a plasma is initially energized and is allowed to relax, the tubes of current shown in Fig. 1 reconnect with one another and finally reaches a minimum-energy helix. The helicity falls during the process and reaches a steady value, while the energy falls through an intermediate state. If energy is continuously supplied to the system, instability occurs to expel the excess energy. If energy is supplied and removed at the same rate, the system twists into a minimum energy state but does not stay there; the state recurs intermittently. The behavior of the system depends on the mechanism of excess entropy removal.

An experimental study of self-organized structures and the conservation of helicity was presented by Urrutia, Stenzel, and Rousculp [Proceedings, Vol. **3**, p. 257]. In the apparatus shown in Fig. 2, several cubic meters of plasma is produced in a 10-G magnetic field, and a negative voltage pulse is applied to the electrode on the left to start a flow of electron current to the opposite end. However, the current does not flow directly from left to right; rather, the current paths form force-free structures which end at various parts of the wall. These current paths tend to form knotted ribbons (Fig. 3). The helicity $\iint \mathbf{A} \cdot \mathbf{B} dS$ integrated over a cross section decays from left to right because of dissipation, but the *normalized* helicity $\iint \mathbf{A} \cdot \mathbf{B} dS / \iint ABdS$ remains constant (Fig. 4), showing that the topology does not change. (Here **B** is the magnetic field and **A** its vector potential.)

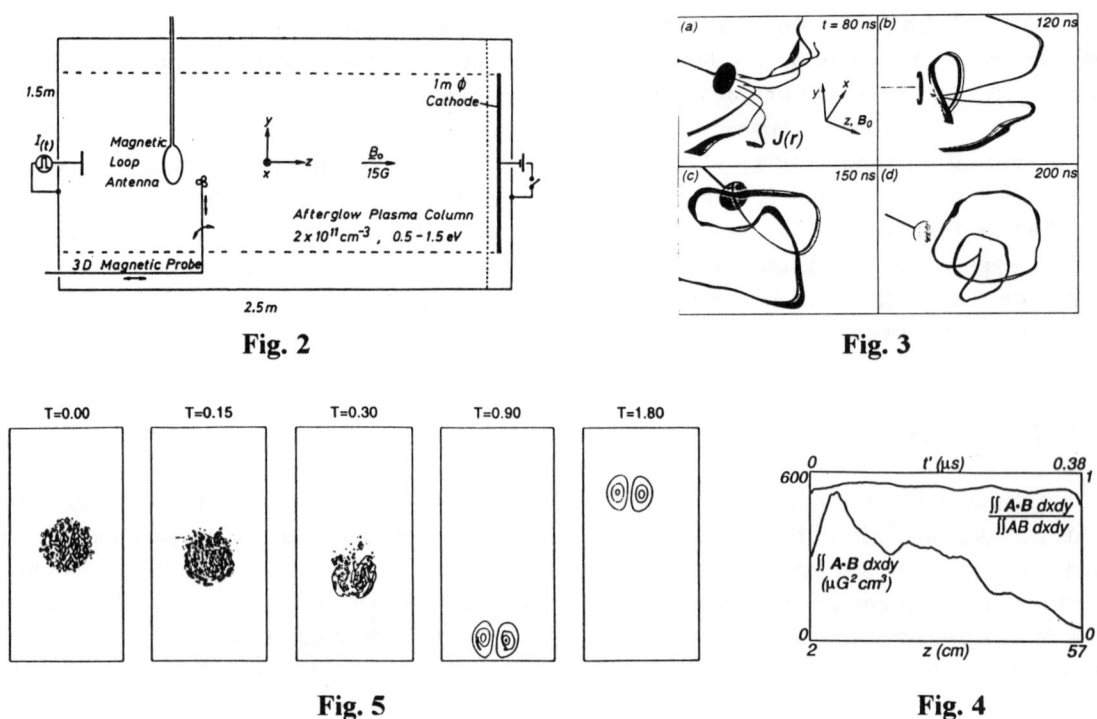

Fig. 2

Fig. 3

Fig. 5

Fig. 4

A different kind of coherent structure is treated in an invited paper by Rasmussen et al. of Denmark. Two-dimensional flows in hydrodynamics and in strongly magnetized plasmas follow the same equations and can be treated together. Fig. 5 shows how a turbulent patch evolves into a dipole vortex structure in a homogeneous medium. These studies apply equally well to convective cells in fusion toruses and to planetary atmospheres.

3. BASIC EXPERIMENTS

Dusty plasmas. The study of plasma comprising or containing charged particulates is a relative newcomer among the traditional topics for basic plasma physics. It has been known for several years that dust in plasmas is important in two different arenas: in the solar system, where dust grains affect the behavior of comet tails and planetary rings; and in plasma processing, where micron-sized contaminants are a major source of defects. More recently, several groups have found dust to provide an easy way to study strongly coupled plasmas, those which have less than one particle per Debye sphere. Since dust particles can have charges of order $10^4 e$ or above, this

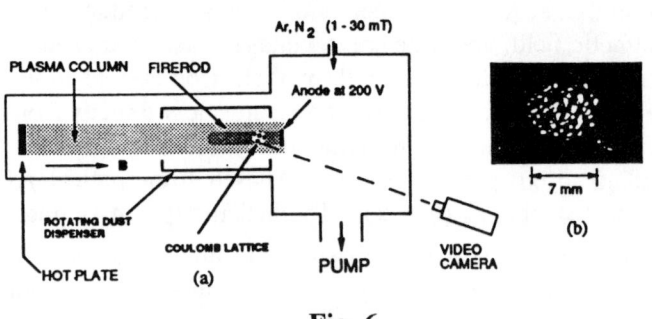

Fig. 6

condition is easily produced. In an invited paper, Merlino reported on an experiment of the Iowa group in which a "solid" plasma, or Coulomb lattice was directed observed. In Fig. 6, a Q-machine plasma is surrounded by a perforated drum containing 10-μm particles which are dropped into the plasma as the drum rotates. By applying a high voltage to an anode, a cone-shaped double layer is formed, and the dust grains, which are charged negative like floating probes, are trapped in the potential maximum. A video camera shows the motions of the dust particles in this "firerod" as they keep their interparticle distances at a fixed value.

Other papers on dust include ones from Belgium and Japan. F. Verheest of Belgium (2-286) points out that, since the charge Q on dust grains does not have to be constant, fluctuations in Q induced by a wave can feed back and cause an instability of the wave. T. Taziuti of Japan (2-282) shows a method for measuring the charge Q employing an electrostatic deflector. The particles are of evaporated silver with a measurable radius, so that their mass is known. The result of $Q \approx 900e$ agrees with the value $Q = CV$, where the capacitance C is given by the radius and V is the floating potential.

Solitons. One of the first nonlinear phenomena to be studied, acoustic solitons have been the subject of many experiments for the past 25 years. Several papers at this conference showed that interesting basic work on solitons is still going on. For instance, Nakamura and Kurahashi of Kanagawa, Japan (3-193) used a double-plasma machine, as is normal for soliton experiments, but added SF_6 to the Ar plasma to give negative F^- ions and also added a separate ion beam source which was insulated from the main plasma by ferrite magnets in its casing (Fig. 7). This allowed them to create both positive and negative solitons and to amplify them by ion-ion interaction with the beam. Both linear waves and solitons could be amplified, but the solitons were amplified only when the component of the ion beam velocity in the soliton direction was above about twice the acoustic speed. Hellberg of the Univ. of Natal in South Africa reported on theoretical work by Gray, Hellberg, Verheest, and Mace (2-327) on solitons in four-component plasmas consisting of electrons and positrons, each with a hot and a cold species. It was found that if all the species had a Boltzmann distribution, only small-amplitude solitons can exist.

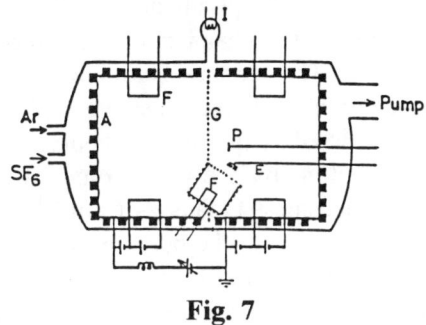

Fig. 7

Double-layers. The discovery of double-layers in the ionosphere triggered numerous experiments in double-plasma devices both in the US and in Scandinavian countries. This activity has greatly subsided or was not well represented at this conference. Two Japanese papers bore on the subject. Tonegawa, Kawamura, and Takayama (3-65) demonstrated the production of a current-free double layer by a LaB_6 hollow cathode discharge. Honzawa, Sekizawa, and Nagasawa (1-281) measured the distribution function of ions trapped in a potential well in a double-plasma and showed how such ion clouds could distort the results from large gridded energy analyzers. A Swedish experiment by Gunnell is described in the next section.

Electron beams. Beam-plasma interactions have continuously been subject to basic experimentation since the late 1950s in many countries. New phenomena are still being discovered. It has been known for some time that as the current of an electron beam in increased, violent disruption of the current eventually occurs, presumably because of the formation of a double layer. These disruptions have been difficult to study because they occur randomly, just as tokamak disruptions do. In an experiment reported by Gunnell, Brenning, and Torvén (1-273), a special double probe is designed to detect plasma oscillations during bursts of ra-

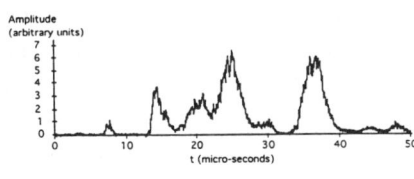

Fig. 8

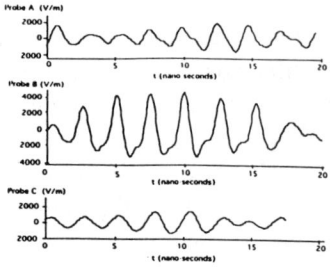

Fig. 9

diation. The probe consists of two thin wires 0.5 mm apart, connected to coaxial cables leading to circuitry giving the difference voltage. To minimize both pickup and disturbance, the wires are formed into a twisted-spiral section before they enter the probe shaft. These probes can be calibrated and respond to the 400-MHz plasma frequencies in the 10^9 cm^{-3} plasma. The plasma is created between a ring anode and ring cathode, and a double layer is formed by drawing current to a central anode. Bursts of radiation are seen near the double layer, as shown in Fig. 8. These bursts consists of localized plasma waves, as shown by the signals on three probes (Fig. 9). Note that the amplitudes are of order kV/m, though the double-layer voltage is only 27 V. The threshold for Langmuir collapse, however, is 6 kV/m, and the bursty nature of the waves is attributed instead to electron trapping. A magnetic mirror was used to stabilize the double-layer position, but it is not clear whether or not a magnetic field is necessary to produce the bursts.

In the experiment of do Prado et al. of Brazil (**3-217**), a planar e-beam is created by biasing the source chamber of a double-plasma device negatively. Only a surface magnetic field is present. The beam relaxation length L is measured, and its behavior attributed to Langmuir turbulence without direct observation of the plasma oscillations. The length L at first falls as $(n_b/n_o)^{-1}$, as predicted by quasilinear theory, then reaches a plateau as modulational instability sets in, forming cavitons by the ponderomotive force of the assumed Langmuir waves. Finally, for larger beam densities, L falls as $(n_b/n_o)^{-1/2}$ as the modulational effects develop. A double plasma was used to generate the beam-plasma interaction also in the paper of Hayashi et al. of Kyushu University (**3-33**), but here the high-frequency waves in the 150-Mhz regime were directly observed. They had a complicated frequency spectrum and formed a standing wave pattern in the experimental region. By launching test waves, it was found that waves below ω_p, with frequency $\omega_p[1-(n_b/4n_0)^{2/3}]$ due to the beam-plasma interaction were generated at the center of the chamber, while Bohm-Gross waves with $\omega > \omega_p$ were generated near the wall.

The surprisingly simple experiment of Varma et al. of India (**3-121**) yielded the most baffling result. Here, a weak beam of only a few nA was projected across a distance L ≈ 50 cm of vacuum along a 78G magnetic field to a grounded grid and collector. As the beam energy was varied from 100-500 eV, the collected current showed peaks and dips, with peaks occurring at energies E_j which follow the relation $E_j = \frac{m}{2}\left[\frac{f_c L}{j+\phi_j}\right]^2$, where f_c is the electron cyclotron frequency, j is an integer and ϕ_j is a small phase correction. This formula had previously been given by Varma, but the fact that the grid and collector currents were anticorrelated probably indicates that the explanation is not so simple.

Waves and instabilities. Though it is difficult to find new problems with linear waves and instabilities, many problems remain with waves that are not quite linear. Tsukabayashi et al. (**3-253**), using a double-plasma device, found that a large-amplitude ion acoustic wave train in a 3-G magnetic field excites a 2nd harmonic wave bunch behind it. We think that the halved wavelength is probably caused by the periodicity of the ponderomotive force of the original wave. Takeda et al. (**3-245**) claim to have seen the Buneman instability in a magnetized deuterium plasma generated by a Ti washer gun. When a capacitor discharge is fired along the field, capacitive probes see UHF signals in the 350-MHz range, in agreement with Buneman theory. Morales et al. (**3-189**) reported on Alfvén wave experiments by Gekelman et al. at UCLA, in which the waves are confined to channels as narrow as an electron collisionless skin depth. In doing the theory for this case, they found the new phenomenon of collisionless diffraction in radiation cones.

4. BASIC THEORY

Chaos. The field of nonlinear dynamics has its growth spurt about a decade ago when bifurcations, attractors, and chaos were discovered with fast computers. Several papers indicated that this field still holds interest for many practitioners. Corso and Rizzato (**2-21**) considered bifurcations in a relativistic wave-particle system, and Pakter et al. (the same people) did the same for a cyclotron maser system (**2-131**). Spatschek and Eickermann (**2-155**) showed that even in weakly ionized plasmas, drift-type instabilities can lead to chaotic convective transport across a magnetic field. Serbeto et al. (**3-229**) studied the chaotic behavior of a relativistic electron beam interacting with a slow transverse electromagnetic wave. Tanaka et al. (**3-249**) computed the chaotic acceleration of electrons by electron cyclotron waves.

Laser-plasma interactions. What has happened to inertial fusion and laser-plasma interactions? A hot topic not so long ago, this field was poorly represented at this conference. Worldwide interest is still strong, but the ICPP is not a major conference for this field. Most of the submitted posters were not given. On the experimental side, there were only a few papers on the fabrication of microballoons. On the theoretical side, there was an invited review by T. J. M. Boyd of the UK and two interesting poster papers. The first, by Barr, Boyd, et al. (**1-77**), considers strongly driven Raman backscattering in sub-picosecond pulses. The advent of short-pulse lasers of extremely high brightness has necessitated a new approach to the theory. The interaction length is now limited by the length of the laser pulse itself, and computations are done in the moving frame. There is mode conversion at the boundaries. Linear dispersion relations no longer hold because of the strong coupling. What is needed is a global treatment which includes all parametric processes at the same time. This paper purports to explain the baffling observations of a blue shift of the scattered light. On a different subject, Sakagami and Nishihara (**1-249**) gave 3D simulations of the Rayleigh-Taylor instability in laser-fusion targets. It was found that, surprisingly, three-dimensional distortions of the surface grew faster than 1D or 2D perturbations. The sphere deforms into a cluster of bubbles. Such effects as the vorticity at the bottom of the bubbles are needed to explain this computational result.

Other topics. Of the large number of theoretical posters submitted, a majority were not given or merely posted in the absence of the author. Of the remainder, we cite only a few random ones of general interest. Simpson and Law of Australia (**3-53**) treated a problem that is important now that industrial uses of gas discharges are becoming more prevalent. These discharges usually contain noble gases possessing a Ramsauer minimum. The motion of electrons in gases with a non-monotonic cross section is not easily treated using normal methods, and this paper tackles this important problem. Mace and Hellberg of South Africa (**2-95**) solved the fundamental problem of the plasma dispersion function (Z-fn.) in a plasma containing superthermal particles. Jiménez-Domínguez of Mexico (**3-181**) found a clever use of the Z-fn. that does not involve plasma physics at all. Noticing that the Hilbert transform of the Voigt profile of a spectroscopic line is identical to the real part of the Z-fn., he was able to use the standard methods for approximating the Z-fn. to evaluate the integrated intensity of a spectral line. Riemann of Germany (**2-139**) considered the problem of the Bohm sheath criterion in the frequently occurring case where there are several ion species in the plasma. He found that the equation giving the critical ion velocity had a singularity for each ion species, but that only the one corresponding to the lightest ion was relevant.

5. PLASMA PROCESSING

Applications of partially ionized plasmas played a major role in this conference for the first time as a strong showing was made by many countries in the poster sessions on this subject.

Plasma sources. In addition to papers on the "standard" methods for producing plasmas for industrial applications, it was refreshing to see novel ideas that are not presented in the major conferences devoted to this field. An invited paper was given by Y. Kawai on the work at Kyushu Univ. in Fukuoka, Japan, on producing plasmas in the 10^{12}-10^{13} cm^{-3} density range, uniform over >20 cm diameter, using electron cyclotron resonance (ECR). A slotted waveguide is used at the standard frequency of 2.45 GHz. Fukao of Japan (3-89) showed a large ECR discharge with linear multipole surface magnetic confinement. The resonance zones at 875 G, where $\omega = \omega_c$, are located at the cusps; and the plasma diffuses into the field-free interior, where the density becomes uniform. Though this large machine requires too much power for practical uses, the concept could be used in small devices using permanent magnets. Chen et al. (3-21) and Krämer et al (3-37) reported on basic studies of the helicon wave discharge, which is one of the leading contenders for the industrial source of the future. The emphasis at the moment is to understand the mechanism of antenna coupling and the propagation and ionization characteristics of helicon waves. For low-damage semiconductor fabrication, it is sometimes desirable to have a low electron temperature and no energetic particles. Matsumoto et al. (3-317) have developed a 6-phase ac discharge that runs off the 60-Hz power lines. The voltage is applied to six meter-long rods inside a large tank and produces a uniform plasma of mid-10^9 cm^{-3} density and with T_e's only 0.3 eV. The extensive numerical modeling which industry supports was missing from this conference, but there were scattered theoretical treatments. Massi et al. (1-293), for instance made a transformer model of an rf inductive discharge at São José dos Campos, and Deutsch and Schwarz of Stuttgart made a circuit model of the sheath in an RIE (Reactive Ion Etching) discharge. Plasma sources for sputtering are widely used in industry. Ferreira et al. (3-297) of Brasília reported on a multidipole source for producing ion beams for sputtering, and Garamoon et al. (3-301) of Cairo showed measurements of a dc magnetron sputtering unit. Urai et al. (3-69) of Tokyo studied the electron orbits in a wire-anode ion plasma source, in which they precess around one or two straight wires in a magnetic field. Choi et al. (3-25), in a UK-Chile-Israel collaboration, studied the ionization processes in a transient hollow cathode discharge. Nogeira et al. (3-329) of Brazil were concerned with optimizing a dc methane discharge between graphite electrodes in regard to how the wall temperature affects the breakdown voltage. The discharge is used for plasma enhanced chemical vapor infiltration to produce carbon composites of importance to aerospace. These examples amply illustrate the international nature of this research field.

Plasma polymerization. The use of plasmas to produce strong plastic films and coatings is usually covered in chemical rather than plasma conferences, but in ICPP'94 there were several papers on plasma polymerization from Brazil. Durrant et al. (3-293) studied how the properties of films made from acetylene (C_2H_2)-Ar-O_2 plasmas depended on the concentrations of O_2 and the CH and CO precursors. Teixera et al. (3-353) studied CH and H formation in Ar-C_2H_2 and N_2-C_2H_2 plasmas excited by both dc and rf and measured the refractive index of the films produced. Da Cruz et al. (3-277) measured plastic films made in SF_6-methanol and SF_6-acetone discharges.

Mota et al. (**3-325**) spectroscopically analyzed the radicals in a dc-excited Ar-HMDSO (hexamethyl disiloxane) plasma, commonly used to make strong Plexiglas films.

Miscellaneous applications. In other applications there were too few papers to form any coherent group. Kitajima et al. (**3-309**) of Tsukuba studied how the production rate of oxide layers on Si depended on the substrate bias in an O_2 plasma. Vasilevskiy et al. (**3-361**) of Nizhniy Novgorod calculated the dissociation rates of metal-organic compounds in an rf He discharge used for PECVD (plasma enhanced chemical vapor deposition) of epitaxial layers. The large field of plasma source ion implantation (PSII) was represented only by a handful of papers. For instance, Mukherjee and John of India (abstract only, p. 206) examined the rarefaction wave that ensues when a high negative voltage on the substrate is turned off. Rossi et al. of Brazil (**3-341**) tacked one of the main problems in PSII--the design of the large power supply, in this case giving a 50kV, 5A, 5µs pulse at 10 Hz. Thermal plasmas--i.e., those at nearly atmosphere pressure--are less well known to the plasma community than to industry, where the use of torches, arcs, and sprays in manufacturing is widespread. Szente gave an invited review of these applications. Bruzzone et al. of Argentina (**3-273**) showed how a dense TiN coating on steel can be formed with a pulsed, Ti-cathode arc in a nitrogen discharge, provided that the object is exposed to a glow discharge first.

6. NOVEL APPLICATIONS

Plasma isotope separation is potentially an important application, but no papers were given on the ion cyclotron resonance method, which has been studied extensively for uranium but not for other elements of medical importance. There was, however, a paper on atomic vapor laser isotope separation (AVLIS) by Matsui (**3-313**) of Japan. Here the problem is to collect the desired ion species rapidly once it has been selectively ionized by the laser. Matsui proposed to add a magnetic field and then cyclotron accelerate the ions to the walls. Using a more classical approach to isotope separation, Simpson et al. (**3-349**) of Australia are developing a high-pressure plasma centrifuge with a cylindrical anode and an axial cathode in a magnetic field. Another type of plasma centrifuge was studied by Dallaqua et al. (**3-281**) of Brazil. Here, the cathode does not extend into the separation chamber but is separated from it by a mesh anode. A radial electric field is nonetheless produced which spin the different species in a magnetic field.

Experiments in fusion-like toroidal geometries were reported by two groups. Mukherjee and Bora (**3-49**) of India used helicon waves to create a plasma in a torus and measured the wave and plasma profiles. Armstrong et al. (**3-153**) of Norway told of a large toroidal device in Oslo which was filled, according to Schrittwieser and Armstrong (**3-225**), by an arc discharge between an internal filament and the limiter. No toroidal drifts were seen, a result that is not yet explained. In another fusion-related basic development, Idehara et al. (**3-285, 325**) reported on the development and use of high-power gyrotrons, including a new world frequency record of 50W at 830 GHz.

Besperstov et al. of Moscow (abstract only, p. 424) have found a way to make fine metal powders by exploding a continuously fed metal wire with a 6 kV capacitor bank in an argon atmosphere. Wire diameters of 0.3 mm produce 30-nm diam particles. Irie et al. (**3-305**) of Waseda Univ. in Shinjuku, Tokyo, are engaged in lightning research, of importance to electrical power companies because of the need to protect high-voltage transmission lines from lightning strikes. It is commonly thought that an electron trail produced with a laser would be an efficient way to trigger lightning strikes so that they would not occur in unexpected places. However, in experiments using a rail gun plasma, it was found that excited neutrals or negative O_2^- ions are more important in initiating lightning than free electrons.

7. CONCLUSION

The distribution of papers in this conference reflects the global change in basic plasma research away from the traditional areas to areas to more immediate utility. A number of papers have been cited as examples; the author apologizes to those authors whose work could not be mentioned in this short review. The organization of this conference at a remote location and in conjunction with a rare display of solar plasma was a formidable and thankless task. The organizers of ICPP'94 are to be congratulated not only for accepting the challenge but also for providing a highly successful forum in which plasma scientists all over the world can show the strength and vitality of their research efforts.

SPACE AND ASTROPHYSICAL PLASMAS*

Reuven Opher
University of Sao Paulo, Instituto Astronomico e Geofisico, Av. Miguel Stéfano 4200, Sao Paulo, SP 04301-904, Brazil

I. INTRODUCTION

Space and astrophysical plasmas have taken an increasingly important part of the ICPP meeting we have had excellent contributions on SPACE PHYSICS (aurora, ionosphere, magnetosphere, solar windmagnetosphere interaction), THE SUN (flares, sunspots, coronal holes, prominences, radio emission, eclipse observations, solar neutrinos), OUR GALAXY (protostellar jets, stellar winds, pulsars, supernova), EXTRAGALACTIC ASTRONOMY (cosmic rays, magnetic fields in galaxies, clusters of galaxies) and COSMOLOGY (large scale structure of the universe, primordial magnetic fields, primordial nucleosynthesis). I present some of the highlights of these excellent contributions (with due apologies to the authors of the other excellent contributions which I did not revue due to the lack of space). At the end of each section I take advantage of my position as reviewer to make some comments on future research which I think should be done.

II. SPACE PHYSICS

I separate the discussion on space physics into the aurora, ionosphere, magnetosphere and the solar wind-magnetosphere interaction, although the subjects are obviously clearly interrelated.

II.1 Aurora

Lopes, Alves and Chian noted that whistler, Langmuir and Alfvén waves, along with 100 eV - 10 keV electrons, are observed to be found in the same region in the aurora. They suggested that a large amplitude whistler wave can nonlinearly generate Langmuir and Alfvén waves through a threewave parametric instability. The threshold for the electric field of the whistler wave to have the instability is $\propto (\Gamma_A \Gamma_L)^{1/2}$, where $\Gamma_A (\Gamma_L)$ is the decay rate of the Alfvén (Langmuir) wave. _Potapenko, Assis, Elfimov and Azevedo_ suggested that large amplitude Alfvén waves can cause anomalous electron precipitation in the aurora. The Alfvén waves do this by distorting the electron distribution over angles and widen the distribution by plasma heating. _Roth, Muschietti and Brown_ discussed the observational evidence of Langmuir wave packets in the aurora (of size ~ 100 - 1000m). They made a simulation injecting an observed bump-on-tail electron beam into a weak Langmuir packet. They calculated the compression of the packet with time. The compression occurs when the range of electron velocities in the positive slope of the electron distribution is greater than $4\omega/kN$, where $\omega(k)$ is the plasma frequency (wave number) and $N = kD$ is the number of waves in the packet of dimension D. _Schukla_ presented numerical simulations to try to explain the highly nonlinear structures observed in aurora.

can we make a self-consistent theory of the aurora which will reproduce the electrons, whistler waves, Alfvén waves, Langmuir waves and nonlinear structures?

VI.2 Ionosphere

Muralikrishna, Abdu and Aquino discussed the observations of a rocket borne Langmuir probe launched into the ionosphere from Natal, Brazil. Two clear depressions were observed at 370 km and at 430 km, evidence for the existence of plasma bubbles at these altitudes. The power spectra of the electron density fluctuations were large in the regions of downward gradients of electron density in the bubbles.

What is the quantitative theory for the formation of the bubbles and the electron density fluctuations?

II.3 Magnetosphere

Uberoi studied the impulse excitation of the magnetosphere. The magnetosphere was found to oscillate for ~100 Alfvén periods. _Gomberoff, Gnavi and Gratton_ discussed observed preferential heating of the minor O^+ ion cyclotron wave (EICW), the O^+ ions are found to be heated from a few eV to ~ 100 eV. Linear theory can not explain the necessary heating of O^+. They made a nonlinear investigation of the decays of EICW whose results indicate that O^+ (as well as $^3He^+$ from a previous study) are preferentially heated.

Can we refine the theory to make quantitative predictions of the enhancements of O^+, $^3He^+$ and other minor ions in the magnetosphere?

© 1995 American Institute of Physics

II.4 Solar Wind-Magnetosphere Interaction

Gonzales analyzed the data obtained from the ISEE-3 satellite during the time when it was at the L_1 libration point of the sun-earth system (at a distance of about 240 earth radii). About 90% of intense geomagnetic storms were preceded by fast shocks with a negative B_Z magnetic field component. It is suggested that this negative B_Z of the earth's magnetosphere creating reconnection which induces the geomagnetic storms. *Goldman* compared Vlasov simulations with observations from the ISEE-1 satellite of the electron foreshock. The observed peaked electron distribution has plateaus above and below the peak. He noted that the Zakharov equations do not include flattening, advection and tail formation of the electron distribution. The Vlasov simulations reproduce well the observed electron distribution. The simulations show the growth of both beam-resonant and backscattered Langmuir waves. The origin of the plateau near the positive slope of the peak is obvious: the positive slope electrons create Langmuir resonant waves which flatten the electron distribution to make the plateau. The second plateau is due to electron acceleration (diffusion) by backscattered Langmuir waves. The Vlasov simulations were started with a beam which was periodically replenished in order to simulate advection.

What is the detailed theory that explain the transfer of energy from the solar wind through the magnetosphere to make the geomagnetic storms? Can a self consistent calculation be made of the electron foreshok to reproduce the observed electron distribuion starting from the solar wind-magnetosphere interaction?

III. THE SUN

The sun is the nearest star and is a plasma physics laboratory where many existing plasma physics phenomena occur which await explanation.

III.1 Flares

Bagala, Rovira, Mandrini, Demoulin and Henoux studied vector magnetograms from Marshall Space Flight Center. Electric currents, inferred from transverse magnetograms, are found at the border of separatrices where also are found flare brightenings of H_α, UV and X-rays. Since in this region the magnetic field is in opposite directions on either side of the border, magnetic reconnection is indicated as the source of the flare brightenings. *E. Opher* used general functions in reconnection theory, instead of the usually used eigenfunctions, to amplify an assumed level of turbulence $\Delta B/B \sim 10\%$ to initiate reconnection. It was found that reconnection proceeded at a rate two orders of magnitude faster than conventional theory. *Roth, and Temerin* noted that in the corona $^3He/^4He \sim 5\times10^{-5}$ while in flares $^3He/^4He$ can reach unity. Similarly, Fe/O is 0.1 in the corona and can reach unity in flares. They found that electromagnetic ion-cyclotron waves (EICW) can simultaneously heat ^3He and heavy ions in a flare. A threshold level of ECIW is found necessary to heat the heaviest ions.

What triggers magnetic reconnection in a flare? What is the preflare level of magnetic turbulence? Can we make a quantitative theory of the heating of the ions ^3He, O, Fe etc by EICW to accurately predict their relative abundance's in a flare?

III.2 Sunspots

Ochi, Sakanaaka, Faria, Azevedo and Assis constructed a two plasma theory to explain observations of sunspots: one plasma describing the central umbral region and the other the outer penumbral region. The outer wall was assumed to be rigid. They found that discrete Alfvén waves reproduce the 230-300 second observed period of the penumbral waves.

Can the calculation of the sunspot region be improved by relaxing the rigid wall assumption?

III.3 Coronal Loops

Elfimov and Azevedo suggested that Alfvén waves in a coronal loop create an axial current sufficient to produce the observed poloidal field ~ 10G. An Alfvén flux necessary to heat the coronal was assumed. *Goodbloed* discussed the difference between the magnetic structure in coronal loops as compared to that in Tokamaks. In coronal loops the perturbations are zero at the ends (i.e., line tying) while in Tokamaks the boundary conditions are periodic. While in tokamaks Alfvén, Fast and Slow Magnetosonic Waves exist independently, they do not in coronal loops because of line-tying. This qualitative difference between tokamaks and coronal loops is responsible for their different behaviour in the onset of instability: In tokamaks we have enhanced MHD activity while in coronal loops we have violent solar flares.

Do the coronal loops show the necessary asymmetry to create an axial current? Can a single coronal loop make a violent solar flare, or do we need at least two coronal loops (i.e., flares occur due to the interaction of coronal loops such as is indicated by the 3D vector, magnetograms of Bagala et al. shown in this conference)?

IIII.4 Prominences

Azevedo, Assis, Tennfors & Hellsten showed that the discrete slow wave spectrum can explain the observed ~ 75 minute period.

Can we refine the theory of discrete slow waves to explain all the characteristics of prominences?

III.5 Radio Emission

Chian, Abalde, Alves, and Lopes showed that transverse electromagnetic radiation can be produced from the interaction of Langmuir waves with Alfvén waves, which is complementary to the commonly investigated process of the production of electromagnetic radiation from the interaction of Langmuir waves with sound waves.

What is the ratio of the nonlinear interaction with Alfvén waves to the nonlinear interaction with sound waves?

III.6 Eclipse Observations

Livingston noted that in the July 11, 1991 eclipse observed from Mauna Kea an ejection of a plasmoid was seen, as well as very fine structure in the corona.

What are the explanations for the creation of a plasmoid and the fine structure in the corona?

III.7 Coronal Holes

Jatenco-Pereira, Opher and Yamamoto found that a flux of Alfvén waves emitted from coronal holes (open field lines) plus the network activity predicted by Parker, can explain the peaks in the temperature in the corona, the increasing velocity with distance of the ejected plasma up to velocities > 400 km/s, as well as the observed Alfvén wave spectrum at 0.3 AU (~ 64 solar radii). Can the predicted Alfvén wave flux be observed at distances < 0.3 AU?

III.8 Solar Neutrinos

Guzzo tried to explain the possible time variation of the different solar neutrino detectors sensitive to different energy neutrinos. He suggested that neutrinos have a nonzero magnetic moment which couples to a time varying solar magnetic field. The coupling of the neutrinos with the solar magnetic field transforms left-handed neutrinos (which are detectable) to right-handed sterile neutrinos (which are undetectable by present detectors).

Can we make a more quantitative analysis, including the magnetic field time variations in the sun?

IV. OUR GALAXY

Our galaxy consists of ~ 10^{11} stars with many fascinating phenomena: 1). The birth of stars involves collapse, which has never been clearly identified. What is seen are outgoing protostellar jets; 2) Stellar winds are ubiquitous. We observe stellar winds up to nine orders of magnitude greater than that of the sun; 3) Pulsars are magnetic rotating neutron stars with electric fields 10^9V/cm at their surfaces; and 4) Supernovae are exploding stars.

IV.1 Protostellar Jets

DalPino made a series of 3D simulations of the propagation of protostellar jets. The simulations show appreciable evidence of instabilities. *DalPino and Cerqueira* found in their simulations that an increasing ambient density with distance causes increasing collimation. *Ferari* studied the conditions for the collimation or destruction of jets. He found that heavy jets are more stable than light ones.

What is more important for the collimation of a jet, the heaviness of the jet or the increasing ambient density? Can we include the magnetic field (which is observed) into the simulations?

IV.2 Stellar Winds

Ferro-Fontan presented analytic solutions for the winds of rotating stars. *Goncalves, Jatenco-Pereira and Opher* studied early-type (hot) stars and noted that observations of an infrared excess and a large H_α blue wing indicate the existence of clumps.

They explained the formation of the clumps by a thermal instability. *Jatenco-Pereira, Lazarian and Opher* studied late-type (cool) stars and found that Alfvén waves can align the grains produced by the stars by the Gold mechanism (i.e., flow of plasma over the grains).

Can we make a self-consistent theory of stellar winds taking into account radiation pressure on grains, molecules and resonant lines in atoms, thermal gradient, MHD waves, rotation and the inhomogeneity due to the presence of a magnetic field?

IV.3 Pulsars

Tsintsadze discussed the condition for the creation of nonlinear Langmuir wave structure, such as filaments and cavitons, by strong electromagnetic radiation (EM) which exists in pulsars. For $(\gamma^2 - 1) = (eA/mc^2)^2 \gg 1$, these structures can always form with a growth rate $(1/2)\omega_p^2/\omega$, where A(ω) is the amplitude (frequency) of EM and ω_p is the plasma frequency. The general condition to create the instability is $(\gamma^2 - 1)/(1 + \gamma^{-2}) > (\omega_p/2\omega)^2$ and the instability has a growth rate $(1/2)(\omega_p^2/\omega)(1 - \gamma^{-2})^{1/2}$.

Is the instability for the creation of nonlinear Langmuir wave structures the explanation for the micropulses observed in pulsar radiation? Can we make a quantitative comparison with observations?

IV.4 Supernovae

Bingham noted that present calculations do not produce a supernova explosion with the small neutrino energy deposition $\Delta E \sim 1\% E$, where E is the total energy of the neutrinos. In order to increase ΔE, neutrino-plasmon scattering is suggested to occur. The plasmons (L) are suggested to be created by the instability $v \rightarrow v' + L$ (a type of stimulated Raman scattering) which has a threshold for the neutrino energy density $W \propto \Gamma_L \Gamma_v n^{-3/2}$, where $\Gamma_L(\Gamma_v)$ is the damping rate of the plasmons (neutrinos) and n is the ambient density. Since Γ_v is very small, W is very small, resulting in all of the neutrinos losing their energy by stimulated Raman scattering and making $\Delta E \sim 100\% E$, which is much too large. It is suggested that a dephasing of the neutrinos occurs due to the scattering (diffusion) of the neutrinos. We then have $\Delta E \sim \Delta\Theta^2 E$ and for $\Delta\Theta^2 \sim 0.1$ we have $\Delta E \sim 10\% E$, which is more reasonable.

Can we make a self-consistent theory? The neutrinos need to escape the central region and deposit their energy in the outer region to create a supernova, so how does the threshold condition $Wn^{3/2}/\Gamma_L\Gamma_v$ and $\Delta\Theta$ behave with distance? Does it have the correct dependence? What about the effects of a magnetic field (since we know we leave a neutron star with a magnetic field $B \sim 10^{12} G$ and with turbulence we might expect equipartition $B^2/8\pi \sim nkT$, as seen in the interstellar medium and in extragalactic jets)?

V. EXTRAGALACTIC ASTRONOMY

Many interesting phenomena occur of our galaxy. For example, many galaxies have active nuclei which are $> 10^4$ more luminous than the center of our galaxy. The highest energy cosmic rays observed ($\sim 10^{20} eV$) probably have an extragalactic origin. WEe observe clusters of galaxies. The origin of the ubiquitous magnetic field in galaxies is also a great problem.

V.1 Cosmic Rays

Medina-Tanco, Jatenco-Pereira and Opher discussed the observational evidence in active galactic nuclei of a high rate of supernova explosions at high density in a small volume (Terlevich et al. 1992). They applied the model of Medina-Tanco and Opher (1993) of cosmic ray acceleration in our galaxy by multiple supernova remnants to active galactic nuclei. Using a volume of $(10 pc)^3$, density $n \sim 10^7 cm^{-3}$ and a supernova rate $\sim 1/week$, they found that cosmic rays are accelerated to energies $10^{17} eV$.

Can this model be made more quantitative and explain cosmic rays $\sim 10^{20} eV$ which are observed?

V.2 Clusters of Galaxis

Jafelice and Friaca discussed the observations of filaments with optical emission lines which indicate cooling flows of galaxies. Calculations followed the cooling condensations until the phase of optical line emission. Magnetic reconnection was found necessary to obtain good agreement with observations.

Can magnetic reconnection be proven to exist in the region of the observed filaments?

V.3 Magnetic Fields in Galaxies

Field discussed the dynamo mechanism in spiral galaxies for creating magnetic fields. The basic dynamo relation for the change of the magnetic field with time consists of three terms: 1) The first term transforms poloidal field to toroidal field due to differential rotation; 2) The second term transforms toroidal field back to poloidal field due to turbulent motion and is proportional to a parameter α; and 3) The third term describes turbulent diffusion and is proportional to a parameter β. For a galactic disk of thickness 2h, the growth rate is $\gamma = 0.8(\alpha_o \Omega/h)^{1/2} - 2.3(\beta/h^2)$ where Ω is the angular velocity of the spiral galaxy and α is assumed to have the form $\alpha = \alpha_o sin(\pi z/h)$. Traditionally it is assumed that $\alpha \sim 10^4 cms^{-1}$ and $\beta \sim 10^{26} cm^2 s^{-1}$. Field (1995) found in a specific model with $\alpha_o = 2.1 \times 10^4$ and $\beta \sim 10^{26}$, that $\gamma < 0$ (i.e., no dynamo!). A general problem with a present dynamo theory is that the turbulent velocities are assumed to be given in order to determine α and β, rather than the turbulent velocities be generated self-consistently from dynamo theory.

Can α and β be determined self-consistently? Even if α and β can produce a dynamo magnetic field in a spiral galaxy, how does the magnetic field diffuse out of the spiral galaxy fast enough to explain magnetic fields in elliptic galaxies and in the intergalactic medium? If the Faraday rotation measurements are confirmed that an appreciable magnetic field exists in high redshift systems (i.e., early times), can a dynamo be sufficiently efficient as to make an appreciable magnetic field in only a few rotations of the galaxy?

VI. COSMOLOGY

What made the large scale structure of the universe and primordial magnetic fields? Are observations consistent with primordial nucleosyntheses of the light elements? These are just a few of the open questions in cosmology.

VI.1 Primordial Magnetic Fields and Galactic Voids.

Miranda, M. Opher and R. Opher studied a model in which the first object to collapse in the primordial universe creates a supernova explosion and the swept up material by the supernova shock subsequently collapses and makes further supernovae. The final cavity produced is ~ 60 Mpc, on the order of the size of observed voids of galaxies. The density gradient in the primordial shock is not parallel to the temperature gradient and produces a magnetic field $\sim 10^{-9}$ G by the "Biermann" mechanism.

Is the distortion of the cosmic background radiation consistent with the model of multiple supernova explosions?

VI.2 Large Scale Structure of the Universe

Oliveira, Opher and Pires found evidence for inhomogeneity in the galaxy distribution for distances $>100 h^{--1}$ Mpc.

What is the structure of the inhomogeneous distribution of galaxies for distances $>100 h^{--1}$ Mpc?

VI.3 Primordial Nucleosynthesis

M. Opher and R. Opher found that at a given temperature plasma fluctuations increase the expansion rate of the universe due to their energy density and change nuclear reaction rates due to their electromagnetic fields. They found that fluctuations predicted by the fluctuation dissipation theorem at frequencies $\omega \sim 0$ create a change in the predicted primordial helium abundance $\Delta Y/Y \sim 10^{-4}$ and for $\omega \sim \omega_p$: $\Delta Y/Y \sim 10^{-3}$.

What other important plasma effects change primordial nucleosynthesis?

VII. FINAL COMMENTS

From the above discussion we note that the contributions on space and astrophysical plasmas at the 1994 ICPP were at a very high level. they indicate that there is extremely interesting plasma physics outside of terrestrial laboratories. In particular, some of the most basic problems in astrophysics await a plasma physic solution, such as the origin of: 1) Fluctuations in the universe that gave birth to stars and galaxies; 2) The primordial magnetic field; 3) Large scale structures of universe; 4) Protostellar and extragalactic jets; 5) Supernova explosions; and 6) Solar Flares. We also have the space physics problems of: 7) Energy transfer from the solar wind through the megnetosphere; and; 8) the time and space structure of the aurora.

Our understanding of space and astrophysical plasmas is rapidly increasing so that we can expect greater insight and perhaps solutions to some of the above problems at the next ICPP.

Acknowledgements: I would like to thank the brazilian agency CNPq for partial support.

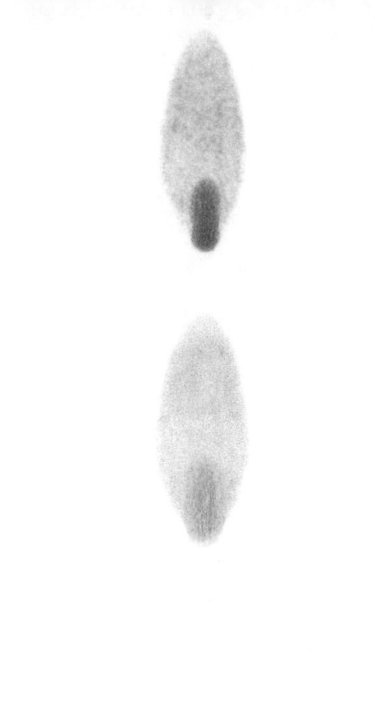

OPENING COLLOQUIUM

SOLAR ECLIPSES

William Livingston
Nation Solar Observatory, NOAO
Tucson, Arizona 85726, USA

ABSTRACT

The occasion of a total eclipse impacts the human observer with a bewildering rapid sequence of phenomena: mid-day cooling, failing light without accustomed color change, shadow-bands transiting the ground, cessation of bird sounds, possible frantic beating of jungle drums, Baily's beads, appearance of flame-like prominences, and most fantastic of all THE CORONA (hottest object in the universe visible to the naked eye; a plasma whose convolutions plainly map out the sun's large scale magnetism). We consider that although the corona is known to be 2-20 (10^6) K, there is a lack of consensus on the heating mechanism, except the energy must be non-thermal and derived from surface and sub-surface convective motions. Theoreticians invoke the Joule dissipation of magnetic fields by Alfvén waves, electric currents in loop structures, or MHD turbulence. Although eclipse experiments to discriminate between these ideas generally fail, the sighting of 'plasmoids' was reported from the CFHT on Mauna Kea at the 1991 eclipse. Next Thursday's experiments include: IR mapping of the coronal spectrum, spectroscopic velocity measurements, and the continued search for waves, nanoflares, and plasmoids.

1) HISTORICAL

Although total eclipses, being spectacular events, have been noted throughout recorded history, descriptions of the event have been meager and uninformative [1,2]. At least until mid-nineteenth century.

The human mind rejects, or fails to see, the unexpected. It is as simple as that. Today when we observe a total eclipse, our attention is immediately drawn to the corona with all its intricacies of rays, helmets, and arcs. Yet Kepler in 1605 was the first to even notice this glow around the sun. Those before him were apparently overwhelmed by the darkness of the moment, perhaps with a sense of foreboding, and saw nothing else.

When awareness of the corona did become a part of eclipse lore, people began to at least devote a few words to it, evoking a pale light or glory; perhaps it was a manifestation of the moon's atmosphere. More likely it was just an atmospheric illusion.

As photography began to replace the eye it was found that corona structure was the same when recorded at sites hundreds of km apart. The corona was therefore solar and not terrestrial. (The moon was ruled out as having no atmosphere).

Polarizing eyepieces revealed that coronal light was radially polarized as would be expected from reflected sunlight. But, curiously, the spectroscope showed no evidence of

Fraunhofer absorption lines. It did, however, show a bright emission line, dubbed 'coronium,' at 5303Å. Over half-century later B. Edlen identified this green corona line as due to Fe XIV in a diluted gas at around $2(10^6)$ K. Doppler blurring by fast moving particles at this temperature explained the absence of a scattered light Fraunhofer spectrum. (Much further out Fraunhofer lines do appear, but the ('F-corona') scattering media proved to be dust particles that gravitate in the ecliptic and not the high temperature plasma closer in to the sun).

2) SCIENCE GLEANED FROM PAST ECLIPSES

A) The Chromosphere

At second contact, when the moon first covers the white-light disk, a reddish arc flashes out around the limb: the chromosphere. Lying just above the 5700K photosphere and below the $2(10^6)$ K corona, this layer is about 5000 km thick. Actually, to call it a layer is misleading because it is very structured, being made up of spicules, plages, and all sorts of magnetically related features.

On the disk, outside of eclipse, the cores of strong absorption lines like H-alpha, Na D, Mg b, Ca K, etc. are all formed in the chromosphere. At the limb these same lines, and many lesser ones, go into emission. Because this limb emission region subtends only a few arc-seconds, it is difficult to observe in the presence of ubiquitous atmospheric seeing.

However, at an eclipse, as second-contact evolves, the chromospheric spectrum is uncovered by the lunar limb independent image tremor. Furthermore, the moving lunar limb serves as a time varying probe of height within the chromosphere. D.H. Menzel devised a 'jumping-film' camera to record this flash spectrum. His technique reached a state of perfection at the hands of a joint Sac Peak/High Altitude Observatory/National Bureau Standards/U.S. expedition at Lae, New Guinea, in 1962 [3]. Some 3500 emission lines were measured in the wavelength range 3200-9100Å with a height resolution of about 100 km.

The moon as a height probe becomes even more significant at sub-millimeter and longer wavelengths where telescope aperture is normally a limiting factor because of diffraction. F. Shimabukuro (at 3.3 mm), T. Allen Clark (at 0.4 mm), and E. Becklin, et al. (at 0.03, 0.5, 0.1 and 0.2 mm) have discovered 'limb-brightening' and an extended chromosphere above the visible-light segment. Combining the flash-spectra and these long wavelength eclipse observations has led to sophisticated chromospheric models applicable even to the stars.

The above models indicate a minimum temperature of 4700K just above the photosphere with a rise to 6000K, and above, as we progress outward. Limb spectra outside eclipse have revealed the CO fundamental (4.6 μm) in emission under good seeing. Since the CO molecule could not exist at 4700K, it has been proposed that the chromosphere bifurcates into hot and cold parts. Alternatively, there may be a very thin CO layer immediately below the 'normal' chromosphere. To better measure the height and thickness of this CO layer, T. Alan Clark recently made use of a deep, near annular partial eclipse (10 MAY 94) to cut off the sun's limb with its CO emission. The geometric cutoff was similar to that at a total eclipse except now the action was scissor-like. By moving along the occulting limb a series of wedge-like segments were imaged on a 256 x 256 indium antimonide array camera. Anomalously cool (3000K) CO radiation was resolved to a height of better than 100 km.

B) The Corona

Once it was understood that the corona was a $2(10^6)$K plasma, the outstanding question was: what heats it? This is a hard question and has not yet been answered. Obviously, there is a violation of the Second Law of thermodynamics here because the underlying photosphere, through which all energy must flow, is at a temperature of only 5700K.

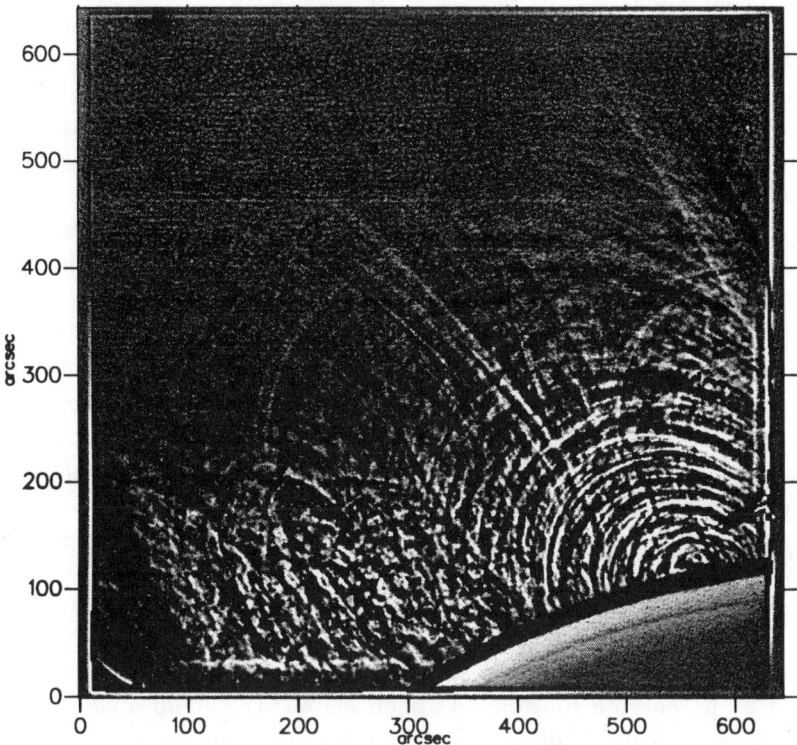

Fig 1. Loops detected by the CFHT 1991 experiment after image enhancement (November & Koutchmy private communication).

Three of the many proposed heating mechanisms might be clarified by eclipse observations. These are: Alfvén wave dissipation, electric current dissipation (coronal loops and so-called 'nanoflares'), and the decay of a turbulent magnetic field.

Pasachoff [4] has searched for intensity oscillations in Fe XIV 5303Å at the 1980 eclipse. Something was seen at the 1% level, presumably from Alfvén waves. But Parker [5] points out that such hydromagnetic waves do not produce density fluctuations to first order. Pasachoff plans to repeat the experiment using improved CCD imaging on 3 NOV 94.

Nanoflares [6] is another possibility and these were searched for at the 1991 eclipse by Zirker and LaBonte using the University of Hawaii 2.24 meter and by Koutchmy, et. al [7] at the CFHT 3.6 meter, both on Mauna Kea. A high-resolution video (shown a this ICPP conference) fails to reveal any nanoflares in visible light at 4471Å. Similarly, nothing resembling nanoflares is found on the various 3.6 meter images.

What the superb CFHT data do show is an extensive field of fine loop-like structures at various levels in the corona. Corrected for image smearing, the FWHM of these threads is about 2 arc-sec. Thus filamentation of the corona occurs at small scales which could foster current dissipation, Figure 1.

Some evidence of the corona's turbulent motions, and by inference its magnetic field, comes from spectra in Fe XIV 5303Å. Livingston and Harvey [8] operated a multi-slit spectrograph to measure flow patterns, including rotation, at several eclipses (7 MAR 70; 29 JUN 73; 16 FEB 80; 11 JUN 83). At a spatial resolution of about 25 arc-sec velocities of up to 12 km s^{-1} were seen, but 0.5 km s^{-1} was the average value. This result is not suggestive of a vigorous turbulent magnetic field although higher resolution might bring it out.

At the 11 JUL 91 eclipse, Vial et al. [9] detected what they call plasmoid activity. During their $3\frac{1}{2}$ minute time sequence, a tiny 2 x 4 arc-sec condensation, which they deduce to be a confined plasma, was ejected from the sun. They invoke a diamagnetic plasma moving in a divergent magnetic field. By observing the plasmoid in different spectrum windows (H-alpha, Fe X 6374Å, white light continuum —Thomson scattering) physical conditions such as temperature and electron density were found.

C) Earth's atmosphere

Veteran eclipse observers become acquainted with shadow bands: faint, moving patterns of bands that appear on the ground as 2nd contact approaches or immediately following 3rd contact. Generally, they can be seen but not photographed —the eye is particularly good at perceiving motion. At each eclipse the band spacing and velocity differ. No detailed explanation is offered, but atmospheric density inhomogeneities must be responsible —the same thing that produces the twinkling of stars at night.

Many meteorological observations are conducted at each eclipse. We will mention that E.O. Hulburt [10] measured sky brightness in and out of eclipse at Bocaiuva, Brazil, at the eclipse of 20 MAY 47.

D) Solar diameter change

Sir Edmund Halley collected and published eye-witness accounts of whether the eclipse of 3 MAY 1715 was total or partial depending on the English village sites of the observers. The northern limit proved to be Darrington, the southern Cranbrook, leading to a solar diameter of 959.63 arc-sec with an error estimate of ±0.1 arc-sec [11]. A change of solar luminosity is accompanied by a diameter change, and so historical sightings of eclipse bounds help to determine irradiance variation over a long time base.

3) PLANNED EXPERIMENTS FOR 3 NOV 1994

- *Photography of white-light E-corona* : O.R. White (Boulder), will deploy the 'Newkirk camera' with its radially graded filter (Chile); O. Matsuura (Sao Paulo) and N. Dziubenko (Kiev) will operate multistation color film cameras (Brazil); S. Koutchmy (Paris) plans the same in Chile; M. Molodensky (Moscow) will photograph through polaroids to deduce 3-D structure (Brazil).

- *Photography of F-corona*: I. Kim (Moscow) and O. Matsuura (Sao Paulo), by use of standard astronomical filters and deep exposures, will try to detect circumsolar dust grains.

- *Photography of Fe XIV 5303Å corona*: J. Zirker and R. Coulter (Sunspot) will make long exposures through $\Delta\lambda=1.8$Å filter of the middle corona out to $3R_\odot$ (Chile).

- *Fabry-Perot imaging*: I. Kim (Moscow) and E. Picazzio (Sao Paulo) will operate in 5303 and Fe XI 7892Å(Brazil); R. Smartt (Sunspot) and I. Kim (Moscow) in 5303 (Chile), all to measure plasma flows.

- *Spectra of coronal lines*: S. Koutchmy (Paris) has an f/6 spectrograph with 4.5Å mm^{-1} set on 5303Å (Chile); R. Cowsik (Bangalore) will take multi-slit spectra in 5303 and Fe X 6374Å imaging on a CCD chip; J. Kuhn and M. Penn (Sunspot) will make an exploratory spectrum map of inner corona from 1.0 to 1.5μm using a NICMOS camera (Chile).

- *Wave detection*: J. Pasachoff (Willamstown) will make high sensitivity CCD pictures through filters which avoid emission lines from 3900 to 4100Å (Chile); O.R. White (Boulder) will acquire rapid pictures with a CCD camera using a Questar (Chile); O. Matsuura (Sao Paulo) and I. Kim (Moscow) plan cinematography through a rotating polaroid (Brazil); V. Kulidzanishvili (Abastumani) and E. Picazzio (Sao Paulo) will make photoelectric measures of linear polarization to 4 $R_\odot$.

- *Prominences and transition layer*: I. Kim (Moscow) and S. Livi (Rio Grande do Sul) employ narrow bandpass filters on H-alpha and He II 4686Å lines in conjunction with other observations in attempt to sense this thin layer.

- *Radio observations*: H. Sawant (San Jose dos Campos) will look for structures at 1600MHz (Brazil).

- *Timing of contacts*: G. Vergara-Gavagnin (Montevideo) in connection with solar diameter programs and solar luminosity variability (Brazil).

4) WHAT TO LOOK FOR ON 3 NOV 1994

The following, amended to reflect the current circumstances, is taken from Lynch and Livingston [12]:

An eclipse begins with little fanfare. At 'first contact' the invisible new moon takes its first bite out of the sun. This will occur at 09:34 AM in Foz do Iguacu with the sun at an altitude of 36° and due east [13]. Except for the disk becoming occulted, nothing very striking happens until about 10 minutes before totality when we become aware that the sunlight is failing. Several unfamiliar conditions then ensue. For one, the sky begins to darken without the accustomed reddening that precedes sunset. Hues that are normal for mid-day continue to prevail even though it is getting dark. Shadows become strange. Openings in tree foliage overhead, which normally cast circular patches of light on the ground, now project crescent shapes. If we examine the shadows of our own hands, finger sharpness depends on the angle we hold them. This is a consequence of the sun becoming a slit-like source of light, broad in one direction, narrow in the perpendicular.

Within 5 minutes of totality shadow bands may begin running across the ground. Shadow bands are low contrast, roughly parallel, ridges of light and dark that would escape our attention except for their movement [14]. The distance between the bands varies from 5 cm to perhaps 2 m; their velocity is 2 to 20 m s^{-1}.

About thirty seconds before totality one has the the growing sense that something is desperately 'wrong'. The sky itself now darkens rapidly as the shadow of the moon sweeps toward us on the horizon. There is a hint of chill to the air.

Heralding the onset of totality is the Diamond Ring. The last sliver of photospheric light combines with the emerging ring of the inner corona to mimic a diamond ring in the sky. The ring part is sometimes peppered by Baily's Beads, sparkling points of photospheric light shining through lunar valleys.

Suddenly totality is upon us (at 10:44). This is the moment of second contact when the moon first completely covers the sun and the solar corona is now revealed in all its splendor! What you are witnessing can be seen in no other way and is unique to this moment. Neither spacecraft nor the most powerful ground-based telescopes can record what your own eyes now behold. Even photographs taken by astronomers at this same eclipse will not do justice to your personal perception of this visual scene. For once the unaided eye reigns supreme.

The corona is a soft, pearly white light encircling the eclipsed sun. Structures in the form of helmets, fans, and rays can be traced out to one or more solar radii. The corona is the tenuous outer atmosphere of the sun at an temperature of 1 - 2 million degrees. This makes it the hottest object in the universe visible to the unaided eye. During sunspot maximum, the inner parts of the corona are fairly symmetrical and as bright as the full moon, or one millionth the intensity of the photosphere. Around sunspot minimum (which we are approaching now), the coronal equatorial rays may be expected to be unsymmetrical and develop to their greatest elongation.

The corona is polarized. This is because sunlight is scattered off free electrons. It is tangentially polarized up to 40% at a distance outward of one quarter the radius of the sun from its edge. Beyond about one solar radius, interplanetary dust particles begin to dominate over electrons as the scattering source and polarization diminishes. These are the same particles responsible for the zodiacal light.

Close in to the occulting moon we may see tiny, red features called prominences. Their color originates primarily from a spectral line of hydrogen which is in emission at 6563Å. At both the second and third contact the chromosphere flashes out briefly as a narrow reddish arc. This is the hot, normally invisible layer overlying the photosphere. Again the red light of hydrogen dominates. Binoculars greatly assist in seeing the chromosphere and prominences.

The planets Mercury, Venus and Jupiter will be near the sun at this time. If one is located on a hill, a view toward the horizon may show the edge of the moon's umbra against the still illuminated distant atmosphere. One could watch the progression of this shadow across the landscape, but this observation would cost precious time that is best devoted to the corona.

At third contact the events preceding second contact take place in reverse order: there is the chromospheric flash, Baily's beads, and then the blossoming diamond ring drives the corona to invisibility. Before you know it, the spectacle is over.

The following is a list of possible eclipse-related phenomena to watch for near totality. You only have a few minutes and a prepared mind will help one to get the most out of the experience. Some are subtle, all have been reported, but few photographed.

- Coronal transients (short-lived, bubble-like enhancements in coronal structure)

- Comets near the sun (which are invisible outside eclipse)

- Edge of the moon opposite sun is visible in partial phase (indicating existence of coronal light which would otherwise go unnoticed)

- Inner parts of the corona seen before totality

- Brighter stars and planets seen outside totality

- Prominences seen several minutes before or after totality

- Moon darker than surrounding sky during totality (Mach effect?)

- Earthshine visible on the eclipsed moon; look for lunar features

REFERENCES

1. Schove, D.J., 1984, "Chronology of eclipses and comets AD 1-1000", Boydell Press, Dover, New Hampshire.

2. Zirker, J.B., 1984, *Total eclipses of the Sun*, Van Nostrand, New York.

3. Dunn, R.B., Evans, J.W., Jefferies, J.T., Orrall, F. Q., White, O.R., and Zirker, J.B. 1968, "The Chromospheric Spectrum at the 1962 eclipse" *Ap. J. Suppl.*, **15**, 275-458.

4. Pasachoff, J.M. and Landman, D.A., 1984, "High-frequency coronal oscillations and coronal heating", *Solar Phys.*, **90**, 325-330.

5. Parker, E.N., 1991, Comment on p. 29, Ulmschneider, P., Priest, E.R., and Rosner, R. (eds.), "Mechanism of chromospheric and coronal heating", *Springer-Verlag* , Berlin.

6. Parker, E.N., 1988, "Nanoflares and the solar x-ray corona", *Ap. J.*, **330**, 474-479.

7. Koutchmy, S., Belmahdi, M., Coulter, R.L., Demoulin, P., Gaizauskas, V., MacQueen, R.M., Monnet, G., Mouette, J., Noens, J.C., November, L.J., Noyes, R.W., Sime, D.G., Smartt, R.N., Sovka, J., Vial, J.C., Zimmermann, J.P., and Zirker, J.B. 1994, "CFHT eclipse observation of the very fine-scale solar corona" *Astron. Astrophys.*, **281**, 249-257.

8. Livingston, W. and Harvey, J., 1982, "Preliminary results from eclipse coronal velocity observations", *Proc. Indian National Sci. Acad.*, **48**, 18-28.

9. Vial, J.-C., Koutchmy, S. and the CFH Team, 1992, "Evidence of plasmoid ejection in the corona from 1991 eclipse observations with the Canada-France-Hawaii Telescope", in Solar Physics and Astrophysics at Interferometric Resolution, L. Dame and T.-D. Guyenne (eds.) ESA SP-344, 87-90.

10. Hulburt, E.O., 1949, "Night sky brightness measurements in latitudes below 45°" *Jnl. Optical Soc. America*, **39**, 211-215.

11. Morrison, L.V., Stephenson, F.R., and Parkinson, J., 1988, "Diameter of the sun in AD 1715", *Nature*, **331**, 421-423.

12. Lynch, D.K. and Livingston, W., 1995, *Color and Light in Nature*, Cambridge Univ. Press, Cambridge (in press).

13. Matsuura, O.T., Boczko, R., and Marques dos Santos, P., 1993, "Information regarding the observations of the eclipse of November 03, 1994, in Brazil", private communication.

14. Codona, J.L. 1991, "The enigma of shadow bands", *Sky & Tel.*, **81**, 482.

FUSION PLASMAS

THE FUNDAMENTALS OF THE ITER PHYSICS

S.Putvinski and the ITER Joint Central Team

The ITER design developed by ITER Joint Central Team and Home Teams since the start of Engineering Design Activity in 1992, is based on our current understanding of the main physics processes in the tokamak plasmas and their extrapolations to the ITER scale. It is shown that our present knowledge is sufficient to choose ITER basic parameters and design specifications. However, because of the far extrapolation from the present day machines to the ITER scale, ignition margins are required. The paper discusses the most important physics issues for ITER and R&D needs in the physics area.

I. INTRODUCTION

International Thermonuclear Experimental Reactor (ITER) is acknowledged by the international fusion community as an important step in the fusion program which should be undertaken by the world community. The ITER agreement [1] between the European Union, the government of Japan, the Russian Federation and the USA stated that the ITER objectives should be accomplished "by demonstration of controlled ignition and extended burn of deuterium-tritium plasmas, with steady-state as an ultimate goal, by demonstrating technologies essential to a reactor in an integrated system, and by performing integrated testing of the high-heat flux and nuclear components required to utilize fusion energy for practical purposes". To accomplish these goals ITER must a) have a capability for a long inductive pulse t > 1000 s under burning conditions, b) provide during the flat top neutron wall load of about 1 MW/m^2, and c) be designed to achieve fluence of at least 1 MWa/m^2 to carry out material tests. In addition the ITER heating system should have sufficient current drive capability.

The lessons learned during Conceptual Design Activity (CDA) phase [2] of the ITER made possible a few important improvements in the ITER design. It was decided to change the divertor configuration from a double null to a single null and to increase space for the divertor by decreasing plasma elongation. To have the plasma current and hence the same plasma energy confinement capability at the lower plasma elongation the plasma minor and major radii were increased. Further optimization of the Toroidal Field (TF) coil design made it possible to increase the toroidal field from 4.85 to 5.7 T. The list of the ITER CDA and Engineering Design Activity (EDA) basic parameters is given in Table 1.

	CDA	EDA
Plasma major radius R_0 (m)	6.0	8.1
Plasma minor radius a (m)	2.15	3.0
Plasma elongation k_{095}	2	1.55
Plasma current I_p (MA)	22	24
Toroidal field (at $R = R_0$) B (T)	4.85	5.7
Fusion power P_{fus} (GW)	1	1.5
Plasma average density n_e (m^{-3})	10^{20}	10^{20}
Plasma temperature T (keV)	10-20	10-20
Maximum He concentration (%)	5-10	10-20

Table 1. ITER CDA and EDA basic parameters

II. DESIGN OF THE BASIC MACHINE

The superconducting magnet system consists of the 24 TF coils, Central Solenoid (CS) and seven Poloidal Field (PF) coils which provide plasma initiation, equilibrium and control. The TF coils have bending free shape and are bucked on the CS which allows more favorable stress conditions and therefore allows an increased toroidal magnetic field and the amount of volt - seconds available for the inductive pulse. All coils will use Nb$_3$Sn superconducting cable in conduit form at a current level of 40-60 kA and a magnetic field up to 13 T at the conductor. The number of toroidal coils is large enough to provide a low ripple (< 2%) at the plasma surface and allows feasible remote maintenance schemes.

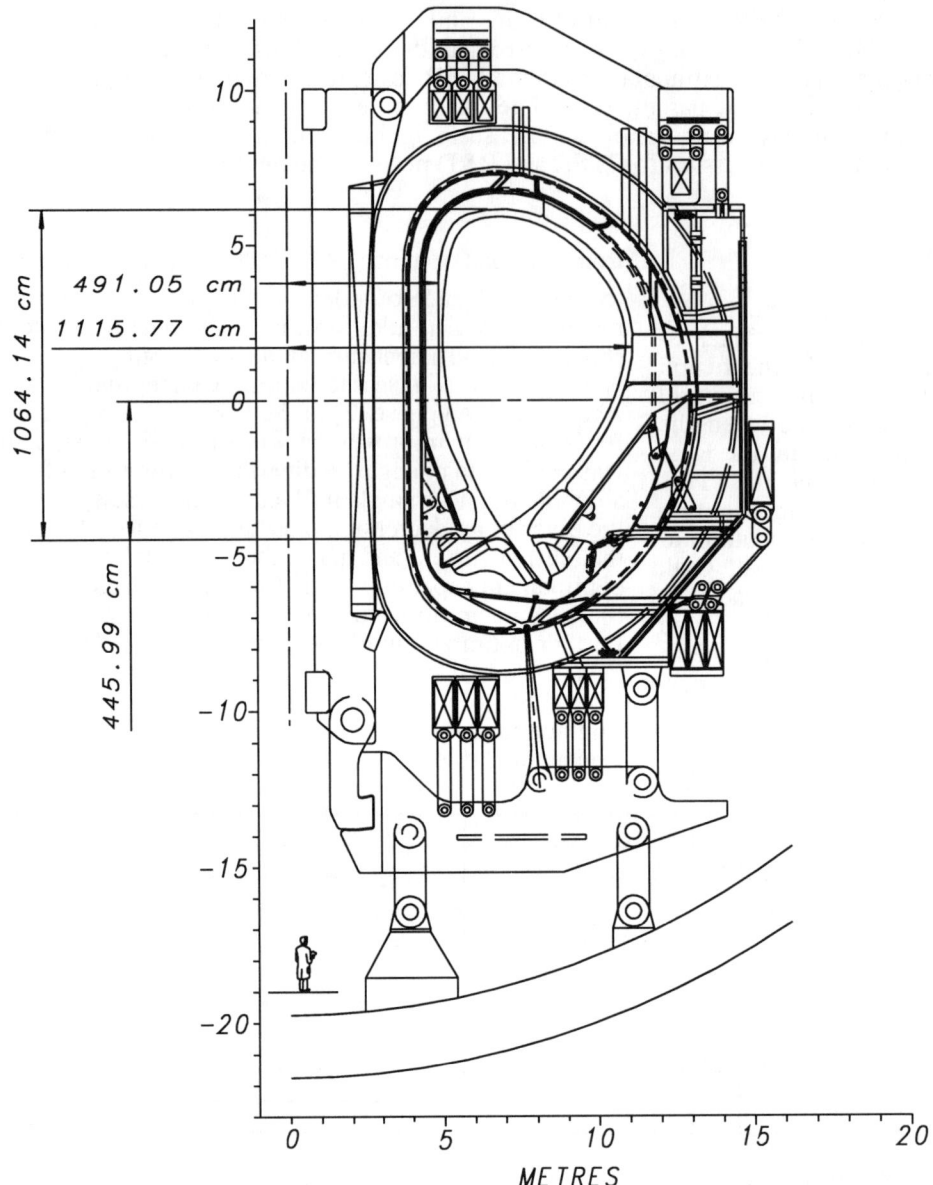

VERTICAL CROSS SECTION

Because of the neutron radiation from the DT plasma the ITER first wall, blanket and Vacuum Vessel (VV) designs are principally different from the VV/first wall of the present day tokamaks. They must provide shielding of the superconducting magnet system from the neutron radiation, must have a powerful cooling system capable of removing 1.5 GW of fusion power and withstand the strong electromagnetic load produced during disruption (up to 1 MPa) and vertical plasma displacements (up to 100 MN of total vertical force). They must also satisfy demanding requirements imposed by safety, reliability and maintenance considerations. These requirements make the design of these elements crucial for the ITER success.

The ITER VV provides high quality vacuum and is designed as a double wall structure formed by two nested toroidal shells (see Fig. 1). The cavities between the shells and internal ribs are filled with insulated steel balls which are directly cooled by the water flowing through the system. The VV is the major safety barrier and is capable of withstanding a large pressure rise due to plausible accidents without losing the confinement.

The first wall and blanket assembly is located between the plasma and VV (see Fig.1). At the basic performance phase this structure is used for shielding and for heat removal produced by the fusion power. However, at the extended performance phase which will be used for nuclear testing, ITER will need to breed the bulk of tritium it will consume and at this phase the blanket will have to breed tritium with a relatively high breeding rate > 0.8. The first wall has a separate cooling system and is composed of the steel tubes embedded in the copper alloy about 2 cm thick. The plasma facing side of the wall is covered by 5 mm beryllium protection. The shielding blanket is fabricated in modules each 1 m wide toroidally and 12 m high poloidally. It is planned to replace, before the extended performance phase the shielding blanket with one which breeds tritium.

The ITER divertor is probably the most challenging element of the plasma facing components. It became clear during the CDA phase that very high heat loads on the divertor target plates in the conventional high recycling divertor concept requires new approaches which allow the reduction of heat loads at the plates [10]. The gas target/radiative divertor concept accepted for ITER relies on a radiation transport of a large fraction of power conducted into the SOL. The neutral gas pressure is contained in the closed divertor geometry by baffles and the circulating in the divertor neutral gas flow which allows a reduction of the total pressure in the gas bag formed by semi-transparent walls. The divertor gas box consists of 96 divertor cassettes mounted on toroidal rails. The dump plates and the side walls have beryllium plasma facing protection and are cooled by water separated from the first wall/blanket cooling contour. The gas bag is pumped through the private flux region and allow flexibility of the target plate geometry. The cryogenic pumps are located inside each of 16 lower ports and are designed to allow the various pumping speeds for impurity, fuel and He ash.

The auxiliary heating system must deliver 50 MW of power to achieve ignition. The heating method has not been chosen yet. ICRH, ECRH and 1 MeV neutral beam are considered as main candidates for the heating system. Further upgrading of the power up to 100 MW is probably necessary for demonstration of reactor relevant steady state operation in ITER.

The detailed design, Research & Development program and manufacturing of the ITER systems require a deep involvement of the four Home Teams and industries, to prepare transition to the construction phase. The ITER team is working toward this goal.

III. ITER PHYSICS BASIS

The design described above is based on a detail optimization of plasma and machine parameters based on the present day understanding of the plasma physics processes. The present knowledge is sufficient to make projections for ITER basic parameters. However, ITER is an experimental machine and must be flexible to perform physics study of the plasma processes in the ignited plasma to provide information for the next step DEMO type of the fusion reactor.

A. Plasma Energy Confinement.

Plasma energy confinement defines the size of the plasma necessary for achieving sustained burn. Two different models were used for extrapolation of the plasma energy confinement of the present day experiments to the ITER scale: Rebut-Lalia-Watkins (RLW) model [3] and empirical scaling ITER-89P [4] for plasma energy L-mode confinement time. In terms of dimensionless parameters they represent GyroBohm and Bohm types of scalings. Both types of scalings describe reasonably well the present day experiments but differ in the extrapolation to the ITER scale. The results of the modeling show the sustain ignition margin 1.34 for L-mode RLW model and only 1.06 for ITER-89P, even with an enhancement factor H=2. Enhancement factor H=2 corresponds to the H-mode of the present day experiments.

Fig.2b shows the ignition margin as a function of the Helium recycling rate, R, which is defined as the ratio of He inboard flux to the outboard flux at the plasma separatrix. R=0.98 is the reference value which corresponds to the experiments at the DIII-D and relevant to the geometry of the ITER pumping system. At a low He density the (low R) ITER has large ignition margins predicted by both types of scalings.

As soon as the fraction of He builds up ITER-89P predictions yield only marginal ignition.

The enhancement factor H=2 is somewhat arbitrary and was used as an example to describe the H-mode

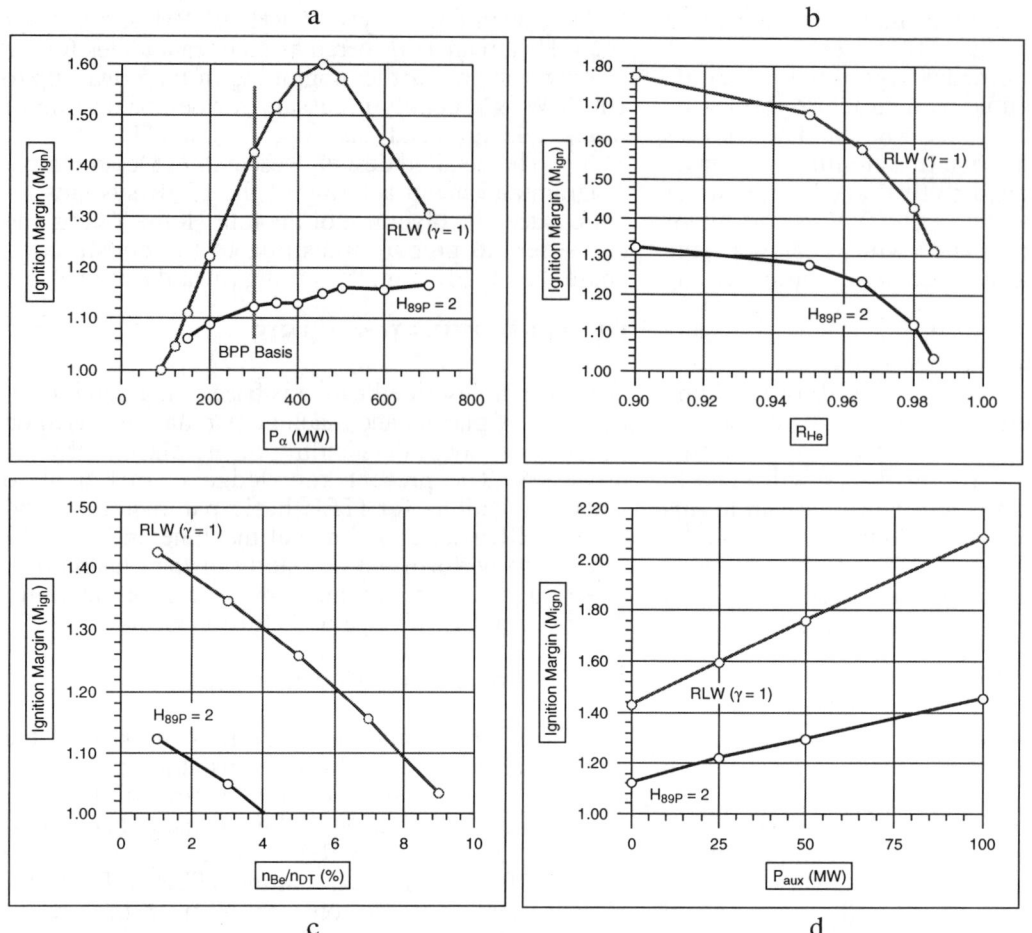

Fig. 2 Ignition margins as a function of fusion power (a), He recycling rate (b), Be concentration (c) and auxiliary power (d). Ignition margin M_{ign} is defined as $M_{ign} = <n_{DT}><T>\tau_E / \{<n_{DT}><T>\tau_E\}_{marginal}$.

enhancement over the scaling obtained on the L-mode data base. Even larger enhancement factors in the H-mode in the present experiments have been demonstrated. However, according to the empirical scalings the triple product $<nT\tau>$ is roughly proportional to $(H/q)^2$ and the search for relevant ITER like discharges with $H/q>2/3$ in DIII-D and JET databases shows that $H/q=0.67$ is close to the maximum attainable value for long pulse operation. At the same time modeling of ITER discharges with the Bohm type energy confinement scaling shows that sustained ignition is no longer possible if H/q falls below 0.6. The ignition under pure L-mode scaling H=1 is not possible in ITER even with the assistance of 100 MW of auxiliary power. It makes the transition to the H-mode a necessary condition for the success of ITER operation.

The extrapolation to the ITER parameters by the power threshold scaling predicts a very high power threshold in ITER:

$$P_{thr}(MW) \approx 300 n_e ,$$

and the path to the H-mode requires operation at low plasma density at least at the transient phase.

However, dependence of the threshold power on the plasma size is not very reliable for the far extrapolation required for ITER. It is known from experiments that the L-H transition is sensitive to plasma edge parameters and plasma wall interaction. The control of the L-H transition is vital for ITER and study of the threshold for the L-H transition was acknowledged as one of the most urgent tasks for the present experiments.

B. Ideal MHD Stability

Linear analysis performed for ITER reference equilibriums and profiles shows comfortable stability margins up to normalized $\beta_N \equiv \beta / (I / aB_0) = 3.4$ for ballooning instabilities and $\beta_N = 4.4$ for global kink modes, for L-mode type profiles and approximately the same values for H-mode type profiles. The limit is much higher than required for achieving the nominal fusion power of 1.5 GW, $\beta_N = 2.2$.

The normalized beta above 2 can be achieved in the present tokamaks. However, fusion power is proportional to $(\beta_N / q)^2$ and the ITER reference scenario requires $\beta_N / q \approx 0.7$. There is an indication from experiments that the maximum normalized beta in the H-mode could be much lower than predicted by ideal MHD analysis. As in the area of the plasma confinement we need from the present experiments demonstration discharges which can assert accepted for ITER $\beta_N / q \approx 0.7$

C. Alpha-Particle Confinement

In the ignited plasma fusion alpha-particles provide about 300 MW of plasma heating power to sustain plasma energy balance. It is essential in the ITER design to predict the sources of alpha-particle loss and regions in the operational parameters that are dangerous to alpha-particle confinement.

The toroidal field ripple, due to discreteness of the TF coils, is the most predictable type of magnetic field perturbation in tokamaks which can affect alpha-particle confinement. The theory of the fast particle ripple loss is well understood and effective numerical codes have been validated in dedicated experiments on tokamaks [5,6,7]. The new element in the ITER EDA plasma configuration is its up-down asymmetry. The analysis of the alpha-particle ripple loss in ITER, done by the Central Team and Home Teams, shows that the main mechanism responsible for alpha-particle ripple loss in ITER is ripple well trapping and the following convective loss of fast particles. In spite of the plasma asymmetry, the total loss fraction does not depend on the direction of the toroidal drift. However, Monte-Carlo analysis of the ripple loss show that at the upward direction of the toroidal drift the heat load of the first wall is concentrated in the small areas of the first wall in between the TF coils. The energy loss fraction for the reference scenario is not very large, < 1% even for the H-mode type of the plasma profiles. The ripple loss is dangerous because they provide high concentration of the heat load at the wall especially in the upward drift direction.

MHD modes and especially Toroidal Alfven modes represent a potential threat to the alpha-particle confinement. TAE modes and associated energetic particle loss have been observed in beam simulation experiments and in ICRH heated plasmas. Detailed stability calculations of TAE modes in ITER have been performed by several groups. The alpha particles contribute in a destabilizing manner for n>5 and overcome background dissipation for n values 10<n<50. There are several unstable eigenmodes for each n, with a total of about $M \approx 500$ unstable modes. Analysis of particle loss caused by TAE modes show that we would expect appreciable alpha particle diffusion. However, issues of finite aspect ratio, elongation and finite beta which may lead to significant stabilization of TAE modes [9] and limit the quantitative reliability of the Candy/Rothenbluth code, while the other codes are restricted to low n. Clearly better calculations of TAE instability are required to obtain definitive predictions by obtaining more parametric scans and improving the physics invoked in existing codes.

Additional loss of alpha particles from the plasma center to the periphery can be caused by fishbone instability and sawteeth oscillations. The source of the fusion alpha-particles in the plasma center is very peaked and even a small feeding of the ripple loss region from the central region can cause a significant increase of the local first wall loads.

D. The ITER Divertor Concept

The nominal alpha heating power is 300 MW with 600 MW possible near the beta limit. Neglecting radiation and other power losses, the peak power density on the target plates is approximately 140 MWm^{-2} for a major radius at the x-point of ~ 7m and a scrape-off thickness of ~ 0.02m at the target plate. Tilting the target plates can reduce this load by up to a factor of 4 resulting in about 35 MWm^{-2}. This peak load, coupled with the requirement to tolerate disruptions, is too large to allow the development of a reasonable engineering design. The reference ITER divertor concept seeks [10] to reduce the peak heat loads to the 5 MW/m^2 level, or less by the use of impurity radiation and other atomic processes in order to spread out the energy from the divertor plasma to the divertor side walls.

Radiative and charge exchange losses have been used to reduce the peak heat loads in early divertor experiments [11, 12]. More recently, radiative losses have resulted in "detachment" (a substantial reduction of the heat and particle flux) of the plasma from the divertor plate in DIII-D[13, 14], JET[15] and other divertor experiments[16]. Detachment is obtained in these experiments by increasing the recycling by intense gas puffing and/or puffing impurity gases such as Neon or Argon. The ITER divertor is designed to operate in a similar fashion without such intensive gas fueling.

The reference divertor concept (Fig. 3) utilizes baffles to contain the recycling neutral hydrogen and impurity gases in the divertor chamber and facilitate momentum exhaust. The radiated power falls on water cooled louvers parallel to the plasma. The louvers are semi-transparent to the neutrals. The louvers ensure that the recycling neutral flux is relatively uniform on the divertor plasma, and that the fast neutrals formed by charge exchange can transfer their momentum to the wall before striking other neutrals. The remaining plasma energy falls on dump targets at the bottom of the divertor chamber.

The key physics issues associated with the ITER divertor include:
1. Control of the location of the radiating region in the divertor,
2. Achievement of a high fraction of radiated power to heating power with an acceptably low impurity content in the main plasma, including retention of sputtered and recycling impurities in the divertor chamber,
3. The effect of transient events such as ELM's,
4. Confinement of neutrals in the divertor chamber and
5. Compatibility of the requirements for optimizing the performance of the plasma and optimizing the performance of the divertor.

Research on divertor physics is a key part of the ITER divertor design process. We need better theoretical understanding of the complicated physics processes in the plasma SOL, development of powerful codes able to provide the projection for the ITER divertor performance and a broad experimental program on the major tokamaks to test different divertor configurations and operational regimes. This task can be done in a close cooperation of all four ITER Home Teams and the ITER Central Team. However, the ITER is experimental machine and divertor physics study is a large part of the ITER experimental program. Therefore the design of the ITER divertor must be flexible and allow optimization of the divertor regimes during ITER operation.

E. Steady State Operation Modes

Demonstration and experimental study of steady state operational regimes in ITER is one of the ITER goals. This requires high beta operation to produce relevant fusion power, high bootstrap operational scenarios to cope with the limited auxiliary power and relatively low q<4-5 to be relevant to the power reactor.

The reversed shear scenario was examined as one of the candidates for ITER steady state operation modes. Modeling shows [17] that MHD stable regimes can be achieved by 100 MW of auxiliary power. The fusion power is 1.5 GW and bootstrap current provides 80% of the total plasma current (see Fig.4). These advanced operational modes are compatible with the ITER Poloidal Field system and divertor operation and require a current drive figure of merit g=0.25 10^{20} A/W-m^2. All three candidate methods for plasma auxiliary heating, i.e. ICRH, Neutral Beam and ECRH, are capable of producing the required current density in the plasma center. The off axis current drive is also required to control q profile. This off axis current could be provided by ICRH or ECRH but low hybrid waves are the best suited to carry out this task due to the highest figure of merit for the current drive.

It is obvious that development of a reliable steady state operational mode will require an extensive experimental program and a large part of this work will be done on ITER. The ITER device must be flexible enough to reach this goal.

III. CONCLUSIONS

The Joint Central Team in collaboration with the Home Teams is carrying out the ITER design which fulfills the objectives together with the cost limitation laid down by the four Parties. Our state of knowledge of plasma processes suffices to determine the ITER basic parameters and design specifications. However, much important physics remains to be studied in the plasma of the ITER scale. ITER will be an experimental machine which will produce not only nuclear tests, but will also yield important physics information required to ensure the successful use of the fusion energy.

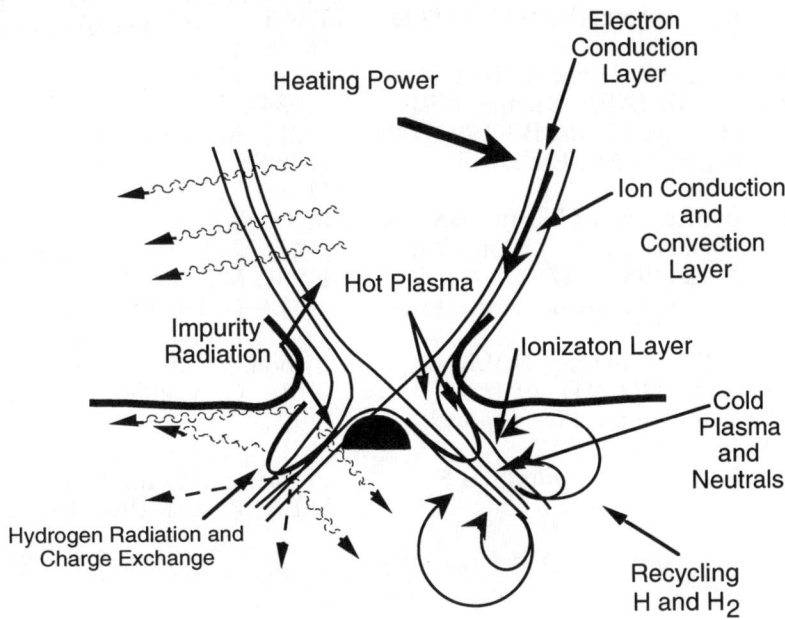

Fig. 3 ITER divertor concept.

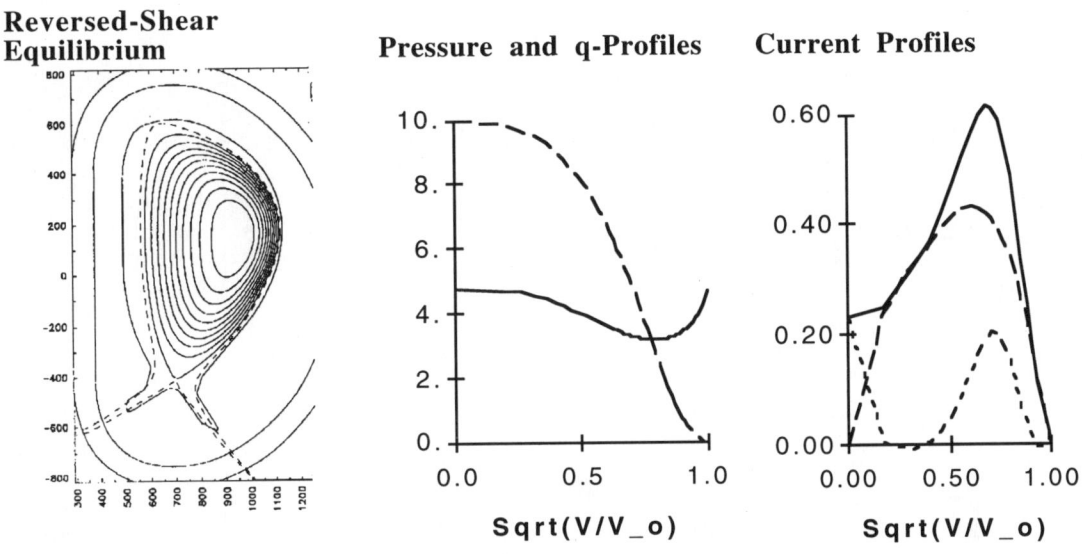

Fig.4 Plasma configuration and plasma profiles for the steady state reversed shear scenario

REFERENCES

[1]. ITER EDA Agreement and Protocol 1, ITER EDA Doc. series No.1.
[2]. ITER Conceptual Design Report, ITER documentation Series No.18, IAEA, Vienna, 1991.
[3]. P.H.Rebut et.al., Phys.Fluids, **B3**(1992)2209
[4]. J.P.Christiansen, et.al., Nuclear Fusion, **32**(1992)291
[5]. P.N.Yushmanov, Review of Plasma Physics, Vol.16 (Kadomtsev, B.B. Ed.), Consultant Bureau, New York (1987) 117.
[6]. K.Tobita, K.Tani, Y.Neyatani, et. al., Phys. Rev. Letters **69**(1992)3060.
[7]. S.V.Putvinskij, B.J.DTubbing, S.V.Konovalov, Nucl.Fusion **34**(1994) 495.
[8]. J.Candy, M.N.Rosenbluth, Plasma Phys. Controlled Fusion **35**(1993)957.
[9]. F.Zonca and L.Chen , Phys. Fluids B **5** (1993)3668.
[10]. M.L.Watkins, P.-H.Rebut, Proceedings of the 19th EPS Conf on Controlled Fusion and Plasma Physics, Insbruck, Austria, Vol. 16C, Part II(1992)731
[11]. M. G. Bell, et al., *J. Nuc. Mat.* **121**, 132 (1984).
[12]. M. Shimada, et al., *Nuc. Fus.* **22**, 643-655 (1982).
[13]. T. Petrie, et al., *J. Nuc. Mat.* **196-198**, 849-853 (1992).
[14]. T. Petrie, APS DPP Annual Meeting, St. Louis, Mo., 1993).
[15]. G. Janeschitz, et al., 20th EPS Converence on Controlled Fusion and Plasma Physics, Lisbon, Portugal, 1992), vol. 17C, Part II, pp. 559.
[16]. G. Matthews, *J. Nuc. Mat.* (**to appear**), (1994).
[17]. W.Nevins et.al., 15th Conference on Plasma Physics and Control Nuclear Fusion Research, IAEA, Vienna, 1994, IAEA–CN–60/E–P–5.

JET Physics in Support of ITER: Results and Future Work

M. Keilhacker and the JET Team*
JET Joint Undertaking, Abingdon, Oxon, OX14 3EA, United Kingdom.

The JET Programme to 1999 concentrates on issues that must be solved before a decision to construct ITER can be taken. The paper discusses three areas representative of the physics support provided: confinement studies, divertor studies and fusion performance. Due to its large size and proximity to reactor conditions, JET provides the most ITER relevant confinement data of all tokamaks. JET has an active divertor programme (Mark I, 1994/95; Mark IIA, 1996; Mark IIGB, 1997/98) aimed at developing a viable divertor concept for ITER, based on, for example, the gas target/radiative divertor concept. High performance ELM-free H- and VH-mode discharges, complemented by quasi-stationary ELMy H-modes in an ITER-like configuration will be developed further in 1994/95 and provide the basis for D-T experiments in 1996 and 1999.

1. INTRODUCTION

The approach to a fusion reactor based on magnetic confinement foresees a series of major facilities such as the Joint European Torus (JET) (start of operation in 1983), the International Thermonuclear Reactor (ITER) (~2005) and a demonstration reactor (DEMO) (~2025). JET is the European representative in the family of "Large Tokamaks" currently in operation and has always had a strong ITER and reactor relevance due to its large size, proximity to reactor plasma conditions and use of reactor relevant technologies (tritium and remote handling).

This paper focusses on the physics support that JET is providing to ITER. The JET Programme is proposed to continue to the end of 1999 and concentrates on those physics issues that must be solved before a decision on ITER construction can be taken. The main objectives of the JET Programme are therefore to develop a viable divertor concept for ITER, to improve plasma performance and to carry out D-T experiments in an ITER-like configuration (see JET Programme Schedule in Fig. 1).

JET is distinguished by its large scale (major radius~3m; plasma volume~80m^3; a plasma current capability up to 6MA), powerful heating and current drive systems (22MW of neutral beam injection; 20MW of ion cyclotron waves; 10MW of lower hybrid waves), efficient fuelling and pumping systems (a centrifuge delivering approximately forty x 2-3mm pellets a second; a divertor cryopump capable of pumping 170m^{3}s^{-1}) and a flexible divertor to control particle and energy exhaust and plasma purity. The divertor configuration will also permit a comparison between carbon fibre composite and beryllium target tiles.

The present paper concentrates on three areas which are representative of the physics support that JET provides for ITER: confinement studies, divertor studies and fusion performance.

2. CONFINEMENT STUDIES

Confinement studies on JET have recently concentrated on changes in transport in the different regimes of confinement often referred to as L-mode (low confinement), H-mode (high confinement) and VH-mode (very high confinement).

First, it is important to define the conditions under which a change in confinement occurs and, since the physics basis for such changes is not yet understood, empirical scalings have been derived from experimental results. For example, the threshold power required to enter the H-mode is available in a multi-device database to which JET has made substantial contributions [1]. Based on these data, dimensionally correct scalings have been constructed. So far, however, it has not been possible to distinguish between two of these scalings (A: $P_{thresh} \sim n_e^{0.75} B_T L^2$ and B: $P_{thresh} \sim n_e B_T L^{2.5}$) which fit the data equally well. This situation must improve since the two scalings give quite different projections to ITER (A→ $P_{thresh} \approx 100$MW; B→ $P_{thresh} \approx 200$MW).

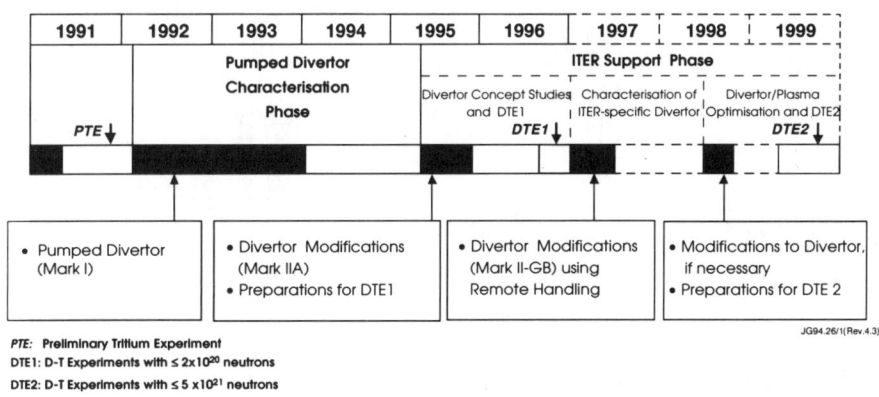

Fig. 1: The JET Programme Schedule to 1999

---

* See Ref [16]

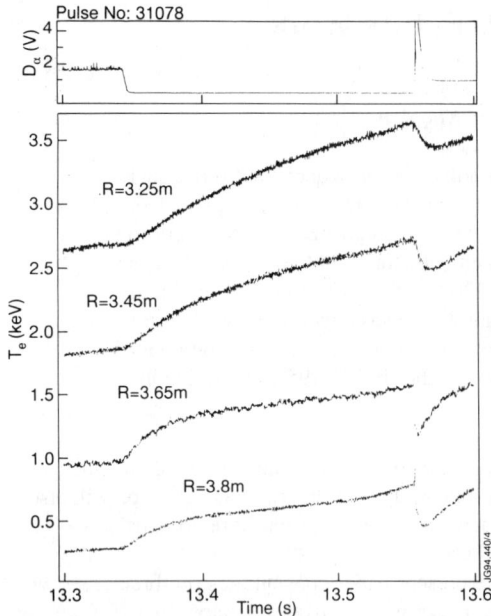

Fig. 2: Evolution of electron temperature (T_e) at the L- to H-mode transition across the plasma radius obtained using a 48-channel ECE heterodyne radiometer

Following the transition from L-mode to H-mode in JET, the evolution of the electron temperature across the whole plasma has been measured using a 48 channel ECE heterodyne radiometer [2]. These results are shown in Fig.2 and confirm the earlier result [3] of a very rapid temperature response over the entire plasma volume. The local transport model used previously to simulate L-mode plasmas [4]:

$$\chi_e = \frac{c^2}{\omega_{pe}^2}\frac{\varepsilon V_{T_e}}{qR} + \alpha_e \frac{|\nabla(nT_e)|}{eBn}aq^2, \quad \chi_i = \chi_i^{neo} + \alpha_i \frac{|\nabla(nT_e)|}{eBn}aq^2$$

is unable to simulate this rapid response, even when a "barrier" of reduced transport is imposed at the plasma edge. A much more non-linear model would be needed.

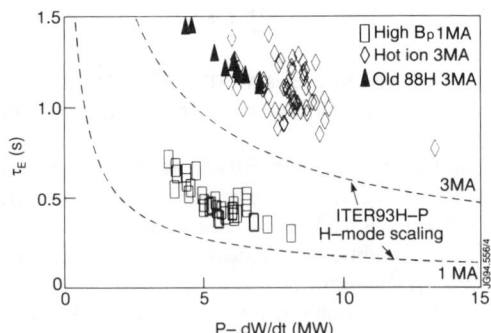

Fig. 4: JET ELM-free H-mode data exhibiting a range of enhancement factors above ITER 93H-P H-mode scaling at both 1MA and 3MA plasma currents

However, by reducing the predominantly L-mode (Bohm-like) contribution to the transport over the entire plasma (the numerical coefficients, α_e and α_i, are reduced from $\alpha_i^L = 3\alpha_e^L = 9.9 \times 10^{-4}$ in L-mode to $\alpha_i^H = 3\alpha_e^H = 5.4 \times 10^{-5}$ in H-mode [5]), the temperature response can be well-modelled (Fig.3). Furthermore, temperature profiles in both L- and H-mode phases of the discharge are then well described by this "global" model. Pedestals in temperature and density form spontaneously when the following edge boundary condition is used:

$$\chi_{e,i} n \nabla T_{e,i}\big|_{r=a} = \beta_{e,i} D T_{e,i} \nabla n\big|_{r=a} \text{ where } D = \frac{\chi_e \chi_i}{\chi_e + \chi_i}$$

The overall confinement of tokamak discharges are often described in terms of a scaling for the energy confinement time as a function of net input power, current, etc. For example, ELM-free H-mode data from JET and other tokamaks have been combined to provide a ITER93H-P H-mode scaling [6]. However, these scalings can be misleading, as shown in Fig. 4 where JET ELM-free H-mode data shows a range of enhancement factors above, for example, the ITER93H-P H-mode scaling at both 1MA and 3MA. Furthermore, there is no observable transition between a "normal" H-mode and the VH-mode with better energy confinement.

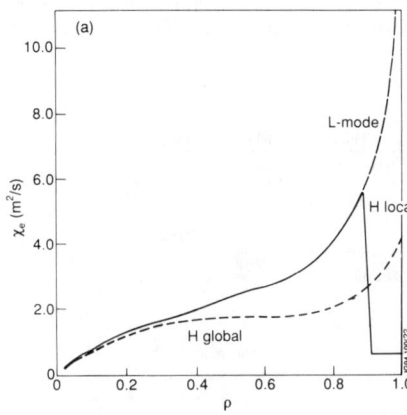
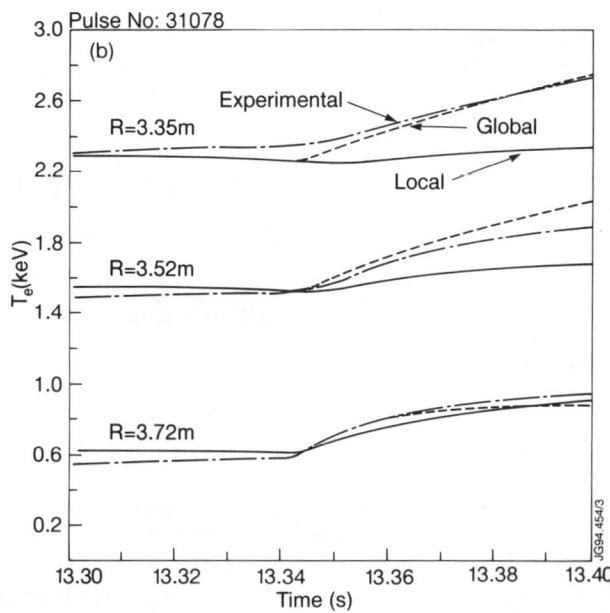

Fig. 3: Using the local and global models for χ_e shown in (a), the modelled temperature response at the L- to H-mode transition is compared with experimental results in (b)

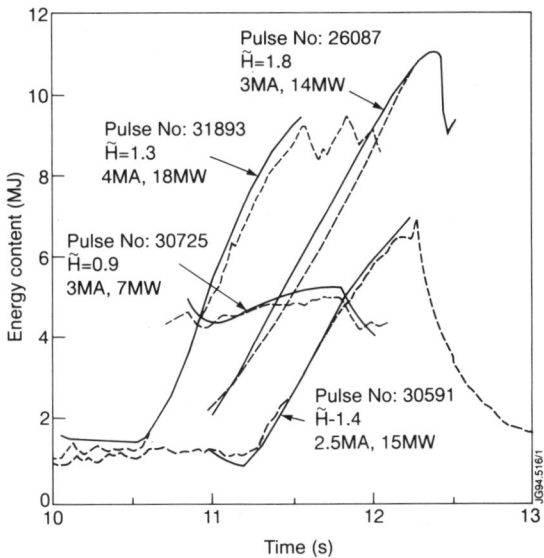

Fig. 5: Modelling (solid lines) of four H-mode discharges with widely different enhancement factors ($\tilde{H}$) over ITER93H-P scaling

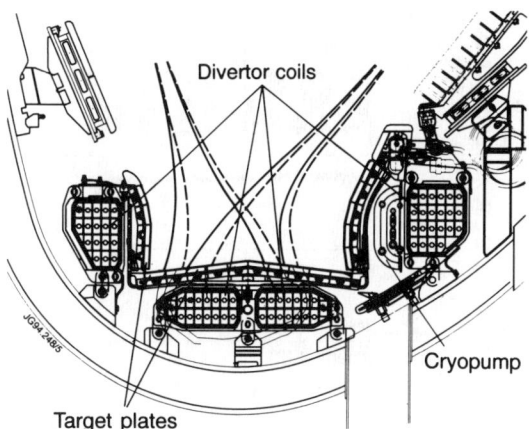

Fig. 6: A poloidal cross section through the JET Mark I pumped divertor, showing two positions of a moderately swept equilibrium

It is reasonable to ask whether the transport is indeed different between the H-mode and VH-mode. Figure 5 shows the result of modelling four quite different H-mode discharges with a range of currents up to 4MA and input powers up to 18MW. Even though the four discharges have widely different enhancement factors over the ITER93H-P H-mode scaling, the energy content is well-matched in each case [5] using the same transport model with the same numerical coefficients described earlier. It is to be concluded that these ELM-free H-modes and VH-modes in JET show no difference in transport properties; the difference is rather in impurity radiation, power deposition profiles and recycling.

It should be noted that the above transport model is predominantly Bohm-like in L-mode and might be Gyro-Bohm-like in H-mode. This distinction is strongly supported by global scaling expressions (Goldston [7] and ITER89-P [8] in L-mode; and ITER93H-P in H-mode [6]) and specific experiments on JET, TFTR and DIII-D. Since Bohm-like and Gyro-Bohm-like models extrapolate quite differently towards ITER, further experiments to clarify this issue are planned on JET.

3. DIVERTOR STUDIES

3.1. JET'S DIVERTOR PROGRAMME

The functions of a divertor, as typified by the Mark I pumped divertor presently installed in JET (Fig.6), are to exhaust power at acceptable erosion rates and heat loads on the plasma facing surfaces (divertor target plates), to control the impurity content of the main plasma by removing the source of impurities from the main chamber, by minimising the production of impurities at the target and retaining them in the divertor region, and finally to remove the He ash. JET has an active divertor programme, consisting of Mark I, Mark IIA, and Mark IIGB (see JET Programme Schedule in Fig. 1). The programme is designed to ensure the success of JET's high performance programme, as well as to contribute to the development of the ITER divertor. These JET divertors use a four-element divertor coil system which makes possible a large variety of magnetic equilibria, covering a broad range of target to X-point heights, connection lengths, and magnetic flux expansion. A large capacity cryopump is used, in connection with gas puffing and/or repetitive shallow pellet fuelling, to induce a flow in the scrape-off layer (SOL) to purge impurities and He ash from the main chamber.

The JET divertor programme will investigate a large variety of divertor concepts relevant to ITER. The Mark I divertor is relatively wide, giving it great flexibility with respect to the variety of plasma equilibria it can accommodate, at the expense of being somewhat open to the escape of recycling neutrals. It will investigate both CFC and Be target tiles.

The Mark II divertors utilize a common "base structure" into which modules of tiles and tile carriers are installed and removed, allowing for relatively quick and inexpensive changes in divertor geometry, carried out fully by remote handling. Mark IIA is a moderate slot divertor which is more closed than Mark I and has improved static power handling capability (to eliminate the need for sweeping) and improved pumping. It allows high power, high current operation on both the bottom domed targets and the vertical sideplates. Following the next JET tritium experiment (DTE1) in late-1996, Mark II will be reconfigured, using remote handling, into the Mark IIGB geometry to provide a specific test of the gas box concept currently favoured by ITER.

3.2. THE "HIGH RECYCLING" DIVERTOR

"Next Step" divertor designs carried out prior to the ITER EDA phase relied upon the "high recycling" divertor concept to fulfil the functions described above. In this concept, a very dense, cool (~10eV) plasma develops adjacent to the targets by operating with a closed divertor, at moderately high SOL densities. In a narrow "high recycling" zone close to the target, the plasma temperature drops to the point that target sputtering becomes insignificant, reducing erosion and impurity production. The heat load is also somewhat reduced because re-ionisation of the recycling neutrals is accompanied by substantial amounts of hydrogen radiation, which is emitted over 4π steradians. Up to 40% of the power flow into the divertor can be re-directed in this way. In general, however, this is not sufficient to reduce the heat flux onto the targets to $\leq 5MW/m^2$, as required. For that reason, other methods of reducing the heat load to the targets are being developed. These

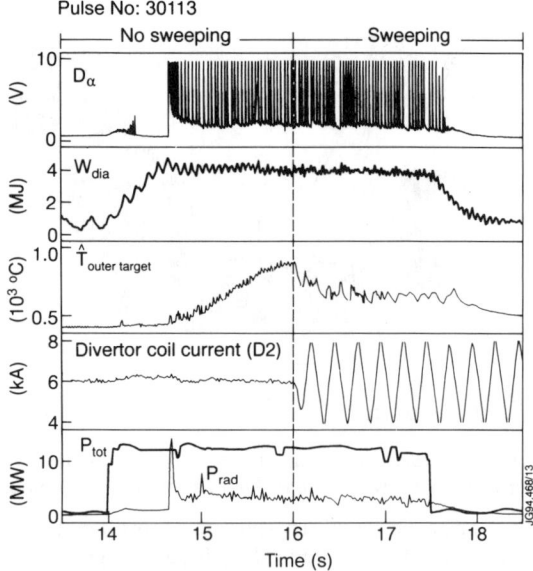

Fig. 7: Time traces for a quasi-steady H-mode pulse. Sweeping was initiated at 16 seconds, resulting in a reduction of the temperature of the outer target plates ($\hat{T}$). There was no effect on the ELMs (D_α), the stored energy (W_{dia}), or the radiated power (P_{rad})

include sweeping of the divertor plasma back and forth across the target plates, and creating a "detached" divertor plasma, where all power flowing into the divertor is dissipated by radiation and charge exchange before reaching the target plates (so-called gas target divertor).

3.3. SWEEPING

Sweeping has been employed very successfully in the JET Mark I divertor [9], where the plasma strike points are swept with an amplitude of up to 10cm at 4Hz by imposing an oscillating component on the divertor coil currents. Figure 7 shows an example where the sweeping was initiated about halfway through the neutral beam heating phase. It can be seen that this had no effect on the plasma stored energy, the ELM signal (D_α), or the radiated power, but that the temperature of the tiles, which had been rising continuously prior to initiation of the sweeping, was reduced to a steady level of about 700°C, well within the limits allowed. Using this technique, up to 140MJ of energy have been exhausted without evidence of tile

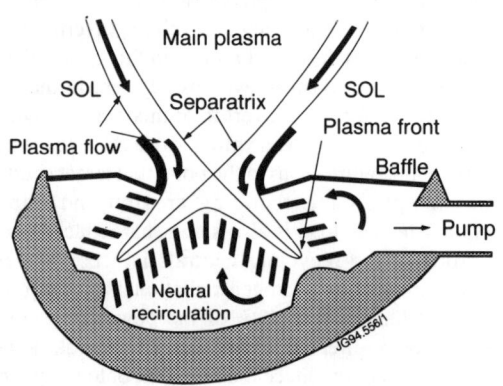

Fig. 8: A schematic of the "gas target" divertor proposed for ITER

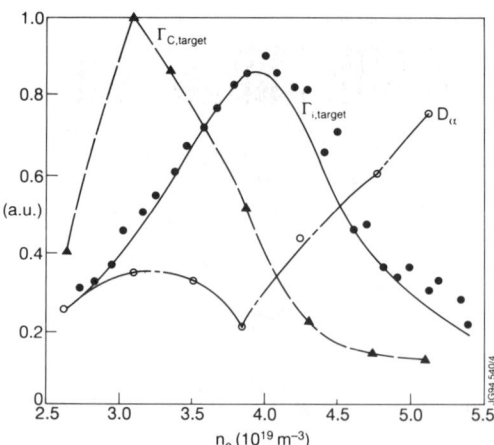

Fig. 9: Hydrogen flux to the target (Γ_i), carbon flux from the target (Γ_C), and D_α measured at the target, as a function of the main plasma density for an ohmic pulse, illustrating detachment as the density increases

overheating. This is a great improvement on the 1991-92 JET campaign, where carbon "blooms" typically limited exhaust energies to 15MJ or less.

Although sweeping has worked successfully in JET, there are doubts about its viability for ITER. The main concern is that lateral sweeping of the X-point is not compatible with the concept of a tightly "closed" divertor. This is required to reduce main chamber recycling to a tolerable level in the presence of high divertor neutral pressures associated with the gas target divertor concept. In addition, there are technical concerns about the difficulty of sweeping with superconducting coils.

3.4. THE "GAS TARGET" DIVERTOR: ITER'S PREFERRED SOLUTION

Figure 8 shows a sketch illustrating the principles of the "gas target" divertor [10]. The baffle at the top is positioned and dimensioned to reduce the neutral pressure in the main chamber several orders of magnitude below that in the divertor, which may be a few tens of millitorr. The neutrals interact with the plasma to remove energy, by radiation and charge-exchange (CX), and momentum, reducing the plasma pressure near the target ("detachment"). Calculations suggest that a recycling impurity such as neon will need to be added to the divertor plasma in small amounts ($\cong 1\%$) to radiate sufficient energy. The structure labelled "transparent wall" is intended to redirect hydrogen and impurity neutrals from the lower part of the divertor to the upper, while removing the energy of energetic CX neutrals before they re-enter the plasma. The divertor chamber is pumped to induce a SOL flow for the removal of He ash and intrinsic main chamber impurities. Several candidate geometries for detached, highly radiating divertors have been suggested, and will be tested at JET in the Mark IIA and IIGB series.

Detachment of the plasma from the target plates tends to occur in most divertor geometries, at a given exhaust power level, when the main plasma density is raised sufficiently. The role of a closed divertor is to promote detachment at moderate densities, where compatibility with H-mode operation can be demonstrated.

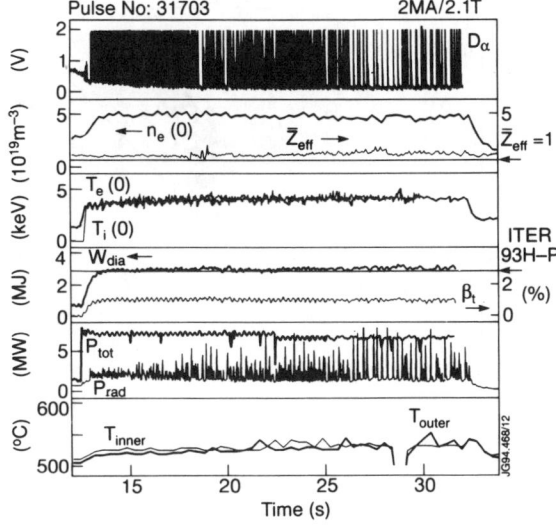

Fig.10: Time traces for an ELMy H-mode pulse. Steady conditions were maintained for 20s (more than forty energy confinement times)

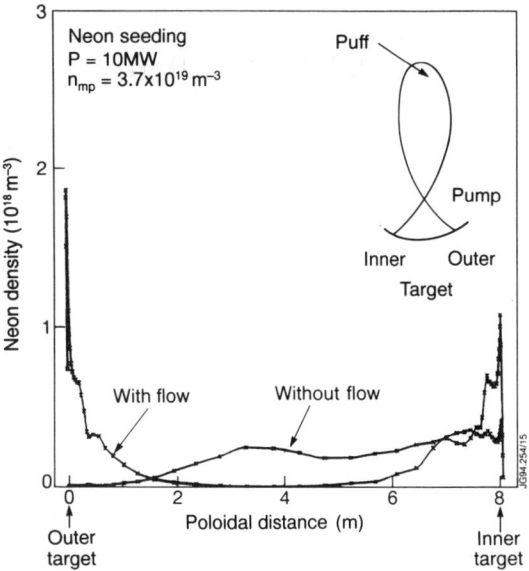

Fig.11: Neon density versus poloidal distance along the separatrix, from EDGE2D simulations of the JET Mark IIGB divertor with and without SOL flow induced by puffing in the main chamber and pumping in the divertor

In JET, detachment and the concomitant reduction of heat flux to the target plates was reported during the 1991-92 campaign [11]. With Mark I, these studies have been extended over a broader range of plasma conditions and equilibrium configurations. Figure 9 shows results from an Ohmic pulse where the density was ramped from $3 \times 10^{19} m^{-3}$ to $6 \times 10^{19} m^{-3}$ over a period of several seconds. As the density increased the particle flux density (Γ_i) to the target first increased (the high recycling regime), and then decreased dramatically as detachment set in. The carbon influx (Γ_c) from the target begins to decrease during the high recycling phase as the divertor plasma temperature falls below that required for significant sputtering, and then plunges further as the temperature and the particle flux decrease during the detached phase. Simultaneously, the recycling light increases as the number of excitations per ionisation increases rapidly at very low temperature.

The pattern of divertor radiation also changes as detachment proceeds. It varies from being localised near the target plates to being primarily in the region of the X-point, depending on the degree of detachment. The precise location of the radiating region and its control is of great importance, for example in connection with determining the H-mode power threshold, and is being studied in detail [12].

3.5. PARTICLE CONTROL AND STEADY-STATE H-MODES

In addition to handling the power exhaust, it is essential to maintain good particle control during a discharge. In the JET Mark I pumped divertor, the cryopump is observed to remove particles at a rate corresponding to several times the beam fuelling rate, facilitating density control [9]. This improved density control has made it possible to produce ELMy H-mode plasmas which are essentially steady state, as shown in Fig.10. In the pulse shown, the density, Z_{eff}, stored energy, recycling signal, plasma temperatures, β, and radiated power all remained constant, while the temperature of the divertor tiles, which are not actively cooled, increased at a very slow rate, remaining very far below their design limits. The length of the pulse corresponds to about forty times the energy confinement time, which was equal to that given by ITER93H-P H-mode scaling. The safety factor, q_{95}, was 3.3, giving an H/q value of 0.6. The demonstration of steady-state H-modes of this quality is a significant step along the road to successful operation of ITER.

3.6. SIMULATION OF THE SOL/DIVERTOR

JET has developed a robust, two dimensional edge code, EDGE2D/NIMBUS, which treats hydrogen and impurity species in the fluid approximation, and the neutrals by a coupled Monte Carlo description [13]. This code is now used routinely for interpretation of edge data as well as for design-related prediction. An example of the latter is given in Fig. 11, which shows some results from an EDGE2D simulation of the JET Mark IIGB divertor [14]. This simulation used the full multi-species version of the code with Neon as the injected impurity. The figure shows that without pumping the Neon resides primarily in the SOL outside the divertor. With divertor pumping and puffing in the main chamber, the resultant SOL flow entrains the Neon in the divertor, and a majority of the radiation shifts to the divertor region, fairly equally balanced between inner and outer legs.

4. FUSION PERFORMANCE IN JET
4.1. HIGH PERFORMANCE OPERATION

The main high performance regimes currently pursued at JET are: (a) quasi-stationary, ELMy H-modes and (b) transient, hot-ion H- and VH-modes. So far, operation in the new pumped divertor configuration has been with currents up to 4MA, but will soon be extended to its full current capability of 6MA. In general, JET H-modes in 1994 are more ELMy than previously. As a result, steady-state conditions are more readily achieved

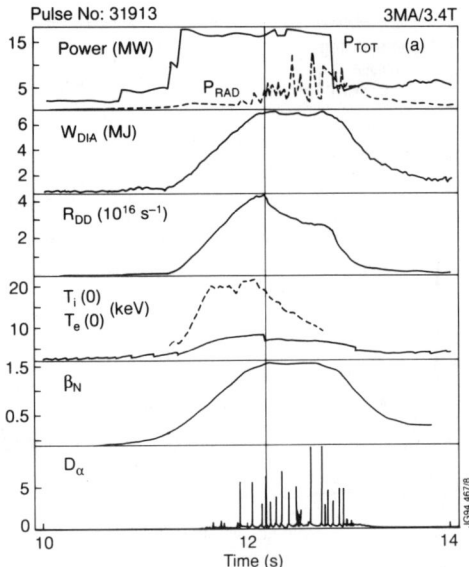
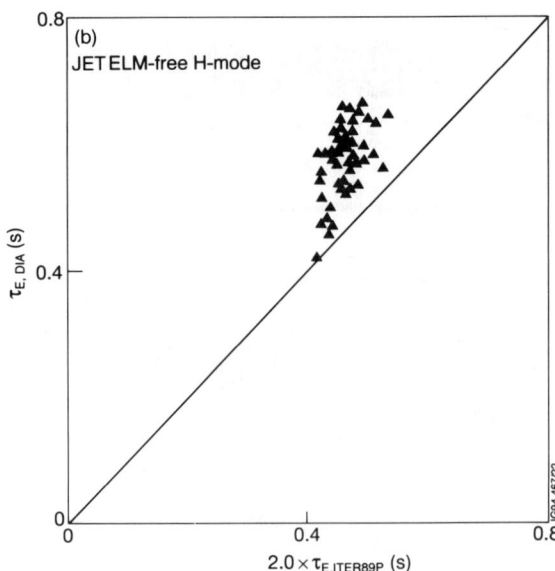

Fig.12:(a) A hot-ion H-mode (at 3MA) which reached a value for the triple fusion product of $5.6 \times 10^{20} m^{-3} keVs$; (b) the experimental energy confinement time ($\tau_{E,DIA}$) as a function of the calculated value of $2.0 \times \tau_E(ITER89\text{-}P)$ for a number of JET hot-ion ELM-free H-modes

(see Fig.10) but performance so far is limited to below that achieved previously in 1991/92. Transiently, performance as measured by the fusion triple product $n_D(0)T_i\tau_E$ has reached $5.6 \times 10^{20} m^{-3} keV$ in 3MA hot-ion H-modes, such as shown in Fig.12(a) [15]. It should be noted that, as shown in Fig.12(b), energy confinement during the ELM-free phase of similar discharges is well above normal H-mode expectations.

So far, recycling in the new JET configuration is higher than in 1991/92 and the performance is correspondingly lower. Only when the full cryopump is now used is recycling reduced to that of the earlier period of high performance. Special attention is being paid therefore to identifying and reducing the source of higher recycling which could be the cold divertor targets or larger surfaces of bare metal in the new JET configuration. In particular, the divertor targets and vacuum vessel have been baked to high temperatures (200°C and 320°C, respectively) and the cryopump used to deplete neutral reservoirs. The "plugging" of the divertor with plasma (by high flux expansion) to reduce the leakage path of neutrals into the main plasma has also been beneficial. This is seen in Pulse No:32867 (Fig.13), which features a clear ELM-free period and a distinct improvement in plasma performance when compared with Pulse No:31913 (Fig.12(a)). As an additional option the use of carbonisation, boronisation or lithium pellet injection is being assessed.

Figure 14 summarises performance achieved so far in 1994 in terms of the fusion triple product and central ion temperature [16]. Comparison is made with the best results from the 1991/92 campaign. At present, 1994 performance is about a factor of 1.5 below the best achieved in 1991/92 and approximately equivalent to that achieved in the two discharges of the Preliminary Tritium Experiment (see Section 4.3).

4.2. THE CONFINEMENT OF ENERGETIC IONS

Even when high fusion performance is achieved, α-particle heating in deuterium-tritium (D-T) plasmas will only be effective if the energetic α-particles are thermalised within the plasma volume. Given the importance of confining energetic particles, significant experiments have already been carried out to simulate in JET deuterium discharges some of the α-particle physics expected in D-T plasmas.

The thermalisation of tritons produced by the fusion process in deuterium discharges allowed the classical behaviour of individual energetic particles to be demonstrated [17], while a minority population of ions heated at the ion cyclotron resonance frequency (ICRF) to energies of a few MeV produced no deleterious collective effects for energetic particle pressures (β)

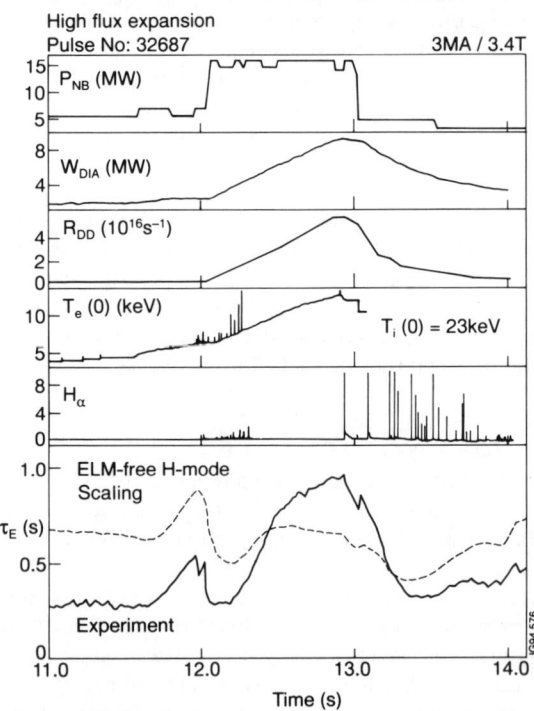

Fig.13: Characteristics of Pulse No:32687 showing the improved performance. In particular, the reaction rate (R_{DD}) reaches $6 \times 10^{16} s^{-1}$.

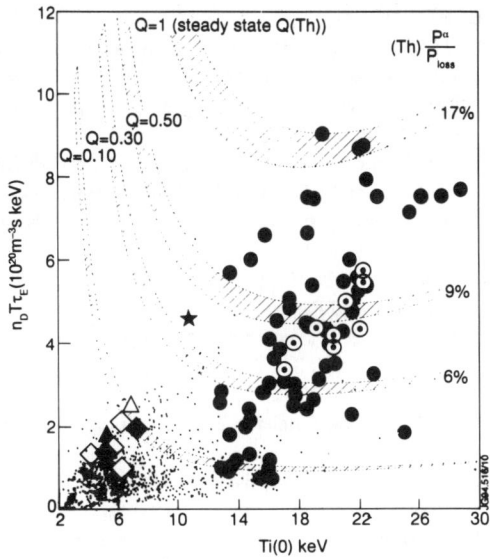

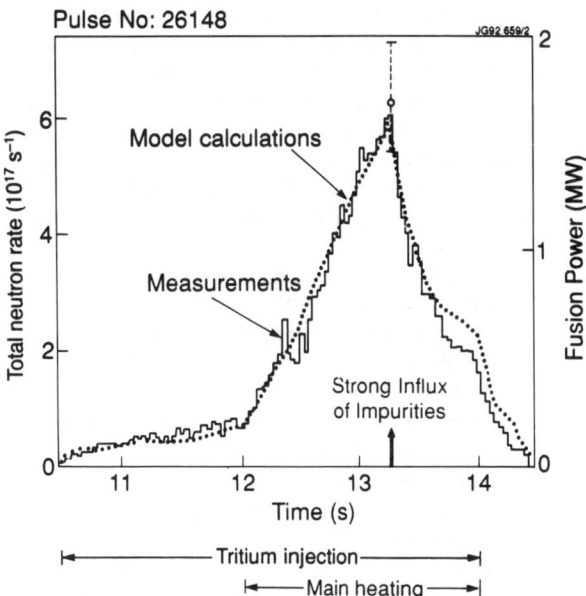

Fig.14: The fusion triple product of $n_D(0)T_i(0)\tau_E$ versus $T_i(0)$, obtained in 1991/92 (●) compared with 1994 quasi-stationary ELMy H-modes (at 2.5MA(♦), 3MA(◊), 3.5MA(▲), and 4MA(△)) and transient H- and VH-modes (at 2.5MA/3MA(o) and 4MA(★))

Fig.16: The total neutron rate and the fusion power as a function of time for the Preliminary Tritium Experiment (PTE)

up to the investigated value of $\beta_{energetic} \approx 8\%$ [18]. Furthermore, by operating JET with only 16 toroidal magnetic field (TF) coils energised (rather than the standard 32 TF coils), the effect of increased TF ripple on energetic particles was investigated [19], showing a one to two order of magnitude reduction in the flux of MeV hydrogen ions. Furthermore, ELM-free H-mode operation (thought to be necessary for ITER) was prevented when only 16 TF coils were energised. This essential new design input to ITER will be augmented further by an improved TF ripple experiment when a continuous variation between 16 and 32 TF coils will be possible early in 1995.

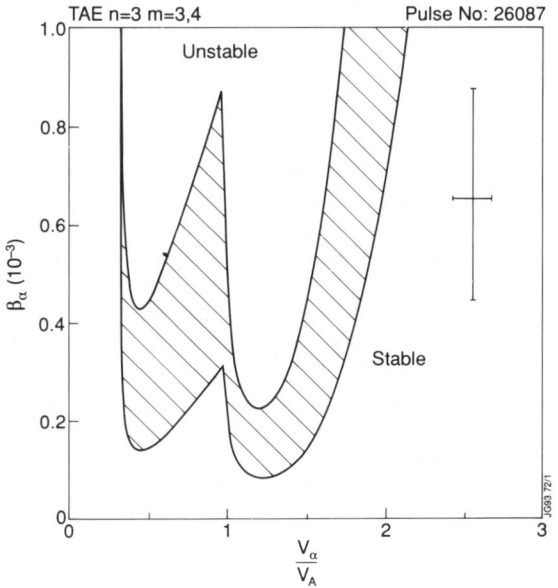

Fig.15: Stability boundaries for the most dangerous TAE modes (n=3, m=3, 4) for JET discharge Pulse No:26087 during the PTE series

Theory predicts that energetic α-particles can lead to the resonant excitation of global Alfvén Eigenmodes (TAE) which can, in turn, induce α-particle losses. These losses can lead to unacceptably high heat loads on the first wall of fusion devices. Figure 15 shows that a high performance deuterium discharge in JET when operated with equal concentrations of deuterium and tritium would be stable to the most dangerous TAE modes (n=3, m=3,4) [20]. However, TAEs could be excited spontaneously in future D-T experiments on JET if operated at a lower density, since β_α would increase, and v_α/v_A would decrease. There are two loss channels. Prompt α-particle losses scale linearly with the amplitude of the applied TAE perturbation, which, for the case above, leads to 1.5% α-particle losses. Stochastic α-particle diffusion losses on the other hand set in above a threshold field perturbation which is fairly high. For the case above, the threshold is about $\delta B_r/B \approx 1 \times 10^{-3}$ at the position of the gap for the considered case of the n=3 TAE mode. In that case, $\delta B_r/B \approx 3 \times 10^{-3}$ would lead to an additional 1% α-particle loss [21]. However, in the perhaps more realistic case that a large number of TAE modes is unstable with n in the range between 3 and 10, the stochastic regions associated with each mode may overlap leading to significantly higher losses.

In the meantime, the stability of TAEs and their interaction with energetic ions is being investigated in active excitation experiments using internal saddle coils and fast magnetic coils to excite and detect TAEs. First results have been reported [22].

4.3. DEUTERIUM-TRITIUM OPERATION

JET started its programme of deuterium-tritium (D-T) operation in 1991, when performance had reached the level which warranted the use of tritium for the first time in a laboratory plasma experiment. In this Preliminary Tritium Experiment (PTE), the activation of the machine structure had to be strictly controlled so that only two PTE discharges were planned and executed with a tritium content limited to 10% (rather than the optimum ~50%). As seen in Fig.16, each of two practically identical PTE discharges produced peak fusion power of 1.7MW averaging 1MW over a two second period [23]. The duration of the high performance phase of this and similar high performance discharges was abruptly terminated by MHD effects followed by an unacceptably high influx of carbon impurities (the so-called carbon "bloom") originating from divertor targets.

D-T operation on JET is next scheduled for the end of 1996 with the DTE1 series of experiments (see Fig. 1), which will have the physics objectives:

- to demonstrate long pulse fusion power production ($Q \approx 1$ with more than 10MW lasting for several energy confinement times) and to study α-particle effects; and
- to make important contributions to D-T physics in an ITER relevant divertor configuration (H-mode threshold, ELM and confinement behaviour, and RF heating studies).

The technical objectives are:

- to demonstrate the ability to process tritium (using the Active Gas Handling System) while supporting a reacting tokamak plasma; and
- to demonstrate a reactor-relevant remote handling operation.

The fusion performance expected during DTE1 has been assessed on the basis of high performance deuterium Pulse No. 26087 obtained during the PTE series of discharges. Assuming that the carbon "bloom" can be avoided (as with the Mark I divertor), TRANSP calculations show that a 50:50 D-T plasma would produce 14MW of fusion power and $Q \approx 1$ ($Q_{thermal} \approx 0.67$). α-particle heating would then lead to a clear increase in the central electron temperature (from 13keV to 16.5keV).

The neutron budget for DTE1 will be limited to 2×10^{20} D-T neutrons (TFTR has produced 1.2×10^{20} neutrons so far) corresponding to 0.6MJ of fusion energy. The programme would then continue with the remote handling exchange of the divertor target structure. If manual intervention became necessary, the experimental programme could still continue after a delay of at most about one year.

More extensive D-T experiments (DTE2) are then planned to follow in 1999.

5. CONCLUSIONS

Following a shutdown that lasted nearly two years, JET with its new pumped divertor configuration is again contributing to ITER and a fusion reactor. The present experimental programme is already addressing the most critical scientific issues (such as confinement and divertor physics and high performance operation) that must be solved before ITER construction can proceed. The proposed JET Programme to the end of 1999 will complement this work. This should ensure that Europe maintains its present lead in fusion research.

ACKNOWLEDGEMENTS

The author is especially grateful for the contributions of M L Watkins and G C Vlases and for the assistance of B E Keen in editing the paper.

REFERENCES

[1] ITER Confinement Data Base and Modelling Group, Proc. of the 15th International Conf. on Plasma Phys. and Controlled Fusion Res. (Seville, Spain, 1994) Paper No: IAEA-CN-60/E-P-3

[2] D V Bartlett, et al, In Proceedings of the 8th International Workshop on ECE and ECRH (Gut Ising, Germany, 1992)

[3] J G Cordey et al, Plasma Physics and Controlled Fusion **36** (1994) A267

[4] A Taroni, M Erba, E Springmann, and F Tibone, in Plasma Physics and Controlled Fusion **36** (1994) 1629

[5] V V Parail, et al, Proc. of the 15th International Conf. on Plasma Phys. and Controlled Fusion Res. (Seville, Spain, 1994) Paper No:IAEA CN-60/A-2-II-3

[6] ITER H-mode Database Working Group, Nuclear Fusion **34** (1994) 131

[7] R J Goldston, Plasma Physics on Controlled Fusion **26** (No 1A), (1984) 87

[8] P N Yushianov, et al, Nuclear Fusion **30** (1990) 1999

[9] JET Team presented by D J Campbell, Proc. of the 15th International Conf. on Plasma Phys. and Controlled Fusion Res. (Seville, Spain, 1994) Paper No:IAEA-CN-60/A-4-I-4

[10] P-H Rebut et al, Fus. Eng. and Design, **22** (1993) 7

[11] G Janeschitz et. al., Proc. 19th EPS Conference, Innsbruck **16C** (1992) II-727

[12] R Reichle et. al., Bulletin of the American Physical Society, (APS, Minneapolis, 1994)

[13] A Taroni et. al., Proc. of Joint Varenna-Lausanne International Workshop on the Theory of Fusion Plasmas, Varenna, Italy, 22-26, August 1994

[14] R. Simonini et. al., Proc. 21st EPS Conference, Montpellier, **17C** (1994) II-771

[15] JET Team presented by P J Lomas, Proc. of the 15th International Conf. on Plasma Phys. and Controlled Fusion Res. (Seville, Spain, 1994) Paper No:IAEA-CN-60/A-2-I-4

[16] JET Team presented by D Stork, Proc. of the 15th International Conf. on Plasma Phys. and Controlled Fusion Res. (Seville, Spain, 1994) Paper No:IAEA-CN-60/A-1-I-3

[17] O N Jarvis et al, Proc. 18th EPS Conference, Berlin **15C** (1991) I-21

[18] G Sadler et al, Proc. 18th EPS Conference, Berlin **15C** (1991) I-27

[19] JET Team presented by B J D Tubbing, Proc. of the 14th International Conf. on Plasma Phys. and Controlled Fusion Res. (Würzburg, Germany, 1992), Vol 1 (1993) 429

[20] W Kerner, D Borba, G T A Huysmans, F Porcelli, S Poedts, J P Goedbloed and R Betti, Plasma Physics and Controlled Fusion **36** (1994) 911

[21] W Kerner et al, Proc. of the 15th International Conf. on Plasma Phys. and Controlled Fusion Res. (Seville, Spain, 1994) Paper No:IAEA-CN-60/D-P-II-4

[22] A Fasoli et al, Proc. of the 15th International Conf. on Plasma Phys. and Controlled Fusion Res. (Seville, Spain, 1994), Post Deadline Paper

[23] JET Team, Nuclear Fusion **32**(2) (1992) 187

Deuterium-Tritium Experiments on TFTR

N.L. Bretz, H. Adler, P. Alling, C. Ancher, H. Anderson, J.W. Anderson, V. Arunasalam, G. Ascione, C.W. Barnes,[1] G. Barnes, S. Batha,[2] G. Bateman, M. Beer, M.G. Bell, R. Bell, M. Bitter, W. Blanchard, C. Brunkhorst, R. Budny, C.E. Bush,[3] R. Camp, M. Caorlin, H. Carnevale, S. Cauffman, Z. Chang,[4] C. Cheng, J. Chrzanowski, J. Collins, G. Coward, M. Cropper, D.S. Darrow, R. Daugert, J. DeLooper, W. Dorland,[5] L. Dudek, H. Duong,[6] R. Durst,[4] P.C. Efthimion, D. Ernst,[7] H. Evensen,[4] N. Fisch, R. Fisher,[6] R.J. Fonck,[4] E. Fredd, E. Fredrickson, R. Fromm, G. Fu, T. Fujita,[8] H.P. Furth, V. Garzotto, C. Gentile, J. Gilbert, J. Giola, N. Gorelenkov,[9] B. Grek, L.R. Grisham, G. Hammett, G.R. Hanson,[3] R.J. Hawryluk, W. Heidbrink,[10] H.W. Herrmann, K.W. Hill, J. Hosea, H. Hsuan, M. Hughes,[11] R. Hulse, A. Janos, D.L. Jassby, F.C. Jobes, D.W. Johnson, L.C. Johnson, M. Kalish, J. Kamperschroer, J. Kesner,[7] H. Kugel, G. Labik, N.T. Lam,[4] P.H. LaMarche, E. Lawson, B. LeBlanc, J. Levine, F.M. Levinton,[2] D. Loesser, D. Long, M.J. Loughlin,[12] J. Machuzak,[7] R. Majeski, D.K. Mansfield, E. Marmar,[7] R. Marsala, A. Martin, G. Martin, M. Mauel,[13] E. Mazzucato, M.P. McCarthy, J. McChesney,[6] B. McCormack, D.C. McCune, K.M. McGuire, G. McKee,[4] D.M. Meade, S.S. Medley, D.R. Mikkelsen, S.V. Mirnov,[9] D. Mueller, M. Murakami, J.A. Murphy, A. Nagy, G.A. Navratil,[13] R. Nazikian, R. Newman, M. Norris, T. O'Connor, M. Oldaker, J. Ongena,[14] M. Osakabe,[15] D.K. Owens, H. Park, W. Park, P. Parks,[6] S.F. Paul, G. Pearson, E. Perry, R. Persing, M. Petrov,[16] C.K. Phillips, M. Phillips,[11] S. Pitcher,[17] R. Pysher, A.L. Qualls,[3] S. Raftapoulos, S. Ramakrishnan, A. Ramsey, D.A. Rasmunsen,[3] M.H. Redi, G. Renda, G. Rewoldt, D. Roberts,[4] J. Rogers, R. Rossmassler, A.L. Roquemore, E. Ruskov,[10] S.A. Sabbaugh,[13] M. Sasao,[15] G. Schilling, J. Schivell, G.L. Schmidt, R. Scillia, S.D. Scott, I. Semenov,[9] T. Senko, S. Sesnic, R. Sissingh, C.H. Skinner, J. Snipes,[7] J. Stencel, J. Stevens, T. Stevenson, W. Stodiek, J.D. Strachan, B.C. Stratton, J. Swanson,[18] E. Synakowski, H. Takahashi, W. Tang, G. Taylor, J. Terry,[7] M.E. Thompson, W. Tighe, J.R. Timberlake, K. Tobita,[8] H.H. Towner, M. Tuszewski,[1] A. Von Halle, C. Vannoy, M. Viola, S. von Goeler, D. Voorhees, R.T. Walters, R. Wester, R. White, R. Wieland, J.B. Wilgen,[3] M. Williams, J.R. Wilson, J. Winston, K. Wright, K-L. Wong, P. Woskov,[7] G.A. Wurden,[1] M. Yamada, S. Yoshikawa, K.M. Young, M.C. Zarnstorff, V. Zavereev,[19] and S.J. Zweben

Plasma Physics Laboratory, Princeton University, Princeton, New Jersey, USA

ABSTRACT

A peak fusion power production of 9.3 ±0.7 MW has been achieved on the Tokamak Fusion Test Reactor (TFTR) in deuterium plasmas heated by co and counter injected deuterium and tritium neutral beams with a total power of 33.7 MW. The ratio of fusion power output to heating power input is 0.27. At the time of the highest neutron flux the plasma conditions are: $T_e(0) = 11.5$ keV, $T_i(0) = 44.$ keV, $n_e(0) = 8.5 \times 10^{19}$ m^{-3}, and $<Z_{eff}> = 2.2$ giving $\tau_E = 0.24$ s. These conditions are similar to those found in the highest confinement deuterium plasmas. The measured D-T neutron yield is within 7% of computer code estimates based on profile measurements and within experimental uncertainties. These plasmas have an inferred central fusion alpha fraction of 0.2% and central fusion power density of 2 MW/m^3 similar to that expected in a fusion reactor. Even though the alpha velocity exceeds the Alfven velocity throughout the time of high neutron output in most high power plasmas, MHD activity is not substantially different from that in comparable deuterium plasmas and Alfven wave activity is low. The measured loss rate of energetic alpha particles is about 3% of the total as expected from alphas which are born on unconfined orbits. Compared to pure deuterium plasmas with similar externally applied conditions, the stored energy in electrons and ions is about 25% higher indicating improvements in confinement associated with D-T plasmas and consistent with modest electron heating expected from alpha particles. ICRF heating of D-T plasmas using up to 5.5 MW has resulted in 10 keV increases in central ion and 2.5 keV increases in central electron temperatures in relatively good agreement with code predictions. In these cases heating on the magnetic axis at $2\Omega_T$ gave up to 80% of the ICRF energy to ions.

1. INTRODUCTION

Experiments with deuterium and tritium have been underway since December 1993 on the Tokamak Fusion Test Reactor (TFTR) using nearly equal amounts of each isotope.[1,2,3,4] The larger fusion cross section for D-T mixtures has resulted in a substantial increase in the neutron yield over pure deuterium and in the production of a substantial alpha population which has the potential to heat the plasma, cause instabilities, or escape. Work has focused on maximizing the performance of D-T plasmas and understanding their heating and transport properties. To date the highest peak fusion power production has been 9.3 ±0.7 MW in a deuterium plasma heated by co and counter injected deuterium and tritium neutral beams whose total power was 33.7 MW. At the time of the highest neutron flux the plasma conditions were: B_T = 5.1 T, I_p = 2.5 MA, $T_e(0)$ = 11.5 keV, $T_i(0)$ = 44. keV, $n_e(0)$ = 8.5x10^{19} m^{-3}, $<Z_{eff}>$ = 2.2, W_{tot} = 6.5 MJ, τ_E = 0.24 s, and $P_{Thermonuclear}:P_{Beam-Target}:P_{Beam-Beam} \approx$ 0.33:0.53:0.14. The inferred central fusion alpha fraction, n_α/n_i, is about 0.2 % and the central fusion power density is 2 MW/m^3 similar to that expected in a fusion reactor. Even though the alpha velocity exceeded the Alfven velocity throughout the time of high neutron output in this and most other high power D-T discharges, MHD activity has not been significantly different than in comparable deuterium plasmas. Some activity has been seen near the TAE frequency, but this appears unrelated to alpha loss. The measured direct loss rate of energetic alpha particles is about 3% of the total as expected from alphas which are born on unconfined orbits and consistent with the first orbit loss of other suprathermal particles. Compared to pure deuterium plasmas with similar externally applied conditions, the stored energy in electrons and ions is about 25% higher indicating small improvements in confinement associated with D-T and consistent with modest electron heating from alphas. ICRF heating powers of up to 5.8 MW has been used on beam heated D-T plasmas resulting in 10 keV increases in central ion and 2.5 keV increases in central electron temperatures in good agreement with code predictions. In these cases heating was done on the magnetic axis at $2\Omega_T$ with and without a degenerate minority $1\Omega 3_{He}$, and up to 80% of the ICRF energy was coupled to ions. These TFTR D-T results have extend experiments performed on JET which achieved a fusion power of 1.7 MW from ~10% tritium admixtures.[5] So far TFTR has had 180 discharges with tritium beams and 70 discharges with trace amounts of tritium introduced from the edge as a gas. Experimental work will continue through the end of 1995.

2. PLASMA PERFORMANCE AND SCALING

The type of plasma used in TFTR to produce the best D-T results are called supershots and are characterized by extensive limiter conditioning, low prebeam densities, and balanced beam injection.[6] These conditions have led to plasmas with centrally peaked densities, high ion temperatures ($T_i>>T_e$), broad T_e profiles, and confinement typically three times L-mode. The best D-T results have been obtained by using extensive Ohmic conditioning and lithium pellet injection[7] to prepare the limiters and walls, and by operating with high toroidal magnetic field and plasma current to maximize the β-limit. Fig. 1 shows the D-T fusion power versus time for several of the highest power plasmas. The 9.3 MW discharge ended in a high-β disruption with a total stored energy of 6.5 MJ corresponding to β_n (=$10^8\beta_t aB_t/I_p$) of 2.0, but many others in the range 6 to 8 MW have produced a steady output for the 0.8 s duration of the neutral beam. The fusion yield is determined by a number of neutron detector systems which measure the instantaneous flux at several sensitivities to accommodate a wide dynamic range, the total flux integrated over the pulse, the poloidal distribution of the flux, and which discriminate D-T neutrons at 14 MeV from D neutrons at 2.5 MeV. Most systems are absolutely calibrated and cross checked for consistency; the resulting accuracy is believed to be ±7%. These measurements have been used in a heating and transport code (TRANSP) to estimate the distribution between thermonuclear, beam-target, and beam-beam reactions. Typically this distribution is in the range (0.3 to 0.4):(0.5 to 0.6):(0.1 to 0.15), respectively. Computer modeling of the fusion power using TRANSP based on measurements of T_i, T_e, and n_e profiles, Z_{eff} from visible bremsstrahlung, and edge T/D ratios from $T\alpha/D\alpha$ spectroscopic measurements typically give an estimate of the D-T neutron rate which is within 25% of the measurement. One source of uncertainty is the role of edge fueling which is dominated by deuterium.

emission spectroscopy[14,] show small to no isotopic changes.[15] Mode activity at TAE frequencies, which can be 2-3 times larger in D-T plasmas, is not correlated with alpha loss. Other MHD activity is not significantly different in D-T plasmas. If the transport dependence on the average isotope mass, $<A>$, is expressed at constant local parameters, the relation $\chi_i^{tot} \sim \langle A \rangle^{-1.8 \pm 0.2}$ is inferred. This is to be contrasted with weak isotopic scaling observed in TFTR with hydrogen and deuterium in L-mode discharges. Plans are being made to do isotope scaling experiments with tritium in L-mode discharges.

In addition to overall studies of transport in plasmas with substantial tritium fractions, experiments have been carried out to examine trace tritium and He ash transport.[16]

3. RF HEATING IN D-T PLASMAS

Plasma heating in the ion cyclotron range of frequencies (ICRF) are being carried out to study RF physics of heating and current drive in D-T plasmas, to enhance the performance of D-T supershots,[17] and to study physics related to the channeling of alpha power to increase the current drive and reactivity.[18] ICRF heating experiments at 43 MHz have been carried out in supershot D-T plasmas to 5.8 MW using out of phase current straps. The location of the $2\Omega_T$ resonance was located in the vicinity of the magnetic axis. The best ICRF heating results so far (which have a 2% ^{3}He minority to reduce eigenmodes) increases $T_i(0)$ from 26 to 36 keV, $T_e(0)$ from 8 to 10.5 keV, and the total stored energy from 3.4 to 4.1 MJ. In these cases heating on the magnetic axis gave up to 80% of the ICRF energy directly to ions in relatively good agreement with code predictions. Plasmas with no ^{3}He have fast particle losses consisting primarily of ~600 keV tritium ions accelerated from the ~100 keV heating beams by ICRF waves confirming computer models of the $2\Omega_T$ heating process. In addition first orbit loss of alphas is increased by up to 50% with 5.5 MW of ICRF power.

Mode converted Ion Bernstein Waves (IBW) have been generated in D-^{3}He-^{4}He plasmas. Up to 80% of the ICRF power at 2 to 4 MW has been coupled to the electrons at the mode conversion surface. Experiments with ±90° in the phasing of the two strap TFTR antennas result in significant differences in the response of the electron temperature indicating that the directivity of the IBW can be controlled. Recent modeling has indicated that, if the mode converted IBW wave is generated off the midplane, then the resulting high poloidal mode number may be able to channel alpha energy into electrons or ions.[19] The present ICRF frequency and maximum B_T does not permit on axis mode conversion studies in D-T alone. However, IBW alpha particle effects in D-T-^{3}He are under investigation.

4. DISCUSSION

The safe and routine operation of tritium plasmas in TFTR has been accomplished during the last year, and a number of issues related to high power D-T plasma operation have been settled. It is especially reassuring that many of the predictions based on experience with deuterium plasmas have been verified while the fear that instabilities and plasma loss related to alphas or other unknown causes have not been apparent. Some of the differences between deuterium and tritium plasmas are due to classical, well understood effects. Other differences are not as well understood, but show a favorable isotopic scaling in going to D-T plasmas which is stronger than expected. Taken together these results imply renewed confidence in plans for ITER and other fusion reactor concepts and raise new questions about how to extrapolate the isotope effect.

ACKNOWLEDGEMENTS

The authors wish to express their thanks to the entire technical and engineering staff who have made these experiments possible, to the many outside collaborators for their valuable contributions, and to Drs. R. Davidson, P. Rutherford, and other staff members who have undertaken many administrative burdens to make this work possible.

This work was supported by US Department of Energy Contract No. DE-AC02-76-CH0-3073.

Table I

	D Supershot	D-T Supershot	D-T High-β_p
I_p (MA)	2.0	2.5	1.5
B_t (T)	5.0	5.1	5.1
P_{NB} (MW)	30.8	33.7	30.9
$n_T(0)/[n_D(0)+n_T(0)]$	--	0.50	0.48
$n_e(0)$ (10^{19} m^{-3})	9.6	8.5	8.0
$n_{Hyd}(0)$ (10^{19} m^{-3})	6.8	6.3	5.8
Z_{eff}	2.6	2.2	2.3
$T_e(0)$ (keV)	11.7	11.5	10.0
$T_i(0)$ (keV)	29.	44.	30.
W (MJ)	5.4	6.5	5.4
dW/dt (MW)	2.1	7.5	0
$\tau_E = W/(P_{tot}-dW/dt)$ (s)	0.19	0.24	0.21
$\tau_E^* = W/P_{tot}$ (s)	0.18	0.20	0.21
$\tau_E/\tau_E^{ITER-89P}$	2.4	2.4	3.2
P_{fusion} (MW)	0.065	9.3	6.7
P_{fusion}/P_{NBI}	0.0021	0.27	0.22
$n_{Hyd}(0)\tau_E^* T_i(0)$ (10^{20}m$^{-3}\cdot$s$\cdot$keV)	3.6	5.5	3.7
β_n	2.1	2.0	3.0
β_n^*	3.4	3.0	4.2
q^*	3.8	3.2	4.7

Comparison of the previous highest performance D supershot obtained during the 1992 experimental campaign with the highest performance D-T supershot and high-β_p shot from the present experimental campaign. $n_{Hyd} = n_H + n_D + n_T$

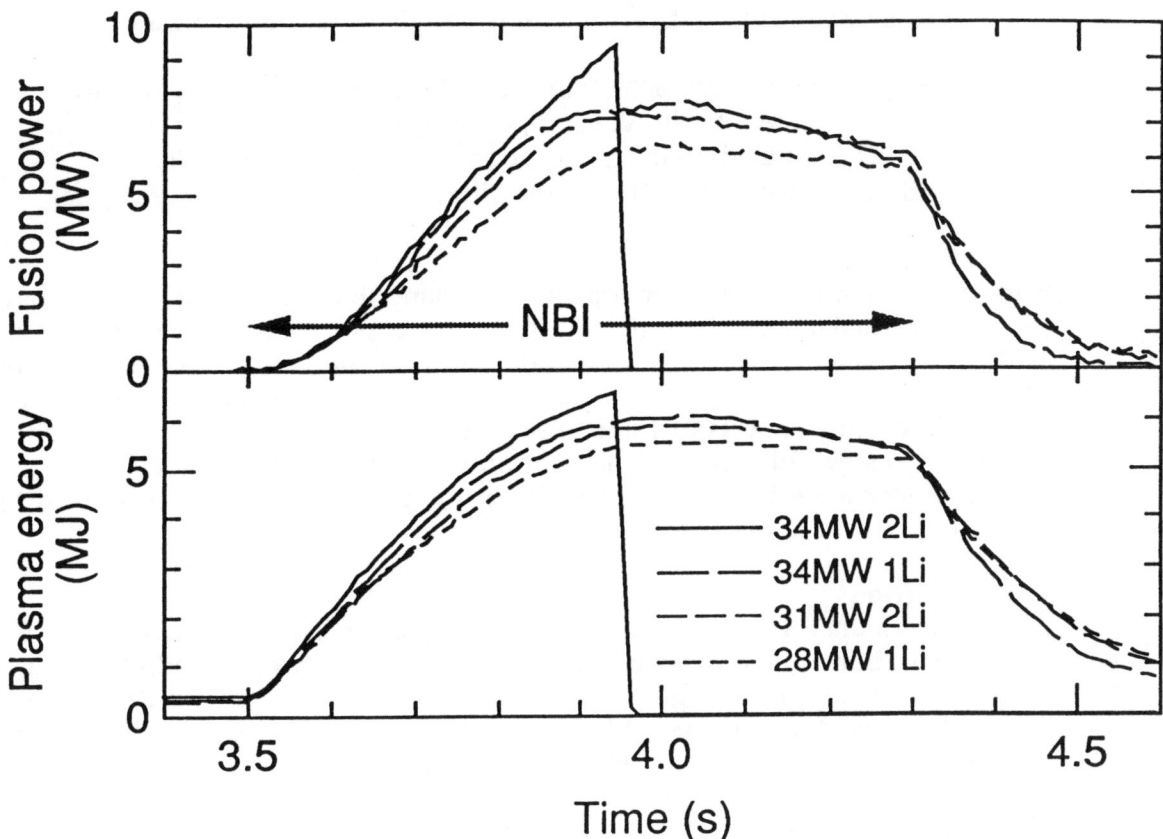

Fig. 1 Evolution of the DT fusion power and the plasma stored energy for a series of plasmas with mixed D and T NBI leading up to the shot which produced the highest instantaneous fusion power. Discharges with D-NBI only were interspersed in this sequence. One or two lithium pellets were injected into the plasma prior to NBI. Plasma parameters: major radius 2.52m, minor radius 0.87m, toroidal magnetic field 5.1T, plasma current 2.5MA.

Observations of MHD activity in JET high performance plasmas

M.F.F.Nave[*], B.Alper, P.Smeulders, S. Ali-Arshad, P. Lomas, F. Marcus,
E. Monticelli, V. Parail, R. Sartori, M.F.Stamp

JET Joint Undertaking, Abingdon, OXON, OX14 3EA, UK

[*] Associação EURATOM/IST, Instituto Superior Técnico, Lisbon, Portugal

Abstract

High performance plasmas were obtained in many JET experiments, both before and after the major shut down of 1992-93. In this paper, a summary is presented, of the MHD activity which, in both the old and the new JET configuration, appears to be most clearly correlated to the collapse of the neutron yield. The paper will discuss aspects of MHD observations in three types of performance limitation; a) slow roll-overs of the neutron rate, which appear to start with the growth of low n modes in the outer 30% of the plasma, b) sawteeth coupled to ELMs, and c) giant ELMs.

1 - Introduction

High performance H-modes have been successfully obtained in several JET experiments. Enhancement factors, $H = \tau_E/\tau_{E(ITER\ 89P)}$, between 2 and 4 have been reached in three types of regimes [1]: Pellet Enhanced Plasma (PEP) H-modes [2], high β-poloidal H-modes [3] and hot-ion H- modes [4-6]. In addition, high performance studies in high plasma current discharges, Ip=3-5 MA, have recently been initiated [7]. While the aim of these experiments is to achieve high confinement in a steady-state, the high performance phase is found to be transient, lasting <2 sec. Thus the understanding of the causes of the loss of the high confinement is at present of primarily importance.

Among a varied phenomenology, possibly indicating a variety of causes which may limit high performance, one finds in all types of regimes, cases where the loss of confinement appears to be associated to MHD activity [1]. The most clear cases where MHD activity has a limiting effect, are those where the collapse of the neutron rate occurs following a giant ELM or a large sawtooth crash. In this paper we will discuss aspects of the MHD phenomena which appear to have a limiting effect in hot-ion H-modes, which is the regime where the highest neutron yields have been achieved in JET [6].

High performance plasmas were obtained both before and after JET's major shut down of 1992-93. During the shut-down, the JET vessel was extensively modified in order to accommodate new installations, which include a pumped divertor, a torus cryopump and new target plates [8]. As a consequence of better designed target plates and better power handling, large Carbon Blooms, which were one of the limitations to high performance in the previous JET configuration [9], are no longer observed. Without the Carbon Blooms, we are able to study more clearly other underlying causes of the limitation of the high confinement.

From the point of view of MHD observations, there is a marked difference between the observation of edge activity in the old and the "new" JET machine. While the previous JET H-modes were nearly ELM-free, the new ones are more

ELMy. This has some advantages, but also some disadvantages. In the new machine configuration, steady-state H-modes in the presence of small ELMs lasting up to 20s were obtained /10/. In high performance plasmas, on the other hand, a common limitation to achieving high neutron yields are now giant ELMs, which have larger amplitudes and occur more often than what used to be observed.

A detailed analysis of MHD observations in hot-ion H modes up to the JET 1992 shut-down can be found in reference /11/. In the present paper a summary of the MHD activity which appears to be most clearly correlated to the collapse of the neutron yield, both in the old and the new JET configuration is presented. Section II starts with a summary of the observed phenomenology. The following 3 sections, describe observations of low m/n modes (section III), sawteeth combined with ELMs (section IV) and giant ELMs (section V). The ELM section will show data on the giant ELM precursors and ELM global effects which were not discussed in reference /11/.

2 - Types of termination of high performance

An overview of the JET discharge with the highest neutron yield in a DD plasma is given in fig. 1. This was obtained in a series of hot-ion H mode discharges run in preparation to the JET preliminary tritium experiments of Nov. 1991 /6/. The time evolution of the D_α emission and temperature signals shown in fig. 1, indicate that the high performance phase coincides with a period free of sawteeth and nearly free of ELMs. An early sawtooth, at 12.4s, and the early, small amplitude ELMs, have no large effect on confinement, and do not prevent the development of a high performance regime. The high confinement phase, H>1, lasts ~900ms, and it finishes with a slow roll-over of the plasma stored energy and the neutron rate. This is also the time when the hot-ion H-mode regime is lost as T_i decreases to the same level as T_e. During the phase of termination of high performance, a slow increase in D_α emission and in impurity level is observed. At the end of this phase a sawtooth crash occurs almost simultaneously with a large ELM.

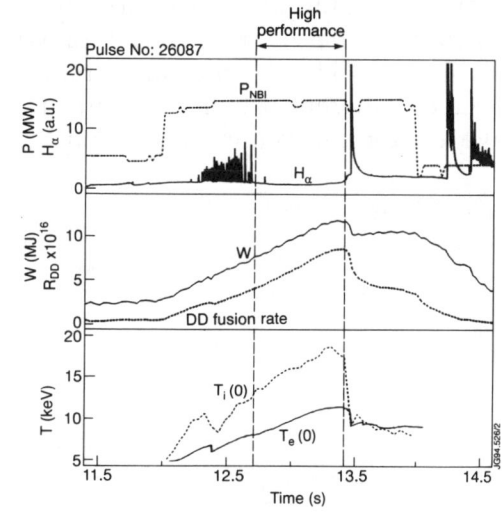

Fig. 1 - Time evolution of the best D-D discharge. The high performance phase, when H>1, is free of sawteeth and ELMs.

Fig. 2 shows the total neutron rate and the CIII line emission for 3 discharges exemplifying different types of termination phases. Case (a) will be referred to as a "slow termination", since a slow increase in CIII emission is observed coinciding with the collapse of the neutron rate. Cases (b) and (c) show fast increases in CIII emission, which in both cases coincide with a large MHD phenomenon. In case (b) the collapse of the neutron rate coincides with an ELM, while in case (c) it coincides with a sawtooth shortly followed by an ELM. In many discharges, e.g. the best DD discharge shown in fig. 1, we observe a combination of a slow roll-over, as in (a), followed by either a large ELM or a sawtooth combined with an ELM.

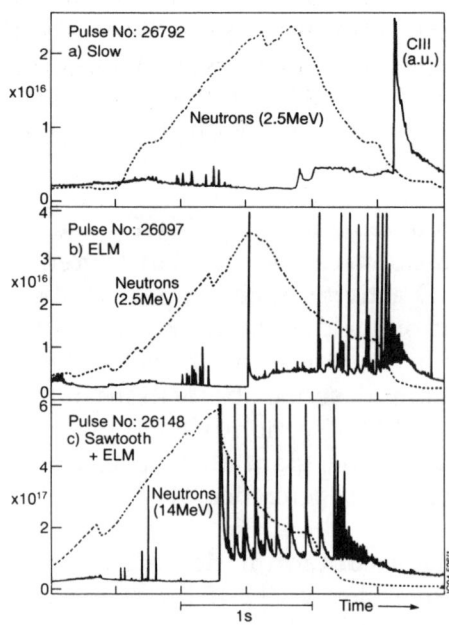

Fig. 2 - Total neutron rate (broken line) and CIII emission line(solid line) for discharges showing different types of termination of the high performance. (Pulse 26148 is a D-T discharge with 10%T.)

3 - Observations of MHD activity during the slow termination

The slow termination starts with a gradual temperature collapse. which is initially observed outside the q=1 region, typically within the outer 30% of the plasma. During this phase, a rich variety of MHD oscillations is observed in all layers of the plasma /11/. In the outer region, where the temperature profile is first affected, coherent oscillations with frequencies of 5-9 kHz are observed. These have low toroidal n numbers, in most cases n=1.

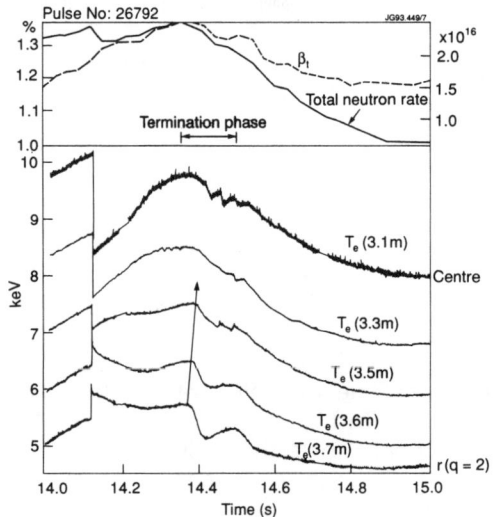

Fig. 3 - Time evolution of β_t, neutron rate and ECE signals for a slow termination, case (a) of fig. 2. The neutron rate roll-over starts with a temperature collapse, observed first around q=2.

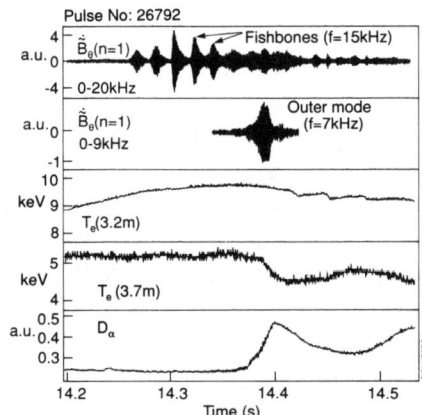

Fig. 4 - n=1 MHD activity observed during the slow termination phase: a) n=1 signal; b) n=1 signal with central modes filtered out; c) ECE traces; and e) D_α emission. The slow temperature collapse coincides with the growth of the slower, outer, n=1 burst.

Figures 3 and 4 show in detail the slow termination example, case (a), of figure 2. The top traces in figure 3 give the total neutron rate and β_t. The roll-over of the neutron rate occurs when the normalized beta toroidal $\beta_N = \beta_\phi / (I(MA)/a(m)B_\phi(T))$ =1.7 (i.e. at 60% of the Troyon limit). It coincides with a slow collapse of the temperature, which starts in a region outside q=1. Figure 4 shows that, in this case, two types of n=1 activity are observed during the slow collapse. Fishbone oscillations with frequencies around 15 kHz (close to the frequency of rotation of the plasma at q=1), and a slower mode with a frequency of 7 kHz. The temperature collapse in the outer region of the plasma, and the slow increase in D_α which is a characteristic signature of the termination phase, is well correlated with the growth of the slow n=1 burst.

The coincidence between the beginning of the slow termination phase and the appearance of n=1 (and sometimes n=2) modes has been observed in many discharges. However the estimated island sizes, in some cases are small, e.g. only a few cms for a m=2,n=1 island in the example above, thus it is difficult to understand how they could have a large effect on confinement /11/. Recent measurements with an infra-red camera show that the slow MHD burst is followed by an increase in the temperature of the target plates. An influx of impurities is observed to follow by a process not yet fully analysed.

4- Sawtooth effects at the edge

The sawtooth observed in high performance discharges have features which extend beyond the q=1 surface. Those observed at the termination of the high performance are often followed by ELM activity in a few tens of μs. A sawtooth combined with an ELM has been only observed for $\beta_N > 1.4$.

Fig. 5, shows a sawtooth collapse at an intermediate β value, $\beta_N = 1.1$, observed with a SXR camera at the top of the machine. The several traces show the SXR emission in lines of sight crossing the mid-plane at a position R. The centre of the plasma is at R~3.1 m. The signals from the core of the plasma show the oscillations of an m=1,n=1 precursor. The sawtooth collapse is clearly observed in the central signal. As the m=1,n=1 precursor grows and the central emission collapses, a fast perturbation with a ballooning character is observed to move towards the edge. This causes a transient bulging of the flux surfaces beyond q=1, on the low field side.

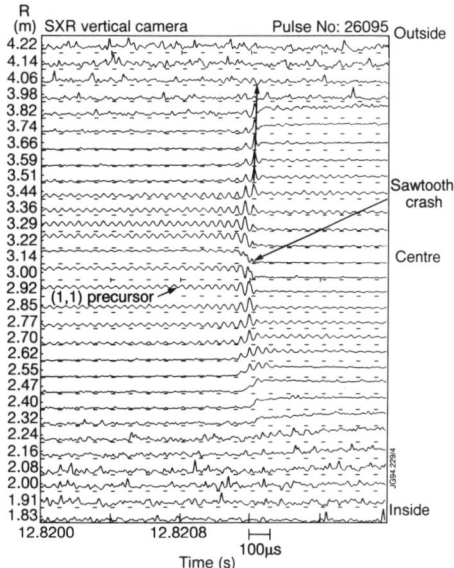

Fig. 5 - Data from the vertical SXR camera for a sawtooth at an intermediate β value, $\beta_N = 1.1$. A perturbation at the time of the sawtooth collapse extends to large values of R.

The radial extent of the MHD perturbation increases as beta increases. Fig. 6 shows the amplitude of the perturbation observed on the low field side for two sawtooth crashes observed in the best DD discharge (shown in fig. 1): a) an early sawtooth, just before the high performance sawtooth-free period, at $\beta_N \sim 0.8$; and b) the sawtooth at the termination phase, at $\beta_N \sim 2.2$. For the sawtooth at the termination phase, the perturbation has a finite amplitude at the edge, corresponding to a displacement on the midplane low-field side of ~5 cm. Modelling indicates that the finite MHD amplitude at the plasma boundary could be explained by coupling of the m=1 precursor to higher poloidal harmonics /12/. The perturbation reaches the edge in 20 μs. An ELM follows in <50 μs.

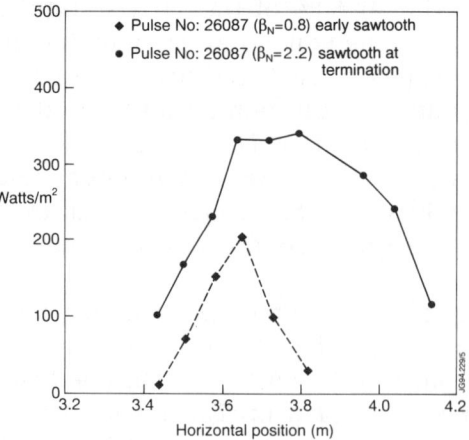

Fig.6 - An illustration of how the sawtooth perturbation grows with β. The magnitude of the perturbation in line integrated SXR emission is plotted as a function of radius on the midplane, low field side.

5 - Giant ELMs

The term giant ELM is applied at JET to bursts of H_α emission with amplitudes above 10^{15} ph/s/sr/cm². The H_α burst consists of a fast spike lasting ~1 ms and a slow decay phase which for giant ELMs lasts from a few tens to a few hundred ms. Although the name "edge localised mode" is appropriate to describe the MHD precursors to the ELM, the effect that the giant ELMs have on the plasma profiles is certainly not restricted to the edge. This is illustrated in figure 7, where one can see that the larger the ELM, the larger is the effect on the temperature and the further in this effect is observed.

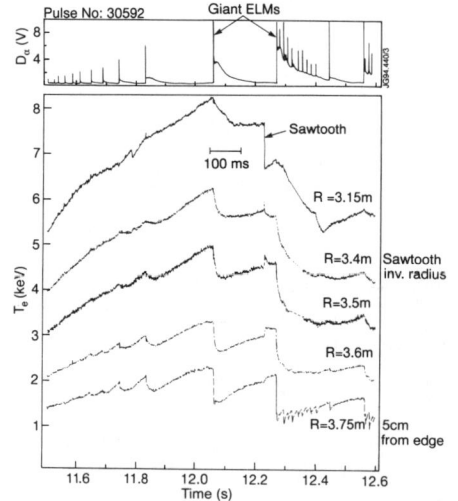 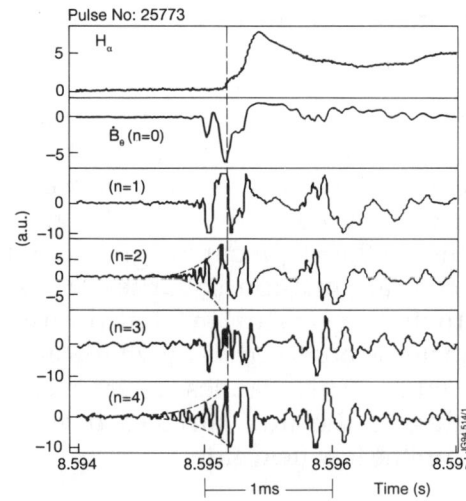

Fig. 7 - D_α signal and ECE signals at different radii showing that giant ELMs affect the whole temperature profile.

Fig. 8 - D_α and magnetic pick-up coil signals showing precursors to a giant ELM.

The giant ELMs observed at the termination of the high performance in JET are similar to DIII-D type I ELMs /13/. Like in DIII-D these ELMs are observed late in the H-mode when the edge pressure seems close to the ideal ballooning limit /7/. Since the highest fusion performance is obtained during an ELM-free period, a lot of effort has been put in trying to extend the ELM-free period. It became quite clear that the ELM frequency depends on the plasma shape, and it has been found that an effective way to increase the duration of the ELM-free period is to increase the plasma triangularity and magnetic shear /7/.

Below, two aspects of the observations of giant ELMs will be described: a) the associated MHD oscillations, and b) global effects observed with the SXR cameras.

5a) Precursors and observed mode structure

The L to H transition is normally attributed to a reduction of fluctuations in the plasma. Nevertheless, immediately following the transition, a large variety of high frequency coherent modes, f >30 kHz, and large toroidal mode numbers n=3-12, start to be observed. These high frequency modes grow throughout the H-mode until interrupted by an ELM. Some of these long-lived, high frequency, coherent modes present a sudden growth before the ELM and are know as ELM precursors /14/. In the case of giant ELMs, low frequency precursor oscillations are also observed. These are often much short lived than the high frequency precursors, being observed in the magnetic pick-up coils only within a few ms of the H_α spike. The giant ELM itself has mainly a low frequency, low n mode structure, as can be seen in fig. 8.

Fig. 8 shows the observed mode structure for a large ELM which terminates an H-mode. The H_α and magnetic pick-up coil data shown here, have been digitized together allowing a precise timing between the events observed in the two sets of data. A long lived high frequency precursor f ~80 kHz was observed both in magnetic and edge reflectometer signals (see fig. 4 in reference /14/). At 0.5 ms before the H_α spike, an n=4 mode with frequency of 15 kHz starts to grow, followed in 0.2 ms by an n=2 mode. The large amplitude components of the ELM itself are n=0 (>40 Gauss), and n=1 (>15 Gauss). The n=0 component correspond to a movement of the plasma. In the new JET divertor configuration, the plasma is in most cases observed to move inwards and upwards, consistent with a loss of β_p.

Measurements with ECE, SXR and density reflectometry diagnostic systems indicate that both high frequency and low frequency precursors are localized within 5 cm from the edge. (Although their growth may sometimes be triggered by an MHD event in the core, as in the sawtooth/ ELM combination discussed in the previous section.)

5b) Global effects

Although the MHD phenomena starts near the edge, it is quite clear from fig. 7, that a giant ELM may provoke a collapse of the temperature in the whole plasma. The JET X-ray diode arrays are particularly suitable to study global ELM effects, as they have the advantage of a high spatial and temporal resolution. Figures 9 and 10, below, show a giant ELM, associated with the termination of high performance, observed with the SXR cameras. For contrast, the figures also show a core MHD phenomena, a fishbone which occured 220ms before the ELM activity.

Figure 9 shows an H_α signal and 4 SXR signals from around the edge, measured with a SXR camera viewing the plasma from the top. In this example, the sampling rate of the SXR data is 10 kHz. The outermost SXR signal shows several spikes corresponding to ELM activity. Preceding the largest ELM there is an n=1 oscillation with a frequency of 3.4 kHz. Further in, the effect of the ELM is observed, first as a fast collapse of the SXR emission, $\Delta t\sim$ 1 ms, followed by a slower increase of the emission. In order to find out how far in radius those effects are observed a tomography inversion of the SXR /15/data was performed. The temporal resolution is too low to determine the extent of the chaotic zone in great detail, however it does allow the study of the slower peak in the emission that follows the ELM.

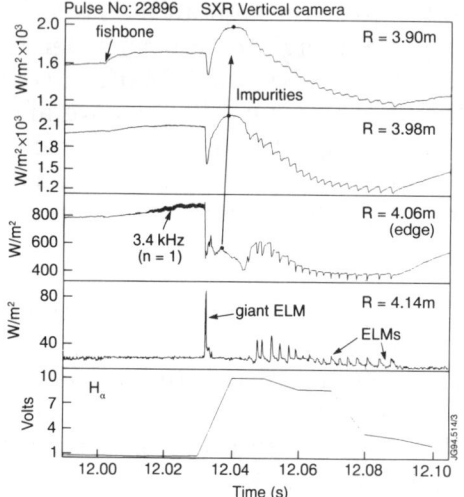
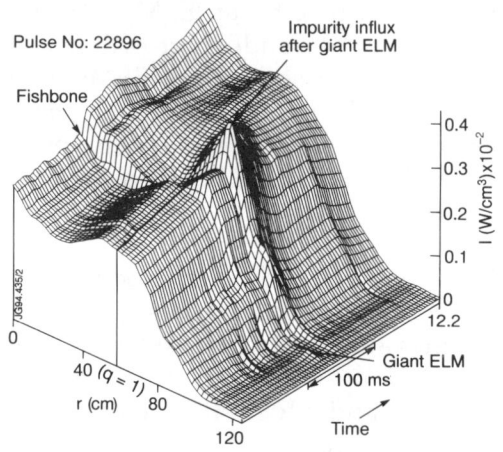

Fig. 9 - D_α and SXR line integrated signals near the edge showing ELM activity at the termination of a high performance phase.

Fig. 10 - SXR intensity profiles (2-camera tomography). 12ms after the giant ELM, a peak in SXR emission is observed at r(q=1).

Figure 10 shows the temporal evolution of the SXR intensity profiles. The SXR emisson increase which follows the giant ELM is observed up to the q=1 radius. This change in the SXR profiles after the ELM is similar to observations in laser blow-off experiments, where the diffusion of impurities injected at the edge is studied /16/. The time from the observation of the ELM at the edge to the peak at q=1 is 12ms. This is a a very short time, shorter than the time for diffusion of impurities in an H-mode, an possibly 5-10 times shorter than in an L-mode /16/. The figure also shows a fishbone, which is a phenomenon occurring around q=1. Its observed effects are similar to those of sawteeth crashes. Here, since the central SXR profile is hollow, we are able to observe an increase in the central SXR emission. Thus the fishbone, similar to a sawtooth, brings to the core of the plasma impurities which were outside the q=1 region, in an anomalously fast time scale /16/.

6 - Summary

Three main types of MHD phenomena are found associated to the termination of high performance in JET experiments: low n number modes in the outer third of the plasma, sawteeth coupled to ELMs, and giant ELMs.

During the phase of termination of the high performance both sawteeth and ELMs lose their local characteristics. The sawtooth, normally thought as a core phenomena, couples to edge activity. Typically, the sawtooth in this phase is followed by an ELM in <50 µs. On the other hand, the large ELMs, although starting with an MHD phenomena close to the edge, are seen to affect the temperature and SXR emission profiles of the whole plasma.

An increase in SXR emission after a giant ELM appears to indicate an impurity influx from the edge to the $q=1$ radius, in a time shorter than the time for impurity diffusion, which is observed with laser blow-off experiments during L-modes.

A much more subtle type of limitation to high confinement in hot-ion H-modes, known as the "slow termination", appears to start with the growth of low frequency modes in the outer 30% of the plasma. These modes have low toroidal mode numbers $n=1-4$, the most common observation being that of $n=1$ modes around the $q=2$ and $q=3$ surfaces.

References

/1/ JET Team (presented by D. Stork), Plasma Phys. and Contr. Fus. **36**, suppl. 7A (1994) A23
/2/ P. Smeulders et al, "Survey of pellet enhanced performance JET discharges", JET-P(94)07, accepted by Nuc. Fus.
/3/ C. D. Challis et al, Nuclear Fusion **33** (1993) 1097
/4/ Thompson, E, Stork, D., de Esch, H.P.L.and the JET Team, Phys. of Fluids B5 (1993) 246
/5/ Balet, B. et al. , Nuc.Fus. **33** (1993) 1345/
/6/ JET Team, Nucl. Fus. **32** (1992) 187
/7/ JET Team (presented by P.Lomas), 15th Int. Conf. on Plasma Phys. and Contr.Fus. Research, Seville,Spain (1994), IAEA-CN-60/A-2-I-4
/8/ JET Team (presented by A.Tanga), Plasma Phys. Control Fusion **36** (1994) B39
/9 D. Stork et al, Proc. 18h EPS Conf. on Contr. Fusion & Plasma Phys., (Berlin) vol. 1 (1991) 357
/10/ JET Team (presented by D.Campbell), 15th Int. Conf. on Plasma Phys. and Contr.Fus. Research, Seville, Spain (1994), IAEA-CN-60/A-4-I-4
/11/ M.F.F.Nave et al., " MHD activity in JET Hot ion H-mode discharges", JET-P(93) 79, accepted by Nuc. Fus.
/12/ B. Alper et al, Procc. 21th EPS Conf. on Contr. Fus. and Plasma Phys., (Montpellier) vol.1, p.202, 1994
/13/ Zhom et al., Proc. 19th EPS Conf. Contr. Fusion & Plasma Phys.,(Innsbruck) vol. I (1992) 243
/14/ Ali-Arshad et al., Proc. 19th EPS Conf. Contr. Fusion & Plasma Phys.,(Innsbruck) vol. I (1992) 227
/15/ R. S. Granetz and P.Smeulders, Nuc. Fus. **28** (1988)462
/16/ D. Pasini et al., Plasma Phys. and Controlled Fusion **34** (1992) 677

Acknowledgements

We would like to thank M. Garriba, S. Clement, J. Wesson, R.Gill, R. Giannella and J. Cordey for useful discussions. M.F.F.Nave is gratefull to Fundação de Amparo a Pesquisa do Estado de São Paulo and Fundação Gulbenkian, Lisbon, for the financial support for attendance to this conference.

NON-INDUCTIVE CURRENT DRIVE AND PROFILE CONTROL IN JT-60U

K. Ushigusa and the JT-60 TEAM

Japan Atomic Energy Research Institute
Naka Fusion Research Establishment
Mukoyama 801-1, Naka-machi, Naka-gun
Ibaraki-ken, Japan

Abstract

Recent studies in JT-60U to establish physics base of a steady-state tokamak reactor are presented. High fusion performance with fusion triple product $n_D(0)\tau_E T_i(0)$ =1.2×10^{21} keVsm^{-3} (equivalent $Q_{DT} = 0.6$) has been achieved transiently in the high β_p H-mode regime. By optimizing the current and pressure profiles, a quasi steady-state high β_p H-mode with $n_D(0)\tau_E T_i(0) = (0.4 - 0.5) \times 10^{21}$ keVsm^{-3} (equivalent $Q_{DT} = 0.25 - 0.36$) was maintained for up to 1.5 s. In order to maintain such high fusion performance discharges for a long time, non-inductive current drives and their capability of profile control were widely examined in the JT-60U tokamak.

High-power (~7 MW) lower hybrid current drive(LHCD) experiments were performed by using a simplified multijunction launcher. A driven current of 3.6 MA has been achieved with a current drive efficiency of $(0.3 - 0.35) \times 10^{20}$ m^{-2}A/W. Neutral beam current drive (NBCD) was also studied by using co/counter and on/off axis tangential beams (~4 MW). The maximum driven current and the current drive efficiency were 0.55 MA and 0.06×10^{20} m^{-2}A/W, respectively. Capability to control the current profile by LHCD and NBCD with a wide range of the plasma current was verified. Fully non-inductively driven discharges ($I_p \leq 1$ MA) with good energy confinement ($\tau_E/\tau_E^{ITER89P} \leq 2.5$) and a high fraction of the bootstrap current ($I_{BS}/I_p = 0.6 - 0.7$) have been obtained with NBCD and/or LHCD. These results emphasized an importance of the profile control to obtain highly integrated tokamak performance.

To establish integrated basis of physics and technology for an advanced steady-state tokamak reactor and to contribute ITER activity, we continue conceptual design of a steady-state tokamak, JT-60 Super Upgrade ($I_p = 10$ MA, $B_t = 6.25$ T, $R_p/a_p = 4.8/1.5$ m, $P_{NBCD} = 65$ MW, $P_{LHCD} = 20$ MW with a pulse length of >2000 s) as well.

INTRODUCTION

Many efforts to obtain physics base of a steady-state tokamak reactor have been devoted in the recent experiments on the JT-60U tokamak ($I_p \leq 5$ MA, $B_t \leq 4.4$ T, R_p =3.1 - 3.5 m, $a_p = 1$ m, $P_{NB} \leq 34$ MW, $P_{LH} \leq 7$ MW, $P_{IC} \leq 6$MW. A steady-state tokamak reactor such as SSTR[1] and ARIES[2] requires 1) non-inductively current driven discharge with a high bootstrap current fraction, 2) good energy confinement (H-factor = $\tau_E/\tau_E^{ITER89P} > 2$), 3) stable high power density (high normalized beta $\beta_N > 3$-3.5), 4) disruption-free discharges and 5) efficient heat and particle exhaust. Recent results in JT-60U have shown significant progress in current drive, confinement, MHD stability to demonstrate physics base of steady-state tokamak reactor. It should be emphasized that the active profile control is very important to realize an integrated non-inductively driven steady-state tokamak performance which satisfies mentioned conditions above.

NON-INDUCTIVE CURRENT DRIVE

1) Lower Hybrid Current Drive (LHCD)

A simplified multijunction launcher[3] has been installed in addition to the previous 24x4 multijunction launcher. A new launcher consists of 4x4 multijunction modules. In each module, the RF power is divided into toroidally 12 sub-waveguides at a junction point through an oversized waveguides. The application of the oversized waveguides to the multijunction module simplifies significantly the structure of the LHCD launcher

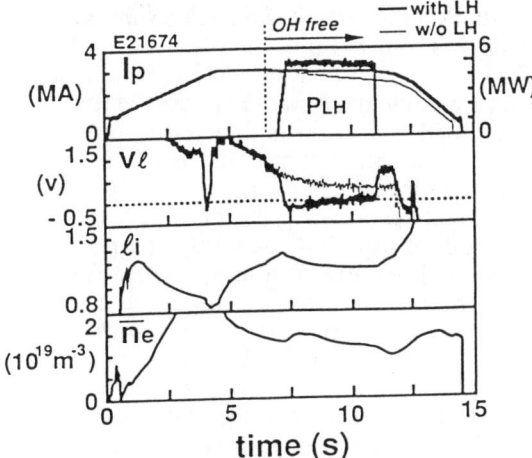

Fig. 1 Typical example of 3MA current sustainment by LHCD with keeping OH coil current constant.

Fig. 2 The current drive efficiency against the driven current by LHCD.

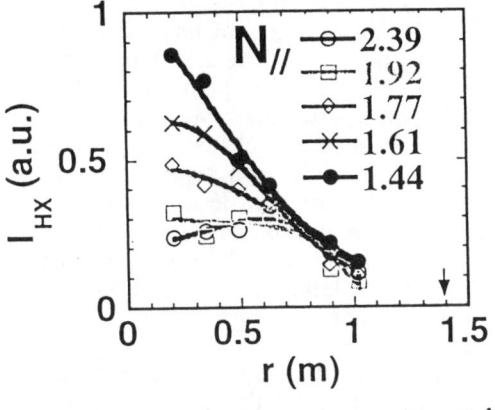
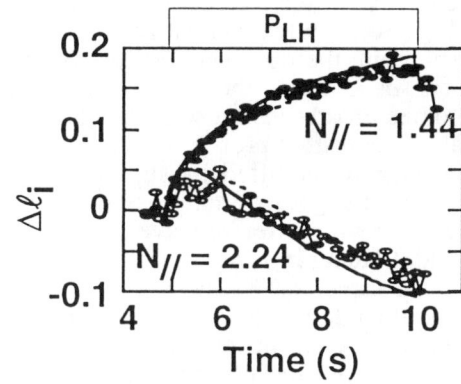

Fig. 3 a); Hard x-ray profiles for various $N_{\parallel}^{peak}$ where I_p = 1.2MA, B_t = 4T, $n_e \sim 10^{19} m^{-3}$. b); Time evolution of the internal inductance with $N_{\parallel}^{peak}$ = 1.44 and 2.24. Solid and dotted lines are the calculated results with the measured temperature profile and assuming that the driven current profile is the same as the hard x-ray profile with Zeff = 5 and 3, respectively.

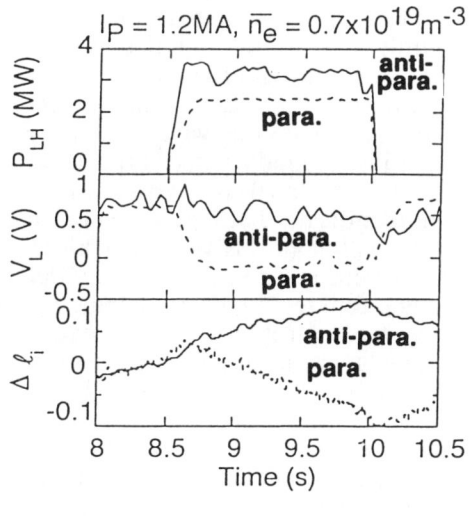
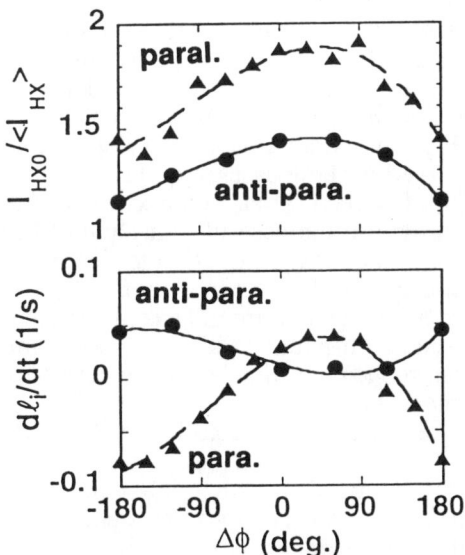

Fig. 4 a); Time evolution of parallel and anti parallel current drive by LH waves
b) Dependence of hard x-ray peaking factor and $d l_i/dt$ on wave phasing

which requires a large number of waveguides for exciting a narrow wave spectrum with a good directivity. Up to now, 7MW of LH power was injected into the plasma and the driven current of 3.6MA was achieved by LHCD alone. Figure 1 shows 3MA current sustained discharge by LHCD where the OH primary coil current was kept constant at t = 6.5 - 12 s and the LH power of ~4.2 MW with $N_{\parallel}^{peak}$ = 1.7 was injected into n_e ~ $1.2 \times 10^{19} m^{-3}$. Plasma current of 3MA is almost sustained with a current drive efficiency of ~ $2.5 \times 10^{19} m^{-2} A/W$. The current drive efficiency is plotted against the driven current in Fig. 2. The current drive efficiency is improved with increasing the plasma current and consistent with the previously reported empirical scaling $\eta_{CD} \propto <T_e(keV)>/(5+Z_{eff})$[2] where $<T_e>$ is the volume average electron temperature. The change in the plasma internal inductance and the hard x-ray profile suggest that the driven current profile becomes broader with increasing the plasma current with the same launched spectrum. Although the reason of outward shift of the deposition profile at high current discharges is not fully understand yet, there may be some dependence of deposition profile on the safety factor which is suggested by a ray tracing analysis.

Capability to control the current profile by LHCD is widely examined by using two multijunction launchers. Figure 3a shows the radial profile of hard x-ray for various $N_{\parallel}^{peak}$ which is excited by 24x4 multijunction launcher. This figure clearly demonstrates that the LH deposition profile can be controlled by the launching $N_{\parallel}^{peak}$. The change in the internal inductance during LHCD is shown in Fig. 3b. Several lines in Fig. 3b show calculated results by assuming that the driven current has the same profile with the observed hard x-ray profile. It was also found that the LH deposition profile depends on the width of the launched wave spectrum; the deposition profile becomes broad when a wide launched spectrum is injected with the same $N_{\parallel}^{peak}$. A peaked deposition profile, which is produced by an injection of low $N_{\parallel}^{peak}$ waves, has been largely modified by superimposing low power wave with high $N_{\parallel}$ component. This result suggests another possibility to control the deposition profile by two wave injections with different wave spectra. The dependence of the deposition profile on the poloidal launching angle was investigated by using two launchers. The wave spectra with the same $N_{\parallel}^{peak}$ and the width were injected from two launchers located on the poloidal angle of 45° and 0°. A peaked deposition profile and increase in the internal inductance were observed in the case of launching from the mid-plane, while the injection from the poloidal angle of 45° shows a broad hard x-ray profile and the decrease in the internal inductance. A ray tracing analysis indicates that a large $N_{\parallel}$ up-shift is expected in the injection from the poloidal angle of 45°. This result indicates that the launcher poloidal location should be optimized to control the deposition profile, and suggests the toroidal effect on the change of $N_{\parallel}$ during the wave propagation.

Current profile control by driving RF current opposite direction to the original OH current (anti-parallel current drive) has been examined. Figure 4a compares parallel and anti-parallel current drive with the same $N_{\parallel}^{peak}$. The internal inductance decreases during LHCD in the case of parallel current drive, while increase in the inductance is observed in the anti-parallel CD. The peaking factor of hard x-ray profile and the dl_i/dt are plotted against the launcher phase of 48x4 launcher for parallel and anti-parallel CD in Fig. 4b. Although a similar dependence of hard x-ray profile on the wave phasing is observed for both cases, a quite different behavior in the internal inductance is found. In spite of peaked (broad) hard x-ray profile is produced, a broad (peaked) current profile is formed in the case of anti-parallel CD. These results indicate that LH waves drive current in opposite direction to the original OH current. Although it is very difficult to produce a peaked current profile by LHCD in high current regime, the anti-parallel current drive can formed peaked current profile in high current regime by reducing edge current. Anti-parallel CD may play an important role on the control of H-mode because it can reduce the edge bootstrap current produced by an edge pressure pedestal after the H-mode transition.

2) Neutral Beam Current Drive (NBCD)

JT-60U has four tangential beams with a beam energy < 100keV for NBCD experiments. Two tangential beams are directed along the plasma current (co beams; 4MW), while other two are against Ip (counter beams; 4MW). Typical example of NBCD experiments is shown in Fig. 5 where the time evolution of the loop voltage is

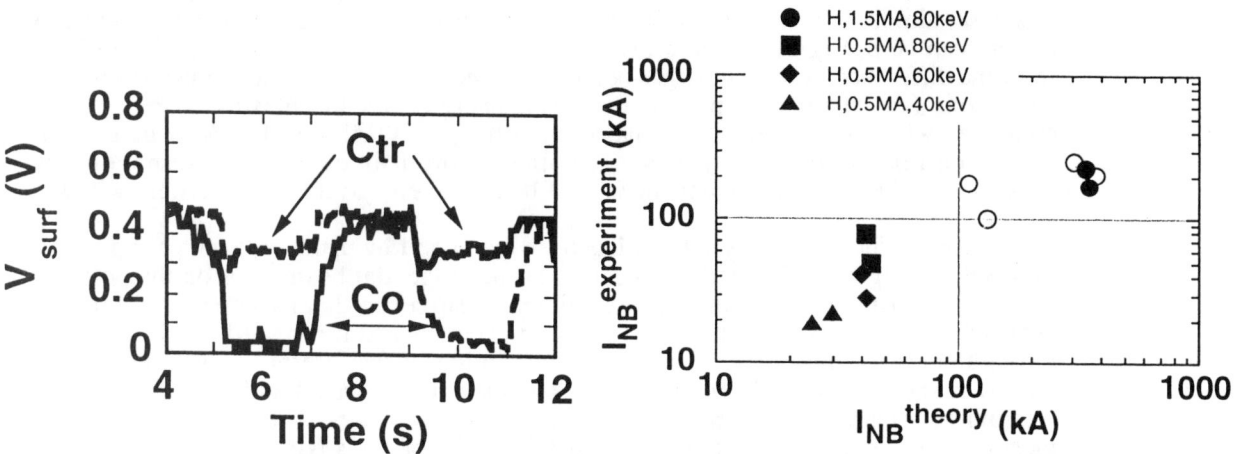

Fig. 5 a) Time evolution of the loop voltage during co- and counter tangential beam injections
b) Comparison of measured beam driven current and the estimated current from ACCOME code for various conditions.

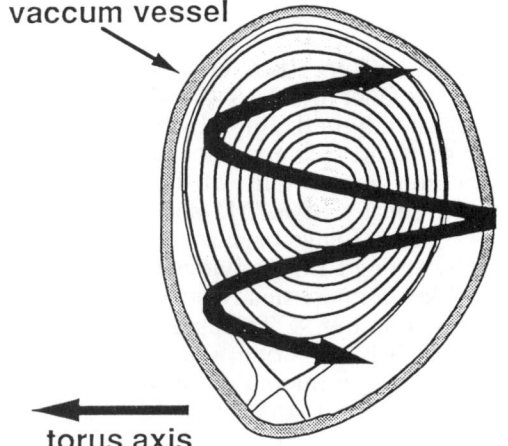

Fig. 6 Equilibrium configuration for NBCD profile control.

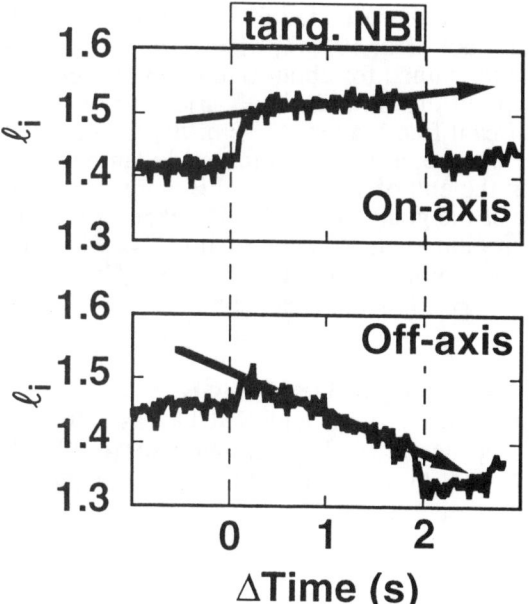

Fig. 7a) Time evolution of the internal inductance for on axis and off axis co-beam injections. P_{CD} = 4MW, I_p = 1MA, B_t = 3T, n_e = $0.6 \times 10^{19} m^{-3}$.
b) Profile of beam driven current estimated by ACCOME code.

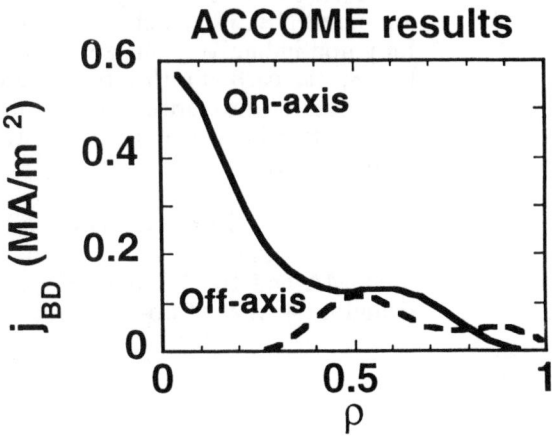

plotted for both co and counter beam injections at I_p = 0.6MA, n_e = 0.3x10^{19}m^{-3}. In this case, almost 90% of the plasma current is driven by NBCD alone. The bootstrap current can be negligible in these low β_p discharges. Beam driven current is estimated from the difference of loop voltages between co and counter beam injections for various conditions (the beam ion species, the beam energy, and the plasma current), and compared with the theoretical estimation from the ACCOME code[5]. Measured beam driven current agrees fairly well with the theoretical estimation; beam energy dependence, the temperature dependence of beam driven current are consistent with the theory.

Current profile control by NBCD has been examined by using co/counter and on/off axis NBCD. Figure 6 shows the projection of the tangential beam lines on the poloidal plane. The plasma position in these experiments was set so that two upward tangential beams (co and counter beams) view the plasma center and downward two beams pass through outer regime of the plasma. Time evolution of the internal inductance during NBCD is shown in Fig. 7a where on axis co-beam (upper trace) and off-axis co-beam (bottom trace) are injected and I_p = 0.6MA, n_e = 0.6x10^{19}m^{-3}. The internal inductance increases slowly in the case of on-axis NBCD, while off-axis NBCD decreases l_i. This result indicates that off-axis NBCD produces broad current profile. Figure 7b shows profiles of beam driven current for on- and off-axis NBCD by the ACCOME code. This result agrees qualitatively with the temporal evolution of l_i. Toroidal rotation profile has been also controlled by tangential beams in JT-60U. By combining of on-axis/co-beam and off-axis/counter-beam injections, a strongly peaked profile of the plasma rotation velocity has been formed. This can be used for studying the velocity shear effects on the confinement.

QUASI STEADY-STATE HIGH FUSION PERFORMANCE

High fusion performance with fusion triple product $n_D(0)\tau_E T_i(0) = 1.2 \times 10^{21}$ keVsm^{-3} (equivalent Q_{DT} = 0.6) has been achieved transiently in the high β_p H-mode regime[6]. A high β_p H-mode is characterized by two transport barriers at the central core region (q < 3) and the plasma edge. The central ion temperature of 41keV, the neutron production rate of 5.6×10^{16} s^{-1}, the stored energy of 8.7MJ and the confinement improvement factor (H-factor = $\tau_E/\tau_E^{ITER89P}$) of 3.7 have been achieved by this mode. This good confinement phase is terminated by strong ELM activities or the βp collapse. By optimizing the current and pressure profiles, a quasi steady-state high β_p H-mode with ELM activities has been obtained as shown in Fig. 8. To suppress the β_p collapse (low (m,n) pressure driven instabilities) in high β_p H-mode, the magnetic axis was shifted to produce broad beam deposition profile. The internal inductance of the target plasma was carefully adjusted to make an appropriate frequency of ELM activities[7]. During the injection of ~30MW of NB power, the stored energy of >7MJ, the central ion temperature of >30keV and the neutron production rate of > 3×10^{16}s^{-1} are maintained for about 1.5s. The confinement improvement factor is 2.2 with dW/dt = 0. A peaked density profile with $n_e(0)/<n_e>$~2.3 is also sustained during ELMy phase. It should be emphasized that fusion triple product of $n_D(0)\tau_E T_i(0) = (0.4 - 0.5) \times 10^{21}$ keVsm^{-3} (equivalent Q_{DT} = 0.25 - 0.36) is maintained for about five times longer than the energy confinement time. From a viewpoint of non-inductive current drive, these discharges are very attractive because significant fraction (35-45%) of the total current is driven by the bootstrap current. With an intense core current drive like an negative ion based NBCD, which will be installed in the end of 1995, these discharges may be fully non-inductively sustained. This is one of the goals of the JT-60U experiments. In Fig. 8, the carbon influx from the divertor region starts to increase at t=7.4s when the temperature of divertor plates at the outer strike point exceeds 800 degree. Control of the divertor heat load is still critical issue to sustain quasi steady-state high fusion performance.

INTEGRATED STEADY-STATE PERFORMANCE

Full current drive by combination of bootstrap current and NBCD was achieved at I_p=1 MA and B_t=3 T (q_{eff}=7.1, q_{95}=5.2) in the ELMy H-mode with P_{NB}~20 MW including 4 MW of co-tangential NB for current drive. Figure 9 shows a typical time traces of plasma parameters of 1MA full CD discharge with including the bootstrap current, beam driven current and surface one turn voltage estimated with the 1.5D time dependent analysis code TOPICS. The fully noninductive current drive condition is sustained for ~0.7 s where the bootstrap current and the beam driven current are 0.74MA and 0.37MA (this corresponds to η_{CD}(NBCD) ~ 0.7×10^{19} m^{-2}A/W), respectively. Discharge shows highly-integrated performance:β_p=2.6, the normalized beta β_N=2.9, H-factor=2.5 and $n_D(0)\tau_E T_i(0)$=0.8 $\times 10^{20}$ m^{-3}skeV. The discharge region (q_{95}, β_N, β_p, H-factor, bootstrap fraction, etc.) obtained in this experiment is almost the same as the design region of SSTR. In this case, the safety factor profile was almost unchanged during the NB pulse because of a high $<T_e>$ of ~3 keV and no MHD mode except ELMs was observed. By keeping OH coil current constant, sustainment of the discharge with NBCD and the bootstrap current was examined to investigate stability of performance. In Fig. 10, the OH coil current is kept constant from t = 6s and 4MW of co-beam power is applied in addition to 19MW of perpendicular NB power. The plasma current increases rapidly after the onset of beam power because a rapid increase in vertical filed, which is due to β_p increase, supplies the volt-second. In the case of solid line in Fig.10, MHD oscillation with m/n = 3/2 mode appears at t > 8s. This mode degrades the core confinement, and consequently reduces the driven current. When the core confinement is improved as shown in the dotted line of Fig.10 (t ~ 7s), the plasma reaches to β_p-limit and β_p-collapse occurs (t ~ 7.4s). Driven current decreases after the collapse. These two examples show that careful control of the current profile and pressure profile is important to obtain steady-state non-inductive discharges with high bootstrap current fraction and without OH coil.

To study the effects of the current profile evolution due to the bootstrap current, fully noninductive current drive plasmas were also produced at Ip=0.5 MA/ B_t=1.5 T in which low $<T_e>$ ~1 keV allows significant current diffusion. Integrated performance with β_N=2.5-3.1, β_p=2.5-3.0 and H-factor=1.8-2.2 was sustained quasi-steadily for ~1 s (t=6.5-7.5 s) which is much longer than τ_E (~0.14 s) under the full noninductive current drive condition (bootstrap current 60% and NB driven current 48%, q_{eff} =8.6). We injected co-tangential off-axis NB (4 MW) for current drive and 8 MW of near-perpendicular NB. The orbit loss of NB power evaluated by the orbit following Monte Carlo code is 35% for the perpendicular NB and 10% for the tangential NB. The small value of the bootstrap fraction of 60% for β_p=3 is caused by the high fraction (57%) of beam component in the stored energy. Time evolution of l_i suggests the current profile nearly reaches a steady state [8]. The estimated safety factor profile has two resonant surfaces with q=2.5 consistently with appearance of the 5/2 mode. In this type of ELMy discharges, the confinement degrades by the appearance of the low (m,n) pressure driven modes. Therefore a more flexible method of current profile control is required to sustain a high bootstrap fraction longer than the current diffusion time.

Similar studies have been performed by using LHCD and the bootstrap current. Figure 11 shows an example of combined LHCD and NB heating with perpendicular beams (no beam current) at Ip = 0.7MA. In LHCD phase, the surface loop voltage becomes negative and all current is driven by LHCD. At the first 0.6s of NB pulse, the surface voltage shows positive value due to the increase in β_p but it becomes again negative at t > 7.6s. Bootstrap current and the LH driven current in combined phase are estimated to ~0.42MA and 0.4MA, respectively. The confinement enhancement factor is ~1.5 without a correction of ripple loss of beam ions (H-factor ~2 with the correction). By changing the launched wave spectrum and the density, the internal inductance just before NB pulse can be controlled. It was found that the energy confinement is improved by increasing the target l_i by LHCD[9]. This result emphasizes again an importance of the profile control in high bootstrap current discharges.

FUTURE PROGRAM

In the end of 1995, 10MW of negative ion-based neutral beam (N-NBCD) system with a beam energy of 0.5MeV will be installed in JT-60U. This system may play an important role in the integrated stead-state high fusion performance as an intense core current driver. Preliminary estimation by ACCOME code indicates that N-NBCD

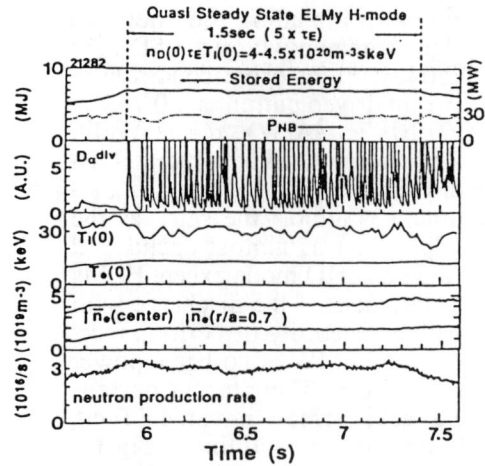

Fig. 8 Typical waveformes of a quasi steady state ELMy H-mode. $P_{NBCD} \sim 30MW$, $I_p = 2.1MA$, $B_t = 4.4T$, $n_e = 4 \times 10^{19} m^{-3}$, H-factor = 2.2, $\beta_p = 1.4$, $I_{BS}/I_p \sim 0.4$.

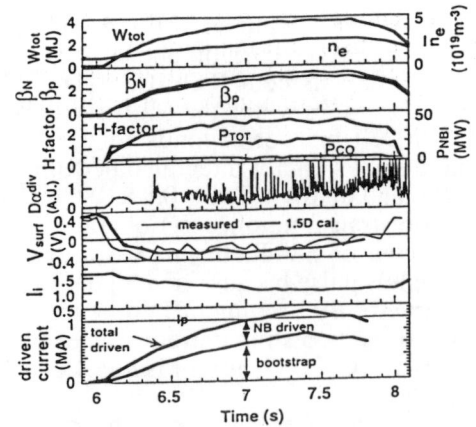

Fig. 9 Time evolution of fully current driven discharge with NBCD ($I_p = 1MA$, $B_t = 3T$, $q_{eff} = 7.1$, $q_{95} = 5.2$, $I_{BS}/I_p \sim 0.74$, $I_{BD}/I_p \sim 0.37$, $\beta_p = 2.6$, $\beta_N = 2.9$, H-factor = 2.5)

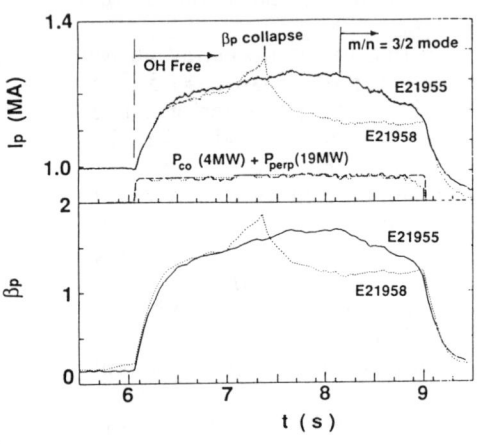

Fig.10 Sustainment of current by NBCD and bootstrap current without OH input.

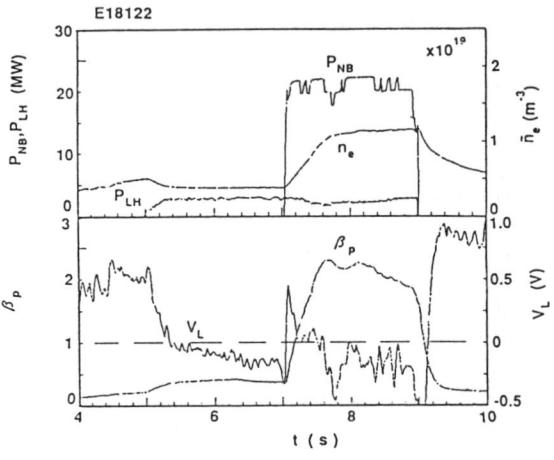

Fig.11 Time evolution of fully current driven discharge with LHCD ($I_p = 0.7MA$, $B_t = 3T$, $I_{BS}/I_p \sim 0.6$, $I_{LH}/I_p > 0.4$, H-factor ~ 2)

Table 1 Main machine parameters of JT-60SU

Toroidal field, B_t at 5 m	6.25 T
Plasma current, I_p	10 MA
Major radius, R_p	4.8 m
Minor radius, a_p	1.3-1.5 m
Aspect ratio	3.2-4
Elongation, κ	1.8
Triangularity, δ	0.4-0.8
Configuration	SN/DN
Number of TF-Coil	18
Field ripple at plasma surface	0.9 %
Total volt·second	170 V·s
Pulse length	>2000 s for full current drive case
Neutral beam power, P_{NB}	65 MW
Beam energy	500-700 keV
Lower hybrid power, P_{LH}	20 MW
Frequency of LHRF	3.7 or 5 GHz
ECH power for break down	1 MW
Frequency of ECH	140 GHz

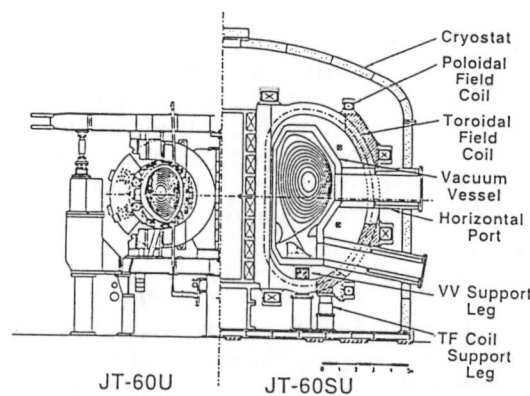

Fig.12 Cross-sectional view of JT-60SU in comparison with JT-60U.

system is enough to realize full CD condition in quasi-steady-state high fusion performance discharges as shown in Fig. 8. Divertor will be modified in the end of 1996 to realize cold dense divertor plasma and reduce divertor heat load in the integrated stead-state high fusion performance experiments with N-NBCD system. All these research activities in JT-60U may establish a physics base of a steady state tokamak reactor. However, there are some issues which are essentially unsolved in JT-60U. The plasma current, the density and the pulse length of the discharge in JT-60U are too low and too short compared with those of steady-state tokamak reactor, motivating us to study a conceptual design of the JT-60 Super Upgrade tokamak[10].

The design study of JT-60SU is based on the experimental results on JT-60U. Wide range of data base in JT-60U experiments shows that the discharge with qeff ~ (5 - 6) is the best from a point view of the confinement improvement factor, the beta limit, current quench speed at disruption, frequency of disruption occurrence, the divertor heat load and the particle recycling flux at the divertor. Therefore the operation of JT-60SU is optimized at a high safety factor. By upgrading N-NBI system (0.5MeV, 10MW) in JT-60U, 65MW of N-NBCD (0.5MeV) power will provide heating and current drive in JT-60SU. In addition to a capability to vary the beam energy of N-NBCD system for controlling beam deposition profile, 20MW of LHCD power will be used for profile control in advanced steady-state performance. Steady-state high fusion performance high β_p ELMy H-mode as demonstrated in JT-60U will be sustained fully non-inductively by N-NBCD and LHCD. The basic design parameters of JT-60SU are listed in the Table I. The toroidal magnetic field at the magnetic axis is set at 6.25T which corresponds to q_{95} = 6.8 at I_p = 6MA and q_{95} = 4.1 at I_p = 10MA. The cross section of JT-60SU is compared with JT-60U in Fig. 12. In addition to contribute ITER activities, an integrated basis of physics and technology for a steady-state tokamak reactor will be established by research studies in the JT-60SU tokamak.

SUMMARY

Significant Progress in non-inductive current drive, profile control, confinement and MHD stability has been made in recent studies in JT-60U to establish the physics base of highly integrated steady-state tokamak performance. Especially following progress should be emphasized; 1) 1MA of fully non-inductively driven discharges with a good confinement (H-factor = 2-2.5), high normalized β (β_N = 3) and high bootstrap current fraction (I_{BS}/I_p =74%) were realized, 2) quais steady-state (duration ~ $5\tau_E$) high fusion performance with high $n_D(0)\tau_E T_i(0)$ (= 3-5x10^{20}m^{-3}keVs) and high bootstrap current fraction (~40%) was demonstrated, 3) non-inductive discharge with large current (3.6MA) was achieved, 4) importance of active profile control to sustain a good confinement and good stability was highlighted. Upcoming installation of the intense core current driver with N-NBI and modification of the divertor with radiative pumped concept in JT-60U will ensure the establishment of physics base of steady-state tokamak operation. Based on these studies, JT-60SU will establish an integrated basis of physics and technology for ITER and a steady-state tokamak reactor.

REFERENCE

[1] Y. Seki et al., in Plasma Phys. and Contr. Nucl. Fusion Res., 1991 (Proc. 13th Int. Conf. Washington), vol. 3, 473.
[2] R.W. Conn et al., in Plasma Phys. and Contr. Nucl. Fusion Res., 1991 (Proc. 13th Int. Conf. Washington), vol. 3, 659.
[3] Y. Ikeda, et al., Fusion Eng. Design **24**(1994)287.
[4] K. Ushigusa et al., Nucl. Fusion **29**(1989)1052.
[5] K. Tani et al., J. Comp. Phys. **98**(1992)332.
[6] M. Mori et al., Nucl. Fusion **34**(1994)1045.
[7] Y. Kamada et al., Plasma Phys. Contr. Fusion **36Suppl(7)A**, 1994, A123.
[8] Y. KAMADA, et al., Nucl. Fusion (1994), in press.
[9] O. Naito et al., Plasma Phys. Contr. Fusion **35**(1993)215.
[10] H. Ninomiya et al., in Plasma Phys. and Contr. Nucl. Fusion Res., 1994 (Proc. 15th Int. Conf. Seville), IAEA-CN-60/F-1-I-1

EVIDENCE FOR MODIFIED TRANSPORT DUE TO SHEARED ExB FLOWS IN HIGH-TEMPERATURE PLASMAS

R.J. Groebner, M.E. Austin,[a] K.H. Burrell, T.N. Carlstrom, S. Coda,[b] E.J. Doyle,[c]
P. Gohil, J. Kim, R.J. LaHaye, C.L. Hsieh, L.L Lao, J. Lohr, R.A. Moyer,[c]
C.L. Rettig,[c] T. Rhodes,[c] G.M. Staebler, R.E. Stockdale, D.M. Thomas,
and J.G. Watkins[d]

General Atomics, P.O. Box 85608, San Diego, California 92186-9784 USA
[a]University of Maryland, USA
[b]Massachusetts Institute of Technology, USA
[c]University of California at Los Angeles, USA
[d]Sandia National Laboratories, Albuquerque USA

ABSTRACT. Sheared mass flows are generated in many fluids and are often important for the dynamics of instabilities in these fluids. Similarly, large values of the E×B velocity have been observed in magnetic confinement machines and there is theoretical and experimental evidence that sufficiently large shear in this velocity may stabilize important instabilities. Two examples of this phenomenon have been observed in the DIII–D tokamak. In the first example, sufficient heating power can lead to the L-H transition (transition from low-mode to high-mode confinement), a rapid improvement in confinement in the boundary layer (narrow region just inside the last closed flux surface) of the plasma. For discharges with heating power close to the threshold required to get the transition, changes in the edge radial electric field are observed to occur prior to the transition itself. In the second example, certain classes of discharges with toroidal momentum input from neutral beam injection exhibit a further improvement of confinement in the plasma core leading to a regime called the VH–mode. In both examples, the region of improved confinement is characterized by an increase of shear in the radial electric field E_r, reduced levels of turbulence and increases in gradients of temperatures and densities. These observations are consistent with the hypothesis that the improved confinement is caused by an increase in shear of the E×B velocity which leads to a reduction of turbulence. For the VH–mode, the dominant term controlling E_r is the toroidal rotation v_ϕ, indicating that the E_r profile is controlled by the source and transport of toroidal momentum. At the edge of the plasma, indirect measurements indicate that the change in E_r is initiated by a change in the V×B velocity of the main ions but at later times is dominated by the ion diamagnetic velocity.

I. INTRODUCTION

It has long been recognized in fluid mechanics that the interaction between sheared velocity fields and turbulence can have important consequences for the stability of various modes. Sufficiently large velocity shear can drive the Kelvin-Helmholtz mode unstable [1]. On the other hand, Rayleigh-Taylor modes can be stabilized by velocity shear [2]. Theoretical and experimental work during the last few years has provided strong evidence that similar mechanisms are important in magnetized plasmas. For this case, the fundamental velocity is not the mass velocity but is rather the E×B velocity, the velocity at which turbulent eddies are convected, where E and B are the electric and magnetic fields. Basic scaling arguments of nonlinear theory show that if the shear (spatial gradient) of the E×B velocity varies sufficiently on the scale of a turbulent eddy, the correlation time of the eddy is decreased by the shear [3]; thus, the eddy is effectively ripped apart by the shear leading to a reduction of the rate of transport due to turbulent processes. Numerous linear stability calculations have shown that sufficient shear in the E×B velocity is a robust mechanism to stabilize flute-like modes [4] (modes whose structures are elongated along magnetic field lines). Self-consistent nonlinear simulations have been performed for several modes of interest and they generally support the results of the linear analyses [5]. An important difference is that for some modes, the curvature (second spatial derivative) of the E×B velocity field is predicted to me much more important for stabilizing turbulence than the shear (first spatial derivative) [6].

There is a significant body of experimental evidence that sheared E×B flows do in fact modify and reduce the transport in magnetic confinement devices [7–9]. The evidence comes from machines with a variety of magnetic geometries and indicates that the effect is a basic phenomenon of magnetized plasmas rather than a peculiarity of a given confinement scheme. The clearest example in tokamaks is the L-H transition, which is a spontaneous transition from a base level of confinement called the L–mode to an increased level of confinement called the H–mode [10]. The transition to H–mode is initiated in a narrow layer at the periphery of the plasma. In this narrow layer, it is simultaneously observed that large shear (both first and second spatial derivatives) develops in the radial electric field E_r, that the level of density fluctuations decreases, and that temperature and density gradients increase. This phenomenology is consistent with, and in some cases has been predicted by, the theoretical work on suppression of turbulence due to sheared E×B flows. Thus, the theoretical paradigm that shear in the E×B flow can stabilize turbulence has become the primary hypothesis for guiding studies of the L-H transition. Additional support is given to the hypothesis by experiments in which the transition has been induced with application of a radial electric field from a probe inserted part way into the plasma [11].

In at least two tokamaks, DIII–D [12] and JET [13], the H–mode discharge has been observed to evolve into a further improved regime of confinement called the VH–mode, which exhibits values of τ_E which are 3-4 times those in L–mode discharges. In DIII–D, the phenomenology of the VH–mode is also consistent with the hypothesis of E×B shear suppression of turbulence [14,15]. Whether this phenomenology is peculiar to DIII–D or is a universal aspect of the VH–mode remains to be seen.

The E×B shear suppression mechanism is of interest for both basic and applied studies of plasma physics. Non-linear analyses show that a fundamental understanding of saturated turbulence in a magnetized plasma requires self-consistent knowledge of both E×B suppression [16] and transport of momentum by the turbulent fields [17]. From a practical point of view, design studies aimed at producing a tokamak which can obtain an ignited thermonuclear burn generally indicate that some enhancement of confinement over L–mode levels is required to build a practical machine. H–mode and VH–mode discharges are candidates for achieving this increased confinement. Thus, if E×B shear suppression is required to produce the H–mode and VH–mode plasmas, it is a rather important subject from the point of view of applied physics as well as basic physics.

The goal of this paper is to examine the present status of H–mode and VH–mode studies in the DIII–D tokamak. In Section II, the phenomenology of the L-H transition and of the production of VH–mode discharges is briefly examined to show that the observations are consistent with expectations from the E×B shear suppression hypothesis. New results are presented for discharges in which the time scale for the transition has been increased. These discharges exhibit changes in E_r prior to the transition and thus provide evidence that E_r causes the transition. Section III discusses the status of knowledge about the origin of E_r from the point of view of the radial force balance equation for toroidal equilibria. At this level of analysis, it is clear that the observed electric fields are a self-consistent result of various physical processes in the tokamak. Recent evidence is presented which indicates that the V×B term of the main ions controls the rapid change in E_r at the transition. Section IV contains a summary and conclusions. In this paper, the term "shear suppression" will mean suppression of turbulence by an E×B profile with non-zero first or second spatial derivatives. For the cases of interest, it is radial variation of E_r which produces this shear in E×B.

II. STUDIES OF L-H TRANSITION AND VH–MODE

The spontaneous form of the L-H transition is generated in tokamaks by application of heating power above a required minimum called the power threshold. After the heating power is applied, the plasma passes through a phase of L–mode confinement during which temperatures and possibly densities rise monotonically with time. The classic signature marking the time of the transition is an abrupt and marked drop in the intensity of the D_α signal emitted from the plasma boundary. This signature is interpreted as indicating that the outflux of particles from the plasma decreases dramatically due to the formation of a transport barrier at the plasma edge which holds the particles in the plasma. The transport barrier is manifested as a dramatic steepening of the electron density n_e, electron temperature T_e and ion temperature T_i profiles in a region with a width of the order of 1 cm just inside the last closed flux surface and a reduction of n_e on the open field lines [9]. These modified boundary conditions allow an improvement of confinement (thermal diffusivity) in the core of the plasma to occur so that the discharge attains a value of τ_E which is roughly twice its value in the L–mode phase.

The phenomenology of the L-H transition in DIII–D is consistent with the expectations of the shear suppression hypothesis as has been discussed elsewhere [18-23]. Basically, at the L-H transition and in the same region of the plasma, there is a simultaneous formation of a negative "well" in E_r [9,18-21], a reduction in the level of electron density fluctuations $\tilde{n}$ [9,18,19,21-23] and an increase in electron and ion pressure gradients [9,18,19]. Due to the short time scale in which these changes occur, it has not been possible in previous work to unambiguously determine if changes in E_r occur prior to other changes, as might be expected from the shear suppression hypothesis. However, during the last year, experiments have been conducted in which the heating power was adjusted to be near the threshold value required to obtain the transition and this technique has slowed down the rate at which the transition occurs. Notably, significant changes in E_r are observed to occur prior to the transition [19,21], with E_r being determined directly from Langmuir probe measurements and by the standard spectroscopic techniques. These measurements of early changes in E_r are reminiscent of data reported several years ago [24] and are the strongest evidence for the causal role of E_r in the spontaneous L-H transition.

Figure 1 illustrates typical waveforms from these discharges with "slow" transitions. The time of the transition is inferred from Fig. 1(d) which displays the D_α emission from the divertor target. The drop in the signal at 1480.4 ms is the classic signature of the transition. This time coincides with a marked reduction of the ion saturation current drawn by Langmuir probes in the divertor target. Figure 1(a) displays the values of E_r as obtained from two radii very near the plasma edge and shows that E_r starts becoming more negative and that the shear in E_r starts to increase several milliseconds before the transition. The level of electron density fluctuations is reduced abruptly at the transition as indicated by an FIR scattering system [Fig. 1(b)] and a microwave reflectometer system [Fig. 1(c)]. The signal from both instruments originates at the plasma edge. As indicated in Fig. 1(e), the electron pressure in the same region increases during the H–mode with a time constant of several milliseconds.

Under appropriate conditions, H–mode discharges evolve into a new regime of enhanced confinement, called the VH–mode, which exhibit τ_E values 50%–100% larger than those obtained in H–mode discharges. The transition to VH–mode confinement occurs gradually over a period which is typically several tens of milliseconds [14,15]. During this time, the energy confinement time is increasing, the transport barrier gradually builds further into the plasma, there is an increase in the toroidal rotation velocity v_ϕ in the core of the plasma, and there is a simultaneous reduction of core microturbulence as determined via FIR scattering [25]. Heat transport coefficients in the plasma decrease during this time. The increasing value of v_ϕ indicates that shear in the E×B velocity is increasing in the region in which transport improves and the turbulence decreases. The changes in the VH–mode occur in the core of the plasma rather than at the plasma boundary as for the L-H transition. The shear suppression hypothesis predicts that if the core E_r shear is destroyed, then the VH–mode should be destroyed. This prediction has been verified with the aid of a magnetic perturbation which decreases the toroidal rotation [18,19,26] with a corresponding decrease in the shear of E_r. Simultaneously, the core turbulence levels rise, the local heat diffusion coefficients increase and the global confinement decreases. Thus,

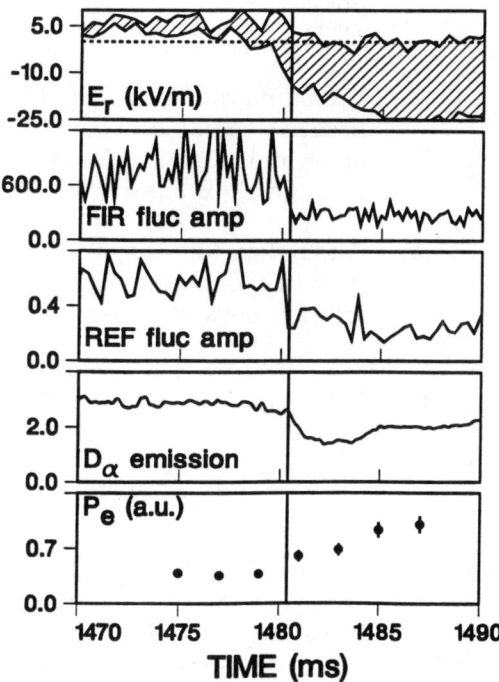

Fig. 1. Waveforms for a "slow" L-H transition, occurring at 1480.4 ms as indicated by vertical line. (a) E_r at R = 2.289 m (lower line) and R = 2.295 (upper line); cross-hatching highlights the shear in E_r; (b) Density fluctuations from FIR scattering; (c) Amplitude of fluctuating signal from 24 GHz microwave reflectometer channel shows reduction in level of density fluctuations at transition; (d) D_α emission from plasma hit spot in divertor; (e) Electron pressure at R = 2.281 m.

the hypothesis of E×B shear suppression of turbulence is capable of describing the phenomenology of the VH–mode as well as of the L-H transition in DIII-D.

III. ORIGIN AND SHAPE OF RADIAL ELECTRIC FIELD

A full understanding of the L-H transition and of the VH–mode requires an understanding of the physics which controls the shape and magnitude of E_r. There has been a large theoretical effort devoted to this issue which has led to the formulation of equations to describe E_r near the plasma boundary in toroidal geometry with neoclassical and anomalous effects included [27–29]. The experimental assessment of theory is very difficult due to the fact that many of the quantities of interest cannot be measured with existing diagnostic capability. The ideas which are being actively pursued can perhaps be divided into four categories:

1. Ion orbit loss [30–32]. Ions are preferentially lost from the plasma edge because ions in the loss cone intersect material surfaces. Thus, the plasma is charged up negatively.

2. Stringer spin-up [33,34]. A large poloidally asymmetric sink or source of particles overcomes the natural damping of poloidal rotation and allows a large

poloidal rotation to develop. The relationship between poloidal rotation and E_r has not been specified in this model.

3. Pressure-gradient drive [35,36]. A toroidal equilibrium naturally develops a negative radial electric field to balance the pressure gradient of the ions. The large and negative E_r generated by the pressure gradient reduces transport through E×B shear. This feedback can cause a transport bifurcation.

4. Anomalous viscosity [27,37,38,39] or turbulent Reynold's stress [17,40]. Transport of momentum can modify the average flow profile of the plasma. The relation between plasma velocity and E_r has been specified for these models.

On the experimental side, a convenient framework which is being used to examine the origin and shape of the E_r profile is the lowest order radial force balance equation

$$E_r = (Z_i e n_i)^{-1} \nabla p_i - v_{\theta i} B_\phi + v_{\phi i} B_\theta \quad , \qquad (1)$$

where i labels the ion species, Z_i is the charge of the ion, n_i is the ion density, e is the electronic charge, p_i is the ion pressure, $v_{\theta i}$ and $v_{\phi i}$ are, respectively, the poloidal and toroidal rotation velocities and B_θ and B_ϕ are, respectively, the poloidal and toroidal magnetic fields. Although this equation hides a great deal of physics, it shows clearly that the plasma's E_r is a self-consistent solution which satisfies several physical processes. These include the sources and transport of toroidal momentum, poloidal momentum, heat and particles as well as the generation of the magnetic equilibrium. Although the sources of several of these quantities are fairly well understood, the transport mechanisms in tokamaks are not understood. Thus, radial transport of heat, particles and momentum is generally considered to be "anomalous". There are theoretical suggestions that neoclassical effects may be important for rotation at the edge of the plasma, particularly in H–mode [30,31,38,41-43]. An underlying assumption of present research is that, for a given experimental situation, one term on the right hand side of the force balance equation dominates and is considered to be responsible for E_r and the present goal of experimental research is to determine which is the dominant term.

For the VH–mode in DIII–D, the dominant term controlling E_r is $v_{\phi i} B_\theta$. At this level of understanding, the E_r profile and its shape are controlled by the input of toroidal momentum from neutral beam injection, which is understood to be a classical process, and the transport of toroidal momentum, which is not understood and is considered to be an anomalous process.

Studies of L-H transition have generally been based on the premise that the main ions control the physics of the edge E_r, and one research goal has been to determine whether it is a change of the pressure gradient or of the V×B term of the main ions which initiates the rapid change in E_r observed at the L-H transition. Although direct measurements of the behavior of pressure gradient and V×B terms for the main ions are not available at the transition, the slow transitions discussed above provide strong evidence that the main ion pressure gradient does not control the change in E_r at the transition [19]. These conclusions are based on measurements which show that the gradient of T_i does not change sufficiently at the transition to account for the E_r change and measurements which show that there are no significant changes in the electron density prior to the transition [19].

More quantitative analysis of the discharge discussed in Fig. 1 supports this conclusion. For this discharge, E_r and T_i profiles are available at the plasma edge every 0.5 ms before and after the L-H transition. These data are obtained from He II, which is present in the plasma as an impurity and these measurements are a diagnostic used to determine the E_r and T_i values for the main ion force balance equation. In addition, a sequence of profiles of n_e and T_e from Thomson scattering are available every 2 ms from 1475 ms to 1487 ms, a range which brackets the L-H transition. Under the assumption that the ratio of n_e to the main ion density n_i is constant through the region of interest, the measurements of T_i and n_e can be combined to make an indirect estimate of the main ion pressure gradient contribution to E_r. This is so because the constant of proportionality between n_i and n_e cancels out in the pressure gradient term of the force balance equation for the main ions.

Figure 1(a) demonstrates that there is a large change in the edge E_r between 1475 ms, the time of the first Thomson laser pulse during the time of interest, and 1481 ms, the time of the first Thomson laser pulse after the transition. Figure 2 is a comparison at these two times of the full E_r profiles and of the estimate for the pressure gradient term in the main ion force balance equation. These data show that although there is a profound change in E_r from 1475 to 1481 ms, there is only a very modest change in the estimate for the main ion pressure gradient term. Thus, it must be concluded that either changes in the main ion pressure gradient do not cause the change in E_r across this transition or that the gradient of n_i changes drastically between these two times with little change occurring in the gradient of n_e. The second possibility appears unlikely, particularly because simple estimates suggest that n_i may have to go negative to provide the required pressure gradient.

The lowest order force balance equation is expected to be valid for timescales much shorter than 1 ms. Thus, the estimate of the main ion pressure gradient term developed above can be used to make an estimate of the temporal behavior of the V×B component of the main ion force balance equation at the L-H transition. Figure 3(a) shows the time history for the total radial electric field E_r, obtained at the location where the most negative value of E_r is observed in the H-mode. At the same location, Fig. 3(b) shows the temporal history of $(en_e)^{-1} \nabla(T_i n_e)$. There is very little change in this term across the transition (1480.4 ms) with the pressure gradient term subsequently becoming significantly more negative. Figure 3(c) shows the difference of E_r and the pressure gradient term and is an estimate of the temporal behavior of the V×B term of the main ions at the L-H transition. With the

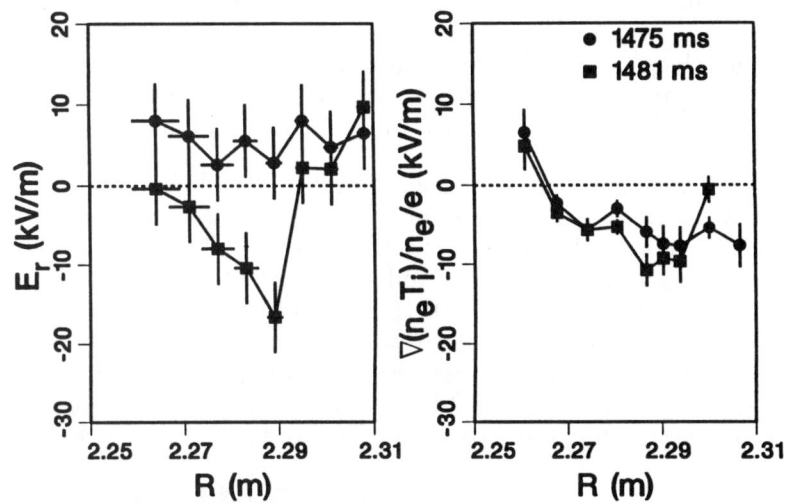

Fig. 2. Profiles at midplane of terms in force balance equation. Data for 1475 ms obtained in L-mode and data for 1481 ms obtained within one ms after transition. (a) E_r (obtained from He II force balance equation) shows significant changes between 1475 and 1481 ms; (b) Estimate of main ion pressure gradient term, based on assumption that n_e and n_i profiles have similar shapes, shows little change between 1475 and 1481 ms.

caveats already discussed, these data indicate that the primary reason that E_r becomes more negative at the transition is that the V×B term of the main ions rapidly becomes less positive. The data suggest that the V×B term changes sign at the transition, but such a conclusion is perhaps premature until more data are examined. Figure 3(c) also suggests that as the H–mode develops the main ion V×B term tends to go more positive again. For the claims made here, it is crucial that the relative timing accuracy of the Thomson Scattering and charge exchange recombination spectroscopy (CER) systems be better than about one millisecond. The absolute timing system accuracy of the CER system is good to within 0.1–0.2 milliseconds and the accuracy of the Thomson system is even better.

These inferences about the V×B term of the main ions are consistent with direct measurements of the main ions in other discharges. The positive value for the V×B term of the main ions has been confirmed with direct measurements of the main ions in discharges with fast L-H transitions [43]. For those discharges, it is known that the V×B term of the main ions is always positive except for a window of ±3.5 ms around

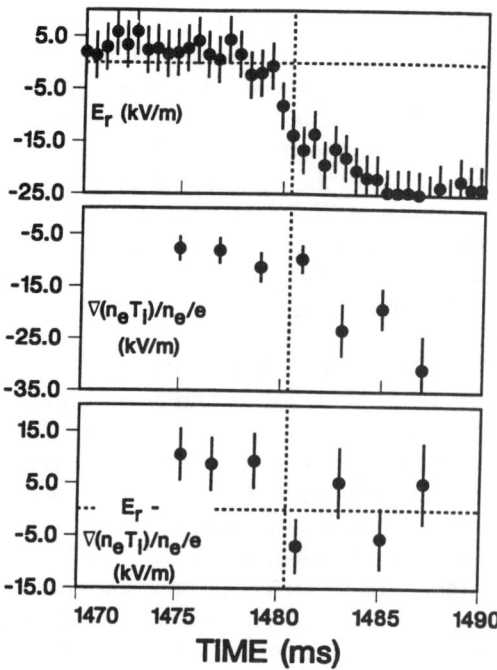

Fig. 3. Temporal study of force balance at location of most negative E_r (R = 2.289 m). (a) E_r starts to become more negative prior to L-H transition; (b) Estimate of main ion pressure gradient term, based on T_i and n_e measurements; (c) Difference of data in (a) and (b). This is estimate of V×B term of main ion force balance equation.

the time of the transition itself. Unfortunately, due to technical limitations associated with the measurements, data for the main ions across the transition are not available. Thus, there are no direct measurements of changes in the main ion V×B term. The direct measurements do show that 3.5 ms after the L-H transition, the pressure gradient of the main ions is the dominant term in the force balance and causes the negative E_r.

At various times in the H–mode, all of the terms of the main ion force balance equation are important in controlling the negative E_r observed in the transport barrier in DIII–D. At the time of the transition itself, the V×B term appears to be important, indicating that the sources and transport of poloidal and, perhaps, toroidal momentum are important processes. Of these quantities, only the source of toroidal momentum in neutral-beam heated discharges can be considered well understood at this time. Shortly after the transition the pressure gradient becomes dominant and thus the edge E_r profile is controlled by the sources of heat and particles, which are understood as classical processes, and the transport mechanisms of heat and particles which are considered to be anomalous.

IV. SUMMARY AND CONCLUSIONS

Sheared mass flows are of general interest in fluid mechanics because of their role in driving or subduing instabilities. For magnetized plasmas, there is a large body of theoretical and experimental evidence which indicates that sheared E×B flows, due primarily to sheared E_r fields, can stabilize a large class of instabilities found in laboratory plasmas. In the DIII–D tokamak, the phenomenology of two modes of improved confinement are consistent with the hypothesis that increased shear in E_r leads to a suppression of turbulence and improved confinement. In both the transition from L–mode to H–mode confinement and the transition from H–mode to VH–mode confinement, there is a strong spatial and temporal correlation between the increase of shear in E_r, a reduction of $\tilde{n}/n$ which is evidence of a reduced level of turbulence and increased gradients in temperature and density profiles.

The mechanisms which control the magnitude and shape of E_r in a tokamak are poorly understood. The core E_r in the VH–mode in DIII–D is controlled by the plasma's toroidal rotation which is in turn controlled by the momentum input from neutral beam injection. Although, this source of momentum input is understood, the transport of toroidal momentum is also important in forming the E_r profile and is not well understood. The situation for the L-H transition is even more difficult, due to the rapidity with which the edge plasma evolves at the transition. However, the time scale for the transition has been increased by operating near the H-mode power threshold. Evidence acquired from such transitions shows that the edge E_r starts to become more negative before changes are observed in turbulent quantities and before the transition occurs. In addition, the available data provides indirect but very strong evidence that it is the V×B rather than the pressure gradient term for the main ions which changes rapidly at the transition. Further studies are required to determine whether it is v_ϕ or v_θ of the main ions which changes at the transition.

ACKNOWLEDGMENTS

The results presented here are the result of the dedication and hard work of the entire DIII–D staff, including the technical, engineering, operations, computer and physics groups. This work was supported by the U.S. Department of Energy under Contract Nos. DE-AC03-89ER51114, W-7405-ENG-48, DE-FG03-89ER51121, DE-FG03-86ER53225, DE-AC05-84OR21400, DE-FG02-91ER54109 and DE-AC03-76DP00789.

REFERENCES

[1] Tajima, T., *et al.*, Phys. Fluids B **3**, 938 (1991).

[2] Kuo, H.L. Phys. Fluids **6**, 195 (1963).

[3] Biglari, H., Diamond, P.H., Terry, P.W., Phys. Fluids B **2**, 1 (1990).

[4] Hassam, A.B., Comments Plasma Phys. Controlled Fusion **14**, 275 (1991).

[5] Hamaguchi, S., Horton, W., Phys. Fluids B **4**, 319 (1992).

[6] Carreras, B.A., *et al.*, Phys. Fluids B **4**, 3115 (1992).

[7] Ritz, Ch. P., *et al.*, Phys. Rev. Lett. **65**, 2543 (1990).

[8] Groebner, R.J., Phys. Fluids B **5**, 2343 (1993).

[9] Burrell, K.H., *et al.*, Plasma Phys. Contr. Fusion **36**, A291 (1994).

[10] Wagner, F., *et al.*, Phys. Rev. Lett. **49**, 1408 (1982).

[11] Taylor, R.J., *et al.*, Phys. Rev. Lett. **63**, 2365 (1989).

[12] Jackson, G.L., *et al.*, Phys. Fluids B **4**, 2181 (1992).

[13] JET Team, Plasma Phys. Contr. Fusion **36**, A23 (1994).

[14] Greenfield, C.M., *et al.*, Plasma Phys. Contr. Fusion **35**, B263 (1993).

[15] Osborne, T.H., *et al.*, Plasma Phys. Contr. Fusion **36**, A237 (1994).

[16] Diamond, P.H., *et al.*, Phys. Rev. Lett. **72**, 2565 (1994).

[17] Diamond, P.H., and Kim, Y-B., Phys. Fluids B **3**, 1626 (1991).

[18] Burrell, K.H., *et al.*, Phys. Plasmas **1**, 1536 (1994).

[19] Burrell, K.H., *et al.*, "H–Mode and VH–Mode Confinement Improvement in DIII–D: Investigations of Turbulence, Local Transport and Active Control of the Shear in the $E \times B$ Flow," in Proc. 15th Intnl Conf. on Plasma Phys. and Contr. Fusion Research 1994 (International Atomic Energy Agency, Vienna) to be published.

[20] Gohil, P., *et al.*, "The Parametric Dependence of the Spatial Structure of the Radial Electric Field at the Plasma Edge in the DIII–D Tokamak," in Proc. 21st Euro. Physical Society Conf. on Contr. Fusion and Plasma Phys., Montpellier, France, 1994 (EPS, Petit-Lancy, Switzerland) to be published.

[21] Moyer, R.A., to be published in Physics of Plasmas.

[22] Doyle, E.J., *et al.*, (1993). *Plasma Physics and Controlled Nuclear Fusion Research 1992* (International Atomic Energy Agency, Vienna, 1993), Vol. 1, p. 235.

[23] Rhodes, T.L., *et al.*, Rev. Sci. Instrum. **63**, 4599 (1992).

[24] Groebner, R.J., Burrell, K.H., Seraydarian, R.P., Phys. Rev. Lett. **64**, 3015 (1990).

[25] Rettig, C.L., *et al.*, Plasma Phys. Contr. Fusion **36**, A207 (1994).

[26] La Haye, R.J., *et al.*, in Local Transport Studies in Fusion Plasmas (Proc. Varenna Workshop, 1993) (Societa Italiana di Fisca, Bologna, 1994) **ISPP-14** p. 283.

[27] Rozhansky, V., and Tendler, M., Phys. Fluids B **4**, 1877 (1992).

[28] Hinton, F.L., and Kim, Y.-B., Nucl. Fusion **34**, 899 (1994).

[29] Yushmanov, P.N., *et al.*, Phys. Plasmas **1**, 1583 (1994).

[30] Itoh, S.-I., and Itoh, K., Nucl. Fusion **29**, 1031 (1989).

[31] Shaing, K.C., Crume, Jr., E.C., and Houlberg, W.A., Phys. Fluids B **2**, 1492 (1990).

[32] Tendler, M. *et al.*, (1993). *Plasma Physics and Controlled Nuclear Fusion Research 1992* (International Atomic Energy Agency, Vienna, 1993), Vol. 2, p. 243.

[33] Hassam, A.B., Antonsen, Jr., T.M., Drake, J.F., and Liu, C.S., Phys. Rev. Lett. **66**, 309 (1991).

[34] McCarthy, D.R., *et al.*, Phys. Fluids B **5**, 1181 (1993).

[35] Hinton, F.L., and Staebler, G.M., Phys. Fluids B **5**, 1281 (1993).

[36] Staebler, G.M., *et al.*, Phys. Plasmas **1**, 909 (1994).

[37] Staebler, G.M., and Dominguez, R.R., Nucl. Fusion **33**, 77 (1993).

[38] Rozhansky, V., and Tendler, M., "Prospects for Improved Confinement in Hot Plasmas," in Proc. 21st Euro. Physical Society Conf. on Contr. Fusion and Plasma Phys., Montpellier, France, 1994 (EPS, Petit-Lancy, Switzerland) to be published.

[39] Itoh, S-I., *et al.*, Plasma Phys. Contr. Fusion **36**, A261 (1994).

[40] Carreras, B.A., *et al.*, "Dynamics of L to H Bifurcation and Ion Pressure Gradient Evolution," to be published in Phys. Plasmas **1**.

[41] Stacey, W.M., and Jackson, D.R., Phys. Plasmas **1**, 909 (1994).

[42] Rogister, A., Phys. Plasmas **1**, 619 (1994).

[43] Hinton, F.L., and Kim, Y.-B., "Poloidal Rotation in Tokamaks with Large Electric Field Gradients," General Atomics Report GA-A21600, April, 1994, to be published in Phys. of Plasmas.

[44] Kim, J., *et al.*, Plasma Phys. Contr. Fusion **36**, A183 (1994).

RECENT RESULTS ON CURRENT PROFILE SHAPING ON TORE SUPRA

Equipe Tore Supra
(*presented by A. Bécoulet*)

Association EURATOM-CEA sur la Fusion,
Département de Recherche sur la Fusion Controlée,
C.E. Cadarache, FRANCE

Abstract

The link between the current profile and the confinement is studied, involving various regimes: high power minority ion cyclotron resonant heating, high power lower hybrid current drive, fast wave direct electron heating and current drive and pellet enhanced performance. It is shown that the electron heat diffusivity decreases when the magnetic shear increases in the confinement zone and/or when it decreases in the plasma centre.

1. Introduction

Tore Supra is a large superconducting Tokamak, with a circular cross-section (major radius R=2.4m, minor radius a=0.8m, B(axis)≤4T). Its lower hybrid current drive (LHCD) system, as well as an active cooling of the plasma facing components (12m^2 graphite inner wall and pump limiters), has allowed long pulse discharges up to a stationary, 62s flat-top, 1MA plasma [1] (3MW/180MJ). LHCD has also demonstrated the capability for coupling up to 6MW and has obtained full current drive discharges up to 1.6MA. In parallel, the ion cyclotron resonant heating (ICRH) system, operating in the 35-80 MHz range, has been able to reach 8MW pulses (using 3 double-loop actively cooled antennas), and a 2.5MW/30s shot in minority heating scenarios. Up to 5MW/2s in direct electron heating scenario (Fast Wave Electron Heating, FWEH) have also been obtained. The directional phasing of the antennas has lead to the demonstration of non inductive current driven by the fast magnetosonic wave (Fast Wave Current Drive, FWCD) [2].

Several evidences of a plasma current profile shaping have been obtained using these two additional heating systems, either separately and together. Systematically, these modifications of the current profile lead to an improvement in the confinement properties. Section 2 of this paper describes the behaviour of the global confinement in Ohmic and L-mode discharges. Then Section 3 reviews the various operating regimes in which a clear link between current shaping and confinement enhancement has been observed on Tore Supra. Finally, at the light of the so-called "steady-state advanced" scenarios which require a very accurate current profile control, Section 4 concludes and gives perspectives for further experiments.

2. Confinement Studies

2.1 Ohmic and L-mode confinement

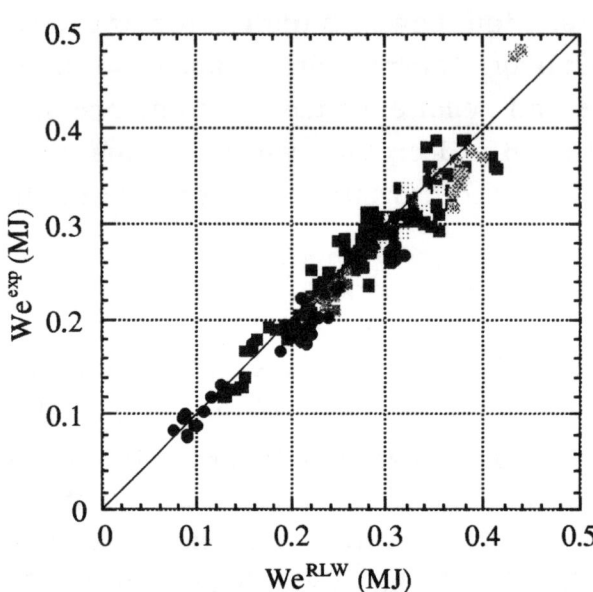

fig1: Electron energy content compared to RLW prediction for ohmic shots (circle) and L-mode shots (square).

On Tore Supra, the global Rebut-Lallia-Watkins (RLW) prediction for the electron energy content (We) is in very good agreement with a wide range of experimental data (fig1), i.e. ohmic (in the non-saturate regime with respect to density), low power minority ICRH and low driven current fraction LHCD plasmas. We note that both ICRH and LHCD lead to discharges where the electron energy content is large or even dominant.

2.2 Ohmic ramp-down experiments

The global RLW prediction has also been tested during ohmic ramp down experiments [3], (Ip= 1.7MA-> 0.7MA at -1.5MA/s). During the ramp-down phase, the value of the internal inductance l_i strongly increases and We exceeds the RLW value, by a factor up to 1.5. During this phase, it is noted that the current profile is altered in the outer part of the plasma ($\rho > 0.6$), and correlatively that the shear in this region is increased.

3. Current Profile Shaping Experiments

A clear link between confinement and current profile has also been established in high power auxiliary heated plasmas with predominant electron heating. Using sufficient LH and/or ICRF waves can lead to significant current profile modifications, due to non-inductive driven current (RF or bootstrap) or simply due to changes in the plasma resistivity by localised power deposition. On Tore Supra, the polarimetry and magnetic measurements give access to the safety factor (q) profile ($0.1<\rho<0.7$), and the internal inductance (l_i, dominated by the edge current profile). Local transport studies lead to a satisfactory determination of χ_e in the gradient zone ($0.4<\rho<0.7$), using both the LOCO [4] and TRANSP [5] codes. Hard-X ray measurements provide quantitative information on the fast electron population. The relationship between current profile and confinement is analysed in four different experimental schemes: monster sawteeth, high-l_i and Lower Hybrid Enhanced Performance (LHEP), FWEH and FWCD, and finally Pellet Enhanced Performance (PEP).

3.1 The monster sawtooth phase

Monster sawteeth have been obtained on Tore Supra using minority ICRH. For a given power, the duration of the sawtooth-free period increases with density, up to a maximum where the stabilising phenomenon disappears. This increase has revealed the role played by the passing (or marginally trapped) fast ions in the stabilising process [6]. During the sawtooth-free period where the electron temperature (T_e) saturates, the central safety factor q0 constantly decreases, while the q=1 surface increases, until the crash. Consequently, the magnetic shear in the gradient zone significantly increases as well as the stored energy. Only a weak increase in l_i is observed.

In order to compensate the broadening of the q=1 surface, presumably linked to the monster sawtooth collapse, LHCD was superimposed to the ICRH power. From the neutral particle analysers, there is no evidence of LHCD power directly coupled to the fast ion population. This is coherent with the predicted LHCD off-axis deposition profile.

Above a power threshold (typically $P_{LH}\approx3-3.5MW$, for $P_{ICRH}\approx4MW$, Ip=1.5MA, Bt=3.7T, $\bar{n}_e=4.10^{19}m^{-3}$), the q0 evolution is frozen, as shown

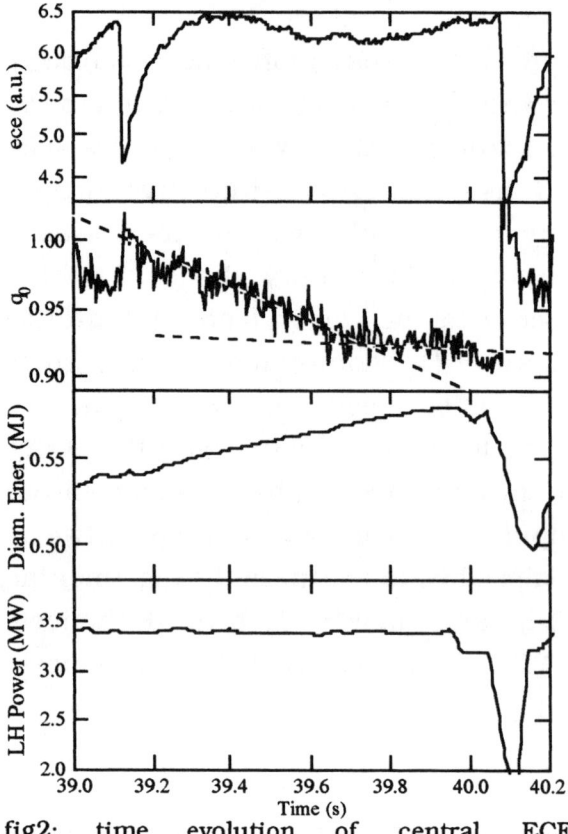

fig2: time evolution of central ECE measurements, q0, the diamagnetic energy and the LH power during combined ICRH/LHCD sawtooth stabilisation experiments.

on fig2. However the stored energy continues to increase, and the resistive evolution of the current profile is certainly not finished (LHCD only drives 20% of the total current in that case). In the best shot, arcing on the LHCD launchers provoked unfortunately a loss of power after 1 s of stabilisation, leading to the sawtooth crash. A doubling of the sawtooth-free period duration has nevertheless been obtained in controlling partially the current profile, but a stationary sawtooth-free period (in particular with a saturation in the stored energy) has not been reached yet.

3.2 The high-l_i LHCD discharges and the Lower Hybrid Enhanced Performance (LHEP)

Steady-state (up to 15s) so-called "high-l_i" regimes have been studied, using LHCD in current ramp-down experiments (Ip= 1.7MA-> 0.8MA at -1MA/s, Bt=4T, $\bar{n}_e=2.10^{19}m^{-3}$, $P_{LHCD}=2.5MW$). An internal inductance l_i value up to 2 is obtained, and a confinement enhancement (with respect to RLW global scaling) is observed. This enhancement factor scales approximately like $l_i^{0.7}$ [7]. Local transport studies have demonstrated a clear correlation between χ_e and the magnetic shear in the confinement zone, coherently with the local RLW predictions [8].

This regime has been extended to the so-called LHEP regime. In that case, the switch-on of LHCD power generates first an increase in the gradient zone shear, followed about one second after by a vanishing (or even a reversal) of the central magnetic shear. Consequently the central electron temperature (T_{e0}) rises up to 8-10 keV. The local transport

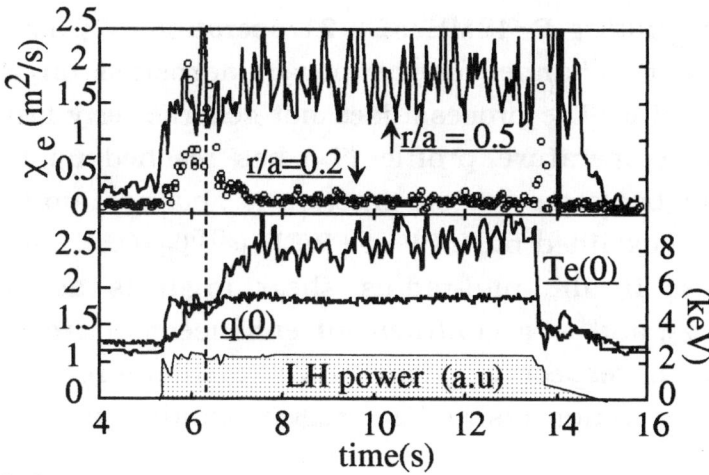

fig3: the LHEP regime. Time evolution of χ_e (at $\rho=0.2$ and 0.5), T_{e0}, q0 and LH power.

analyses exhibit a strong decrease of the central χ_e (fig3). The strong pressure gradient, combined with this shear lowering leads the plasma core ($\rho<0.25$) into the second stability zone (stable with respect to ballooning modes, TRANSP[5] analysis). Up to now, this regime has only been triggered in nearly zero-loop voltage discharges, where LHCD also controls the central current fraction. The LHEP phase has been sustained on time scales larger than the current diffusion time (up to 8s).

3.3 Fast Wave Electron Heating (FWEH) and Current Drive

In the high-l_i and LHEP regimes, the modifications of the shear in the confinement zone are mainly triggered by non-inductive fast electron current drive generated by LH waves in that region. The agreement with simulations and hard-X ray measurements is fairly good [9]. An alternative way to achieve that goal is to generate a large bootstrap current fraction, which also peaks around mid-radius.

This has been performed on Tore Supra through the direct coupling of the fast magneto-

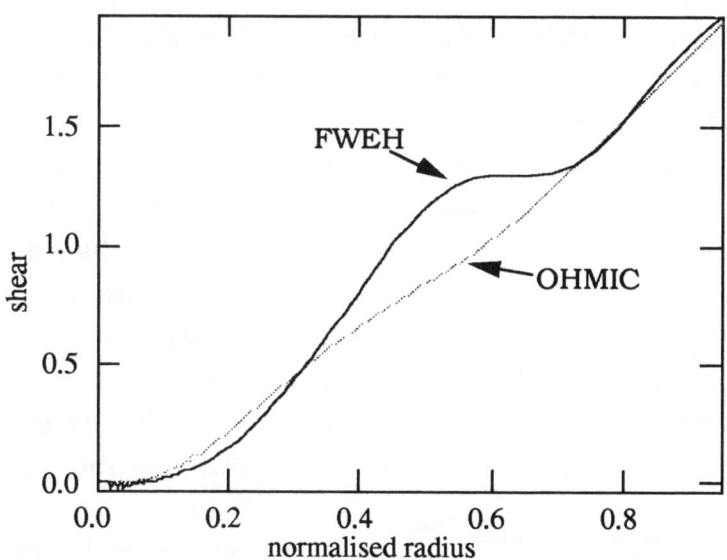

fig4: comparison of magnetic shear profiles during ohmic and FWEH phase.

sonic wave to the electrons, during F=48MHz, Bt=2T operations [2]. The electrons are the only damping channel for the power, deposited on a strongly centrally peaked profile. The process does not generate any fast electron tail, but a peaked temperature profile (T_{e0} has reached up to 5keV). Consequently, a large bootstrap current fraction is obtained: up to 45% of a 400kA plasma are sustained by 4MW of FWEH ($\epsilon^{0.5}\beta_p$=0.5). The consecutive strong increase in the mid-radius shear (fig4) is again correlated to the drop of χ_e, and to a confinement enhancement factor reaching more than 2 (with respect to global RLW prediction). The situation lasts for the pulse duration (2s). FWEH regime exhibits almost no modification of l_i.

Phasing the currents which flow in the antenna straps up to ± 90° lead to co- and counter-FWCD. In co-current, up to 80 ±40 kA of FWCD has been driven, on a 400kA discharge. The bootstrap current was 90 ±20 kA for a 2.5MW total FWCD power. The current drive efficiency (30kA/MW, $\gamma \approx 0.02 \; 10^{20}$ A/W/m^2) is in perfect agreement with simulations. The polarimetry angles and the time evolution of l_i are well reproduced, using the CRONOS [10] code, when assuming a current profile slightly broader than the source term derived from codes. The experimental current profile is in fact close to the temperature profile, which is consistent with the fact that the FWCD is carried by thermal electrons. In counter-current operation, the FW driven current fraction is smaller and not in agreement with predictions (even including the poloidal effects). No satisfactory explanation of this effect is up to now available, and further experiments are planned on this discrepancy.

3.4 The Pellet Enhanced Performance (PEP)

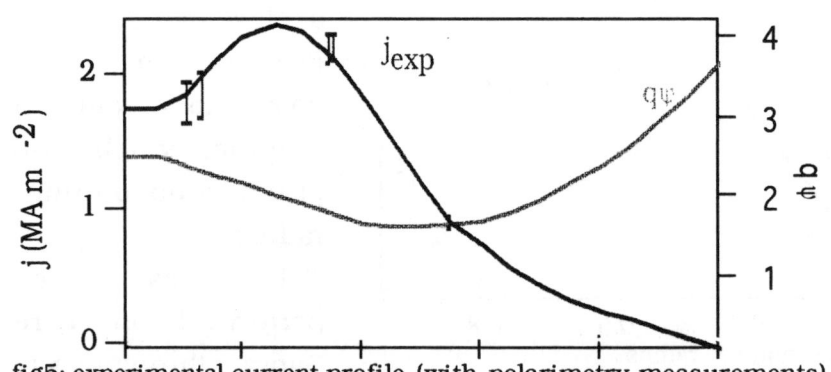

fig5: experimental current profile (with polarimetry measurements), bootstrap current profile and q profile during the PEP phase.

Finally, the PEP regime has been reproduced on Tore Supra with a clear evidence of a shear reversal in the plasma core [11]. A fast pellet is launched in the plasma with a two-stage injector (pellet velocity 3.3km/s) on a 3MW ICRF heated plasma. The central density reaches $2 \cdot 10^{20}$ m^{-3}, and the triple product $n_i(0)T_i(0)\tau_E = 1.1 \cdot 10^{20}$ m^{-3}.keV.s. During a 300ms sawtooth-free period following the pellet ablation, a modification of the current profile is correlated to the decrease of heat conductivity. The central shear reversal (fig5) results from the central temperature drop and the bootstrap current driven by the strong density gradient ($\approx$ 15% of the total current). In this regime, the electron and ion energy contents are comparable, showing that the confinement enhancement due to a shear modification probably does not require hot electron modes.

4. Discussion and perspectives

The magnetic shear profile appears to play the key role in the confinement of Tore Supra plasmas. The global confinement can be improved by increasing the shear value in the gradient zone ($0.4 < \rho < 0.7$). In the FWEH experiments, a large bootstrap current fraction has been achieved (45% of the total current), leading to a significant enhancement factor ($H = W_e/W_{eRLW} \approx 2$). Steady-state improved confinement has been demonstrated in the LHEP regime, with a strong electron heating (T_{e0} = 8-10 keV), in which the additional central performances are linked to the low or negative shear. A triple product $n_i(0)T_i(0)\tau_E = 1.1 \cdot 10^{20}$ m^{-3}.keV.s has been reached during PEP. The results on the shear modifications are summarised on fig6, where the shear profile, normalised to the corresponding ohmic profile, is shown for high-l_i, LHEP, monster sawteeth and FWEH regimes.

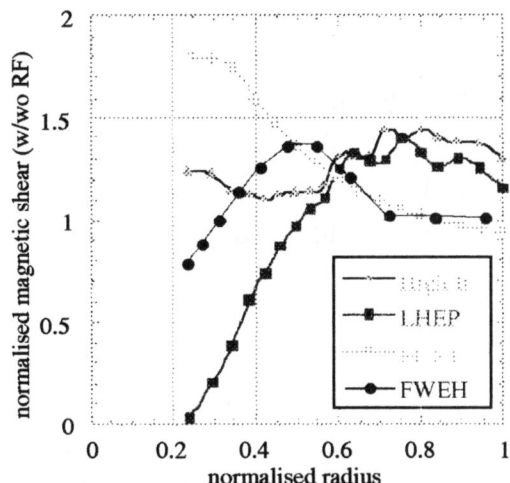

fig6: shear profile normalised to ohmic in various improved confinement regimes.

The local RLW model [8] correctly describes the qualitative influence of the shear on the local χ_e. However, the comparison between predictions and various experimental situations (ohmic [3], L-mode, improved confinement regimes) showed a quantitative agreement only when the T_e gradient is much larger than the critical temperature gradient (∇T_{ec}) which appears in the RLW theory. This suggests that ∇T_{ec}, given in [8], is too low. A slight modification of ∇T_{ec}, including a shear dependence, gives a better fit for all regimes described above [2]. Supplementary studies are requested to fully validate such an assumption.

Both LHCD and ICRF have supplementary potentialities for tailoring the current profile: the synergism between LHCD and Bernstein wave close to the ion-ion hybrid layer [12], mode conversion heating and CD [13], which present presumably a high efficiency and a good localisation. Increasing coupled powers and pulse duration, and/or combining some regimes presented above, give confidence in the possible exploration of the high-beta advanced tokamak regime. This regime should combine current profile control (shaping and feedback of shear reversal profiles) to the long time scale stability (in particular with respect to MHD), in high power-low current discharges.

References

[1] D. Van Houtte and Equipe Tore Supra, Nucl. Fus, **33** (1993) 137.
[2] Equipe Tore Supra, presented by B. Saoutic, 21th EPS Conference on Controlled Fusion and Plasma Physics, Montpellier (FRANCE), June 27- July 1, 1994.
[3] L. Guiziou et al., 21th EPS Conference on Controlled Fusion and Plasma Physics, Montpellier (FRANCE), June 27- July 1, 1994.
[4] G. R Harris et al., Nucl. Fus, **32** (1992) 1967.
[5] R.V. Budny et al., Nucl. Fus, **32** (1992) 429.
[6] B. Saoutic et al., 10th Topical Conference on Radio Frequency Power in Plasmas, Boston (USA) April 1-3, 1993.
[7] G.T. Hoang et al., Nucl. Fus, **34** (1994) 75.
[8] P.H. Rebut et al., Phys Fluids B **3** (1991) 2209.
[9] J. P. Bizarro, Y. Peysson et al., Phys Fluids B **5** (1993) 3276.
[10] X. Litaudon, private communication.
[11] X. Garbet et al., 15th IAEA International Conference on Plasma Physics and Controlled Nuclear Fusion Research, Seville (SPAIN) 1994.
[12] C. Gormezano, 10th Topical Conference on Radio Frequency Power in Plasmas, Boston (USA) April 1-3, 1993.
[13] J. Wilson et al., 15th IAEA International Conference on Plasma Physics and Controlled Nuclear Fusion Research, Seville (SPAIN) 1994.

Appendix:

Equipe Tore Supra and collaborations :

G. Agarici, J. P. Allibert, J.M. Ané, R. Arslanbekov, S. Balme, B. Bareyt, V. Basiuk, P. Bayetti, L. Baylor [1], B. Beaumont, R. Becherer, A. Bécoulet, M. Benkadda [2], G. Berger-By, S. Bério, D. Bessette, P. Bibet, J.P. Bizarro [3], G. Bon-Mardion, P. Bonnel, J. Boscary, J.M. Bottereau, F. Bottiglioni, S. Brémond, R. Brugnetti, R. V. Budny [4], Y. Buravand, H. Capes, J.J. Capitain, J.L. Carbonnier, J. Carrasco, C. Chamouard, P. Chappuis, D. Chatain, E. Chatelier, M. Chatelier, J.H. Chatenet [5], D. Ciazynski, F. Clairet, L. Colas, J.J. Cordier, A. Côté [6], J.P. Coulon, B. Couturier, J.P. Crenn, P. Cristofani, C. Deck, P. Decool, B. De Gentile, H. Demarthe, J. C. De Haas, C. De Michelis, P. Deschamps, C. Desgranges, P. Devynck, L. Doceul, C. Doloc, M. Dougnac, H.W. Drawin, M. Druetta [7], M. Dubois, J.L. Duchâteau, T. Dudok de Wit, L. Dupas, D. Edery, D. Elbèze, D. Escande, J.L. Farjon, I. Fidone, J.R. Ferron [8], M. Fois, C. Forrest [8], C.A. Foster [1], D. Fraboulet, D. Frigione [9], V. Fuchs [6], M. Fumelli, B. Gagey, L. Garampon, X. Garbet, E. Gauthier, A. Géraud, F. Gervais [10], P. Ghendrih, C. Gil, G. Giruzzi, M. Goniche, R. Gravier, B. Gravil, M. Grégoire, D. Grésillon [10], C. Grisolia, A. Grosman, D. Guilhem, B. Guillerminet, R. Guirlet, L. Guiziou, J. Harris [1], G. Haste [1], P. Hennequin [10], F. Hennion, P. Hertout, W.R. Hess, M. Hesse, G.T. Hoang, J. Hogan [1], T. Hutter, J. Idmtal, R. Isler [1], C. Jacquot, B. Jager, F. Jequier, E. Joffrin, J. Johner, J.Y. Journeaux, P. Joyer, S. M. Kaye [4], C. Klepper [1], F. Kazarian, K. Kupfer [8], H. Kuus, D. Lafon, P. Laporte, J. Lasalle, L. Laurent, C. Laviron, G. Leclert [11], P. Lecoustey, A. Ledyankine, C. Leloup, Y.Y. Li [12], P. Libeyre, M. Lipa, X. Litaudon, T. Loarer, P. Lotte, J.F. Luciani [5], T. Lutz [13], R. McGrath [13], P. Magaud, A. Mahdavi [8], R. Magne, G. Martin, A. Martinez, E.K. Maschke, R. Masset, M. Mattioli, G. Mayaux, Y. Michelot, P. Mioduszewski [1], J. Misguich, P. Mollard, P. Monier-Garbet, D. Moreau, F. Moreau, J.P. Morera, B. Moulin, D. Moulin, M. Moustier, R. Nakach, F. Nguyen, R. Nygren [13], P. Ouvrier-Buffet, L. Owen [1], M. Pain, J. Pamela, F. Parlange, G. Pastor, R. Patris, M. Paume, J. Payan, A.L. Pecquet, B. Pégourié, Y. Peysson, D. Piat, J.M. Picchiottino, P. Platz, C. Portafaix, M. Prou, L. Qualls [1], A. Quemeneur [10], J.M. Rax [14], H. Renner [15], G. Rey, P. Riband, B. Rothan, P. Roussel, S.A. Sabbagh [16], R. Sabot, F. Samaille, A. Samain, B. Saoutic, J. Schlosser, A. Seigneur, J.L. Segui, F. Smits, K. Soler, P.G. Sonato [17], Y. Stephan [18], M. Talvard, J. M. Theis, C. E. Thomas [1], S. Tobin [13], G. Tonon, A. Torossian, A. Truc [10], E. Tsitrone, B. Turck, S. Turlur, T. Uckan [1], G. Urquijo, J.C. Vallet, J. Valter, D. VanHoutte, A. Verga [2], D. Vézard, H. Viallet, J. Weisse, R. White [4], T. Wijnands, M. Zabiego, XL. Zou.

---

[1] Oak Ridge National Laboratory, Oak Ridge, Tennessee, USA.
[2] Equipe Turbulence Plasma, Institut Méditerranéen de Technologie, Marseille, France.
[3] Instituto Supérior Tecnico, Association EUR-IST, Lisboa, Portugal.
[4] Princeton Plasma Physics Laboratory, Princeton University, New Jersey, USA.
[5] Centre de Physique Théorique, Ecole Polytechnique, France.
[6] Centre Canadien de Fusion Magnétique, Varennes, Québec, Canada.
[7] Laboratoire de Traitement du Signal et d'Instrumentation, CNRS, St Etienne, France.
[8] General Atomics, San Diego, California, USA.
[9] C.R.E. Association EUR-ENEA, Frascati, Italy.
[10] Physique des Milieux Ionisés, Ecole Polytechnique, Palaiseau, France.
[11] L.P.M.I., Université de Nancy I, Vandœuvre les Nancy, France.
[12] Institute of Plasma Physics, Hefei, China.
[13] Sandia National Laboratories, Albuquerque, New Mexico, USA.
[14] Université Paris XI, Orsay, France.
[15] Max Planck Institut fur Plasmaphysick, Association EUR-IPP, Garching, Deutschland.
[16] Department of Applied Physics, Columbia University, New York, New York, USA.
[17] C.S.Gas Ionizzati, Association EUR-ENE-CNR, Padova, Italy.
[18] CISI Ingénierie, Centre d'Etudes de Cadarache, France.

Physics Research with Small Tokamaks

P.K. Kaw
Institute for Plasma Research
Bhat, Gandhinagar 382 424
INDIA

Abstract

In this paper we review the contributions that small tokamaks have made to physics research in general and to fusion physics in particular. It is argued that although most spectacular and newsworthy results are obtained from large tokamaks like JET, TFTR, JT-60 etc., they are often based upon ideas and concepts that have originated in the community of small tokamaks. This latter group are therefore the 'unsung heroes' who form the true backbone of tokamak research.

We start our discussion with the definition : 'What is a small tokamak ?' Tokamaks are toroidal magnetic cages produced by external toroidal fields and internal plasma currents. The device complexity and cost typically goes as some power of the plasma current I. A small tokamak is one in which the complexity is small and so the current is restricted, $I \lesssim I_c$. Note that this definition means that the physical size of a machine is irrelevant and may sometimes be deceptive. Thus for a circular plasma

$$I \sim B_\theta a \sim \frac{\epsilon}{q} B_T a \lesssim I_c$$

implies that $(B_T a) \lesssim$ critical value. Thus we may have a physically large machine ('a' large) with small B_T classified as a 'small tokamak' (e.g. Macrotorr) or a physically small machine with large B_T classified as a medium-size or large tokamak (e.g. Alcator C-Mod). The only thing that we must ensure from the size is that for an interesting and relevant tokamak $(\Delta a)_n \ll a$ shere $(\Delta a)_n$ is the penetration depth of neutrals so that the important atomic physics phenomena are dominant only in the edge of the machine. Quantitatively, we shall take the working definition that tokamaks with I less than 300-500 Kamps. may be classified as small tokamaks.

A limitation on I limits the cost and complexity of a tokamak but also limits its parameter space of operation. However, it should be noted that the typical objective of a small tokamak is <u>not</u> pushing parameters (this is best left to the 'large' machines). Instead, one finds that small tokamak groups use their flexibility and ingenuity to
(a) do support work for large devices,
(b) explore new territories, and
(c) elucidate physical phenomena.
There are around 50 small tokamaks operating in the world today. We shall now present examples from some recent work to illustrate how small tokamaks are contributing to the above mentioned topics. The choice of the material is highly subjective (simply meant to illustrate a few points) and I apologize to those groups whose work could not be included because of lack of time and space.

1 Support Studies for Large Tokamaks :

Large tokamak research, which is now culminating in the final design of the proposed ITER device, has uncovered several areas of concern and much of small tokamak research is directed to these areas. We now cite a few examples.

1.1 β limit studies :

Experiments and theoretical studies have shown that the plasma pressure one may load on a given toroidal field is limited by MHD stability considerations. The critical β (ratio of plasma pressure to magnetic pressure) is given by the so called Troyon formula, $\beta_c = g(I/aB_T)$ where g is a geometrical factor, which may be optimised by shaping the magnetic surfaces and the pressure and current profiles. Interestingly, even the energy confinement of the tokamak plasma is strongly influenced by surface shaping and profile optimisation. A new small tokamak, the TCV device, totally dedicated to these studies, presented some preliminary results[1] at IAEA (1994). Fig. 1 illustrates the enormous variety of plasma shapes that can be explored with

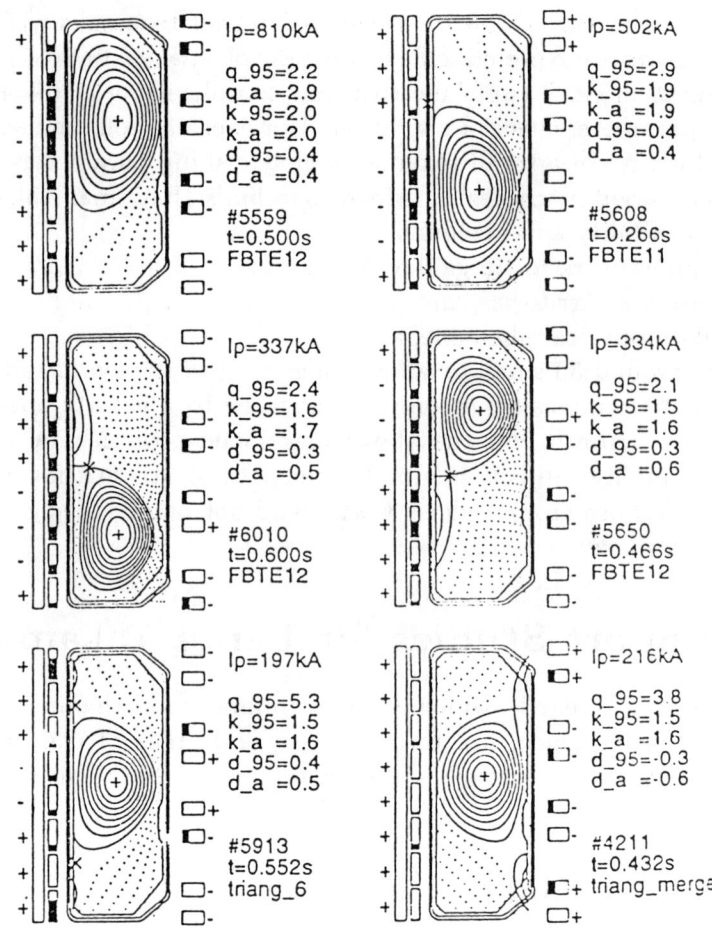

Figure 1: Various plasma shapes produced in TCV (Lister et al [1])

this device (because of the large vacuum vessel and the versatile poloidal field coil system). Experiments/theory have also indicated that whereas the internal MHD modes may be completely stabilized by shaping and profile optimisation, the plasma may still be unstable to low n external kink modes ; these are stabilized by the presence of passive conducting shells in the close vicinity of the plasma. One of the worries has been the possible influence of toroidal gaps (necessary for diagnostic and auxiliary heating) on the stabilizing properties of conducting shells. The HBT-EP experiment[2] presented some excellent data indicating that narrow gaps do not compromise the stabilizing properties of conducting shells. Fig. 2 illustrates how the lifetime of a discharge could be significantly enhanced by the insertion of retractable shells; this is further corroborated by observation of significant reduction in MHD fluctuations when the shells are inserted.

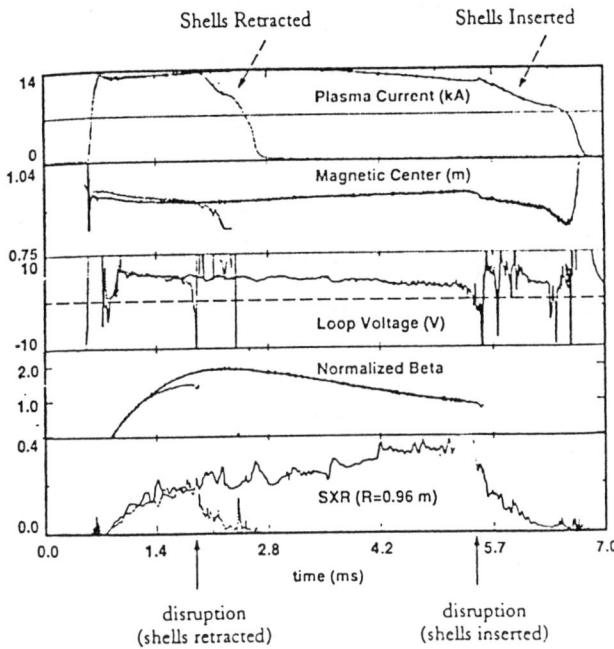

Figure 2: HBT-EP plasma evolution with conducting shells in retracted and inserted positions (Ivers et al [2])

1.2 Disruption Control :

A major area of concern for all large tokamaks is how to work in an interesting parameter space (e.g. close to 'density limits') and still prevent plasma disruptions, which if allowed to take place too frequently, can severely stress the machine components and compromise its integrity. Many small tokamaks have developed active methods of control involving the use of feedback coils producing relevant magnetic perturbations at the boundary, and the use of ECRH heating. We present results from the DITE experiment[3] using resonant feedback coils (Fig.3) and the JFT-2M experiment[4] using ECRH methods (Fig.4); both illustrate how a higher density parameter space may be explored with the help of these controls. Similarly, recent work on Compass[5] and Doublet D-III D[6] devices has identified the importance of error fields in generating locked modes and disruptions. It has been pointed out that ITER size machines can tolerate only very small error fields. The use of correction coils, which can reduce the error fields to acceptable levels, has thus become an integral part of the design of all large devices - thanks to the fundamental work on small tokamak experiments.

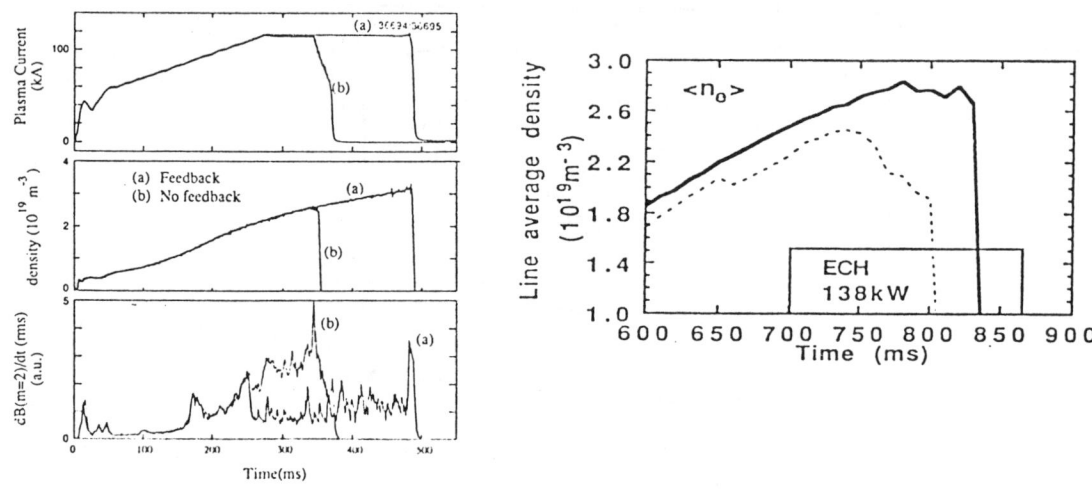

Figure 3: Extension of density limit in DITE by resonant feedback coils (Morris et al [3])

Figure 4: Extension of density limit in JFT-2M by ECRH (Hoshino et al [4])

1.3 Divertor Studies :

The physics of tokamak divertors is crucial from the point of view of particle control (fuel recycling, impurity influx reduction, helium ash removal etc.) and the handling of the heat flux coming out at the tokamak edge. Small tokamaks are making important contributions to this physics area e.g. tokamaks JFT - 2M and T deV have demonstrated[7,8] that biasing of divertor plates with respect to plasma chamber walls can (i) control the asymmetry of the particle/heat flux between the inner and outer divertor plates, (ii) help in the pumping of helium by additional ExB drifts in the divertor region. Figs 5 and 6 illustrate these phenomena. Controls like this are bound to play an important role in the utilisation of divertor in large tokamaks.

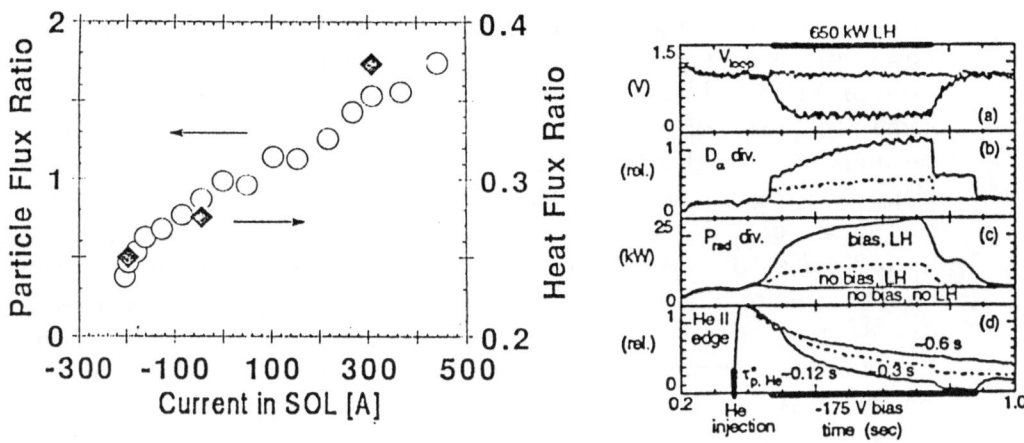

Figure 5: Particle/Heat flux control by divertor plate biasing in JFT-2M (Tamai et al[7])

Figure 6: Improved Helium pumping in T dev by biasing of divertors (Decoste et al[8])

2 Exploration of New Territories :

Large and medium size tokamaks are typically conservative and too inflexible to try out new ideas for the first time and to explore totally new territories. They excel in producing spectacular and impressive results using ideas which have been tested out on small tokamaks first. We now illustrate this with a few examples.

2.1 The START Experiment :

It has been apparent for some time now that due to several theoretical considerations[9], a low aspect ratio tokamak ($A \lesssim 1.5$) may have several advantages over its higher aspect ratio cousins. Yet, such is the conservatism of the tokamak community, that aspect ratios have hovered around 3, and no experiments were tried in the proposed regime till recently. The START experiment (Small Tight Aspect Ratio Tokamak experiment) has therefore, come as a breath of fresh air[10]. The stated advantages of a low aspect ratio tokamak are : a higher MHD stable β, higher bootstrap current, natural high elongation, natural divertor etc. Fig.7 ilustrates some of the features mentioned above. Using induction and compression methods, plasma currents upto 250 Kamps. have been reported[10]. The experiment has also demonstrated some totally unexpected features. Thus the confinement sems to be (Fig.8) superior to what comes out of neo-Alcator scaling (perhaps the effect of shaping or perhaps the plasma is in an Ohmic H-mode). The plasma also exhibits unexpectedly good resilience to disruptions as one approaches the density limit. It appears that the plasma undergoes a series of internal reconnection events ('minor disruptions') as the density limit is approached so that the current drops softly to zero rather than have a hard crash. If this turns out to be a genuine reproducible characteristic of such low aspect ratio discharges, it could be a major advantage over conventional tokamaks. The START experiment has been so successful that there are already proposals for megaampere size devices and even reactor studies based on this concept.[11]

An interesting variation on this concept is the small CDX-U experiment[12]. Here a low aspect ratio tokamak configuration was created through 100%

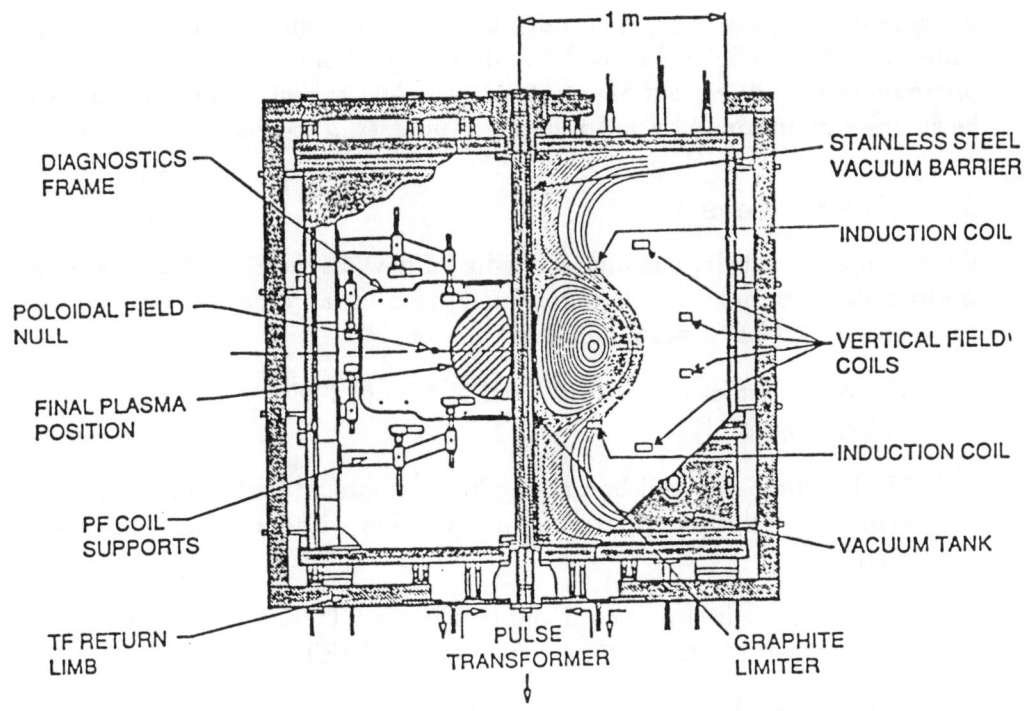

Figure 7: The START Configuration (Sykes et al[10])

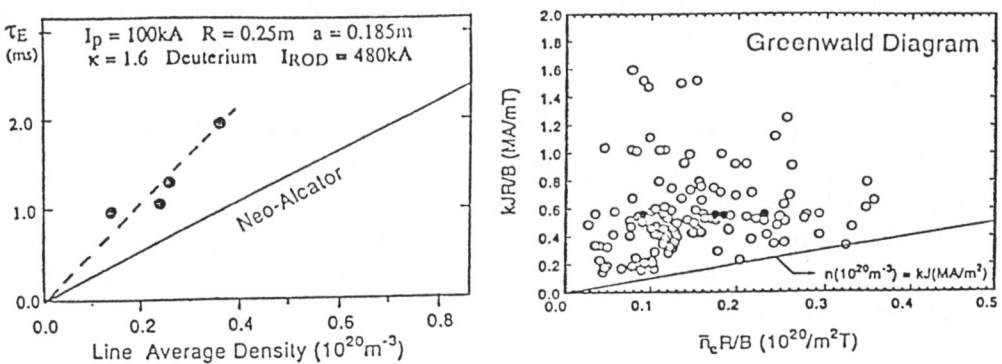

Figure 8: τ_E vs. $\bar{n}_e$ for START for discharges shown in Greenwald diagram. (Sykes et al [10])

pressure-driven ('bootstrap') currents in a toroidally magnetized plasma with additional vertical fields, heated by ECR heating. The experiment is still very preliminary in nature, but illustrates how a low aspect ratio tokamak may be steadily maintained by self-generated 'bootstrap' currents.

2.2 New Ideas :

We now present a brief but impressive list of new discoveries/ideas first studied in small tokamaks and now routinely used in large devices and forming an essential part of tokamak folk-lore!

1. H-modes first discovered in ASDEX[13] (note ITER <u>must</u> operate in H-mode to ignite !)

2. Wall carbonizing and boronizing first explored[14] in Textor (no impressive results with auxiliary heating possible without using these methods)

3. AC tokamaks first proposed and tested[15] in STOR-M (most spectacular results[16] with 2 Mamps. current plasma in JET)

4. Lower Hybrid start-up and current drive first explored in devices like JIPP-T II[17] (most impressive recent results[18] with 3.5 Mamp LHCD in JT-60 U)

5. Steady state 70 min LHCD tokamak experiment[19] in TRIAM-1 (to be continued on a much more impressive scale in the proposed TPX and JT-60 SU experiments[20])

The list can go on as new ideas continue to be explored. To conclude this section, we present Fig. 9 taken from the results of T - deV experiment[21], which illustrates how a compact torus may be used to fuel a tokamak plasma, an idea first tested on the Caltech tokamak. We look forward to seeing more impressive results using the technique and its adoption by the large/medium-size tokamak community.

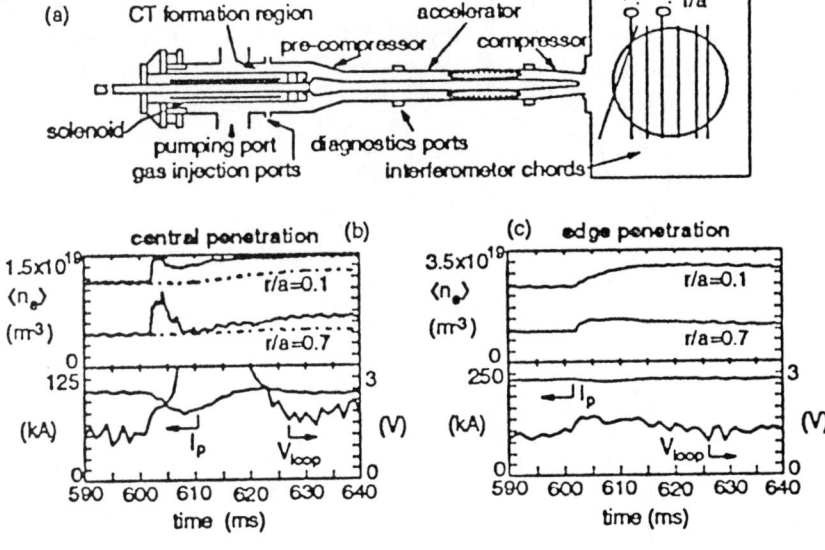

Figure 9: Compact toroid injection into Tdev (Decoste et al [21])

3 Elucidation of Physical Phenomena :

Perhaps the most lasting contribution from small tokamaks is in elucidation of physical phenomena taking place in these devices. Here I shall illustrate this aspect by only one example viz. the study of the physics of L - H transition in tokamaks. The understanding is still far from complete but has already led to new discoveries and experimentally verifiable predictions for novel operating scenarios. H-modes were first discovered[13] in the divertor

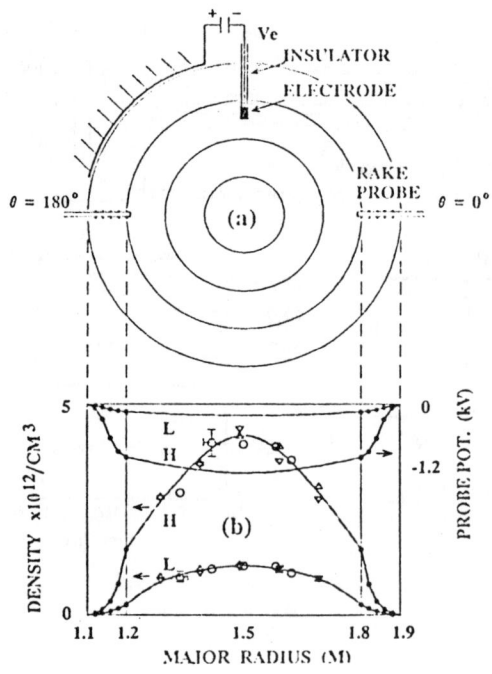

Figure 10: Edge profiles of n_e and ϕ in L and H modes produced by a biased probe (Taylor et al [24])

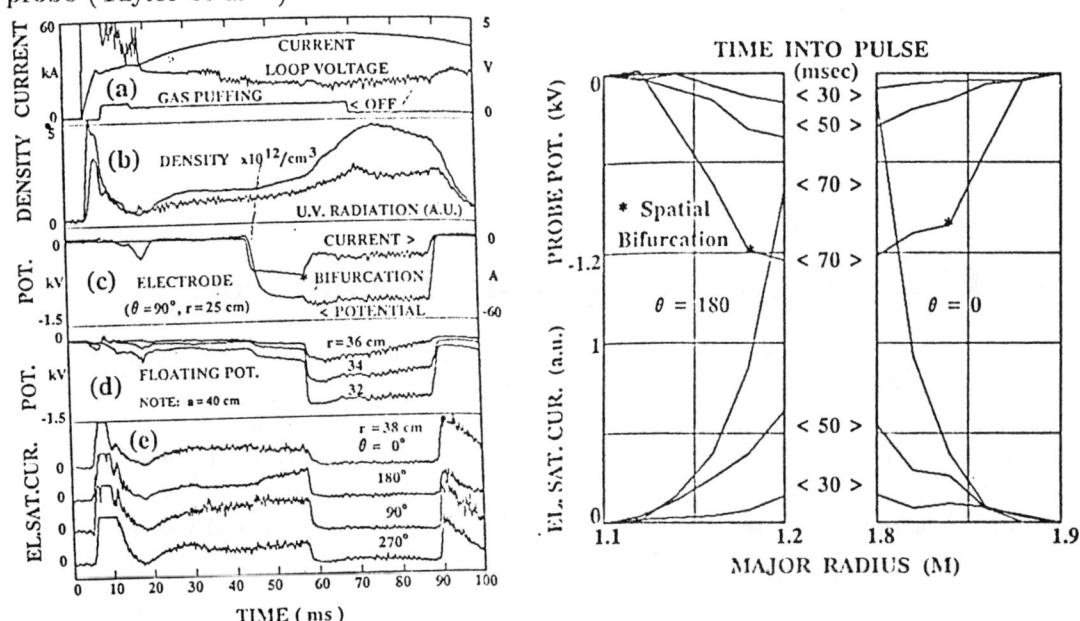

Figure 11: Characteristic H-mode behaviour induced by biased probe in Ohmic Limiter plasma (Taylor et al [24])

Figure 12: Edge Gradient sharpening behaviour seen between 50 and 70 msec in CCT (Taylor et al [24])

tokamak ASDEX in 1982. It is well known that tokamaks need auxiliary heating to reach fusion temperatures and that confinement which is determined by anomalous and turbulent processes, degrades with auxiliary power input P, giving us the L - mode scaling $\tau_E^L \propto P^{-0.5}$. ASDEX discovered that if the auxiliary heating power P exceeded a threshold P_{th}, the plasma suffered a bifurcation phenomenon and found itself in a better confined state with reduced fluctuation level at the edge, sharper gradients of density and temperature at the edge and significant improvements in particle and energy confinement times e.g. $\tau_E^H/\tau_E^L \sim 2$, $\tau_p^H/\tau_p^L \sim 4 - 10$ etc. Theoretical speculation[22,23] led to the concept that perhaps a radial electric field layer was generated at the boundary which could then explain the improved observed performance.

At this point an impressive and rather simple experiment was carried out[24] in the CCT device. As illustrated in Fig. 10, a biased probe was introduced in a limiter tokamak with Ohmic heating (no divertors, no auxiliary input power but instead an external charging system which could produce radial electric field layers). The plasma was found to bifurcate into a better confined state (Figs. 10-12) with chracteristics very similiar to the H-mode. Experiments on CCT and other small tokamaks have established the importance of 'transport-barriers' created by toroidal or poloidal velocity shear associated with radially lcoalized electric field layers in the plasma. These electric field layers may be created by major external intervention(such as in the biased probe experiment of CCT) or by self-consistent processes in a turbulent plasma in which the external injected power provides only a minor stimulating event. A paradigm involving the bifurcated states of a nonlinear stationary turbulent plasma in which the fluctuation level, the electric fields and rotation speeds and the net heat and particle transport are all tied to each other, is being developed and refined by further experiments and theoretical calculations. Many features are understood and many need further work. But it is no exaggeration to say that the CCT experiment triggered major new theory, simulation and experimentation on the L - H transition. This has led to discovery of many new confinement modes such as the VH modes seen[25] first in D-III D (apparently associated with transport barrier in the core), the observation of migrating transport barriers[26] etc.

A most impressive recent result is that of the discovery of Ohmic H-modes. It has been most thoroughly studied in the TUMAN experiment[27]. This experiment has no divertor, no auxiliary heating, no biased probes and

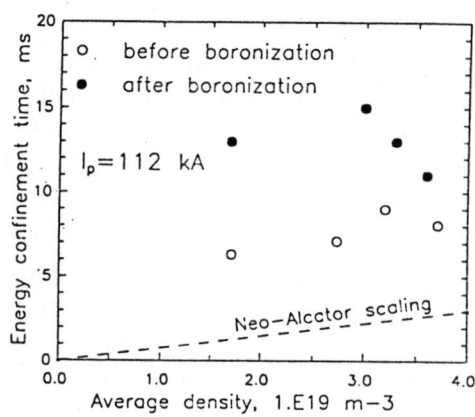

Figure 13: τ_E vs. $\bar{n}_e$ TUMAN in Ohmic H regime (Andrejko et al [27])

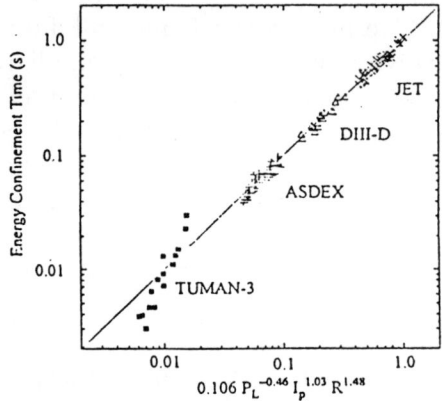

Figure 14: τ_E versus predictions of JET/D III D scaling for H - modes (Andrejko et al [27])

yet observes the plasma bifurcating into the better confined H mode (Figs 13 and 14). The TUMAN group also finds a simple scaling law which is satisfied by both Ohmic H-mode and the auxiliary heating D-III D, ASDEX and JET plasma experiments. This seems to suggest that very similar physics must be operating in Ohmic H discharges also. Obviously much more work needs to be done on this topic.

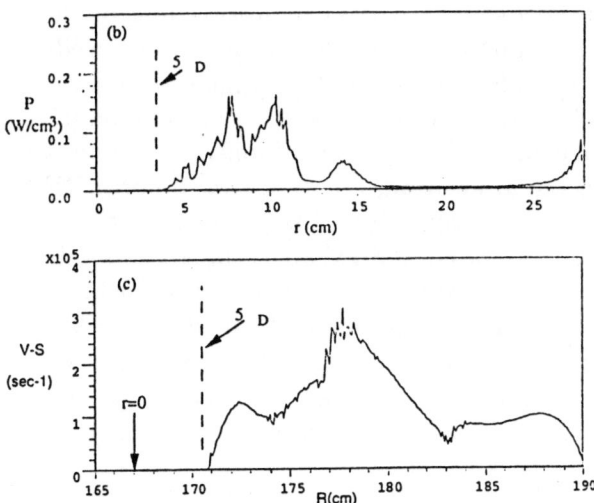

Figure 15: Calculated RF power profile and velocity shear profile due to ion Bernstein Modes in PBX-M (Ono et al [28])

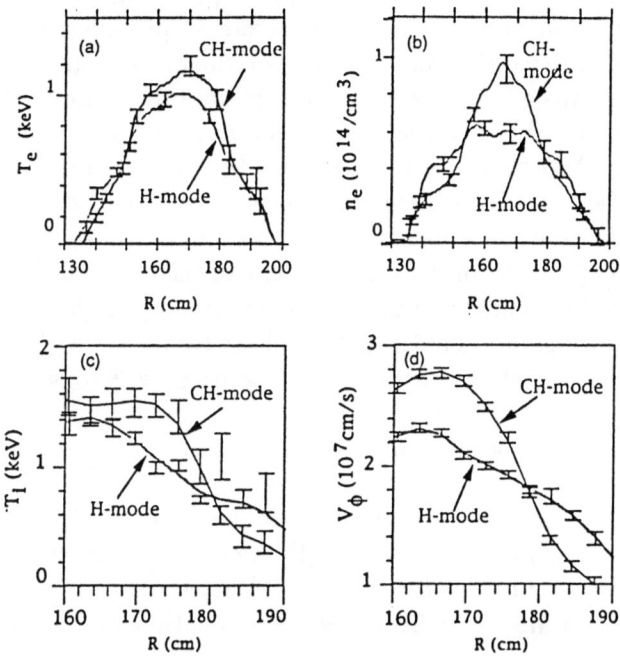

Figure 16: Gradients of n_e, T_e, T_i and V_ϕ measured in CH - mode in PBX - M (Ono et al [28])

We now illustrate how even a partial understanding of the complex set of phenomena described above has led to an interesting prediction for performance improvement and its apparent experimental verification[28]. The PBX-M group has considered the possibility of creating a velocity shear layer in the core of PBX plasma by the injection of ion-Bernstein modes, which mode convert at a given radial location, creating non-uniform RF fields and associated ponderomotive forces[29] (Fig. 15). An experiment was carried out in PBX-M and demonstrated the existence of a CH-mode in which much sharper density and temperature gradients could be sustained in the velocity shear region (Fig.16). This experiment illustrates how we may be able to control pressure profiles in a fusion-grade plasma using external RF control and our understanding of the importance of velocity shear layers and their relevance to formation of transport barriers.

4 DISCUSSION

I hope that this brief discussion has convinced you that 'small' tokamaks are making a significant contribution in (a) supporting large tokamaks (b) exploring new ideas and (c) elucidation of tokamak physics. In the mad rush to be 'useful' and 'relevant', there is a danger that most small tokamak programs will lean towards (a). There is thus an imperative need that the peer group pressure helps this community to maintain a good balance between (a), (b) and (c). As one looks towards the future, one may dream that a multi-billion dollar world-wide tokamak programme will devote at least a few percent of its resources to research with small tokamaks. As demonstrated above, the payback on this investment in terms of understadning and generation of new ideas and techniques, is remarkably high. Thus the small tokamak community should continue to look at topics in (a) (of which there will be no dearth) but should also look at some new things in (b) and (c) such as :
(i) investigation of scaling and engineering advantages of large aspect ratio tokamaks
(ii) fundamental studies in Tokamak turbulence using modern techniques of nonlinear physics (such as those in TEXT[30], current filamentation studies[31] in RTP, 'bubbly' transport studies[32] in ADITYA etc)
(iii) fundamental studies in collisionless reconnection of crucial interest to

solar physics, magnetospheric physics and astrophysics
(iv) tokamaks as nonlinear irreversible systems exhibiting coherent patterns in driven states far from equilibrium and acting as negentropy machines[33]
(v) investigation of direct conversion systems which are likely to be crucial for advanced fuel tokamaks[34]
(vi) non-fusion applications of high grade tokamak plasmas such as in radiation sources, fusion torch applications, radioisotope sources etc.

Finally, I would like to propose that the 'small tokamak' community develop a personality of its own by meeting regularly and promoting one international meeting such as the IAEA Technical Committee meeting on Research using Small Tokamaks, which is held once every year.

REFERENCES

1. J. Lister et al., IAEA Conf. on Plasma Physics and Controlled Fusion, Seville (1994), Paper IAEA-CN-60/A-5-I-2.

2. T.H. Ivers et al., ibid IAEA-CN-60/A-3/5-P-15

3. A.W. Morris et al., Phys. Rev. Letters **64** (1990) 1254.

4. K. Hoshino et al., ibid IAEA-CN-60/A-5-II-3.

5. J.T. Seoville et al., Nucl. Fusion **31** (1991) 875.

6. A.W. Morris et al., Proc. of 18th Europeon Conf. on Controlled Fusion and Plasma Physics Berlin (EPS, Petit-Lancy, Switzerland 1991) **15c** II, 61.

7. H. Tamai et al., ibid IAEA-CN-60/A-1-II-3.

8. R. De Decoste at. al., ibid IAEA-CN-60/A-4-II-5.

9. Y.K.M. Peng and D.J. Strickler, Nucl. Fusion **26** (1986) 769.

10. A. Sykes et al., ibid IAEA-CN-60/A-5-II-5, Nucl. Fusion **32** (1992) 691.

11. R. Buttery et al., ibid IAEA-CN-60/F-I-3-1.

12. M. Ono et al., Proc. of 14th IAEA Conf. on Plasma Physics and Controlled Fusion, Wurzburg 1992 (IAEA, Vienna 1993) **1**, 693.

13. F. Wagner et al., Phys. Rev.letters **49** (1982) 1408.

14. F. Waelbrocek et al., Proc. 9th Int. Vacuum Congress, Madrid 1983 (Int. Union for Vac. Science, London) p.693.

15. H. Kuwahara et al., Proc. of 11th IAEA Conf. on Plasma Physics and Controlled Fusion, Kyoto 1986 (IAEA, Vienna 1987) **1**, 413.

16. M. Keilhacker et al., Proc. of 14th IAEA Conf. on Plasma Physics and Controlled Fusion, Wurzburg 1992 (IAEA, Vienna 1993) **1**, 15.

17. K. Toi et al., Phys. Rev. Letters **52** (1982) 2144.

18. Y. Ikeda et al., ibid IAEA-CN-60/A-3-I-1.

19. S. Itoh et al., Proc. of 13th IAEA Conf. on Plasma Physics and Controlled Fusion, Washington, D.C. 1990 (Vienna 1991) **1**, 733.

20. H. Ninomiya et al., ibid IAEA-CN-60/F-1-I-1.

21. R. Decoste et al., ibid IAEA-CN-60/A-4-II-5.

22. S.I. Itoh and K Itoh, Phys. Rev. Letters **60** (1988) 2276

23. K.C. Shaing and E.C. Crome, Jr., Phys. Rev. Letters **63** (1989) 2369; S. Itoh and

24. R.J. Taylor et al., Phys. Rev. Letters **63** (1989) 2365.

25. K.H. Burrell et al., ibid IAEA-CN-60/A-2-I-5.

26. Y. Koida et al., ibid IAEA-CN-60/A-2-I-3.

27. M. Andrejko et al., ibid IAEA-CN-60/A-2/4- P-5.

28. M. Ono et al., ibid IAEA-CN-60/A-3-I-7.

29. G.G. Craddock and P.H. Diamond, Phys. Rev. Letters **67** (1991) 1535.

30. R.V. Bravenec et al., Phys. Fluids **B4** (1992) 2127.

31. N.J. Lopes Cardozo et al., Phys. Rev. Letters **73** (1994) 256.

32. R. Jha et al., ibid IAEA-CN-60/A-4-II-4-1 ; Phys. Rev. Letters **69** (1992) 1375.

33. P.K. Kaw, Current Sci. **59** (1990) 1065 ; Current Sci. (to be published 1995).

34. M.N. Rosenbluth and F.L. Hinton, Plasma Physics and Controlled Fusion **36** (1994) 1255.

Tight Aspect Ratio Tokamak experiments and prospects for the future

A Sykes, Y-K M Peng[1], J.W.Connor, M Gryaznevich, T C Hender, J Hugill,
I Jenkins, R Martin, A W Morris, C Ribeiro[2], D C Robinson, M J Walsh
and H R Wilson

UKAEA Government Division Fusion, Culham Laboratory, Abingdon, Oxon UK OX14 3DB
(1) ORNL, Tennessee, USA
(2) University of Sao Paulo, Brazil

Abstract

The present status of experimental results from low aspect ratio tokamaks is described, together with plans for physics experiments at the mega-amp level. Further development of the concept, and its potential for a materials/component test facility or ultimately a fusion power plant, are indicated.

1 Introduction

Following the initial interest in the low aspect ratio ($A = R/a \leq 1.6$) tokamak [1], first experimental investigations were made by fitting a central rod (to provide a toroidal field) in two quite different existing devices: the spheromak at Heidelberg [2] and the Rotamak at Lucas Heights [3]. As shown in Fig 1, these provided 'tokamak' discharges having high current for a short time (spheromak) and low current for a relatively long time (Rotamak).

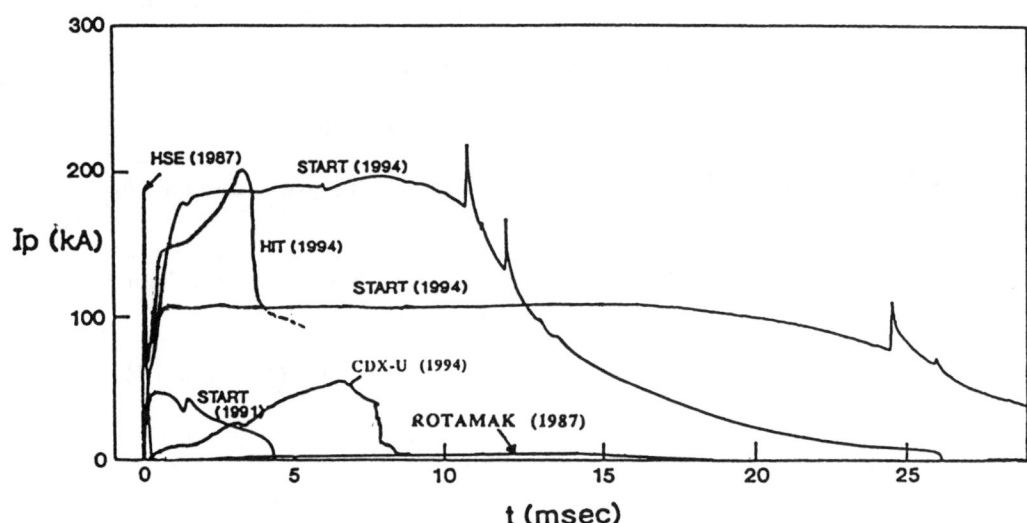

Fig 1 Plasma current vs time, for low A experiments

The first purpose-built device to produce standard 'spherical tokamak' discharges was START (Small Tight Aspect Ratio Tokamak) at UKAEA Culham Laboratory, which became operational in 1991, and results from START and plans for future development are described in section 2.

More recently, other low aspect ratio devices have become operational: CDX-U at Princeton, TS-3 at Tokyo University, and HIT at Seattle, and early results from these devices are summarised in section 3.

Many theoretical advantages are claimed for the low aspect ratio concept and experimental results so far appear to justify these and indeed to suggest further advantages. It is important to test out these properties in a well-diagnosed, additionally heated, relatively long pulse machine, and designs for several such machines are presently being considered. An outline design for MAST (Mega-Amp Spherical Tokamak) proposed by Culham is described in section 4.

If the present optimism is confirmed, the spherical tokamak should offer significant advantages over the conventional tokamak, due to improved plasma properties and also to its simpler, more compact construction. Steady-state scenarios for the spherical tokamak have recently been proposed, and possibilities for a low-A materials test facility, and ultimately a fusion power plant, are discussed in section 5.

2 The START experiment

2.1 Description of device

START, at UKAEA Culham Laboratory, became operational in January 1991 [4]. A cross-section of the device, showing the original and present layouts, is given in Fig 2.

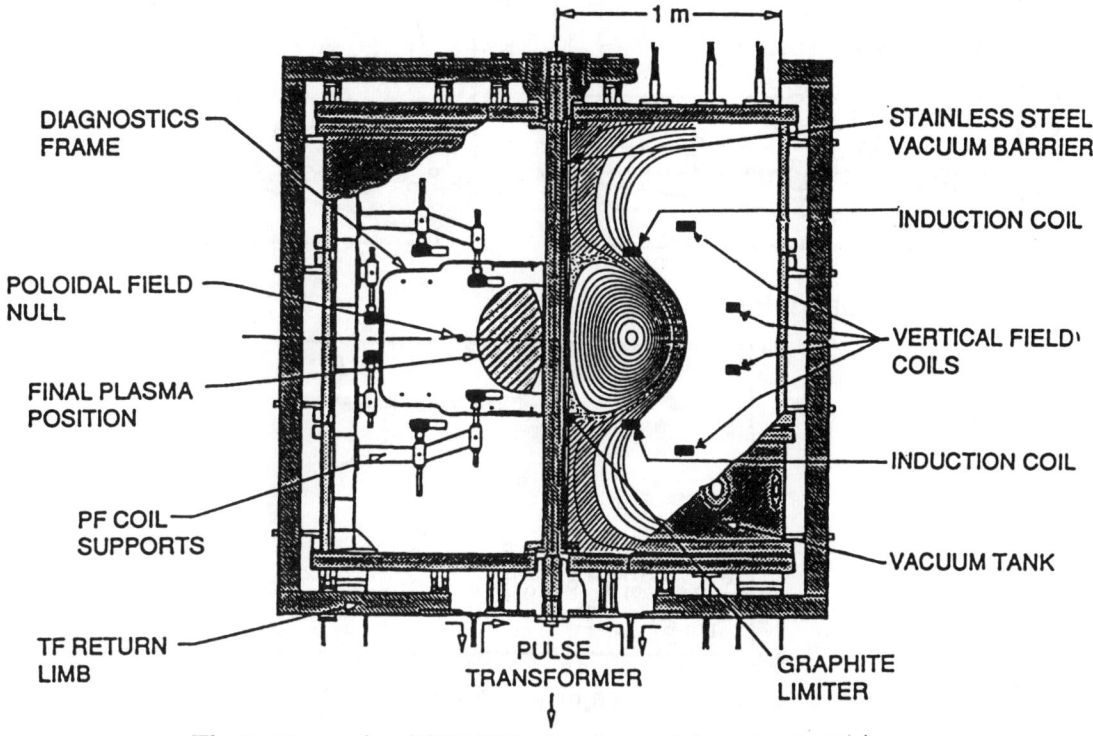

Fig 2 schematic of START, showing original (on left) and present (right) coil configurations

A feature of START is the use of in-vessel poloidal field coils. These are necessary to give adequate field control since the thick aluminium tank has a long time constant, and in addition the inboard pair (Fig 2) are used for plasma induction, currents of over 230kA having been obtained by the process of induction around these coils followed by compression to the low-A configuration. Use of the central solenoid alone (presently operated at up to 40mVs) can produce plasma currents of 160kA by direct induction. A combination of induction-compression followed by a current ramp using the central solenoid can produce even higher currents [5]. A second feature of START is that the central TF rod is part of the single-turn secondary of a large pulse transformer located under the vacuum tank. The combination of these features gives a very compact centre column and enables study of plasmas at aspect ratios as low as 1.25.

Each of the toroidal and poloidal field systems and the central solenoid are powered by capacitor banks, limiting the maximum pulse length to ~ 50ms. As shown in Fig 1, plasma conditions may be held approximately constant for 10 - 20ms (compared to a plasma energy confinement time τ_E ~ 2ms) enabling studies of plasma properties to be made. These properties are outlined in the following sections.

2.2 Operating space

Operation is in hydrogen or deuterium, using boronisation by glow-discharge in trimethyl-boron. Location of START data on conventional operating space diagrams is given in Fig 3.

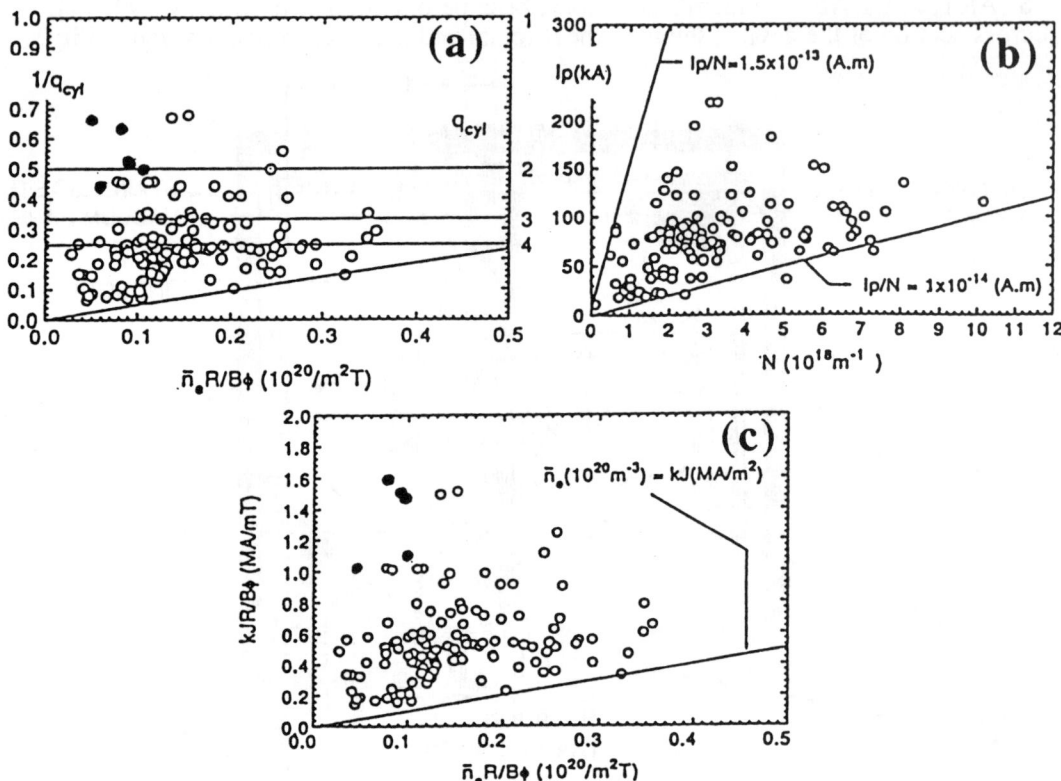

Fig 3 operating regimes for START (a) Hugill diagram (b) I/N diagram (c) Greenwald diagram

START can access relatively low values of q_{cyl} (= $5a^2 (\frac{1+k^2}{2}) B_T/RI_p$) as shown on the Hugill diagram (Fig 3a), and has a high Murakami number for an ohmic device. START data slightly exceeds the usual I/N ~ 1×10^{-14} Am boundary (Fig 3b where N = n_e nab) but is well represented by the Greenwald scaling (Fig 3c), showing the beneficial effect of plasma elongation.

2.3 Confinement time

Plasma energy confinement at low A is difficult to predict from present scaling laws. First results from START showed that τ_E exceeded the neo-Alcator scaling, and moreover showed no evidence of saturation at higher density [6]. Confinement in START appears to be better represented by the semi-empirical Rebut-Lallia (RL) [7] or Connor-Lackner-Gottardi (CLG) scalings. This latter is an adaptation of the Lackner-Gottardi scaling [8]:

$$\tau_E = 0.12\, I_p^{0.8}\, R^{1.8}\, a^{0.4}\, \frac{k}{(1+k)^{0.8}}\, \bar{n}_{e20}^{0.6}\, q_{cyl}^{0.4}\, P_{tot}^{-0.6}$$

to include a more appropriate treatment of the tight aspect ratio geometry obtaining

$$\tau_E = 0.21\, I_p^{0.8}\, R\, a^{1.2}\, \left(\frac{n_{e20}}{P_{tot}}\right)^{0.6}\, \frac{q_{cyl}^{0.4}}{f(\varepsilon)(2\varepsilon)^{0.8}}\, \frac{k^{1.4}}{(1+k^2)^{0.8}}$$

where $f(\varepsilon) = \dfrac{\langle q_{95}\rangle^{0.4}}{q_{cyl}^{0.4}} \approx 1.05 + 0.11\,(\varepsilon + 0.35)^{6.4}$

Recent results showing the dependence of confinement time on density for fixed plasma size and current are compared with various predictions in Fig 4. Radiated power, not included in the evaluation of the experimental confinement time, becomes a significant fraction of the input power for these discharges, so that the START points shown are underestimates, especially for the higher density cases.

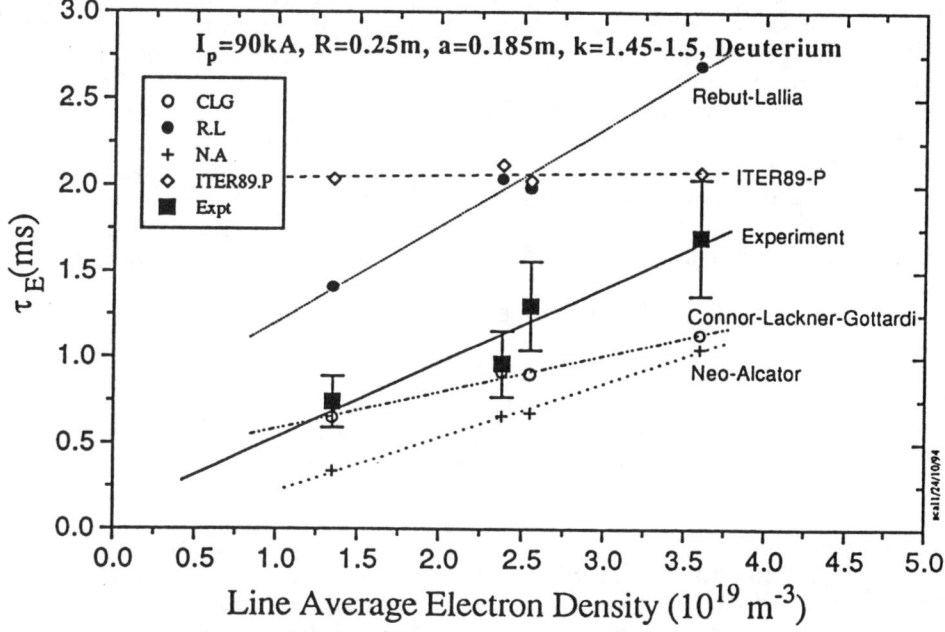

Fig 4 Comparison of recent START data, for fixed plasma size and current, with predictions of common scaling laws

2.4 Plasma exhaust

An equilibrium modelling of a typical START plasma was shown in Fig 2. The vertical field provided by the poloidal coils is approximately uniform, and the equilibrium illustrates two features: a high 'natural' plasma elongation [5] (which can reach k ~3 or higher for flat or hollow current profiles), and the 'natural divertor' whereby most of the exhaust plume is led away from the centre post.

Probe measurements of the scrape-off-layer (SOL) in START [13] indicate that this is significantly wider than that in conventional tokamaks, supporting recent theoretical predictions [14]. Together with the long connection length (several metres) and plume expansion observed in Fig 1, these features suggest that the spherical tokamak may have significant advantages over the conventional tokamak in the key area of plasma exhaust and divertor loading.

2.5 MHD

In general, the MHD behaviour observed in START is very similar to that observed in conventional aspect ratio tokamaks [6]. However, current terminating disruptions are not observed. START does suffer from 'internal reconnection events' (IREs) in which positive current spikes occur, and the IRE has similarities to the first stages of a current terminating 'major' disruption. The IRE occurs typically once per discharge, often during the current ramp-down, as in the examples shown in Fig 1. If the operating space boundaries are approached, IREs become more frequent: near the high density boundary, interferometer signals indicate that IREs act to reduce line average density; near the low-q boundary, to reduce the current. In this latter case a succession of IRE's can occur, and since there is no feedback control of the vertical field on START the plasma is driven onto the central limiter as the current/field ratio decreases. As the plasma size falls, the aspect ratio increases and a current quench can occur.

Three separate features may contribute to the observed resilience to the 'major disruption'. Firstly, at low A there is high q_a and high shear: $q_a/q_{cyl} \sim A/(A-1)$ [15]. Secondly, an IRE is observed to further increase plasma elongation, indicative of a reduction in internal inductance l_i. Since the external inductance of a low-A plasma, already small, is further reduced by increased elongation [16] the total plasma inductance is considerably reduced. This accounts for the large positive current spike which has the consequence that the inward plasma motion observed in conventional tokamaks is reduced or possibly reversed. In START this implies reduced contact with the central (and only) limiter. Finally, passive stabilisation may be important. The centre column of START itself may act as a stabilising conducting shell, and the conducting steel coil cases of the Induction coils (Fig 2) also have a stabilising effect.

3 The CDX-U, TS-3 and HIT experiments

CDX-U has a similar vacuum vessel to START but uses a multi-turn TF system, relies on a central solenoid for plasma induction and uses external poloidal field coils. Although the more complex centre column imposes a minimum aspect ratio of 1.5, higher than that of START, CDX-U has the potential advantage of a steady state TF system.

Initial plasmas studied in CDX-U have low density (to optimise induction) but interesting information on the MHD behaviour has been reported [9]. Using ECR pre-ionisation it is found that plasma current production is relatively insensitive to the toroidal magnetic field and plasma currents of over 50kA have been achieved with $q_a \sim$ 3.5, $q_{cyl} \sim 1.7$ and $I_p/I_{TF} \sim 0.35$. Plasmas having q_{cyl} down to unity have been obtained but with a significant enhancement of $m = 1,2/n = 1$ internal MHD activity. Plasmas with low q_a are more likely to terminate abruptly, although inadequate plasma position control may be partly responsible. Contact with upper and lower limiters is observed during the termination and, from START experience, this may indicate a sudden elongation following an IRE.

TS-3 at Tokyo University has recently been modified to operate as an Ultra-Low-Aspect-Ratio Tokamak "ULART" at aspect ratio as low as 1.05 [9]. Discharge lengths are necessarily very short, due to the absence of central solenoid or non-inductive current drive, and TS-3 is studying the connection between spheromak and ULART regimes, for example the stabilisation of tilt modes.

HIT (Helicity Injection Tokamak) has demonstrated very effective current drive of up to 200kA [10,11]. So far, plasma loop voltage has been ≥ 10V compared to ~ 1V for CDX-U and START plasmas and further development is required to see if the plasma current is HIT can be maintained - for example, by a conventional solenoid - or indeed if the helicity injection technique can be more generally applied.

4 Future development

Existing experiments (START, CDX-U, TS-3, HIT) are continuing to yield useful information. A 300kA device, GLOBUS-M should shortly begin construction at the Ioffe Institute, St Petersburg and larger 1MA devices are under consideration, including MAST at Culham which is described in the next section.

Key parameters of START are compared with those of MAST, a component test facility (CTF), and power plants STR in Table 1.

parameter	START	MAST	CTF	STR[1]	STR[2]
R(m)	< 0.3	~ 0.7	0.7	2.8	3.1
A = R/a	> 1.2	> 1.3	1.6	1.3	1.3
Ip (MA)	< 0.25	1 - 2	7 - 10	34	27
$B_{\phi 0}$(T)	< 0.5	< 0.6	2.3	1.4	1.1

Table 1 Key parameters of present and future applications of low-A devices

4.1 MAST (Mega Amp Spherical Tokamak)

This is essentially a scaled up version of START. It retains the internal coils and the induction-compression facility, but will have a multi-turn TF system capable of long pulse operation. As in START, the vacuum tank is distant from the plasma, especially in the vertical direction, so that plasma exhaust features can be investigated (Fig 6).

An important feature of the MAST programme will be the study of neutral beam additional heating and current drive, which appears the most straightforward method at the relatively low toroidal field envisaged for most spherical tokamaks. Preliminary investigations will be made on START in 1995 using a 0.5 MW 40keV beam, to determine the heating efficiency and to evaluate the effect of auxiliary heating on confinement.

4.2 Steady state scenarios for a spherical tokamak

Although it is not entirely certain whether a tokamak power plant has to be steady state, it would clearly be desirable if it were. Furthermore, for a large-volume component test facility to be most effective, it should be operared continuously. It is therefore of interest to determine what steady-state scenarios exist and whether these can be tested on MAST.

Conventionally, it is concluded that externally driven current requirements are minimised by obtaining high bootstrap fractions, which in turn implies high βp plasmas at large aspect ratio [17]. However, the large-aspect-ratio approximations used in these estimates can be entirely misleading at low aspect ratio and a full calculation has to be performed. At higher beta, a significant diamagnetic current contribution occurs, and a model equilibrium having 70% total pressure driven currents is described in [12].

Two types of steady-state scenario can be tested on MAST using beams of the type to be investigated on START. The beam current and profile are calculated consistently with the full numerical equilibrium using a Monte Carlo particle deposition code coupled to a Fokker Planck solver; equilibrium with monotonic q-profiles, which are ballooning (first regime) and external kink stable, are obtained.

The first, more modest, test (Fig 5) has low plasma and central rod currents (0.5 and 1.5MA), and 3.6MW of beam power at 27 keV. The pressure profiles are within the first region of stability, with Troyon factor $\beta_n = 2.5$ and energy confinement enhancement factor $H_{ITER-89P} = 1.5$, $H_{RL} = 1.7$, $H_{CLG} = 2.7$.

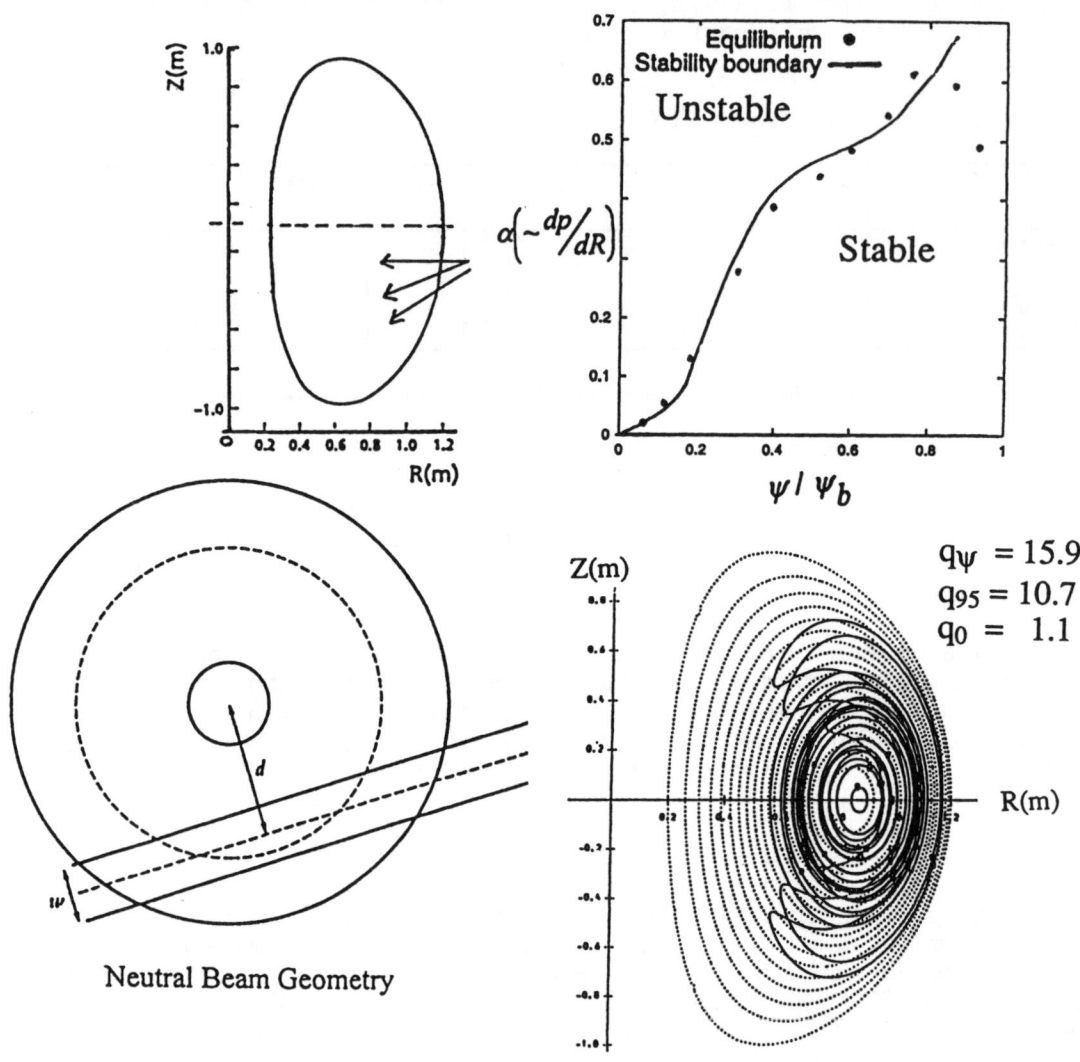

Fig 5 Steady state equilibrium in MAST. The total plasma current (0.5MA) comprises 40% bootstrap, 4% diamagnetic and 56% neutral beam contributions.

A second example tests the practicality of the 'second regime' of ballooning stability at high q_0 (2.8 in this case).. Here 5MW of beam power at 40keV is used, with $\beta_N = 5$, $H_{ITER-89P} = 3.1$, $H_{RL} = 1.9$, $H_{CLG} = 2.7$. Simulations show a full access to second stability across the entire radius.

Hence MAST should allow the important concepts of steady state operation, and access to (and steady state operation in) the second regime of stability, to be evaluated.

5. ST materials test facility, or power plant.

The steady state scenario outlined above, makes possible a compact materials test facility capable of providing the necessary neutron fluence required for component testing for future DEMO reactors. One design [12] is compared in Fig 6 with the START and MAST experiments. This example has a water-cooled central copper conductor, and uses the induction-compression technique to provide an initial current which is then driven by the neutral beams.

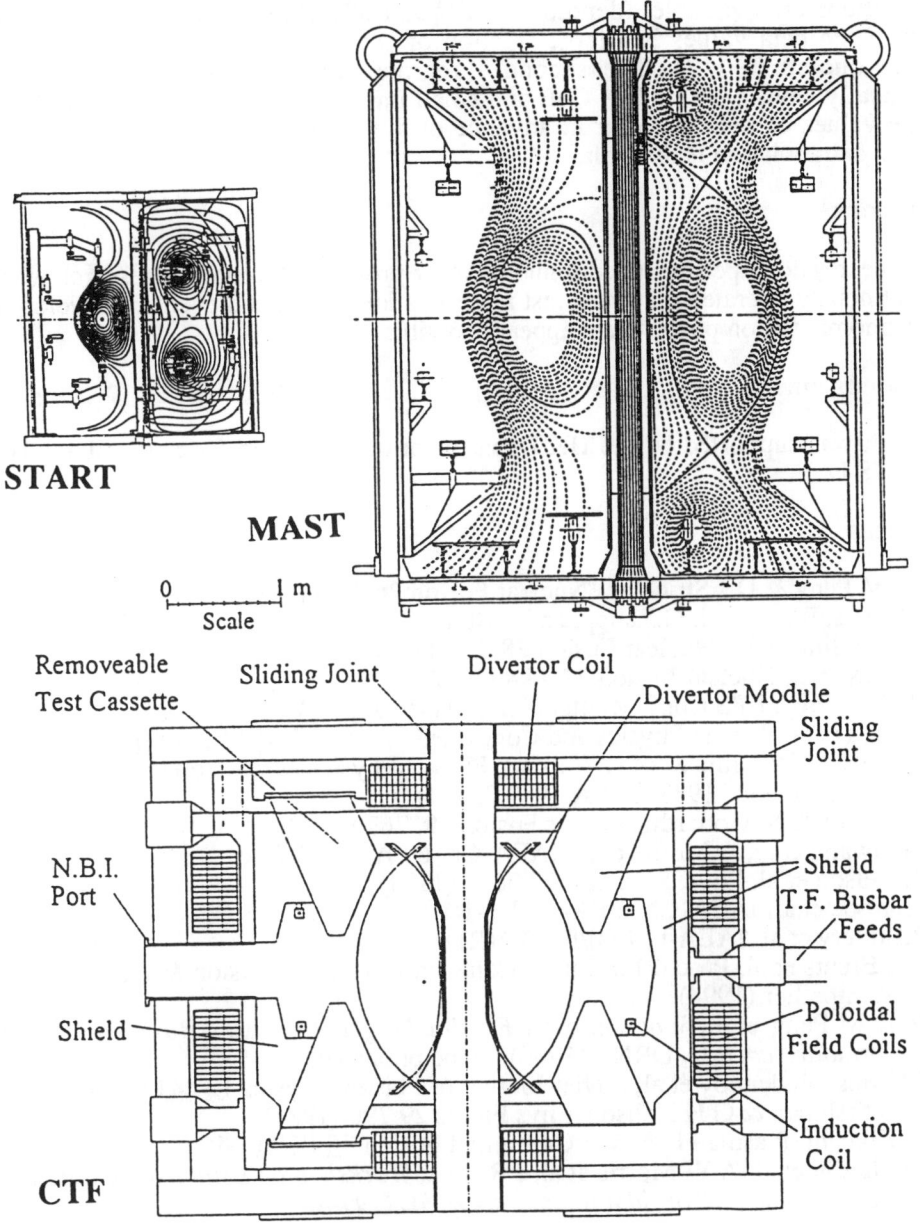

Fig 6 comparision of START, MAST and CTF design (to same scale). The plasmas in the MAST figure show a 'naturally' diverted plasma (left) and a double null diverted plasma using X-point coils (right).

The physics assumptions required for this design are quite modest, indeed similar to the first-stability region test proposed for MAST, with normalised beta $\beta_n = 2.9$ but more advanced confinement enhancement factors of $H_{ITER-89P} = 3.6$, $H_{RL} = 2.4$.

5.2 Fusion power plant

Whereas the function of the materials test facility is just to provide the required neutron fluence, so that a driven system is acceptable, a power plant has a substantial additional requirement - to give a competitive cost-of electricity (CoE). At present, this appears quite possible for the spherical tokamak power plant [12], by making use of the very high beta values theoretically predicted [18]. Both first regime, and second regime stable options are available, (denoted $STR^{(1)}$ and $STR^{(2)}$ in Table 1) and associated with the high beta values is a large fraction of pressure-driven current (78% in the example quoted in [12]. The second - regime option is especially attractive economically having both a low CoE and small unit size. However operation at such extreme values of beta is so far in advance of any present experimental results that confident extrapolation is difficult.

Conclusions

Tight aspect ratio experiments continue to show promise. A route for research leading to a continously operated materials test facility is identified, and eventual development to an economic fusion power plant appears possible.

Acknowledgement

This work was supported by the UK Department of Trade and Industry, EURATOM, and the US DoE.

References

[1] Y-K M Peng & D.J.Strickler, Nuclear Fusion **26**, 769 (1986)
[2] H.Bruhns, R.Brendel, G.Raupp, and J. Steiger, Nuclear Fusion **27**, 2178, (1987)
[3] G.A. Colling et al, Nuclear Fusion **28**, 255 (1988)
[4] A.Sykes et al, Nuclear Fusion **32**, 694 (1992)
[5] A.Sykes, Proc. EPS Conf, Montpellier 1994 (to appear)
[6] A.Sykes et al, Plasma Physics and Controlled Fusion **35**, 1051 (1993)
[7] P H Rebut, P P Lallia & M L Watkins, Plasma Physics & Controlled Fusion Research, Nice 1988 (IAEA 1989,**2**, 191).
[8] K.Lackner & N. Gottardi, Nuclear Fusion **30**, 767 (1990)
[9] Y.S.Hwang et al, A.Morita et al, IAEA-CN-60/A-5-II-6-2(R)
[10] B. Nelson et al, Phys. Rev. Lett.72 3666 (1994)
[11] T.Jarboe et al, IAEA-CN-60/A-%-II-6-1(R)
[12] R.Buttery et al, IAEA-CN-60/F-I-3-1(R)
[13] S.K.Erents et al, Proc. 21st Europ. Conf. on Controlled Fusion & Plasma Physics, Montpellier (1994)
[14] Y-K M Peng et al, *Workshop on Establishing the Physics Basis for Compact Toroidal Reactors*, ORNL (1994), to be published
[15] S G Bespoludennov et al, Soviet Journal of Plasma Physics <u>12</u> 441 (1986)
[16] S P Hirshman & G H Neilson, Phys Fluids <u>29</u> 790 (1986)
[17] M Kikuchi, Plasma Physics & Controlled Fusion <u>35</u> B39 (1993)
[18] S C Jardin et al; A W Morris et al; J Ramos, *Workshop on Establishing the Physics Basis for Compact Toroidal Reactors*, ORNL (1994).

CONTINUUM DAMPING IN PLASMA PHYSICS

Z. Sedláček
Institute of Plasma Physics, Academy of Sciences of the Czech Republic
P.O. Box 17, Za Slovankou 3, CZ-182 00 Prague 8, Czech Republic

Abstract

An attempt is made at a unified description and interpretation of non-dissipative damping phenomena of waves and oscillations in Vlasov or inhomogeneous plasmas as a leaking phenomenon in a suitably defined transform space. Phase mixing phenomena connected with the presence of a continuous spectrum are thus converted into a dispersive wave propagation and scattering in the transform space.

1 INTRODUCTION

In Plasma Physics, the non-dissipative damping phenomena of waves and oscillations in inhomogeneous or high-temperature collisionless plasmas are of unceasing interest. To document this it is enough to look at the enormous extent of work done recently on the continuum damping of toroidal Alfvén eigenmodes or on the Alfvén-wave heating of high-temperature plasma.

Phenomena of this kind are found in many other branches of physics. They are always connected with the presence of a *continuous spectrum* of eigenfrequencies (hence the name "continuum damping") and accompanying them *singular continuum eigenmodes*. The solution of the initial-value problem, expressed as an integral superposition of these continuum eigenmodes, may exhibit damping in time, fully understandable from the mathematical point of view but somewhat puzzling in view of the fact that the physical systems in question are energy-conserving and the operators involved are Hermitian.

The Laplace transform and Green's function techniques make it possible to express the solution in an alternative form of a contour integral in the complex frequency plane which in the end may be reduced to the sum over the singularities of the Laplace transform. The complex poles generated by zeros of the dispersion function located on "unphysical sheets" of its Riemann surface give rise to "modes" exponentially damped in time (called collective modes, quasi-modes, virtual modes, resonances, etc.), a still more puzzling and easily misinterpreted fact, as such solutions are usually encountered in dissipative systems.

Equivalency of the normal-mode analysis and the Laplace transform approach may rigorously be proved. Another fact to be stressed is that the complex zeros of the dispersion function on unphysical sheets are not part of the spectrum, there are no corresponding orthonormal eigenmodes and there is no expansion theorem which would be useful, for example, in non-linear perturbation theories. The situation that such an all-important phenomenon like Landau damping has not its own eigenmode should therefore be considered as unsatisfactory.

There were attempts in the past to resolve the situation: by Trocheris [1] who introduced a complex velocity space and quite recently by Crawford and Hislop [2] who employed spectral deformation techniques. But these theories do not solve the problem completely: the continuous spectrum persists in both and also they are rather formal.

Here we present another attempt at a unified description and interpretation of continuum damping phenomena which usually find their explanation in terms of what is known under the generic name of *phase mixing*. At the basis of our approach is the idea of conversion of phase mixing into propagation. The possibility of this conversion rests upon the observation that the problems under consideration permit description in terms of very similar integral equations consisting of a multiplication operator perturbed by an integral operator. We Fourier transform this original integral equation, thus converting it into a generalized wave equation which describes dispersive wave propagation and scattering in the transform space. All the familiar concepts of wave transmission, reflection and trapping in the "scattering center" are then available to interpret and disentangle the complex physical process of phase mixing. There is no damping in the true sense, only an intricate redistribution of the amplitudes of the excitation in the transform space: exponential damping phenomena like the Landau damping of plasma oscillations then correspond to the transient leaking of a wave-packet imperfectly trapped in the scattering center, exactly as the so called radioactive decay states in quantum mechanics. An enormous advantage of the transform space description is that it completely unfolds this redistribution process, separating various decay modes and clearly showing their causes.

In addition to transient processes there may also be stationary states in the transform space, scattering states of a monochromatic wave for example. They correspond to singular continuum eigenmodes known under the names of Van Kampen-Case [5, 6] and Barston [7]. A wave-packet perfectly trapped within the scattering center would correspond to a genuine *discrete eigenmode*.

In Sec. 2 we consider the simpler problem of Vlasov plasma oscillations. We derive the Vlasov-Poisson equation in the Fourier transformed velocity space,[1] a uni-directional wave equation perturbed by an integral operator, and show its basic properties. In particular we analyse the "free-propagation" which corresponds to the free-streaming, a simple shift in the Fourier transformed velocity space outside the scattering region. Then we solve the initial value problem for this equation by the Laplace transform technique and reduce the solution to the form of an eigenfunction expansion. This time all zeros of the dispersion equation, including those in the lower half of the complex frequency plane (Landau zeros among them), become part of the discrete spectrum and well-behaved eigenfunctions may be defined, by means of the residues at the poles of the Green's function. The eigenfunctions are also derived directly from the general solution of the basic eigenmode equation to show that it is the boundary condition imposed at infinity of the Fourier transformed velocity space that determines in a crucial way the character of the resulting solution and makes the whole difference between a singular (monochromatic wave) and regular (wave-packet) solution.

---

[1] Apart from an earlier application of this space in the numerical solution of the Vlasov equation, it was used for the first time by Best [3, p. 278].

In Sec. 3 we take up the problem of linear electrostatic oscillations of a cold inhomogeneous plasma and its mathematically equivalent counterpart, linear Alfvén oscillations in inhomogeneous incompressible magnetofluids (MHD), which was originally solved by normal-mode analysis [7, 8, 9] and by Laplace-transform and Green's-function techniques [8, 9]. As the transform space we use the wave-number space.[2] We derive the basic equation, the Fourier-transformed integro-differential equation for the transverse component of the electric field [9, 10], discuss the properties of free propagation in the wave number space, which is this time dispersive, and use them to calculate the asymptotic formula for the evolution of a freely propagating wave-packet.

2 VLASOV PLASMA OSCILLATIONS

Consider a spatially homogeneous collisionless plasma of electrons with the equilibrium distribution function $F(v)$ normalized to unit density and with immobile ions forming a neutralizing background. The linearized Vlasov equation for the perturbation distribution function $f(x, v, t)$ is first Fourier transformed with respect to the spatial coordinate x

$$\frac{\partial f_k(v,t)}{\partial t} - ikv f_k(v,t) + \frac{eN}{m}\frac{\partial F(v)}{\partial v} E_k(t) = 0, \qquad (2.1)$$

the perturbation electrostatic field $E_k(t)$ being given by the Poisson equation. The set of Vlasov and Poisson equations is now Fourier transformed with respect to velocity (q being the Fourier variable corresponding to v) to find that the electric field is given by the value of the Fourier transform of the distribution function $\varphi_k(q,t)$ (or, in other words, of its spectrum) at the zero point of the q-axis. With this result the Fourier transformed Vlasov-Poisson equation may be written as

$$\frac{\partial \varphi_k}{\partial t} - k\frac{\partial \varphi_k}{\partial q} + \frac{\omega_p^2}{k} q \Phi(q) \varphi_k(0, t) = 0, \qquad (2.2)$$

where ω_p is the electron plasma frequency.

The Fourier transform $\Phi(q)$ of a smooth equilibrium distribution function $F(v)$, like a Maxwellian, decays very rapidly as $q \to \pm\infty$. Thus, asymptotically for $q \to \pm\infty$, the Fourier transformed Vlasov-Poisson equation becomes a simple advective equation

$$\frac{\partial \varphi_k(q,t)}{\partial t} - k\frac{\partial \varphi_k(q,t)}{\partial q} = 0 \qquad (2.3)$$

which is the Fourier transform of the free-streaming part of the Vlasov equation. To the free-streaming solution $f_k(v,t) = f_k(v,0)\exp(ikvt)$, a typical phase mixing process consisting in creation of finer and finer structures in the velocity space as time proceeds, there corresponds a simple uni-directional shift, the "free-propagation", in the Fourier transformed velocity space $\varphi_k(q,t) = \varphi_k(q+kt,0)$. The q-axis may therefore be divided into three regions: the "channels", the two semi-infinite regions of free-propagation far from the origin where $\Phi(q)$ is negligible and the "scattering center" around the origin, where $\Phi(q)$ is substantial so that the full equation (2.2) must be considered.

---

[2]The idea of propagation in the wave-number space may be found in a paper by Bondeson [4] in the context of shear Alfvén wave turbulence.

The following picture of the behaviour of a disturbance in the Fourier transformed velocity space thus emerges: if the system is initially disturbed far from the origin, on the positive half of the q-axis say (the "input channel"), the disturbance propagates without changing its form through the channel towards the scattering center (assuming $k > 0$). Once it crosses the origin it starts to interact with the scattering center, is being distorted and propagates with a changed form out of the scattering center into the "output channel". The process of distortion in fact consists of partial trapping in and leaking from the scattering center: an exponential tail is created behind the propagating disturbance. The electric field of the disturbance, given by its value at the origin, therefore decays in time exponentially, and this is the effect that is called Landau damping.

The eigenfunction expansion corresponding to the Vlasov-Poisson equation in the Fourier transformed velocity space (2.2) can most easily be extracted from the Laplace transform solution of the initial value problem. Defining the Laplace transformation with $-i\omega$ as the transform variable we get for the transformed equation (2.2)

$$\frac{\partial \varphi_{k\omega}(q)}{\partial q} + \frac{i\omega}{k}\varphi_{k\omega}(q) - \frac{\omega_p^2}{k^2} q\Phi(q)\varphi_{k\omega}(0) = -\frac{1}{k}\varphi_k(q,0), \qquad (2.4)$$

where $\varphi_k(q,0)$ is the initial condition. This first order ordinary differential equation may easily be solved to find $\varphi_{k\omega}(q)$. The result still contains $\varphi_{k\omega}(0)$, an unknown quantity which is, as we have seen above, proportional to the Laplace transform of the electric field. It is calculated by imposing a boundary condition on the solution at one point, we assume this point to be infinity and require that $\varphi_{k\omega}(q)\exp(i\frac{\omega}{k}q) \to 0$ for $q \to \infty$. In the end we find

$$\varphi_{k\omega}(q) = -\frac{1}{k} e^{-i\frac{\omega}{k}q} \int_0^q \varphi_k(s,0) e^{i\frac{\omega}{k}s} ds$$

$$+ \frac{1}{k} \frac{e^{-i\frac{\omega}{k}q}\left[1 + \frac{\omega_p^2}{k^2}\int_0^q s\Phi(s)e^{i\frac{\omega}{k}s}ds\right]}{1 + \frac{\omega_p^2}{k^2}\int_0^\infty s\Phi(s)e^{i\frac{\omega}{k}s}ds} \int_0^\infty e^{i\frac{\omega}{k}s}\varphi_k(s,0)\,ds. \qquad (2.5)$$

It is seen that this result consists of two distinctly different terms, corresponding to the splitting of the system into two subsystems: the first term describes a shift of the *negative* part of the spectrum into minus infinity without any change of form, whereas the second term describes a more complicated process going on with the *positive* part of the spectrum. To analyse this process we close the Bromwich integration path in the Laplace transform inversion formula by a large semi-circle Γ_n with radius R_n in the lower half of the complex ω-plane and use the theorem of residues

$$\varphi_k(q,t) = \sum_n \psi_{kn}(q)\left\langle \psi_{kn}^\dagger(q), \varphi_k(s,0)\right\rangle e^{-i\omega_n t} - \lim_{R_n \to \infty} \frac{1}{2\pi}\int_{\Gamma_n} (\text{integrand})\, e^{-i\omega t} d\omega. \qquad (2.6)$$

The inversion has the form of an expansion in terms of discrete eigenfunctions

$$\psi_{kn}(q) = e^{-i\frac{\omega_n}{k}q}\left[1 + \frac{\omega_p^2}{k^2}\int_0^q s\Phi(s)e^{i\frac{\omega_n}{k}s}ds,\right], \qquad (2.7)$$

with expansion coefficients given by the scalar product $\langle\cdot,\cdot\rangle$ of the adjoint eigenfunctions

$$\psi_{kn}^\dagger(q) = H(q) e^{-i\frac{\omega_n^*}{k}q}, \qquad (2.8)$$

where $H(q)$ is the Heaviside step-function, with the initial condition $\varphi_k(s,0)$. The summation is over all zeros ω_n of the function occurring in the denominator of the integrand, encircled by the contour,

$$1 + \frac{\omega_p^2}{k^2} \int_0^\infty q\Phi(q) e^{i\frac{\omega}{k}q} dq = 0. \tag{2.9}$$

In the upper half of the complex ω-plane the function on the left-hand side is equivalent to the standard form of the dispersion function of the Vlasov plasma oscillations. In the lower half it gives its analytic continuation (see, e. g., [11, 12]). The sum in (2.6) therefore comprises the contributions of all the complex zeros of the dispersion function, including not only those in the upper half-plane and corresponding to unstable discrete eigenmodes, but also those in the lower half-plane which correspond to damped eigenmodes. For entire analytic equilibrium distribution functions an infinity of them exists. Thus, in this reformulation of the classical Vlasov problem, the Landau damped "mode" acquires the status of a genuine eigenmode. Also it is seen that the continuum Van Kampen-Case eigenmodes play no role in this expansion.

The integral over the large semi-circle may or may not converge to zero as its radius grows without any limit, depending on the initial condition, as Ecker and Frömling [13] have demonstrated. A deeper analysis reveals that the discrete eigenmodes are never complete, an additional invariant subspace of the Vlasov-Poisson operator always exists, and the contribution of the large semicircle just corresponds to the projection of the initial condition onto this additional invariant subspace. In some sense one can say that this is what remains of the continuum eigenmodes in this limit.

The discrete eigenfunctions can of course be derived directly, as a solution of the basic Fourier-transformed Vlasov Poisson equation (2.2). We assume harmonic time-dependence, $\varphi_k(q,t) = \exp(-i\omega t)\varphi_{k\omega}(q)$, and substitute it into eq. (2.2) to obtain a first order ordinary differential equation identical to the homogeneous part of eq. (2.4). The solution is easily found to be

$$\varphi_{k\omega}(q) = \varphi_{k\omega}(0) e^{-i\frac{\omega}{k}q} \left[1 + \frac{\omega_p^2}{k^2} \int_0^q s\Phi(s) e^{i\frac{\omega}{k}s} ds \right]. \tag{2.10}$$

If this general solution is restricted by imposing the zero boundary condition at $+\infty$ as before, the dispersion relation (2.9) for the discrete eigenvalues is again found. The unstable eigenmodes, if they exist, are acceptable on the whole q-axis because they are bounded at $+\infty$ due the dispersion relation and decay exponentially at $-\infty$. On the other hand, the stable eigenmodes, while again bounded at $+\infty$, suffer from the "exponential catastrophe" at $-\infty$, but this does not matter since we need them, as we have seen, only on the positive half-axis.

The adjoint discrete eigenfunctions can be obtained in the same way from the adjoint Fourier-transformed Vlasov-Poisson equation with the boundary condition this time at $-\infty$: $\varphi_{k\omega}(q) e^{i\frac{\omega}{k}q} \to 0$ for $q \to -\infty$. A straightforward solution of this equation then gives the adjoint discrete eigenfunctions $\psi_{km}^\dagger(q)$, identical to (2.8). It can be verified that the discrete eigenfunctions and their adjoints form a bi-orthonormal set on the positive half of the q-axis.

The set of adjoint eigenfunctions is never complete: a function always exists, which is orthogonal to all of them. By a well-known theorem of functional analysis the linear manifold of all such functions, the orthogonal complement of the invariant subspace of the adjoint Vlasov-Poisson operator that is spanned by all its adjoint discrete eigenfunctions, generates an invariant subspace of the Vlasov-Poisson operator.

For an arbitrary real $\omega = \sigma$ the general solution (2.10) yields another class of acceptable solutions bounded over the whole q-axis:

$$\psi_{k\sigma}(q) = e^{-i\frac{\sigma}{k}q}\left[1 + \frac{\omega_p^2}{k^2}\int_0^q s\Phi(s)e^{i\frac{\sigma}{k}s}ds\right]. \qquad (2.11)$$

This solution describes scattering states of a monochromatic wave in the Fourier-transformed velocity space. These are the singular eigenmodes corresponding to the continuous spectrum σ comprising the whole real axis of the complex ω-plane. It can be verified [14] that the inverse Fourier transformation of (2.11) gives the Van Kampen-Case continuum eigenmode.

All possible eigenmodes of the Fourier-transformed Vlasov-Poisson equation (2.2) can thus be extracted from a single formula (2.10) just by imposing an appropriate boundary condition.

3 COLD INHOMOGENEOUS PLASMA OSCILLATIONS

Here we discuss another well-known problem, the electrostatic oscillations of a cold inhomogeneous plasma [7, 8, 9]. Planar geometry is assumed with the density gradient in the x-direction, the plasma frequency profile being $\omega_p(x)$. The wave propagates in a direction perpendicular to the gradient with the wave number $K = \sqrt{k_y^2 + k_z^2}$. For the present purpose it is necessary to use as the basic equation the integro-differential equation for the transverse component of the electrostatic field [9]

$$\frac{\partial^2 E(x,t)}{\partial t^2} + \omega_p^2(x)E(x,t) = \int_{-\infty}^{\infty} Q(x,x')E(x',t)\,dx', \qquad (3.1)$$

with

$$Q(x,x') = -\frac{K}{2}\int_{x'}^{\infty} e^{-K|x-u|}\,d\omega_p^2(u). \qquad (3.2)$$

It is seen that the integro-differential equation (3.1) is similar to the Vlasov equation (2.1). We can therefore expect analogous results if (3.1) is Fourier transformed with respect to the variable x, the variable q playing this time the role of the wave-number in the x-direction. The "free-oscillation" part, the left-hand side of (3.1), is easily transformed by means of the convolution theorem so that, in the end,

$$\frac{\partial^2 E_q(t)}{\partial t^2} + \int_{-\infty}^{\infty} \Omega_{q-q'}E_{q'}(t)\,dq' = \int_{-\infty}^{\infty} Q_{qq'}E_{q'}(t)\,dq'. \qquad (3.3)$$

Here

$$\Omega_q = \frac{1}{2\pi}\int_{-\infty}^{\infty}\omega_p^2(x)e^{-iqx}dx, \qquad Q_{qq'} = -\frac{1}{2\pi}\varepsilon_p(q-q')a_q b_{q'}, \qquad (3.4)$$

$$\varepsilon_p(q) = \int_{-\infty}^{\infty}\frac{d\omega_p^2(u)}{du}e^{-iqu}du, \qquad a_q = \frac{K^2}{K^2+q^2}, \qquad b_q = H_q, \qquad (3.5)$$

H_q being the Fourier transform of the Heaviside function.

This equation may again be interpreted as an equation describing wave propagation in the wave-number space accompanied by scattering. As before, the propagation properties are characterized by its "homogeneous part", translation invariant in the wave-number space. The propagation is this time dispersive and since the equation is of second order in time, propagation in either direction is possible, but it can be shown that there is no reflection. The scatterer is again localized around the origin of the wave-number space: the better so, the smoother the density profile as may be inferred from Eq. (3.5).

To analyse the "free-propagation", described by the homogeneous part of Eq. (3.3), we first note that it admits monochromatic wave solution $E_q(t) = C \exp[-i(\omega t + \kappa q)]$, where κ is the wave-number in the wave-number space (q-space), with the dispersion relation decomposable into two branches,

$$\omega^2 - \omega_p^2(\kappa) = 0 \Rightarrow \begin{cases} \omega = \omega_p(\kappa) \\ \omega = -\omega_p(\kappa) \end{cases} \tag{3.6}$$

corresponding, for $\kappa > 0$, to propagation in the direction of the negative and positive q-axis, respectively. The formula for the dispersive propagation of a general wave-packet in the wave-number space is then obtained as an integral superposition of these monochromatic wave solutions

$$E_q(t) = \int_{-\infty}^{\infty} C_1(\kappa) e^{-i[\omega_p(\kappa)t + \kappa q]} d\kappa + \int_{-\infty}^{\infty} C_2(\kappa) e^{-i[-\omega_p(\kappa)t + \kappa q]} d\kappa. \tag{3.7}$$

It is easy to analyse the dispersion process asymptotically for large times. The present problem is very similar to that of waves on water of constant depth [15]. Indeed, the group velocity in the wave-number space can be defined as usual by $d\omega/d\kappa$. For monotonic density profiles like those analysed in [16] the group velocity has a single maximum at $\kappa = 0$ and we can expand around this point:

$$\omega_p(\kappa) = \omega_p(0) + c_0 \kappa - \gamma \kappa^3 + \ldots, \tag{3.8}$$

where

$$c_0 = \left.\frac{d\omega_p(\kappa)}{d\kappa}\right|_{\kappa=0} \tag{3.9}$$

is the maximum group velocity and

$$\gamma = -\frac{1}{3!} \left.\frac{d^3\omega_p(\kappa)}{d\kappa^3}\right|_{\kappa=0}. \tag{3.10}$$

Assume further a very narrow initial wave-packet in the wave-number space corresponding to the initial condition $E_q(0) = \delta(q)$. This means that the initial condition in the original x-space is very wide: $E(x, 0) = 1/2\pi$, $\dot{E}(x, 0) = 0$. The evolution of the wave-packet generated by this initial condition and propagating in the direction of the positive q-axis is given by the second term in Eq. (3.7). The approximation (3.8) and substitution $s = -\sqrt[3]{3\gamma t}\kappa$ converts this integral into the standard Airy integral

$$E_q(t) \approx e^{i\omega_p(0)t} \frac{1}{2\sqrt[3]{3\gamma t}} \operatorname{Ai}\left(\frac{q - c_0 t}{\sqrt[3]{3\gamma t}}\right). \tag{3.11}$$

Thus, after a sufficiently long time, an initially narrow wave-packet far away from the scattering region will be dispersed into a steadily widening oscillatory formation with a steep, exponentially increasing front, slowly decreasing tail and steadily diminishing maximum amplitude. Both the widening and the damping are governed by the $t^{-\frac{1}{3}}$ law. This is the remarkable "excitation front" propagating in mode-number space that Cally [17] observed in his numerical calculations. Further discussion and numerical analysis of this result will be found in [18, 19] where graphical comparison is made with exact solutions for a specific plasma frequency profile. The correspondence is excellent.

References

[1] Trocheris, M.: Nucl. Fusion **5**, 299 (1965).

[2] Crawford, J. D., Hislop, P. D.: Ann. Phys. (N. Y.) **7**, 349 (1989).

[3] Best, R. W. B.: *Proceedings of the 7th International Conference on Phenomena in Ionized Gases,* Vol. II, Beograd 1966.

[4] Bondeson, A.: Phys. Fluids **28**, 2406 (1985).

[5] Van Kampen, N. G. : Physica **21**, 949 (1955).

[6] Case, K. M.: Ann. Phys. **7**, 349 (1959).

[7] Barston, E. M.: Ann. Phys. (N. Y.) **29**, 282 (1964).

[8] Sedláček, Z.: J. Plasma Phys. **5**, 239 (1971).

[9] Sedláček, Z.: J. Plasma Phys. **6**, 187 (1971).

[10] Uberoi, C., Sedláček, Z.: Phys. Fluids B**4**, 6 (1992).

[11] Roos, W. B.: *Analytic Functions and Distributions in Physics and Engineering.* Wiley 1969.

[12] Gnavi, G., Gratton, F.: IEEE Trans. Plasma Sci. **12**, 223 (1984).

[13] Ecker, G., Frömling, G.: Z. Naturf. **29**a, 1863 (1974).

[14] Sedláček, Z., Nocera, L.: J. Plasma Phys. **48**, 367 (1992).

[15] Lighthill, M. J.: *Waves in Fluids.* Cambridge University Press 1978.

[16] Cally, P. S., Sedláček, Z.: J. Plasma Phys. **48**, 145 (1992).

[17] Cally, P. S.: J. Plasma Phys. **45**, 453 (1991).

[18] Sedláček, Z., Cally, P. S.: J. Plasma Phys., in press.

[19] Cally, P. S., Sedláček, Z.: J. Plasma Phys., in press.

LONG WAVELENGTH MODES AND ANOMALOUS TRANSPORT IN TOKAMAKS

A. Hirose[1]

Plasma Physics Laboratory, University of Saskatchewan
Saskatoon, Saskatchewan S7N 0W0, Canada

Abstract

Stability of long wavelength modes in tokamaks with $(k_\perp \rho)^2 = 0.001 \sim 0.01$ has been investigated in terms of kinetic analysis. In electrostatic limit, the ion acoustic drift mode persists due to the suppression of ion Landau damping by the ion magnetic drift. The mode is further destabilized by finite β. For modes propagating in the ion diamagnetic drift, an η_i driven ballooning mode is predicted in the MHD second stability regime.

1. Introduction

In light of the mixing length ansatz, longer wavelength modes are more harmful to plasma confinement. Long wavelength drift type modes are generally believed to be stabilized by the ion Landau damping as the mode frequency approaches the ion acoustic transit frequency $\omega_s = k_\parallel c_s$. In toroidal geometry, the ion magnetic drift effectively suppresses ion Landau damping and the drift mode with $(k_\perp \rho)^2 \simeq 0.01$ (thus $\omega \simeq \omega_s > \omega_*$) remains unstable [1]. In general, electromagnetic corrections due to a finite β ($= 8\pi p/B^2$, the ratio plasma pressure/magnetic pressure) have a stabilizing influence on the electrostatic drift type modes with $|\omega| \simeq \omega_*$ (the diamagnetic frequency). For example, the toroidal η_i (ion temperature gradient) mode is stabilized by a modest β [2, 3]. Stabilizing role of magnetic fluctuations on the η_i mode may qualitatively be seen from the untrapped electron density perturbation in the limit $|\omega| \ll k_\parallel v_{Te}$,

$$n_e = \left(\phi - \frac{\omega - \omega_{*e}}{ck_\parallel} A_\parallel \right) \frac{e}{T_e} (1 - \sqrt{\epsilon}) n_0, \qquad (1)$$

where ϕ ($A_\parallel$) is the scalar (vector) potential, ω_{*e} is the electron diamagnetic frequency, and $\sqrt{\epsilon}$ is the fraction of trapped electrons. Since the η_i mode propagates in the ion diamagnetic drift ($\omega_r < 0$), the magnetic fluctuation $A_\parallel$ (which is proportional to $\beta\phi$) effectively enhances the adiabatic electron density perturbation which is evidently stabilizing. However, for long wavelength modes propagating in the electron diamagnetic drift with a frequency exceeding the diamagnetic frequency ($\omega > \omega_*$), such as the ion acoustic drift mode [1], magnetic fluctuations can be destabilizing because the electron response becomes less adiabatic,

---

[1]In collaboration with L. Zhang, M. Elia and A.I. Smolyakov.

$n_e/n_0 < e/T_e$. As will be shown, the long wavelength drift mode in tokamaks is strongly destabilized by finite β.

The MHD ballooning mode in tokamaks has been studied almost exhaustively in the past [4]. Kinetic [5, 6] and two-fluid [7] analyses have essentially confirmed the MHD stability properties including the existence of the second stability regime. One exception is the prediction of an η_i driven second mode which coexists with the MHD branch [5]. Recently, it has been shown that a kinetic η_i driven ballooning mode prevails in the MHD second stability regime [8]. The η_i driven mode is characterized by a broad eigenfunction in the θ space and a sufficiently large shooting distance is required to reveal the mode. The instability originates from the ion magnetic drift resonance which is absent in ideal MHD formulation. As will be shown, two-fluid approximation is able to predict the mode at least qualitatively.

2. Finite β Effects on the Ion Acoustic Drift Mode

In toroidal geometry, the ion resonance condition is given by

$$\omega + \frac{Mc}{B^3}\left(\tfrac{1}{2}v_\perp^2 + v_\|^2\right)(\nabla B \times \mathbf{B})\cdot \mathbf{k} - k_\| v_\| = 0. \tag{2}$$

For modes propagating in the electron diamagnetic direction ($\omega > 0$), the resonance is forbidden and long wavelength drift modes having frequencies close to the ion acoustic transit frequency remain unstable. Electromagnetic corrections (finite β effects) are expected to be destabilizing as explained in the Introduction. In collisionless tokamak discharges, destabilization is provided largely by magnetic drift resonance of trapped electrons.

In order to confirm the absence of ion Landau damping and finite β destabilization of long wavelength drift modes, we solve the semilocal electromagnetic dispersion relation [9] for low frequency modes in a low β tokamak discharge,

$$\left\{k^2 + 2\left(\frac{\omega_{pe}}{c}\right)^2 F_{e2U} + 2\left(\frac{\omega_{pi}}{c}\right)^2 F_{i2}\right\} D_{ES} = 2\left(\frac{\omega_{pe}}{c}\right)^2 \left(F_{e1U} + \sqrt{\frac{\tau m}{M}} F_{i1}\right)^2, \tag{3}$$

where D_{ES} is the kinetic electrostatic dispersion relation,

$$D_{ES} = F_{e0U} + F_{e0T} - 1 - \tau(1 - F_{i0}), \tag{4}$$

and the moments F's are defined by

$$F_{in} = \left\langle \left(\frac{v_\|}{v_{Ti}}\right)^n \frac{\omega + \widehat{\omega}_{*i}}{\omega + \widehat{\omega}_{Di} - k_\| v_\|} J_0^2(\Lambda_i) \right\rangle, \tag{5}$$

$$F_{enU} = \left\langle \left(\frac{v_\|}{v_{Te}}\right)^n \frac{\omega - \widehat{\omega}_{*e}}{\omega - \widehat{\omega}_{De} - k_\| v_\|} \right\rangle_U, \quad F_{eoT} = \left\langle \frac{\omega - \widehat{\omega}_{*e}}{\omega - \widehat{\omega}_{De}} \right\rangle_T, \tag{6}$$

$\langle \cdots \rangle$ denotes the average over the velocity with Maxwellian weighting, $J_0(\Lambda_i)$ is the Bessel function with $\Lambda_i = k_\perp v_\perp / \Omega_i$,

$$\hat{\omega}_{*s} = \omega_{*s}\left[1 + \eta_s\left(\frac{m_s v^2}{2T_s} - \frac{3}{2}\right)\right], \quad \hat{\omega}_{Ds} = \frac{m_s c}{eB^3}\left(\frac{1}{2}v_\perp^2 + v_\parallel^2\right)(\nabla B \times \mathbf{B}) \cdot \mathbf{k}, \qquad (7)$$

are the energy (velocity) dependent diamagnetic (magnetic drift) frequency, respectively, $\langle \cdots \rangle_{U,T}$ is the average for untrapped and trapped electrons, U ($v_\parallel^2 < \epsilon v_\perp^2$), T ($v_\parallel^2 > \epsilon v_\perp^2$). The norms of the differential operators for a simple trial eigenfunction, $\phi(\theta) = 1 + \cos\theta$, $|\theta| < \pi$, are

$$\begin{cases} \langle k_\parallel^2 \rangle_\theta \simeq \dfrac{1}{3(qR)^2}, \\ \langle k_\perp^2 \rangle_\theta = k_\theta^2\left[1 + \dfrac{s^2}{3}\left(\pi^2 - \dfrac{15}{2}\right) - \dfrac{10}{9}s\alpha + \dfrac{5}{12}\alpha^2\right], \\ \langle \omega_{Ds} \rangle_\theta = 2\epsilon_n \omega_{*s}\left(\dfrac{2}{3} + \dfrac{5}{9}s - \dfrac{15}{12}\alpha\right), \end{cases} \qquad (8)$$

where s is the shear parameter and $\alpha = q^2 R\beta/L_p$ is the ballooning parameter with L_p the pressure gradient scale length.

Figure 1 shows the mode frequency and growth rate (both normalized by ω_{*e}) as functions of $b_\theta = (k_\theta \rho)^2$ for various β values. The frequency of unstable modes can exceed the diamagnetic frequency particularly at low β. The case $\beta = 0.2$ % is nearly electrostatic and there exists a threshold in b_θ, $b_\theta > 0.002$, for instability. Longer wavelength modes in the region $b_\theta < 0.002$ are destabilized by β. Resonant destabilization occurs at $\beta \simeq 1.5$ % for the case shown. The resonance condition numerically found is well described by $\omega_0 = ck_\parallel k_\perp/k_{De} \simeq \omega_*$, or $\beta \simeq (\epsilon_n/2q)^2/(1 - \sqrt{\epsilon})$, which emerges from two-fluid approximation as well.

The quasilinear ion and electron thermal fluxes induced by general electromagnetic fluctuations may be evaluated from

$$q_i = -\sum_k \mathrm{Im}\, F_{iq} \frac{cT_i k_\theta}{eB}\left(\frac{e\phi_k}{T_i}\right)^2 n_0 T_i, \qquad (9)$$

$$q_e = \sum_k \mathrm{Im}\, (F_{eqU} + F_{eqT}) \frac{cT_e k_\theta}{eB}\left(\frac{e\phi_k}{T_e}\right)^2 n_0 T_e, \qquad (10)$$

where the thermal flux moments F are defined by

$$F_{iq} = \left\langle \frac{Mv^2}{2T_i} \frac{\omega + \hat{\omega}_{*i}}{\omega + \hat{\omega}_{Di} - k_\parallel v_\parallel} J_0^2(\Lambda_i)\left|1 - \frac{v_\parallel}{v_{Te}} \frac{D_{ES}}{F_{e1U} + \sqrt{\tau m/M}F_{i1}}\right|^2 \right\rangle, \qquad (11)$$

$$F_{eqU} = \left\langle \frac{mv^2}{2T_e} \frac{\omega - \hat{\omega}_{*e}}{\omega - \hat{\omega}_{De} - k_\parallel v_\parallel}\left|1 - \frac{v_\parallel}{v_{Te}} \frac{D_{ES}}{F_{e1U} + \sqrt{\tau m/M}F_{i1}}\right|^2 \right\rangle_U, \qquad (12)$$

$$F_{eqT} = \left\langle \frac{mv^2}{2T_e} \frac{\omega - \widehat{\omega}_{*e}}{\omega - \widehat{\omega}_{De}} \right\rangle_T. \tag{13}$$

In Fig. 2, the imaginary parts of the total electron thermal flux moment $F_{eq} = F_{eqU} + F_{eqT}$ and that of ions F_{iq} are shown for $\beta = 0.2$ % and 2 %, together with the mode frequency and growth rate. The thermal flux moments Im F_{eq} and Im F_{iq} at long wavelength region, $b_\theta < 0.005$, are strongly enhanced from the electrostatic limit ($\beta = 0.2$ %). The increase in the ion thermal flux is more significant than that in the electron thermal flux.

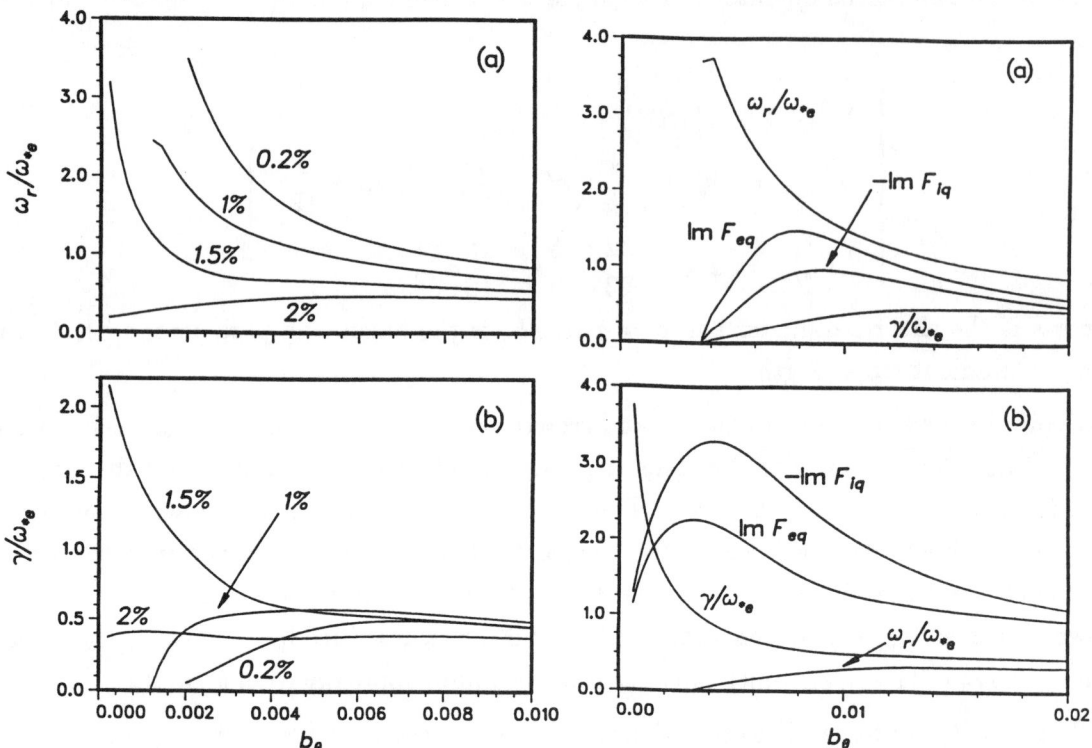

Fig. 1 ω_r/ω_{*e} and γ/ω_{*e} of the drift mode vs. the finite radius parameter at various β in %. $L_n/R = 0.4, r/R = 0.25, T_i = T_e, s = 1, q = 2, \eta_e = \eta_i = 1$.

Fig. 2 Mode frequency, growth rate, Im F_{eq}, and Im F_{iq} vs. b_θ. (a) $\beta = 0.2\%$, (b) $\beta = 2\%$. Other parameters are unchanged from those in Fig. 1 except $q = 1$.

3. η_i Driven Ballooning Mode

The importance of the ion temperature gradient on the ballooning mode has earlier been noticed by Cheng [5]. It has recently been shown that an ion temperature gradient η_i induces a kinetic ballooning mode which persists above the MHD unstable window in the ballooning parameter [8]. The analysis in Ref. [8] was fully kinetic and involved somewhat complicated velocity space integrations. Destabilization is caused by the ion magnetic drift resonance because the ballooning mode propagates in the ion diamagnetic drift, $\omega_r < 0$. The fluid approximation for the ion density perturbation developed by Weiland and co-workers [10] appropriately retains the resonance effects in terms of two-pole approximation. In this section, the two-fluid approach [7] to the ballooning mode is revisited.

For modes in the intermediate frequency regime $k_\| v_{Ti} \ll |\omega| \ll k_\| v_{Te}$, the electron and ion density perturbations based on two-fluid approximation are

$$n_e = \left(\phi - \frac{\omega - \omega_{*e}}{ck_\|}A_\|\right)\frac{e}{T_e}n_0, \tag{14}$$

$$n_i \simeq \frac{\left(\omega + \frac{5}{3}\omega_{Di}\right)[\omega_{*e} - \omega_{De} + (\omega + (1+\eta_i)\omega_{*i})(k_\perp \rho_s)^2] + \left(\frac{2}{3} - \eta_i\right)\omega_{Di}\omega_{*e}}{\left(\omega + \frac{5}{3}\omega_{Di}\right)^2 - \frac{10}{9}\omega_{Di}^2}\frac{e\phi}{T_e}n_0 \equiv Q_i\frac{e\phi}{T_e}n_0. \tag{15}$$

The corresponding electron current perturbation along the magnetic field is

$$J_{\|e} = \frac{n_0 e^2}{k_\| T_e}\left\{(\omega_{*e} - \omega)\phi + \frac{(\omega - \omega_{*e})(\omega - \omega_{De}) + \eta_e \omega_{*e}\omega_{De}}{ck_\|}A_\|\right\}. \tag{16}$$

From the charge neutrality $n_e = n_i$ and Ampere's law $\nabla^2 A_\| = -4\pi J_{\|e}/c$, we obtain the following mode equation,

$$\frac{d}{d\theta}\left\{[1 + (s\theta - \alpha\sin\theta)^2]\frac{d\phi}{d\theta}\right\} + \frac{\beta}{4}\left(\frac{qR}{L_n}\right)^2 V(\theta, \Omega)\phi = 0, \tag{17}$$

where $\Omega = \omega/\omega_{*e}$,

$$V(\theta, \Omega) = (\Omega - 1)(\Omega - f) + \eta_e f - \frac{(\Omega - 1)^2}{Q_i}, \quad f(\theta) = 2\epsilon_n\omega_*[\cos\theta + (s\theta - \alpha\sin\theta)\sin\theta].$$

The growth rate and mode frequency (normalized by the Alfven frequency $\omega_A = V_A/qR$) found from shooting code analysis of Eq. (17) are shown in Fig. 3 (solid lines). For comparison, the ideal MHD growth rate (dotted line) and eigenvalue found from exact kinetic analysis (dashed lines) are also shown. The two-fluid approximation agrees with the kinetic analysis at least qualitatively. In particular, the two-fluid approximation does predict the persistence of the η_i driven ballooning instability in the MHD second stability region. However, it underestimates the critical α at the first instability boundary. The eigenfunction $\phi(\theta)$ revealed from both kinetic and two-fluid analyses for modes in the MHD second stability regime extends to $\theta \simeq 30$ indicating that a sufficiently large shooting distance must be chosen.

The critical η_i for the kinetic instability in the MHD second stability regime is approximately unity. It should be noted that the η_i driven ballooning mode is not the toroidal η_i mode with electromagnetic corrections. The toroidal η_i mode is predominantly electrostatic and subject to finite β *stabilization*. As shown in Refs. [2, 3], the mode is stabilized as α approaches the first MHD unstable boundary. In contrast, the η_i driven ballooning mode is *destabilized* by β. The magnitude of the mode frequency $|\omega_r| \simeq V_A/2qR$ is significantly larger than that of the toroidal η_i mode which is of order ω_{Di}.

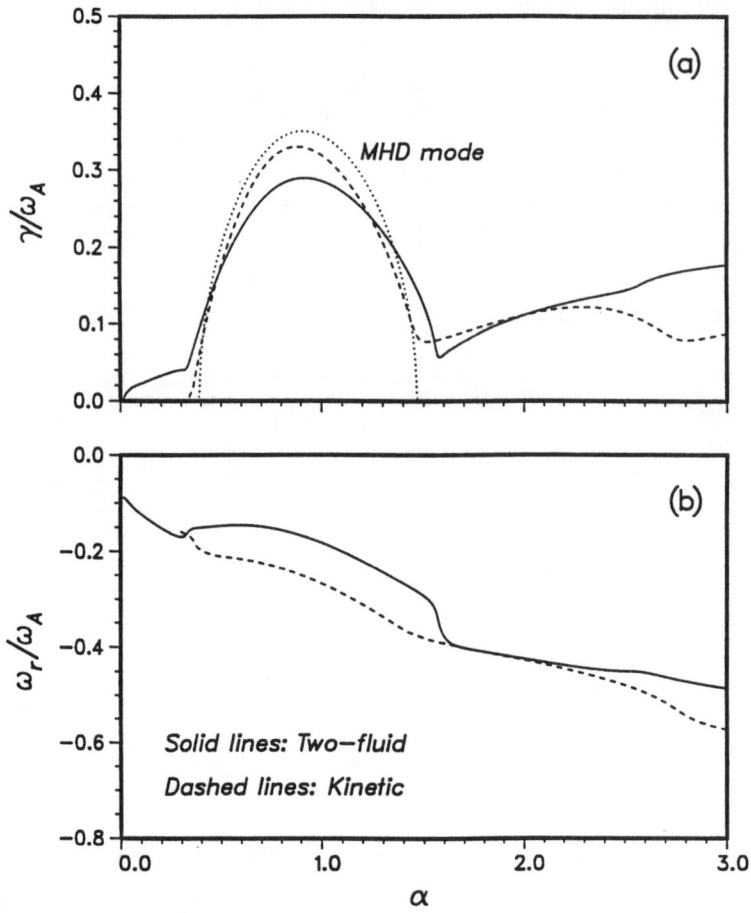

Fig. 3 Growth rate (a) and mode frequency (b) normalized by $\omega_A = V_A/qR$. Solid lines from two-fluid model, dashed lines from kinetic analysis, and dotted line from ideal MHD. $s = 0.4, b_\theta = 0.01, \eta_i = \eta_e = 2, L_n/R = 0.175$.

There are several stabilizing agents for the ballooning mode, such as the trapped electrons, ion acoustic transit effect, and magnetosonic ($B_\parallel$) perturbation. However, they are unable to suppress the η_i driven ballooning mode [11]. The effect of the trapped electrons is to reduce the parallel current carried by the untrapped electrons which effectively enhances the stabilizing Alfven magnetic field line bending. The reduction factor in the parallel electron current numerically found is approximately $1 - 0.6\sqrt{\epsilon}$. The ion transit frequency $k_\parallel v_{Ti}$ remains smaller than the ballooning mode frequency $\omega_r \simeq -V_A/2qR$ by a factor $\sqrt{\beta}$ and implementing the ion transit effect is not expected to modify significantly the eigenvalue. In Ref. [11], a perturbative method has been developed to assess the stabilizing role of the ion transit frequency $k_\parallel v_{Ti}$. Fig. 4 shows the growth rates found from kinetic analysis with and without $k_\parallel v_{Ti}$ where

$$\delta = \frac{\langle k_\parallel \rangle_\theta v_{Ti}}{\omega_{*i}},$$

with $\langle k_\parallel \rangle_\theta$ the norm of the operator $k_\parallel = -i(1/qR)\partial/\partial\theta$ based on the eigenfunction without the ion transit effect. At small α ($\propto \beta$), the ion transit effect is ignorable. Reduction in the growth rate can be seen at large α. However, the ion transit effect does not suppress the η_i driven ballooning mode.

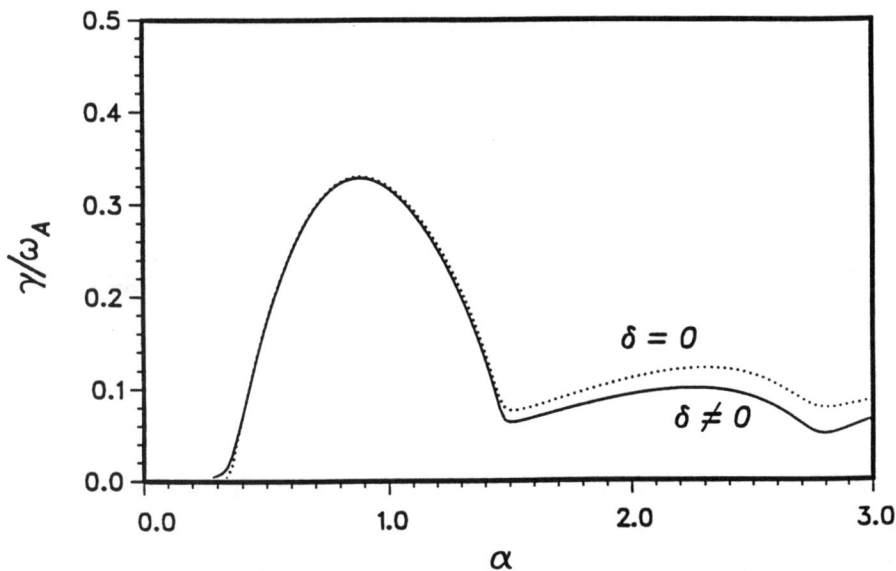

Fig. 4 Growth rates found from kinetic analysis with ($\delta \neq 0$) and without ($\delta = 0$) the ion transit frequency. The parameters are the same as in Fig. 3.

4. Discussions and Conclusions

Finite β has significant destabilizing effects on the long wavelength drift mode in tokamaks. Modes in the regime $(k_\perp \rho)^2 \simeq 10^{-3}$ persists in realistic discharges with modest β without suffering ion Landau damping even though the mode frequency is close to the ion transit frequency ($\omega \simeq k_\parallel v_{Ti} \simeq k_\parallel c_s > \omega_{*e}$). The quasilinear ion and electron thermal fluxes are enhanced by β from those in electrostatic limit. For the ion acoustic mode, parallel ion dynamics is dominant and the turbulence level at saturation may be estimated from

$$\gamma \simeq \sqrt{\frac{ek_\parallel E_\parallel}{M}}, \tag{18}$$

which yields $e\phi/T_e \simeq (r/R)^2$. A resultant thermal diffusivity is Bohm-like,

$$\chi \simeq q^2(1 + \beta/\beta_c)(cT/eB)(r/R)^2, \qquad (19)$$

where β enhancement found in the present investigation is implemented. This predicts the following scaling for the energy confinement time,

$$\tau_E \propto \frac{R^2(aBn)^{1/3}I_p^{2/3}}{P^{2/3}}. \qquad (20)$$

For modes propagating in the ion diamagnetic drift ($\omega < 0$), a kinetic ballooning mode driven by an ion temperature gradient has been predicted in the MHD second stability regime. The mode can also be revealed within relatively simple two-fluid approximation although in detail significant deviations from exact kinetic analysis have been found.

Acknowledgements

Helpful discussions with Drs. J. Weiland, H. Nordman and D. Sigmar are acknowledged with gratitude.

This research has been sponsored by the Natural Sciences and Engineering Research Council, and by the Office of National Fusion Program, Canada.

References

1. A. Hirose, Phys. Fluids B **3**, 1125 (1991); A. Hirose, A.I. Smolyakov, and O. Ishihara, Nucl. Fusion **33**, 735 (1993).
2. B.G. Hong, W. Horton, and D.I. Choi, Plasma Phys. Controlled Fusion **31**, 1291 (1989).
3. J. Weiland and A. Hirose, Nucl. Fusion **32**, 151 (1992).
4. For a review of MHD ballooning mode, see, for example, O.P. Pogutse and E.I. Yurchenko, in *Reviews of Plasma Physics*, (Consultants Bureau, New York, 1986), vol. 11, p. 65.
5. C.Z. Cheng, Phys. Fluids **25**, 1020 (1982).
6. C.Z. Cheng, Nucl. Fusion **22**, 773 (1982).
7. H. Nordman, B. Jhowry, and J. Weiland, Phys. Fluids B **5**, 3465 (1993).
8. A. Hirose, L. Zhang, and M. Elia, Phys. Rev. Lett. **72**, 3993 (1994).
9. A. Hirose, Phys. Fluids B **3**, 1599 (1991).
10. A. Jarmen, P. Andersson, and J. Weiland, Nucl. Fusion **27**, 941 (1987).
11. A. Hirose, L. Zhang, and M. Elia, submitted to Phys. Plasmas.

FILAMENTATION IN TOKAMAKS

N.J. Lopes Cardozo, C.J. Barth, C.C. Chu, J. Lok, A. Montvai, A.A.M. Oomens, M.Peters, F.J. Pijper, M. de Rover, F.C. Schüller, M.F.M. Steenbakkers, and RTP team

FOM-Instituut voor Plasmafysica 'Rijnhuizen',
P.O. Box 1207, 3430 BE Nieuwegein, The Netherlands

ABSTRACT.

The relevance of a nest of toroidal flux surfaces as a paradigm of the magnetic topology of a tokamak plasma is challenged. High resolution Thomson scattering measurements of electron temperature and density in RTP show several hot filaments in the plasma centre and sharp gradients near the sawtooth inversion radius and structures outside the sawtooth region under central ECH. In ohmic plasmas, too, the pressure and temperature profiles show significant bumps. These measurements give evidence of a complex magnetic topology. Transport in a medium with spatially strongly varying diffusivity is considered. It is shown that macroscopic transport is determined by the microscopic structure: a transport theory must predict this structure and the diffusivity in the insulating regions, while the 'turbulent' diffusivity is irrelevant. A numerical approach to equilibria with broken surfaces is presented.

1. INTRODUCTION

It is often assumed that the topology of the magnetic field in a tokamak is that of a nest of toroidal flux surfaces. Indeed, this topology is guaranteed if a scalar pressure and zero flow are assumed. The major part of theory of tokamak plasmas has been developed within this idealisation. This includes the Grad-Shafranov magnetic equilibrium, which describes the shape of the flux surfaces for given distributions of current and pressure as function of the flux coordinate, and the transport of heat and particles, which is normally treated as a 1-D problem (from surface to surface).

The Grad-Shafranov equation appears to give a good description of the tokamak equilibrium, at least as far as the global parameters: major and minor radius, Shafranov shift, and plasma shape are concerned. Transport, on the other hand, has proven to be enigmatic. Neither the absolute value nor the radial dependences of the transport coefficients are understood on the basis of theory.

In this paper we investigate the possibility that the topology of the magnetic field is essentially more complex than the ideal nest of flux surfaces, in such a way that transport could be strongly affected while the global characteristics of the Grad-Shafranov equilibrium are retained.

A fundamental problem of perfect toroidal flux surfaces is that it is known from mathematical theory that toroidal surfaces on which the field line winding ratio (q) is rational, are topologically unstable and can very easily be destroyed [1]. The KAM theorem [2] states that in this case the topology changes into a mix of 'good' surfaces, magnetic islands and regions of stochastic field: see Fig 1.

The perfect flux surfaces can be described by an unperturbed, integrable Hamiltonian. Degeneration of the surfaces is produced by an infinitesimal non-integrable perturbing Hamiltonian. In the case of a tokamak, anything that breaks the symmetry of the system could play this role: the discrete coils, localised momentum or power input, a poloidal asymmetry of the heat flux, etc.

A second fundamental problem of ideal equilibria is that they do not satisfy Ohm's law [3]. Further, the class of possible equilibria widens up if the fluid velocity is not put to zero. Typically, helical equilibria appear to be favoured when plasma flow is allowed [4].

The experimental evidence is that the tokamak plasma does have non-zero flow, and that magnetic structures do form. Hence, while the paradigm seems quite successful in describing the global plasma geometry, there is little ground to believe that the conditions for its validity are fulfilled.

© 1995 American Institute of Physics

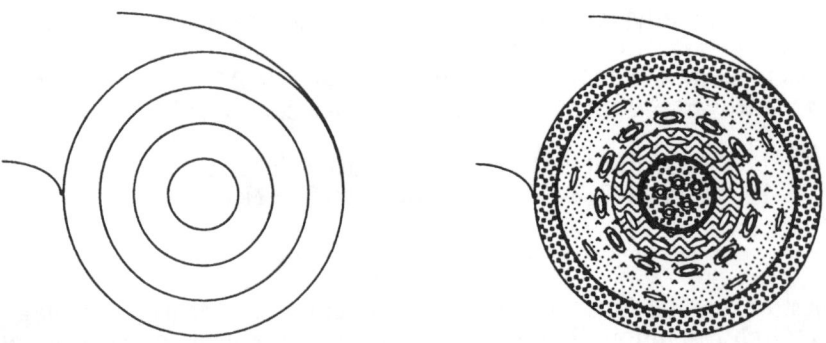

Figure 1. The standard theory of tokamak plasmas assumes a simple magnetic topology: perfect flux surfaces which form a nested set. Under perturbation, the flux surfaces at which q is rational degenerate to form magnetic islands embedded in a region of stochastic field.

In this paper we briefly review experimental evidence for deviations of the ideal topology (Sec.2); we present recent measurements of the pressure in the RTP experiment in Rijnhuizen which show small scale structures, including filamentation in the plasma center (Sec.3); we discuss transport in tokamak plasmas in the framework of the evidence that the tokamak equilibrium has a complex magnetic topology (Sec.4); and we present a numerical approach to equilibria with complex topology (Sec.5). A brief general discussion is given in Sec.6.

2. REVIEW OF EXPERIMENTAL EVIDENCE FOR SMALL SCALE PLASMA STRUCTURES

Evidence for small scale structures in the <u>plasma edge</u> is widely available, thanks to the relatively easy diagnostic access. To mention two typical results: in TFTR filamentation of H_α emission was demonstrated [5]. Measurements with a CCD-camera showed a regular structure of parallel, narrow bands of high emissivity. These were aligned with the field lines. The poloidal extension was short (3-5 cm) while the parallel wavelength was much longer ($\approx$ 100 cm). Similarly, in JET broadband (10 - 60 kHz range) magnetic fluctuations measured with Mirnov coils showed a correlation length along the field lines of $2\pi R$ or longer, and a very limited poloidal extent: m = 10 to 100. The magnetic fluctuations correlated with the fluctuations of the visible light emission [6].

In the <u>plasma core</u> measurements of small scale structures are very difficult to make. Well known departures from the ideal equilibrium are the large magnetic islands associated with disruptions and the sawtooth instability. An example of a long-lived small structure is the so-called 'snake', a m=n=1 structure of high density and pressure, which was first observed in JET [7]. It occurs frequently after pellet injection, but also generatio spontanea is reported. The snake is thought to be a closed flux tube, and the excellent confinement properties confirm the hypothesis that inside magnetic islands transport is strongly reduced. The same conclusion was reached based on analysis of transport in large m=2 islands in RTP [8]. In TEXTOR it was discovered that a thin, m=n=1 helical beam of 30 MeV runaway electrons could survive a short period of strong ergodization induced by pellet injection [9]. The surviving beam is exceptionally stable and shows vanishingly small radial and poloidal diffusion.

Evidence for the common occurrence of even smaller magnetic structures in tokamak discharges is derived from the ablation of pellets injected into the plasma, from correlation reflectometry and from detailed measurements of the T_e and pressure profiles using high resolution Thomson scattering.

The H_α emission of a pellet that travels through the plasma is a measure of the ablation rate. In Tore Supra, this intensity was found to dip when the pellet crossed rational q-surfaces [10]. The hypothesis is that on a rational surface the pellet is in contact with only a small plasma volume, which cools rapidly. This mechanism requires that the rational surfaces have a finite width, i.e. that they are degenerated.

Measurements in JET [11] with 4 reflectometer channels tuned to slightly different radial positions in the plasma, showed that the radial correlation length of density fluctuations was very short (<1cm) in Ohmic and H-mode plasmas, while in L-mode it was much longer (up to 5 cm or more) and increasing with the input power. Studies with toroidally separated channels showed that the observed density fluctuations are due to unsmooth surfaces of constant density that rotate toroidally. This could be interpreted as evidence for the existence of small magnetic structures.

Bumpy T_e profiles were measured a.o. in JET with the LIDAR Thomson scattering diagnostic [12]. Flat spots in the profile were associated with q-surfaces of simple rationality, and interpreted as evidence for magnetic islands. Similar observations are reported from TFTR, but attempts to measure T_e-structures with ECE did not support their existence [13].

At RTP, a research programme is built around ultra high resolution Thomson scattering measurements of the electron temperature and pressure profiles, taken during intense local heating with ECRH or during pellet injection. These measurements show ample evidence of small scale structures in the plasma [14] and will be described in more detail in Sec.3.

Flow velocities are measured on many tokamaks; the ohmic plasma usually rotates toroidally at a significant speed, and beam injection can bring the velocity up to the ion sound speed. It is important to notice that when large MHD modes are present in the plasma, these are normally coupled, so that the toroidal speed of the modes is uniform over the plasma. The fluid velocity, however, has a radial profile, and hence the fluid flows with respect to the magnetic structure.

Closely related with the flows is the question if poloidal asymmetries are observed. To give two examples: i) in TEXTOR it was observed that the heat pulses induced by the sawtooth instability have different amplitude and phase velocity at the low and high field sides of the plasma [15]; ii) in RTP asymmetric T_e profiles were measured with off-axis ECH[16].

Summarizing, there is overwhelming evidence that a tokamak plasma does not satisfy the conditions required for the simple magnetic topology of nested perfect flux surfaces. The plasma generally has a non-zero fluid velocity, there are poloidal asymmetries, there are often large scale magnetic structures, and the body of evidence is growing that small scale structures are also common: in the edge, where detection is easy, they are commonly observed, and they are found in the center when an effort is made to detect them.

It is of great importance that the cases in which departures from the paradigm topology are demonstrated do not show inferior confinement. This may be regarded as evidence that the surfaces are probably always broken. Conversely, in those cases where good surfaces may be expected, such as inside an m=n=1 flux tube (snake and runaway snake, magnetic islands, filaments (Sec.3)) the measured confinement is superb.

3. EVIDENCE FOR PLASMA FILAMENTATION IN RTP

RTP (major/minor radius 0.72m/0.165m, toroidal field 2.2 T, boronised vessel) is equipped with Electron Cyclotron Heating (ECH) (2×180 kW at 60 GHz, 500 kW at 110 GHz), a pellet injector, and an extensive set of diagnostics, including magnetics, 19-channel FIR interferometer, 20-channel heterodyne radiometer, 80-channel 5-camera Soft X-ray tomography, etc. Crucial for the present study is the Thomson scattering system, which measures T_e and n_e along a vertical chord through the center of the plasma at 100 radial positions simultaneously, with a spatial resolution of 1.7 mm. Details of the diagnostic are given in [14,17]. Much attention has been paid to the statistical measuring error, which is determined in two independent ways.

T_e and n_e profiles were measured in ohmic and ECH heated hydrogen plasmas, for a variety of plasma parameters: q_a=3.2-16, $\overline{n_e}$ = 1-2×10^{19} m^{-3}. Typically, the global energy confinement time is $\tau_E \approx$ 1- 5 ms. In heated discharges an ECH pulse (central deposition) of >50 ms duration was applied in the current flattop. The Thomson scattering measurement was carried out 5 ms before the end of the ECH pulse. Striking observations are:
- hot filaments in the plasma core (Fig 2a,b)
- steps of T_e, apparently associated with the sawtooth inversion radius (Fig 2c)
- quasi periodic structures outside the sawtooth region with ECH, and in ohmic discharges (Fig 2d)

Neither in the ohmic phase nor during ECH is MHD activity observed on the Mirnov coils

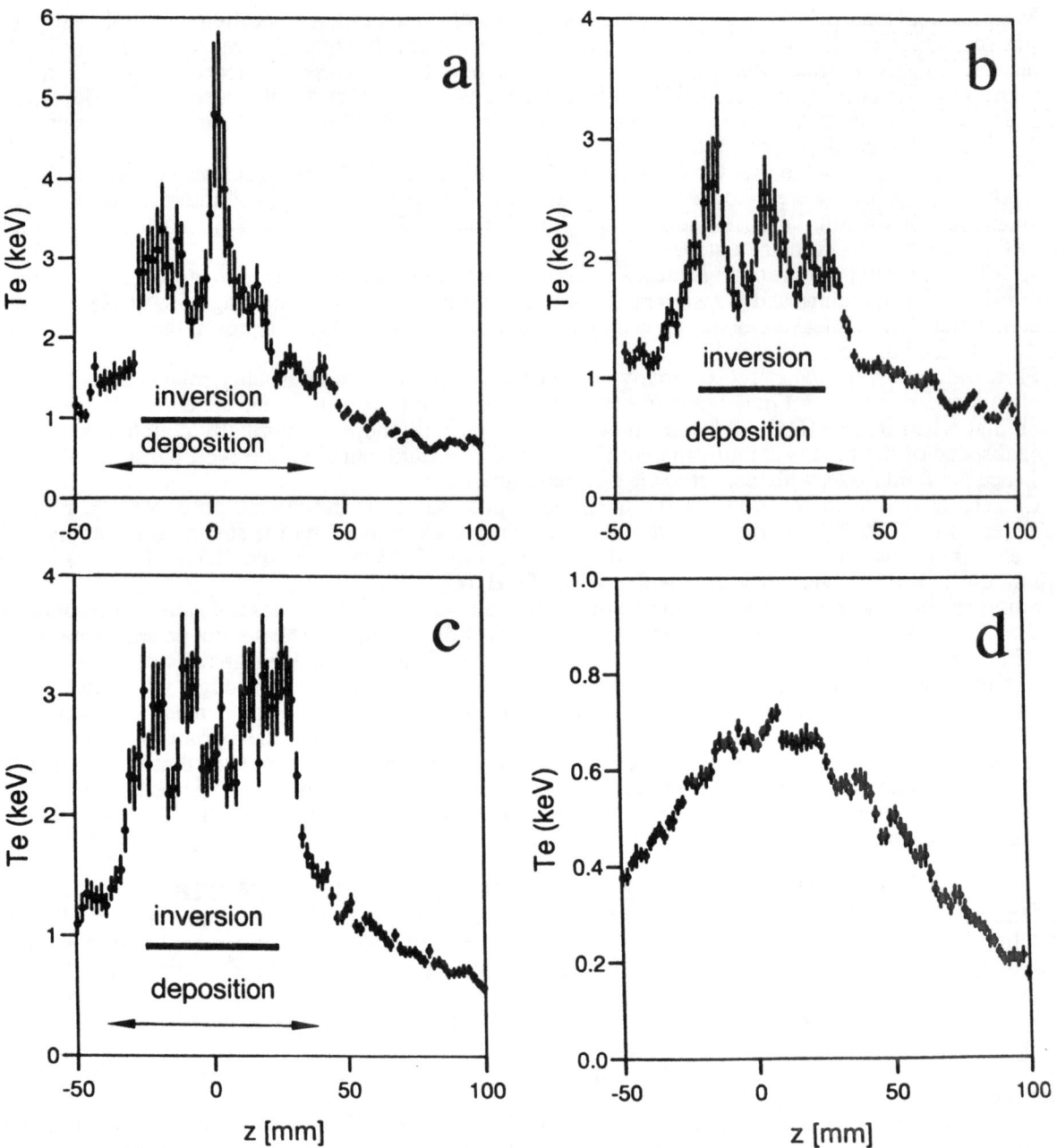

Figure 2. Examples of measured T_e profiles. The corresponding profiles of n_e show very little structure, and are therefore omitted. a) Under central ECH several peaks of high T_e are observed; b) the hot filaments are still there 300 μs after switching off ECH; c) Under central ECH, near the sawtooth inversion radius a step in T_e is observed, which is especially pronounced in high q_a discharges; d) high density ohmic discharge showing a 'bumby' T_e profile.

The profiles of n_e are rather flat and smooth: the electron pressure and T_e profiles exhibit the same features.

The hot filaments have a typical diameter of 5-10 mm. They appear only in the ECH deposition region, and are interpreted as closed flux tubes which are strongly heated by the intense ECH power. The diffusivity inside the tubes must be very low to allow the high T_e values. An estimate obtained by assuming that a tube absorbs a fraction of the ECH power proportional to its surface relative to the deposition region, gives χ_e(inside) ≈ 0.01 m^2/s for 6 mm diameter. This is one to two orders of magnitude lower than the diffusivity in the ambient plasma, and approaches the neoclassical limit. The single pulse Thomson scattering measurement does not allow an investigation of the dynamics of the filaments. However, it was measured that the hot filaments are still alive up to ≈ 500 µs after switching off the ECH (Fig 2b), which demonstrates the long life of the magnetic structures and confirms the excellent confinement inside them.

The extraordinarily steep gradients (up to 0.5 MeV/m) near the sawtooth inversion radius are particularly pronounced in high q_a discharges. They represent a transport barrier: a layer with a width of a few millimeter in which the diffusivity drops by two orders of magnitude, to a value close to neo-classical. Inside this barrier, the magnetic topology appears to be complicated, with several flux tubes (the hot filaments) embedded in a sea of ergodic field.

Whereas the filamentary structures such as presented in Fig.2a-c are manifest, the structures in the ohmic profile (Fig.2d) may give rise to doubt as to their statistical significance. One way of testing this is by fitting a cubic spline with free knots to the profile, and computing the improvement of the fit upon adding an extra knot [18]. The probability that the N-knot curve is not a better model than the (N-1)-knot curve can be computed (the F-test) and is plotted in Fig.3a for 4 examples of a series of high density ohmic discharges with very similar conditions. Clearly, up to 20 -25 knots are significant, with a confidence level of 95%. With this number of knots, the profiles are well fitted (Fig3b). In particular, structures with a typical size down to 1 cm are shown to be statistically significant.

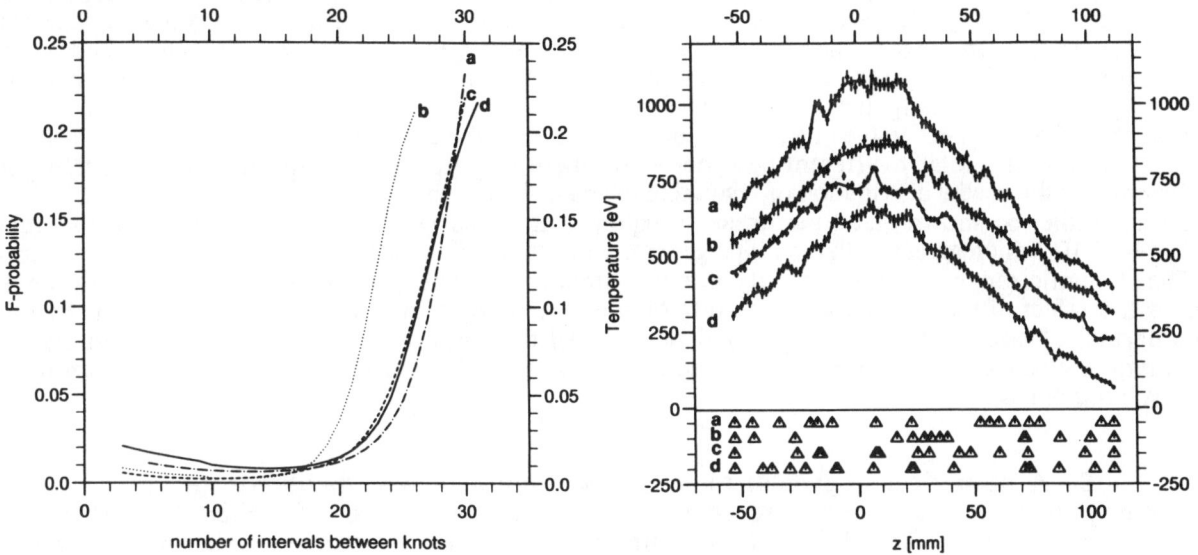

Figure 3. The statistical significance of the structures in ohmic profiles is tested by free knot cubic spline fitting. Four ohmic discharges out of a series of 10 were taken as examples for this figure. a: The F-test shows that up to 20-25 knots are supported by the data. b: The corresponding fitted profiles feature structures with a typical size of down to 1 cm. The number of intervals between knots used were 24, 19, 21, and 22 for curves a-d. The positions of the knots are indicated by triangles. The curves a-c have been shifted up by 300, 200, and 100 eV, resp.

In summary, detailed measurements of the T_e profile give evidence of a complex magnetic topology. In the center several flux tubes are found, probably embedded in a sea of ergodic field; this region is confined by a transport barrier near the sawtooth inversion radius; outside the sawtoothing region, the T_e-profile is definitely unsmooth, indicative of a structured magnetic topology.

4. Consequences for transport; Transport in an inhomogeneous medium

Clearly, the measurements presented above prompt a discussion of the relevance of the nest of ideal flux surfaces as the paradigm topology of a tokamak plasma. To narrow down the question to the transport problem: does it make sense to use a 1-D transport model, i.e. to apply flux surface averaging? Inside the transport barrier, in the region where filaments are produced this is very questionable indeed. However, outside that radius, the profiles do show a high degree of up-down symmetry. Hence, the dominant topology of the plasma appears to be toroidal shells. The measured profiles featuring small structures could represent snapshots of a T_e profile with broadband T_e-fluctuations. As a consequence, the 1-D transport coefficients which are determined by a power balance analysis could be regarded as representing the spatially and temporally smoothed transport. As such, they are useful numbers, characterizing the profiles and confinement of the plasma. The real problem comes when these numbers are interpreted in terms of a transport model. This is illustrated by the discussion below.

The measurements presented above show that there can be thin layers in the plasma where the electron thermal diffusivity (χ) drops by several orders of magnitude. This we take as an incentive to investigate the generic properties of transport in a medium in which the diffusivity is a strongly varying function of the spatial coordinate. We consider a plasma slab with a sequence of conducting and insulating layers. Such a slab would feature a stair-step T_e profile. The question to be addressed is whether the macroscopic transport properties are influenced by the microscopic structure or not.

From elementary calculus it follows that if measurements of T_e are taken that do not resolve the structure, a power balance analysis yields a (macroscopic) χ^{pb} that is related to the local (microscopic) χ by $\chi^{pb} = \langle \chi^{-1} \rangle^{-1}$, where $\langle . \rangle$ denotes spatial averaging. An obvious but important consequence is that if χ varies by more than an order of magnitude, the measured value χ^{pb} is almost entirely determined by the insulating regions. Consequently, a theory that aims at predicting the macroscopic χ^{pb} needs to give the value of χ in the insulating layers, and the fraction of the space that is occupied by these layers. Interestingly, the 'turbulent' value of χ, i.e. the one that governs the conducting zones, is of no interest any more. This implies a totally different perspective on the transport problem: not the turbulent transport coefficient, but the spatial distribution of insulating layers is the nut to crack.

It can further be shown that a perturbative experiment yields $\chi^{pert} = \langle \chi^{-0.5} \rangle^{-2}$. This leads to a ratio $\chi^{pert}/\chi^{pb} \approx 2\text{-}10$, which agrees with the range generally found in tokamak plasmas [19].

The dynamic range of the diffusivity is much smaller for ions than for electrons, which strongly reduces the effect of the microscopic structure on the macroscopic ion transport. The reverse is true for suprathermal electrons. It has been pointed out in [20] that in a system with good and bad regions suprathermal electrons can have better confinement than thermal electrons, even if their transport in the stochastic regions is faster.

5. A numerical approach to magnetic equilibria with broken topology

In an attempt to come to a numerical scheme that can handle the situation of broken flux surfaces, the following approach was adopted. The plasma current is represented by a large number of small current elements. The global profiles of plasma current density and pressure are prescribed. In an iteration procedure, the direction of the current elements is adjusted to make pressure balance. The procedure is described in more detail in [21]. At each step in the iteration, a Poincaré plot can be made to show the structure of the field.

The iteration does not represent a dynamical evolution of the equilibrium. The numerical set up is purely a vehicle to reach a magnetic equilibrium without presupposing the existence of flux surfaces. In physical terms, the current filaments may be thought of as small scale magnetic perturbations.

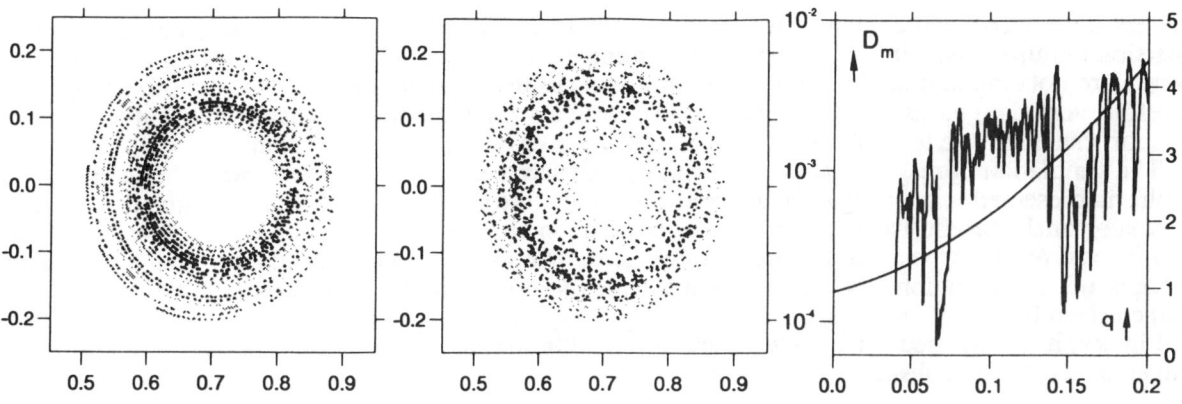

Figure 4. Poincaré plots of the magnetic field in a plasma in which the plasma current is represented by a large number of small current elements. Pressure balance is achieved by - iteratively - adjusting the direction of these current filaments. The system starts out with near perfect flux surfaces (a: 1st iteration) (a), but converges to a mildly stochastic state (b: 8th iteration). The variance of the radial position of a fieldline in this state is plotted as a function of radius in (c). The unperturbed q-profile is plotted for reference.

First results show that the system starts out with nearly perfect flux surfaces, but quickly moves to a mildly stochastic situation. The system stays quite close to the Grad-Shafranov type equilibrium, with nested toroidal regions as the basic topology, but the surfaces do break up and form layers of stochastic field. The solution converges quickly and after three or four iterations it does not change any more in a statistical sense. Fig.4 shows Poincaré plots after the 1st and 8th iteration. Clearly visible are a few surfaces that remain intact, while in other places the surfaces are completely destroyed.

A measure of the destruction of the surfaces is given by the magnetic diffusivity D_M=var(r), where r is the radial coordinate of a field line and the variance is taken following the field line along its path around the torus. D_M is plotted in Fig.4c, together with the unperturbed q-profile. The variation in stochasticity seen in the Poincaré plot is indeed borne out by D_M, which makes excursions over more than one decade.

It is remarked that this model is still very crude. No equilibria with poloidal variation of the pressure have yet been modelled. Still, the first results show interesting parallels with the experimental data, such as the persistence of regions with very low stochasticity, in which steep pressure gradients could develop.

6. Summary and discussion

The filaments of high T_e and pressure in the plasma center cannot be reconciled with the concept of nested flux surfaces. The working hypothesis is that the filaments are closed flux tubes. It is not clear whether they are <u>formed</u> by the intense heating, or that they pre-existed and are only made visible by their high temperature. From the experiments it is clear that once they have formed, their life time is ≈ 0.5 ms. This corresponds to their energy confinement time (using the diffusivity obtained from a local power balance), as well as to their skin time. Also, once they have formed, they will act as current channels. In fact their existence may be the result of some instability whereby a current perturbation is amplified by the heating proces. A similar mechanism is invoked in [22] to predict filamentation of ohmic plasmas. If the hot filaments are current channels, this will ergodize the surrounding field. Thus, the central part of the plasma would be an ergodic bath in which several flux tubes are immersed, surrounded by the layer of low conductivity.

The transport barrier near the sawtooth inversion radius is a quite reproducible observation. The fact that χ drops so sharply could only be measured thanks to the ultra high spatial resolution of the measurement. Interestingly, on the basis of simulations of sawteeth, a similar drop of χ -also located between the sawtooth inversion and mixing radii - was found in [23]. Further evidence for this transport barrier was obtained from the evolution of the pressure profile during pellet injection in RTP[24]. Transport barriers have also been demonstrated near q=3 in JT-60 [25], and for impurities, near the sawtooth inversion radius in JET [26]. Thus, there is mounting evidence that the transport coefficients in a tokamak plasma are not smooth functions of the minor radius. To describe the transport by smooth coefficients is therefore very dangerous, for the reasons given in Sec.4. The simple fact that a few thin layers of low transport can dominate the global confinement is often overlooked in transport studies.

The simulation of magnetic equilibria with broken surfaces presented here is really in its infancy. The equilibria presented may be regarded as Grad-Shafranov like equilibria, with added magnetic fluctuations. The many small current filaments, produce a perturbation of the field which may be more realistic than the often used Fourier expansion of a perturbing field. The latter typically produces chains of islands at surfaces of simple rationality. The current filaments, on the other hand, produce a topology that is characterized by 'fuzzy' surfaces.

Finally, it is reiterated that structures are observed in plasmas that do not have degraded confinement with respect to the established scaling-laws.

Acknowledgement

Many thanks are due to Geert Verhaag, Peter Busch and Ben Grobben, who made a great effort to get the Thomson diagnostic and its software running, and to the technical staff of RTP. Tim Luce introduced us to the technique of free knot spline fitting. This work was performed under the Euratom-FOM association agreement, with financial support from NWO and Euratom.

References.

[1] B.V. Chirikov *Phys Rep* **52** (1979) 265
[2] V.I. Arnold, Soviet Math Dokl **3** (1962) 1008
[3] D.Montgomery and X. Shan, *Comments on Plasma Phys. Contrlled Fusion* **16** (1994) 35
[4] Y.Z. Agim and D. Montgomery, *Plasma Phys. and Controlled Fusion* **33** (1991) 881
[5] S.J. Zweben and S.S. Medley, *Phys. Fluids. B* **1** (1989) 2058
[6] M. Malacarne and P.A. Duperrex, *Nucl Fusion* **27** (1987) 2113
[7] A. Weller et al, *Phys Rev Lett.* **59** (1987) 2303
[8] B.Ph. van Milligen et al, *Nucl. Fusion.* **35** (1993) 1119
[9] R. Jaspers et al, *Phys Rev Lett* **72** (1994) 4093
[10] M.A. Dubois et al, *Determination of the safety factor profile in Tore Supra from striations observed during pellet ablation*, Report EUR-CEA-FC-1446 (1992)
[11] P.Cripwell and A.E. Costley, *Proc. 18th EPS Conf. on Controlled Fusion and Plasma Physics, Berlin (1991) Part 1 p17*; A.E. Costley et al, *Proc. 19th EPS Conf. on Controlled Fusion and Plasma Physics, Innsbruck (1992) I-199*
[12] M.F.F. Nave et al, *Nucl. Fusion* **32** (1992) 825
[13] M.C. Zarnstorff et al, *Proc. Varenna Workshop on Local Transport in Fusion Plasmas* (1993)
[14] N.J. Lopes Cardozo et al, *Phys Rev Lett* **73** (1994) 256
[15] A. Krämer-Flecken A. et al, *Nucl. Fusion* **33** (1993) 921
[16] J.A. Konings et al, *Proc. Varenna Workshop on Local Transport in Fusion Plasmas* (1993)
[17] C.C. Chu et al, *Proc. 21st EPS Conf. on Controlled Fusion and Plasma Physics, Montpellier (1994)*
[18] D.L.B. Jupp, *SIAM J. Numer. Anal* **15** (1978) 328
[19] N.J. Lopes Cardozo et al, *Plasma Physics and Controlled Fusion* **32** (1990) 983
[20] C.C. Hegna C C and J.D. Callen, *Phys. Fluids B* **5** (1993) 1804
[21] A. Montvai et al, *Proc. 21st EPS Conf. on Controll Fusion and Plasma Physics, Montpellier (1994)*
[22] M.M.G. Haines and F. Marsh, *J. Plasma Physics* **27** (1982) 427
[23] G.M.D. Hogeweij et al, *Proc. 21st EPS Conf. on Controlled Fusion and Plasma Physics, Montpellier (1994)*
[24] A.A.M. Oomens et al, *Proc. 15th Int. Conf. on Plasma Phys. and Contr. Nucl. Fusion Res., Sevilla*, (1994) post dead line paper
[25] Y.Koide et al, ibid, paper IAEA-CN-60/A-2-I-3
[26] R. Giannella et al, *Plasma Phys. and Controlled Fusion* **34** (1992) 687

Experimental Results from the TUMAN 3 Tokamak.

V.E.Golant, M.V.Andrejko, L.G.Askinazi, V.A.Korneev, S.V.Krikunov, B.M.Lipin,
S.V.Lebedev, L.S.Levin, K.A.Podushnikova, G.T.Razdobarin, V.A.Rozhansky',
V.V.Rozhdestvensky, M.Tendler'*, A.S.Tukachinsky, S.P.Jaroshevich

A.F Ioffe Phys.-Tech. Institute, Russian Academy of Sciences, 194021, St.Petersburg, RUSSIA
' *State Technical University of St.Petersburg, 195251, St. Petersburg, RUSSIA*
**Alfven Lab., Royal Institute of Technology, EURATOM-NFR Association, Stockholm, 10044, SWEDEN*

The "TUMAN-3" Tokamak programme concentrates on issues of improved confinement. In 1989 the transition from an ordinary Ohmic regime into an improved confinement mode was achieved. The signatures of the H-mode in auxiliary heated tokamaks have been observed in this regime. The crucial role of the boundary radial electric field was found in the experiments with internal bias probe. Other techniques were demonstrated to disturb the boundary plasma which led to H-mode triggering: short increase of working gas puffing, minor radius magnetic compression and pellet injection.

The novel scaling of the energy confinement time in the Ohmic H-mode was obtained, which differs dramatically from the scaling for the ordinary Ohmic regime. There were found a strong dependence of τ_E on plasma current and a weak dependence on density.

The maximum value of τ_E was 10 times longer than in the ordinary Ohmic region. The τ_E scaling for the Ohmic H-mode is consistent with the scaling proposed for devices with powerful auxiliary heating. The result shows that H - mode physics is universal in toka–maks with different geometries and heating methods.

I. THE TUMAN-3 DEVICE

TUMAN-3 is a small tokamak with a circular cross section [1]. It has a metallic vacuum vessel and limiters. Some machine and plasma parameters are listed in Table 1.

	Range	Typical OH
R_0, m	0.44-0.64	0.53-0.55
a_1, m	0.11-0.24	0.22-0.24
B_T, T	0.44-1.2	0.40-0.55
I_p, MA	0.04-0.2	0.09-0.12
$<n>$, 10^{19} m^{-3}	< 5.2	1.0-2.0
T_{e0}, keV	0.3-0.8 (1.0-ICRH)	0.4-0.5
T_{i0}, keV	0.09-0.17 (0.4-ICRH)	0.12-0.15

Table 1. TUMAN-3 parameters

An essential feature of the TUMAN-3 design is the small aspect ratio, A = R/a = 2.3, which allows to increase Ohmic power input due to neoclassical enhancement of the resistance. In TUMAN-3 neoclassical correction for resistance is about a factor of two. The small aspect ratio also provides for a relatively high plasma current at a low longitudinal magnetic field. Thus in the steep torus, the ratio of the input power to B_T is significantly larger compared with a conventional tokamak. For TUMAN-3 it is about $I_p/B_T \approx 0.2$ MA/T at q^{cyl} = 2.5. TUMAN-3 was designed to carry out magnetic compression experiments. In order to achieve minor radius compression by B_E ramp-up during a short time scale, the toroidal magnet has a low inductance. It consists of 24 single-turn coils connected in series. For the a-compression experiments, $B_T^C/B_T^0 \leq 3$, τ^C = 3.5 ms. The poloidal field system is capable to maintain plasma currents up to 200 kA during 80 ms. It can also produce a second fast current ramp - up/down to control j (r) profile. The highest rate of current ramping is 25 MA/s. Plasma equilibrium is maintained in a variety of discharge scenarios and provided by the feedback position control system. The auxiliary heating system includes an ICRH generator with power up to 1.5 MW (8.5 MHz) and a 200 KW gyrotron (37 GHz) of ECRH near the plasma boundary. A pneumatic gun for pellet injection provides the possibility of accelerating a solid pellet at room temperature up to a velocity of 300 m/s to cause density and temperature disturbances in order to study transport phenomena. Diagnostic equipment installed in the device enables the measurement of a variety of plasma parameters in the core and boundary regions.

Many experiments were carried out during the TUMAN-3 operation. The most important are listed below:

1. Studies of Ohmic heating peculiarities in a small aspect ratio tokamak [1,2,3]
2. Minor/major radius compression and combined compression [4,5,6]
3. Fast current ramp-up/down [7,8,9];
4. ICR and ECR heating [10,11]
5. Ohmic H-mode studies [12,13,14,15]

During recent years the main efforts were focussed on the Ohmic H-mode studies. The main results obtained in these experiments are reviewed below.

II. OHMIC H - MODE IN TUMAN-3

The H-mode has been discovered in 1982 while carrying out experiments with powerful heating provided by neutral beam injection on ASDEX [16]. Thereafter possibilities to initiate H-modes by means of different auxiliary heating methods were demonstrated in a number of experiments. Experiments on DIII-D demonstrated that the ordinary Ohmic regime constitutes a degraded confinement mode and could be improved upon entering Ohmic H-mode [17]. Hence, the peculiarities of a heating scheme are not a cause for the confinement degradation in L-mode and its improvement after transition into H-mode. However, in the majority of experiments the L-H transition resulted in divertor configurations.

Below, results obtained on the TUMAN-3 tokamak during the Ohmic H-mode studies in circular limiter configuration are presented. At first, Ohmic H-mode appeared either spontaneously or after a short increase of the puffing rate [13]. The traces of the main plasma parameters in such a mode of operation are shown in Fig.1.

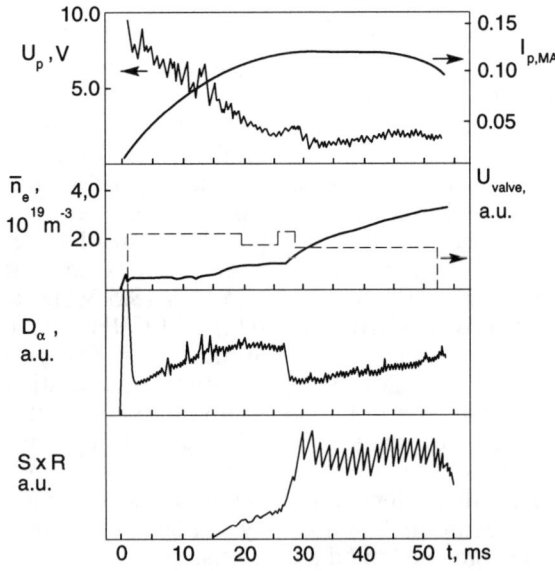

Figure 1. Temporal behaviour of the loop voltage, plasma current average density, D_α radiation and Soft X Ray emission in the 0.115 MA Ohmic H-mode.

The signatures of the TUMAN-3 H-mode are clearly demonstrated in this and in the following figures: (i) sharp drop of D_α radiation (at 27th ms), accompanied by the monotonic increase in plasma density, thereby suggesting the enhancement of the particle confinement time in H-mode by approximately a factor of 3 compared to the standard reference Ohmic shot, (ii) rapid formation of the steep gradient zone at the periphery during the transition from OH to H-mode and the relatively slow rise of n in the bulk plasma, see Fig.2,

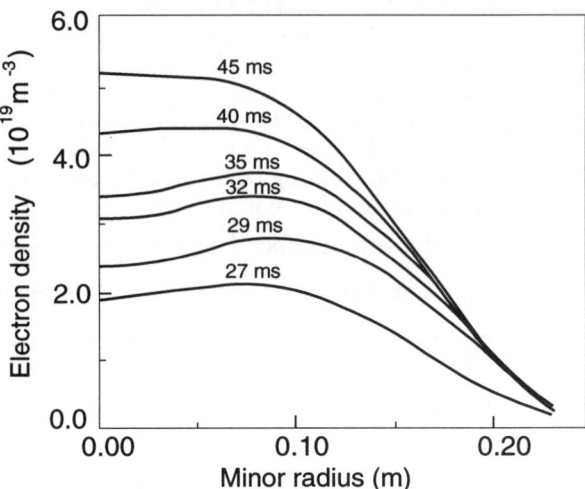

Figure 2. Density profile evolution in the 0.115 MA Ohmic H-mode.

(iii) notable changes in the electron temperature profile, see Fig. 3, widening a thermal transport reduction barrier in the gradient region and an increase in energy confinement time [13], (iiii) a slow increase in loop voltage during the last stage of H-mode caused, apparently, by impurity accumulation in the bulk plasma due to the improvement in particle confinement.

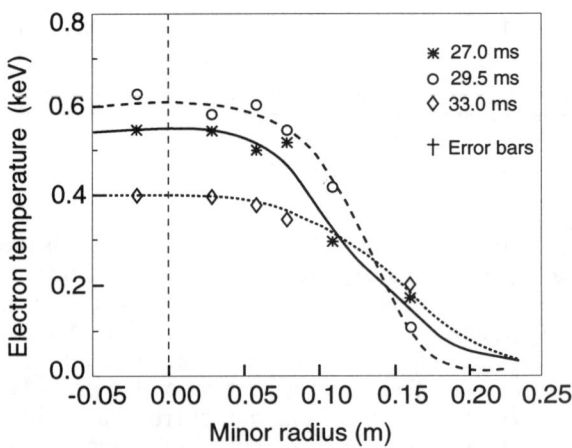

Figure 3. Electron temperature profile evolution in the 0.115 MA Ohmic H-mode.

The density profiles measured by the interferometer and shown on Fig. 2. do not provide sufficient space resolution near the edge. Therefore density profiles in the vicinity of the LCFS (Last Closed Flux Surface) have been measured by electrostatic probes. After the transition into H-mode the density rises everywhere inside LCFS and drops outside [18]. An example of the ion saturation current measured at radial positions r=a-5mm and r=a+5mm is shown in Fig.4.

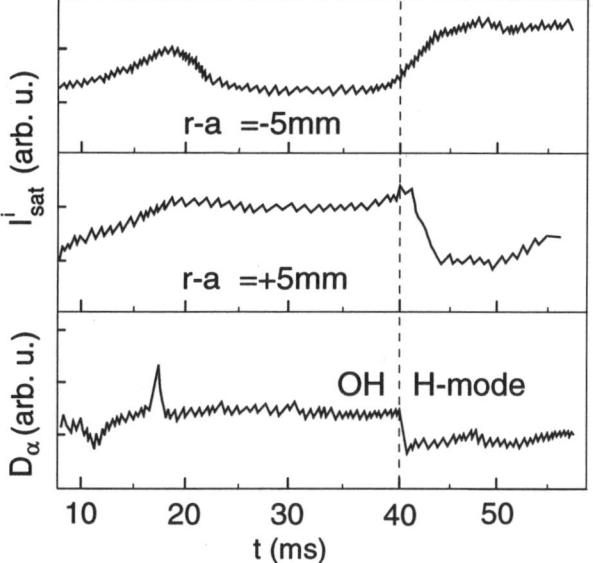

Figure 4. Ion saturation current to the probe, measured 5 mm inside and 5 mm outside LCFS through H-mode transition.

The e-folding length at the LCFS resulting from these data is 12 mm before and 3 mm after the transition suggesting the formation of the transport barrier in this zone. Using multi-electrode probes it was found that the transition was accompanied by the suppression of the fluctuation-driven particle flux[13]. Suppression of the high frequency fluctuations has also been observed at the low field side by the reflectometer adjusted to boundary density, Fig.5 [19].

These observations appear consistent with the explanation of the transport reduction in H-mode due to suppression of the turbulence by sheared rotation [20]. Yet, the contradiction with the model has to be mentioned. Indeed, the high field side interferometer measurements showed a moderate increase in the high frequency part of the spectrum, thereby suggesting strong poloidal asymmetry in the turbulence damping.

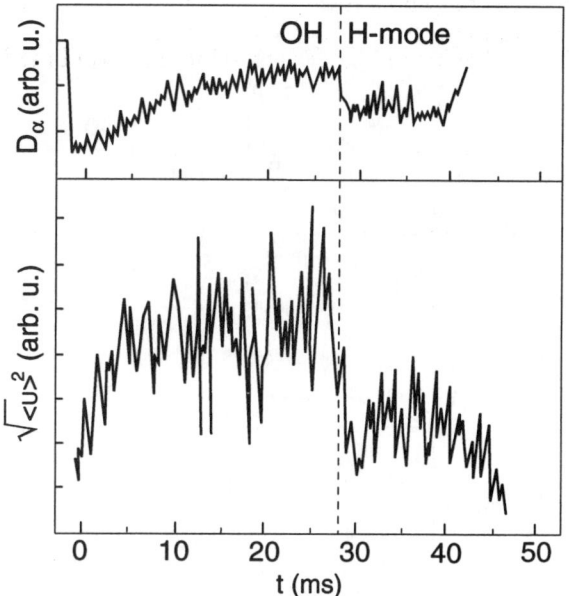

Figure 5. Average amplitude of high frequency turbulence (f_{f1}>200kHz) measured by an RF reflectometer (f_{RF}=16.5GHz) looking from the low field side.

III. METHODS OF H-MODE INITIATION.

First of all, increase of the puffing rate is not the only method of the H-mode triggering, demonstrated on TUMAN-3. There were realized also other methods of boundary plasma perturbations which led to initiation of the H-mode. Among these are: biasing of the plasma edge, minor radius compression and LiD pellet injection.

In more detail, in TUMAN-3 the minor radius magnetic compression is provided by means of fast increase of the longitudinal magnetic field in the quasi-stationary stage of the Ohmic regime. In these experiments the initial toroidal field was equal to 0.45 T and the compression factor B_T^C/B_T^0 was equal to 2 [13]. Time scans of plasma parameters during the shot with the H-mode initiated by a-compression are shown in Fig.6.
Enhancement in particle and energy confinement is borne out by an increase of the density and SXR emission. The density increase exceeds the adiabatic law scaling factor. Indeed, this factor for a 2-fold compression is 1.4, whereas the density rises by more than a factor of 2. Furthermore, the time scale of the density growth is longer than the compression time, $\tau_p^* \approx 10$ ms. D_α emission drops just after the compression starts thereby indicating some improvement in particle confinement resulting from the triggered H-mode.

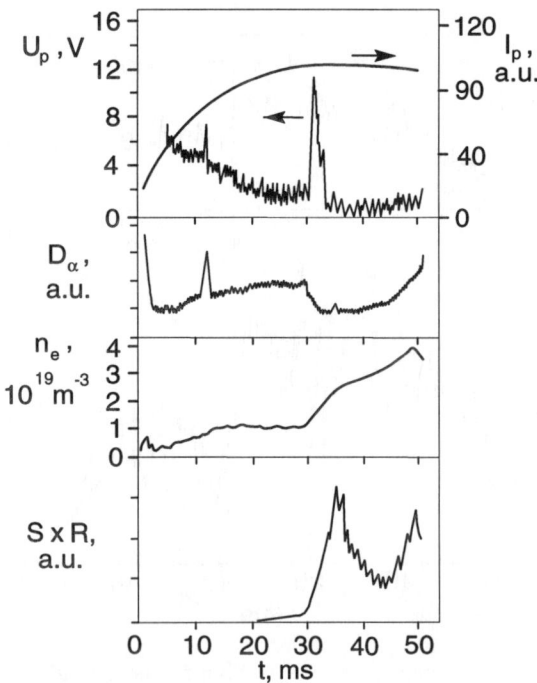

Figure 6. Temporal behaviour of some plasma parameters in the shot with H-mode initiated by a-compression.

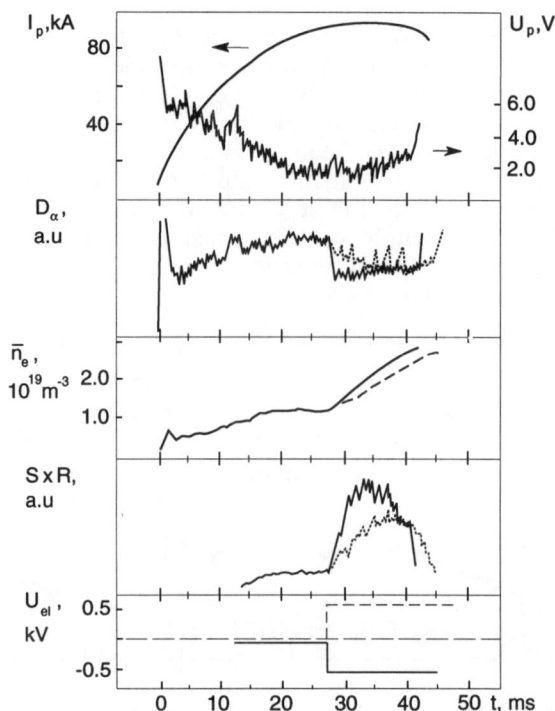

Figure 7. Evolution of the loop voltage, plasma current, D_α radiation, averaged density and SXR emission in Ohmic H-mode caused by negative (solid) and positive (dashed lines) biasing of the plasma edge.

SXR-emission growth continues during 6 ms which is also longer than the compression time. It is comparable with the energy confinement time in compressed plasma. The data demonstrate the option to trigger the H-mode by minor radius magnetic compression. They show that the confinement improvement due to the a-compression experiments may be caused by transition into H-mode. In order to elucidate the role of the radial electric field in the L-H and H-L transitions the experiments with boundary biasing using an internal bias probe have been performed. A molybdenum electrode with a surface of 3cm^2 perpendicular to the magnetic field was inserted from the low field side to a distance of 4 cm inside the LCFS. After obtaining a stationary stage of the discharge both electrodes were biased to positive and negative voltages with respect to the vessel. The limiter was kept at the vessel potential. Transition to the H-mode occured at a bias < -0.25 kV (H$^-$ mode) for negative electric fields and > 0.5 kV (H$^+$ mode) for positive electric fields (see Fig.7.) [21]. The improvement in confinement was enhanced in the case of negative biasing compared to the case of positive bias. The dN/dt (where N is the total number of particles in the tokomak) is larger for the H$^-$ mode than for the H$^+$ mode, whereas D_α emission is reduced.

For the particle confinement time we estimate $\tau_p^- =$ 7.9 ms, $\tau_p^+ = 4.3$ ms and $\tau_p^{OH} = 2.5$ ms for the H$^-$, H$^+$ and and reference Ohmic discharges, respectively. The main effect of the improved confinement occurs in the H$^-$ mode and is demonstrated by the enhanced increase in the intensity of soft X-ray radiation, indicating enhanced energy confinement time in the H$^-$mode. Note that in the auxiliary heated TEXTOR plasma, particle confinement has been found to be mostly improved in the H$^-$ mode as well, whereas in contrast to our results, the strongest enhancement in energy confinement has been reported for the H$^+$ mode [22].

In general, the features of the H$^-$ mode obtained by negative bias are found to be very similar to those of the spontaneous Ohmic H-mode. Hence, it suggests that the natural sign of a radial electric field in the Ohmic H-mode is negative. Evidence for this suggestion was found in the experiment with positive bias applied in the Ohmic H-mode, see Fig. 8.

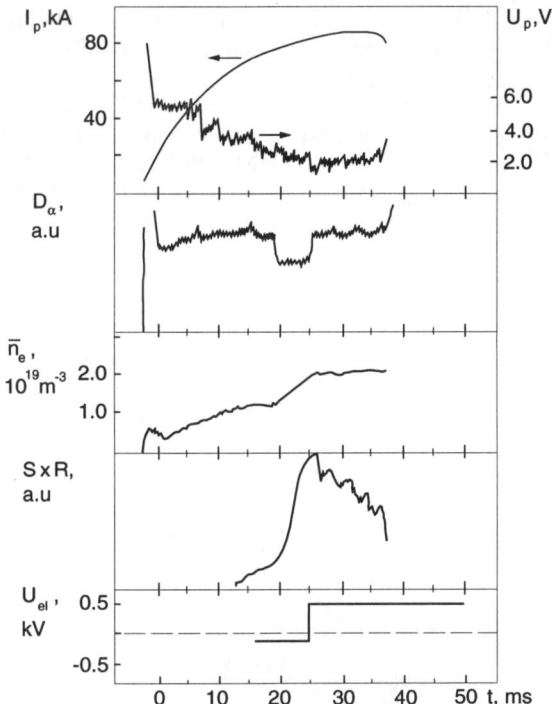

Figure 8. Evolution of the loop voltage, plasma current, D_α radiation and averaged density in the shot with Ohmic H-mode termination by positive biasing of the edge.

After switching on the positive voltage the increase in both plasma density and the SXR intensity is terminated and the D_α radiation increases to the Ohmic heating level. The Ohmic H-mode has been degraded to a normal Ohmic discharge (H-L transition). Thus it can be inferred that the spontaneous Ohmic H-mode is characterized by a naturally emerging negative E_r at the edge and when this field is cancelled by the external positive E_r the transition to the regime with degraded confinement occurs.

None of the active methods for H-mode initiation described above are feasible for the next generation of tokamaks with hotter plasmas. This is due to the low efficiency of gas puffing in larger tokamaks with divertors, technical difficulties with compression experiments and the impossibility of immersing any electrodes in a hot plasma during a long pulse. It is therefore, highly desirable to find a practical alternative to trigger an H-mode using a more suitable method. Pellet injection, which is regarded as an effective means of fuelling plasmas, is shown to offer such an option. Pellet injection causes rather steep density and temperature gradients. If the pellet is sufficiently large and the penetration depth is small, the region of steep density gradient occurs close to the LCFS. According to the relationship $E_r = T_i/e[\, d\ln n/dr + (1-k)\, d\ln T_i/dr]$, where k is a proportionality coefficient dependent on collisionality [23]. More importantly, it has been shown in [24] that there is an important causal relationship between the density profile gradient and the electric field value on the LCFS. Steeper profiles cause stronger electric fields and vice versa.

Hence, in the pellet injection experiments, the density dependence of the radial electric field impacts most significantly. An improved confinement regime (PCH mode) has been found experimentally by LiD pellet evaporation near the edge [25]. These experiments have demonstrated the possibility of triggering H-mode by pellet injection. This could be a promising way to provide H-modes in the next generation of tokamaks.

IV. ENERGY CONFINEMENT IN OHMIC H-MODE

The Ohmic H-mode was first observed on DIII-D [17] and an enhancement factor of 1.2 over Neo-Alcator scaling has been reported. The first Ohmic H-mode experiments on TUMAN-3 resulted in an enhancement factor of 2-4 [13,14]. Hence, the question emerged whether energy confinement in Ohmic H-mode yields Neo-Alcator scaling with some enhancement factor or an expression with another parametric dependencies. The issue of τ_E dependence on the main plasma parameters is addressed below. The experiments were carried out employing boronization which provided for the substantial increase in the density and current ranges [15]. Study of the confinement before and after boronization offered a unique opportunity to draw conclusions of the τ_E dependence on the input power in Ohmic regimes, maintaining narrow range for parameters such as R, a, I_p, <n>, B_T, A_i.

In the boronized vessel the maximum plasma current was increased up to 160 kA. Measurements of the energy confinement time in Ohmic-H mode with Ip = 63, 112 and 155 kA, revealed a strong dependence of τ_E on plasma current, see Fig.9. The other plasma parameters remained almost unchanged in this scan, <n> = (1.7-3.5) 10^{19} m^{-3}, q^{cyl}(a) =2.4-3.1.

The dramatic dependence of τ_E on I_p contradicts the Neo-Alcator scaling, usually assumed for Ohmic regimes. In contrast, it is consistent with the scalings empoloyed for the H-modes in auxiliary heated plasmas. The study of the τ_E dependence on input power and density was performed in the regime with a plasma current of 112 kA. The scaling on the input power benefited from the substantial decrease in loop voltage after boronization maintaining plasmas with similar global parameters.

147

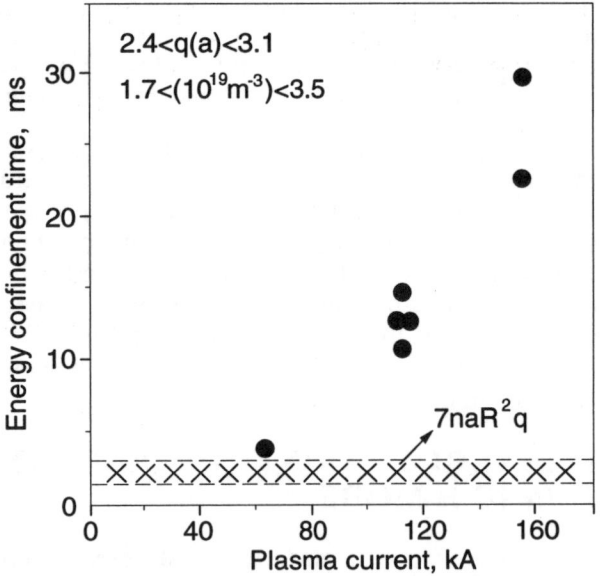

Figure 9. Energy confinement time in Ohmic H-mode in the boronized vessel as a function of plasma current.

Indeed, after boronization the input Ohmic power P^{OH} drops by a factor of two. The decrease in the input power in combination with small changes in density and temperature profiles suggests the significant rise in energy confinement

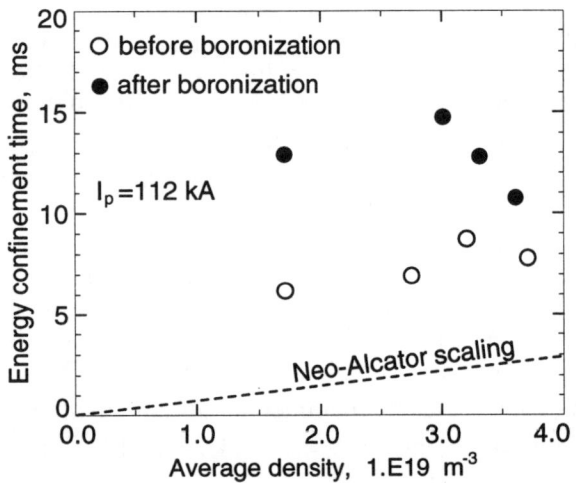

Figure 10. Energy confinement time as a function of density for the regime with current 112 kA before and after boronization.

Note that solid circles (boronized vessel) are significantly above open circles (before boronization).

Moreover, the τ_E dependence on the density is weak or negligible as shown in Fig.10. The Neo-Alcator scaling is also shown for comparison. Although the data base is not sufficient to draw definite conclusions on the exact P^{OH} and $<n>$ dependencies, the contradiction with the Neo-Alcator scaling results clearly. In contrast, the similarity with the ELM- free H-mode scalings in auxiliary heated plasma emerges. Indeed, the comparison of the results from Ohmic H-mode confinement studies on TUMAN-3 with the DIII-D/JET scaling for ELM-free H-mode plasma [26,27] shows remarkable agreement.

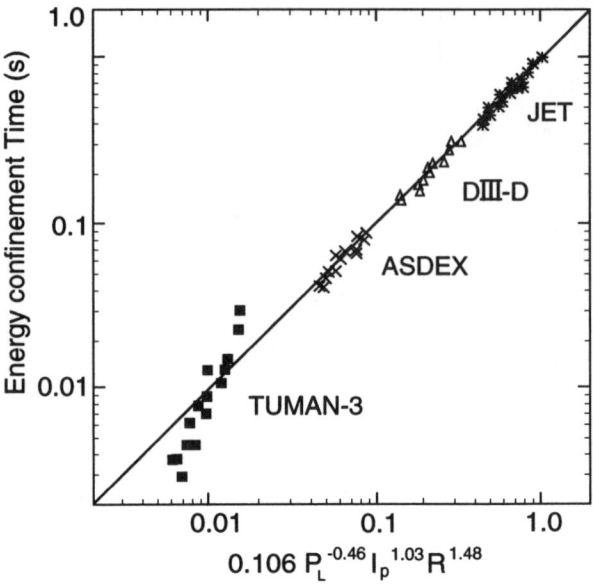

Figure 11. Energy confinement time versus predictions of the DIII-D/JET scaling for ELM-free H-mode.

TUMAN-3 points lie near the linear approximation for three tokomaks.

In summary, this scaling was proposed for plasma in large devices with strong auxiliary heating (JET, DIII-D, ASDEX). Fig. 11. shows that the majority of the TUMAN - 3 discharges are near the linear dependence for these three tokamaks. These results provide evidence that H-mode physics has universal nature in tokamaks with different geometries and heating methods.

ACKNOWLEDGEMENTS

This work was partly supported by The International Science Foundation (Grant N R2T000).

REFERENCES

1. Vorob'ev G.M., et al, Sov. J. Plasma Phys., v.9 (1983), P.65
2. Golant V.E. et al, Sov J. Plasma Phys., v.8 1982), p.1167
3. Golant V.E., et al, JETP Lett., v.39 (1984), p.108
4. Vorob'ev G.M., et al, Proc. 10th Eur. Conf. on Contr. Fusion and Plasma Phys. (Moscow, 1981), v.1, H-2A
5. Golant V.E., et al, Proc. 11th Eur. Conf.

6 Askinazi L.G., et al, Proc. 12th Eur. Conf. on Contr. Fusion and Plasma Phys. (Budapest, 1985), v.9F, Part I, p. 363
7 Bender S.E., et al, Proc. 11th Eur. Conf. on Contr. Fusion and Plasma Phys. (Aachen, 1983), v. 7D, Part. I, p. 111
8 Askinazi L.G., et al, "Voprosi Atomnoi Nauki i Tehniki" (in Russian), v. 3 (1988), p. 53
9 Sakharov N.V., et al, Plas.Phys. and Contr. Fusion, v.35 (1993), p.411
10 Askinazi L.G., et al, Plas. Phys. and Contr. Fusion, v 31 (1989), p.3
11 Arbuzov A.N., et al, Proc. Int. Workshop on Strong Microwaves in Plasma (Suzdal, 1990), H-18
12 Arbuzov A.N., et al, Proc. 17th Eur. Conf. on Contr. Fusion and Plasma Heating (Amsterdam, 1990), v 14B, part 1, p 299
13 Akatova T. Yu., et al, Proc. 13th IAEA Conf. on Plas. Phys. and Contr Nucl. Fusion Res. (Washington, 1990), v.1, p. 509
14 Andrejko M.V., et.al. Proc. 14th IAEA Conf. on Plas. Phys. and Contr. Nucl. Fusion Res. (Wuerzburg, 1992), v.1, p. 485
15 Andrejko M.V., et al, Plasma Phys. and Contr. Fusion, v. 36 Suppl. 7(A), (1994), p. A165.
16 Wagner F., et al, Phys. Rev. Lett., v.49 (1982), p. 1408
17 Schissel D.P., et al, Proc. 16th Eur. Conf. on Contr. Fusion and Plasma Phys. (Venice, 1989), v.13B, Part 1, p. 115
18 Askinazi L. G., et al, Proc. 18th Eur Conf. on Contr. Fusion and Plasma Phys., (Berlin, 1991), v. 15C, Part I, p.401
19 Bulanin V.V., et al, Proc. 18th Eur. Conf.on Contr. Fusion and Plasma Phys., Berlin, 1991, v. 15C, part IV, p.321
20 Biglari H., et al, Phys. Fluids, B2 (1990), p.1
21 Askinazi L. G., et al, Nuclear Fusion, v.32 (1992), p.271.
22 Van Nieuwenhove R., et al, Proc. 18th Eur. Conf. on Contr. Fusion and Plas. Physics (Berlin, 1991), v.15C, part I, p.405
23 Rozhansky V.A. and Tendler M., Phys. Fluids, B4 (1992), p.1877
24 Tendler M., Daybelge U., Rozhansky V., Proc. 14 IAEA Conf., Wuerzburg, Germany, 1992 paper IAEA - CN -56/D-4-8
25 Askinazi L.G., et al, Phys. Fluids, B5 (1992), p.2420.
26 Schissel D.P., et al, Nuclear Fusion, 31 (1991), p.73
27 Simonen T C et al 1991 Preprint General Atomics GA-A20711

EXPERIMENTAL PROGRAMME OF HL - 1 TOKAMAK.

Qian, ShangJie and HL-l Group
Southwestern Institute of Physics,
P. O. Box 432, Chengdu,
Sichuan 610041, P R China.

The experimental programme of HL-1 was divided into two parts. In the first period, the expansion of the HL-1 operation regime connected with wall conditioning and MHD studies was successfully finished in 1989. The lowest q(a) of 1. 8 has been achieved and the maximum plasma density, with energy confinement time of its design value, has been successfully obtained. The reproducibility of the plasma is quite good for detailed research. The program of the second part was to address the improvement of plasma confinement with the emphasis on boundary plasma studies. The typical improvement factors of confinement are: particle confinement time is increased by a factor of 2 to 3 and plasma energy confinement time is increased by 1. 7. The ECRH, LHCD at several kilowatts and Pellet Injection experiments have been carried out on the HL-1 Tokamak. The improvement of confinement has been observed in all investigated cases except during ECRH and the LHCD/plasma biasing combined experiments. In the latter case, the plasma confinement results to be poor for unknown reasons. The results of inert-gas-filled discharge cleaning and molecular injection using the Laval Nozzle are also reported in this paper.

I. INTRODUCTION

The HL-1 Tokamak is a Chinese-made fusion plasma machine built in the Southwestern Institute of Physics. China Nuclear Energy Authority. in 1984 [1]. Since 1985 it has been operated for 7 years until the shut-down in October 24, 1992. It is being modified to HL-lM, that is scheduled to be in operation this year. The main purposes of the modification are the improvement of accessibility for more auxiliary plasma heating, pellet injection, pumping limiter and their combined experiments with much more detailed diagnostics. Installation of a new vacuum chamber and the upgrade of the poloidal field power supply are almost finished now. The variety of additional equipment of 1 MW LHCD. 1 MW ICRH. 1 MW NB and 8 shots Pellet Injector is being prepared on schedule. The next machine. HL-2 is under design and will be ready for the next National Five Years Plan that starts in 1996. The parameters of HL-1, HL-lM and HL-2 are given in Table 1. Some results from the HL1[2] are given below.

II. WALL CONDITIONING AND RECYCLING CONTROL

In order to control the recycling on the wall and impurities in the plasma, various wall conditioning methods and their combinations have been used on the HL-1 with an inert-gas-filled discharge cleaning. The methods used on HL-1 are listed in Table 2. The Electron Resonance Discharge Cleaning gives the highest efficiency but Alternative Discharge Cleaning is the easiest to use regularly with reasonable efficiency. In order to increase the efficiency, a variety of rare gases: Neon, Xenon and Krypton were mixed with the Hydrogen to form the different working gases. The best result was obtained using the H+Kr mixture discharge cleaning. The water yield is 3-fold higher than that using pure Hydrogen and the sum of residual gas partial pressures, measured immediately after shut-off of discharge cleaning (AC-DC), decreases by a factor of 4 in comparison to that using Hydrogen only. The recycling coefficient is obviously decreased. Titanium vaporization and wall carbonization were also used for wall condition control.

Besides wall conditioning and coating, a movable pumping limiter located at a radius of 18-21 cm, acting as a main limiter, is effectively used to control the edge plasma and to reduce the cold gas at the periphery. The measured particle exhaust efficiency was 6.5 to 12 % of the total particle outflux when the slot of pumping limiter was facing the ion drift side. The recycling rate is decreased by 15 % and the fueling efficiency is increased by 30 % in comparison to that without a pumping limiter.

According to HL-1 experimental results, stable operation could be easily obtained at a recycling rate of about 0. 8. Considering the control of impurities by using a graphite limiter and complete wall cleaning, the higher fueling efficiency at a lower recycling rate provided a smaller cold gas periphery, lower radiation loss and better confinement.

III. PLASMA FUELING

Gas puffing is used as a regular plasma fueling method. For normal gas puffing from the wall, the fueling efficiency is as low as 10-20% due to a higher gas launch angular divergence of about 40 degrees. Most of the gas puffed into the chamber is absorbed at the clean wall in terms of the gas source for recycling and forms a cold gas periphery which causes the plasma edge to cool down.

© 1995 American Institute of Physics

Sometimes it leads to discharge termination. By means of puffing gas directly at the Last Closed Flux Surface (LCFS) through two pipes mounted at the graphite limiter with a reduced puffing rate for the same discharge condition, the fueling efficiency is somewhat increased because of the immediate ionization of gases at the radiating layer. The maximum plasma density rising rate obtained is 7.10^{20} m^{-3}s^{-1} without any plasma instability for D+He discharges.

A new collisionless molecular beam provided by the Molecular Beam Injector (MBI), which consists of the Laval Nozzle and skimmer with the opening located 7cm behind the LCFS, has been installed on the HL-1 Tokamak. When the molecular beam with velocity of 100m/s is injected during the plasma current ramp, the ratio of current rise rate to the density rise rate reaches its maximum of $9.5.10^{-19}$kAm^3 which is approximately 3 times higher than that obtained using normal gas puffing. An obvious rapid density rise due to beam injection at the plateau of density $1,5.10^{19}$m^{-3} easily leads to the higher density of 5.10^{19}m^{-3}. The observation of H_α indicates that the injected molecular beam probably penetrates into a plasma up to a radius of 8cm which is nearly the q = 2 location. This method is suitable for fueling the peripheral plasma.

A pneumatic single shot hydrogen pellet injector has been developed in SWIP and pellets were successfully launched into the HL-1 tokamak to increase the plasma fueling efficiency and to peak the density profile. The pellet diameter of 1.1 mm. with a length of 1.1--2.5 mm was accelerated up to a speed about 300 - 500 m/s. The density profile measured with the multi-channel HCN interferometer shows a rise of the peaking factor, the ratio of local density at r/a= 0 to that at r=0.9, ranging from 1.3 to 1.5. Meanwhile, the fueling efficiency observed is above 50% which is 3 to 5 times higher than that obtained using normal gas puffing with an opening at the chamber wall.

IV. PLASMA SURFACE INTERACTION
The typical time evolution of the main parameters of the HL-1 Tokamak for a high density ohmic discharge that ends with a MARFE is shown in Fig. 1. The evolution of the Abel inverted brightness of H_α signal shows the detachment lasting for about 200ms. The brightness of H_α at the limiter ports always exceeds that at a port regardless where the gas is introduced. The radial profiles of the deposited impurities, mainly Ni, Cr, Fe and Ti on the graphite sample facing the plasma after exposure to 74 shots, were obtained from analysis by SIMS cross calibrated by means of AES. Both the H_a observation and the SIMS analysis suggest that the plasma limiter interaction plays the key role for recycling and impurity issues.

V. MHD INSTABILITY AT THE CURRENT RISE PHASE
By means of gas puffing and plasma current programming, the MHD stability has been significantly improved. It was concluded that the m/n=4/1 mode does not result in disruptive instability, the m/n=3/1 mode does not induce the disruption by itself but perhaps it can trigger the development of the m/n=2/1 mode which is the key problem for stable operation of HL-1. It is worth attempting to control the plasma profile with both gas puffing and plasma current programming to peak the profile during the plasma current rise phase for suppression of the m/n=3/1 and 2/1 modes. The experimental observation shows that the most stable discharge could be obtained with the ratio of plasma current rise rate to density rise rate in the regime of $(20-30) \times 10^{-13}$ kAcm^3 (Fig. 2). Therefore the plasma current profile during rise phase is optimized by a combination of plasma ohmic heating and boundary cooling due to gas puffing. With this method and careful wall surface condition control, the highest line-averaged plasma density of $0.72.10^{20}$m^{-3} and the lowest q (a) of 1.8 have been successfully achieved.

VI. SAWTOOTH ACTIVITIES
Sawtooth oscillations were observed on the HL-1 with soft x-ray, hard X-ray, microwave and ECE diagnostics, as well as employing loop voltage and H_α measurements. Most of the measurements was carried out by using soft X-ray diagnostics.

Sawtooth oscillations appeared within a broad range of discharge parameter (n = 0.6 - 11 $\times 10^{19}$ m^{-3}) in the HL-1 (table 3). Up to now, only a portion of the experimental data has been analyzed in detail, either experimentally or theoretically. Their period and crash times are consistent with the prediction of both the Kadomtsev and the turbulent models, specially for a sawteeth of discharges with $2.5<q(a)<3.5$.

According to the observation of a crash event, obtained with a soft X-ray tomograph (Fig. 3), it is believed that cold bubble convection near the q= 1 surface causes the crash of a monster sawtooth in an ohmically heated plasma.

From a recent theoretical study of resistive kink modes [13], which determines a critical safety factor $q_c(0)$. When $q(0) < q_c(0)$, sawtooth disruption could be triggered by m/n = 1/1. Besides, relaxation of a sawtoooth-like soft disruption produced by the m/n= 2/1 mode, was also commonly observed (Fig. 4).

VII. IMPROVEMENT OF CONFINEMENT

The relative aspects of the improved confinement have been studied in great detail employing the mentioned above methods: the fueling and recycling control and the MHD activities including sawtooth oscillations and disruptions. Focusing on the boundary plasma, serious studies on the improvement of plasma confinement have been carried out on HL-1 by the following means : ECRH, Pellet Injection, Molecular Beam Injection (MBI), Biased Electrode, Biased Pump Limiter, Programmable Puffing Control in IOC and edge plasma fluctuation suppression due to LHCD. The typical improved confinement mode is the H-mode induced by a biased electrode.

VIII. THE H MODE INDUCED BY A BIASED ELECTRODE

The H-mode has been induced on the HL-1 Tokamak by means of the radial electric field produced by a positively biased electrode which is inserted 2 to 4cm inside the limiter. A pulse of 100ms, 0—600V with respect to the chamber wall, is applied to the head of the electrode during the plasma current plateau. As the biasing voltage is higher than 350V, the electrode current could be over the threshold and the L-H mode transition takes place on a time scale of several hundred microseconds. The time evolution of diagnostic signals during the L-H mode transition is given in Fig. 5. The data shows that at the L-H mode transition, the H_α or D_α signal drops abruptly with a decrease in the emission of impurities and the electrostatic fluctuation significantly suppressed on a timescale of 200 to 300 microseconds. The fluctuation of edge plasma is reduced by a factor of 10. The results of the plasma biasing show that the line averaged density increases by a factor of 1—3, $H_\alpha(D_\alpha)$ emission intensity decreases by 20—40 %, and the plasma particle confinement time increases by a factor of 2—3. The edge plasma parameters are clearly effected as follows: the edge plasma density and the temperature clearly drop, the e-folding length reduces from ~14 cm of the L mode to ~4 cm of the H mode. The profile of plasma floating potential, measured by single, double and triple Lamgmuir probes, suggests that the the gradient of the plasma potential is greatly increased. This increase probably leads to the suppression of fluctuation and the improvement of confinement.

The diamagnetic measurement shows that the diamagnetic flux increases by a factor of 2—3 during the L-H mode transition, thereby implying an irnprovement of plasma energy confinement. The experimental data shows the increase of plasma energy confinement is approximately 30 % which is roughly in agreement with the data from laser scattering and is much lower than that of particle confinement. However, the observed smoothly increased intensities of impurities C, O, Cr, Ni and Mo, hard X-ray and soft X-ray signals show an increase of radiation loss, indicating that the impurities are gradually accumulated on the same time scale as particle confinement. The H mode is degraded and finally terminated due to the accumulation of impurities.

Most of the confinement improvement mode has similar features to those described above. These features will not be repeated in the following description.

IX.. BIASED LIMITER.

The plasma behavior of the bias electrode and the biased limiter is very similar except that the improvement of plasma confinement is more difficult to obtain and is more temporal in the latter case. The particle flux in the limiter throat decreases by 60 % during the L-H transition. The bias current in the limiter must be 80—120A for the Hydrogen shot as compared to a current of 30—40A in the bias electrode experiment. For a shot with a limiter current held at 40 —80A . the plasma confinement could not be improved.

X. PELLET MODE

For the given target plasma (B_t=2 - 2.4T, I_p=90—110kA, $T_,$=O.5—O.7keV. n = (1—1. 4) $\times$ $10^{19}m^{-3}$ and a = O. 2m), the sudden rise (in less than 0. 5ms) of plasma averaged line density of about 50%, has been clearly observed due to pellet injection. A pellet penetration of 10cm has been observed by means of the H_α line detectors, in reasonable agreement with a theoretical estimation of 8—13cm which is far away from the q = 1 location. The density profile from the multi-channel HCN interferometer shows an increase of the peaking factor from the normal value of 1.3 to 1.5. In other experirnents, it could penetrate closer to or inside the q= 1 surface that is normally at r = 2.0—3.O cm, depending upon the discharge condition using larger pellets of 1. Omm^3.

A sudden rise (in less than 0. 5ms) of plasma line averaged density and loop voltage, accompanied by simultaneous drops of plasma temperature and magnetic/electric fluctuation at the edge, have been clearly observed. The period of the sawteeth is prolonged to about 5 ms, an eventual suppression of sawteeth oscillation for 12ms and after that the giant sawtooth appear with a phase which is exactly inverse to the phase before pellet injection.

Usually, the energy confinement time is prolonged by 50% and particle confinement time is increased by 100%.

X.ECRH MODE

The gyrotron of 75.0 GHz, 30ms pulse-width, with an output of horizontally polarized Gauss mode HE11, is launched into the plasma from the low field side at the discharge plateau. The experiments indicated that the content of plasma energy increased from 0.85 kJ to 0.99 kJ. The typical experimental data shows that the intensities of the ECE signal, the signal of diamagnetic measurement and the CIII, OII and the H_α radiation from the plasma edge, were obviously increased with the decrease of loop voltage. For example, the intensity of OVIII radiation rose by 23 %.

In contrast to most of the improved confinement mode, the edge plasma density and electron temperature were increased from $1.1\times 10^{12} cm^{-3}$ to $1.2\times 10^{12} cm^{-3}$ and from 24 to 31 eV respectively, with the observed drop of the line averaged plasma density approximately from $1.4\times 10^{13} cm^{-3}$ to $0.8\times 10^{13} cm^{-3}$ The plasma energy confinement time decreases by 30 % with an increase of the central plasma temperature of 30% given by the ECE. It is reasonable to summarize that the heating takes place near the center for shot number 12550. There is no clear correlation between resonance location and confinement obtained.

The intensity of the soft X-rays has increased mostly during heating. The sawteeth were changed significantly and in some cases were suppressed completely. The effects of ECRH on sawteeth are quite sensitive to the location of resonance at the position nearby the q=1 surface. When the resonance was located at 1.7 cm of the low magnetic field side, namely inside the q=1 surface (approximately 2.0 cm) the sawteeth were suppressed entirely and appeared again after the heating pulse. For the discharges with the resonance located outside the q = 1 surface, the sawteeth periods were prolonged and the effects diminished gradually with the resonance moving towards the plasma edge. It has been observed that the inversion of sawteeth took place before and after the heating pulse in the same shot. The obvious effect on the sawteeth seems to be caused by the influence of the magnetic shear in the vicinity of the q=1 surface, resulting from the localized change of the current profile due to thermal electrons produced by launched microwaves.

XI. IOC MODE

When the plasma density rises to about $5\times 10^{19} m^{-3}$, which is the normal operating density limit, the gas puffing rate is reduced suddenly and the L-H mode transition takes place. The line averaged density can remain constant or rise somewhat. The plasma peaking factor is kept roughly constant after the reduction of the gas puffing, indicating that the plasma density profile is almost unchanged during IOC. The observation of the inversion radius by soft X-rays gives an expansion of the q=1 surface from 3.5 to 4.5 cm during IOC mode. It is estimated assuming that the plasma current profile is unchanged. This transition could be obtained with a clean wall and low recycling rate of about 0.6. During the transition of the L-H mode, the sputtering produced impurities and desorption-resulting gasinput from the limiter and wall are reduced, which have been proved by diagnostics using spectroscopy. The energy confinement time, after the abrupt reduction of gas puffing, has increased by 40%.

XII. LHCD EFFECT ON CONFINEMENT

The Chinese-made LHCD experimental system, with an output power of 200kW and a pulse width of 100ms at a frequency of 2.45GHz, has been installed on the HL-1 Tokamak and the current drive has been obtained with an antenna of 2×8 multijunction grill. The difference of phase angle is fixed at +90 degrees and —90 degrees.

At the plateau, the efficiency of the current drive is 0.8A/kW. The experimental optimal density is $(0.9 - 1.2)\times 10^{19} m^{-3}$. The maximum particle confinement is improved by a factor of 3 and the improvement is approximately proportional to the power of the low hybrid wave at a line averaged density lower than $3\times 10^{19} m^{-3}$ with a power threshold of 25kW. The improvement of confinement reaches its maximum at a density of $1.10^{19} m^{-3}$ which is nearly equal to the optimum density for a non-induced current drive as shown in Fig. 6, in contrast to the edge heating due to the RF field in the front of the antenna. Soft X-ray data shows that the q=1 surface has expanded. The steepening of the electron temperature at the edge probably contributes to the improvement of confinement. In addition, the hard X-ray radiation and synchrotron microwave radiation at 8mm and 3cm increase. The increase of the hard X-rays from the central region and the decrease of that from the limiter and wall indicate that the suprathermal or runaway electrons are concentrated at the plasma axis after the L-H transition.

XIII. THE SYNERGY OF THE PLASMA BIASING AND LHCD.

In order to compare the H mode induced by a positively biased electrode and that by the launching of LHCD, a comBINED experiment has been performed. When the H mode is induced by biasing, the feed-in of LHCD degrades the H mode back to a normal ohmic discharge. The density and temperature of the edge plasma increase dramatically with an increase of its fluctuation and a decrease of the bulk plasma density, indicating that the H mode is terrninated. According to the effect of the edge electric field on the improvement of plasma confinement, it seems reasonable to assume that the polarization of the edge electrostatic field induced by the LH wave is opposed to the field induced by biasing, but this has not been proved experimentally. The exact reason is unknown to us at this time.

XIV. DISCUSSION

The experimental e-folding length of the edge plasma density and temperature for the H-mode reduces to about 1/3 of that for the L-mode due to the L-H transition. The steepeness of the gradient edge plasma floating potential, on the same time scale as the H_α or D_α fluctuation drops, is greatly increased. Its negative edge radial electric field and associated $E \times B$ rotation probably leads to the fluctuation suppression and turbulence driven loss reduction which seems to be the reason for the confinement improvement. Normally, the H-mode was terminated due to the accumulation of impurities. The contribution of the lower frequency MHD fluctuation suppression and sawteeth activities to confinement improvement remains unclear. The cause of the termination of the bias induced H-mode by the concominant LHCD is also unclear.

The HL project is progressing now with a better possibility to make many more contributions to the world-wide fusion efforts. The HL-1M is scheduled to be operated this year with more powerful diagnostics, auxiliary heating, current drive and pellet injection. It is capable of studying compound effects in detail.

ACKNOWLEDGMENTS

The authors would like to express their deepest gratitude to all of the staff for their data and helpful discussions, especially with Prof. Li, Zhenwu and his student Yan, Longwen.

REFERENCES

1. HL-1 Tokamak Group, IAEA-CN-47/A-VII-1 (1986)
2. S. J. Qian & HL-1 Group, International Conference on Plasma Science and Technology '94 June 16-20, Chengdu.
3. Y. Z. Zheng, et al., IAEA-CN-56/A-7-10(1992)
4. Y. Z. Zheng, et al., Effects of Plasma Edge Control and LHCD on Confinement in the HL-1 (1994) to be published
5. S. J. Qian, et al., Journal of Nuclear Materials 176(1990) 968
6. H. R. Yang, et al., Journal of Nuclear Materials 196 & 198(1992) 908.
7. L. B. Ran, et al., Proc. 19th EPC Conf. on Contr. Fus. and Plasma Phys. Vol 16c. Part 1 I-223(1992)
8. L. H. Yao, et al., Journal of Nuclear Materials 196 & 198 (1992)52.
9. C. H. Luau, et al., Proc. 1991 IAEA TCM on Research Using Small Tokamaks, (1991)82.
10. G. D. Li, et al., Journal of Nuclear Materials 196 & 198(1992).
11. L. W. Yan, et al., Proc 1991, IAEA TCM on Research using Small Tokamaks, (1991)111.
12. X.R. Duan, et al., Chn Phys Lett 11(1994)157
13. B.R. Shi, et al., A Critical Magnetic Shear Theory for Tokamak Resistive Internal Kink Mode and Application (1994) to be published.

Table 1

The Design Parameters of HL-1, HL-1M and HL-2

parameter	HL-1	(*)	HL-1M	HL-2
major radius, cm	102	102	102	150
plasma minor radius, cm	20	up to 20	25	0.45
plasma elongation	1	1	1	1.7
toroidal field, T	5	0.3~3.5	3.0	3—5
plasma current, MA	0.4	0.225	0.3—0.4	1.0—2.0
plateau of plasma current, s	0.08	1.0	1.6	5—10
plasma density, $10^{19} \times m^{-3}$	10	7	7	8
plasma temperature, keV	0.88	2	10	
energy confinement time, ms	40	15—38	30	80—150
effective Z	<3	<1.6		
safe factor at limiter	2.5	~1.8		
auxiliary heating power MW	0	0.3	2	10—20

(*) The experimental parameters

Table 2
The methods of wall conditioning used on HL-1

methods	TDC[1]	ECR-DC[2]	AC-DC[3]	PDC[4]	DDC[5]
gas pressure, 10^{-4}torr	8	2	2	5	4
discharge current, kA	4	RF,1.5kW	1—5kW	>10kW	100kW
discharge voltage, v	20	15	<20	3	
toroidal field, G	850	850	850	1000	18000
electron density cm^{-3}	4×10^{12}	1×10^{11}	1×10^{11}	7×10^{12}	7×10^{12}
electron temperature ev	3—5	4—7	2—8	>10	
single molecular layers per hour	0.1	0.2	0.15		

[1]Taylor discharge cleaning, [2] Electron cyclotron resonance discharge cleaning, [3] Alternative current discharge cleaning, [4] Power discharge cleaning, [5] Disruptive discharge cleaning.

Table 3
Sawtooth events observed on the HL-1

type	sawteeth period τ_s(ms)	number of occurrence	diagnostics
normal	1—4	frequently	Sx,Hx,MW,ECE H_α,V_l^*
compound	$2\tau_s$	much less	Sx
giant	11	a few	Sx
monster	45	once	Sx,B_θ
sawteeth-like	2—10	commonly	Sx,Hx,B_θ,V_l^*

* Sx=soft x-ray; Hx=hard x-ray; MW=microwave; V_l=loop voltage

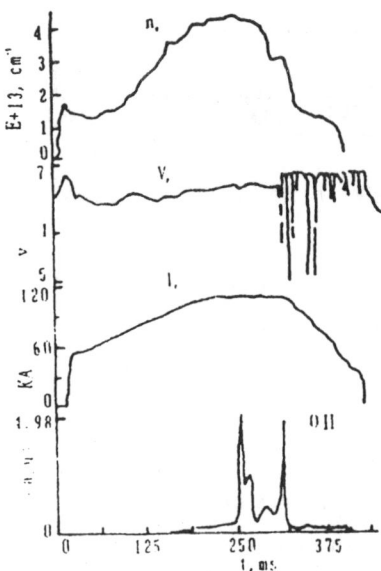

Fig. 1 O I line intensity, shows MARFE and DP

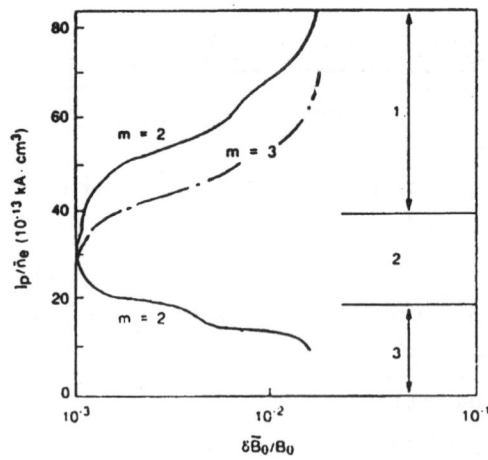

Fig. 2 I_p/n_e versus perturbation of magnetic field $\delta B_0/B_0$. 1: Discharge region with strong Mornor oscillation; 2: Stable dischage region; 3: Discharge region with disruption;

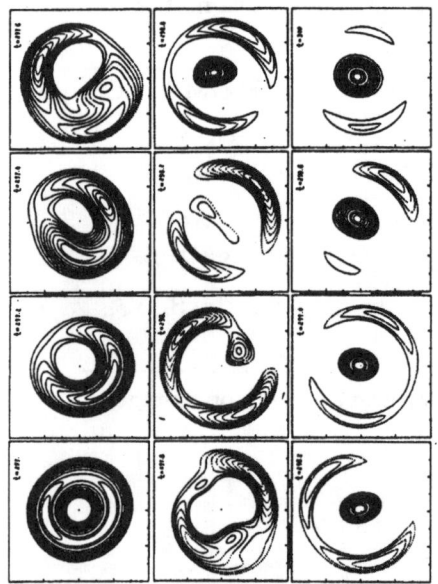

Fig. 3 Tomography of the crach process for a monster sawtooth from ohmically heated plasma during 297—300ms

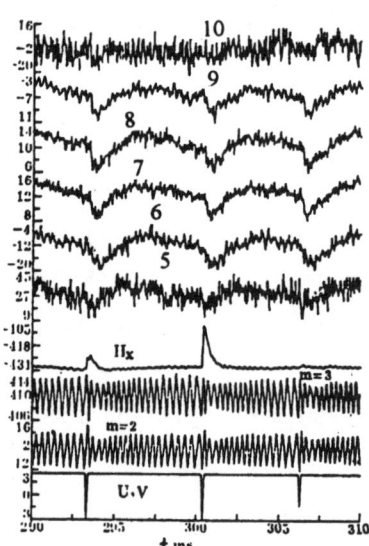

Fig. 4 Sawtooth like relaxation Curve 1—10 are the loop votage, m=2 and m=3 modes, hard X-ray, and soft X-rays at −5.5, −3.5, 2.5, 3.5, 5.5 11.5cm away from R=105cm, respectively

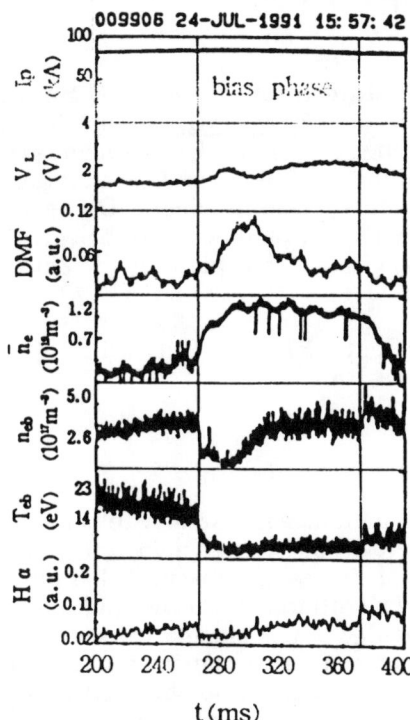

Fig. 5 Time evolution of the diagnostics signals during L-H transition induced by bias electrode.

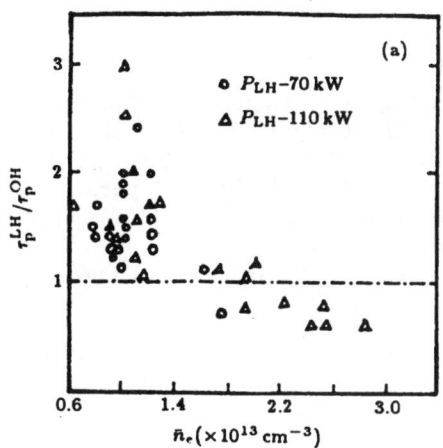

Fig 6 Dependence of T_p^{LH}/T_p^{OH} on the line-average density (T_p^{LH} and T_p^{OH} represent the particle confinement time during LHCD and ohmic discharge period, respectively) (a) anticurrent drive ($\triangle \varnothing = -\pi/2$), (b) current drive ($\triangle \varnothing = \pi/2$).

HIGH Z MATERIAL OPERATION IN FTU

F.Alladio, M.L.Apicella, G.Apruzzese, R.Bartiromo, M.Borra*, G.Bracco, G.Buceti, P.Buratti, C.Centioli, M.Ciotti, V.Cocilovo, I.Condrea*, F.Crisanti, R.DeAngelis, F.DeMarco, B.Esposito, C.Ferro, G.Franzoni*, D.Frigione, L.Gabellieri, E.Giovannozzi, G.Granucci+, M.Grolli, A.Imparato, H.Kroegler, M.Leigheb, L.Lovisetto, G.Maddaluno, A. Mancuso G.Mazzitelli, P.Micozzi, A.Moleti, F.Orsitto, L.Panaccione, D.Pacella, M.Panella, V.Pericoli, L.Pieroni, S.Podda, G.B.Righetti, F.Romanelli, S.E.Segre**, E.Sternini, A.A.Tuccillo, O.Tudisco, F.Valente, V.Vitale, R.Zagorski[†], V.Zanza, M.Zerbini

Associazione EURATOM-ENEA sulla Fusione, Centro Ricerche Frascati,
C.P.65 - I-00044 Frascati, Rome, Italy

ABSTRACT

High Z materials, as possible plasma facing components for a fusion reactor, are being re-evaluated due to their advantages connected to the low sputtering yield and small erosion. A data base of experimental results related to their use on present Tokamak is missing. FTU has carried out a series of experiments with different medium and high Z limiter materials, Ni, Mo and W, to improve the knowledge of the plasma behaviour in these conditions. The results on plasma operation, core parameters, impurity content and corresponding radiation losses are presented, and compared with those obtained with a low Z dominant impurity (siliconization). A simple model coupling the SOL and the plasma core is able to reproduce the experimental results, indicating that sputtering and self-sputtering are the main physical mechanisms of impurity production.

INTRODUCTION

It is well recognised that wall materials have largely influenced the performance of magnetically confined plasma in each experimental device. Particularly, adoption of low Z materials, such as carbon and beryllium, has led to good performances in the present large Tokamaks [1,2]. Nevertheless the choice of materials for the divertor plates of a fusion reactor is still a debated question. Two main problems have to be solved in the Next Step of fusion research: how to maintain a steady state ignited plasma avoiding plasma pollution from the first wall, and how long is the life time of the various structures subjected to high fluxes of ions, neutrals and neutrons and high heat loads. The main effort has been done, up to now, to optimise the plasma core performance using low Z materials, carbon or beryllium, as first wall. This kind of materials could not guarantee an acceptable plasma dilution; Be is satisfactory only at low plasma edge temperature (Te_{edge} < 10eV) and C is good only if chemical sputtering can be reduced (e.g. by some doping). Moreover large erosion is foreseen for low Z materials, due mainly to sputtering. From a preliminary assessment both Be and C will survive only 1000 full power ITER discharges assuming a plate temperature of 10-30 eV[3]; at lower temperature C is even worse if the chemical sputtering is not reduced.

---

*   ENEA Guest
**   Dipartimento di Fisica, II^a Università di Roma
\+   Istituto di Fisica del Plasma CNR Milano
†   Institute of Plasma Physics and Laser Microfusion, Warsaw, Poland

Thus a re-evaluation of high Z refractory materials (such as Tungsten) as divertor plates for a reactor is required [4]. In fact, mainly due to the bad experience made in some previous experiments [5], systematic investigations on high Z materials have been lacking with consequent poor data base for them. Such materials could be used for divertor targets because of their low sputtering rates and the high probability of redeposition of the eroded particles[6]. Both effects are expected to be most efficient in a high density divertor condition. The sputtering yield is strongly non linear with the edge temperature and can be orders of magnitude lower than for low Z materials. Considering erosion due to neutrals and ions sputtering with an edge temperature of 10 eV, it has been evaluated that tungsten could survive for 10^6 full power ITER discharges [3]. It has been also evaluated that, to avoid a plasma core degradation due to the high W radiation intensity, a maximum impurity concentration of the order $n_w/n_e \sim 10^{-5}$-10^{-6} is allowed.

For these reasons we have carried out on FTU experiments with different medium and high Z materials limiter: Inconel(the usual FTU limiter material), Molybdenum and Tungsten. The results have been compared with the ones obtained with the machine just siliconized.

EXPERIMENTAL RESULTS

FTU is a compact high field tokamak characterised by a high power density (>1 MW/m^3) and able to operate in a wide density range (0.3-3 10^{20} m^{-3}) [7]. Up to now mainly ohmic experiments have been performed with a maximum power input of 2 MW. The vacuum chamber consists of stainless steel AISI 304 and the circular cross section of the plasma is defined by a poloidal limiter. Previous experience in FTU [8] has demonstrated that it is necessary to reduce the level of oxygen in order to minimise the production of metallic impurities by sputtering [4,9]. A week of baking at T=130°, in conjunction with glow discharge cleaning in hydrogen, is usually sufficient to guarantee a very low level of oxygen, down to a concentration of 0.2% of the electron density. After cooling at liquid nitrogen temperature, the machine remains clean and with a high reproducibility of the discharges, even after repetitive major disruptions [10]. The limiter does not cover the full poloidal angle, but consists of two halves encompassing a total poloidal angle of 220° symmetrical with respect to the equatorial plane. Both the two limiter halves are composed by an external structural support and by an internal set of mushrooms facing the plasma (Fig. 1). In all our experiments the limiter support was made of Inconel or stainless steel. The materials used for the mushrooms were Inconel, TZM (molybdenum) and TZM covered by a coating of 2 mm of tungsten. It is worthwhile to note that in a typical plasma discharge the power load on the limiter is about 0.5 MW; since the limiter mushrooms surface is about 0.1 m^2 this means that the average surface power load is of the order of 5 MW/m^2. Therefore even in ohmic regime the power deposition on the materials is reactor relevant.

Plasma Operations.

As already noted in previous experiments, using high Z materials as first walls, [4,5] plasma operation can be a critical issue and could need a peculiar care in controlling some parameters.

Our overall experience in FTU has shown that the first 20-40ms are critical for the evolution of the plasma temperature profile. This is the phase following the break-down with a fast current rise ($\dot{I}p \sim 5\ 10^6$A/s). Using a standard discharge pre-programming (same rate of Ip, same gas fuelling, etc.) we obtained substantial differences between discharges dominated by light impurities (C, O, Ne, Si) and the ones dominated by heavy impurities (Ni, Mo, W). In the former ones the temperature profile is peaked immediately after the break down phase and the central temperature evolves with an overshoot followed by the saw-tooth onset. In the latter ones the saw-tooth activity can be absent. In

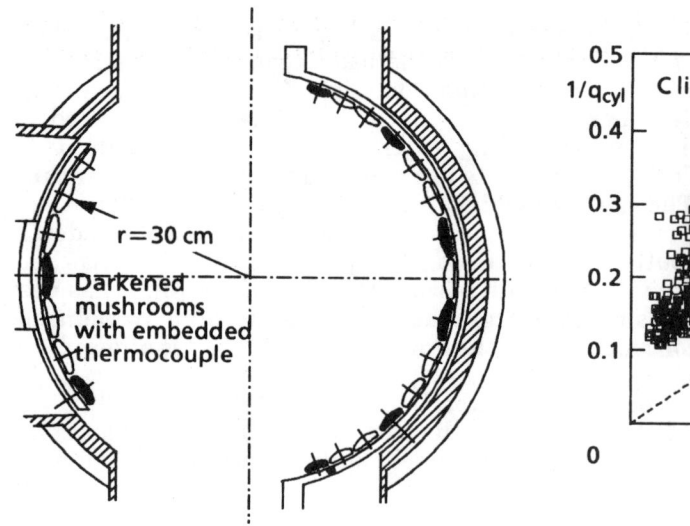
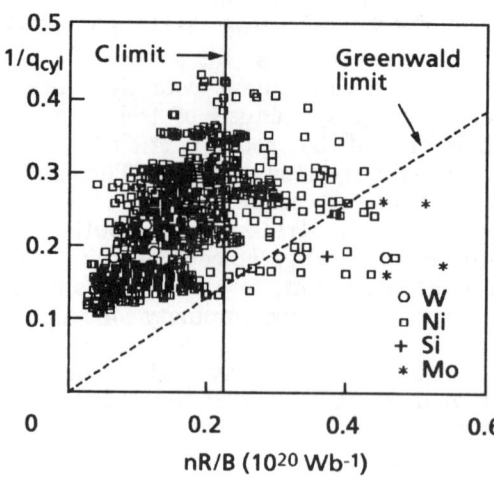

Figure 1 - FTU limiter

Figure 2 - Hugill plot for FTU for different plasma facing materials

this case the central temperature has no overshoot and the temperature profile can be hollow. A careful interplay in using gas puffing just after the break-down and the plasma current rise rate to control the I_p/n_e rate allow to reproduce the "normal" electron temperature profile evolution. All the FTU experience agrees with sputtering and self sputtering being the only mechanism of impurity generation. When cooling of the plasma edge (by strong gas puffing or Ne puffing) reduces the sputtering yield [9] and thus limits high Z impurity concentration in the plasma bulk, with the consequent radiation losses, "normal" temperature profile evolution is achieved.

Bulk Plasma Characteristics

The plasma behaviour in presence of different materials for the limiter has been compared in discharges with a magnetic field of 6 T, plasma current varying between 0.5 and 0.8 MA, an input power < 1.3 MW, and with density ranging from 0.3 to 3 10^{20} m^{-3}. Experiments with tungsten limiter have been somewhat limited, due to a smaller number of shots available.

In Fig. 2 the Hugill plot shows the operational space of FTU. When plasma was dominated by carbon (due to a failure of inner rings supporting magnetic pick-up coils and the exposure of organic insulator to the plasma), it was not possible to operate at line average density larger than 10^{20} m^{-3} at all currents. Just after siliconization [11], when both the vacuum chamber and limiter surfaces were coated with silicon, density limit was found to approach the Greenwald limit. When medium or high Z impurity (Ni, Mo and W) is the main FTU plasma impurity, the density limit was found to exceed substantially the usual Greenwald limit. Actually, only by gas puffing, we have achieved about a factor 2 above the latter limit in steady state discharges [12].

In FTU the approach to the density limit is always characterised by the onset of a Marfe instability: the critical density for the Marfe onset is lower for silicon dominated than for metal dominated plasma [13]. On the other hand, global parameters, such as energy confinement time, average electron temperature, total thermal energy, do not change for discharges with identical macroscopic parameters, but with different dominant impurity.

IMPURITY BEHAVIOUR

In Fig. 3 the relative concentration of Mo is plotted versus the average plasma density. Mo concentration has been deduced by the analysisin the region between 3 and 6 Å (where Mo L transitions are localized), as measured by a rotating flat X rays crystal spectrometer. The same diagnostic has been employed to derive a normalized concentration of tungsten (exploiting M transitions) Fig. 4. Both concentrations are strongly decreasing with density. Absolute values for the Mo density have been calculated: at high density n(Mo) is $<10^{15}$ m^{-3}, i.e. below the critical level required to maintain an ignited state in a reactor. The behaviour of MoI line, as measured by visible spectroscopy looking at the limiter surface, is shown in Fig. 5 as a function of density. As density increases, the line intensity decreases, indicating a corresponding decrease of the impurity source. A similar measure of WI line has not been possible, since its intensity was always below the background level. For Ni and Cr lines, which are the main

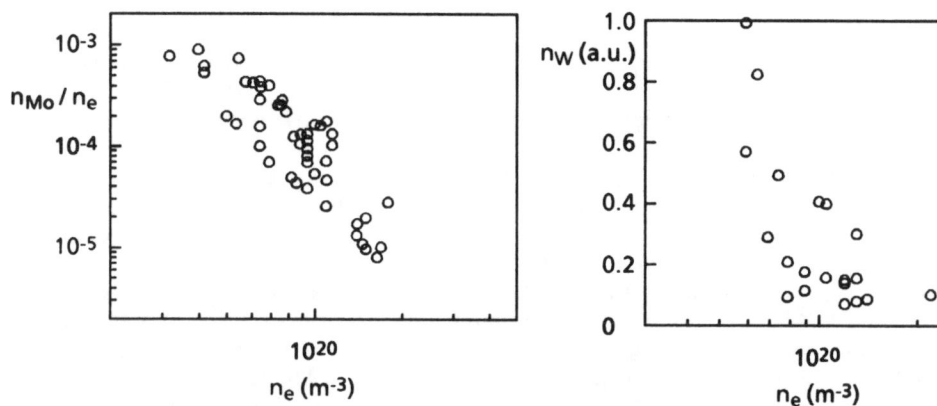

Figure 3 - Relative Mo concentration versus line electron density

Figure 4 - Normalized W concentration versus line electron density

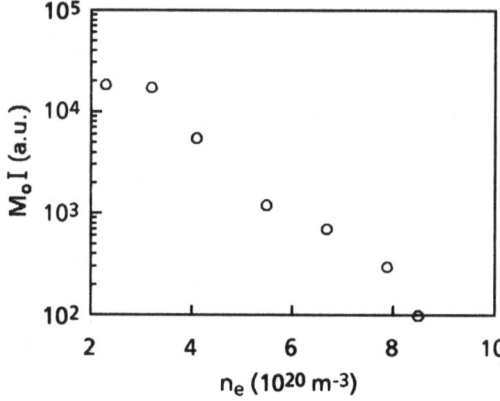

Figure 5 - MoI line (3902Å) intensity versus line electron density

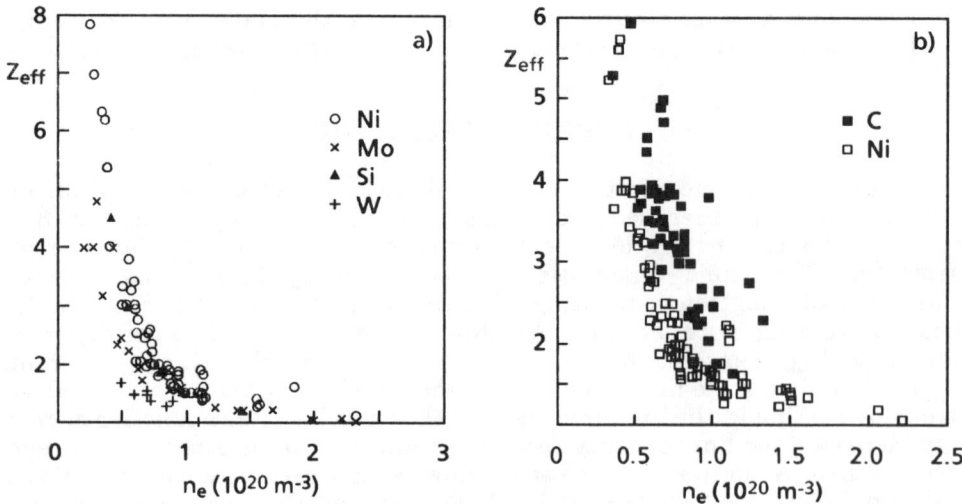

Figure 6 - a,b) Z_{eff} ($I_p = 0.5$ MA) versus electron density for discharges with different plasma facing materials

components of the Inconel limiter, a similar behaviour with density is observed [14]: however even at the highest density their intensity is well above the background emission, consistent with the different thresholds for the sputtering yield.

At high density and in a purely metallic machine, oxygen lines are the most prominent in the VUV spectrum. When approaching the density limit, and with a Marfe extending to the whole poloidal section, the predominant lines in the VUV spectrum are the deuterium Lyman series, while metallic lines are completely absent. When silicon dominates, oxygen lines are replaced by silicon lines. In Fig. 6 Z_{eff} obtained from bremsstrahlung measurements, is plotted versus electron density for discharges with Ip=0.5 MA. As usual Z_{eff} decreases with increasing density, but, while discharges dominated by Ni and Si impurities have the same value of Z_{eff} the same density, in Mo and W discharges Z_{eff} is smaller and approaches unit more quickly. In carbon dominated discharges, Z_{eff} was systematically higher over the whole density range.

In Fig. 7 typical spatial profiles of radiation losses, as determined by a multichannel bolometer system, are shown for the different impurities. As expected, the radiation emissivity for Si is concentrated at the boundary, while for Ni, Mo, W is shifted towards the centre. The low central emission, especially for Mo, is due to the high power density characteristic of a compact high field Tokamak, as FTU, and therefore to the high value of electron temperature at the centre ($T_e(0) \approx 3$keV) obtained at low electron density (4×10^{19}m^{-3}). In Fig. 8 the total radiated power is plotted versus the electron density for Si, Ni, Mo, W discharges. The main feature is the different behaviour of Mo and W with respect to Ni; this can be explained from the different sputtering threshold among the materials and has been reproduced by code simulation as discussed in the next session. In Fig. 9 the total radiated power is plotted versus the relative Mo concentration, showing again a strict correlation between the two quantities.

The poloidal limiter is equipped with thermocouples and from the temperature rise during a shot is possible to estimate conductive/convective losses [15]. We checked that heat loads on the limiter are complementary to radiation losses, i.e. increasing heat loads are detected when radiation losses are decreased.

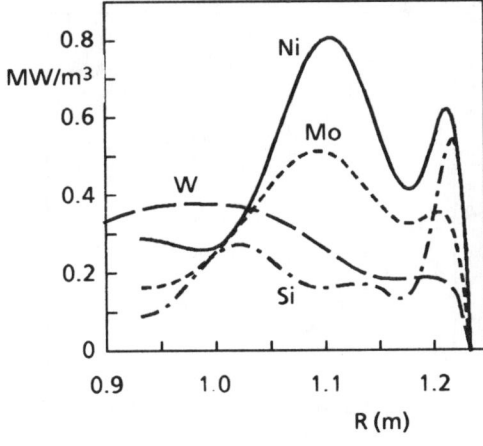

Figure 7 - Radial bolometric profiles

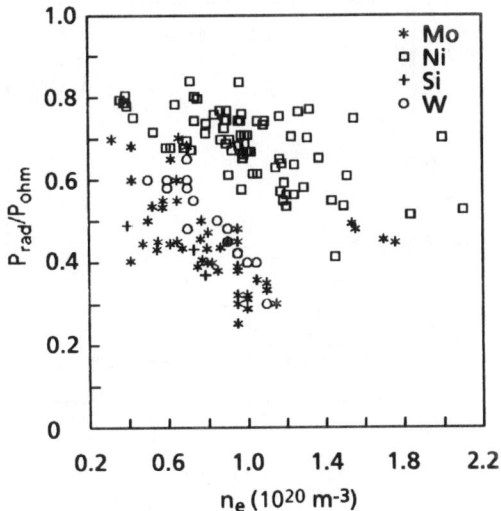

Figure 8 - Total radiation losses

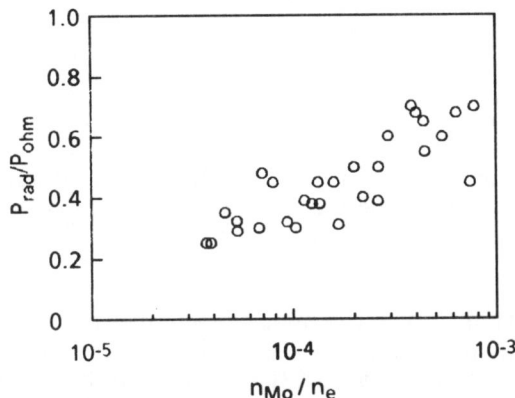

Figure 9 - Total radiative losses versus the relative Mo concentration

SIMULATION WITH A COUPLED EDGE-CORE MODEL

Since medium and high Z impurities loose power from the plasma bulk, the plasma core and the SOL become strongly linked. Therefore we have compared our experimental results with a self consistent analytical model [16] that couples the SOL parameters and the transport of the impurities generated at the limiter to the bulk parameters. The model is based on a 0-D description of the plasma transport at the centre and a 1-D model for the SOL dynamics. Steady state is considered and all the profiles ($T_{e,i}$, n_i) are described by a generalised parabola. The FTU scalings [17,18] for edge electron density and central temperature are adopted. Only one impurity is considered and the corresponding cooling rate is calculated assuming corona equilibrium. The impurity profile is assumed to be consistent with a flat Z_{eff} profile. Impurity generation is only due to physical sputtering by deuterium and by self-sputtering. Equal ion and electron temperature are assumed in the SOL. The model inputs are the plasma current and the geometry of the plasma shape

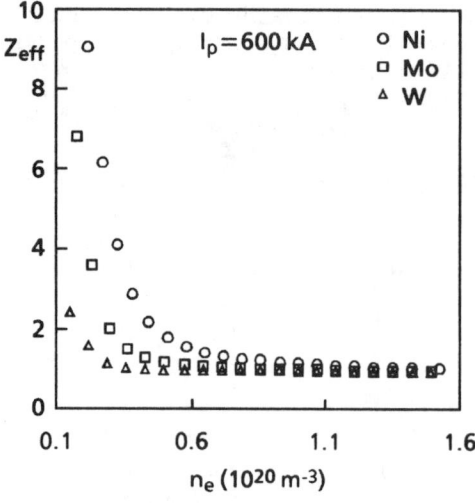
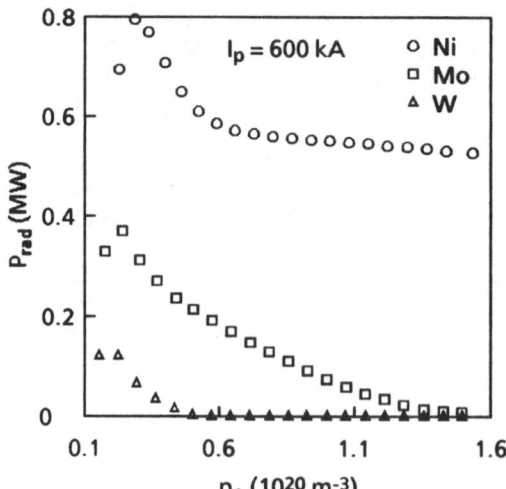

Figure 10 - Computed Z_{eff} versus electron density for different prevalent impurity

Figure 11 - Computed P_{rad} versus electron density for different prevalent impurity

and of the poloidal limiter. In Fig. 10 and 11 the computed value for Z_{eff} and radiative power losses are plotted as a function of the electron density for Ni, Mo and W. The agreement with experimental results is good not only qualitatively but also in absolute values.

The code confirms that on FTU the main mechanism of impurity production is the sputtering and the self-sputtering [19,20] so that operation is possible until the total sputtering yield is kept below 1. Due to different total sputtering yield between Mo-W and Ni, at the same bulk plasma parameters much less Mo-W are produced. For instance at $n_e = 3.2 \times 10^{19} m^{-3}$ the Mo density at the last closed magnetic surface is 2.5 times less than the corresponding Ni density. The lower Mo concentration means smaller value of Z_{eff} and smaller radiative losses. Furthermore, the decrease of the plate temperature as density increase reduces Mo production much more than for Ni, so that Z_{eff} close to 1 is reached more quickly. It is to be noted that even if Z_{eff} is close 1, the contribution of Ni to the radiation losses is significant even at high density (~40% of the total input power) while Mo-W radiation is predicted to disappear at high density.

CONCLUSIONS

Experimental results in FTU using medium Z (Ni) and high Z (Mo, W) as limiters materials, have been compared with those obtained using a silicon coated limiter (low Z). A few conclusions can be draw from this comparison.

Density limits and onset of Marfe are strictly correlated with the presence of low Z impurities. Regime of high density ($>2 \times 10^{20} m^{-3}$) can only be achieved with fully metallic plasma facing materials. In these conditions a density limit substantially higher than the Greenwald value can be achieved using only gas puffing.

- The main production mechanism for metallic impurities is physical sputtering from background ions and self-sputtering. Due to the different sputtering yield for Ni and Mo or W, Z_{eff} and radiation losses in Mo and W are lower.
- The correlation between the influx of impurities from the limiter and the impurity concentration in the plasma core for high Z ions suggests that no accumulation in the plasma centre is occurring, at least in standard saw-toothing ohmic discharges.
- The relative concentration of Mo in the plasma core at medium and high density is below the critical limit needed to obtain and maintain ignition in a Tokamak.
- A model coupling the SOL dynamic to bulk plasma has been developed and it reproduces both qualitatively and quantitatively the experimental FTU results with different first wall materials. This means that the most significant physic is included in this simple model.
- Bulk parameters such as energy confinement time, average temperature, total kinetic energy, show no variation in presence of different first wall materials.

It can be concluded that these results show the possibility of using high Z materials as plasma facing components (e.g. divertor plates) in a fusion reactor.

REFERENCES

[1] Jackson,G.L., et al., Phys. Rev. Lett. 67, 3098 (1991)
[2] Janeschitz,G.,and JET Team, Proc. 14th IAEA Conf. on Plasma Physic and Contr, Wurzburg 1992
[3] Janeschitz,G.,et al., Proc. 15th IAEA Conf. on Plasma Physic and Controlled Nuclear Fusion Research, Seville 1994
[4] Tanabe,T.,Noda,N.,Nakamura,H.J.,J. Nucl. Mat. 196-198 (1992) 11
[5] Bol,K., et al.,Proc. 7th IAEA Conf. on Plasma Physic and Controlled Nuclear Fusion Research, Innsbruck 1978, Vol.I,11,1979
[6] Fussmann,G., et al.,Proc. 15th IAEA Conf. on Plasma Physic and Controlled Nuclear Fusion Research, Seville 1994
[7] Andreani,R.,et al., 1991 Fusion Technol., 218
[8] Alladio,F., et al.,Proc. 14th IAEA Conf. on Plasma Physic and Controlled Nuclear Fusion Research, Wurzburg 1992,Vol.I,141,1993
[9] Stangby,P.C.,McCracken,G.M., Nucl. Fusion, 30 (1990) 1225
[10] Apicella,M.L.,et al.,Proc. 20th Eur. Conf. on Controlled Fusion and Plasma Physics, Lisboa (1993) 17 619
[11] Apicella,M.L.,et al.,Proc. 21th Eur. Conf. on Controlled Fusion and Plasma Physics, Montpellier (1994)
[12] Frigione,D., Pieroni,L., and FTU team, Proc. 20th Eur. Conf. on Controlled Fusion and Plasma Physics, Lisboa (1993) 17 271
[13] Alladio,F.,et al.,Proc. 21th Eur. Conf. on Controlled Fusion and Plasma Physics, Montpellier (1994)
[14] Condrea,I.,De Angelis,R.,Gabellieri,L.,Proc. 20th Eur. Conf. on Controlled Fusion and Plasma Physics, Lisboa (1993) 17 623
[15] Ciotti,M.,Ferro,C.,Franzoni,G.,Maddaluno,G., and FTU TEAM Proc. 20th Eur. Conf. on Controlled Fusion and Plasma Physics, Lisboa (1993) 17 627
[16] Zagorski,R.,Romanelli,F.,Pieroni,L.,submitted to Nucl. Fusion
[17] Bracco,G., private comunication
[18] Pericoli Ridolfini,V.,Zagorski,R., et al., 1994 11th PSI Conf. to be published in J.Nucl.Mater.
[19] Bartiromo,R.,Condrea,I.,De Angelis,R.,Leigheb,M., 1994 11th PSI Conf. to be published in J.Nucl.Mater.
[20] Ferro,C.,Zanino,R.,1994 Contr. Plasma Physics 34 337.

RECENT RESULTS ON HEATING AND CONFINEMENT IN TEXTOR

P.E. VANDENPLAS[1], R. KOCH[1], A.M.MESSIAEN[1*], K.H. FINKEN[2], J.ONGENA[1*],
V. PHILIPPS[2], U. SAMM[2], B. UNTERBERG[2], G. VAN OOST[1], G. VAN WASSENHOVE[1], J. WINTER[2], P. DUMORTIER[1],
F. DURODIE[1], H.G. ESSER[2], H. EURINGER[2], B. GIESEN[2], E. HINTZ[2], M. LOCHTER[2], M.Z. TOKAR[2], G.H. WOLF[2],
G. BERTSCHINGER[2], G. FUCHS[2], D.L. HILLIS[3], F. HOENEN[2], P. HÜTTEMANN[2], L. KÖNEN[2], H.R. KOSLOWSKI[2],
A. KRÄMER-FLECKEN[2], G. MANK[2], D. PETTIAUX[1], A. POSPIESZCZYK[2], B. SCHWEER[2], H. SOLTWISCH[2], H.F. TAMMEN[4],
G. TELESCA[2], R. UHLEMANN[2], R. VAN NIEUWENHOVE[1*], M. VERVIER[1], G. WAIDMANN[2], R. R. WEYNANTS[1].

(1) Laboratoire de Physique des Plasmas - Laboratorium voor Plasmafysica
Association "Euratom-Etat belge" - Associatie "Euratom-Belgische Staat"
Ecole Royale Militaire - B-1040 Brussels - Koninklijke Militaire School
(2) Institut für Plasmaphysik, Forschungszentrum Jülich, GmbH
Association "Euratom-KFA", D-52425 Jülich, FRG
(3) Oak Ridge National Laboratory, P.O.Box X, Oak Ridge, TN 37831, USA
(*) Researcher at NFSR, Belgium
(4) FOM-Instituut voor Plasmafysica "Rijnhuizen", P.O. Box 1207, 3430 BE Nieuwegein,
The Netherlands

1. Introduction

There has been a considerable amount of experimental activity on TEXTOR preceding the June 93 shut-down of the machine. The machine is presently being commissioned after upgrading for longer pulse operation (10 s). The main line of experimentation was the high auxiliary power and high density operation aiming at reactor-relevant tests of new coatings [1] and of the concept of edge radiative cooling [2]. However, a number of other RF experiments, which will also be discussed here, were performed mostly in presence of NBI, namely (i) Third harmonic D heating, (ii) ^{3}He pumpout experiments, (iii) High Z limiter tests. Finally, as a last point, we shall discuss the years-long experience gained on TEXTOR in using unshielded antenna(s).

TEXTOR is a medium size machine with plasma radius $a_p<0.48$m, major radius $R_0\approx1.75$m, plasma current $I_p\leq500$kA and toroidal magnetic field $B_T\leq2.8$T. The auxiliary heating system of TEXTOR is composed of : (1) two parallel injection neutral beam lines, one of co-injection (NBI-co) the other of counter-injection (NBI-counter), with injection energies up to 55keV and a maximum power of 1.7 MW-10s each and (2) two radiofrequency (RF) generators of 2MW-3s each (25-38MHz). Each generator is connected to an antenna pair. Antenna pair 1 was constantly operated without Faraday shield. The transmission line of antenna pair 1 was equipped with an automatic real-time tuning system [3].

In 1994, the RF system has been upgraded for 10-20s operation with automatic tuning on both lines. The injectors are also be upgraded to deliver 2MW at 60kV.

2. I-mode with radiating boundary

(i) <u>On the one hand</u>, improved confinement operation (I-mode) has been one of the main auxiliary heating research subjects of TEXTOR. The improved confinement operation (I-mode) on TEXTOR obtained with boronized walls and low edge radiation, has been characterised by Ongena et al. [4]. The I-mode is reached (usually) without transition, but transition from I to L-mode is sometimes observed. The I-mode can be obtained in a well-conditioned machine with co-injection alone, with co+counter and with any of these beam combinations together with ion cyclotron resonance heating (ICRH). In addition, balanced injection, in presence or not of ICRH, can produce discharges with a highly peaked density profile reminiscent of the "supershot" regime of TFTR [17]. Counter-beam alone or ICRH alone or both together remain in the L-confinement mode. The enhancement factor f_H with respect to the ITER L89-P scaling $f_H = E_{dia}/E_{ITER\ L89-P}$ attains a mean value of 1.65 for stationary conditions in the line density range $1.5 < \bar{n}_{eo} < 4 \times 10^{13}$ cm^{-3} for discharges having an energy contribution due to the fast particles population less than 40 % of the MHD energy measurement (see ref. [4] for a complete discussion of the properties of these discharges). For larger values of $\bar{n}_{eo}$, f_H decreases progressively.

(ii) <u>On the other hand</u>, the concept of radiative edge, or "cold plasma mantle", was first investigated in TEXTOR using neon puffing [5]. In later experiments, this technique was compared with wall siliconisation and di-silane (Si_2H_6) puffing [2]. We refer the reader to the papers [6,7] for pure silicon wall operation. These experiments showed that the light impurities neon and silicon are valuable for generating a cold plasma mantle capable of radiating a significant fraction of the total heating power [2,4]. The thickness of the radiation belt, of the order of 20cm, was observed to be similar with neon and silicon. However, the control of the radiation level in the silicon case, is more difficult because Si sticks to the wall. Part of the Si is released by sputtering. Therefore, the maximum radiated power fraction $\gamma = P_{rad}/P_{tot}$ (P_{rad} and P_{tot} are respectively the radiated power and the total power coupled to the plasma) reached with Si were only about 75%. In contrast, neon does not stick to the walls but is recycled, which means that in the absence of a sink the neon level cannot be controlled. In TEXTOR, the necessary sink is provided by the pumping capability of the ALT-II toroidal belt limiter system and the radiation level can be feedback-controlled using the neon VIII line intensity [2,15]. Feedback control of the radiation is an essential ingredient of the cold plasma mantle concept in order to avoid the 100% radiated power limit, which leads to detachment, or at high auxiliary power to MARFEs, quite often followed by disruption. The use of controlled neon injection has allowed to radiate in stable and stationary discharges more than 90% of the heating power.

Recent experiments have not only proven the viability of edge cooling with high auxiliary heating power, but they have extended the range of the high confinement (I-mode) regime to higher densities [8]. The basic improvement gained by operating with a radiating belt can be described as follows. Comparing a discharge without a strong edge radiative layer with another one with it (neon or silicon), and for identical heating power, one can note two main differences : (1) the radiated power fraction increases from about 20% to typically 50-90% with radiating belt, (2) much higher densities can be obtained in the latter case for a similar value of the electron temperature. Therefore a larger energy content is obtained with a (strong) radiating boundary. The gain in energy results from the density increase, the density profile being more peaked than without radiating layer, while the bulk temperature profile is almost unchanged, or slightly more peaked. The plasma edge temperature is, on the other hand, reduced in presence of radiation cooling. For shots with equal central electron temperature, the neutron yield is also identical. Since neutral beam injection nearly all the neutrons are produced by beam-target reactions, the rate of which is proportional to the number of deuterium ions in the plasma core, this is indicative of a maintained plasma purity, in agreement with earlier findings at moderate auxiliary power [2]. It is thus noteworthy that in this improved confinement mode, there is no sign of central impurity accumulation.

The significant gain in confinement resulting from the operation at high density with radiating boundary is illustrated by the figures 1 and 2. On Fig. 1 is shown a density scan of the plasma energy obtained without (boronized wall conditions : $\gamma \cong 20\%$) and with (siliconized wall conditions : $\gamma \cong 60\%$) radiation cooling for shot heated by balanced injection in the "supershot" regime. The figure illustrates the decrease of f_H at large $\bar{n}_{eo}$ in absence of radiation cooling which is not observed in presence of it and the ability at large $\bar{n}_{eo}$ with radiation cooling to maintain a significative value for f_H. A record shot is shown on Fig. 2.

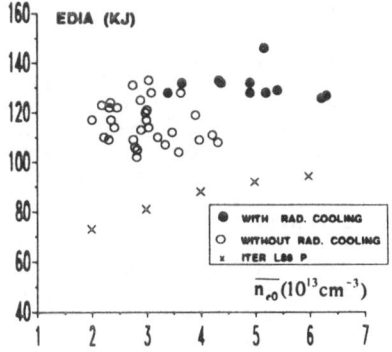

Fig. 1 - Density scan of the diamagnetic energy obtained for siliconized (closed symbols) and boronized (open symbols) conditions (400 kA, 1.5 MW NBI-co + 1.5 MW NBI-counter).

In this shot a total stored energy content of 200kJ is obtained with only 4MW of total heating power (with no correction of the beam power for shine through, etc.). This can be compared with the earlier record of 200kJ [9] (no neon puffing) which was obtained with more than 6MW of additional power. In the case of Fig.2a, the density is about twice higher. The weak density dependence of the usual confinement scaling can certainly not explain such a difference in heating power. The enhancement factor f_H is also shown on the figure. The diamagnetic energy value corresponds to a Troyon factor of 2 which is close to the toroidal β_t limit of TEXTOR (as explained in [4] this operational β limit is characterized by the appearance of MHD activity eventually followed by loss of energy content). Note that the ratio of the enhancement factor to the edge q value f_H/q_a attains 0.59 in this shot, value which is close to the value required for the reactor. Near 1.4s, there is a sudden loss of confinement linked to the large toroidal β value and afterwards, the energy follows lower confinement regimes. Let us also remark that for this shot the density exceeds 10^{14} cm^{-3} in a large fraction of the small radius and that the q = 2 layer is situated in the radiating zone (see fig. 2b).

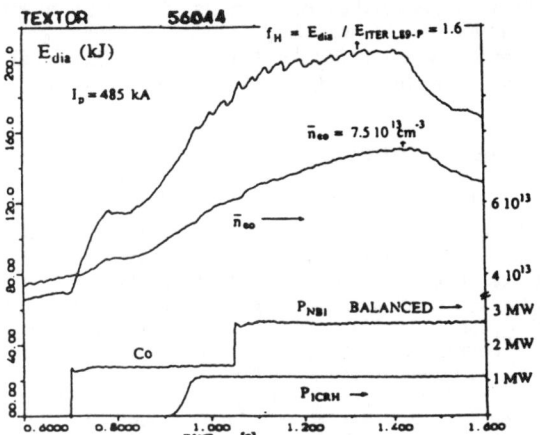

Fig. 2a - Record shot with total additional power of nearly 4MW (siliconised wall and neon puffing). The traces of the diamagnetic energy E_{dia}, of the central line-averaged density $\bar{n}_{e0}$ and of the beam and ICRH power are shown.

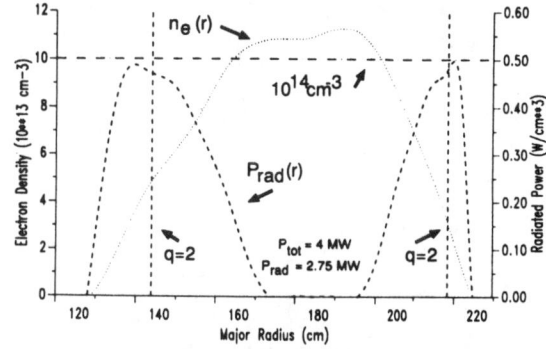

Fig. 2b - Electron density and radiated power density profiles for the shot of fig. 1a at t 1.3 - 1.4 s. The location of the q = 2 surface (from TRANSP [20] simulation) is indicated.

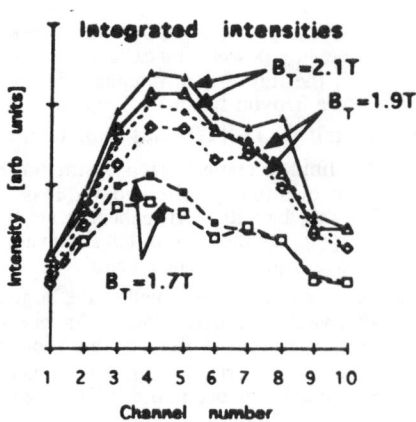

Fig. 6 - Profiles of the global intensity (i.e. spectral intensity integrated over the wavelength range corresponding to the range 11-40 keV of fast ^{3}He$^+$ ions) of the CXS signal for three shots at different B_t values and 2 time intervals.

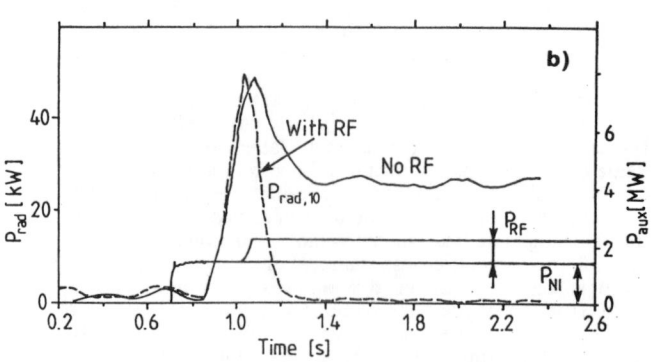

Fig. 7 - b) Central radiated power inside a minor radius r = 10 cm as a function of time for the shots shown in Fig. 7a). The NBI and RF powers for the same shots are also indicated.

5. High-Z limiters

In some discharges with a high-Z test limiter, heated with NBI (D$^\circ$ → D$^+$) a centrally peaked radiation profile was observed as shown on Fig. 7a. Remarkably, this central radiation decreases as soon as ICRH is added to NBI and disappears afterwards, enabling to maintain a sufficiently clean core plasma in a steady state with good confinement. The comparison of the time evaluation of the amount of radiated power inside a minor radius equal to 10 cm is shown in Fig. 7b for the case of pure NBI heating and for the case where ICRH is applied during the NBI pulse. The radiation profiles are also compared in Fig. 7a. A similar RF-induced expulsion of W from the centre was also observed in an ohmic target plasma. No accumulation of W in the centre was found in discharges with edge radiative cooling using neon and heated with NBI and/or ICRH [15].

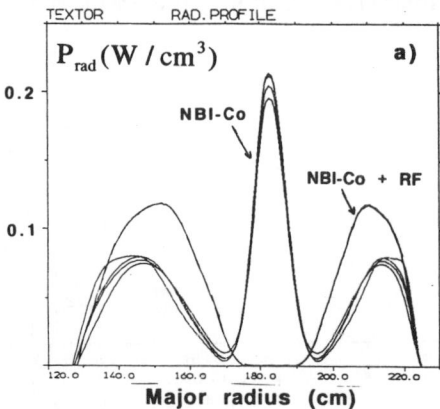

Fig. 7 - a) Radiation power profiles (bolometry) for a shot with NBI-co alone (#55815) and another one with combined NBI-co + RF (#55816) taken at various times (1.6, 2.0, 2.4 s) during the heating pulse. Tungsten test limiter at r = 44.5 cm, toroidal belt limiter at 46 cm.

6. Experience with unshielded antenna

The TEXTOR RF system has been constantly operated with at least one unshielded antenna [16] for about two years. At the same time as heating was performed in a wide variety of conditions, the parameter range over which successful operation of the unshielded antenna is obtained was thus automatically extended, thereby proving the viability of unshielded antennas under a very wide variety of experimental conditions.

7. Conclusions

In conclusion, the last experiments on TEXTOR (before its upgrading to TEXTOR-94) have allowed to extend the improved confinement mode (I-mode) to the high density region using the radiating belt concept. At the same time it was shown that in this regime, stable long-pulse operation with high auxiliary power could be achieved with up to 90% of the power radiated in the outer plasma region. This technique could help to solve the power exhaust problem of the reactor. The high performance (I-mode) conditions were also achieved with third harmonic D heating and balanced injection. In this case a high neutron production is obtained even at high density. First experiments of ^{3}He pump-out indicate that the expected interaction between the RF and the fast He ions is indeed taking place and that preferential expulsion of fast He ions is thus possible. Expulsion of central high-Z impurities by the RF has been observed in high-Z limiter tests. Finally we stress that all these experiments were constantly conducted using one unshielded antenna pair without any problem.

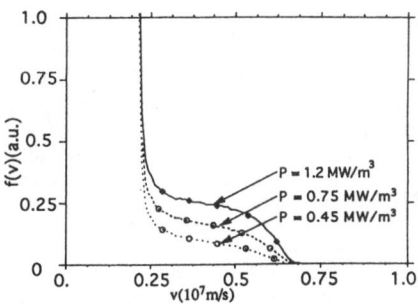

Fig. 4b - Velocity distribution profile the beam ions with 3rd harmonic heating, for 3 different values of the ICRH power density ($n_e = 9 \times 10^{13}$ cm^{-3}, $T_e = 1.5$ keV, $T_i = 1.7$ keV).

With 1.6 MW 3rd harmonic ICRH combined with the same amount of NBI power we were able to reach the β_{limit} of TEXTOR. (Troyon factor = 2.2, $B_t = 1.7$ T, $I_p = 350$ kA).

In some instances, the addition of a modest amount of RF power causes a spectacular increase in energy (see Fig. 5a). This is correlated to a sudden increase of the density peaking (see Fig. 5b) and can be interpreted as a triggering by the RF of an improved confinement regime of the supershot type. The abrupt loss of confinement occurring at about 2.2 s is to be related to the attainment of a toroidal β value close to the β limit of TEXTOR (Troyon factor close to 2).

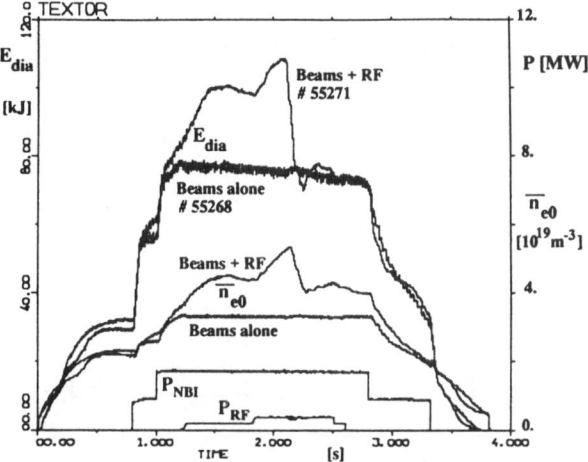

Fig. 5a - A comparison of two shots, one with balanced injection only, the other with, in addition, a small amount of RF power. The diamagnetic energy, E_{dia}, the line averaged density $\bar{n}_{eo}$ and the auxiliary power are shown ($I_p = 350$ kA, $B_t = 1.7$ T, $\omega = 3\omega_{CD}$).

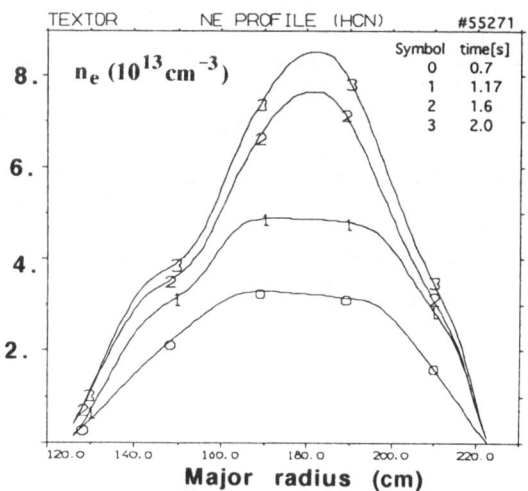

Fig. 5b - Density profiles at four different times during the shot of Fig. 5a before and during the RF heating phase.

4. Helium pump-out experiments

An RF-based technique for pumping the helium ash out of a reactor was proposed by C.S. Chang [12,13]. This method has been tested on TEXTOR using ^{3}He neutral beam injection [14,20]. Off-axis heating by the RF at the 2nd harmonic of ^{3}He induces a drift of the fast ^{3}He ions by changing their $\bar{B} \times \nabla \bar{B}$ drift. In the case of high field side (HFS) heating this would result in an RF induced outward drift of the fast ^{3}He. In case of low field side (LFS) heating an inward flux is induced. The number of fast ions is measured using charge exchange spectroscopy (CXS) with a deuterium beam. The interpretation of the signals requires an adequate modelling including corrections for the light emitted by ^{3}He$^+$ coming from injected ions before second ionisation. The complete interpretation is not yet ready. Nevertheless, for high density discharges, the light emitted by the injected ^{3}He$^+$ is weak and the global intensity in each channel is proportional to the number of fast ^{3}He^{++} ions along the line of sight. When comparing HFS ($B_T = 1.7$ T) and LFS ($B_T = 2.1$ T) heating (solid and dashed lines in Fig. 6) the light intensity changes by a factor 2 indicating a change by a factor 2 in the density of fast ^{3}He ions.

Fig. 3 shows the dependence of the enhancement factor f_H on density for shots heated by co-injection together with ICRH. The large increase of this factor at the highest densities is its most striking feature. The strong density dependence of f_H contrasts with the earlier I-mode behaviour [4] which showed only a weak density dependence of the ITER L89-P type. Note that the largest densities are reached only with large radiated power fractions, and that there is no systematic dependence, at the same density, on the radiated power fraction (γ). This implies that, contrary to what might have been feared, not only the radiating boundary does not deteriorate the confinement, but it gives access to the high density domain where the contribution to the energy content of the fast particle population is very small.

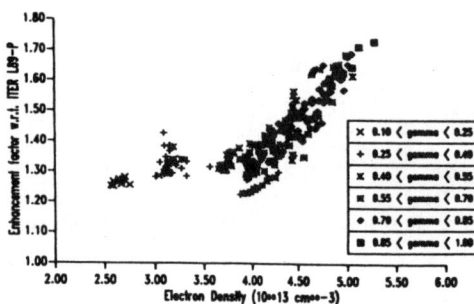

Fig. 3 - Enhancement factor f_H of the diamagnetic energy relative to the ITER L89-P scaling [17] for shots with radiative cooling by neon puffing. f_H is plotted versus the central chord line-averaged density $\bar{n}_{eo}$. The statistics is made for shots with $\approx$ 1.45 MW of NBI-co + 1 to 2 MW of ICRH. The data discrimination is relative to $\gamma = P_{rad}/P_{tot}$, the ratio of the total radiated power to total input power (OH + additional). Plasma current I_p=350kA, minority H heating, 32.5MHz, 2.25T.

It should be stressed that, in addition to the above confinement results, these experiments at high density and large heating power provide a proof of the compatibility of the radiating boundary concept with high confinement regimes. This concept might be useful to relieve the loading of the divertor (or limiter?) of the reactor by radiating a large fraction of the power all over the plasma outer surface. The obtained heating power flux (> 0.1 MW/m^2) and radiated power density ($\sim$ 1 MW/m^3) are relevant to reactor conditions.

Interpretation and modelling of these results have been made by the use of different codes [18,19]. From these analysis it appears that the reason for the confinement improvement is probably linked to (i) the slightly more peaked current density profile j(r) appearing in the modelling due to the reduced edge electron temperature and the hollow Z_{eff} profile, or (ii) the reduced conductive and convective heat flow near the boundary which leads to a reduced edge temperature and a possible decrease of the edge turbulence.

3. Third harmonic heating

Heating scenarii combining both NBI (neutral beam injection D $\rightarrow$ D$^+$) and third harmonic heating have been experimentally studied on TEXTOR; two scenarios were investigated : ICRH + pure Co injection and ICRH + balanced injection. A large increase in neutron production when ICRH is added to the NBI heating is observed in these experiments. For a given NBI power the neutron yield increases in proportion to the RF power. This is shown on Fig. 4a for the case of a plasma heated by balanced injection with a variable amount of RF power. The diamagnetic energy also increases linearly with the RF power. These experimental results are compared with the predictions from SWHAP (wave absorption code) and BECHSI (Fokker Planck code) codes [10]. From this analysis we conclude :
(i) The neutron production in these experiments is mainly due to reactions between fast particles and plasma target. A modest change in the neutron production is due to the change in the plasma target electron temperature (< 10 %). Most of the increase in the neutron production is due to the ICRH induced change in the velocity distribution function of the fast ions. The deformation of the distribution function is shown in Fig. 4b. The shape of the tail is independent of the RF power but its amplitude is proportional to that power. This explains the linear dependence of the neutron yield on the RF power.
(ii) When fast ions of the tail can be confined, the heating efficiency of the 3rd harmonic heating scenario is at least as efficient as the minority heating scenario. Low energy increases measured during the first series of experiments [11] can be explained by the deconfinement process of the fast ions.

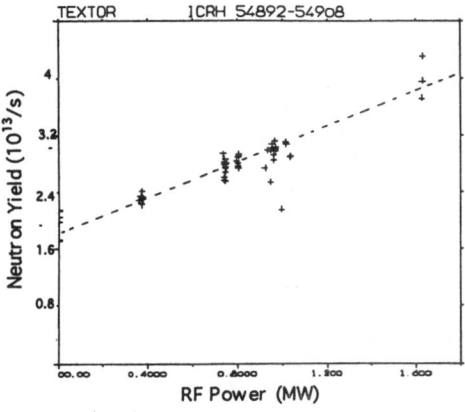

Fig. 4a - Neutron yield as a function of ICRH power. (Co + Cou injection (D$^\circ \rightarrow$ D$^+$), P_{NBI} = 1.6 MW, $\omega = 3\omega_{CD}$, I_p = 350 kA, B_t = 1.7 T).

References

[1] J. Winter et al., 20th EPS on Contr. Fus. and Plasma Phys., Europhysics Conf. Abstracts Vol 17C, Part I (1993) 279.

[2] U. Samm et al., Plasma Phys. and Contr. Fusion $\underline{35}$ supplement (12)B (1993) B167.

[3] F. Durodié and M. Vervier, Europhysics Topical Conf. on Radiofrequency Heating and Current Drive of Fus. Devices, Europhysics Conf. Abstracts Vol 16E (1992) 80.

[4] J. Ongena et al., Nucl Fus. $\underline{33}$ (1993) 283.

[5] U. Samm et al,, Plasma Phys. and Contr. Nucl. Fusion Res., (Proc. 14th Int. Conf. Würzburg, 1992), Vol. 1, IAEA, Vienna (1993) 309.

[6] J. Ongena et al., 20th EPS on Contr. Fus. and Plasma Phys., Europhysics Conf. Abstracts Vol 17C, Part I (1993) 127.

[7] J. Winter et al., Phys. Rev. Lett. $\underline{71}$ (1993) 1549.

[8] A.M. Messiaen et al.,, Nucl. Fus. $\underline{34}$ (1994) 825.

[9] A.M. Messiaen et al., Plasma Physics and Controlled Fusion, $\underline{35}$ Supplement A (1993) A15.

[10] G. Van Wassenhove et al., 21st EPS Conference on Controlled Fusion and Plasma Physics, Montpellier, France. Europhysics Conf. Abstracts $\underline{18B}$, Part II, 964 (1994).

[11] G. Van Wassenhove et al., (1992) Europhysics Topical Conference on Radiofrequency Heating and Current Drive of Fusion Devices, Brussels, Vol. $\underline{16E}$, 141.

[12] C.S. Chang, Phys. Fluids $\underline{B3}$ (1991) 259.

[13] C.S. Chang, J.-Y. Lee and H. Weitzner, Phys. Fluids $\underline{B3}$ (1991) 3429.

[14] K. Finken et al., Phys. Rev. Lett. $\underline{73}$ (1994) 436.

[15] G. Van Oost et al., 21st EPS Conference on Controlled Fusion and Plasma Physics, Montpellier, France. Europhysics Conf. Abstracts $\underline{18B}$, Part II, 1020 (1994).

[16] R. Van Nieuwenhove et al., Nucl. Fus. $\underline{31}$ (1991) 1770.

[17] J.D. Stachan et al., Phys. Rev. Lett. $\underline{58}$ (1987) 1004.

[18] J. Ongena et al., Workshop "Transport in Fusion Plasmas", Aspenäs, Göteborg, Sweden, June 1994, to be published.

[19] A. Messiaen et al., Plasma Phys. and Contr. Nucl. Fusion Res. (Proc. 15th Int. Conf. Sevilla, 1994) paper IAEA-CN-60/A-2-III-3.

[20] R.J. Hawryluk, in Physics of Plasmas Close to Thermonuclear Conditions (Proc. Course Varenna 1979), $\underline{1}$, CEC, Brussels (1980) 19.

CHARACTERISTICS OF PELLET INJECTED DISCHARGES IN TEXTOR

K.H. Finken[1], K.N. Sato[2], H. Akiyama[3], G. Fuchs[1], R. Jaspers[1], S. Kogoshi[4], H.R. Koslowski[1], G. Mank[1], H. Sakakita[2], M. Sakamoto[2], M. Sander[1], H. Soltwisch[1]

[1]Institut für Plasmaphysik, Forschungszentrum Jülich GmbH, Assocition Euratom-KFA, 52425 Jülich, FRG
[2]National Institute for Fusion Science, Nagoya 464-01, Japan
[3]Kumamoto University, Kumamoto 860, Japan
[4]Science University of Tokyo, Noda 278, Japan

Abstract

Pellets injected into the TEXTOR tokamak lead to a density profile peaking which is strongest at low plasma current and weakest at high current independent of B_T. After the injection two types of density oscillations are excited, the first type follows immediately the injection and the second one is excited with a delay of more than ten milliseconds. The oscillations are also observed in runaway discharges; the synchrotron light from the relativistic electrons drops after the pellet injection and is subsequently modulated due to a trapping of the runaways in magnetic islands. First Faraday measurements have been performed indicating that the distribution of the plasma current is not measurably modified by the pellet.

I. Introduction

Pellet injection into tokamak discharges presents several interesting features: The density is increased, the density profile is more peaked than before the injection and the plasma energy is better confined, especially at high plasma densities. The pellet injection typically leads to an increase of 30% - 100% of the pre-pellet density. It is a major perturbation and often excites modes in the plasma which provide information about the current profile in the discharge. The improvements after the pellet injection are attributed to a shift of the fuel ionization to the central plasma, while in normal gas fuelled discharges the ionization zone is at the boundary. This paper treats the aspect of the profile peaking after the pellet injection, especially its dependence on the plasma current and the safety factor q(a). In a second section some special features of the mode excitation in the plasmas are described. Then, observations of the pellet injection into runaway discharges and its influence on the runaways are described. The high energetic runaway electrons are measured by synchrotron radiation and can be seen as probes for the central plasma. Finally, changes of the current distribution after pellet injection are analyzed.

II. Experimental Set-Up

Pellet injection discharges have been studied in the TEXTOR tokamak for some years already. TEXTOR is a medium size limiter machine with a major radius of 1.75 m, a minor radius of 0.46 m, a plasma current of up to 500 kA and a pulse length of up to 4 s. Up to October 1992 a single shot gas gun type injector delivering a maximum pellet velocity of 900 m/s was employed. The pellet size is 1.4 mm in length and in diameter; it is guided via a four meter tube to the equatorial midplane of the tokamak. The propulsion gas is differentially pumped at three locations along the pellet path, and the gas separation is supported by valves closing after the passage of the pellet.

Since December 1992, the single shot injector was replaced by a nine shot gas gun type injector. Also, this injector provides cylindrical deuterium pellets of 1.5 mm diameter; the length is adjustable between 1 mm and 2 mm and the pellet velocity amounts to about 1200 m/s. After leaving the barrel the pellets fly through a 4 m evacuated channel into the discharge. The propellant gas is diffused by a muzzle break immediately behind the barrel and separated via two electro-magnetic valves, which close the gas path in about 1 ms. The injection time of the individual pellets is freely programmable and all kinds of pellet combinations are possible.

III. Profile Peaking

The pellet injection into ohmic discharges leads to a pronounced plasma peaking. The peaking is stronger for high density discharges than for low ones. Characteristic for the peaked, improved plasma state is the suppression of the sawtooth oscillations. The sawtooth suppression and the improvement of the central confinement may indicate that the changes due to pellet injection depend on the central magnetic field structure i.e. on the q-profile. A modification of the central q-profile has been deduced in several publications [1,2].

Here, further investigations of the peaking are performed. It is investigated whether the confinement improvement depends on the volume enclosed by the q=1 surface. Figure 1 shows the density evolution for the time around the pellet injection (t=1.5 s) for the conditions a) I_p = 300 kA, B_T = 2.25 T, q(a) = 4.53 and b) I_p = 500 kA, B_T = 2.25 T and q(a) = 2.72. The initial density on axis is in both cases $4 \cdot 10^{19}$ m^{-3} $\leq n_e \leq 5 \cdot 10^{19}$ m^{-3}. In the first case the target plasma profile is more peaked than in case b). After the pellet injection the peaking increases further and the improved state persists for several hundred milliseconds. The sawtooth activity is rather weak before the pellet injection and stops completely after the injection. The target density profile in the case of high plasma current (b) is rather flat and the sawtooth activity is clearly seen during this phase. After the pellet injection the plasma density starts to peak and the sawtooth-activity stops for a few periods. Then after about 50 ms, the activity starts again and is even stronger than before. With the occurrence of the sawtooth the plasma density decays quickly and returns to the initial value after less than 100 ms. These measurements indicate

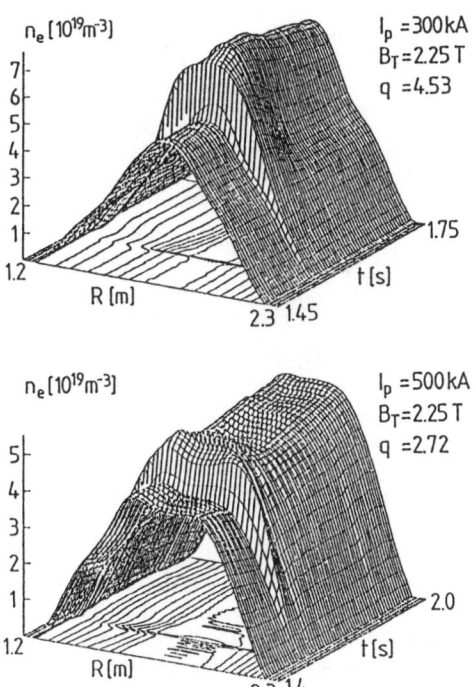

Fig. 1: Density profile evolution after pellet injection for low and high plasma current. I_p, B_T and $q(a)$ are indicated in the plots.

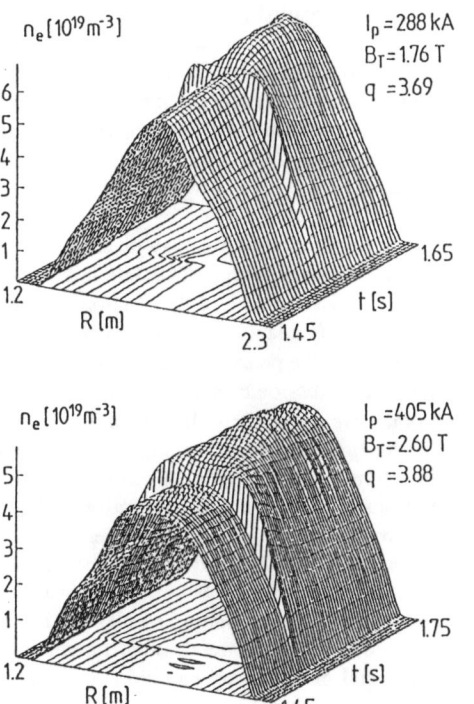

Fig. 2: Density profile evolution after pellet injection for low and high plasma current at nearly constant $q(a)$. I_p, B_T and $q(a)$ are indicated in the plots.

that the confinement after pellet injection is linked to the plasma current and little to the volume spanned by the $q=1$ surface.

In a second run it was tested whether the plasma peaking remains similar for constant values of $q(a)$ or whether it is linked to the value of the current alone. Fig 2 gives the density distribution for a) $I_P = 288$ kA, $B_T = 1.8$ T, $q(a) = 3.7$ and for b) $I_P = 405$ kA, $B_T = 2.6$ T and $q(a) = 3.9$. Despite the fact that the plasma in both cases has nearly the same $q(a)$ value, the target density is more peaked for the lower plasma current. The density increase due to the pellet injection is similar in both cases. The peaking of the density remains again higher for the lower current case. For nearly equal plasma currents but different q-values as shown in figs 1a and 2a, the peaking behavior is very similar. In fig. 2b a sawtooth activity is observed whereas it is not prominent in 2a. The plasma current in fig. 2b is in between the examples shown in figs. 1; the profile peaking and the sawtooth activity are also at a medium level. The results show that the profile peaking and the tendency for sawtooth activity depends more strongly on the value of I_P than of $q(a)$. In the future it has to be investigated whether the q-profile shows a similar tendency.

IV. Excitation of Plasma Modes

The excitation of fast and large density oscillations has been observed in the core region of pellet-injected TEXTOR plasmas [3,4]. The oscillations occur more likely with deeper pellet penetration into the plasma. The frequency is in the range of 0.7 kHz up to 2 kHz.

The excitation of modes has also been observed on other tokamaks [1,2,5], but TEXTOR shows some peculiarities. In TEXTOR two types of oscillations are observed: One type of oscillation occurs immediately after the pellet injection and persists for about ten milliseconds. This type is called the "first oscillation". A "second oscillation" occurs some 10 ms after the pellet injection. It has been observed that pellet injection can lead to the excitation of both oscillation types, of only one type or of none. Fig. 3 a - d shows the characteristic changes of the density oscillations after the pellet injection as a function of the target plasma temperature; (a)$T_{eo} = 1.1$ keV, (b) = 0.95 keV, (c) = 0.88 keV and (d) = 0.8 keV, respectively. The pellet velocity amounts to about 680 m/s and the plasma current to 340 kA. Although the density changes slightly from (a) $n_e = 2.5 \cdot 10^{19}$ m^{-3} to (d) $n_e = 3.8 \cdot 10^{19}$ m^{-3}, the overall penetration varies from shallow (a) to deep (e). It is observed that shallow pellet penetration doesn't excite any modes (a), then the excitation increases with deeper penetration (b) and finally both the first and second oscillation are excited (c,d). With deeper penetration they separate (d). The same characteristics of mode excitation has been verified by varying the pellet speed with the same target plasmas: A low pellet speed results in non or single oscillation excitation and a high pellet speed prefers double mode excitation.

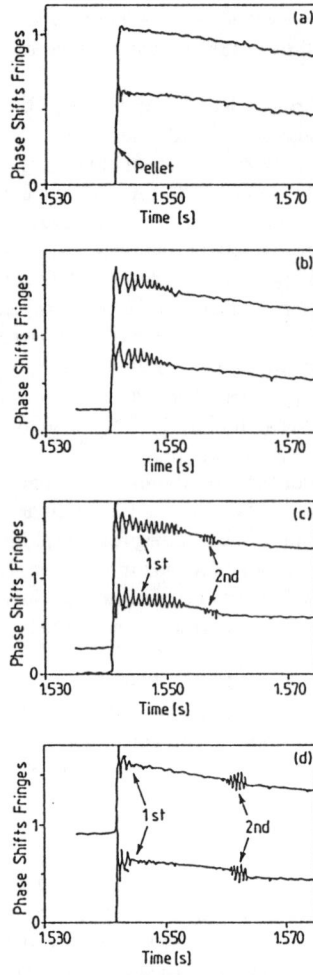

Fig. 3: Excitation of plasma oscillations for different plasma temperatures; (a): $T_{eo} = 1.1$ keV, (b): $T_{eo} = 0.95$ keV, (c): $T_{eo} = 0.88$ keV and (d) $T_{eo} = 0.8$ keV.

The oscillations are density modes around the q=1 surface. A time evolution of the density profile is shown in Fig. 4. Immediately after the pellet injection the density increases. The initial distribution is inhomogeneous with an initial maximum at the high field side. The density oscillation shows up for about 10 ms and resembles the snake-phenomenon observed at JET [6]. The mode features are prominent in the interferometer signals (9 channels) and in soft x-ray mode signals (40 channels). The mode data are best fitted by assuming a poloidal mode number m=1 and a toroidal mode number n=1. The position of the mode agrees with the q=1 surface within the experimental errors.

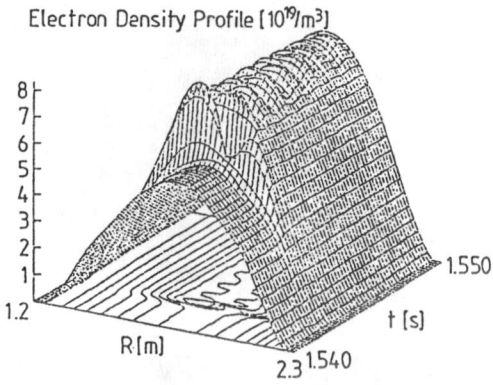

Fig. 4: Density profile with oscillation after pellet injection.

V. Pellet Injection into Runaway Discharges

Runaway discharges are of special interest in TEXTOR because they allow an unique insight into the discharge: It was found that runaway electrons gain an energy up to the range of 25 MeV - 30 MeV. In this energy range the highly relativistic electrons ($\tau \approx 50$) emit synchrotron radiation in the middle IR spectral range which is extremely directed forward (opening angle of the light beam in the velocity direction amounts to about $1/\tau$) [7,8]. The combination of toroidal motion of the electrons with superimposed gyration around the guiding center leads to the observed opening angle of the synchrotron radiation of 100 mrad. From this value it can be estimated that the observed radiation is integrated only over a path length of less than 20 cm. The observation of the synchrotron radiation in toroidal direction with an IR-scanner thus provides a good spatial resolution.

The pellet is injected into the runaway discharge at about 2.5 s after the start of the discharge. This late injection allows a development of the acceleration of the electrons and establishes sufficient synchrotron radiation. The line averaged density trace in fig. 5a shows that the pellet is well accepted by the plasma. Fig. 5b shows the emission of hard X-rays which are created when the runaways are lost and hit the walls. The intensity increases gradually as the runaways gain power and at the time of pellet injection a strong burst is observed. This loss of the runaways is shown in a high time resolution in fig. 5c. The emission is strongly structured and the largest loss of the runaways is observed about 50 ms after the pellet injection. After about 100 ms the X-ray emission becomes low again. The structured emission of the X-ray signal is well correlated in time and frequency with the m=2 mode activity which is plotted as fig 5d. The end of the X-ray emission coincides with the locking of the mode. After about 50 ms the mode starts rotating again and the frequency increases.

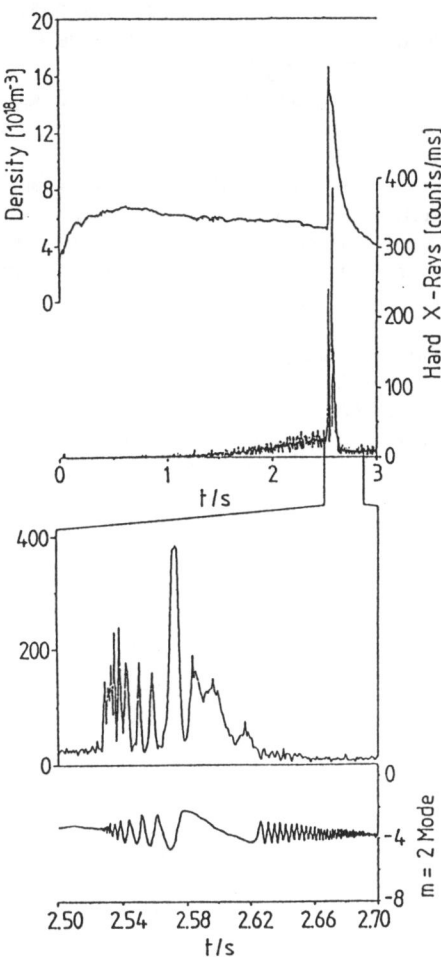

Fig. 5: From top to bottom: Density and hard X-ray traces for a pellet injected discharge; then X-ray and magnetic m=2 mode signals with high time resolution.

After its approach, the pellet causes a sudden drop in the synchrotron light intensity. The loss occurs on a time scale of less than half a millisecond and is accompanied by an X-ray burst. From discharge to discharge both the loss time of the runaways (relative to the pellet injection time) and the decrease of the synchrotron intensity varies widely. For the given example the loss fraction is relatively small and amounts to about 50%.

The modes, seen in the X-ray and magnetic signal, show up in the synchrotron light as a modulation in the IR-TV pictures. It is consistent to assume that the modulation is a time modulation and not a spatial modulation. This interpretation is possible as the scanner translates the incoming radiation by a set of two sweeping mirrors into a TV-norm picture. Fig 6 shows an example of the modulated synchrotron light. The modulation in time is confirmed by operating the scanner in a line scan mode. All signals show consistently that the modulation frequency varies in time.

Fig. 6: Modulation of the synchrotron light after pellet injection into a runaway discharge.

Often the modulation frequency at first decelerates and then again accelerates and finally the modulation depth decreases to zero. The modulation is present in the synchrotron light and not in the thermal background light and this allows a distinction between thermal and non thermal radiation. For observing the runaway electrons the pellet unfortunately has to be injected at the end of the discharge and a further relaxation of the runaways is terminated then.

The observations are interpreted as follows: During the pellet injection turbulences and/or ergodization of the magnetic field are created which lead to a rapid loss of the runaways from a major part of the plasma. The remaining runaway electrons are tied to an island winding around the $q=1$ surface. These runaway electrons, however, are well confined and survive all perturbations during the injection phase. When the island rotates toroidally, the forward directed synchrotron light falls into the detector for a short phase and misses it during other phases of the rotation. The observed modulation is thus similar to that of a pulsar. The mode structure of the island with $n=1$ $m=1$ is the same as discussed above and the good confinement properties of the island resemble the 'snake' phenomenon observed on JET.

VI. Faraday Rotation Measurements

The analysis of the sawtooth signals shows an inversion radius of 10 cm before the pellet injection and of 8 cm after the injection. The sawtooth inversion radius is often attributed to the $q=1$ surface and a change of the radius then would imply a change of the profile of the safety factor $q(r)$. To investigate this point the first Faraday rotation measurements have been performed on pellet-injected TEXTOR plasmas with the HCN laser system. Fig. 7 shows the angle of the Faraday rotation. Within the limits of error the Faraday rotation observed in these discharges is explained by the density rise. A redistribution of the plasma current due to the pellet injection was not found so far.

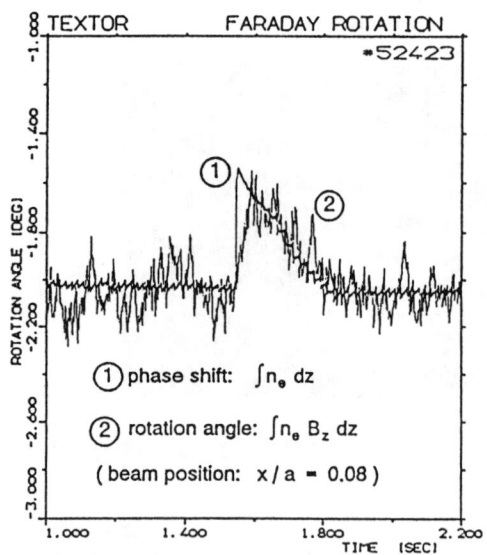

Fig. 7: Faraday rotation measurement in pellet injected plasma; 1) phase shift of interferometer and 2) rotation angle of polarization of FIR laser.

VII. Summary

Pellet injection experiments on TEXTOR have been performed to investigate the plasma peaking, the excitation of modes, the influence on relativistic runaway electrons and on the change of the plasma current distribution. It was found that the plasma peaking after the pellet injection mainly depends on the plasma current and not on the safety factor at the boundary. For lower plasma currents the peaking is stronger than for high plasma currents. Two types of oscillations were excited, the first one immediately after the injection and sometimes a second one with some delay. The excitation of the oscillations depends on the penetration depth of the pellets. The oscillations are also visible in the light of synchrotron radiation which is emitted from the plasma under runaway conditions. The analysis of the runaways gives insight into a loss channel during the injection and into characteristics of the oscillation which resembles the snake phenomenon on JET. Up to now, Faraday rotation measurements do not indicate any redistribution of plasma current during the injection of pellets.

References

[1] R. Yoshino et al., Nucl. Fus., **29** (1989) 2231
[2] M. Hugon, B.Ph. van Milligan, P. Smeulders et al., Nucl. Fus., **32** (1992) 33
[3] K.N. Sato, S. Kogoshi, H. Akiyama et al., Controlled Fusion and Plasma Physics (Proc. 18th Eur. Conf., Berlin 1991) Vol. 15C, I-333 (1991)
[4] K.N. Sato, H. Akiyama, K.H. Finken et al., Controlled Fusion and Plasma Physics (Proc. 20th Eur. Conf., Lisboa 1993) Vol. 17C, I-239 (1993)
[5] J. Parker, M. Greenwald, R. Petrasso et al., Nucl. Fus., **27** (1987) 856
[6] A. Weller, A.D. Cheetham, A.W. Edwards et al., Phys. rev. Lett., **59** (1987) 2303
[7] K.H. Finken, J.G. Watkins, D. Rusbüldt et al., Nucl. Fus., **30** (1990) 859
[8] R. Jaspers, N.J. Lopes Cardozo, K.H. Finken et al., Phys. Rev. Lett., **72** (1994) 4093

Rotation and Ion Temperature Measurements in Tokamaks and Stellarators

K.Ida, Y. Miura[a], K.Kondo[b], J.V.Hofmann[c], F.Sano[b], S.Hidekuma, H.Yamada, H.Iguchi,
T.Matsuda[a], H.Zushi[b], T.Obiki[b], K.Watanabe, J.Geiger[c], N.Nakajima, F. Rau[c]

National Institute for Fusion Science, Nagoya, 464-01, Japan
(a) Japan Atomic Energy research Institute, Ibaragi 311-01, Japan
(b) Plasma Physics Laboratory, Kyoto University, Uji, Kyoto, 611, Japan
(c) Max-Planck-Institute fur Plasmaphysik, Garching bei Munchen, D-85748, Germany

Radial structures of plasma rotation and ion temperature are experimentally studied in JIPP T-IIU and JFT-2M tokamak, CHS and Heliotron-E heliotron/torsatron and Wendelstein 7AS stellarator devices. The profiles of the radial electric field are derived from the rotation profiles. The mechanism determining the rotation profile is studied and the relation between plasma rotation (radial electric field) shear and transport barrier is discussed. Plasma rotation profiles are determined from the balance of momentum deposition and the perpendicular and parallel viscosities. The parallel viscosity, which predominantly damps the toroidal velocity in heliotron/torsatron and stellarator devices, is found to be neoclassical. On the other hand, the perpendicular viscosity, which is dominant in dictating the toroidal rotation in tokamaks, is anomalous. The transport analysis based on density and electron and ion temperature profiles shows that the sheared plasma rotation (radial electric field) improves particle and electron and ion heat transport both in bulk and edge plasma regions of tokamaks.

I. INTRODUCTION

A radial electric field has been considered to possibly reduce the ripple loss and to prevent the degradation of confinement in stellarators and helical devices. After the transition from low confinement mode (L-mode) to high confinement mode (H-mode) (L/H transition) was found in ASDEX[1], a spontaneous bifurcation of the radial electric field was theoretically proposed to cause it[2,3] and refreshed the motivation of research. The phenomenon, that associated with the L/H transition, the radial electric field suddenly changes at the plasma periphery (few cm), was observed in DIII-D[4], JFT-2M[5], ASDEX[6] and Wendelstein 7AS[7]. Thus the radial electric field has been considered to have an important effect on confinement and anomalous transport in the plasma, and the experimental research on the mechanisms determining the radial electric field in a toroidal plasma flourished recently. Since the radial electric field is coupled with plasma rotation, research on the radial electric field and ion transport is accompanying the research on the radial profile of plasma rotation and ion temperature.

II. CHARGE EXCHANGE SPECTROSCOPY

In order to obtain a detailed radial profile of the electric field and to analyze ion transport, high spatially resolved measurements on radial profiles of toroidal and poloidal rotation velocities and ion temperature are required. Charge exchange spectroscopy[8] has the advantage of providing multi-chord data when using optical fibers and thus gives a high spatial resolution up to 5mm (in JFT-2M CXS). Charge exchange collisions between fully stripped impurity ions of the plasma and fast hydrogen atoms from neutral beam result in excited ions with one more electron in an upper levels. For carbon impurities this process is expressed as

$$H^0(NB) + C^{6+} \rightarrow H^+ + C^{5+}(n,l)$$
$$\rightarrow H^+ + C^{5+}(n',l') + h\nu \quad : \quad n' < n$$

For our measurements we use the C^{5+} (n=8-7) transition at $\lambda = 529.05$nm. The emission cross section for this transition at an impact energy of 40 keV is smaller than that for the n=7-6 transition (343.37nm) by a factor of three. However, the n=8-7 transition (529.05nm) has the advantage of lower attenuation through our optical fibers, a larger Doppler shift, an absence of interfering impurity lines nearby and it is used in many tokamaks and helical devices. Principally, in this kind of measurement the emission originates from a volume defined by the intersection of the neutral beams and the line of sight. However, radiation from the plasma periphery due to charge exchange of impurities with thermal hydrogen atoms ("cold component") is superimposed. Due to the low temperature in the plasma periphery this emission exhibits a smaller Doppler broadening and can thus be distinguished from the more central ("hot") radiation. In order to correctly determine and eliminate this cold component, two sets of optical fiber arrays looking on and off the neutral beam line are used[9].

Multi-chord measurements can be done relatively easily by combining a spectrometer with an optical fiber array and a two dimensional detector. Figure 1 shows a schematic diagram of a multi-chord measurement. An optical fiber array is arranged parallel to the entrance slit of a Czerny-Turner spectrometer and a two dimensional detector is arranged at the exit plane. This results in a spectral dispersion in the horizontal direction and a spatial resolution in the vertical direction. The size of our CCD in the direction of wavelength is 6mm (243 channels), corresponding to a spectral range of 2.4nm which is simultaneously measured with an 1 m spectrometer equipped with a 2160 groves/mm grating. A frame transfer CCD with 60 Hz interlaced scan is used to avoid a mechanical shutter for charge transfer. The signal of the CCD is a standard video signal in NTSC and has 262+1/2 lines per frame. One line has 256 data sampled with 5MHz clock and 12 bit dynamics. Since one frame has 64K data, 16 frames (267ms) can be stored with 1M word memory. The spectra of each optical fiber are obtained simultaneously

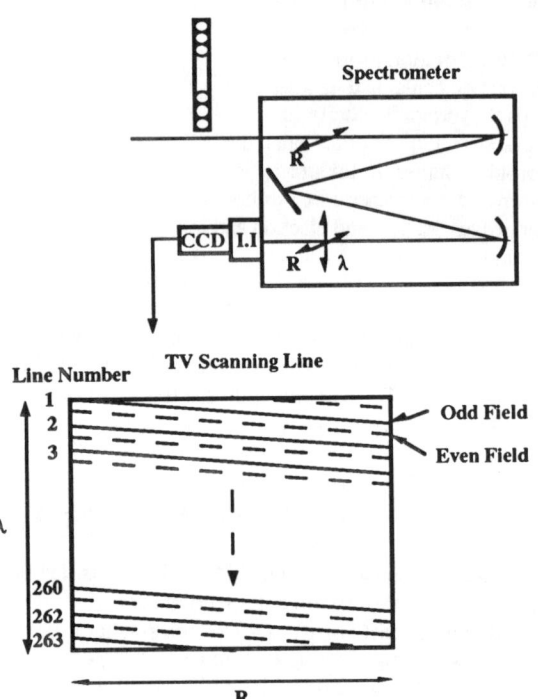

Fig.1. Schematic diagram of the multi-channel measurement using optical fiber array. Solid lines show interlace scans in the odd frame and dashed line show scans in the even frame.

III ION TEMPERATURE AND THERMAL DIFFUSIVITY PROFILES

Here, we discuss the ion temperature and ion diffusivity profiles in JFT2M[10] and Heliotron-E[11] as a comparison of ion transport between tokamak and torsatron/heliotron. Heliotron-E is an axially asymmetric torsatron/heliotron with l=2, m=19, major radius R=2.2m, magnetic field B=1.9T, magnetic axis shift -4cm. Figure 2(a) shows the central ion temperature $T_i(0)$ as a function of electron density n_e for two type of discharges. When the neutral beam is injected into the target plasma produced by electron cyclotron heating (ECH) with low electron density below

$1 \times 10^{19} m^{-3}$, $T_i(0)$ increases in time up to 0.8keV, which is called high T_i mode, as n_e is increased, while $T_i(0)$ normally decreases as n_e is increased. Figure 2(b) shows ion temperature profiles for the L-mode discharge with high density ($n_e = 8 \times 10^{19} m^{-3}$) and high T_i mode discharge with low density ($n_e = 2.3 \times 10^{19} m^{-3}$) with neutral beam injection of 3.2MW (1.1MW to electrons 0.63MW to ions).

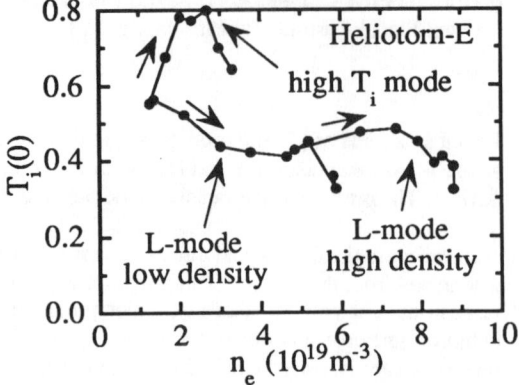

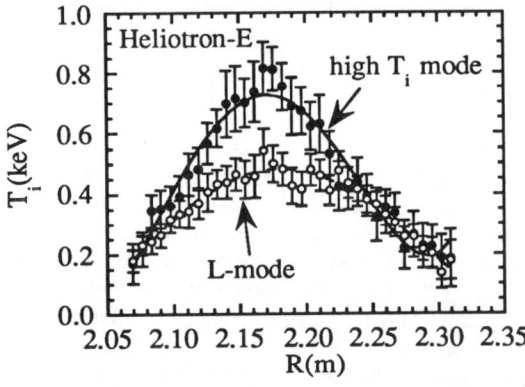

Fig. 2. (a) Central ion temperature as a function of line averaged electron density and (b) ion temperature profiles for L-mode and high T_i mode in Heliotron-E torsatron/heliotron.

JFT-2M is a tokamak with major radius R=1.31m and minor radius a=0.33m. It has two tangential neutral beams; one is parallel (co-injection) and the other is anti-parallel (counter-injection) to the plasma current. We can switch the neutral beams from co-injection to counter-injection during discharge to induce a positive or negative radial electric field. Such series of experiments were done under the conditions of a toroidal field of B_t = 1.3 T, a plasma current of I_p = 240 kA, deuterium working gas, a limiter configuration, an elongation of κ = 1.2 and a NBI power of 0.5-0.6 MW. Figure 3 shows ion temperature profiles for L-mode discharge (co-injection NBI) and high T_i mode discharge (counter-injection NBI) with a line averaged electron density of 2.4 and $3.4 \times 10^{19} m^{-3}$ and neutral beam injection power of 0.71 MW (0.38 MW to electron and 0.24 MW to ions) and 0.63 MW (0.34 MW to electron and 0.22MW to ions).

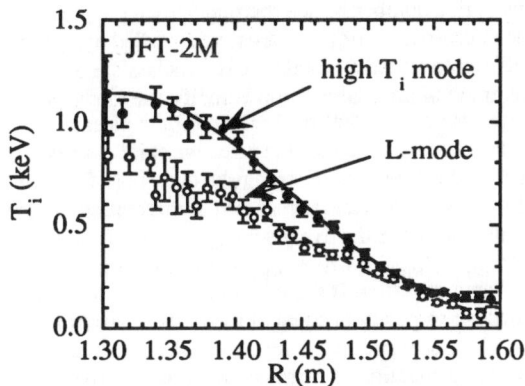

Fig. 3. Ion temperature profile of L-mode and high T_i mode in JFT-2M tokamak and Heliotron-E torsatron/heliotron.

Figure 4 shows ion thermal diffusivity profiles for high T_i mode in JFT-2M and Heliotron-E. The ion thermal diffusivity at the outer half of plasma minor radius in Heliotron-E is comparable to that in JFT-2M . In the tokamak, the thermal diffusivity tends to decrease sharply towards the plasma center, whereas, in Heliotron-E it gradually decreases. The difference in the χ_i profiles between tokamak and torsatron/heliotron is may be due to the difference of the q profiles (magnetic field shear or poloidal field strength).But more detailed analyses is necessary to verify this conclusion. It should be noted that the electron thermal diffusivity in Heliotron-E is much larger [$\chi_e(0.5) = 4 m^2/s$] than that [$\chi_e(0.5) = 0.7 m^2/s$] in JFT-2M.

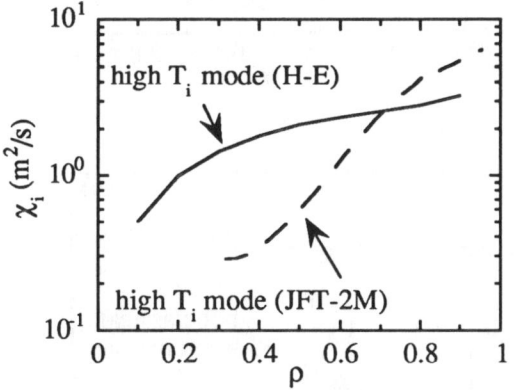

Fig.4. Radial profiles of ion thermal diffusivity in the high T_i mode discharges in JFT-2M and Heliotron-E.

IV PLASMA FLOW AND VISCOSITIES

Plasma rotation results from the balance of poloidal and toroidal momentum input and parallel and perpendicular viscous damping as

$$F_\theta = \langle \nabla_\| \Pi \rangle_\theta + \langle \nabla_\perp \Pi \rangle_\theta \quad \text{poloidal direction } (\theta):$$

$$F_\phi = \langle \nabla_\| \Pi \rangle_\phi + \langle \nabla_\perp \Pi \rangle_\phi \quad \text{toroidal direction}(\phi)$$

and the radial electric field is determined with radial force balance as

$$e n_i Z_i E_r = \partial p_i / \partial r - e n_i Z_i (v_\theta B_\phi - v_\phi B_\theta)$$
$$\text{radial direction(r)}$$

Here, F_θ and F_ϕ are external (and internal) forces in the poloidal and toroidal directions of a toroidal plasma, respectively. The force due to the neutral beam injected in the toroidal direction (tangential direction) contributes to F_ϕ and the forces due to bipolar fluxes such as electron and ion losses work as F_θ and F_ϕ, respectively. With respect to the viscous forces, we consider two viscosities: one is the viscosity coupled to a velocity gradient in the direction parallel to the velocity and the other is the viscosity associated with a velocity gradient in the direction perpendicular to the velocity (radial direction). These viscosities are called parallel and perpendicular viscosity, and the forces are given as $\langle\nabla_\|\Pi\rangle_{\theta,\phi}$ and $\langle\nabla_\perp\Pi\rangle_{\theta,\phi}$. There are off-diagonal terms in these viscosities and the plasma rotations are not independent of temperature or density gradients (which are considered to be part of the internal forces). Taking the standard diagonal terms, the viscous forces can be expressed by the viscosity coefficients $\mu_\|^{\theta,\phi}$, $\mu_\perp^{\theta,\phi}$,

$$\langle \nabla_\| \Pi \rangle_{\theta,\phi} = n_i m_i \mu_\|^{\theta,\phi} V_{\theta,\phi} \quad \text{parallel viscosity}$$

$$\langle \nabla_\perp \Pi \rangle_{\theta,\phi} = -n_i m_i \mu_\perp \nabla^2 V_{\theta,\phi} \quad \text{perpendicular viscosity}$$

When the plasma flows in a nonuniform magnetic field, the parallel viscosity acts on it. The plasma momentum is converted to thermal energy due to the change of magnetic field strength (transit time magnetic pumping: TTMP), and the plasma velocity is damped. In tokamaks, the poloidal rotation is considered to be damped by this mechanism[12], since the magnetic field strength in tokamaks is proportional to the inverse of the major radius. However, in heliotron/torsatron, both poloidal and toroidal rotations are damped by TTMP and no large velocities are expected. Figure 5 shows typical plasma flows measured in a tokamak (JIPP TIIU) and a torsatron/heliotron (CHS). Roughly speaking the toroidal rotation is peaked at the plasma center and the poloidal rotation velocity increases towards the plasma periphery in both devices. In our tokamak, the toroidal rotation velocity is large enough to change the direction of the radial electric field (see two vectors for co-NBI in tokamak point towards positive electric field) and poloidal rotation velocity is small except for the plasma edge in the H-mode.On the other hand, in torsatron/heliotron, the toroidal rotation velocity is too small to affect the radial electric field and it is mainly determined by poloidal rotation.

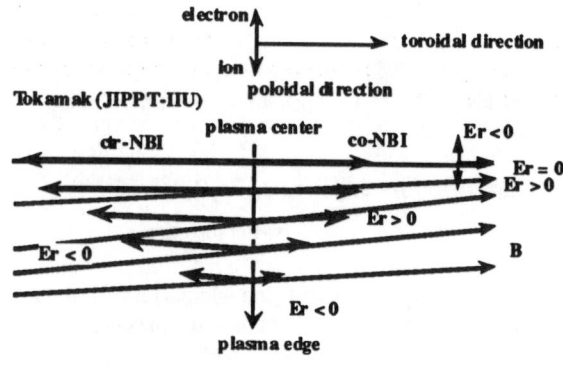
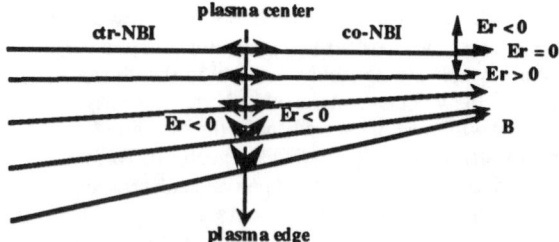

Fig.5 Plasma flow vectors measured in JIPP TII-U tokamak and CHS torsatron/heliotron. The length of the vectors show the magnitude of plasma flow with the same scale between for tokamak and torsatron/heliotrons.

IV-1. PARALLEL VISCOSITY

In heliotron/torsatron devices, it is difficult to produce toroidal rotation large enough to significantly change the radial electric field because of parallel viscosity. For example, the Compact Helical System (CHS) is a heliotron/torsatron device (poloidal period number l = 2, and toroidal period number m = 8) with a major radius (R) of 95 cm and an average minor radius (a) of 20 cm. The magnetic field ripple near the center of the helical coils (R=90-95cm) is negligible, however, it increases sharply for R >95 cm and reaches 8 % at R = 101.6 cm. Therefore the magnetic field ripple at the plasma center can be modified from zero to 8% by shifting the magnetic axis R_{ax} from 89.9cm to 101.6cm and the damping of toroidal rotation due to TTMP can be studied[13].

Figure 6(a) shows toroidal rotation velocity profiles as a result of a major radius scan (R_{ax}=89.8, 94.9, 97.4cm), which is controlled by the vertical field strength in CHS heliotron/torsatron. The plasma is produced initially by electron cyclotron heating (ECH) in hydrogen gas and sustained with tangential NBI (absorbed power of 0.5 MW in the direction parallel to the helical current). The line-averaged density reaches about 2×10^{13} cm^{-3} after NB injection. We define the modulation of the magnetic field strength γ as $\gamma^2 = <(\partial B/\partial s)^2>/B^2$, where s is the length along the magnetic field line and <> is a flux surface average operator. According to neoclassical theory[12,14], the viscosity coefficient of parallel viscosity can be expressed as $\mu_\parallel = <B\nabla\Pi>/(Bm_i n_i v_\phi) \propto \gamma^2 (R/M) v_{th}$, with the modulation of the magnetic field strength γ, the major radius R, the toroidal period number M and thermal velocity v_{th}. The parameter dependence of the viscosity is studied by changing the field ripple to check whether it is neoclassical or not. The damping of toroidal rotation velocity due to charge exchange loss can be neglected except for the plasma periphery. Here we introduce the effective parallel viscosity μ_{eff} as an indication of how strong the damping of central velocity is by parallel and perpendicular viscosities, $\mu_\parallel$ and $\mu_\perp$, in the plasma. Effective viscosity μ_{eff} is defined as $\mu_{eff}^{-1} = v_\phi(0) m_i n_e(0)/f_{NBI}(0)$, where $f_{NBI}(0)R$ is the torque due to neutral beam injection. If there is no perpendicular viscosity this effective viscosity is equal to the parallel viscosity, $\mu_{eff} = \mu_\parallel$. Figure 6(b) shows the inverse of the effective viscosity as a function of magnetic field ripple. The effective parallel viscosity μ_{eff} shows the γ^2 dependence as predicted by the neoclassical theory in the region where the neoclassical parallel viscosity becomes dominant, $\gamma > 0.2$. When the modulation of B decreases below 0.2, the neoclassical parallel velocity becomes small and anomalous perpendicular viscosity becomes dominant.

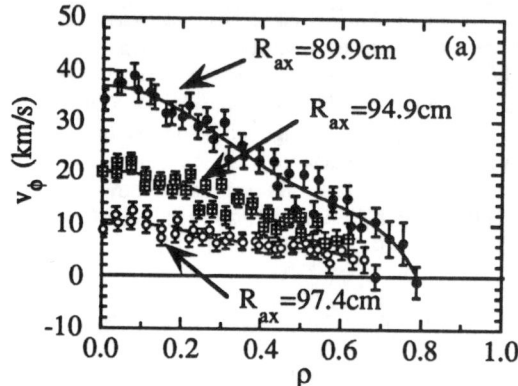
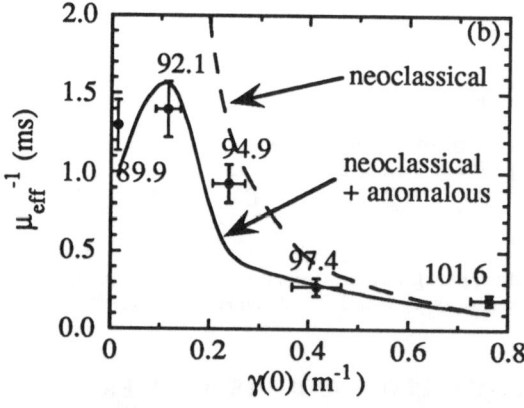

Fig.6 Radial profile of toroidal rotation velocity and inverse of effective viscosity as a function of magnetic field modulation strength γ with the prediction with neoclassical parallel viscosity and neoclassical parallel viscosity + anomalous perpendicular viscosity ($\mu_\perp$=2m^2/s) in CHS torsatron/heliotron.

In order to check confirm CHS results, the measured toroidal rotation velocity is compared with neoclassical prediction in the Wendelstein 7-AS by changing the mirror ratio[15]. The central toroidal rotation velocity is measured using charge-exchange spectroscopy on a single chord and the measurement is line-of-sight integrated within the beamwidth. In order to check that this toroidal rotation is due to the momentum input of the neutral beam, the direction of the neutral beam is switched from co to counter relative to the direction of the bootstrap current during a discharge. Figure 7 shows the difference of toroidal rotation velocities between co and counter-injection phase as a function of the mirror ratio measured and predicted by neoclassical theory. Comparison of measured toroidal rotation velocity with neoclassical theory shows reasonable agreement in cases of high ripple. When the magnetic field ripple is small, measured toroidal rotation is larger than that predicted, because of anomalous perpendicular viscosity like in CHS.

injection is slightly smaller than that for the other one, which is due to the density peaking at the counter-injection phase. The $v_\phi = 0$ line is determined from the spectra of two sets of optical fiber arrays viewing the plasma from opposite directions (on and off the neutral beam line). These transport analyses also show the existence of an offset momentum in the counter direction. It is still open to question which mechanism produces this offset momentum in counter direction. However, preliminary transport analysis based on the assumption that the offset momentum is driven by the ion temperature gradient, was reported[16]. The offset of toroidal rotation will be described later. The perpendicular viscosity on axis in JFT-2M is $0.1 m^2/s$ which is relatively small compared to the viscosities measured in a heliotron/torsatron described later. Therefore, toroidal rotation and radial electric field profiles can easily be controlled by the tangential neutral beam injection.

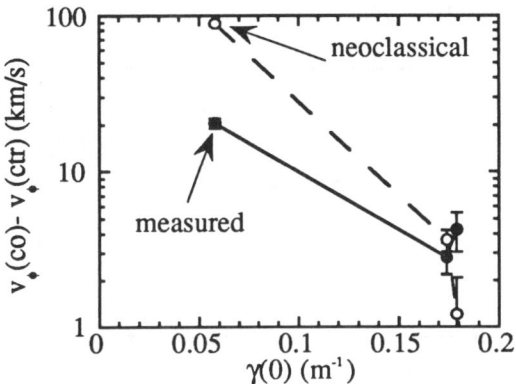

Fig.7 Comparison of central toroidal rotation velocity measured and expected with neoclassical theory in Wendelstein 7AS stellarator.

IV-2. PERPENDICULAR VISCOSITY

Figure 8(a) shows a time evolution of the toroidal rotation velocity at the plasma center in this experiments. Due to an offset of co- and counter-rotation between 560 and 610ms, the perpendicular viscosity is determined from the temporal evolution of the toroidal rotation after the switch from co- to counter- or counter- to co- injection. In tokamaks, the toroidal magnetic field ripple is small enough that the damping of toroidal rotation by TTMP (transit time magnetic pumping) can be neglected but the radial diffusion of toroidal momentum due to the perpendicular viscosity is the dominating effect.

Figure 8(b) shows the radial profile of perpendicular viscosity from a transport analysis for toroidal rotation purely damped by perpendicular viscosity. Similar profiles of perpendicular viscosity are obtained both for discharges switched from co- to counter-injection and counter -to co-injection. The perpendicular viscosity for counter- to co-

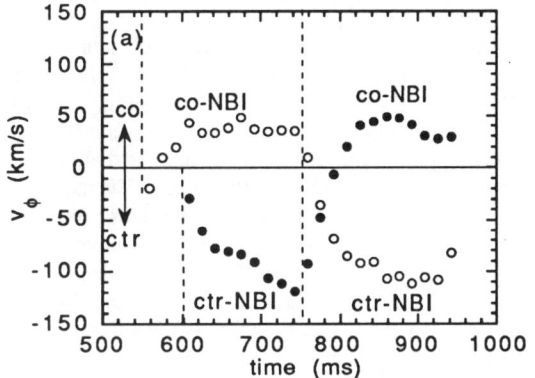

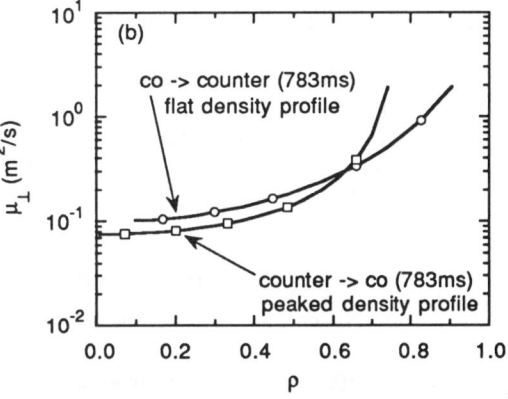

Fig.8 (a) Time evolution of central toroidal rotation velocity and (b) radial profiles of perpendicular viscosity determined at the transient phase of co to counter and counter to co injection in JFT-2M tokamak.

In general, the anomalous perpendicular viscosity is dominant in toroidal rotation and neoclassical parallel viscosity is dominant in poloidal rotation in tokamak. It should be noted that the perpendicular viscosity can be also important even in the poloidal rotation when there is strong poloidal rotation velocity shear such as observed in the plasma periphery in H-mode[17]. The poloidal rotation profile measurement shows that the magnitude of perpendicular viscosity for poloidal rotation at the plasma edge is comparable to the neoclassical parallel viscosity due to the large velocity shear[18].

V. EFFECT OF A RADIAL ELECTRIC FIELD ON PLASMA CONFINEMENT

Since the observation of density peaking in the Alcator-C pellet injection experiments[19], the improvement of core confinement has been recognized to be associated with a peaked electron density Many improved confinement modes, such as the super shot in TFTR[20], IOC-mode in ASDEX[21], improved L-mode in JFT-2M[22] are related to a density peaking. A density peaking model due to a radial electric field has been proposed to explain these observations.

The radial electric field is controlled by changing the direction of NBI from co- to counter- relative to the plasma current in the JFT-2M tokamak as shown in Fig. 9(a). The radial flux of particles is estimated from the beam fueling and time derivative of electron density measured with Thomson scattering and FIR interferometry. The radial particle flux is estimated from the particle source (beam fueling near at the plasma core) and time derivative of measured electron density $\partial n_e/\partial t(t)$. In co-NBI phase, the outward particle flux is balanced by beam fueling. In ctr-NBI phase, however, the particle flux is inward and the density profiles keeps peaking at the plasma center as shown in Fig 9(b). In the nearly steady sate phase (140ms after the change of beam injection direction) of co and counter NBI, the electron density profile in counter-NBI phase is much peaked at the plasma center than that in co-NBI as shown in Fig.9(c).

This experiment indicates that the improved particle transport is related to the negative electric field. This density peaking can be explained by the inward pinch model[23]. According to this mode, the radial flux is determined by the temperature and density gradient, radial electric field, and drift type fluctuations as :

$$\Gamma = -D\left(\frac{\partial n}{\partial r} + \frac{\alpha \partial T}{\partial r}\frac{n}{T} - \frac{eE_r}{T}n + \frac{\omega B_t}{(m/r)T}n\right)$$

where D is a diffusion coefficient, α is a numerical coefficient of order unity ω and m/r denote the frequency and wave number of the mode for drift-type micro-turbulence. The positive radial electric field produce outward flow, while the negative radial electric field produce inward flow, which is consistent with the measurements in JFT-2M. The diffusion coefficient D required to reproduce the time evolution of measured electron density is 0.02m²/s. The magnitude of the diffusion coefficient is consistent with the low diffusion coefficient measured in JFT[24,25].

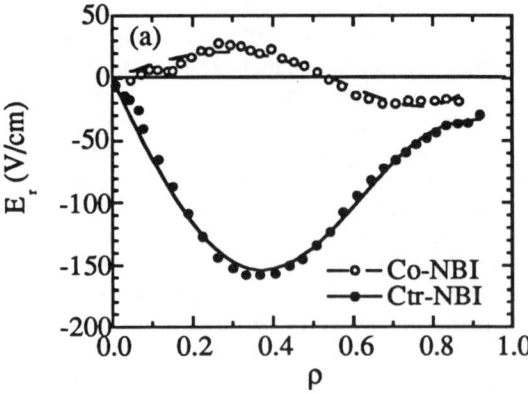
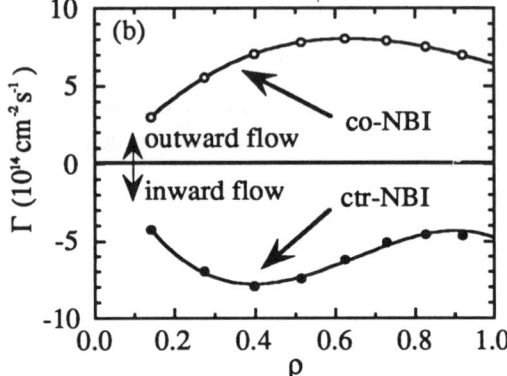
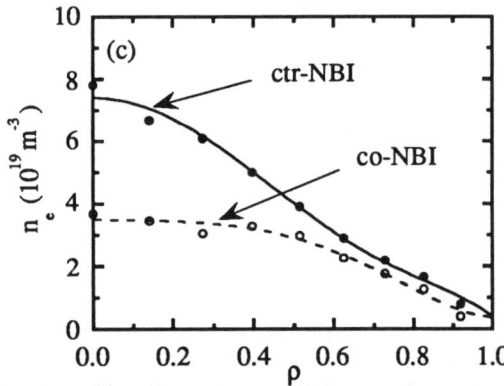

Fig.9 Radial profiles of (a) radial electric field and (b) particle flux, and (c) electron density profiles for co-NBI and counter-NBI in JFT-2M tokamak.

The energy as well as particle confinement is observed to be improved by a negative electric field. Figure 10(a) shows the time evolution of central ion and electron temperatures with co and counter injected neutral beams in the JFT-2M tokamak. The co-NBI is active from 550 ms to 750 ms with an absorbed power Pabs of 0.49 MW, and the ctr-NBI is on from 750 ms to 950 ms with P_{abs} = 0.56 MW. The central ion temperature increases from 0.7keV to 1 1keV in the counter injection phase. The increase of the volume averaged ion temperature, $<T_i>$, as well as of the central ion temperature, indicates the improvement of global energy confinement, which is also supported by the measured energy confinement time of 18 ms for the co-NBI phase and 24 ms (1.5 times) for the ctr-NBI phase. Figure 10(b) shows the effective thermal diffusivity estimated at the coinjection phase (t=690ms) and counter injection phase (t=890ms). The effective thermal diffusivity for the counter injection phase is reduced by the factor of three at the plasma center and 30-50% at the plasma edge relative to the co injection phase. The experiment in JFT-2M shows that the negative electric field due to counter injected neutral beams causes an inward particle pinch and the peaked density profile reduces the heat transport. The mechanism of improved confinement in ctr-NBI have similarities to that in H-mode in terms of E_r and E_r shear[26,27].

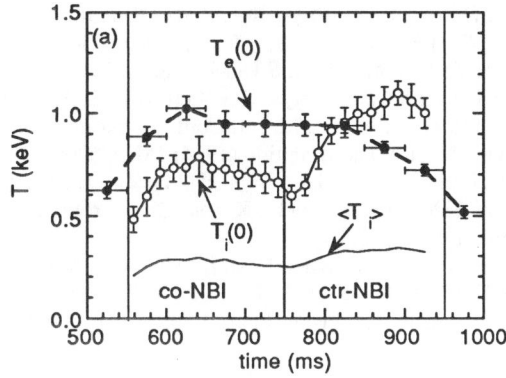

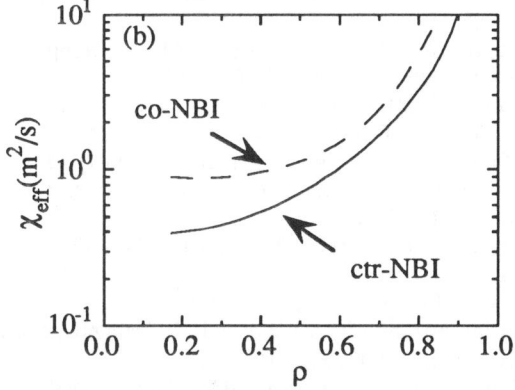

Fig.10 (a)Time evolution of electron and ion temperature and (b) radial profiles of effective thermal diffusivity for co-NBI and ctr-NBI in JFT-2M tokamak.

To establish a causal link between electric field and density peaking, density peaking parameter $n_e(0)/<n_e>$ and ion temperature peaking, $T_i(0)/<T_i>$, where $<>$ is volume averaged, are plotted as a function of central toroidal rotation velocity (positive and negative velocities yield positive and negative radial electric, respectively) as shown in Fig. 11. If the toroidal rotation velocity and radial electric field and density peaking or temperature peaking take place simultaneously, the time trance should be on one line. When there is causality between these two parameter, the time trace becomes circular and the direction of rotation (clockwise or counter clockwise) shows which is first. The measured data clearly shows that the change of toroidal rotation velocity and radial electric field causes the particle inward pinch and the reduction of thermal diffusivity.

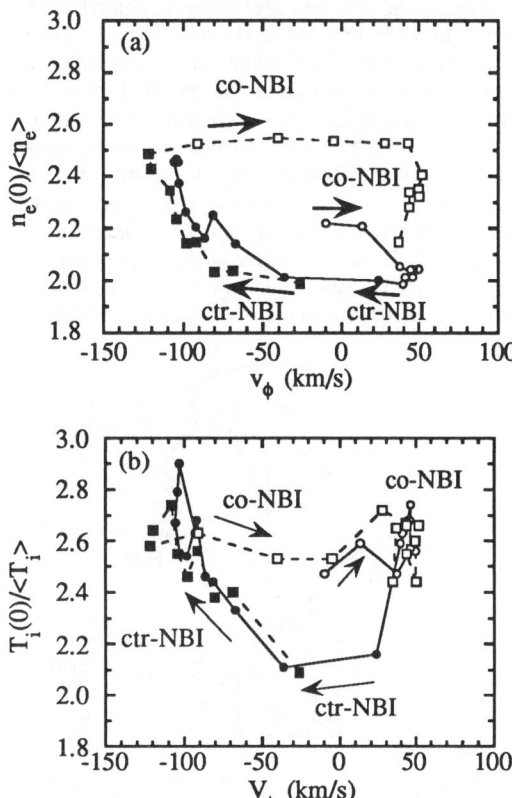

Fig.11 (a)Density peaking parameter and (b)ion temperature peaking parameter as a function of central toroidal rotation velocity for co-NBI and counter-NBI in JFT-2M tokamak

VI. DISCUSSIONS

The mechanism determining the radial electric field and the effect of the radial electric field on transport is summarized in Fig 12. The radial electric field is connected to plasma rotation through the radial force balance. Therefore, the radial electric field in the plasma is determined by the balance between poloidal or toroidal driving forces such as beam injection, bipolar losses due to ECH or NBI, and damping force such as parallel and perpendicular viscosities and charge exchange losses due to collisions with neutrals in the plasma. From measurements of toroidal rotation in helical devices (heliotron/torsatron and stellarator), the parallel viscosity is found to be close to the neoclassical predictions. In tokamaks, where there is almost no parallel viscosity in the toroidal direction, the experimentally determined perpendicular viscosity is found to be anomalously high. We emphasize the importance of understanding the viscosity of plasma rotation, because surface plasma rotation stabilizes ideal kink mode by wall stabilization and increases β-limit in tokamak[28].

It has been experimentally confirmed that the gradients of the radial electric field and/or plasma rotation affect the particle and heat transport. A particle pinch (inward flow) is experimentally found to be triggered by a negative electric field shear driven by external momentum input of a counter injected neutral beam. The thermal transport is reduced associated with a strong negative electric field shear. These observations indicate the importance of off-diagonal terms between momentum transport and particle/heat transport in the transport matrix. These off-diagonal terms in the transport matrix enable us to improve the confinement by the direct (momentum injection) or indirect (heating) control of the electric field. A quantitative study of diagonal and off diagonal terms in the future will improve our understanding of anomalous transport and improved plasma confinement.

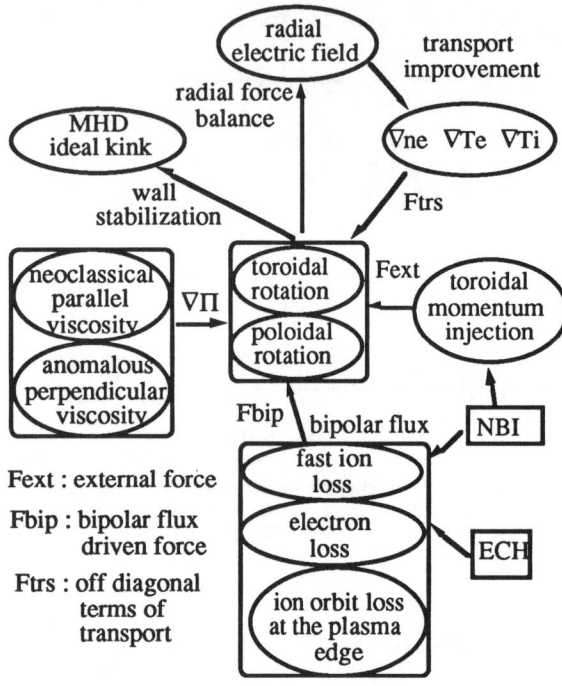

Fig.12 Physical mechanisms determining the radial electric field in a toroidal plasma.

ACKNOWLEDGEMENT

This paper is dedicated to the memory of Dr. Hikosuke Maeda (Japan Atomic Energy Research Institute). The authors acknowledge Drs. K.Itoh, S.-I.Itoh (Kyushu Univ.), A.Fukuyama (Okayama Univ.), H.Idei, H.Sanuki for their useful discussions. The authors also thank Drs. T.Hamada, K.Toi, K.Kawahata, and NBI group of JIPP T-IIU experiments, and Drs. K.Matsuoka, S.Okamura, ECH and NBI groups of CHS experiments, machine operation group of JFT-2M experiments, Mr. T.Hamada and H.Funaba for the transport analysis in Heliotron-E, and Mr. M.Kojima and Mr. C.Takahashi for data acquisition for these experiments.

[1] F.Wagner, G.Becker, K.Behringer, D.Campbell, A.Eberhangen, et al., Phys. Rev. Lett. **49**, (1982) 1408.
[2] S.-I.Itoh and K.Itoh, Phys. Rev. Lett. **60**, (1988) 2276.
[3] K.C.Shaing and E.C.Crume Jr, Phys. Rev. Lett. **63**, (1989) 2369.
[4] R.J.Groebner, K.H.Burrell, and R.P.Seraydarian, Phys. Rev. Lett. **64**, (1990) 3015.
[5] K.Ida, S.Hidekuma, Y.Miura, T.Fujita, M.Mori, et. al., Phys. Rev. Lett. **65**, (1990) 1364.
[6] A.R.Field et al., Nucl. Fusion **43**, (1992) 1191.
[7] V.Erckmann, F.Wagner, J.Baldzuhn, R.Brakel, R.Burhenn, et. al., Phys. Rev. Lett. **70**, (1993) 2086.
[8] R.J.Fonck, M.Finkenthal, R.J.Goldston, D.L.Herndon, R.A.Hulse, et. al.r, Phys. Rev. Lett. **49**, (1982) 737.
[9] K.Ida, S.Hidekuma, Rev. Sci. Instrum. **60**, (1989) 867.
[10] K.Ida, S.-I. Itoh, K.Itoh, S.Hidekuma, Y.Miura, H.Kawashima, etal., Phys. Rev. Lett. **68**, (1992) 182.
[11] T.Obiki, et al., Porc. 15th Int. Conf. on plasma phys. and Contl. Nucl. Fusion, Seville (1994) IAEA-60/A-6-I-2.
[12] K.C.Sgaing, J.D.Callen, Phys. Fluids. **26** (1983) 1526.
[13] K.Ida, H.Yamada, H.Iguchi, K.Itoh, CHS Group, Phys. Rev. Lett. **67**, (1991) 58.
[14] N.Nakajima, er al., Nucl. Fusion. **29** (1989) 605.
[15] J.V.Hofmann,K.Ida, et al., 21th EPS Conference on Contl. Fusion and Plsma Phys. Montpellier, (1994) Or-22
[16] K.Ida, K.Itoh, S.-I.Itoh, Y.Miura, National Institute for Fusion Science report NIFS-241 (1993).
[17] M.Tendler and V.Rozhansky, Comments Plasma Phys. Controlled Fusion **13**, (1990) 191.
[18] K.Ida, K.Itoh, S.I.-I.Itoh, Y.Miura, JFT-2M Group, A.Fukuyama, Phys.plasma **1**, (1994) 116.
[19] M.Greenwald, D.Gwinn, S.Milora, J.Paker, R.Parker, S.Wolfe, et al., Phys. Rev. Lett. **53**, (1984) 352
[20] J.D.Strachan et al., Phys. Rev. Lett. **58**, (1987) 1004.
[21] F.X.Soldner, E.R.Muller, F.Wagner, et al., Phys. Rev. Lett. **61**, (1988) 1105.
[22] M.Mori, N.Suzuki, et al., Nucl. Fusion **28**, (1988) 1891.
[23] S.-I. Itoh, J.Phys. Soc. Jpn. **59** (1990) 3431.
[24] D.Pasini, M.Mattioli, A.W.Edwards, R.Giannella, et al., Nucl. Fusion **30**, (1990) 2049.
[25] R.Giannella, et al., in Controlled Fusion and Plasma Heating (Proc. 18th European Conf. Berlin), Vol 15c, Part I, p.197, European Physical Society, Petit-lancy, Switzerland, (1991).
[26] K.H.Burrell, T.N.Carlstrom, E.J.Doyle, Phys. Fluids **B2** (1990)1405.
[27] K.Ida, S.Hidekuma, M.Kojima, Y.Miura, S.Tsuji, et. al., Phys. Fluids. **B4**, (1992) 2552.
[28] R.D.Stambaugh and D-III-D Team, Porc. 15th Int. Conf. on plasma phys. and Contl. Nucl. Fusion, Seville (1994) IAEA-60/A-1-I-4.

Theoretical and Experimental Studies of the Scrape-Off Layer Biasing in a Divertor Tokamak

M.Tendler*, J.-L. Lachambre & the T de V TEAM

Centre Canadien de Fusion Magnetique, 1804 Montee Ste - Julie, Varennes, Quebec, Canada.
* *Permanent Address Alfven Lab., Association Euratom - NFR, Royal Institute of Technology, Stockholm, Sweden*

In this review paper the effects of an electric field are addressed. The focus is on open field lines in a tokamak. Electrically biasing the divertor plates of TdeV with respect to the vacuum vessel sets up a radial electric field and a radial current between the separatrix and the wall. It has been demonstrated that the resulting $\mathbf{E} \times \mathbf{B}$ flow and the $\mathbf{j} \times \mathbf{B}$ force produce fundamental phenomena, affecting confinement on closed field lines, parameters of the SOL and the performance of a divertor.

I. INTRODUCTION.

At present, plasma edge physics is emerging as a major domain of tokamak physics research. On one hand the edge plasma interfaces with the core plasma and on the other it interfaces with material boundaries. Power and particle exhaust are widely recognized as the most difficult issues faced by a viable fusion reactor. Yet, the essential issues in reactor regimes with improved confinement concern transport not only on the closed field lines inside a separatrix, but also on open field lines within the scrape - off layer (SOL). Therefore, to reconcile conflicting requirements such as achieving the high recycling regime or detachment in front of the divertor plates and the equalization of the power loads with improved confinement regimes appears to be a very difficult problem calling for unconventional solutions. In other words, it may require additional control knobs, for example, the biasing of divertor plates with respect to the first wall.

Certain aspects of biasing appear easy to discern. The poloidal component of the $\mathbf{E} \times \mathbf{B}$ drift will move the SOL plasma into a divertor chamber favoured by the direction of the drift. The toroidal torque provided by the $\mathbf{j} \times \mathbf{B}$ force will clearly set up a toroidal motion. However, biasing experiments carried out on different tokamaks have demonstrated many remarkable and surprising effects. The impact of biasing on confinement on closed field lines inside the separatrix, the changes of the SOL parameters resulting from it and the ensuing changes in the divertor performance have been demonstrated. Biasing one divertor plate with respect to another drives a current parallel to the magnetic field in the SOL /1/. In contrast, biasing the separatrix strike point relative to the grounded vacuum vessel results in the radial component (normal to flux surfaces) of the electric field being controlled externally. Here, we aim to address only the latter case termed unipolar biasing. This scheme has been successfully implemented on the double null configuration of TdeV /2/. Therefore, this experiment is chosen as a reference here.

II. THE IMPACT OF BIASING ON CONFINEMENT.

Recently, it has been demonstrated on DIII - D and JFT - 2M, that negative biasing of the SOL reduces dramatically the power threshold needed to trigger the L - H transition /3/. Furthermore, TdeV has reported changes in the central plasma impurity levels consistent with changes in confinement /2/. Moreover, ASDEX /4/ and later JET /5/ have shown that the power thresholds for obtaining H modes are significantly reduced when the ion grad - B drift was directed towards the target in a single null configuration with the penalty of enhanced asymmetry between heat loads deposited on the inner and outer plates. The ion grad- B drift causes the parallel flow in a tokamak, in a similar way to the toroidal flow caused by the $\mathbf{j} \times \mathbf{B}$ force produced by unipolar biasing. In this paper we show that the two effects originate from the same physical phenomena.

Addressing the link between biasing and improved confinement, the following insights appear fundamental. Rotation of a Navier - Stokes fluid resulting in stabilization due to decorrelation of turbulence by sheared flows is a well known effect in fluid mechanics /6/. However, since a sheared flow also provides free energy which can drive the Kelvin - Helmholtz instability, the effect is not frequently observed in fluids. On the other hand, in magnetized plasmas, the stabilization of the Kelvin - Helmholtz instability is easily provided by shear in the magnetic field, thereby facilitating the suppression of turbulence by $\mathbf{E} \times \mathbf{B}$ shear (laboratory frame) in fusion devices based on magnetic confinement /7/. To this end, at present, a large body of evidence obtained on many tokamaks shows that the formation of a transport barrier at the

© 1995 American Institute of Physics

L - H transition is consistent with the hypothesis that increased shear in the edge radial electric field E_r leads to a suppression of turbulence, thereby improving confinement. This explanation of the H - mode confinement improvement has lead to a fundamental contribution to the physics of fluids. Indeed, it has amplified the world wide effort put into studies of the origin and the role of the natural ambipolar electric fields and also encouraged the search for methods to impose electric fields on a plasma by external biasing.

An important area of research is the origin of the electric field E_r at the plasma edge (in particular, the change in E_r occuring during the L -H mode transition) and a search for options to control it artificially. Experimentally, E_r is normally derived from the measured pressure, poloidal and toroidal velocity profiles employing the radial component of the momentum balance equation

$$E_r = \frac{1}{en}\frac{\partial p}{\partial r} - B_\phi V_\theta + B_\theta <V_\parallel> \quad (1)$$

where $<V_\parallel>$ is the flux surface average parallel (or toroidal) velocity, a point often overlooked in the literature. According to neoclassical theory, parallel viscosity couples the parallel and perpendicular flows, resulting in a large parallel flow so that its poloidal component cancels the poloidal $\mathbf{E} \times \mathbf{B}$ and diamagnetic drifts, yielding the poloidal rotation $V_\theta \approx 0$ if the radial temperature gradient is ignored. Hence, the parallel flow plays a major role in a tokamak and is a natural feature of a tokamak due to the grad - B drift inherent to toroidal geometry. Indeed, the vertical current associated with the grad - B drift is shortcircuited by the Pfirsch - Schluter current, thereby affecting the distribution of the electric field. The parallel flows associated with this current (termed here as the Pfirsch - Schluter flow) must undergo cross - field transport within the tokamak plasma. This is essentially angular momentum, the transport of which is grossly neglected within the framework of neoclassical theory.

In general, it is well known that the cross - field transport of energy and particles is largely anomalous in a tokamak. Yet, the anomaly of the cross - field transport of angular (parallel and/or toroidal) momentum is seldom studied. Parallel flows in a tokamak may be imposed externally by an unbalanced neutral beam injection and by biasing, as shown below. In contrast to auxiliary neutral beam heating, the external $\mathbf{j} \times \mathbf{B}$ torque created by biasing affects only the toroidal momentum balance, resulting in acceleration of plasma toroidally. The toroidal component of momentum balance becomes dominant, thereby permitting the testing of model equations and discrimination between different models. To this end, it is important to emphasize that for internal electrode biasing experiments neoclassical theory predicts that the poloidal velocity must remain neoclassical because the parallel component is unchanged and only the toroidal rotation velocity can be amplified due to the $\mathbf{j} \times \mathbf{B}$ torque. Exactly the opposite - fast poloidal rotation and slow toroidal rotation - is observed experimentally /8,15/. When the electric current in TUMAN 3 is 200 A the volume integrated $j_r B_\theta$ force is about 1.2 Newton, which corresponds to momentum input from 2 MW neutral beam injected power on D III - D at 20 kV. Therefore, a credible model has to provide a universal approach to both the natural and the bias - probe - induced H modes. This requires the use of both anomalous and neoclassical transport of angular momentum.

Hence, in /10/ it was predicted that the off - diagonal terms describing the anomalous inertia and viscosity, which provide for the anomalous transport of angular momentum beyond the effects of the neoclassical parallel viscosity (magnetic pumping) are crucial to explain the L - H transition. Having established that the anomalous radial transport of the Pfirsch - Schluter flows has to be included at the edge in order to account for the steep gradients in density and temperature profiles, one obtains from the parallel component of ion momentum balance

$$<\mathbf{B}\cdot nm_i \frac{d\mathbf{V}_i}{dt}> = -<\mathbf{B}\cdot\nabla\cdot\overleftrightarrow{\pi}> \text{ (NEO)}$$
$$-<\mathbf{B}\cdot\nabla\cdot\overleftrightarrow{\pi}> \text{ (AN)} \quad (2)$$

Considering only the Pfirsch - Schluter flow (i.e. retaining for simplicity only the Ohmic H - mode and no biasing) and the neoclassical parallel viscosity,

$$V_{i\parallel} = -2q(V_0 + V_{pi})\cos\theta \quad (3)$$

where V_0 and V_{pi} are the electric and diamagnetic drift velocities, we obtain the following fundamental equation

$$A\frac{d^2 V_\theta}{dr^2} - B\frac{dV_\theta}{dr} + C(V_\theta - V_\theta^{(NEO)}) = 0 \quad (4)$$

where the coefficients A,B and C are strong functions of the anomalous transport of the parallel momentum. Their exact form is given analytically in /10/. The fundamental conclusion is that if the first two terms, which appear in Eq.(4) due to the anomalous transport of parallel momentum, are of the same order as the third term, which depends on the neoclassical parallel viscosity, then the H mode may unfold. In contrast, if the third term dominates inside the separatrix in a tokamak, the discharge remains in the L regime.

The boundary condition on the separatrix, which must be specified in order to obtain a solution, also affects improved confinement. In more detail, it has been shown in /10/ that the solution of Eq.(4), providing for large values of dV_θ/dr and d^2V_θ/dr^2, can be obtained only if the electric field on the separatrix differs significantly from its neoclassical value, which is determined by a combination of the density and temperature gradients. Therefore, the shear in the electric field at the edge depends dramatically on the value at the separatrix, or very close to it (quantitavely, within one poloidal Larmor radius). It may be affected either directly by biasing the separatrix from the inside or the outside /2/, or indirectly by establishing a steep density profile on a separatrix /11/ within the scale length of the poloidal Larmor radius. The latter can be achieved by injecting small pellets on the separatrix /8,9/ or due to spontaneous changes in recycling. The resulting squeezing of the ion banana widths increases the steepness of the electric field profile because the ion diamagnetic flow exceeds contributions from poloidal and toroidal rotation. As a consequence, a strong velocity shear layer and a well localized potential maximum are formed.

In conclusion, the negative well in the electric field profile on the separatrix is crucial for the L - H transition and may be provided either by establishing a steep density profile ($L_{n_i} < \rho_{\theta i}$) on the separatrix /12/ or by imposing an electric field there using the unipolar biasing. The negative well is deepened by a reversal of the natural electric field within the SOL, which is always positive, for the case of negative biasing. This result is illustrated by the reduction of the H - mode power threshold by negative SOL biasing as demonstrated on DIII - D and JFT - 2M and by orienting the ion grad - B direction towards a single null as reported on ASDEX and JET. Addressing the observation on TdeV concerning the lower impurity content due to negative biasing of the SOL, one must bear in mind that a given change in the electric field profile affects impurities more dramatically than main ions because the $\mathbf{E} \times \mathbf{B}$ drift approaches the poloidal sound speed $\Theta\, c_{SI}$ earlier for impurities because of the mass difference. However, the quantitative theory of impurity transport in a strong radial electric field in a tokamak remains to be developed.

In summary, the boundary condition comprising a well defined bell shaped maximum of the space potential on the separatrix (or the Last Closed Flux Surface) and the impact of off - diagonal terms such as anomalous inertia and viscosity which provide a wide barrier region of large shear within the separatrix are asserted to be the prerequisites for improved confinement.

III. THE EFFECT OF BIASING ON THE SOL PLASMA.

Changes in confinement should therefore be explained by modifications in the radial transport of parallel flow resulting due to biasing. Bearing this in mind, it is important to explain how the $\mathbf{j} \times \mathbf{B}$ force applied either to the central or to the SOL plasma with no parallel component can affect the parallel motion. Note that the standard neoclassical theory implicitly neglects any effect. However, the toroidal angular momentum, resulting from the $\mathbf{j} \times \mathbf{B}$ force, has to be balanced by a force resulting from the transport of the angular momentum. This force is provided by anomalous viscosity and inertia, friction with the primary neutrals (arriving directly from the wall, at rest in the laboratory frame of reference, not the charge - exchanged neutrals), perpendicular viscosity due to Coulomb collisions, and surface dissipation on collecting surfaces and side walls. In particular, anomalous damping mechanisms are essential on closed field lines because the neoclassical contribution (occuring due to Coulomb collisions) to the damping of toroidal momentum cancels upon averaging over magnetic surfaces and the other effects listed above can be ignored. On the other hand, since $V_\parallel \approx V_\phi$, almost the same terms emerge in the parallel component, where the force provided by anomalous viscosity and inertia has to be balanced either by the poloidal component of the pressure gradient on open field lines /13/ or by the more subtle non - linear neoclassical parallel viscosity on closed field lines (during the L - mode; note that the remaining anomalous viscosity and inertia are of the same order in the H - mode) because the pressure gradient cancels on closed field lines upon averaging over the poloidal direction. Thus, it has been asserted in /13/ that the poloidal pressure gradient dominates on open field lines and therefore has to balance the $\mathbf{j} \times \mathbf{B}$ force. Moreover, it has been demonstrated recently that by applying a biasing voltage from an electrode across closed magnetic surfaces, thereby forcing the plasma to rotate toroidaly due to the torque provided by the $\mathbf{j} \times \mathbf{B}$ force, the L -H transition is provoked and likely results from enhanced parallel flow in accordance with the previous Section.

Addressing the radial current, one would expect little current to flow across magnetic surfaces. However, early experiments on TdeV (see Fig.1) have demonstrated that unexpectedly large currents have been measured to flow across the SOL. This low resistance across magnetic field lines appears contrary to expectations. It can be illustrated by a model for current continuity developed in /14/ and shown in Fig.2. The cross - field current is assumed to be related to the electric field by a mobility type relationship: $j_r = e\mu_\perp n E_r$. Only the radial variations of plasma parameters are retained in the model. The current flows from the biased plates parallel to magnetic field through a sheath, then moves radially across field lines past the plates and finally out along field lines to the grounded wall through the sheath as shown in Fig.2. The radial profiles of the parallel and perpendicular currents are computed from the model as shown in this figure. Employing the experimental radial profiles of plasma density and temperature, both the voltage - current and electric field profiles were demonstrated to yield a perpendicular mobility $\mu_\perp =$ 0.006 m^2/Vsec. Note that boundary conditions at the sheath are important in establishing the radial electric field in the SOL. To this end, it has to be kept in mind that the plasma facing material in the region between r_p and r_w (see Fig.2) is actually a floating conductor on TdeV, contrary to the assumption employed in the model that it is an insulator. On the other hand, the SOL plasma has to be grounded somewhere because otherwise the bias will provide a potential drop on electric field in the vacuum region between the SOL plasma and the wall. Therefore, it would have very little impact on plasma phenomena.

However, this model does not answer the question concerning the physical processes governing the cross - field mobility, and how is it possible that the SOL plasma encounters a very low cross - field resistance. In order to answer these questions, the damping of the toroidal momentum has to be assessed.

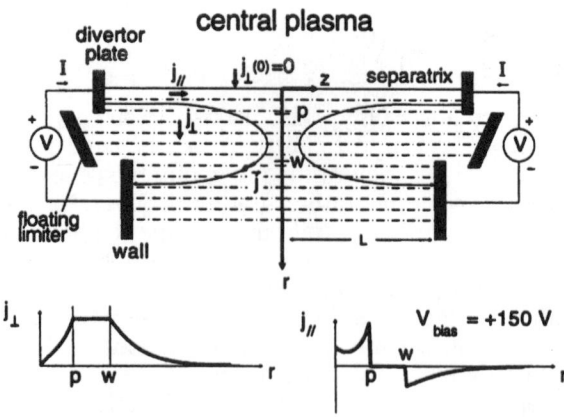

Fig 2. Plasma biasing modelling geometry from Ref.14. The SOL field lines are presented as straight lines bounded at both extremities by three types of electrodes; the neutralization plates, the insulated limiters and the vessel walls. Typical perpendicular and parallel currents are also shown.

Indeed, for a given electric field profile within the SOL (i.e. accounting for the obvious sheath limits imposed by the ion saturation current) the possibility of driving a current across the SOL is primarily limited by the damping of the source of toroidal momentum driven by the torque due to the $\mathbf{j} \times \mathbf{B}$ force. Here, it is assumed that the direct non - ambipolar cross - field transport due to magnetic fluctuations is negligable and the configuration is toroidally axisymmetric. A much smaller direct cross - field conductivity, although amplified due to the Simon effect, is neglected.

In order to quantify the cross - field current, only the damping of the toroidal momentum is addressed. A few damping mechanisms are listed and quantified below. The first to mention are three volume dissipation processes:

the anomalous dissipation of toroidal momentum due to turbulence via anomalous inertia and anomalous viscosity,

charge - exchange processes with primary neutrals, inertia and viscosity due to Coulomb collisions. Note that the Coulomb term cancels on closed field lines upon averaging, but not on open field lines.

In addition, the surface dissipation, as opposed to the volume dissipation discussed above, of the toroidal momentum near a limiter or a divertor plate has to be considered within the SOL. For example, on TdeV, the outermost flux surface makes contact with the grounded guard limiters in front of the first wall.

These effects in the axisymmetric SOL are described by the following equation for the radial current density

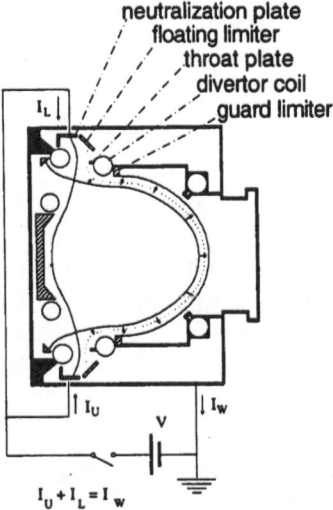

Fig 1. Schematic cross-section of TdeV with plasma biasing electric connections.

$$j_r = \frac{m_i n_i}{B_\theta} (u_i \cdot \nabla) u_{i\phi} +$$
$$+ \frac{1}{B_\theta} \nabla \cdot \overleftrightarrow{\pi} \text{ (AN)} + \frac{1}{B_\theta} \nabla \cdot \overleftrightarrow{\pi} \text{ (COUL)} +$$
$$+ \frac{m_i n_i}{B_\theta} \nu_{ch.-ex.} (u_{i\phi} - V_{n\phi}) +$$
$$+ \frac{m_i}{aB_\theta} (n_i u_{i\theta} u_{i\phi}) |_{C.S.} + \frac{\Theta(r - r_w)}{B_\theta} F$$

The first two terms are due to the dissipation of the toroidal momentum caused by turbulence; the next two are due to classical collisions; and the last two are due to the surface dissipation on the collecting surfaces and side walls, rising very steeply and described by a Heavyside function, Θ. Since the measured SOL widths from both the density and the temperature profiles support that the cross-field particle and energy transport in the SOL are governed by a turbulence, we have good reasons to believe that momentum transport there is also dominated primarily by the first two terms. The notion stems from the ubiquitous basic laws of kinetics and is particularly true in exploiting the present results for extrapolation to future devices. However, significant interaction of the SOL plasma with side structures, such as guard limiters, impeding the flow in a divertor chamber, might amplify the importance of the last two terms. It can be shown that impediments to flow in the divertor throat and elsewhere significantly increase the offset in the voltage-current characteristics. This appears to be the case on TdeV, in particular during positive biasing, since the width of the SOL increases significantly (see below). Note that the magnetization current has to be subtracted from the radial current because it is divergence-free and cannot be extracted from the plasma. Yet, in most cases of practical interest, the contribution of the magnetization current is small.

The scaling law for mobility obtained empirically the TdeV is /14/

$$\mu_\perp = 0.51 \, n^{0.84} \, I_p^{-1.39} \, B_\phi^0$$

while the assumption that anomalous inertia and viscosity dominate yields /13/

$$\mu_\perp \sim n^0 \, I_p^{-1} \, B_\phi^{-1}$$

and that invoking friction with stationary neutrals /15/

$$\mu_\perp \sim n^1 \, I_p^{-2} \, B_\phi^0$$

Surface dissipation on collecting surfaces results in large offsets in the voltage-current characteristics, thereby diminishing the value of empirically derived scalings. In general, offsets obscure experimental scalings. Obviously, all three of these mechanisms are at work at present on TdeV. In a more open divertor configuration, the first mechanism is likely to predominate. Therefore, it is reasonable to employ it for extrapolation to the future devices /16/.

Another surprising effect demonstrated on TdeV is the steepening of the density profile with negative bias and the broadening with positive bias (see Fig3).

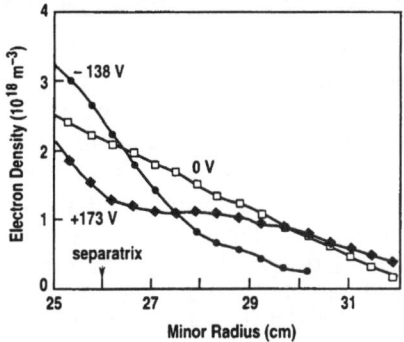

Fig 3. SOL electron density profiles for unipolar bias of the outboard divertor plates for three bias voltages, −138, 0, and +173, on TdeV.
From Ref. 2.

Recall that the shear mechanism for turbulence suppresion does not depend on the sign of the electric field and, indeed, the measurements carried out on TdeV demonstrated that a large bias of either polarity reduces the density fluctuations. Yet, the value of the cross-field current is found to be very large. It is limited either by the ion saturation current at the plates or at obstacles attached to the wall, or by the volume capacity of the SOL and the active divertor. Thus, $I^{(MAX)}/(en)$ may even reach a value of the order of the chaotic particle flow given by Fick's law $u_r = -D \, d(\ln n)/dr$ if the poloidal Larmor radius is roughly the width of the SOL /13/. We suggest therefore that the impact on density profiles borns out the approach to 1. The effect demonstrates that the radial ion flow consists of ambipolar diffusion driven by turbulence weakly affected by biasing, and an equally important ion radial flux providing for the large biasing current.

The fast poloidal rotation expected to result from the significantly enhanced $E_r \times B_\phi$ drift is not observed in the SOL on TdeV. On the contrary, measurements carried out with the Gundestrup Mach probe show that the toroidal rotation at a given radius is much greater than the poloidal rotation. To understand this puzzle, recall that, for a given polarity of the biasing voltage and a given direction of the toroidal magnetic field, the $\mathbf{E} \times \mathbf{B}$ flow, resulting from biasing, enhances or counteracts the sonic flow in the SOL, which is determined by the sheath condition. If the stagnation point of poloidal flow is at the tokamak midplane due to the symmetry of the configuration (the case of a double null configuration on TdeV) particle balance implies that the poloidal component of the parallel flow has to counterbalance the $\mathbf{E} \times \mathbf{B}$ velocity there.

Hence, although the $\mathbf{E} \times \mathbf{B}$ drift amplifies the flow in the poloidal direction, the parallel flow is affected too, thereby providing for a damping of poloidal motion at the midplane of TdeV yielding locally $V_\theta \approx 0$ ($\ll V_\phi$). Bear in mind that, according to neo-classical theory, parallel viscosity couples the parallel and perpendicular flows, resulting in a large parallel flow with a poloidal component canceling the poloidal $\mathbf{E} \times \mathbf{B}$ and diamagnetic drifts and therefore yielding a net poloidal rotation of about zero. In reality, however, the toroidal velocity never exceeds significantly the poloidal rotation velocity, in contrast to predictions of standard neoclassical theory. Hence, the poloidal flow velocity remains slow on open field lines and fast on closed field lines. The conclusion yields the scaling $V_\phi \sim 1/B_\theta$ confirmed on TdeV at the midplane with an offset due to the natural ambipolar electric field and the diamagnetic drift. More importantly, the toroidal rotation of the main plasma ions, when measured at the midplane very close to the separatrix on TdeV, is a very strong function of the biasing voltage. It is linear for small biasing voltages. Measurements indicate also that the condition $V_\theta \approx 0$ is also verified at the midplane. Yet, saturation is seen to occur in the toroidal rotation at the midplane for larger voltages of both polarities, strongly suggesting that the non-linear off-diagonal terms are responsible for the damping of toroidal momentum under these conditions.

Finally, in order to reconcile the parallel and toroidal components of momentum balance, the parallel gradient of the pressure $\nabla_\parallel p$ has to balance the $\mathbf{j} \times \mathbf{B}$ force, because both these terms are affected by almost the same terms which are responsible for dissipation of angular momentum. Indeed, qualitatively, both the electron density and the neutral pressure increase significantly in the "active" divertor chamber (favoured by the direction of the $\mathbf{E} \times \mathbf{B}$ drift). However, experimentally, this theory is confirmed only with negative polarity of biasing, but not with positive polarity on TdeV. The reason for the asymmetry may be due to significantly enhanced friction with the guard limiters on TdeV (due to dissipation on the side walls) because the SOL is much wider with positive polarity, thereby invalidating the ideal poloidal force balance. Yet, a more open configuration foreseen for the upgrades will be beneficial for maintaining this balance with both polarities.

At any rate, the improvement in the active divertor performance is quite spectacular (see Fig.4), facilitating access to a high recycling regime in the divertor chamber with significantly enhanced retention of both heavy and light impurities.

Therefore, biasing offers the option of maintaining a high density and low temperature plasma in the diveror chamber favoured by the $\mathbf{E} \times \mathbf{B}$ drift, which can be controlled externally. This option is particularly attractive because the operation in H-mode creates a large asymmetry in the distribution of the heat load between inner and outer divertor plates, resulting in the overloading of one of the target plates in the SN configuration. On the other hand, the question of synergy of divertor biasing with the improved confinement remains to be explored.

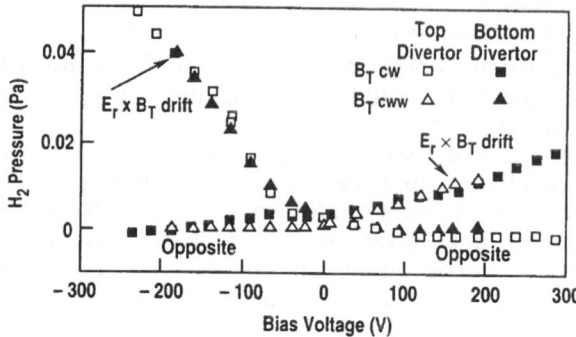

Fig 4. Bias induced neutral pressures in the two divertors of TdeV for two directions of the toroidal magnetic field B_T. The divertor in the direction of the $E_r \times B_T$ drift and the divertor in the opposite direction are indicated. From Ref. 2.

The electron density and the neutral pressure (see Fig.4), in the active divertor, increase linearly with biasing voltage with, however different slopes depending on the biasing polarity. There is an approximately three-fold asymmetry, demonstrating that the negative bias is more favorable in this respect. The probable reason for this has been described above. Quantitatively, at $V_{bias} = -200$ V , the neutral pressure has increased by a factor 5 and the divertor line density by about 40 %.

In summary, divertor plates as electrodes are efficient in generating radial fields and currents within the SOL, in lowering the H-mode power threshold, thereby facilitating regimes with improved confinement, in building up divertor pressure and in driving toroidal rotation. They can be implemented in future reactor alleviating a stringent requirement for an auxiliary power by lowering the H-mode threshold and facilitating successful radiative divertor operation at a lower central plasma density consistent with current drive. In contrast, inactive divertors cannot perform these functions. These and other attractive features justify future work in this area. On the other hand, both neoclassical and anomalous processes have to be included in order to understand the natural and applied electrostatic potential effects on confinement and divertors in tokamaks.The transport of angular momentum appears to be fundamental in the explanation of these experiments. Furthermore, the level of understanding of the physics fundamentals governing the performance of biasing experiments has made sufficient progress to guide present day experiments and to extrapolate the trends to future devices. Yet, there are still many outstanding issues to be addressed both experimentally and theoretically.

ACKNOWLEDGEMENTS

The authors wish to thank the entire TdeV team, in particular Drs. R.Decoste and B.C.Gregory, and Dr.V.Rozhansky, St.Petersburg Technical University for helpful discussions.

The work of one of the authors (M.T.) was supported by the European Communities under an association contract between Euratom and Sweden.

The Centre canadien de fusion magnetique is a joint venture of Atomic Energy of Canada Ltd., Hydro-Quebec and the Institut national de la recherche scientifique, in which MPB Technologies Inc. and Canatom Inc.also participate. It is funded by AECL, Hydro-Quebec and INRS.

REFERENCES.

1. G.M.Staebler, Divertor Bias Experiments, Invited paper, 11th Int.Conf.Plasma Surface Interactions in Controlled Fusion Devices, 23 - 27 th May 1994, Ibaraki, Japan to be published in J.Nuc.Mat.and the references therein.
2. B.C. Gregory, Natural and applied electrostatic potential effects on confinement and divertors in tokamaks, Invited paper, Tokamak Concept Improvement Workshop, Villa Monastero, Varenna, Italy, 29 th August - 3 September 1994 and the references therein.
3. M.J.Schaffer et al. in Controlled Fusion and Plasma Physics (Proc.18th Eur. Conf.Berlin 1991) vol. 15C, Part III, European Physical Society, Geneva (1991) 241 and T.Shoji et al. in Plasma Physics and Controlled Nuclear Fusion Research 1992 (Proc. 14 th Int. Conf. Wurzburg, 1992) (IAEA, Vienna (1993) Vol. I, p.693
4. F.Wagner et al.Nuc.Fus.**25**, 1490 (1985)
5. D.Stork et al. in Plasma Physics and Controlled Nuclear Fusion Research 1994 (Proc. 15 th Int. Conf. Seville, 1994) paper IAEA - CN - 60/A-1-I-3
6. L.Landau & E.Lifshitz Fluid Mechanics, Mir, Moscow, 1988, p.162
7. H.Biglari, P.Diamond & P.Terry Phys.Fluids **B2**,1(1990)
8. L.Askinazi, V.Golant, S.Lebedev, L.Levin, V.Rozhansky & M.Tendler Phys. Fluids **B 5**, 2420 (1993)
9. M.Tendler et al. in Plasma Physics and Controlled Nuclear Fusion Research 1992 (Proc. 14 th Int. Conf. Wurzburg, 1992) (IAEA, Vienna (1993) Vol. II, p.223
10. V.Rozhansky & M.Tendler Phys. Fluids **B4**, 1877 (1992)
11. J.Kim et al. Phys.Rev.Lett.**72**,2199 (1994)
12. F.L.Hinton & Y. - B. Kim Phys.Plasmas to be published
13. V.Rozhansky & M.Tendler Phys.Plasmas **1**, 2711 (1994)
14. J. - L. Lachambre et al. Nuc.Fus. **34**, 1431 (1944)
15. R.Weynants & R.J.Taylor Nuc.Fus.**30**,945 (1990)
16. " Preparation for the 21 st Century - Scientific Program " Internal Report CCFM R1 410e October 1993.

REVIEW ON GAMMA 10 EXPERIMENT

T. Tamano, T. Cho, M. Hirata, H. Hojo, M. Ichimura, M. Inutake, K. Ishii, A. Itakura,
I. Katanuma, R. Katsumata, Y. Kiwamoto, A. Mase, Y. Nagayama,
Y. Nakashima, T. Saito, M. Shoji, E. Takahashi, Y. Tatematsu,
K. Tsuchiya, N. Yamaguchi, K. Yatsu

Plasma Research Center, University of Tsukuba
Tsukuba, Ibaraki 305, Japan

Physics of tandem mirror plasmas have been investigated in GAMMA 10. MHD stability of anisotropic plasma in the average minimum B system is studied in detail. Formation of potentials for axial confinement is performed by ECRH alone without creating sloshing ions. In hot ion modes by ICRH, where $T_i / T_e > (m_i / m_e)^{1/3}$, ion temperatures of about 10 keV are achieved and thermonuclear fusion neutrons are detected for the first time in a tandem mirror. Details of the plasma confinement in the hot ion modes indicate that axial confinement is consistent with classical potential confinement. This implies that radial confinement is better than axial. Density fluctuations due to drift waves are well suppressed by radial electric fields naturally formed as the axial potentials build up.

I. DESCRIPTION OF GAMMA 10

The GAMMA 10 is a tandem mirror consisting of four cell regions: A central cell confining a main plasma, anchor cells attached to both sides of the central cell, plug/barrier cells located next to the anchor cells, and end cells. (See Fig. 1.) The central cell is an axi-symmetric mirror cell with a nearly uniform magnetic field near the mid-plane and with strong mirror throats at its ends. The mid-plane magnetic field is varied from 3 to 5.7 kG and the mirror throat magnetic field is set to about 30 kG. The anchor cells are formed by a set of baseball

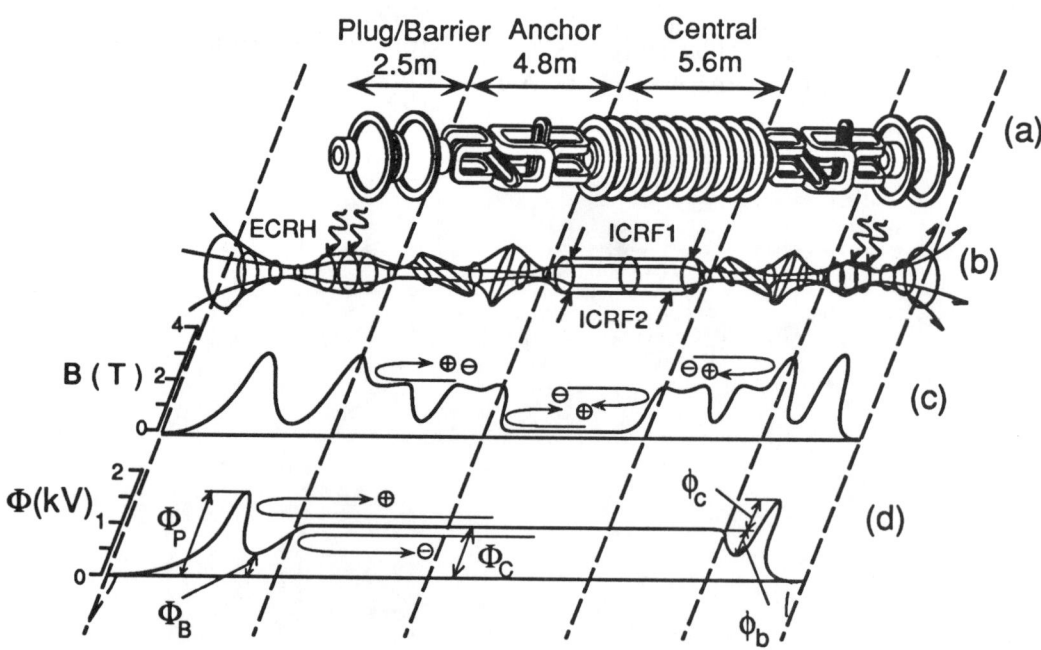

Fig. 1: A schematic of the GAMMA 10. Magnetic field strength and a typical tandem mirror potential are also illustrated.

coils, creating minimum B regions for MHD stability. The coils are designed so that the magnetic fields are axi-symmetric at both ends of the anchors. Thus, the GAMMA 10 is called an axi-symmetrized tandem mirror system. The plug/barrier cells are another axi-symmetric mirror cells attached to the anchor cells. These cells are used to create barrier and plug potentials. The magnetic fields in the barrier and plug regions are about 5 kG and 10 kG. The end cells are separated from the plug/barrier cells by another mirror throats with a magnetic field of about 30 kG. The end cells have end plates at the very end. These end plates are segmented into 4 sections azimuthally and 5 sections radially. They are normally floating, but sometimes grounded or biased.

The length of the central, anchor and plug/barrier cells are 5.6 m, 4.8 m and 2.5 m, respectively. The total length of the GAMMA 10 is about 27 m. The radius of the vacuum chamber in the central cell is about 0.5 m. A fixed limiter with a radius of 0.18 m is placed near the mid-plane of the GAMMA 10. A movable limiter is located in the anchor cell and it limits a plasma radius down to about 0.1 m.

II. MHD STABILITY

Ions are heated by ICRH. Two types of ICRF antennas, called the "RF1" and "RF2" antennas, are located in the central cell. The RF1 antennas excite ICRF fast waves (~9.9 MHz) propagating towards the anchor cells. In the non-axi-symmetric anchor regions, the fast waves are converted into ICRF slow waves that are absorbed by anchor cell ions. The RF2 antennas excite ICRF slow waves (4–10 MHz) which propagate towards the mid-plane and heat central cell ions. Since the GAMMA 10 is an average minimum B system, enough pressure in the anchor cell (minimum B region) must be established before the central cell pressure builds up. Figure 2 shows beta values of the central and anchor cells, $\beta_{\perp C}$ and $\beta_{\perp A}$, for various plasma shots. Here, $\beta_{\perp A}$ is defined as the average beta values of both (east and west) anchors. A clear threshold exists for the ratio of the central beta to the anchor beta with a negligibly small plug/barrier beta. No plasma is sustained above the threshold.

The observed plasma pressure in the central cell region often greatly exceeds the MHD (flute interchange) stability limit expected from the simple theory for an isotropic plasma as is shown by the dashed line. This discrepancy is explained by considering the anisotropic pressure of the central cell plasma. The threshold for the anisotropic plasma is shown by the solid line which agrees well with the experimentally observed pressure limit. More details are available in Ref. [1].

Note that the beta values for the east and west anchors are often unbalanced. However,

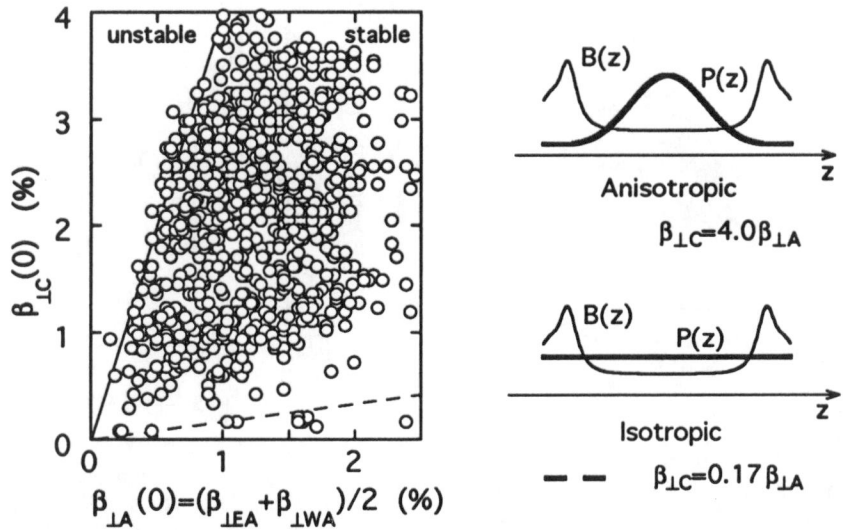

Fig. 2: Central beta values $\beta_{\perp C}(0)$ vs. anchor beta values $\beta_{\perp A}(0)$. Pressure distributions used for evaluating the stability limits are shown in the right.

the same stability limit is given for the same average anchor beta. In our knowledge, this is the first time that the principle of average minimum B is cleanly demonstrated in an experiment since pressure of a minimum B region cannot be varied independently in other types of fusion devices.

Excitation of Alfvén ion cyclotron (AIC) modes and relaxation of the pressure anisotropy due to the modes are observed with a strong ICRH, i.e., with a strong temperature anisotropy A (ratio of perpendicular to parallel temperature $T_\perp / T_\parallel$). Frequencies of the excited modes are slightly less than the applied cyclotron frequency at the mid-plane of the central cell. Magnetic fluctuations are observed by both magnetic probe measurement near the plasma edge and microwave reflectometry looking the core region. The on-set condition for the mode agrees very well with the theoretical prediction. Fluctuation amplitudes increase with an increase with the anisotropy parameter $\beta_\perp A^2$. Relaxation of the anisotropy is also observed. However, plasma confinement is not noticeably affected [2].

III. POTENTIAL FORMATION

Two methods are proposed for generating a plug potential for confining ions in a tandem mirror. One is neutral beam injection to generate sloshing ions [3]. This was successfully demonstrated in the GAMMA 6 [4]. The other is ECRH in the plug region. The original idea for this method is to enhance the plug potential by increasing electron temperature of the plug/barrier cells. Recently, it is found that a plug potential can be made by applying ECRH alone. Recent GAMMA 10 experiments have mostly been carried out by the ECRH method.

Detailed studies are carried out on the potential formation with ECRH alone. The plug potential Φ_P is determined from signals of an analyzer measuring energies of ions lost to the end. This diagnostic system is called ELA (End Loss Analyzer). The central and barrier potentials, Φ_C and Φ_B are measured by heavy ion beam diagnostics using a gold beam (called beam probes). The potential differences $\Phi_P - \Phi_C$ and $\Phi_B - \Phi_C$ are called the ion confining plug potential ϕ_c and the thermal barrier potential ϕ_b. The scaling law between ϕ_c and ϕ_b is theoretically predicted by the Cohen's weak and strong ECH theories [5–7]. This scaling has been checked against GAMMA 10 results as shown in Fig. 3.

The results support the Cohen's strong ECH theory. Electron velocity distribution functions in each cells are measured by x-ray diagnostics. The measured electron distribution function in the plug region shown in Fig. 4 confirms the plateau-shaped distribution function expected from the strong ECH theory [8].

Measured ϕ_c for a given central plasma density n_c increases nearly linearly with the

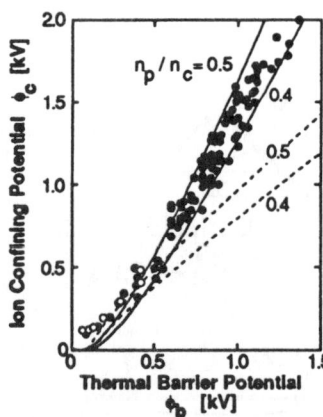

Fig. 3: Ion confining potential ϕ_c is plotted against thermal barrier potential ϕ_b. The solid line and broken lines indicate predictions of the strong and weak theories, respectively.

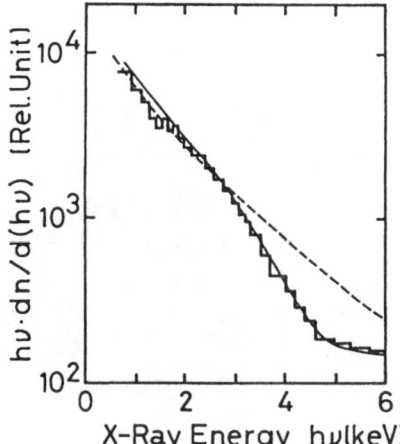

Fig. 4: Electron distribution functions measured by x-ray diagnostics in the plug region matches to a plateau distribution (solid line). The dashed line corresponds to a Maxwellian distribution.

plug-ECRH power P. Figure 5 shows measured $\phi_c + \phi_b$ as a function of plasma density for given ECRH powers. The barrier potential depth ϕ_b and ϕ_c are studied as a function of n_c. Applying the Boltzman relation for the barrier potential depth and the theoretical formula for strong ECRH [5] to the GAMMA 10 geometry, we obtain

$$\phi_b + \phi_c \approx T_{ec}\{0.665\,(n_p/n_c)^{2/3}\,(n_c/n_b)^{50/63}\}, \qquad (1)$$

where T_{ec}, n_p and n_b are the central electron temperature and the plug and barrier densities. Since measurements indicate that the central pressure $n_c T_{ec}$ is nearly proportional to P for a given n_p/n_c, the formula above gives

$$\phi_b + \phi_c \propto n_c^{-13/63}\,n_b^{-50/63}\,P. \qquad (2)$$

Measured values of $\phi_b + \phi_c$ after minor corrections due to a small variation in n_b are plotted against n_c in Fig.5. The result is well explained by the relation above as shown by the dashed curves. This supports a mechanism similar to that employed in the strong ECRH theory for the potential formation. However, detailed mechanisms are still under investigation.

Another important subject regarding the potential formation is sustainment of the potentials. It has been pointed out that the barrier potential might be eliminated as cold ions are trapped in the barrier region due to collisions. Recently, our effort has been directed towards this problem. So far, we finds that the potentials are maintained for about 55 ms, longer than the collision time. This sustainment time is presently limited by the power supply capability for plasma heating.

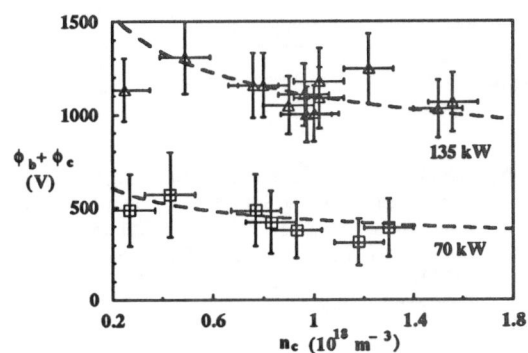

Fig.5: Values of $\phi_c + \phi_b$ are plotted against central cell density n_c for ECRH 80 kW and 140 kW.

IV. PLASMA CONFINEMENT

Recently, energy and particle confinement of ions and electrons contained by electrostatic potentials is closely examined in a hot ion mode regime (~ 3 keV) where a ratio of ion to electron temperature (~ 100 eV) is larger than the cubic root of the ratio of ion to electron mass, $T_i/T_e > (m_i/m_e)^{1/3}$. In this regime ions and electrons are tightly coupled and ions lose most of their energy to electrons before leaving the central mirror cell. This provides a good test base for potential confinement of tandem mirror plasmas.

Discharge duration of GAMMA 10 plasmas is extended to 100 ms and confining potentials are sustained for 55 ms, both the longest in the tandem mirror history, and a steady state condition is established. Increases in plasma density and ion and electron temperatures are observed when the confining potentials are formed, as indicated by the diamagnetic signals, the soft x-ray signals and the line density shown in Figs. 6 and 7. A new diagnostic technique of modulating ICRH power for a short period (5 ms at 70 ms) is applied, and energy and particle confinement times of both electrons and ions are directly and clearly demonstrated. The recovery times represent a characteristic confinement time of 7 – 10 ms for the discharge. Without the potentials, the characteristic confinement time is much shorter (~ 2 ms) as seen in the top figure of Fig. 6.

Details of the energy confinement times are also evaluated from the power balances. Since the main loss mechanism of ion energy is electron drag, some charge exchange processes and end-losses, the ion energy confinement time τ_{Ei} is evaluated by dividing the stored ion

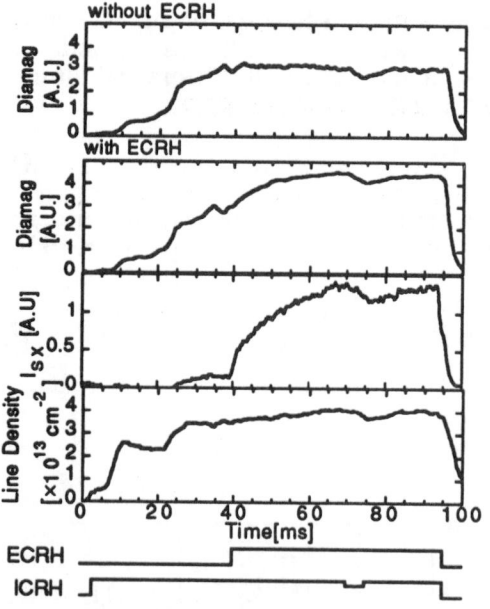

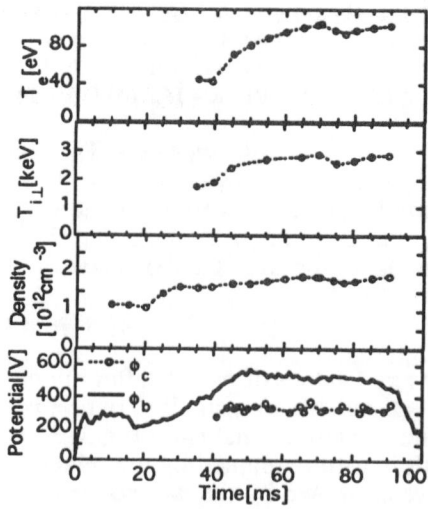

Fig. 7: Electron temperature T_e perpendiular ion temperature $T_{i\perp}$, plasma density confining and potentials are shown as a function of time.

Fig. 6: Characteristics of the diamagnetic signal, the soft x-ray signal and the line density as a function of time. Sequences of ICRF and ECRH are shown in the bottom. Top without ECRH. Next three with ECRH. The ICRH power is slightly modulated for a short period.

energy density by the loss power density. The evaluated ion energy confinement time is plotted as a function of plasma radius r in Fig. 8. The main energy input to the electrons is the energy dragged from ions. The electron energy confinement time τ_{Ee} is evaluated by dividing the electron stored energy by the dragged energy. This is also plotted as a function of radius in Fig. 8. The energy confinement time in the core is 2.3 ms and it is about the same throughout the radius. Particle confinement times for both ions and electrons are evaluated by using two methods [9, 10]: i) Particle balances using end loss fluxes and ii) the Pastukhov's relations between particle and energy confinement times [11–14]. These particle and energy confinement times in the core region are all consistently explained by classical processes of potential confinement.

Figure 9 shows energy confinement times τ_E estimated from measured ion losses at the ends for the cases corresponding to Fig. 5. Energy confinement times are also estimated from the Pastukhov formula by using measured ϕ_c and they are shown by dashed lines in Fig. 9.

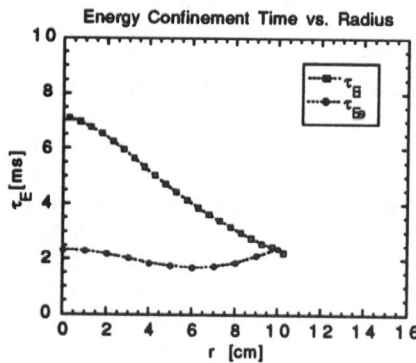

Fig. 8: Evaluated ion and electron energy confinemnt times as a function of plasma radius.

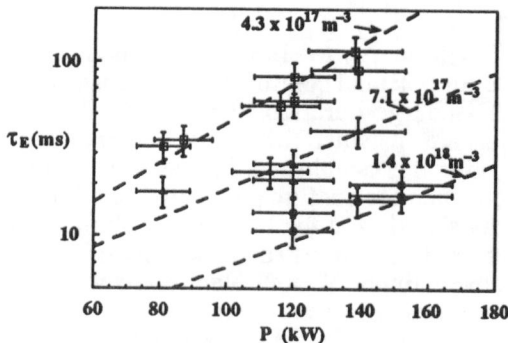

Fig. 9: Energy confinement time τ_E as a function of ECRH power P.

Both agree, and τ_E increases almost exponentially with P for a given n_c.

V. HIGH ION TEMPERATURE EXPERIMENT

A diamagnetic flux mainly due to ion pressure increases almost linearly with the ICRF power radiated from the antennas. When the magnetic field strength of the central cell region is varied from 0.30 to 0.57 T, the diamagnetic flux is about the same for a given radiated ICRF power as is shown in Fig. 10. The plasma density, and the ion and electron temperatures increase when the potentials are formed. This indicates that the ion pressure is nearly proportional to the magnetic field strength. Accordingly, the highest ion temperature is obtained at the maximum magnetic field of 0.57 T. Ion and electron temperatures of up to about 10 keV and 130 eV, respectively, are achieved with a plasma density of about 2×10^{12} cm^{-3} (Fig. 10). The dependence on the magnetic field is approximately consistent with the steady state conditions for both energy and particle balances involving classical processes only.

When the ICRF power at the generator is further increased, the power radiated from the antennas tends to saturate. Microwave reflectometry is used to investigate applied ICRF waves in a plasma. By scanning the microwave frequencies, radial distributions of the ICRF waves are obtained. It is found that, at low ICRF power, enough ICRF waves are detected in the core region. On the other hand, at high power, the waves tend to be located mainly in the peripheral region. Accordingly, the radial temperature profiles for both ions and electrons are broadened. Eventually, the radiated power ceases to increase beyond 70 kW in either magnetic field. One possible reason for the change in the wave distribution is plasma diamagnetism. Although the magnetic field changes only by a few per cent because of the diamagnetism, the parallel wavelength varies significantly near the ion cyclotron resonance condition. Therefore, wave propagation in the radially non-uniform magnetic field due to the diamagnetism needs to be examined.

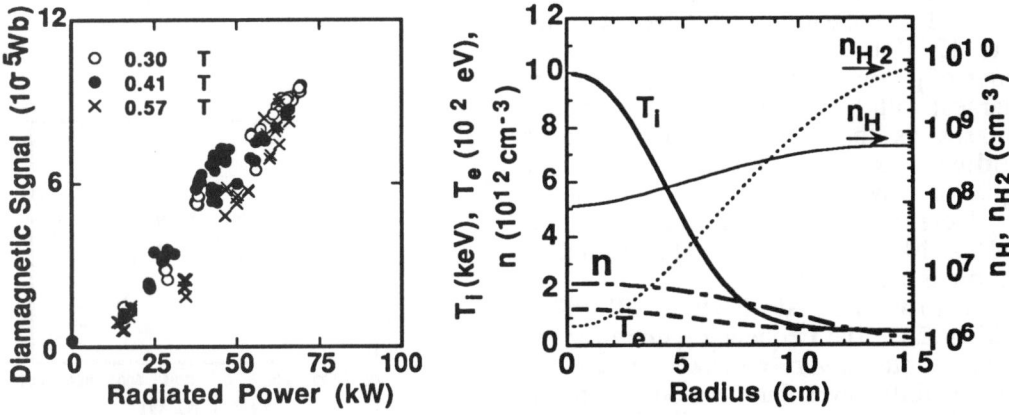

Fig. 10: (Left) Diamagnetic flux vs. ICRF power radiated from the antennas for various central cell magnetic fields
(Right) Radial profiles of density, electron and ion temperatures at 0.57 T.

A mixture of deuterium and hydrogen plasmas is formed (Fig. 11). The central cell ICRF is changed to 4.5 MHz at 0.57 T, which is resonant for deuterium, but not for hydrogen. The anchor cell ICRF (9.9 MHz at 0.6 T) remains unchanged so that hydrogen is heated in the anchor cell for MHD stability. Thermonuclear fusion neutrons are detected for the first time in a tandem mirror. During the discharge, the neutron count rate decreases gradually. This is to be expected because the deuterium to hydrogen density ratio changes during the discharge. The neutron count rate R is approximately proportional to the diamagnetic flux $W_{dia}^{2.9}$. Since the diamagnetic flux is almost proportional to the ion temperature for the hot ion mode, the counting rate and its dependence are, within the uncertainty of the mixture ratio, consistent with the expected thermonuclear fusion reaction rate.

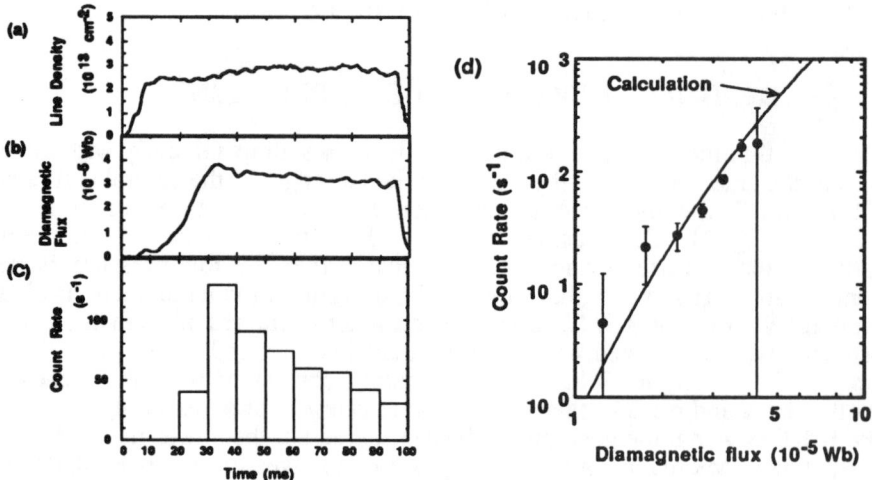

Fig. 11: (Left) Line density, diamagnetic flux and neutron count rate for the mixture discharge is plotted as a function of time.
(Right) Measured and calcurated neutron count rates are plotted against the diamagnetic flux. They agree well.

VI. DRIFT INSTABILITY

Density fluctuations due to drift waves have been identified and their effects on plasma confinement have been studied. The dispersion of the drift wave mode is checked by the Fraunhofer diffraction method. The segmented end plates are biased and effects of radial electric fields on the drift mode have been studied [15].

Figure 12 plots the fluctuation level as a function of plasma potential. The fluctuation is the largest with no radial electric field and it decreases as the electric field increases regardless its sign. When the density fluctuations are large, plasma confinement time decreases. In normal operations of GAMMA 10, the radial potential indicated by the arrow is naturally formed.

A rotational mode due to $E_r \times B$ drift is also identified. The frequency spectra for the ritational mode is very narrow compared to that of the drift waves. Furthermore, the radial distribution of fluctuation amplitudes of this rotational mode differ from that due to the drift mode, therefore, an optimum distribution of the radial electric field may exist.

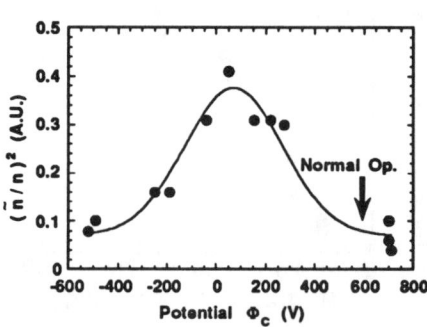

Fig. 12: Fluctuation levels are plotted against the plasma potential.

VII. REMARKS

Good wall conditions of the GAMMA 10 are necessary in order to obtain the results discussed in this paper. Cryo-systems, wall baking, ECRH discharge cleaning and wall conditioning by plasma discharges are all effective for obtaining good conditions. Particular attention is paid for the effects of charge exchange processes. These processes are evaluated by use of an improved neutral particle code (DEGAS code), including the effects of excited hydrogen molecules [16]. The result confirms that charge exchange processes can be small in the estimation of ion confinement time near the axis under a good condition.

The past work has concentrated on the study of plasma confinement in the central cell. The results show that the confinement of central cell plasma is consistent with the classical processes of potential confinement. Plasma confinement in the plug/barrier cells and the end cells are still under investigation. Rough estimates indicates that hot electron confinement in the plug/barrier cells is better than that in the central cell.

VIII. CONCLUSIONS

MHD stability of the GAMMA 10 tandem mirror plasmas in the average minimum B system is well verified. Electrostatic potentials for axial plasma confinement is established by ECRH alone. Observed potentials agree with those predicted by the strong ECRH theory. Axial confinement of the plasmas is consistent with the classical potential confinement, implying that radial confinement is much better than axial, owing to suppression of drift waves by radial electric fields naturally formed when the axial potential is created.

ACKNOWLEDGMENTS

The authors are grateful to the present and former members of the GAMMA 10 Group for their great contributions.

REFERENCES

1. M. Inutake, A. Ishihara, R. Katsumata, M. Ichimura, A. Mase, K. Ishii, Y. Nakashima, Y. Nagayama, N. Yamaguchi, H. Hojo, I. Katanuma and T. Tamano, *Proceedings of Invited Paper, International Conference on Open Plasma Confinement Systems for Fusion, Novosibirsk, 1993,* edited by A.A. Kavantsev (World Scientific, Singapore, 1994), p. 51.
2. M. Ichimura, R. Katsumata, M. Inutake, A. Ishihara, H. Hojo, I. Sasaki, K. Ishii, A. Mase, Y. Nakashima and T. Tamano, *Proceedings of Invited Paper, International Conference on Open Plasma Confinement Systems for Fusion, Novosibirsk, 1993,* edited by A.A. Kavantsev (World Scientific, Singapore, 1994), p. 69.
3. G. I. Dimov, V. V. Zakaidakov, and M. E. Kishinevski, *Fiz. Plazmy* **2** (1976) 597 [*Sov. J. Plasma Phys.* **2** (1976) 326].
4. K. Yatsu, S. Miyoshi, H. Tamai, K. Shida, K. Ishii, and A. Itakura, *Phys. Rev. Letters* **43** (1979) 627.
5. R.H. Cohen, *Phys. Fluids* **26** (1983) 2774.
6. R. H. Cohen et al., *Nuclear Fusion* **20** (1980) 1421.
7. R. H. Cohen et al., *Nuclear Fusion* **23** (1983) 1301.
8. T. Cho et al., *Physical Review Letters* **12** (1990) 1373.
9. M. Hirata et al., *Nuclear Fusion* **31** (1991) 752.
10. S. Miyoshi et al., *Nuclear Fusion, Supplement* **2** (1991) 539.
11. R. H. Cohen et al., *Nuclear Fusion* **18** (1978) 1229.
12. T. Cho, *Nuclear Fusion* **28** (1988) 2187.
13. T. Cho et al., *Nuclear Fusion, Supplement* **2** (1989) 501.
14. V. P. Pastukhov, *Nuclear Fusion* **14** (1974) 3.
15. A. Mase, A. Itakura, M. Inutake, K. Ishii, J. H. Jeong, K. Hattori and S. Miyoshi, *Nucl. Fusion* **31** (1991) 1725.
16. D. B. Heifetz, D. E. Post, M. Petravic, J. C. Weisheit and G. Bateman, *J. Comput. Phys.* **46** (1982) 309.

FIELD REVERSED ION RINGS

R.N. Sudan and Yu. A. Omelchenko

Laboratory of Plasma Studies, Cornell University, Ithaca, NY 14853

In typical field-reversed ion ring experiments, an intense annular ion beam is injected across a plasma-filled magnetic cusp region into a neutral gas immersed in a ramped solenoidal magnetic field. Assuming the characteristic ionization time is much shorter than the long ($t > 2\pi/\Omega_i$) beam evolution time scale, we investigate the formation of an ion ring in the background plasma followed by field reversal, using a $2\frac{1}{2}$-D hybrid, PIC code FIRE, in which the beam and background ions are treated as particles and the electrons as a massless fluid. We show that beam bunching and trapping occurs downstream in a ramped magnetic field for an appropriate set of experimental parameters. We find that a compact ion ring is formed and a large field reversal $\zeta = \delta B/B > 1$ on axis develops. We also observe significant deceleration of the ring on reflection due to the transfer of its axial momentum to the background ions, which creates favorable trapping conditions.

INTRODUCTION

Compact magnetic field-reversed configurations (FRC) are extremely attractive as the basis for the ultimate economic fusion reactor of economical size. Field reversed ion rings which are FRC with the dominant component of the azimuthal current provided by energetic ions with gyroradius of order the major radius [1] have been shown theoretically to have favorable stability [2]. A new ion ring project FIREX has been launched at Cornell University to experimentally achieve the formation of field-reversed ion rings with large reversal factors $\zeta = \delta B/B > 1$ (δB is the diamagnetic self-field on axis and B is the external magnetic field) by injecting a strong proton beam (with the number of particles $N_b \sim (2 \div 5) \times 10^{17}$ and ~ 1 MeV proton energy) through a magnetic cusp followed by a ramped solenoidal magnetic field (Fig. 1). Previous experiments at Cornell [3,4,5] achieved a $\zeta \leq 0.05$ by the injection of $\lesssim 10^{16}$, 400 keV protons into neutral hydrogen at a pressure of ~ 100 mT. We describe below 2d simulations of the injection and formation of ion rings in a background plasma.

MODEL

The previous computer models of ion ring evolution in magnetized plasmas [6], [7] either did not describe the initial injection stage followed by the beam trapping and formation of a ring or did not include the background plasma response in the high conductivity limit. In order to conduct a more thorough study of ion ring dynamics we have developed a $2\frac{1}{2}$-D hybrid, PIC code FIRE. In this code we assume the background plasma to be preformed by rapid beam induced ionization and treat both the beam and background ions as single particles, experiencing slowing-down collisions with free electrons and neutral atoms, respectively. We neglect electron inertia, assume quasi-neutrality $n_e = \sum Z_i n_i$, and treat the electrons as a collisional fluid. We consider the axisymmetric cylindrical geometry ($\partial/\partial\theta = 0$), and represent the magnetic field as $\mathbf{B} = \nabla\psi \times \nabla\theta + rB_\theta \nabla\theta$, where ψ is the flux function and $\nabla \theta = \hat{\theta}/r$. Since we are interested in slow plasma motions, we totally neglect displacement current in Maxwell's equations and arrive at the following system of equations employed by FIRE's field solver:

$$\frac{\partial B_\theta}{\partial t} = -c(\nabla \times \mathbf{E})_\theta \qquad (1)$$

$$\frac{\partial \psi}{\partial t} = \frac{c}{4\pi e n_e}[rB_\theta, \psi] + \eta \Delta_* \psi + S_i \qquad (2)$$

$$S_i = \frac{cr}{\sigma} j_{i\theta} - \frac{1}{en_e}(\mathbf{j}_i \cdot \nabla)\psi$$

$$\mathbf{E}_\perp = -\frac{en_e \mathbf{v}_{e\perp}}{\sigma} - \frac{(\mathbf{v}_e \times \mathbf{B})_\perp}{c} \qquad (3)$$

$$E_\theta = -\frac{1}{cr}\frac{\partial \psi}{\partial t} \qquad (4)$$

$$v_{e\theta} = \frac{1}{n_e e}\left(\frac{c}{4\pi r}\Delta_* \psi + j_{i\theta}\right) \qquad (5)$$

$$\mathbf{v}_{e\perp} = \frac{1}{n_e e}\left(\mathbf{j}_{i\perp} - \frac{c}{4\pi}\nabla(rB_\theta) \times \nabla\theta\right) \qquad (6)$$

Here $\Delta_* = r^2 \nabla \cdot \frac{1}{r^2}\nabla$ and $[rB_\theta, \psi] = (\nabla rB_\theta \times \nabla \theta) \cdot \nabla \psi$ is a Poisson bracket; $\mathbf{j}_i$ is the total ion current found by PIC method, and $\mathbf{v}_e$ the electron fluid velocity; σ is the electron conductivity determined by the characteristic frequencies of elastic electron collisions with plasma ions and neutral atoms, ν_{ei} and ν_{en}, $\sigma = \omega_{pe}^2/4\pi(\nu_{ei} + \nu_{en}); \eta = c^2/4\pi\sigma$.

We have typically run FIRE for 50,000 time steps ($\sim 5\mu sec$) on a grid with 100 and 40 nodes in the axial and radial directions, respectively with the typical numbers of beam and background particles, 10,000 and 25,000, which required approximately 2 hours of CPU time of the NERSC (National Energy Supercomputer Center) Cray-2 computer. The numerical algorithm implemented in FIRE has proven to be remarkably stable and allows long runs with energy conservation errors less than 1%.

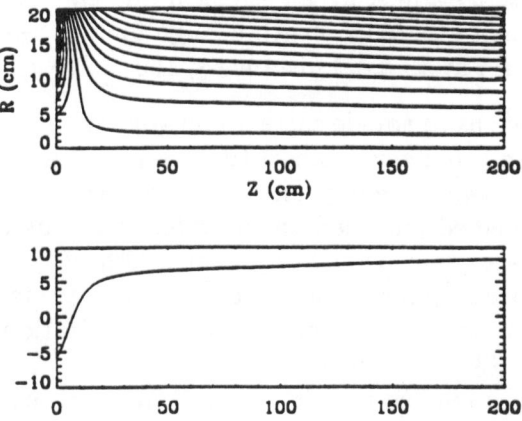

Figure 1: Applied mangetic field: contour lines of ψ (*above*); $B_z(kG)$ on axis (*below*).

RESULTS

We have studied the self-consistent $2\frac{1}{2}$-D dynamics of the intense ion proton beam, injected into a preformed mixture of hydrogen neutral gas and plasma immersed in a strong applied magnetic field ($B \sim 1$ Tesla), forming a narrow (~ 20 cm) cusp-like region followed by a smooth solenoidal part (~ 2m) (Fig. 1). The effect of the radial component of the cusp magnetic field acting on the particles results in the conversion of a substantial fraction of their axial energy (initially ~ 1 MeV) into the azimuthal energy of particle gyration around the center axis with radii $R \simeq 15$ cm, leading to the formation of a cylindrical layer of rotating ions advancing into the solenoidal field. The crucial point of our study was to find conditions favorable for the following stages to be achieved:

(1) fast beam trapping by the self-field,
(2) formation of a compact ring,
(3) achieving field reversal on axis with $\zeta > 1$,
(4) trapping the ring with magnetic mirrors.

FIRE has allowed us to test a broad range of experimental parameters which are important for the design of efficient beam trapping equipment and producing compact field reversed configurations. Below we discuss the most important system characteristics together with some results obtained while investigating them.

1. *External magnetic configuration*, which can be varied by choosing (a) the cusp geometry and (b) the applied solenoidal field. We have tried a number of different cusp configurations to ensure that the beam particles are emitted from the appropriate flux surface to achieve optimum ring formation downstream. The beam dynamics have been also found to be extremely dependent on the strength and configuration of the downstream solenoidal magnetic field. In particular, we have discovered that a ramped magnetic field in the axial direction can not only prevent beam particles from escaping radially to the walls, but also may be regarded as a basic feature of the beam trapping equipment. Our simulations have shown that by adjusting its strength one can efficiently slow down the head particles and reduce the beam axial length enough for the beam self-field to develop and trap the bulk of the beam at an axial distance ≤ 150 cm (Fig. 2).

2. *Beam parameters* such as the diode current and voltage temporal profiles, the beam angular spread and injection geometry. Using FIRE we have tested a number of realistic voltage and current profiles which are expected to be generated in FIREX experiments. We have found that the expected pulse lengths of order of 150 nsec capable of delivering $\sim (3 \div 5) \times 10^{17}$ protons are good enough for about 70% of the beam particles to be trapped by the beam generated perturbations, and large field reversal factors $\zeta \geq 1.5$ (Fig. 3) be achieved. In order to minimize the ring current neutralization in the cusp region by the electron return current [8] the beam was injected close (~ 1 cm) to the conducting cylindrical wall. We have also noticed that it is important to adjust the radial position of the ion emission surface, so that it is located at a radius greater than the mean beam ion gyro-radius at the exit of the cusp region. In this case the trajectories of single particles

wiggling in the poloidal plane around the gyro-radius may reduce the beam axial length, and as a result improve the beam trapping efficiency.

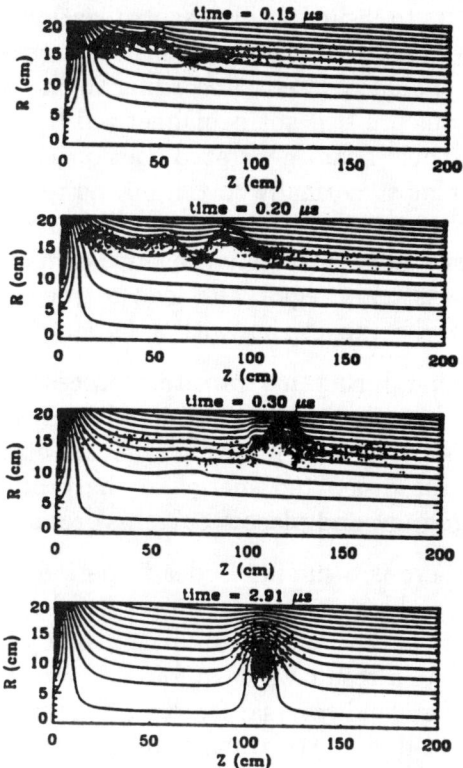

Figure 2: Magnetic field and beam temporal evolution. Field reversal on axis occurs on the diffusive time scale $\sim 2\mu$ s.

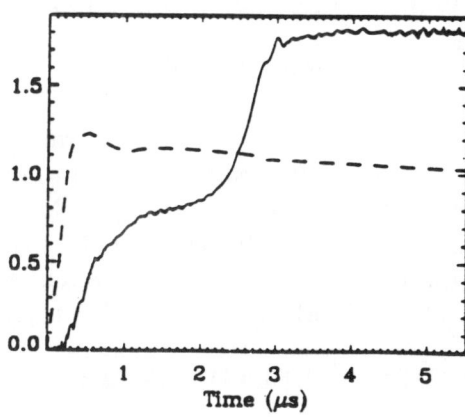

Figure 3: Time histories ($N_b = 4 \times 10^{17}$): magnetic field reversal factor ζ (*solid line*) and ring average axial position $< z_b >$ (m) (*dashed line*).

3. *Background plasma characteristics.* One of the most important system parameters is the plasma density. Since 1 MeV ion beams are expected to have current densities of order $j_b \sim 1 kA/cm^2$, ($n_b \sim 0.5 \times 10^{13} cm^{-3}$), the background plasma must also be dense enough ($n_e \sim 0.5 \times 10^{15} cm^{-3}$) to insure prompt neutralization of the beam space charge by plasma electrons. Therefore, assuming the gas ionization degree $n_b/n_0 \sim 0.1$, a neutral fill with pressure $p \sim 0.2$ Torr should be used. According to the results of our simulations at this pressure, background ions are still not demagnetized by their slowing-down collisions with hydrogen atoms and their beam generated motion can give rise to compressional Alfvén waves propagating almost normally to the downstream applied magnetic field [9]. We have demonstrated that it is these perturbations, which are generated as a result of a resonant beam-plasma instability, that may prevent beam current neutralization by the plasma electrons in the downstream region and lead to the beam bunching and trapping followed by the formation of a compact ring with axial length ~ 15 cm (Fig. 2).

We have further found that during this transitory process approximately half the initial beam kinetic energy is dissipated through the collisional decay of the induced return current [10]. We have performed a series of runs injecting different numbers of beam particles and observed that the ratio of the dissipated energy to the product $I_\theta^{net} I_\theta^p$ ($I_\theta^{net}, I_\theta^p$ are the peak values of space-averaged net and plasma currents) remains constant within 10%.

In a real experiment there is a minimal "dead" period ($\sim 10\mu s$) needed to activate the magnetic mirrors to trap the ring after it is reflected by the ramped solenoidal field. Therefore, an important part of our studies concerned the dynamics of a well-shaped ring in the background plasma. As a result we have found a very interesting and encouraging phenomenon. Rings with large field reversal factors $\zeta > 1$ are significantly decelerated on reflection in the ramped field due to the transfer of axial momentum to the background plasma ions through the generation of Alfvén waves. The average reflected ring speed is at least by an order of magnitude less (Fig. 3) than that expected from single particle motion in a solenoidal magnetic field ramped to increase axially at a rate $\Delta B/B = 0.2$/meter. This provides favorable conditions for experimental ring trapping.

CONCLUSION

Summarizing the results obtained with the code FIRE, we conclude that by tailoring properly the main experimental parameters in the range of interest strong ion rings with $\zeta > 1$ can be created on the equipment under development at Cornell. We intend to further improve our model and develop a 3-D hybrid, PIC code to investigate the stability of three dimensional ring reversed magnetic field configurations, which is an important issue for the practical realization of the ideas underlying the FIREX project.

ACKNOWLEDGEMENTS

We are grateful to our colleagues from the experimental group D.A. Hammer, J.B. Greenly and B. Podulka for their considerable input. The bulk of this work has also been reported at the BEAMS' 94 Conference at San Diego in June 1994. This work was supported by U.S. DOE Contract No. DE-FG02-93ER54221.

REFERENCES

[1] R.N. Sudan, "Particle Ring Fusion", in *Unconventional Approaches to Fusion*, B. Brunelli and G.G. Geotta, editors, Plenum. Publ. Corp., New York, 1982.
[2] J. Finn and R.N. Sudan, Nucl. Fusion **22**, 1443 (1982).
[3] P.L. Dreike, J.B. Greenly, D.A. Hammer and R.N. Sudan, Phys. Rev. Lett. **46**, 539 (1981); Phys. Fluids **25**, 59 (1982).
[4] P.B. Pedrow, J.B. Greenly, D.A. Hammer and R.N. Sudan, Appl. Phys. Lett. **47**, 225 (1985).
[5] J.B. Greenly, D.A. Hammer, P.B. Pedrow and R.N. Sudan, Phys. Fluids **29**, 908 (1986).
[6] A. Mankofsky, A. Friedman and R.N. Sudan, Plasma Phys. **23**, 521 (1981).
[7] P.M. Lyster and R.N. Sudan, Phys. Fluids B **2**, 2661 (1990).
[8] B.V. Oliver, D.D. Ryutov and R.N. Sudan, Phys. Plasma **1**, 3383 (1994).
[9] H.L. Berk and L.D. Pearlstein, Phys. Fluids **19** (1976).
[10] R.V. Lovelace and R.N. Sudan, Phys. Rev. Lett. **27**, 1256 (1971).

RECENT STUDIES OF THE LARGE HELICAL DEVICE

Atsuo Iiyoshi, Kozo Yamazaki and the LHD Group

National Institute for Fusion Science
Furo-cho Chikusa-ku Nagoya 464-01 Japan

ABSTRACT

The Large Helical Device (LHD) is presently under construction (1990-1997) in NIFS, Japan. The magnetic configuration is chosen as an optimized heliotron with double helix coils, succeeding Heliotron-E, ATF and CHS. Target plasma parameters of the LHD with fully superconducting coil systems are; $n\tau_E T \sim 10^{20} m^{-3} \cdot s \cdot keV$, $\bar{\beta}=5\%$ and steady operations with $R_p=3.75m$, $a_p=50-65cm$ and $B_0=3-4T$. The major objectives of the LHD project are to demonstrate good plasma confinement, high beta achievement and steady-state divertor operation in the helical system.

Various R&D results have been obtained on superconducting coils, negative ion source and special diagnostics for LHD. In our new site (Toki-city, Gifu-prefecture), the construction of the main experimental hall of the LHD Building was finished, and the winding of the helical coil has been started there. The construction of LHD will be completed in 1997.

1. INTRODUCTION

There is a growing interest in helical systems from the view-point of demonstrating a steady state helical system reactor. They are characterized by several merits such as steady-state operation without disruption, built-in divertor configuration, simple plasma control and small circulating power. For the demonstration of these advantages, the Large Helical Device (LHD) [1,2] is presently under construction (1990-1997) in NIFS, Japan.

The major objectives of LHD projects are
    (1) good plasma confinement, high temperature and high beta achievements,
    (2) steady-state operation with divertor,
    (3) complementary studies with tokamak researches, and
    (4) development of reactor technology.

For these purposes, world's largest superconducting helical coil system was adopted for LHD.

Figure 1 shows a schematic drawing of LHD machine with superconducting helical and poloidal field coils. The major radius of the main torus R_c is 3.9m, and the helical coil minor radius a_c is 0.975m. Three sets of poloidal coils are installed for positioning and shaping of the plasma column.

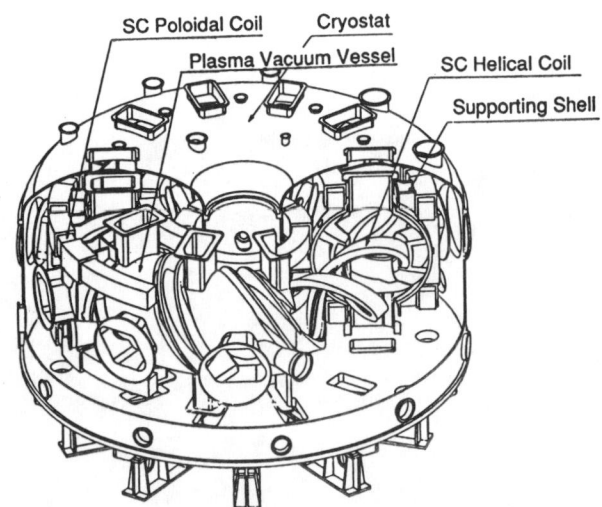

Fig.1 Schematic drawing of LHD

The magnetic configuration is an optimized heliotron with double helix coils, succeeding Heliotron-E, ATF and CHS. Target plasma parameters of the LHD with fully superconducting coil systems are; $n_0\tau_E T_{i0} \sim 10^{20}$m^{-3}·s·keV, $\bar{\beta}=5\%$ and steady operations with $R_p=3.75$m, $a_p=50$-65cm and $B_0=3$-4T. Figure 2 shows $n_0\tau_E$ - T_{i0} diagram for various fusion machines including target region for LHD. This $n_0\tau_E T_{i0}$ value of LHD is lower than the present front of Tokamak researches, however, the long pulse operation of LHD (Fig.3) leads to the new regime aiming at the next-generation device ITER and future commercial reactors. Different from Tokamak operations, helical systems do not require high-power current-drive and give rise to the easy access to steady-state reactors.

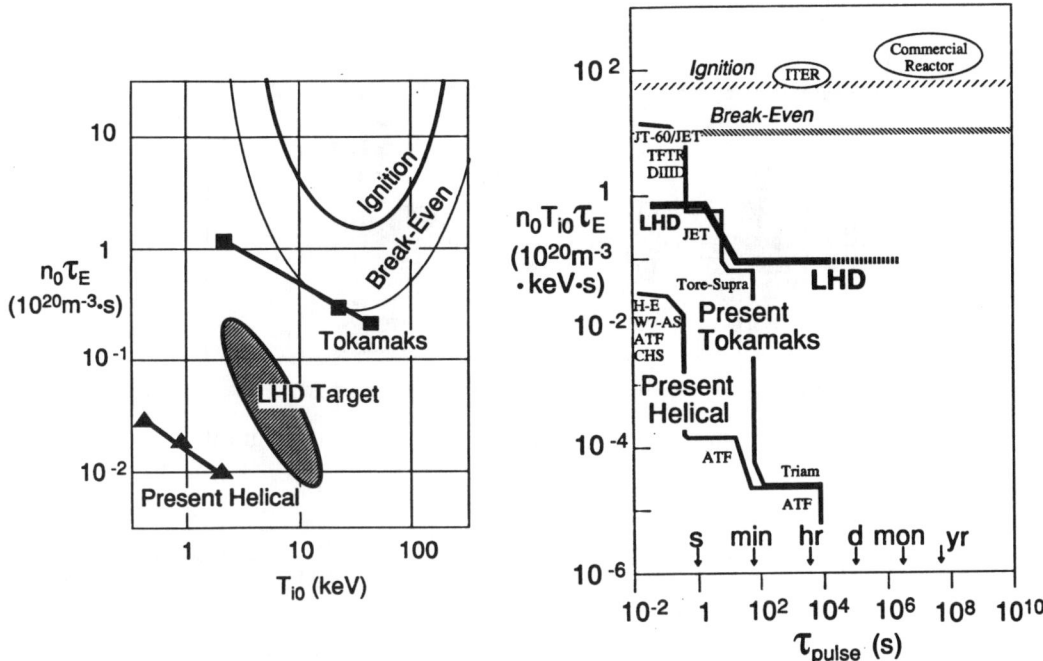

Fig.2 LHD target region in $n_0\tau_E$ - T_{i0} diagram

Fig.3 Long pulse operation regime

2. LHD BASIC DESIGN AND OPTIMIZATION

The configuration of LHD is based on the Japanese original Heliotron concept with continuous helical coils and several poloidal field coils. The continuous coil design is superior to the modular coil design because the clean magnetic field configuration with divertor separatrix can be produced. This is very important in the real experiment to clarify the points of physics. The basic optimization of LHD magnetic coil configurations were performed with respect to physics [3] and engineering [4].

The physics aspects, such as high-β achievement, good particle orbit, appropriate divertor-coil clearance and reliable transport scaling, are taken into account to choose standard LHD configurations. Figure 4 shows a typical example of these optimizations with equilibrium-beta, stability-beta and divertor-coil clearance. The helical pitch number m and the pitch parameter $\gamma (= ma_c/\ell R_c)$ of the helical coil is optimized for LHD configuration. The final LHD system is determined by $m=10$ and $\gamma=1.25$, and positive pitch modulation $\alpha=0.1$. The standard configuration of LHD (Fig.5) is 15cm inward shifted for the improvement of plasma parameters. The major plasma and machine parameters are summarized in Table I and II.

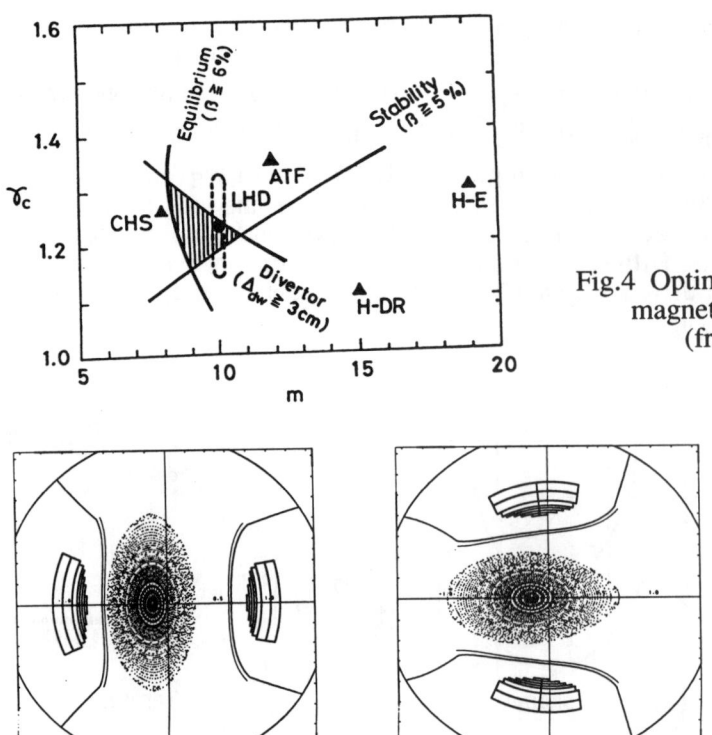

Fig.4 Optimization of magnetic configuration (from ref.[3])

Fig.5 Magnetic surfaces of LHD standard configuration

Table I. LHD configuration parameters

	Phase - I	Phase - II
Machine Configuration		
ℓ	2	
m	10	
γ (picth parameter)	1.25	
α (modulation parameter)	0.1	
major radius	3.9 m	
magnetic field strength	3.0 T	4.0 T
stored magnetic energy	0.9 GJ	1.6 GJ
Plasma Configuration		
plasma major radius	3.75 m	
plasma minor radius	0.6 m	
plasma volume	30 m^3	
ι(0)/ι(a)	0.4 / 1.3	
Heating System		
ECH Power	10 MW	10 MW
NBI Power	15 MW	20 MW
ICRF Power	3 MW	9 MW

Table II. Major parameters of LHD superconducting magnets

	Phase - I	Phase - II
Helical Coil		
cooling type	pool-boiling	
superconducting material	NbTi	
major radius	3.9 m	
magneto-motive force	5.85 MA	7.8 MA
coil current density	40A/mm^2	53.3A/mm^2
cooling temperature	4.2 K	1.8 K
maximum field	7.2 T	9.6 T
Poloidal Coil		
cooling type	forced-flow	
superconducting material	NbTi	
OV major radius	5.55 m	
magneto-motive force	-4.5 MA	
maximum field	5.0 T	
IS major radius	2.82 m	
magneto-motive force	-4.5 MA	
maximum field	5.4 T	
IV major radius	1.80 m	
magneto-motive force	5.0 MA	
maximum field	6.5 T	

3. PHYSICS ANALYSIS ON LHD

The detailed design parameters of the LHD plasma are optimized in terms of particle transports, equilibrium, stability, bootstrap current and divertor property. Several new numerical codes have been developed for three dimensional analysis on MHD equilibrium, stability and transport.

First of all, the equilibrium of helical systems is studied. The good magnetic surfaces are sometimes destroyed by the error field and plasma beta effects (Fig.6). For this analysis, the HINT code has been developed and clarified the island formation due to beta effects[5]. The equilibrium β-limit obtained by this HINT code is shown in Fig.6. The scenario to suppress beta-induced islands by means of several compensation coils is also demonstrated by this HINT code.

MHD stabilities have been extensively studied by VMEC/STEP code. Figure 7 shows typical results of Mercier mode β-limits in LHD as a function of magnetic axis position[6]. The shaded area denotes unstable region and the 15cm inner shift can permit the achievement of the average beta value of 5%. These results are confirmed by the global mode stability analysis with STEP code.

In helical system, bootstrap currents might deteriorate the plasma stability and make the steady-state operation difficult due to possible current disruption. However, in the standard LHD configuration the bootstrap current is about 150k A for $B_0=3T$, which corresponds to the prediction by the LHD scaling for 25MW heating and low density case. The further reduction of bootstrap current is easier in helical systems than in tokamaks, because its currents strongly depend on the magnetic configuration, especially ellipticity and axis-shift, and are simply controlled by the poloidal field coils. Figure 8 shows the typical dependence of bootstrap current on central beta value β_0[7]. The circular points denote the self-consistent equilibrium calculation including bootstrap current in comparison with simple analytical scaling (solid line) and the current-free equilibrium analysis (square points)

The transport analysis on these LHD configurations has been carried out by using the global confinement scaling such as the LHD scaling[8], the neoclassical transport and the theoretical anomalous transport such as drift wave turbulence. The prediction of the LHD plasma confinement based on the empirical LHD scaling is shown in Fig.9. Recently, self-consistent electric-field formation due to fast ion loss during ICRF heating has been analyzed and the importance of radial electric field on LHD is clarified[9].

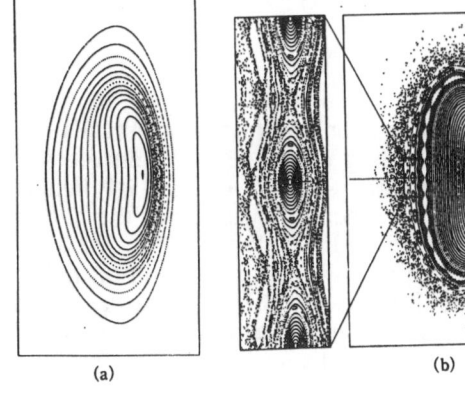
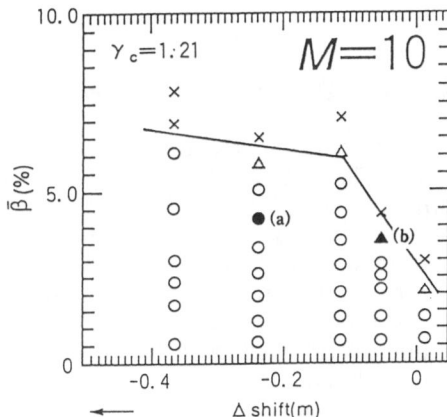

Fig.6 Equilibrium β-limit due to magnetic surface destruction (from ref.[5])
(a) inner shift case (averaged beta $\bar{\beta}= 4.5\%$), and (b) marginal limit case ($\bar{\beta}=3\%$)

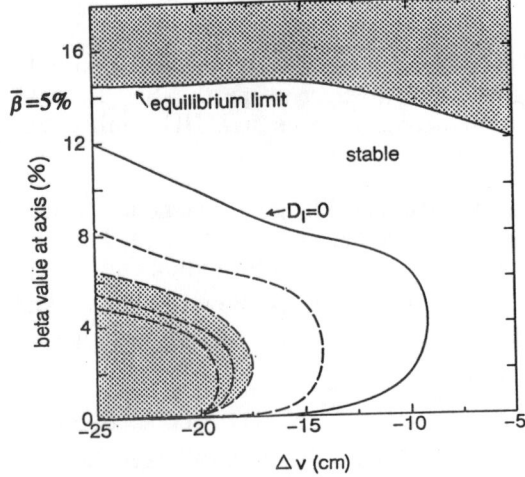
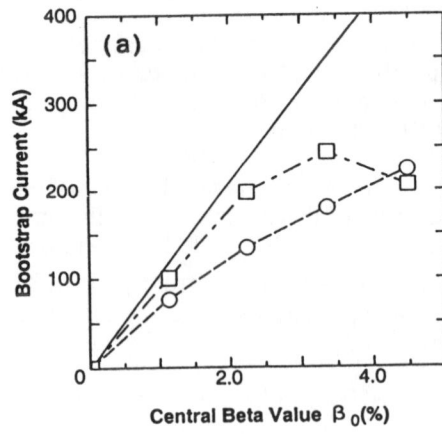

Fig.7 Mercier stability limits on LHD (from Ref.[6])

Fig.8 Bootstrap current in LHD (from Ref.[7])

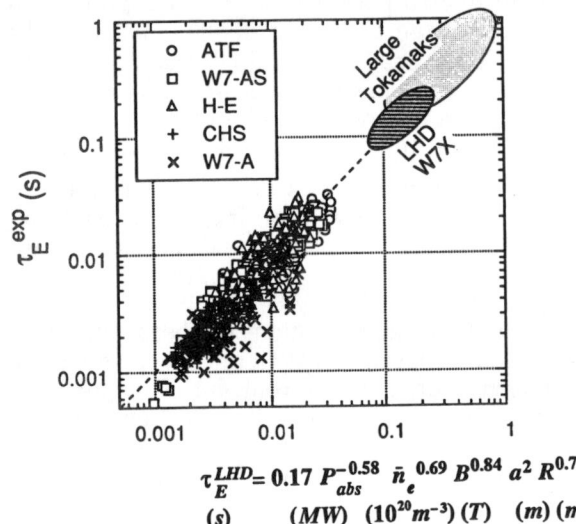

Fig.9 Plasma confinement prediction based on LHD scaling[8]

$$\tau_E^{LHD} = 0.17\, P_{abs}^{-0.58}\, \bar{n}_e^{0.69}\, B^{0.84}\, a^2\, R^{0.75}$$
$$(s) \quad (MW) \quad (10^{20} m^{-3})\, (T) \quad (m)\, (m)$$

4. STEADY-STATE OPERATION SCENARIO

The steady-state operation with high quality plasma is one of the major objectives of LHD. The LHD will be operated by using superconducting coils energized during daytime and the plasma will be maintained by means of steady-state 3~10MW RF heating methods, which in turn requires an active cooling of the vacuum vessel. For the density and impurity controls, helical divertor (Fig.10) is utilized on LHD. We will execute a first divertor experiment in helical system, which will give us fruitful information on helical divertor compared with the tokamak divertor. In addition to the helical divertor, a local island divertor has been recently proposed on LHD. By mean of these excellent divertors and elaborate plasma control techniques, new regimes with enhanced plasma performance will be explored.

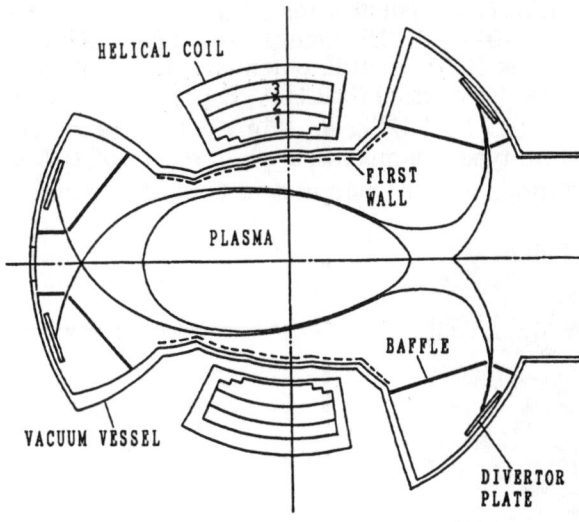

Fig.10
Divertor concept on LHD

5. R&D RESULTS FOR LHD

The various R&D for constructing SC magnet system, heating systems and diagnostic systems have been carried out in the past four years. After elaborate research and development of NbTi superconducting conductors[11], the final design of 13.0 kA (17.3kA in Phase 2) helical coil conductor with 12.5mm x 18.0mm cross-section was decided. The construction of the coil winding machine (Fig.11) for this pool-boiling type helical coil was completed and its winding has been started this year. It will take more than one and half years. Two sets of forced-flow type superconducting poloidal coils (Inner Vertical coils and Inner Shaping coils) was constructed, and the largest poloidal coil set (Outer Vertical coil, 5.55m major radius) is now under construction. Figure 12 shows the cross-section of conductor and the coil configuration for IV coils. The lower half of cryostat, the liquid helium refrigerator and the poloidal coil power supply were also completed last fiscal year. All components for main torus have been designed and constructed on schedule.

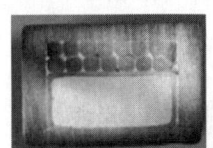
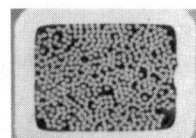

Fig.11 Helical coil winding machine and the superconducting conductor

Fig.12 Inner vertical coil and the superconducting conductor

Various R&D tests for ~20MW NBI system using negative ion source, ~10MW ECH(83GHz/166GHz) gyrotron and ~10MW ICRF antenna system have been performed. Several diagnostic equipments such as HIBP electric potential measurement and YAG Thomson scattering systems have also been tested for LHD. Especially, R&D researches for negative ion source with 45A, 125 keV H⁻ beam (Fig.13) have been successfully conducted[12]. Using 1/3-scale proto-type equipments, a world record of 16A extraction of negative ion beam has been performed with the current density of 45mA/cm^2 (Fig.14).

Fig.13 R&D facility for negative ion source and its power supply.

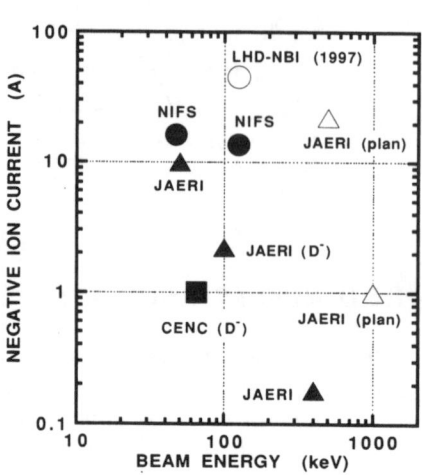

Fig. 14 Progress of negative ion beam development

6. SCHEDULE AND SUMMARY

In our new site (Toki-city, Gifu-prefecture), the construction of the main experimental hall of the LHD Building (Fig.15) was finished last fiscal year, and the above-stated winding machine and the lower cryostat were installed there.

Fig.15 Photograph of LHD Experimental Building.

The construction of the largest superconducting poloidal coil is also started in this experimental hall. Both the machine and the building constructions are on schedule.

In summary, the further understanding of LHD physics has been developed, and the steady-state operation scenarios have been clarified. After various R&D on superconducting coils, heating systems and diagnostic equipments, we have come to the stage of the fabrication of the real machine. The construction of the LHD machine will be completed in 1997.

REFERENCES

[1] A.Iiyoshi, M.Fujiwara, O.Motojima, N.Ohyabu, and K.Yamazaki, Fusion Technol., **17** (1990) 169.
[2] O.Motojima et al., Fusion Engineering & Design **20** (1993) 3.
[3] K.Yamazaki, N.Ohyabu, M.Okamoto et al., *in Plasma Physics and Controlled Nuclear Fusion Research 1990* (International Atomic Energy Agency, Vienna, 1991), Vol.2, P709.
[4] K.Yamazaki, O.Motojima, and M.Asao, Fusion Technol. **21** (1992) 147.
[5] T.Hayashi, A.Takei, and T.Sato., Phys. Fluids **B4** (1992) 1539.
[6] K. Ichiguchi, N.Nakajima, M.Okamoto, Y.Nakamura, and M.Wakatani, Nucl. Fusion, **33** (1993) 481.
[7] K.Watanabe, N.Nakajima, M.Okamoto, Y.Nakamura, M.Wakatani, Nucl. Fusion **32** (1992) 1499.
[8] S.Sudo et al., Nucl. Fusion **30** (1990) 11.
[9] S.Murakami, M.Okamoto et al., *in Proc. 15th International Conf. on Plasma Physics and Controlled Fusion Nuclear Research (Seville, Spain, 1994)* IAEA-CN-60/D-P-I-14.
[10] N. Ohyabu et al., Nucl. Fusion **34** (1994) 387.
[11] J.Yamamoto, O.Motojima et al., *in Proc. 15th International Conf. on Plasma Physics and Controlled Fusion Nuclear Research (Seville, Spain, 1994)* IAEA-CN-60/F-P-3.
[12] A.Ando et al., Phys. Plasmas **1** (1994) 2813.

CONFINEMENT PHYSICS OF NON-AXISYMMETRIC TORUS

M. Wakatani, Y. Ishii, T. Matsumoto, M. Yokoyama,
A. Takayama, and Y. Nakamura

Plasma Physics Laboratory, Kyoto University, Gokasho, Uji, Japan 611

Abstract

MHD stability of heliotron/torsatron type non-axisymmetric torus is analyzed for current carrying plasmas. For highly peaked current profiles the current-driven ideal internal kink mode is destabilized and sawtooth type oscillation is expected even in heliotron/torsatron. The L to H transition similar to tokamak case might be triggered by the sawtooth, when the plasma current exceeds a threshold value. An L to H transition model based on the poloidal viscosity is shown. ELM(edge localized mode) type oscillation is expected by the nonlinear interchange mode in the edge plasma.

1. Introduction

Recently the scaling law of energy confinement time called LHD scaling[1] is established based on Heliotron E[2], Heliotron DR[3] and CHS[4]. It is also shown that the LHD scaling is very similar to the gyro-Bohm scaling in ATF[5]. Based on this scaling law LHD(Large Helical Device) is designed and construction has been started. However, in order to design a compact heliotron/torsatron reactor, it is necessary to realize an improved confinement with about a factor two compared to the LHD scaling.

By including net toroidal plasma current the L to H transition has been observed in CHS[6], although the confinement improvement is not significant. After the transition ELM (edge localized mode) similar to that observed in tokamak H-mode plasmas appears. In this paper we study three subjects related to the L to H transition in the net current carrying heliotron/torsatron.

The first subject is the effect of the net plasma current on the MHD stability as shown in Section 2. In the Heliotron E type configuration the highly peaked plasma current may produce two resonant surfaces for the $(m, n) = (1, 1)$ mode, where $m(n)$ is the poloidal (toroidal) mode number. Then the current-driven internal kink mode is destabilized only at the inner $\iota = 1$ surface for low beta plasmas with $\beta(0) < 1\%$, where ι is the total rotational transform $\iota = \iota_h + \iota_J$ and $\beta(0)$ is the central beta value. Here ι_h is the rotational transform by the external helical magnetic field and ι_J is the one by the plasma current. When $\beta(0) > 1.5\%$, the outer $\iota = 1$ surface also contributes to the instability and the growth rate increases substantially. This result suggests that the sawtooth oscillation appears in the central region of current carrying low beta plasma in Heliotron E, which is consistent with the experimental data[7], and the heat pulse is observed in the edge plasma region.

The next subject is the mechanism of L to H transition in heliotron/torsatron as discussed in Section 3. In tokamaks many experimental results show that the poloidal shear flow driven by the radial electric field plays an essential role at the transition. It is assumed that the poloidal flow is determined by the balance between the momentum loss due to the ion orbit loss near the edge region and the poloidal viscosity[8][9]. When the poloidal viscosity is plotted as a function of the poloidal Mach number, the peak of the poloidal viscosity may correspond to a transition from a low flow velocity state to a high flow velocity one[8]. In heliotron/torsatron configurations such as Heliotron E and CHS, the most important ingredient to make a peak of poloidal viscosity is the reduction of neutral particle density or of momentum loss due to the change exchange reaction as well as the reduction of ion collision frequency. Since there is no L to H transition in the currentless plasmas of CHS, the heat pulse produced by the sawtooth type oscillation due to the plasma current is expected to reduce the neutral density or ion collision frequency at the edge plasma.

The last subject is the theoretical model for the edge localized model (ELM) as demonstrated in Section 4[10]. It is assumed that the L to H transition is triggered by the increase of poloidal shear flow which might be generated by the balance between the ion orbit loss and the poloidal viscosity. When the poloidal shear flow exceeds a critical threshold value, the nonlinear interchange mode shows the ELM-type relaxation oscillation based on the two dimensional slab plasma model[11]. This oscillation is related to the change of the poloidal velocity profile due to the Reynolds stress and the repetition frequency depends on what degree the poloidal velocity shear exceeds the threshold value. When the poloidal velocity shear increases, the repetition frequency becomes high. Thus the ELM appeared after the L to H transition is qualitatively explained by the nonlinear interchange mode in the serape-off layer (SOL). It is considered that the plasma current effect is negligible for the appearance of ELM, since the nonlinear interchange mode is destabilized in the SOL region. In order to confirm this scenario of the ELM we need data showing the poloidal shear flow layer just outside the last closed flux surface in CHS. This is a future experimental subject.

2. MHD Stability of Current Carrying Heliotron/Torsatron

In order to find an optimized magnetic configuration of heliotron/torsatron, MHD equilibrium and stability studies of current carrying plasmas are valuable, since the net toroidal plasma current may control the rotational transform, magnetic shear and magnetic well substantially. Here our concern is in the highly peaked current profile case which produces two resonant surfaces for the rotational transform $\iota = n/m$, where m and n are poloidal and toroidal mode numbers of helical MHD mode, respectively. For the Heliotron

E, the $(m, n) = (1, 1)$ mode may be most dangerous.

By solving the reduced stellarator MHD equations including dissipative processes such as resistivity and thermal conductivity in the large aspect ratio limit of Heliotron E, linear growth rates of $(m, n) = (1, 1)$ mode are obtained as a function of $\beta(0)$ in Fig.1. Although the dissipation is included, numerically obtained growth rates do not depend on the resistivity. For low beta plasmas with $\beta(0) \lesssim 1\%$ the $m = 1$ ideal internal kink mode is destabilized at the inner resonant surface and the central current carrying region shifts in one direction in the nonlinear stage (see Fig.2). For finite beta plasmas with $\beta(0) \gtrsim 1.5\%$ the mushroom like perturbation appears which manifests the existence of interchange drive(see Fig.3). For Fig.3, first the internal kink perturbation grows at the inner resonant surface and subsequent expansion of the perturbation to the outer resonant surface produces the interchange like perturbed state. Since the evolution of the $m = 1$ internal kink mode is essentially same as that in tokamaks[12], it is expected that the sawtooth type oscillation may appear in the Heliotron E with a highly peaked plasma current. This result seems consistent with the observation of soft-x ray and electron cyclotron emission showing the sawtooth oscillations[7]. The consequence of the unstable $m = 1$ internal kink mode is the heat pulse propagation to the plasma edge associated with the sawtooth crash. This may change the edge plasma parameters immediately.

3. L-H Transition in Heliotron/Torsatron

For the L to H transition in tokamaks several theories have been already developed and compared with experimental results. Recent results in W7-AS[13] and CHS[6] showing the L to H transition phenomenon similar to that in tokamaks suggest the same physical mechanism governs the L to H transition in both tokamaks and stellarators. In tokamaks the poloidal flow generation is closely related to the L to H transition. Shaing has developed a theoretical model to explain the poloidal flow generation based on the balance between the poloidal viscosity and the momentum loss due to orbit loss near the plasma surface[8]. Since the poloidal viscosity depends on the plasma flow velocity nonlinearly, it is possible to have two types of solution ; one is low flow velocity state and the other is high velocity state with $M_p > 1$, where M_p is the Mach number of poloidal flow. Shaing also has shown that the nonlinear poloidal viscosity depends on the spectrum of the magnetic field[9]. This means that the poloidal flow generation shows different behavior in stellarator/heliotron compared to tokamaks.

Figure 4 and Figure 5 show poloidal viscosity $\langle \mathbf{B}_p \cdot \nabla \overleftrightarrow{\Pi} \rangle$ as a function of poloidal Mach number is CHS and Heliotron E, respectively[14]. When the normalized ion collision frequency ν_{*i} is not large, reduction of neutral density producing the charge exchange momentum loss at the plasma edge makes two peaks of poloidal viscosity. The first peak with $M_p \simeq 1 \sim 2$ corresponds to that in tokamaks and the second peak appeared at $M_p \simeq |\bar{m} - \bar{n}q|/\bar{m}$ is generated by the stellarator field, where q is a safety factor and

$\bar{m}(\bar{n})$ is poloidal (toroidal) mode number of equilibrium magnetic field. It is considered that when the first peak is clearly seen, transition from the low flow velocity state to the high flow velocity one is possible. Thus the L to H transition might be realized in heliotron/torsatron as well as tokamaks, when the poloidal viscosity shows the clear peak.

The remaining question is that the L to H transition was not observed in currentless plasmas of both CHS and Heliotron E but it is obtained for current carrying plasmas in CHS[6]. One possible explanation is that a current-driven instability appears in the latter case, which may make a difference in the plasma edge. When the current profile becomes highly peaked, the sawtooth type oscillation may produce rapid heat flow from the central region as shown in Section 2. This heat flow may reduce the neutral density at the plasma surface and produce the large poloidal flow as expected from Fig.4 for CHS.

4. ELM-Type Oscillation of Nonlinear Interchange Mode

During the H mode in tokamaks sometimes ELM occurs. The similar ELM is also seen in the H-mode of CHS[6]. It is not clear whether all ELM in both tokamaks and CHS are excited by the same mechanism or not. Here it is demonstrated that the two-dimensional nonlinear interchange mode produces the ELM-type oscillation when the strength of shear flow exceeds a threshold value. The interchange mode in the SOL plasma of heliotron/torsatron is described by the volticity equation and the pressure evolution equation[10][11]. The numerical solution of these nonlinear coupled equations in the $x - y$ plane is shown in Fig.6, where total kinetic energy and kinetic energies of several harmonics are plotted as a function of time. A relaxation oscillation is clearly seen when the background velocity in the y direction is given by $V_y = (2\pi\Phi_0/d)\sin(2\pi x/d)$ and $\Phi_0 > \Phi_{oc}$ where d is the width of SOL plasma and Φ_{oc} is a critical value. The time evolution of the averaged velocity in the y direction is shown in Fig.7. From Fig.6 and Fig.7 it is seen that the disappearance of the velocity shear makes a rapid increase of fluctuation energy. Since this enhanced fluctuation generates velocity shear due to the Reynolds stress and the fluctuation becomes suppressed. Thus the interaction between the shear flow and the interchange drive in the SOL plasma through the Reynolds stress produces the relaxation oscillation which may be the origin of the ELM in CHS or tokamaks appeared in H-mode plasmas.

5. Concluding Remarks

In order to improve confinement in heliotron/torsatron devices, the L to H transition has been also studied in Heliotron E and CHS. In CHS the L to H transition similar to that in tokamaks was observed for the current carrying plasmas. A theoretical model to explain the L to H transition and the ELM in the H-mode of CHS has been proposed. For the transition the current-driven internal kink mode plays a role. For the L to H

transition the orbit loss is necessary to produce the large poloidal shear flow in the edge plasma region.

The poloidal flow is given by the balance between the orbit loss and the poloidal viscosity. The behavior of poloidal viscosity depends on the magnetic spectrum of heliotron/torsatron; however, possibility to obtain a large poloidal flow exists for the edge plasma with a substantially low neutral density.

The ELM may be triggered by the nonlinear interchange mode in the SOL plasma when a substantial shear flow exists.

We acknowledge K.C. Shaing (Oak Ridge National Laboratory) and K. Toi (National Institute of Fusion Science) for useful suggestions.

References

(1) S. Sudo et al., Nucl. Fusion 30 (1990) 11.

(2) K. Uo et al., in 8th IAEA Conf. on Plasma Phys. and Cont. Fusion Research (Brussel, 1980) vol.1, p.217.

(3) S. Morimoto et al., Jap. J. Appl. Phys. 28 (1989) L1470.

(4) K. Nishimura et al., Fusion Technology 17 (1990) 86.

(5) M. Murakami et al., in 14th IAEA Conf. on Plasma. Phys. and Cont. Fusion Research (Würzburg, 1992) vol.2, p.391.

(6) K. Toi et al., in 14th IAEA Conf. on Plasma Phys. and Cont. Fusion Research (Würzburg, 1992) vol.2, p.461.

(7) M. Wakatani et al., Nucl. Fusion 23 (1983) 1669.

(8) K.C. Shaing and E.C. Crume, Jr., Phys. Rev. Lett. 63 (1989) 2369.

(9) K.C. Shaing, Phys. Fluids B5 (1993) 3841.

(10) A. Takayama, M. Wakatani and H. Sugama, submitted in J. Phys. Soc. Jpn.

(11) H. Sugama and M. Wakatani, J. Phys. Soc. Jpn 59 (1990) 3937.

(12) B.B. Kadomtsev, Sov. J. Plasma Phys. 2 (1976) 533.

(13) V. Erkmann et al., Phys. Rev. Lett. 70 (1993) 2086.

(14) M. Yokoyama, M. Wakatani and K.C. Shaing, to be published in Nucl. Fusion.

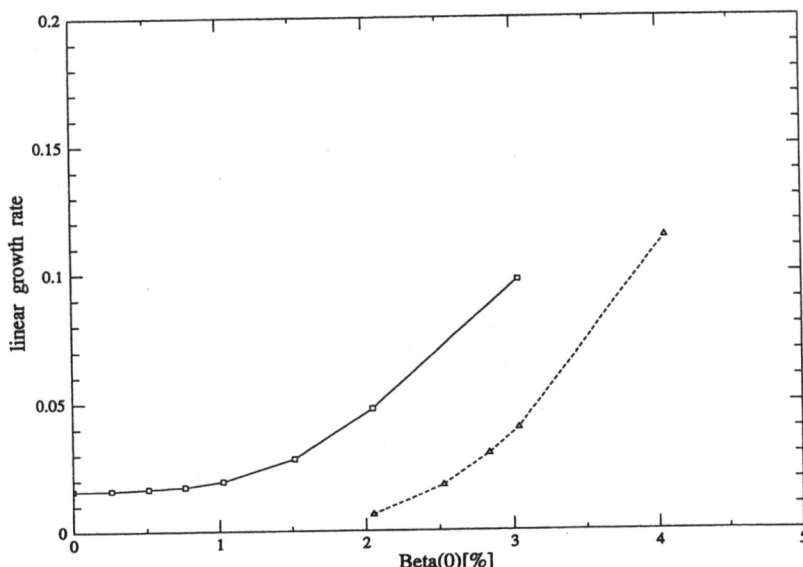

Fig.1 Growth rate versus central beta value $\beta(0)$ for the pressure profile of $p = p_0(1-r^2)^2$. The upper curve corresponds to the highly peaked current profile and the lower curve corresponds to the broad current profile under the almost same total current.

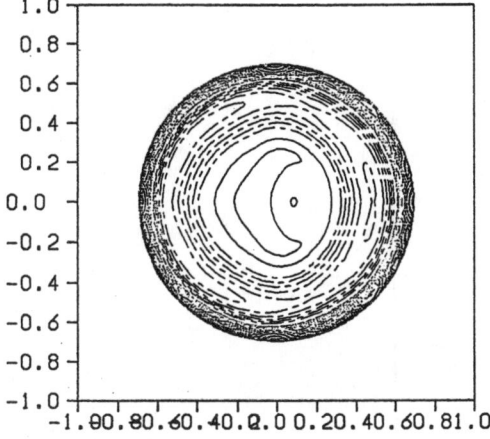

Fig.2 Contours of stream function at the saturated state for $\beta(0) = 1\%$ with the highly peaked current profile.

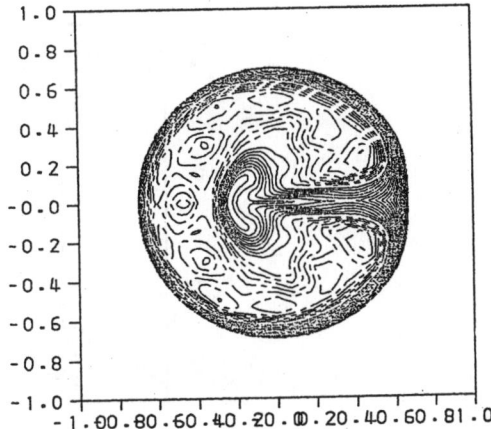

Fig.3 Contours of stream function at the saturated state for $\beta(0) = 1.5\%$ with the highly peaked current profile.

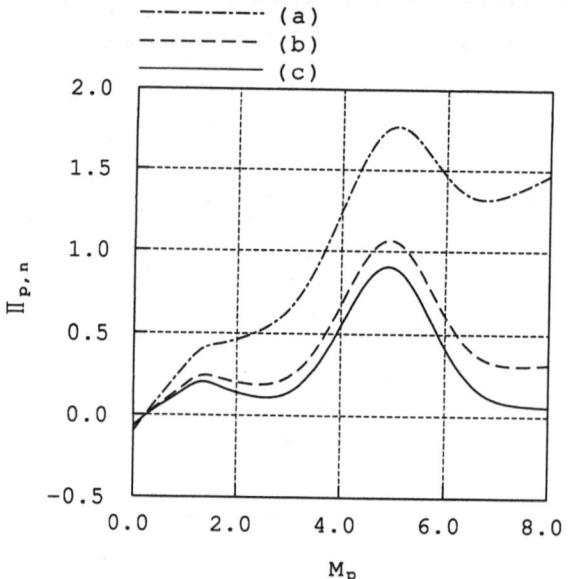

Fig.4 Poloidal viscosity $\Pi_{p,n}$ versus Mach number of poloidal flow M_p in CHS for $(a)\nu_{*i} = 1.0$, $\nu_{eff}/(v_T/Rq) = 5.0$, $(b)\nu_{*i} = 1.0$, $\nu_{eff}/(v_T/Rq) = 1.0$ and $(c)\nu_{*\nu} = 1.0$, $\nu_{eff}/(v_T/Rq) = 0.1$. Here $\nu_{*i} = \nu_i Rq/(v_T \epsilon_t^{3/2})$, ν_i is an ion collision frequency and ν_{eff} is a collision frequency corresponding to charge exchange reaction.

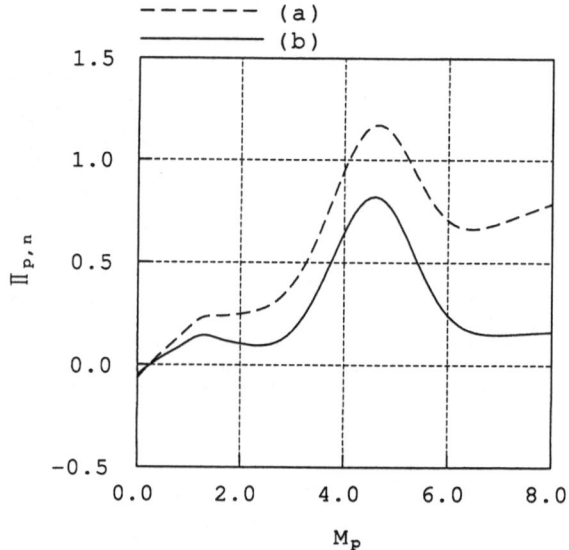

Fig.5 $\Pi_{p,n}$ versus M_p in Heliotron E for $(a)\nu_{*i} = 1.0$, $\nu_{eff}/(v_T/Rq) = 5.0$ and $(b)\nu_{*i} = 1.0$, $\nu_{eff}/(v_T/Rq) = 1.0$.

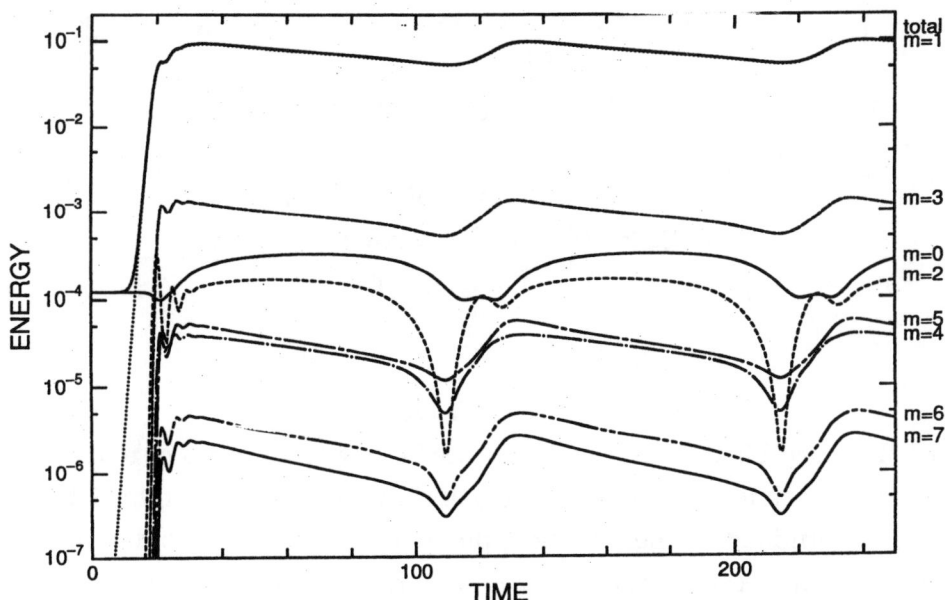

Fig.6 Time evolution of kinetic energy of dominant Fourier modes for $\Phi_0 = 7.0 \times 10^{-3}$, ν(viscosity) $= \chi$(thermal diffusivity) $= 9.0 \times 10^{-3}$. Total kinetic energy is also shown. Time is normalized by $(g\kappa)^{-1/2}$.

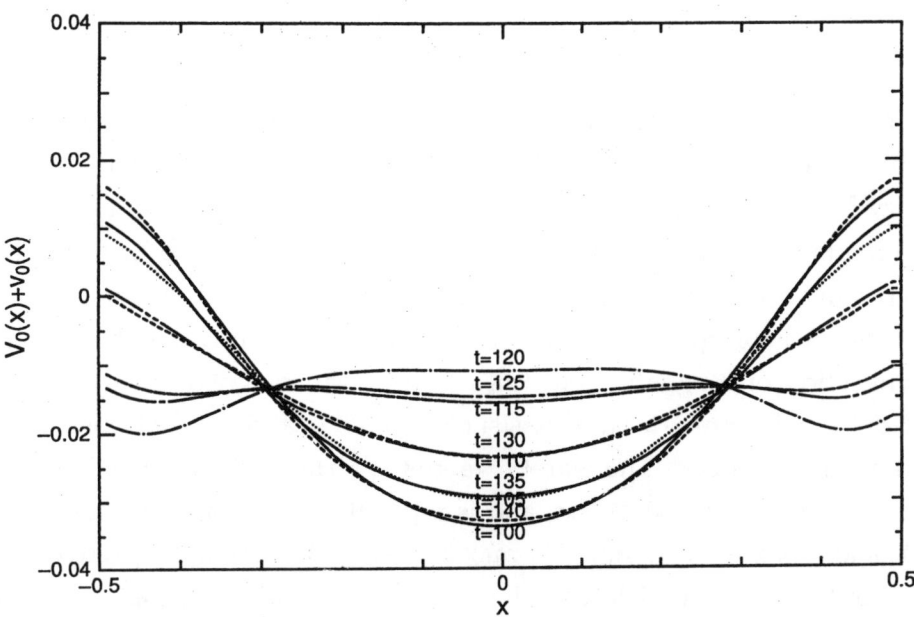

Fig.7 Radial profile of poloidal shear flow $V(x, t)$ for $100 \leq t \leq 140$. The stabilizing phase is at $t \simeq 100$ or $t \simeq 140$ and the destabilizing phase is at $\simeq 120$ in Fig.6. Here x denotes the inhomogeneous direction of background plasma.

REVERSED FIELD PINCH PHYSICS

S. Ortolani*

Istituto Gas Ionizzati CNR
EURATOM-ENEA-CNR Association
Padova, Italy

ABSTRACT

The Reversed Field Pinch (RFP) profiles belong to the class of magnetoplasma systems which tend to exist naturally and can be produced in the laboratory in states determined by the ratio of applied volt seconds to toroidal flux. This, in turn, determines the ratio of toroidal current to toroidal flux, I/Φ, which, theoretically, completely determines the magnetic field profiles. In recent years, significant progress has been made in the understanding of the physics underlying the relaxation process and the corresponding magnetic field and plasma dynamics. Recent results from experimental and theoretical studies are reviewed.

INTRODUCTION

The mean magnetic field profiles of many plasma confinement systems reveal properties of self-organization which can be described by Taylor's relaxation theory [1,2,3]. In particular, the RFP magnetic distribution is, under certain constraints, a minimum energy state described by $\nabla \times \underline{B} = \mu \underline{B}$ with uniform μ. The plasma confined in the RFP relaxes towards this minimum energy state in the presence of dissipative mechanisms which allow the breaking and reconnection of magnetic field lines. The mean magnetic field profiles are toroidally axisymmetric and consist of a poloidal and toroidal component of similar amplitudes. The safety factor $q = rB_\phi / RB_\vartheta$ is < 1, decreases monotonically with the radius, and is typically $\approx (2/3)(a/R)$ on the magnetic axis (a and R are the minor and major radii of the torus, respectively). The value of the magnetic field line shear, $(1/q)(dq/dr)$, is large, especially in the outer region.

---

* Present address: ITER San Diego Joint Work Site, USA

According to the theory, the final state is completely determined by the ratio of applied volt seconds to toroidal flux. This, in turn, determines μ (i.e. the ratio of toroidal current to toroidal flux, I/Φ,) which completely determines the magnetic field profiles.

The equilibrium RFP distribution is such that there exists an infinity of rational surfaces closely spaced together in the proximity of the toroidal field reversal radius. Therefore, a high probability of growth of a large number of unstable perturbations accompanied by a high occurrence of magnetic island overlapping is predictable. This is likely to influence plasma transport. On the other hand, these perturbations are essential for relaxation to occur. In fact, the existence of some level of turbulence is necessary to disrupt the internal symmetry and to allow the configuration access to a lower energy state.

The relaxation process is central to RFP physics. However, it is relevant also for other configurations as the Spheromak and the Multipinch [2]. Furthermore, manifestations of this relaxation process can be quasi-continuous in systems characterized by well-developed turbulence or they can be violent, as in the case of more ordered structures such as, for example, tokamak disruptions and solar flares [3].

Research is being presently carried out actively in many experiments [4-10]. In this paper we will briefly review the proposed interpretations for the mechanism of relaxation and discuss some examples of recent advances in the understanding of the magnetic field dynamics.

MECHANISMS OF RELAXATION

In the RFP configuration there is a large poloidal current in the outer plasma region at about the toroidal field reversal region. Power is required to sustain this poloidal current in plasma regions where resistive dissipation is usually high. Ordinary field diffusion would lead to a loss of the configuration on a resistive diffusion time scale. However, contrary to classical expectation, the profiles are observed to persist for longer times preserving the global mean field distribution and in particular the toroidal field reversal and correspondingly large poloidal currents in the outer region. The existing interpretations can be divided into two broad categories:

- Resistive MHD models [3] where an additional driving force in Ohm's law arises self-consistently from instabilities which can give rise to coherent fluctuations in plasma velocities and magnetic fields, which can, in turn, induce an electromotive force parallel to the magnetic field, thus sustaining the current density distribution despite resistive diffusion. The parallel Ohm's law is in this case written as $E+<vxb>=\eta j$. This description is usually referred to as the MHD dynamo although the term has been improperly taken from the astrophysical dynamo [3].

- Models which postulate the existence of a stochastic magnetic field. This concept is at the basis of the so called Kinetic Dynamo Theory (KDT) [11] and Tangled Discharge Model (TDM) [12]. Although the KDT and TDM models apply to different collisional regimes, in both cases the transfer of the field-aligned current from the interior of the plasma to the colder resistive edge region is accomplished directly by electrons streaming along stochastic magnetic field lines. This mechanism provides the sustainment of the profiles against resistive diffusion.

However, irrespective of the model used, heat transport in stochastic magnetic fields can be a dominant process in relaxing magnetic profiles. In fact, magnetic fluctuations can lead to a mixing of the magnetic field lines in the plasma in such a way that a single field line no longer traces out a flux surface, but rather can fill a large fraction of the plasma volume. In this case, changes in particle velocities are due to collisions and to random magnetic field fluctuations. The rapid transport of thermal energy along these stochastic field lines may result in an anomalously rapid loss of energy from the core of the plasma to the edge compared to what would be expected from perpendicular thermal conduction alone. At least in the Reversed Field Pinch, this process may contribute significantly to the experimentally measured confinement properties. Correlation between magnetic field fluctuations and plasma transport has been experimentally demonstrated by recent measurements in MST [13,14,15].

EXPERIMENTAL MEASUREMENTS OF MAGNETIC FIELD DYNAMICS

In MST [5] measurements of the MHD dynamo term have been made in the plasma edge [16]. In Fig. 1a the measured electric field is compared to the resistive term and in Fig. 1b, to the difference between the resistive term and the applied electric field. The corresponding radial profiles are shown in Fig. 2. These results indicate that the MHD dynamo description of the profile relaxation [3] holds both between and during sawtooth events within experimental error bars.

New detailed measurements of magnetic field fluctuations have been made in the EXTRAP T1 experiment [17]. In Fig. 3 is shown an example of contour plots of constant poloidal magnetic field fluctuation energy as a function of time and toroidal mode number. The dominant modes are resonant near the axis. The amplitudes of these modes are seen to oscillate in a quasi-cyclical fashion. For the same discharge, in Fig. 4 is shown the time evolution of the experimentally measured μ profile. A correlation can be seen between the profile dynamics and the m=1 activity. This is particularly evident at about the three time instants, marked on the figure, where the m=1 power peaks and the spectrum is broadened with a predominant double-peaked structure.

MHD MODELING AND SIMULATIONS

3D-MHD simulation studies continue to provide an essential support for the interpretation of experimental measurements. Recently, a comparison between the fluctuation spectra from a 3D-MHD code [18] and those measured in the EXTRAP T1 experiment has been made [8]. In particular, the study aimed at analyzing the effects of different plasma resistivity profiles. These results, shown in Fig. 5, clearly demonstrate that higher amplitude fluctuations are observed, both experimentally and in the code, when the resistivity is larger in the outer plasma region. This is associated with the need of a larger fluctuation induced electric field to drive the parallel (mostly poloidal) current in the higher resistivity edge region. These results also point to the importance of developing profile control techniques and current drive schemes to improve the confinement properties of the RFP.

In a paper at this Conference [19] the role of strong current sheets generated intermittently during the sustainment phase of an RFP configuration is discussed. It is worthwhile to note that the dynamics of these current sheets is likely to play a role in the generation of fast electrons and anomalous ion heating observed in experiments [20,21].

Finally, it is noteworthy that the recently proposed dynamics for the tokamak sawteeth collapse [22] based on a two-stage reconnection process (which eventually allows q on axis to remain less than 1) is similar to the double reconnection mechanism which can explain the sustainment of the RFP configuration with a single helicity mode [23].

In summary, the RFP is an excellent configuration for physics studies on relaxation phenomena, MHD instabilities and plasma transport. The fundamental properties of the relaxation process have been well established and understood. Plasma confinement in these magnetic profiles is probably limited by the relatively high intrinsic level of fluctuations. However, careful control of the profiles remains to be explored and optimized. The results of recent experimental and theoretical studies have consolidated the MHD interpretation of the plasma relaxation process.

REFERENCES

[1] J.B. Taylor, *Proc. 5th Int. Conf. on Plasma Physics and Controlled Nuclear Fusion Research*, Tokyo 1974, IAEA Vienna, vol. I, 161, 1975.

[2] J.B. Taylor, *Reviews of Modern Physics*, **58**, 741, 1986.

[3] S. Ortolani, D.D. Schnack, *Magnetohydrodynamics of plasma relaxation*, World Scientific Publishing, 1993.

[4] V. Antoni, et al. Comparison of different RFX operation modes, *15th Int. Conf. on Plasma Physics and Controlled Nuclear Fusion Research* Seville 1994, IAEA-CN-60/ C-1-I-2.

[5] J.S. Sarff, et al. Fluctuations and transport in the Reversed Field Pinch: characterization and reduction, *15th Int. Conf. on Plasma Physics and Controlled Nuclear Fusion Research* Seville 1994, IAEA-CN-60/ C-1-I-5.

[6] Y. Yagi, et al. Fluctuations and confinement in the TPE-1RM20 Reversed Field Pinch, *15th Int. Conf. on Plasma Physics and Controlled Nuclear Fusion Research* Seville 1994, IAEA-CN-60/ C-1-I-3.

[7] K. Hattori, et al. Poloidal and toroidal divertor experiments on the TPE-2M Reversed Field Pinch, *15th Int. Conf. on Plasma Physics and Controlled Nuclear Fusion Research* Seville 1994, IAEA-CN-60/A6/C-P-7.

[8] P. Nordlund, et al. Loop voltage anomaly and resistivity profile effects in the EXTRAP T1 Reversed Field Pinch, *15th Int. Conf. on Plasma Physics and Controlled Nuclear Fusion Research* Seville 1994, IAEA-CN-60/A6/C-P-6.

[9] Zhang Peng, et al. Reversed Field Pinch and Ultra Low Q discharges in SWIP-RFP device, *15th Int. Conf. on Plasma Physics and Controlled Nuclear Fusion Research* Seville 1994, IAEA-CN-60/A6/C-P-9.

[10] P. Greene and S. Robertson, Locking of kink modes in a Reversed Field Pinch, *15th Int. Conf. on Plasma Physics and Controlled Nuclear Fusion Research* Seville 1994, IAEA-CN-60/ C-1-I-2.

[11] A.R. Jacobson and R.W.Moses, *Physical Review* A **29**, 3335, 1984.

[12] M.G. Rusbridge, *Plasma Physics*, **19**, 499, 1977.

[13] S. Hokin, et al. Magnetic fluctuation induced transport and edge dynamo measurements in the MST Reversed Field Pinch, *15th Int. Conf. on Plasma Physics and Controlled Nuclear Fusion Research* Seville 1994, IAEA-CN-60/A6/C-P-8.

[14] G. Fiksel, et al. *Physical Review Letters* **72**, 1028, 1994.

[15] M.R. Stoneking, et al. *Physical Review Letters* **73**, 549, 1994.

[16] H. Ji, et al. *Physical Review Letters* **73**, 668, 1994.

[17] S. Mazur and P. Nordlund Toroidal phase locking phenomena and modal interactions with the mean fields in the EXTRAP T1 Reversed Field Pinch, *21st EPS Conference on Controlled Fusion and Plasma Physics* Montpellier, France 27 June - 1 July 1994.

[18] D.D. Schnack, et al. *Computer Physics Communications* **43**, 17, 1986.

[19] S. Cappello and D. Biskamp MHD studies of stationary turbulent dynamics in Reversed Field Pinches *This Conference.*

[20] M. Erba et al. *Nuclear Fusion* 33, 1577, 1993.

[21 S. Ortolani, *Plasma Physics and Controlled Fusion,* **34**, 1903, 1992.

[22] D. Biskamp and J.F. Drake *Physical Review Letters* **73**, 971, 1994.

[23] E.J. Caramana, R.A. Nebel and D.D. Schnack, *Physics Fluids*, **26**, 1305, 1983.

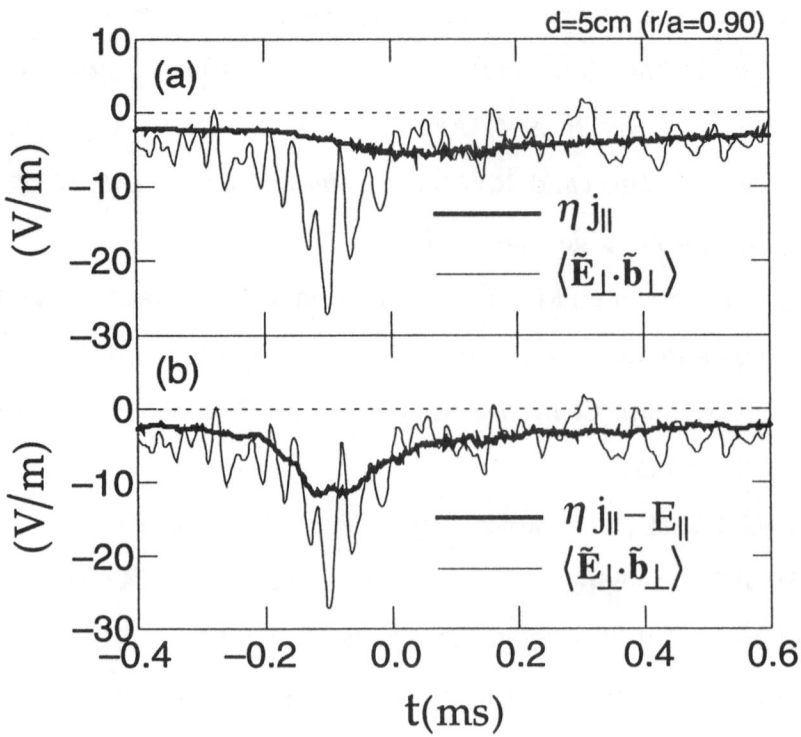

Fig. 1. Comparison of MHD dynamo electric field $\langle \tilde{E}_\perp \cdot \tilde{b}_\perp \rangle$ to (a) $\eta j_\parallel$ and (b) $\eta j_\parallel - E_\parallel$ during one sawtooth crash. Rapid oscillations in $\langle \tilde{E}_\perp \cdot \tilde{b}_\perp \rangle$ indicate experimental uncertainty.[16]

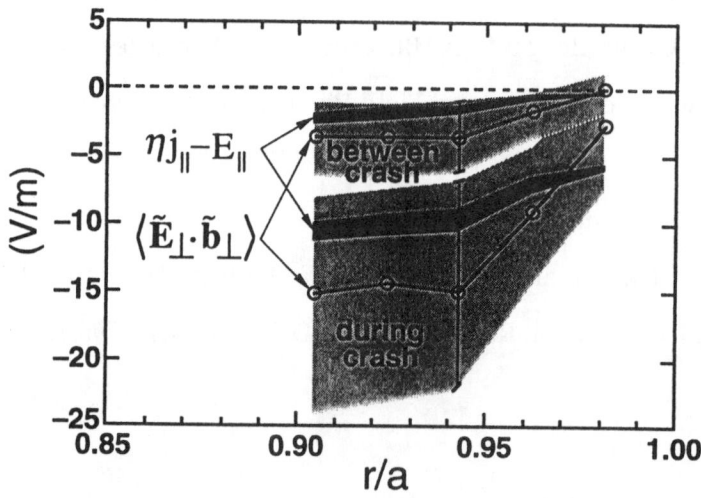

Fig. 2. Edge radial profiles of MHD dynamo electric field $\langle \tilde{E}_\perp \cdot \tilde{b}_\perp \rangle$ and $\eta j_\parallel - E_\parallel$ between and during the sawtooth crashes. The shaded region (blackened region) indicates the measurement uncertainty of $\langle \tilde{E}_\perp \cdot \tilde{b}_\perp \rangle$ ($\eta j_\parallel - E_\parallel$). Here the measurement uncertainty is determined from the ranges of rapid oscillations of Fig. 1.[16]

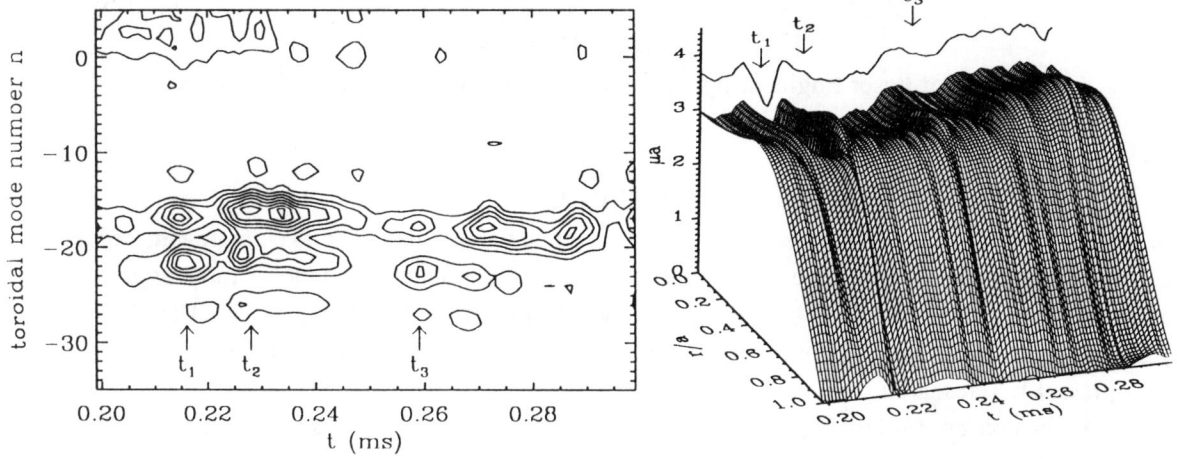

Fig. 3. Contour plot of the time evolution of the $m = 1$ helical spectrum for a $\Theta \approx 2.2$ discharge.[17]

Fig. 4. Time evolution of the associated experimental $\mu a = \mu_0 a \mathbf{j} \cdot \mathbf{B}/B^2$ profile (same discharge as in Fig. 3).[17]

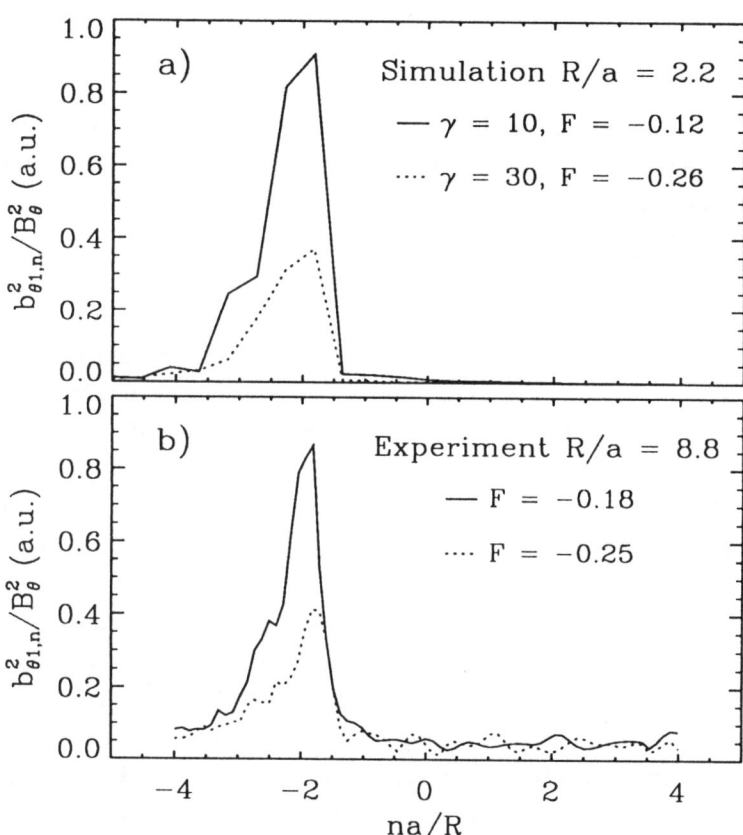

Fig. 5. Comparison of normalised helical Fourier mode spectra of $m = 1$ fluctuations in the edge poloidal field from the MHD simulations and the T1 experiment. a) Simulation spectra: $\Theta = B_\theta(a)/\langle B_\phi \rangle = 1.7$, $\eta(r) = (1+9(r/a)^\gamma)^2$ with $\gamma = 30$ ($\cdots$) and $\gamma = 10$ (—). Narrowing the resistivity profile the reversal decreases from $F = B_\phi(a)/\langle B_\phi \rangle = -0.26$ to $F = -0.12$. b) Experimental spectra: $\Theta = 1.7$, $F = -0.25$ ($\cdots$) and $F = -0.18$ (—). [8]

REVIEW ON LASER ICF PHYSICS
K. Mima and H. Takabe
Institute of Laser Engineering, Osaka University, Suita, Osaka 565, JAPAN

ABSTRACT

The recent developments of the laser ICF physics are reviewed in this paper. The high gain target design is important at present since the ignition experiments are planned and the ignition target parameters have to be determined by the fusion yield scaling for high gain implosion. The hydrodynamic instabilities and implosion symmetry are discussed in details, which are the key issues in the present ICF research.

A new concept for igniting ICF plasmas, so called "fast ignition" was proposed by LLNL group recently. This concept is interested because the symmetry requirement will be reduced. The key physics discussed here on this concept is the relativistic laser plasma interactions. Finally, the present status of National Ignition Facility program and the related research works in U.S. are presented.

§1. INTRODUCTION

Since the middle of '80s, implosion experiments of high density compression and hot spark formation have been carried out both for the direct and indirect drive [1,2,3]. In the high density compression experiments, imploded are pellets which consist of CDT plastic shell, DT cryogenic shell and DT gas contained in a plastic capsule by Gekko XII laser at ILE, Osaka Univ. [1], NOVA laser at LLNL[2], OMEGA laser at LLE, Univ. of Rochester[3], and so on. As the results of the experiments in recent ten years, the imploded core plasmas are compressed to 100~600 times solid density and the symmetry requirements for the formation of hot spark became clear. Although the high density compression is successful and the compressed densities agree with the results of 1D simulations, the neutron yield is significantly degraded in comparison with the one dimensional simulation results. The mechanism of the neutron yield degradation is attributed to the turbulent mixing between central hot spark and cold main fuel. Therefore, the critical issue of the next step of ICF research is to clarify the symmetry and stability conditions required for hot spark formation and igniting DT fuel.

Since the stability of implosion is sensitive to the imploding shell isentrope, we compare two typical MJ laser target designs which are (1) low implosion velocity and low isentrope target and (2) high implosion velocity and high isentrope target [4]. The 1D implosion simulations show that the high implosion velocity is achieved with low inflight aspect-ratio by high intensity laser irradiation. In this implosion, we can ignite the DT fuel with small laser energy and keep the e-foldings of Rayleigh-Taylor instability lower. Although the gain is saturated at a few MJ laser energy to be less than 100, the high velocity implosion is considered to be appropriate for the ignition experiment in the near future.

In order to relax the symmetry and stability requirements, a new ignition concept was recently introduced[5], which is so called "fast ignition". In this concept, a compressed DT fuel is directly heated by an intense short pulse laser. Since the central hot spark is not necessary in this concept, the conditions for implosion symmetry is significantly reduced. The critical physics of this concept is the coupling between an intense short pulse laser and the high density core plasmas. Namely, the critical issues are the propagation of intense short pulse laser and the generation and transport of MeV electron and ion in coronal plasmas. Although 2D particle simulations have been carried out to investigate the channeling of the laser pulse into a dense plasma region and the absorption processes, experimental investigations are still premature and expected to be carried out in near future.

In this paper, also reviewed are the National Ignition Facility (NIF) program and the related research results[6]. Recently, the engineering design action for NIF has been approved by the Department of Energy of US. According to the present plan, the NIF will be completed before the end of 2002. The laser for NIF will deliver 1.8 MJ pulse with 0.35 μm wavelength and have 192 beams which are focused into a hohlraum or on a pellet directly. The NOVA experiments have been carried out to clarify the implosion physics related to the NIF target design.

In this paper, the high gain target design is presented in §2. In §3, presented is the theoretical analysis on the mechanisms of the degradation of neutron yield in the Gekko XII experiments. The section 4 is devoted to the review of the fast ignition concept. Finally, in §5, we show the recent results of NOVA experiments and the NIF program. The summary is given in §6.

§2. ICF SCENARIO AND HIGH GAIN TARGET

There are two approaches in the laser inertial confinement fusion. They are direct drive and indirect drive which use electron heat conduction and x-ray heat conduction respectively to produce a high shell ablation pressure for the implosion as shown in Fig.1.

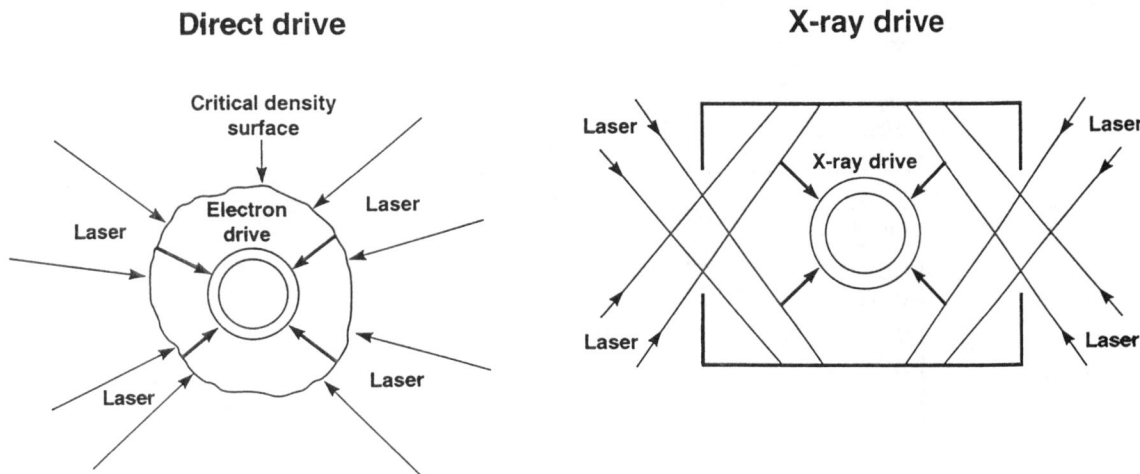

Fig.1. Schematic diagram of laser implosions by direct drive (a) and indirect drive (b).

In the direct drive (Fig.1 left), laser beams are illuminated directly on a capsule surface, which are absorbed on the critical surface. The absorbed energy is transported toward the ablation surface by electron conduction. The energy coupling between laser and core plasma is efficient in comparison with the indirect drive. However, the irradiation symmetry and ablation surface stability are critical in this scheme.

In the indirect drive (Fig.1 right), laser energy is converted into x-rays in a hohlraum. The laser beam nonuniformities are relaxed inside the cavity and the hydrodynamic instabilities are reduced because the mass ablation rate is higher for x-ray driven ablation. However, the energy coupling is not high in particular when the driver energy is not as high as a few Mega jules. Namely, 1～2 MJ laser or ion beam energy is needed for the ignition.

Target designs for 1 MJ laser pulse have been carried out in order to clarify relations between high gain target and that for ignition experiment [7]. The minimum implosion velocity for ignition is 3×10^7 cm/sec. The required ignition energy decreases with increase of implosion velocity, v, which is scaled by v^{-5} both for direct and indirect drive. The required ablation pressure P_a normalized by the solid DT Fermi pressure P_0 is estimated as follows [8],

$$P_a/P_o = v^5(\rho_o/P_o A_{inf})^{2.5} \alpha^{-1.5}$$

where A_{inf}, α and ρ_o are inflight aspect ratio, isentrope and solid DT density. Because of the stability requirement, A_{inf} can not be large. Therefore, the fast implosion needs high ablation pressure and high isentrope. In the table 1, shown are typical design parameters for slow and fast implosions. Since the transition from sub-ignition to ignition occurs around 100 kJ ~ 300 kJ laser energy for the fast implosion, the gain for 1 MJ input energy is higher. In order to increase the implosion velocity by factor 2, we have to increase the laser intensity and the isentrope by factor 3. Note also that we use DT instead of plastic as an ablator in fast implosion in order to reduce the initial Fermi pressure P_0, because the required ablation pressure is lower in that case.

In conclusion of this section, the fast implosion is attractive for the ignition demonstration experiments with direct irradiation.

Table 1. Comparison of target and plasma parameters between slow velocity implosion and high velocity implosion.

Implosion Mode		slow (high gain)	fast (medium gain)
Implosion Velocity (cm/sec)		3×10^7	6×10^7
Isentrope		2	6
In-flight Aspect Ratio		100	70
Core Structure		Isobaric	Isochoric
Laser Intensity (W/cm^2)	foot	1×10^{13}	3×10^{13}
	peak	10^{15}	6×10^{15}
Ablator		CH	DT
Pellet Gain	1 MJ	5	25
	4 MJ	150	60

§3. TURBULENT MIXING IN DIRECT DRIVE IMPLOSION

In laser implosion, pellets have been compressed successfully to a few hundreds times solid density [1,2,3]. However, the neutron yield was far below than that of the corresponding one dimensional numerical simulation. In order to quantitatively explain the degradation of the neutron yield, the turbulent mixing between pusher and fuel is investigated in the implosion. The turbulent mixing models have been developed by Young [10], Takabe [9,11] and Haan [12]. In the analysis of CD shell implosions by Gekko XII, we use the turbulent diffusion coefficient calculated by the k-ε type model [13] in the hydrodynamic code. The k-ε type model by Gauthier [13] was extended to two temperature cases when it is installed in the 1D hydrodynamic code.

In the simulation [9], the turbulence is assumed to be generated in the acceleration phase and reach a level of $\xi = k/(1/2\, u_o^2)$, where u_o is the implosion velocity and $k = 1/2 \langle v_1^2 \rangle$ is the kinetic energy of the turbulence.

In Fig.2, the experimental neutron yields normalized by 1D simulation neutron yields are compared with those calculated by the turbulent mixing model. The normalized neutron yield decreases with increase of the convergence ratio which is defined by the ratio of initial target radius to the compressed plasma radius. The experimental results agree with the curves of $\xi = 10\%$ and 20% as shown in Fig.2.

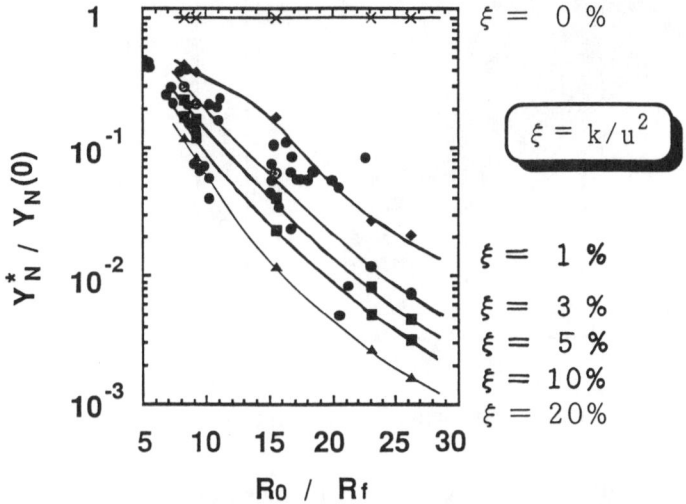

Fig.2 Convergence ration dependence of experimental neutron yields normalized by the neutron yield of 1D implosion code without mixing. Curves indicate reductions of neutron yields for various turbulence level.

The degradation mechanism of the neutron yield by the turbulent mixing is investigated in Figs.3(a) and (b). In the figures, flow trajectories near maximum compression are shown for the case with and without the mix model, where $\xi = 20\%$ is assumed at the starting point of the mixing calculation ; t = 2.87 nsec. By comparing Fig.3(a) and (b), we found that the central spark is collapsed by the mixing.

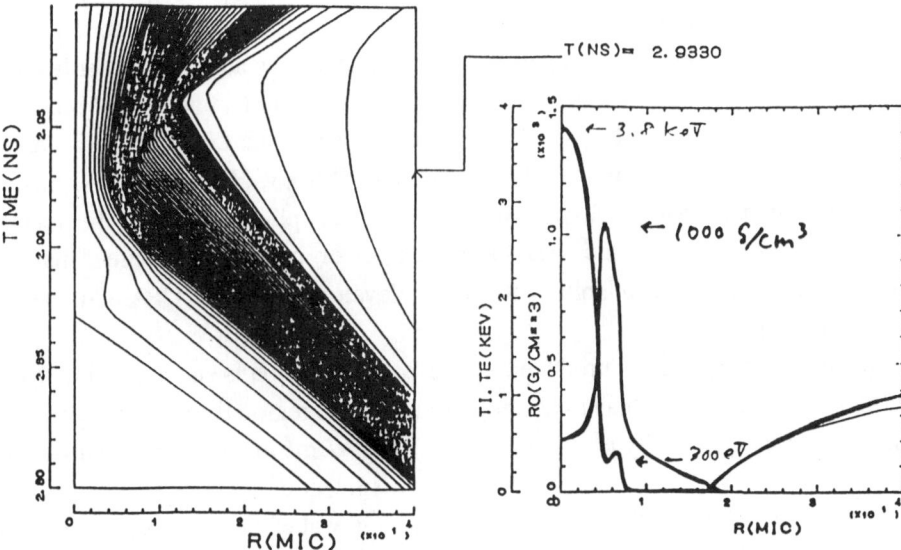

Fig.3. (a) Flow diagrams, density and temperature profiles at the maximum compression of 1D simulation without turbulent mixing.

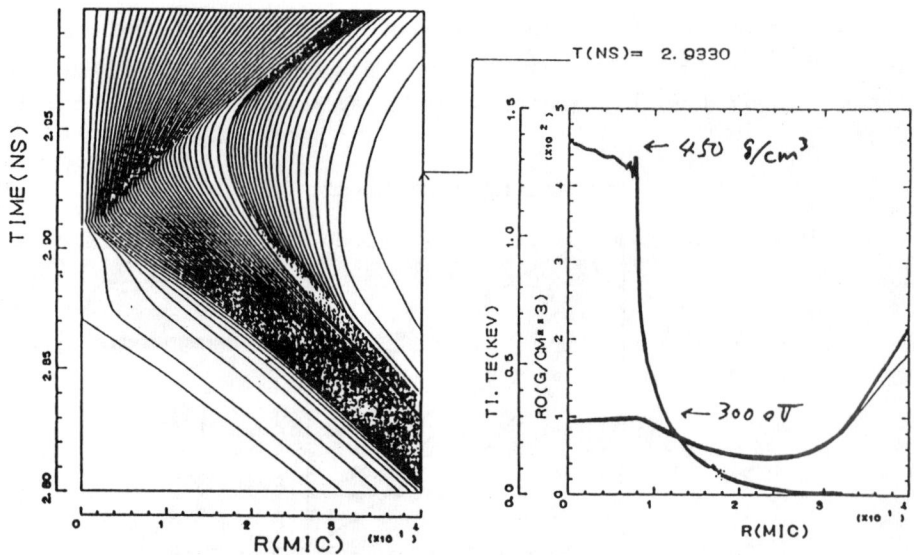

Fig.3. (b) Same as (a) with turbulent mixing.

Therefore, the central spark plasma temperature decreases from 3.5 keV to 0.3 keV. However, the average density and the core plasma radius are 400~500 g/cm^3 and 8~9 μm for both cases. Namely, the high density compression can be achieved with relatively poor implosion symmetry, but the central hot spark is not formed.

In conclusion of this section, the turbulent mixing between pusher and fuel is crucial in the anomalous degradation of neutron yield of the present experiments. Therefore both for direct and indirect drive, it is important in future to reduce the fluctuation level on the interface between pusher and fuel by increasing the irradiation symmetry and improving target surface finish.

§4. FAST IGNITION

As it is discussed in the previous section, the most difficult problem for the demonstration of ignition is the stable formation of the hot spark. The fast ignition concept is introduced in order to detour the above difficulty. Namely, the hot spark is generated externally by the additional laser heating in this concept. Since the imploded high density plasma disassembles within a few tens of picosecond, the heating have to be completed within 10 ps. The total heat capacity of a DT plasma with areal mass density $\rho R = 0.3$ g/cm^2 and the density $\rho = 200$ g/cm^3, is 400 J/keV. Therefore, 2 kJ has to be deposited in the super dense plasma. Taking into account the coupling efficiency, the injected laser pulse energy will be 10 kJ in a few psec. The laser peak power in this case is higher than 10^{15} W. Such a ultra intense laser became available by the recent development of the CPA (Charped Pulse Amplification) technology.

Since the high density plasma produced by the laser implosion is surrounded by a coronal plasma, the ultra intense laser pulse has to be sent into the dense region through the corona, as shown in Fig.4. The focused laser intensity is approximately 10^{20} W/cm^2 in which the electron quiver energy is the order of 1 MeV. Therefore, the relativistic nonlinear effect is very strong to cause the laser hose and modulational instabilities [17] and excite strong wakefield [18]. A part of the laser pulse energy is converted into MeV hot electrons in the corona plasma and/or scattered out by the modulational instability.

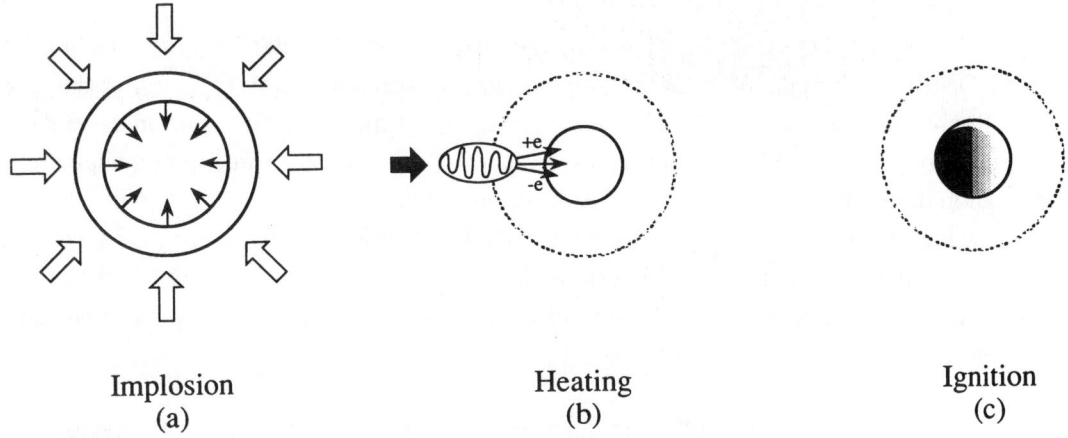

Implosion (a)　　Heating (b)　　Ignition (c)

Fig.4. Illustration of fast ignition processes.

§5. LASER FFSION EXPERIMENTS AND NEW FACILITIES [19]

Recently, many reports on hohlraum target plasma dynamics have been published from LLNL, USA and Limeil, France. Those reports disclose the following radiation hydrodynamics in hohlraum and implosion non-uniformities of capsule [2, 20, 21, 22, 23, 24].

(1) X-ray flux on capsule from laser spots and hohlraum walls; space-and time variation of hohlraum wall albedo which is related to laser pulse shape.
(2) Laser entrance hole effects for x-ray irradiation symmetry.
(3) Motion of hohlraum plasmas which fill the cavity and change the x-ray emission position; plasma colliding and stagnation cause additional x-ray emission and the hohlraum plasma pressure induces perturbations on capsule hydrodynamics.
(4) Laser plasma instabilities; SBS, SRS and filamentation which scatter laser beams and generate hot electrons in the cavity.
(5) Capsule radiation hydrodynamics which includes Rayleigh-Taylor instabilities in acceleration and deceleration processes, turbulent mixing between pusher and hot spark, drive x-ray intensity nonuniformity effects on compressed core shape, and so on.

The experimental and theoretical research results of the above issues have been summarized as a data base for designing target and driver laser for the national ignition facilities (NIF of USA [23] and a new project of France [24]). For examples, Nova experimental results for the hohlraum laser plasmas are compared with the requirements, for NIF as shown in the table 2 [2, 19, 22]. Since the required laser energy for ignition is sensitive to the implosion velocity, achievements of high radiation temperature and high implosion velocity ($\geq 4 \times 10^7$ cm/sec) have been key issues. The LLNL and LANL research groups claim that the requirements have been almost achieve. Another important issues are the drive symmetry and Rayleigh-Taylor instabilities. At present, the improvement of time dependent drive symmetry is a current subject.

According to the Nova results for the implosion campaigns test, the NIF is designed for igniting a capsule by the indirect drive implosion. Recently, the Key Decision 1 was announced by the secretary of US DOE to proceed to the engineering design phase of NIF. An ignition project by a large glass laser system similar to NIF was also recently approved as a national project in France, which will be constructed at Limeil-Valenton. Those projects plan to complete the facilities until 2002 and start ignition experiments.

As for the direct drive approach, two medium size lasers are under constructions in USA, which are OMEGA-Upgrade; 60 beam/30 kJ glass laser system at Laboratory for Laser Engineering, Univ. of Rochester and NIKE; 5 kJ KrF laser at NRL. The experiments will start in the end of 1994 ~ the beginning of 1995. Those experimental results will provide more precise understandings on the irradiation nonuniformity effects on the direct drive implosion.

In Japan, the ignition facility KONGO [25]; the 0.3 MJ of 3ω glass laser system has been designed for demonstrating ignition and burn by direct drive implosion. The 10 kJ KrF laser system Super-Ashura at Electro Technical Laboratory in Japan is under construction to do laser plasma experiments related to direct drive implosion.

Table 2. U.S. Nova results for the implosion campaigns for testing NIF physics.

	Issue	Requirements of NIF	Results of Nova Experiments	Present Status
Hohlraum Physics	Hohlraum Radiation Temperature	300 eV	295 (empty Au cavity, 1 ns pulse) 215 (shaped pulse)	Gas filled hohlraum is under study.
	Hohlraum Plasma Dynamics	Gas fill	Low Z liner which prevents wall motion	Gas filled plasmas are under study.
	Drive Symmetry Implosion Average Time Swings	~1% ±15%	Ni-lined and CH-lined ~1% ±25% measured but not adjustable	Symmetry in gas filled hohlraums in progress
	Laser Plasma Condition	$L \simeq 3\sim 4$ mm, f/8 $I_L = 1\sim 2\times 10^{15}$ W/cm^2 $n/n_{cr} = 0.07\text{-}0.5$ $T_e = 3\text{-}4$ keV	$L \simeq 2$ mm, f/4 or f/8 $I_L = 1\sim 2\times 10^{15}$ W/cm^2 $n/n_{cr} = 0.1$ $T_e = 3$ keV	Continuing activity to qualify scattering mechanisms under NIF conditions
Capsule Physics	Implosion Velocity	4×10^7 cm/sec	3.5×10^7 cm/sec	
	Convergence Ratio (CR)	25~35 (hot spark)	24	In progress on higher CR
	Hydro-instability Acceleration Deceleration	6~7 e-foldings	4~5 e-foldings	High e-foldings with high CR remains to be done
	Pusher density	700~1200 g/cc	140~170 g/cc	complete (Nova limit reached)
	Hot spot density	45~75 g/cc	20~30 g/cc	

More new laser facilities for studying laser fusion and related plasma phenomena have been planned or under construction in UK, Russia, China, Italy and so on. The above laser facilities like NIF and KONGO are expected to contribute to the laser fusion research development in near future.

ACKNOWLEDGEMENT

One of the author (K. Mima) is grateful to Dr. M. Cray of Los Alamos National Laboratory, and Dr. M. Tabak of Lawrence Livermore National Laboratory for providing useful information on Nova experimental results and fast ignition concept.

REFERENCES

(1) H. Azechi, et al., Laser and Particle Beams **9**, 193 (1991).

(2) M. Cable et al., to be published in Phys. Rev. Lett. (1994).
(3) R.L. McCrory, et al., Nature **335**, 225 (1988).
(4) H. Takabe, et al., ICENES '93 proceedings.
(5) M. Tabak, et al., Proceedings of '94 European Conference on Laser Interaction with Matter, to be published.
(6) J. D. Lindl, et al., Proceedings of IAEA '94 on Plasma Phys. and Thermonuclear Fusion Research, Seville, Spain, to be published.
(7) H. Takabe, et al., Proceedings of IAEA Tech. Committee Meeting on Driver for ICF, Osaka 1991, pp.160-165.
(8) J. D. Lindl, et al., in International School of Plasma Physics Piero Caldirola; Inertial Confinement Fusion, edited by A. Caruso and E. Sindori (Societá Italiana de Fisica, Bologna, 1988).
(9) H. Takabe, et al., Proceedings of 15th IAEA Conf. on Plasma Physics and Cont. Nucl. Fusion Research, Seville, Spain, 26 Sept.-1 Oct., 1994, IAEA-CN-60/B-P-9 to be published.
(10) D.L. Young, Physica **12D**, 32-44 (1984).
(11) H. Takabe and A. Yamamoto, Phys. Rev. **A44**, 5142 (1991).
(12) S.W. Haan, Phys. Fluids **B3**, 2349 (1991).
(13) S. Gauthier and M. Bonnet, Phys. Fluids **A2**, 1685 (1990).
(14) M. Tabak, et al., "Ignition and High Gain with Ultra-powerful Lasers", LLNL Quarterly Report '94, p.94.
(15) S.C. Wilks, et al., Phys. Rev. Lett. **69**, 1383 (1992).
(16) J. Denavit, Phys. Rev. Lett. **69**, 3052 (1992).
(17) P. Sprangl, J. Krall and E. Esarey, to be published in Phys. Rev. Lett. (1994).
(18) C.D. Decker, W.B. Mori and T. Katsouleas, to be published in Phys. Rev. Lett. (1994).
(19) M. Cray, private communication.
(20) R.L. Kauffman, et al., Phys. Rev. Lett. 73, 2320 (1994).
(21) T.R. Dittrich, et al., Phys. Rev. Lett. 73, 2324 (1994).
(22) L.J. Suter, et al., Phys. Rev. Lett. 73, 2328 (1994).
(23) M.M. Sluyter, Proceedings of 15th IAEA Conf. on Plasma Phys. and Thermonuclear Fusion Res., Seville, Spain, 26 Sept. - 1 Oct., 1994, IAEA-CN-60/B-1-1-2, to be published.
(24) CEL-V laser team, ibid, IAEA-CN-60/B-1-1-3, to be published.
(25) S. Nakai, et al., ibid, IAEA-CN-60/B-1-1-1, to be published.

GENERATION OF MEGAGAUSS MAGNETIC FIELDS IN LASER-TARGET PLASMAS

T. J. M. Boyd

Physics Department, University of Essex,
Wivenhoe Park, Colchester CO4 3SQ, UK.

Abstract. Many laser-target plasmas are themselves sources of strong, if highly localized, magnetic fields. These fields derive from a number of distinct mechanisms, among them thermoelectric sources, resonance absorption, hot electron fluxes and the inverse Faraday effect. This review will focus on the best documented of these and on thermoelectrically generated fields in particular since these have been characterized in different parameter regimes. Contemporary concern with sub-picosecond laser plasma interactions for which irradiances can be in excess of $10^{18}\,W\,cm^{-2}$ has led to renewed interest in the generation of very intense magnetic fields in the range of tens to hundreds of megagauss. Fields of this magnitude would influence the interaction physics significantly, notably with regard to electron thermal transport but also through parametric processes and resonance absorption.

INTRODUCTION

Self-generated magnetic fields have been a topic of recurring interest in the physics of laser-target interactions for more than twenty years. It is an interest that has waxed and waned with changes in wavelength of the incident laser light and with pulse lengths ranging from hundreds of picoseconds to sub-picosecond pulses on which attention is focussed at present. Two decades ago magnetic fields as large as a few megagauss were predicted and, after some false starts, were in the end detected and characterized [1-7]. Fields one or two orders of magnitude greater still have been forecast under the extreme conditions of sub-picosecond pulse irradiation. If indeed fields of tens or even hundreds of megagauss do occur they have the potential to strongly influence the interaction physics in experiments under these conditions. This talk is of necessity part review, part preview. The review focuses in the main on thermoelectrically generated magnetic fields since these have been best characterized experimentally and modeled in simulations. In passing it is perhaps worth drawing attention to the role of simulation in providing insight into a problem of some complexity, the more so since control experiments – other than by means of target composition or configuration – are not practicable. The part that is preview deals with predictions, from theory and simulation, of ultra-intense magnetic fields for which there is as yet no direct evidence from experiment though attempts to detect them are being made at present. If the presence of fields of the order of tens of megagauss or more is confirmed then their effect on a number of aspects of interaction physics may become a key factor in determining the absorption and transport physics in the target. From the early days of laser fusion physics there has been an awareness that magnetic fields of sufficient strength and extent could affect target performance in a number of ways by disrupting implosion symmetry and reducing thermal transport. Later studies showed that the effects of strong laser generated fields on absorption physics might be beneficial giving rise to enhanced resonance absorption and reducing stimulated Raman scattering through Raman choking, to mention just two examples.

As interest in magnetic field generation in laser target plasmas grew, mechanisms capable of producing fields of megagauss intensities proliferated. The sources included resonance absorption itself [8,9], thermal instabilities [10], hot electron fluxes from the focal spot [11] among others, all of them, at least in theory, alternatives to the favoured thermoelectric mechanism as a source of large dc toroidal magnetic fields. By contrast comparatively little attention was paid to the axial (poloidal) fields due in part to the relatively modest fields measured in the few experiments in which they were reported.

Thermoelectric generation of magnetic fields was first proposed by Biermann as a source of magnetic fields in stars. The mechanism demands non-aligned gradients in electron density and temperature to operate effectively, conditions realized in many laser-target irradiation experiments. The simplest of models suffices to show the effect, though to characterize the field generated and its topology requires fuller analysis. Adopting a fluid description and neglecting electron inertia allows the electric field needed to maintain charge neutrality to be written

$$E = -\frac{1}{eN_e}\nabla p_e - \frac{u \times B}{c} \tag{1}$$

in which N_e and p_e denote electron density and pressure respectively and u, B, the flow velocity and magnetic field. With no magnetic fields present to begin with in the plasma, using (1) and Faraday's law gives

$$\frac{\partial B}{\partial t} = -\frac{ck}{eN_e}(\nabla N_e \times \nabla T_e) \tag{2}$$

where T_e is the plasma electron temperature and k Boltzmann's constant. If one supposes that the dominant density gradient lies along the direction of incidence of the laser on the target while the corresponding gradient in temperature would be expected to be radial then for an electron temperature of $1 keV$ and scalelengths for density and temperature of the order of 10 microns, (2) suggests that magnetic field will be generated at a rate of about $0.1 MG, psec^{-1}$. Thus for $100 psec$ pulses, typical of many experiments a decade ago, fields of several megagauss could be expected.

OBSERVATIONS OF TOROIDAL MAGNETIC FIELDS

Early indications of large toroidal magnetic fields in laser-target plasmas were first found in experiments using magnetic probes in the form of small induction coils set close to the target. Such probes were not capable of diagnosing the interaction region directly so that real progress on characterizing magnetic fields in laser target plasmas began some years later with Faraday rotation measurements. The earliest of these [1] used a tightly focused ($\sim 35\,\mu m$) $1.06\,\mu m$ Nd laser beam with an irradiance of $10^{15} W\,cm^{-2}$ and reported magnetic fields of several megagauss. Later measurements by Raven and his collaborators [3-5] examined the effects of target size and composition on magnetic field generation again with a $1.06\,\mu m$ Nd laser beam with focal spot diameter $\sim 30\mu m$ and irradiances of several times $10^{16} W\,cm^{-2}$. A range of target plasmas was diagnosed using simultaneous Faraday rotation and interferometry. The Faraday rotation probe beam in these experiments was a fourth harmonic enabling magnetic fields to be measured right up to critical density and allowing magnetic field profiles to be determined. A wide variety of targets, plane solid surfaces, microballoons and wires of varying sizes was used and interesting variations in the magnitude of the self-generated magnetic field observed. Magnetic fields of a few megagauss were observed with large plane targets but with small disks there was no evidence of magnetic field generation.

Fig 1. shows the profile determined by Raven and others [5] along with computed field profiles using the model to be discussed in the next section. The observations of the thermoelectrically generated magnetic field were recorded in an axial scan along the edge of the focal spot. The observed field is seen to increase from zero at the refractive cut-off to a peak of just over $3MG$ at a point on the axis corresponding to a density of $0.2N_c$ where N_c is the critical density. From the maximum it drops more or less linearly to zero at a density of approximately $0.1N_c$. With the electron density measured by interferometry this experiment established that the peak field was generated in the underdense plasma.

Experiments using shorter pulses (20 psec) with intensities up to $3 \times 10^{17} W\ cm^{-2}$ were carried out by Burgess, Luther-Davies and Nugent [6]. The maximum density probed in their experiments was about $0.2N_c$ and close to this the magnetic field increases rapidly from the axis to a peak value of about $1.5MG$. These results confirm some features of those reported in [2,3] but unlike these it appears that the interaction physics may now be dominated by fluxes of suprathermal particles. Indeed a significant feature observed by Burgess, Luther-Davies and Nugent was the formation of a fast ion jet emerging from the plasma within the focal spot along the axis of the magnetic field. These results suggest that the magnetic field convects through the corona with a velocity determined by the suprathermal particles. This feature is of particular interest since it appears to presage recent observations of collimated plasma flow normal to the target surface in picosecond laser-plasma interactions [12].

THERMOELECTRIC MAGNETIC FIELDS AND THEIR GENERATION, EVOLUTION AND MORPHOLOGY

Determining the evolution of a laser generated magnetic field and its morphology is fraught with difficulty reflecting the mix of physics that has to be taken into consideration. The general expression for the evolution of the magnetic field has been given - apart from the fourth term - by Braginskii [13] in the form

$$\begin{aligned}
\frac{\partial \boldsymbol{B}}{\partial t} =\ & \boldsymbol{\nabla} \times (\boldsymbol{u} \times \boldsymbol{B}) - \frac{c^2}{4\pi} \boldsymbol{\nabla} \times [\boldsymbol{\eta} \cdot (\boldsymbol{\nabla} \times \boldsymbol{B})] \\
& - \frac{ck}{eN_e} (\boldsymbol{\nabla} N_e \times \boldsymbol{\nabla} T_e) + \frac{c}{e} \boldsymbol{\nabla} \times \left(\frac{1}{N_e} \boldsymbol{\nabla} \cdot P_R \right) \\
& + \frac{c}{4\pi e} \boldsymbol{\nabla} \times \left(\frac{1}{N_e} \left[\frac{1}{2} \boldsymbol{\nabla} B^2 - (\boldsymbol{B} \cdot \boldsymbol{\nabla}) B \right] \right) \\
& - \frac{c}{e} \boldsymbol{\nabla} \times \left[\boldsymbol{\beta} \cdot \boldsymbol{\nabla}(kT_e) - \alpha\, \hat{\boldsymbol{B}} \times \boldsymbol{\nabla}(kT_e) \right] .
\end{aligned} \qquad (3)$$

In (3) u is the fluid velocity, η, the resistivity tensor and N_e and T_e are the electron density and temperature respectively. P_R is the electromagnetic stress tensor. The quantities α and β are thermal transport coefficients that depend on $\Omega_e \tau_{ei}$, where Ω_e is the electron cyclotron frequency and τ_{ei}, the electron-ion collision time. Values of α and the components of β have been tabulated by Braginskii [13]. The various terms appearing on the right-hand side of (3) represent both the mechanisms generating a magnetic field together with those physical processes that determine its morphology. The first two terms are the familiar hydromagnetic contributions describing convection and diffusion. The third term is the thermoelectric source identified earlier. As we have seen for a plasma with a density gradient along the laser axis and a temperature gradient in the radial direction this source term generates an azimuthally directed magnetic field, the peak appearing at the edge of the local spot. The fourth term in (3), which involves the radiation pressure, is a true source and has to be added to the classical Braginskii equation when considering laser-generated magnetic fields.

The fifth term in (3) is the Hall term with components deriving from the magnetic pressure and from the curvature of the field. The final term stems from electron-ion collisions in a plasma with a temperature gradient. The presence of temperature gradients results in an imbalance of the frictional forces on electrons giving a resultant thermal force of order $N_e \nabla(kT_e)/\Omega_e \tau_{ei}$. Neither the Hall term nor the thermal force term of themselves generate magnetic fields though they play an important role determining the morphology of the fields. This is particularly true of the Hall term. The thermal force term serves to smooth fine-scale effects in the structure of the magnetic field.

Equation (3) describing the evolution of the magnetic field is highly nonlinear and is susceptible only to numerical analysis. In general this is complicated in that not only must all the terms on the right-hand side of (3) be retained but some like the thermal force term in which the coefficients α and β each depend on Ω_e/τ_{ei} need careful treatment in the numerical scheme. To date no study has successfully included all the effects represented in (3). It has been suggested for example [14] that the Hall term could be ignored but in fact this term contributes in an important way to determining the morphology of the field. Nevertheless to make progress some simplification is needed. A step in this direction is to discard the diffusion term in (3) on the grounds that diffusion times are typically much larger than times over which the plasma diagnostics operate [15]. Under conditions for which magnetic fields are generated by the thermoelectric source we may ignore the radiation source. However none of the remaining terms can be discarded in general though the thermal force term may be simplified in some circumstances. For example the term describing convection of the magnetic field by the heat flux can lead to strong amplification at high densities, in the region of the ablation surface. However if our concern is principally with field generation in the under-dense corona rather than at super-critical densities there are grounds for discarding the Nernst term. Boyd and Cooke [15] found in simulations that this effect had only a slight influence on the morphology of the field over the parameter range of interest. At higher densities the field is shifted to larger radii.

Sample results from a fluid code incorporating such a reduced version of (3) are shown in figs. 1 and 2 for conditions characterizing the experiments by Raven *et al.* [3,5]. Fig. 1 shows the computed magnetic field profile as a function of z at a radius of 24 μm corresponding to the peak field. Agreement between computed and measured profiles is reasonable as far as the magnitude and position of the peak magnetic field is concerned, not least when the uncertainty of the position of the critical surface is taken into account. The enhancement of the computed field over that measured experimentally at low plasma densities is more apparent than real and arises from the local density profile chosen to satisfy the boundary conditions. Neither the measured fields nor those from the simulation are trustworthy in the high density region, below $z \sim 10\mu m$. Fig. 2 shows the radial dependence of the magnetic field measured at an axial distance corresponding to a density of about $0.16 N_e$ along with computed profiles for two positions on the axis. The measured

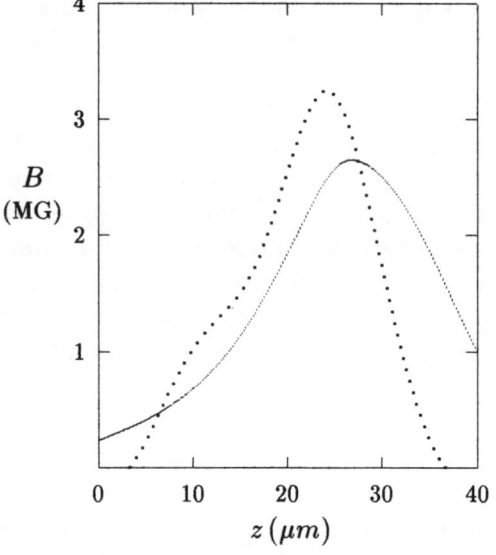

Fig. 1. Experimental [5] ($\cdots$) and computed B(z) [15]. The computed profile is for $r = 24$ μm, corresponding to peak field.

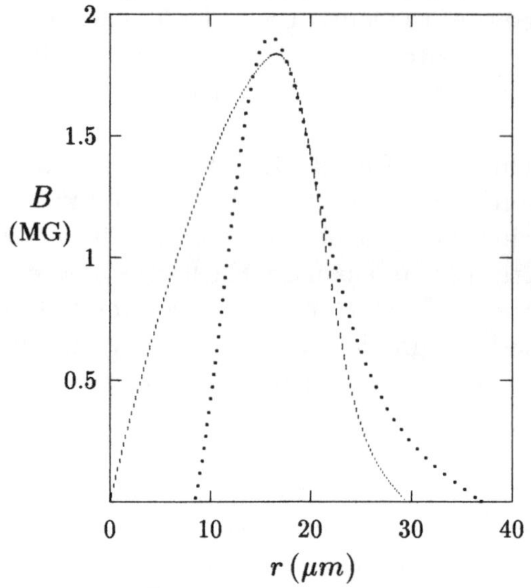

Fig. 2. Experimental [3] ($\cdots$) and computed B(r) [15]. The computed profile is for $z = 28$ μm.

field shows field reversal at a value of r below about $10\mu m$ but this may not be real since experimental error is large in this region on account of the small angular values in the Faraday rotation measurements. Overall, agreement between measured and computed fields is encouraging.

MAGNETIC FIELDS GENERATED IN SHORT PULSE INTERACTIONS

As we have already remarked there is no shortage of mechanisms capable in principle of generating megagauss magnetic fields in addition to the thermoelectric source discussed thus far. However experimental evidence for these competing sources is scant. Whatever uncertainties remain over the evolution and structure of thermoelectrically generated fields in long pulse experiments (tens of *psec* and above) – and they are many – are as nothing beside our lack of understanding of other mechanisms likely to dominate magnetic field generation in the short pulse regime ($O(1psec)$ and below). Differences between the thermoelectrically generated fields observed in [1-5] and those characterized in high irradiance, $20psec$ experiments [6] have already been highlighted. Most noteworthy is the fact that the fields observed by Burgess, Luther-Davies and Nugent correlate strongly with the laser pulse decaying in some tens of picoseconds whereas in the experiments of Raven and others, the peak fields were generated after the duration of the pulse. No less significant was the presence of a jet of fast ions emitted from the plasma along the axis of the self generated field, a feature suggested by other experiments and confirmed by simulation [16]. The simulations showed that in the presence of a self-generated magnetic field fast ions were emitted from the target and could explain the collimated jets of fast ions observed experimentally. Recent results reported by Bell and others [12] have drawn attention to collimated plasma flow from the target surface. Their work involves plasma irradiances up to about $10^{18} W\, cm^{-2}$ and they suggest that the collimation observed arises from a large (tens of megagauss) magnetic field pinching the plasma. Such large fields were found in a two-dimensional MHD simulation with an initial density profile characterized by a very short scalelength of $2 \mu m$. Magnetic inhibition of the heat flow was neither Braginskii-like, $(\omega \tau_e)^{-2}$ – where $\omega \tau_e$ is the Hall parameter – nor Bohm-like, $(\omega \tau_e)^{-1}$ but arbitrarily taken to be $(\omega \tau_e)^{-3/2}$ a choice dictated by the need to get agreement with the observed plasma temperatures. The magnetic fields found by Bell *et al* [12] in turn depend sensitively on this assumption.

Characterizing fields generated in sub-picosecond experiments with irradiances in excess of $10^{18} W\, cm^{-2}$ is clearly taxing, requiring diagnostics on a matching time scale. Thus far attempts at detecting large magnetic fields in short pulse experiments have not been successful. Realistic modeling of field generation in this regime is no less of a challenge. MHD-based simulations are unrealistic given that the appropriate scalelengths are comparable with electron mean free paths and the particle distributions are decidedly non-Maxwellian. At high irradiances particle collisions can be neglected and a particle-in-cell (PIC) code used. PIC codes have their own limitations on both time and length scales but at high irradiances are well suited to modeling the absorption physics and to providing information on magnetic fields generated through resonance absorption [9]. Wilks and others [17] have reported a 2D PIC simulation of the absorption of ultra-intense laser light in which the electron dynamics are strongly relativistic. Amongst other findings they observed intense toroidal magnetic fields, of magnitude $\sim 250\, MG$, formed when p-polarised light of intensity $I_L \lambda_L^2 (\mu m) = 5.6 \times 10^{18} W\, \mu m^2\, cm^{-2}$ is incident on a preformed overdense plasma. These fields attributed to electron heating at the light plasma interface, if confirmed experimentally, have the potential to affect the interaction physics significantly. Recent theoretical work has reported estimates of magnetic fields as large as $1000 MG$ [18]. These fields, however, are local in the sense that they are contained within a skin depth of the plasma-vacuum interface and do not correspond to the field structure observed by Wilks *et al* in the overdense plasma. To describe such non-local effects needs a kinetic formulation to include relativistic electron drift in the direction of propagation of the incident light. Boyd and Tatarinov [19] have shown that with full relativistic dynamics included, the magnetic field associated with the average electron current appears to conform to that observed by Wilks *et al* [17] in both polarity and magnitude, fields of several hundred megagauss being generated at the edge of the focal spot.

In the discussion thus far we have concentrated solely on azimuthal magnetic fields to the exclusion of axial fields. Nevertheless axial magnetic fields of the order of a megagauss have been observed by Briand *et al* [7] and attributed to dynamo action, an interpretation not generally accepted. Whatever the source of the field observed in this particular investigation it is important to recognize that the presence of an axial field is not *ipso facto* a unique signature of dynamo action.

Whatever uncertainties beset the generation of axial magnetic fields by linearly polarized laser light, fields in the megagauss range again appear a possibility if the incident light is circularly polarized. With this polarization an axial field is expected as a consequence of the inverse Faraday effect. This effect is well documented in solids and small fields have been detected in low density plasmas using a microwave driver. By contrast with the azimuthal fields driven thermoelectrically or through resonance absorption or otherwise, the inverse Faraday effect is in principle straightforward, relying as it does on the first order response of electrons to a circularly polarized field. Steiger and Woods [20] were the first to draw attention to large magnetic fields expected from, at the time, unattainable incident intensities. With irradiances in excess of $10^{18} W\, cm^{-2}$ now commonplace, Boyd and Tatarinov [19] have re-examined magnetic field generation by the inverse Faraday effect using full relativistic dynamics as well as incorporating the finite pulse length of

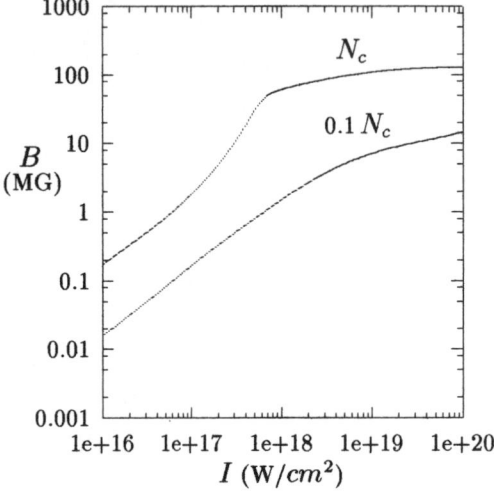

Fig. 3. Magnetic fields generated by the inverse Faraday effect in short-pulse (1 psec) interaction of circularly polarized laser light with uniform plasmas at densities N_c and $0.1\, N_c$ [19].

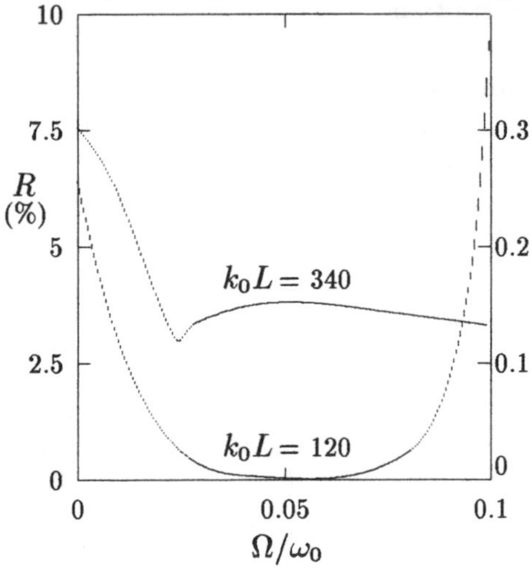

Fig. 4. Raman reflectivity R as a function of magnetic field for two different scale lengths. The left-hand vertical scale refers to $k_0 L = 340$; the right-hand to $k_0 L = 120$ [22].

the incident light in the model. Magnetic fields expected on the basis of this analysis are shown in Fig. 3, with intensities well in excess of $10\, MG$ for $I_L > 10^{19} W\, cm^{-2}$ close to critical density. Attempting to detect IFE generated fields would require discriminating such fields from thermoelectrically generated fields. This might be achieved by using a pre-formed plasma with a broad density maximum and by carrying out a null experiment using linearly polarized light. The same severe constraints on diagnostic response noted earlier make the detection and characterization of these fields a challenge.

EFFECTS OF MAGNETIC FIELDS ON INTERACTION PHYSICS

Until the very large magnetic fields predicted in the sub-picosecond regime have been detected and characterized it is premature to speculate about the influence such fields might have on the interaction physics. From the outset there has been concern in the laser fusion community about the way in which electron thermal transport would be affected by megagauss fields. Electron thermal transport will be inhibited by self-generated magnetic fields when $(\lambda/r_L) > 1$ where λ is a mean free path and r_L the electron Larmor radius. From the ratio

$$\frac{\lambda}{r_L} \sim 20 \left(\frac{n_c}{Zn_e}\right) \left(\frac{10}{\ln \Lambda}\right) T_e^{3/2}(keV) \, B(MG)$$

in which Z is the atomic number and Λ the usual plasma parameter, taking $Z \sim 10$, $T_e = 1\,keV$ there would be increasingly severe inhibition of electron thermal transport for magnetic fields over $1\,MG$. In the long pulse experiments already discussed we saw that the peak field was generated in the underdense plasma with the result that thermal electrons at the critical surface were not likely to suffer serious inhibition by the magnetic field. In the case of sub-picosecond plane target experiments by contrast, transport from the laser spot is likely to be severely reduced.

The role played by self generated fields in parametric instabilities and more widely in interaction physics is not easy to characterize in general. Two examples in which magnetic fields may act beneficially are resonance absorption [21] and in the choking of stimulated Raman scattering [22]. The results of a PIC simulation of Raman reflectivity as a function of magnetic field are shown in Fig. 4. Raman emission is reduced significantly as the magnetic field strength is increased and vanishes for some critical field. This behaviour reflects the increasing value of the parameter $\tau = (wL/c)^{1/3}(\Omega_e/\omega)^{1/2}$ where ω is the frequency of the wave involved and L is the local density scale length. As τ increases the absorption of the scattered wave by mode conversion grows. Minimum reflectivities occur where $\tau \sim 0.8$ and mode conversion is a maximum. For the choice of scalelength in Fig. 4 corresponding to an initially steep profile, Raman emission is effectively switched off over a range of magnetic field strengths before recovering to about its unmagnetized value. What the effect of the large fields predicted for sub-picosecond laser plasma interactions proves to be is a problem for the future, once the fields themselves have been detected and characterized.

ACKNOWLEDGEMENTS

I am indebted to Andrew Mackwood and Ricardo Ondarza for assistance in preparing this paper and to Greg Tallents and Alexey Tatarinov for helpful discussions.

REFERENCES

[1] Stamper, J. A. and Ripin, B. H., Phys. Rev. Lett. **34**, 138, 1975.

[2] Stamper, J. A., McLean, E. A. and Ripin, B. H., Phys. Rev. Lett. **40**, 1177, 1978.
[3] Raven, A., Willi, O. and Rumsby, P. T., Phys. Rev. Lett. **41**, 554, 1978.
[4] Raven, A. and Willi, O., Phys. Rev. Lett. **43**, 278, 1979.
[5] Raven, A. *et al*, App. Phys. Lett. **35**, 526, 1979.
[6] Burgess, M. D. J., Luther-Davies, B. and Nugent, K. A., Phys. Fluids **28**, 2286, 1985.
[7] Briand, J. *et al*, Phys. Rev. Lett. **54**, .18, 1985.
[8] Woo, W. and DeGroot, J., Phys. Fluids **21**, 2072, 1978.
[9] Boyd, T. J. M., Gardner, G. A. and Humphreys-Jones, G. J., Jour. de Phys. **C7**, 551, 1979.
[10] Tidman, D. A. and Shanny, R., Phys. Fluids **17**, 1207, 1974.
[11] Kolodner, R. and Yablonovitch, E., Phys. Rev. Lett. **43**, 1402, 1979.
[12] Bell, A. R. *et al*, Phys. Rev. E **48**, 2087, 1993.
[13] Braginskii, S. I., in *Reviews of Plasma Physics*, ed. M. A. Leontovich (Consultants Bureau, N. Y. , 1966), Vol 1, 205.
[14] Colombant, D. G. and Winsor, N. K., Phys. Rev. Lett. **38**, 697, 1977.
[15] Boyd, T. J. M. and Cooke, D., Phys. Fluids **31**, 651, 1988.
[16] Forslund, D. W. and Brackbill, J. V., Phys. Rev. Lett. **48**, 1614, 1982.
[17] Wilks, S. C. *et al*, Phys. Rev. Lett. **69**, 183, 1992.
[18] Sudan, R. N., Phys. Rev. Lett. **70**, 3075, 1993.
[19] Boyd, T. J. M. and Tatarinov, A. V., (to appear, 1994).
[20] Steiger, A. D. and Woods, C. H., Phys. Rev. **A5**, 1467, 1972.
[21] Barr, H. C., Boyd, T. J. M., Gardner, G. A. and Rankin, R., Phys. Fluids **28**, 16, 1985.
[22] Barr, H. C., Boyd, T. J. M., Gardner, G. A. and Rankin, R., Phys. Rev. Lett. **53**, 462, 1984.

PLASMA BASED 14 MeV NEUTRON SOURCES.

E.P. Kruglyakov.

Budker Institute of Nuclear Physics, 630090, Novosibirsk, Russia.

ABSTRACT

The utmost urgency to develop powerful 14 MeV neutron sources is now well understood among fusion community of engineers and material scientists. The time of influence of 14 MeV neutrons upon the fusion reactor components is estimated at 10-20 years and corresponds to a fluence of $1.5 - 4.5 \ 10^{22}$ n/cm^2. As yet, no reliable data are available on the behaviour of reactor component materials exposed to such a flux. This creates a great deal of uncertainty in the performance and endurance predictions of future fusion reactor main components. The first wall flux from ITER will be approximately 1 MW year/m^2 by the time that the choice of materials for DEMO will have to be made. A comparison of the desired flux from DEMO of 10-20 MW year/m^2 shows quite clearly that ITER will provide too small a flux and too late in time to contribute usefully to the materials selection process. On the other hand, before DEMO is built, suitable materials for the first wall and other critical reactor elements must be selected. Thus, DEMO cannot be constructed without a wide programme of material study (end-of-life, resistivity increase from neutron damage, H, He, dpa production rates, activation of nuclei, etc.) as well as the welds and brazes to be used in the reactor. To realize this programme a high-flux neutron source of DT-neutrons, covering a sufficiently large test zone, should be rapidly constructed.

At present, a number of proposals for 14 MeV neutron sources are known. Among these are: sources based on the acceleration of ions, interaction of ions with targets, volumetric plasma neutron sources based on tokamaks and mirrors, "exotic" schemes using μ-catalysis and lasers. In this paper the advantages and shortcomings of various approaches are discussed, from both the technical feasibility and economic points of view.

I. INTRODUCTION

At present the urgency to develop powerful 14 MeV neutron sources is now well understood among fusion community engineers and material scientists. In the Report of the Fusion Evaluation Board (chaired by U.Colombo) prepared for the Commission of the European Communities [1] the following recommendations are asserted: "The problem of the need for a powerful source of high energy neutrons for material testing should be addressed with the utmost urgency. Such a source should be made an integral part of the ITER programme". In typical schemes of fusion reactors the characteristic value of a 14 MeV neutron flux is 1-3 MW/m^2 ($0.5-1.5 \ 10^{14}$ neutrons/cm^{2}s). The time of influence of such a flux upon the reactor components is estimated as 10-20 years and corresponds to a fluence of $1.5-4.5 \ 10^{22}$ n/cm^2. No reliable data are available on the behaviour of reactor component materials exposed to such a flux. This creates a great deal of uncertainty in the performance and endurance predictions of future fusion reactor main components. The first wall fluence from ITER will be approximately 1 MW year / m^2 [2] by the time that the choice of materials for DEMO will have to be made. A comparison of the desired flux from DEMO of 10-20 MWyear/m^2 shows quite clearly that ITER will provide too small a flux and too late in time to contribute usefully to the materials selection process. On the other hand, before DEMO is built, suitable materials for the first wall and other critical reactor elements must be selected. Thus, DEMO cannot be constructed without a wide programme of material study (end-of-life, resistivity increase from neutron damage, H, He, dose per annum dpa production rates, activation of nuclei, etc.) as well as the welding and brazing technology to be used in the reactor. To realize this programme a high-flux neutron source of DT-neutrons, covering a sufficiently large test zone, should be rapidly constructed. In addition, such a source could be used in investigations to create new, low activated materials for the first wall and breeding blanket components.

Thus, one can understand that the material issues related to the fusion nuclear environment are very serious and require decades to be resolved. The most appropriate time to have the required neutron source for testing is the beginning of the next century (approximately the year 2005), In this case an inte-

© 1995 American Institute of Physics

grated neutron loading of 10 MWyears/m^2 (possibly even 20 MWyears/m^2) will be achieved during the ITER technological phase and before the DEMO design works.

At present a lot of proposals for a 14 MeV neutron sources are described. Here, an attempt is made to compare different approaches from the viewpoints of feasibility, present day state of the art and also economics.

II. COMPARISON OF DIFFERENT NEUTRON SOURCES

A. ACCELERATOR BASED NEUTRON SOURCES.

Among the present existing proposals for high power neutron sources there are only two main approaches, accelerator-based and plasma-based. The accelerator type includes the following concepts.

1. An accelerator with relatively high proton energy (around 1 GeV) producing spallation reactions in the proton beam-heavy target (U, Pb, etc.) interaction, this produces a high flux of energetic neutrons. [3].

2. The use of accelerated deuterons (E=35-40 MeV) in D-Li stripping reactions ^{7}Li (d,n) Be for the production of a neutron flux with an average energy of 14 MeV [4].

At present the final choice has been made in favour of a D-Li source and the design of the International Fusion Material Irradiation Facility (IFMIF) will evolve under the International Energy Agency (IEA) Fusion Materials Agreement. It is planned to complete the conceptual design by early 1997. Typical parameters, selected to meet the materials development mission, are as follows [5]:

Deuterium Beam Current	250 mA
Beam Energy	30-35 MeV
Beam Spot Size	10x10 cm
Target	Lithium Jet with Back Plate
Test Volume	Approximately 1 litre (flux equivalent of 2 MW/m^2)

The facility should have a deuteron current of 250 mA. This will significantly exceed those of the Fusion Material Irradiation Test (FMIT) facility [4] and the Energy Selective Neutron Irradiative Test (ESNIT) [6] for material testing purposes. From the technical viewpoint there are no doubts that the IFMIF can be designed in a reasonable time period. Such a source can produce a lot of important data but it should be noted that the IFMIF cannot solve all the problems of structural material and component testing for future fusion reactors. Beside providing a test zone volume which is too small, this approach has other disadvantages. In particular, the stripping reaction cannot produce monochromatic neutrons of 14 MeV which are prevalent in the D-T reaction.. In addition, the energy spectrum has a pronounced dependence on the angle relative to the deuteron beam.

Fig.1 shows examples of stripping reaction spectra [7] for accelerated deuteron energies of 35 and 45 MeV. The tail of the neutron distribution function for the energy range above 14 MeV may influence the irradiation effects. In other words, the broad neutron spectrum may lead to significant flaws when testing unknown materials.

Fig. 2 shows the behaviour of the activation cross-section in dependence on the neutron energy in the vicinity of 14 MeV. The fast growth of activation cross-section with increasing neutron energy is rather typical for many reactions (see Ref. [8]). Thus, if the cross-section of a process is unknown it is rather dif-ficult to point out a cross-section value for the 14 MeV energy using a broad spectrum of neutrons. From this consideration it follows that, besides an acceleration-based neutron source, it is necessary to

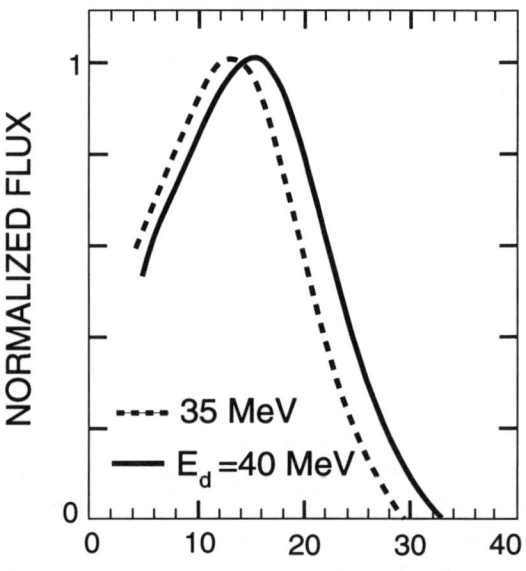

Fig.1. Neutron spectra of stripping reaction for D-Li source [7].

construct a monochromatic 14 MeV neutron source. Before a description of volumetric plasma type neutron sources, one should mention a source which occupies an intermediate position between accelerator and plasma-based NSs.

B. MUON-CATALIZED NEUTRON SOURCE.

The concept for such a source is based upon the scheme proposed in Ref. [9] for a hybrid reactor.

The modern version of the muon-catalized NSs consists of a 12 mA, 1.5 GeV accelerator of deuterons.

The deuteron beam is directed onto a 50 cm long, rotating carbon target inside a strong solenoidal magnetic field to produce negative pions in a vacuum.

The negative muons, originating from pion decay, are guided along the magnetic field to a 50 cm long D-T cell with a pressure about 10^3 atmospheres. The energy cost to form one muon is equal to 8 GeV.

However, due to the catalytic character of muon interactions with the D-T mixture the efficiency of neutron production, resulting from the fusion of mesic molecules, can be high enough.

Experiments have demonstrated that one hundred 14 MeV neutrons can be generated by a single muon. Thus, one can estimate that such an accelerator could produce 200 kW, 14 MeV neutrons.

It is necessary to note that, even for these moderate source parameters, the concept has not yet been studied in detail.

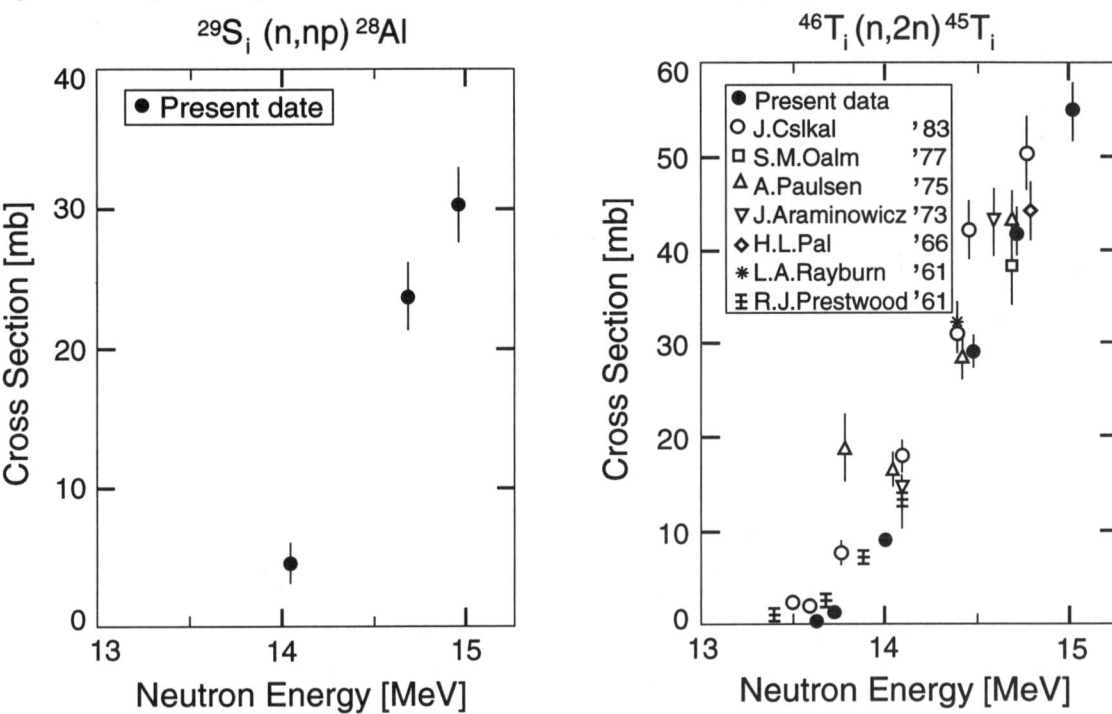

Fig.2. Examples of behavior of activation cross sections for potential elements of structural materials of fusion reactor in the vicinities of 14 MeV [8].

C. PLASMA-BASED NEUTRON SOURCES

The general requirements determined by the International Energy Agency (IEA) specification for volumetric plasma type neutron sources can be formulated as follows:
1. 14 MeV uncollided neutron flux P_n should be above 2 MW/m^2
2. Test zone volume exceeds 30 litres.
3. Flux gradient less than 10 % / cm.

It is obvious that to accumulate a flux dose of 10-20 MWyr/m^2 before the DEMO design is working then the source must start to operate not later than the year 2005. That means a source based on very reliable, well established technologies. An application of new technologies is not warranted.

Discussions concerning plasma type neutron source have started 10 years ago. One of the most important features of a plasma type neutron source is the fact that the neutron spectra is similar to the fusion reactor. Historically the first proposals were based on different mirror system concepts [11-22]. Another important parameter of each project is the ratio of uncollided neutron flux to power consumption. The ratios of these values are shown in Fig.3 for different mirror based concepts. During the last few years a lot of proposals for neutron sources were based on tokamaks [22-29]. Fig.4 demonstrates the same values as Fig.3 for tokamak based neutron sources. One can conclude from these figures that power consumption is three to four times less for mirror based sources than for tokamak ones.

Discussing the Next Step of the Fusion Programme (be it ITER or NET) the Colombo Commission [1] came to the conclusion that: "apart from the Next Step project, these should include a large neutron source for material testing..." It is obvious that the cost of the source should not exceed 5-10% of the ITER capital cost. More expensive projects would lead to a revision of the total ITER concept. The main operational costs are for the electric power and the tritium consumption. Assuming that 1 MW year of electrical energy costs 0.7M$ and 1MW of 14 MeV neutron flux for one year requires about 70 grams of tritium, one understands that most large scale devices needing high power or tritium cannot be candidates for a neutron source for material and component testing. In addition, the total world supply of tritium does not exceed 5 kg. Thus, a neutron source can only be selected among devices with a low tritium consumption. Mirror-based NS has several strong advantages for the choice of fusion reactor materials, subcomponent and component testing. Such a source can be constructed in steps, starting with a relatively inexpensive, small, low flux facility. Subsequently, power, length, and/or a stronger magnetic field may be added to increase the flux or the useful area and volume. This possibility arises from the following mirror properties: Confinement is independent of radius (small is as good as big one). Drive power is proportional to source length. The initial system can be small and inexpensive, then lengthened by additional solenoidal coils and drive power. The fact that mirrors can operate at beta values around unity without loss of equilibrium was demonstrated experimentally (2XllB, TMX, GAMMA-10 & GDT).

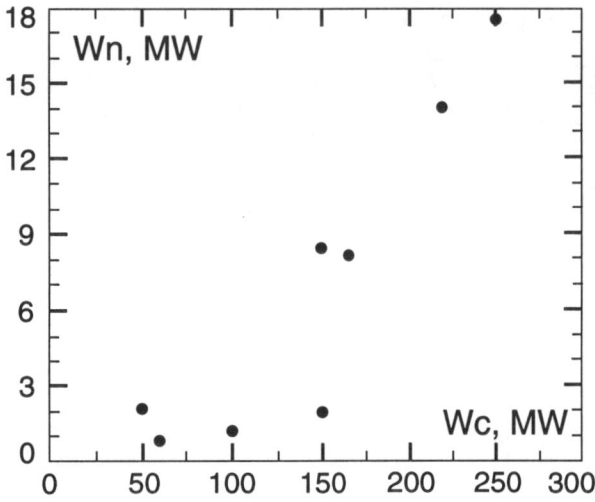

Fig.3 Total neutron flux versus power consumption. Mirror based NSs [11-22]

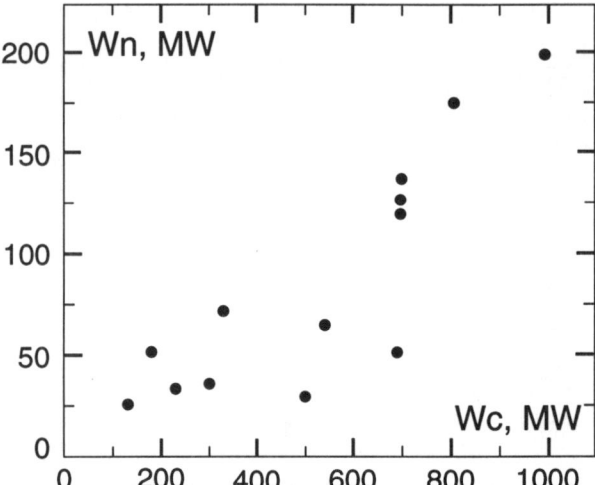

Fig.4 Total neutron flux versus power consumption. Tokamak based NSs [22-29]

A high beta allows a high neutron flux from a small volume. The majority of mirror-based concepts have a low tritium and power consumption. In addition, one should mention the following advantages of mirrors:
1. They are not subject to disruptions or other catastrophic events.
2. They allow steady state operation.
3. Their simple linear-cylindrical geometry provides convenient removal and installation of test assemblies and source components.
4. The magnets and vacuum wall of a mirror NS will have a long lifetime because they are much further from the plasma than the test samples and so receive one to several orders of magnitude less neutron flux than the samples.
5. The physics of the proposed neutron sources is based upon fairly moderate extrapolations from experimental data.

As to the tokamak-based NS, from the physical point of view, this could be built with the existing plasma parameters. At present, technology is mostly developed for low Q (neutral beam-plasma) neutron sources. This can achieve $2MW/m^2$ in a large testing zone from 15-20 m^2 up to several hundred m^2. Large test zones would make possible full scale reactor component testing. However, in the case of relatively large scale NSs their capital and operational cost would be prohibitive. For example, the capital costs of two NSs, presented in Ref. [24], were estimated at 0.48 and 0.45 of ITER's direct cost. Naturally such projects are unrealistic, not only for their costs but also for their large tritium consumption of over 10 kg/year. This value is twice the world's annual tritium production. In addition, the direct cost of tritium is around $ 30,000/gm. One of the most important problems of tokamak-based NS sources is the tritium. Even the compact concept of a tokamak NS, such as the TRINITY version [26], requires 2.5 kg/year of tritium. However, recently new NS candidates have appeared on the scene. These are based on the advances in super compact tokamaks, particularly the START (Small Tight Aspect Ratio Tokamak) spherical tokamak [30], the TUMAN-3 [31] etc. At present START is the only device where $R/a < 2$ (the TUMAN-3 aspect ratio is slightly larger than 2). The benefits of a tight aspect ratio have been demonstrated experimentally. A β value equal to 20% has been reached and vertical stability is also improved. It has been demonstrated that MHD stable, steady state regimes exist with tight aspect ratio tokamaks. START has reached elongations of $b/a > 1.8$ without feedback. In over 20,000 discharges so far, "hard" (current terminating disruptions) have not been observed provided that $R/a < 1.8$.

At present there exists only one spherical tokamak project as a material test facility (MTF) [27, 29]. Perhaps a new project, the compact spherical tokamak GLOBUS [32] with $R/a < 1.6$, can be considered as an NS option.

In general, the plasma parameters of compact tokamaks are presently rather moderate (both electron and ion plasma temperatures are less than 1 keV, discharge duration around 10 ms, etc.). Nevertheless, careful studies show that this concept can be the basis of a neutron source design. However, from an engineering point of view the project has some shortcomings. Indeed, the tokamak will only be a compact if there is no central column shielding. The next issue is whether the centre column should be single or multi-turn. A multi-turn column requires insulators. This would shorten the column lifetime. Thus one should use a monolithic column. This implies a very complicated low voltage, high current power supply. In addition there is the uncertainty concerning the mechanical properties and resistive characteristic of the central copper column under the neutron flux. The next disadvantage is common to any tokamak NS. The source must be sufficiently powerful in order to study the unknown properties of the test materials and reactor components. However, the neutron flux effects on the first wall are completely unknown. Lastly, the minimum wall area in the tokamak NS case can hardly be less than 15 m². This means that even a super compact tokamak will consume around 2 kg/year of tritium.

The plasma-type NSs can best be understood by reviewing mirror-type NSs. Two projects should be mentioned separately: The Livermore project [14] and the Novosibirsk one [18,19]. The Livermore concept uses a warm tritium plasma target and the injection of 200 keV deuterons. The project is based upon the experimental results of the 2XIIB facility and also upon very moderate extrapolations from the parameters achieved there.

The Novosibirsk project, on the other hand, is based upon the axisymmetric Gas Dynamic Trap (GDT) concept [33]. This uses the plasma parameters previously obtained in mirror experiments and incorporates the present day technology level.

Parameters of the mirror-based NS proposals mentioned above are presented in Table 1. It is important to note that these projects can be started on a small scale by designing initially for a small testing zone (under 1 m²). Thus the construction and operating cost can be minimized.

	LLNL-87	GDT-2	GDT-3
Construction Cost (M$)	200	250	250
Annual Cost (M$/yr)	91	67	55
Power Consumption (MW)	100	60	50
Peak Neutron Wall Load (MW/m²)	10	3.9	2.0
Test Zone Radius (m)	0.06	0.06	0.07
Axial Length of Test Zone	0.35	0.6	2.5
Tritium Consumption (kg/yr)	0.1	0.075	0.12
Plasma Density (cm^{-3})	$1.3 \cdot 10^{15}$	$3.7 \cdot 10^{14}$	$2 \cdot 10^{14}$
Electron Temperature (kV)	0.19	0.52	1.1
Neutral Beam Energy (keV)	200(D)	240(T)	80(D); 94(T)
Neutral Beam Power (MW)	47	20	15

The GDT-based concept of an NS, being an axisymmetric system, is the most appropriate if the testing zone area is to be increased. At present it is the most advanced project among many others. The prototype of the gas dynamic trap is operating at Novosibirsk. A full scale model of an NS (so called Hydrogen Prototype) is under construction. More detailed information can be found in Ref. [21]. The oblique injection of tritium (or deuterium and tritium) beams is used in this source, forming the sloshing ions population. As a result the 14 MeV neutron flux peaks at the vicinities of fast ions turning points and rapidly drops off longitudinally. The GDT NS has a high mirror ratio (20% or even higher). In this case the spatial separation of the high neutron flux range and the mirror positions, significantly simplifies the problem of shielding the magnetic coils. Due to the small plasma radius (10 cm) and the large vacuum chamber radius (1m), the neutron and heat wall loading are low enough over the major part of the facility. Thus, only a small part of the first wall must be changed from time to time. Such a situation is unreal in any tokamak-based neutron Finally, it is necessary to mention a new approach to the problem of NS construction. In Ref.[34] a laser initiated NS is proposed. The laser driver should have a wavelength of 0.3 - 1 μm, an efficiency of 2%, a pulse repetition rate of 10 Hz and a pulse energy of 200 kJ. In case of a target with a Q = 1 it will produce a neutron flux in the order of 200 kW. Present lasers with a pulse energy of 50 kJ have an efficiency around 1% and a repetition rate of 10^{-4} Hz.

III. CONCLUSIONS.

The urgency for the design and construction of a dedicated neutron source is obvious. The accelerator-based IFMIF project could be the first facility for material testing but it cannot solve many of the problems. A plasma type neutron source is source system. The main GDT NS parameters are presented below in Table 2.

	GDT-2	GDT-3
NBs, Energy/Power		
D Beams, kV/MW	-	80/8.5
T Beams, kV/MW	240/20	94/6.5
Injection Angle, deg	20	40
Mirror-To-Mirror Length, m	10	10
Magnetic Field, T		
At The Center	1.25	1,8
In The Mirrors	25	26
Electron Temperature, keV	.52	1.1
Warm Plasma Density, cm	$3.7 \cdot 10^{14}$	$2 \cdot 10^{14}$
Plasma Radius (Midplane), cm	6	7
Max Neutron Flux, MW/m	3.9	2.5
Total Power Consumption, MW	60	50

required. At present both the mirror type and the compact tokamak-based NSs can be built. A mirror NS can be constructed in steps starting with a small, relatively inexpensive facility with a testing zone area less than 1 m^2. The most compact tokamak (like START) can be constructed beginning with a testing zone area around 15 m^2. For a neutron flux of 2 MW/m^2 the minimum tritium consumption in the tokamak case would be 2 kg. The final choice will depend upon the level of expense which the plasma and material scientific communities are prepared to pay for neutron source construction. In addition, the problem of tritium consumption deserves a serious consideration. If the neutron source consumes tritium at a rate near that of world production then the construction of such a source is unrealistic.

REFERENCES

1. Colombo U., Jaumotte A., Kennedy E., et al, Report of the Fusion Programme Evaluation Board prepared for the Commission of the European Communities, Bruxelles, (1990).
2. ITER Documentation Series, IAEA, Vienna, (1991).
3. Kley W. and Bishop G.R., EUR-9753 EN, Commission of the European Communities, (1984).
4. Lawrence G.P., Varsamis G.L., Krakowski R.A. et al, In: Proc. of the Intern. Fusion Materials Irradiation Facility (IFMIF) Workshop, vol.2, p.149, San Diego, (1989). Perlado J.M., ibid, vol.2, p.273.
5. Shannon T.E., Cozzani F., Crandall D.H. et al, Proc. of the Intern. Conference on Plasma Physics and Controlled Nuclear Fusion Research, IAEA-CN-60/F-2-ll-5, Seville, Spain, (1994).
6. Noda K., Sugimoto M., Kato Y. et al, ESNIT International Advisory Meeting, JAERY, Japan, p.13, June 1993.
7. Ohno H., Noda K., Sugimoto M., et al ibid, p.1.
8. Ikeda Y., Konno C., Oishi K., et al, Activation Cross Section Measurements for Fusion Reactor Structural Materials at Neutron Energy from 13.3 to 15.0 MeV Using FNS Facility, JAERY 1312, Japan, March 1988.
9. Activation Cross Section Measurements for Fusion Reactor Structural Materials at Neutron

Energy from 13.3 to 15.0 MeV Using FNS Facility, JAERY 1312, Japan, March 1988.
10. Petitjean C., Atchinson F., Heidenreich G. et al, Fusion Technology, vol.25, p.437, (1994).
11. Mirnov V.V., Nagornyj V.P., Ryutov D.D., Report INP 84-40, Institute of Nuclear Physics, Novosibirsk, 1984; Kotelnikov I.A., Mirnov V.V., Nagornyj V.P., Ryutov D.D. In: Plasma Physics and Controlled Fusion Research, v2, p.309, Vienna, IAEA, (1985).
12. Badjer B. et al, TASKA-M, Low Cost, Near Term Tandem Mirror Device for Fusion Technology Testing, Kernforschungszentrum Karlsruhe, Report KfK 3680, 1984.
13. Doggett J.N. et al, A Fusion Technology Demonstration Facility (TDF), LLNL Report UCRL 90824, 1984.
14. Coensgen F.H., Casper T.A., Correll D.L. et al, LLNL Rep. UCRL-97280, Rev.1, 1987.
Coensgen F.H., Casper T.A., Correll D.L. et al, Journal of Fusion Energy, v.8, p.237, (1989).
15. Kesner J., Horne S.F., Pastukhov V.P., Rep. PFC/JA-87-8, Plasma Fusion Center, MIT, Cambridge, Massachusets, 1987.
16. Golovin I.N., Zhiltsov V.A., Panov D.A. et al, Voprosy Atomnoy Nauki i Tekhniki, T.S., No 3, p.3, (1988).
17. Kawabe T., Yamaguchi H., Mizuno N., et al, Journal of Nuclear Materials, v.191-194, p.1387, (1992).
18. Ryutov D.D., Plasma Physics and Controlled Fusion, v.32, p.999, (1990).
19. Kotelnikov I.A., Ryutov D.D., Tsidulko Yu. A. et al, Rep. INP 90-105, Institute of Nuclear Physics, Novosibirsk, 1990.
20. Isanov A.A., Kotelnikov I.A., Kruglyakov Eh. P., Proc. of Symposium on Fusion Technology, vol.2, p.1394, Roma, Italy, Sept. 1992.
21. Kruglyakov Eh. P., Proc. of International Conference on Open Plasma Confinement Systems for Fusion, p.349, World Scientific, (1994).
22. Kruglyakov Eh. P., Plasma Type High Volume 14 MeV Neutron Sources, IEA Workshop on Fusion Neutron Sources, Moscow, July 12-16, 1993.
23. Kawabe T., Characteristics of Mirror Based Neutron Source-FEF, ibid.
24. Peng M., Initial Design Boundaries and Parameters of Small Tokomak VNS Envelope, ibid.
25. Astapkovich A.M., Glukhih V.A., Mineev A.B. et al VNS on the Basis of the High Boot - strap Fraction of Current, ibid.
26. Azizov E.A., Kovan A.I., Koltchenko S.V., et al, The Compact Volumetric Neutron Source on the Tokamak Based ibid.
27. Hender T.C., Tight Aspect Ratio Tokamak Neutron Source, ibid.
28. Abdou M.A., Peng Y.M., Berk S., Intern. Conf. on Plasma Physics and Controlled Nuclear Fusion Research; IAEA-CN-60/F-ll-6, Seville, 1994.
29. Buttery R., Counsell G., Cox M., et al, IAEA-CN-60/F-I-3(1), ibid.
30. Sykes A. et al, Nuclear Fusion, vol 32, p.694 (1992).
31. Andrejko M.A., Ashkenazi L.A., Golant V.E., Tendler M., et al, Intern. Conf. on plasma Physics and Controlled Nuclear Fusion Research; IAEA-CN-60/A-2/4-P-5, Seville, 1994.
32. Golant V.E., Gusev V.K., Sakharov N.V., et al; Project of Physical Tokamaks GLOBUS-M, Report of Ioffe Physico-Technical Institute, Sankt-Petersburg, 1994; Fusion Technology (to be published).
33. Mirnov V.V., Ryutov D.D., Sov. Tech. Phys. Lett., v.5, p.279, (1979).
34. Basov N.G., Subbotin V.I., Feoktistov L.P., Vestnik Rossiyskoy Akademii Nauk, vol. 63, p.878, (1993).

Review of the Z-pinch

M. G. Haines

Blackett Laboratory, Imperial College, London SW7 2BZ

Abstract

The dense Z-pinch is undergoing a renaissance of research following the development of large terawatt pulse-power generator sources and the theoretical modelling of radiative collapse and of conditions for controlled fusion. The issue of macroscopic stability has been made clearer with the construction of a universal diagram delineating regimes in parameter space for the importance of various physical effects. Of particular interest is the regime in which large ion Larmor radius effects are important, since under conditions for fusion with end losses and also for a strong radiative collapse the ion Larmor radius is about 20% of the pinch radius; this coincides with a minimum in the growth rate of m=0 and m=1 modes in two kinetic models for a parabolic density profile. At Imperial College Z-pinch experiments have commenced on the MAGPIE generator (2.4MV, 336kJ, 200ns). In common with other solid fibre experiments on terawatt and smaller machines there is a tendency at low line density for the ablating coronal plasma to expand, consistent with the triggering of lower hybrid turbulence due to the high drift velocity which prevents the current density from being sufficient to provide an inward pinching force to overcome the pressure gradient. At higher line density, when the ion Larmor radius is small, bright spots appear on the axis with a lifetime of about 10ns. The connection of these with MHD m=0 instabilities will be discussed. At later times there can be an abrupt disruption of the discharge accompanied by a hard X-ray burst, and, for CD_2 fibres a neutron pulse also. One possible explanation of this phenomenon is end losses of plasma, leading to a fall of line density below a critical value which triggers a high anomalous resistivity accompanied by the onset of runaway electrons and accelerated ions. A similar triggering would occur if the line density is lowered in a m=0 neck region. Besides the development of a more complete set of diagnostic tools for the better understanding of these phenomena, attention is now being focussed on pinch load designs that will minimise the initial expansion and control the instabilities and disruption.

1. Introduction

One of the intriguing questions in plasma physics is whether the Z-pinch can be useful as a fusion device or even reactor; if so it has the potential for being more compact and less costly than either tokamak magnetic confinement or inertial confinement.

When we calculate the conditions that a Z-pinch has to satisfy for fusion we find two coincidences of nature that fortuitously occur[1,2]. One is that the ion

Larmor radius becomes a significant fraction of the pinch radius, thus allowing the possibility that the stability of the Z-pinch might not after all be so problematic. The second condition is that the current required is close to the Pease-Braginskii value[3,4] for hydrogen, above which a dramatic collapse[5,6,7] to a density 10^4 to 10^5 times solid is theoretically possible.

To achieve these experimental conditions a pulsed power generator capable of delivering several terawatts is required. At Imperial College we have built and tested the MAGPIE generator (Mega-Ampere Generator for Plasma Implosion Experiments) which has a nominal maximum voltage of 2.4MV, an energy store of 336kJ, line impedance of 1.25Ω, and can deliver currents in excess of 1.5MA into a high impedance Z-pinch load with a rise time of 200ns. Figure 1 is a schematic drawing of MAGPIE. Preliminary experiments mainly into 33μm diameter, 3 cm long carbon fibres at 60% of full voltage are being reported separately at this conference[8]. Some of these results will also be quoted here. The machine has been modified and tested to 80% of full voltage recently. During the next year cryogenic fibres of hydrogen, deuterium and possible argon and neon also will be employed, the latter being of potential interest for X-ray lasing schemes.

At the 1989 ICPP conference I gave a review of the basic philosophy and underlying physics of the dense Z-pinch[9]. In this review I will outline some of the current, important issues that confront both experimenter and theorist in this field. The first of these is the pinch formation and energy coupling, the second is stability including non-linear effects, and the third concerns the physics of what we call a disruption, when a hard X-ray burst associated with an intense electron beam and a large negative value of dI/dt (the rate of change of current) occurs.

2. Pinch Formation

There are many experimental ways of forming a Z-pinch. The earliest experiments employed a vessel filled with gas with a cylindrical wall made of an electrical insulator such as pyrex or quartz. In these experiments, which used capacitor banks directly, the gas characteristically formed a current shell close to the insulating wall which pinched inward driving a cylindrical, converging shock wave ahead which ionized the gas. The shock reflected from the axis and later hit the driving current shell. The pinch radius then expanded but accelerated inwards under $\underline{J} \times \underline{B}$ force. This inward acceleration at this time led to the generation of a m=0 Rayleigh-Taylor instability[10]. A slug model[11] in which the shocked gas has a uniform pressure can be a good representation of the initial pinch formation, and is probably better than a snowplough model which ignores the shock. The slug model predicts a pinch radius of about one third of the wall radius.

When a modern pulsed-power generator is employed, to match the 100ns current rise time a narrower and shorter tube vessel is employed. For typical filling pressures this leads to a lower line density $N\left(=\int_0^a 2\pi nr\,dr\right)$, and we have found that such a pinch compresses to about one third of the tube radius and is fairly stable. Under these conditions the ion Larmor radius a_i is about 20% of the pinch radius a, and while a m=0 instability does grow, it saturates to give a modulation to the pinch radius of about 30%. This experiment[12] was carried out at 150kA current in hydrogen. It would be interesting to extend this to higher currents.

Capillary discharges are an extreme form of compression pinch, but usually so extreme that little current flows in the highly inductive gas region close to the axis. Instead current flows in the adjacent capillary walls which forms a dense, cold, high Z plasma with a radially expanding shock. However the recently reported capillary discharge in argon which gave lasing action at 46.9nm[13] actually had a relatively large bore of 4mm, and is probably more like the compression pinch mentioned above[12]. Indeed, since for laser action it is generally accepted that a high degree of axial homogeneity is required, further exploration of the large ion Larmor radius regime and perhaps the resistive regime (see section 3) might be profitable.

A gas embedded pinch triggered by a focussed laser which ionizes a narrow channel on the axis[14], or alternatively uses field emitting sharp electrodes[15], can lead to a similar problem of continuous accretion of cold plasma, though at Imperial College the line density, measured by laser holography, soon saturated in time. An interesting feature, suggestive that the Kadomtsev profile condition[16] was satisfied was that only the m=1 mode was observed, leading to a rapid transition to a helical (minimum energy) pinch configuration[14].

There is a problem of insulator breakdown, however, when extremely high voltages (>1MV) are applied. Most generators require to be coupled to a relatively low inductance load, and loads must be designed so that at least initially, the plasma current flows at large radius. Hence injected hollow gas puffs or wire array loads in vacuum are usually employed for high voltage pulsed-power Z-pinch experiments, the purpose of which is the study of X-rays. Gas-puff Z-pinches are usualy characterised by m=0 MHD instabilities, and X-ray emission from many randomly occurring dense bright spots of short (~1ns) duration[17].

If the pulsed-power generator is designed for a high impedance load, which is the case for MAGPIE, the load can be a fibre. Pioneering work several years ago by Hammel[18] at LANL used cryogenic deuterium fibres with currents up to 700kA. Similar experiments with a lower voltage and rate of rise of current were carried out at NRL[19]. In these experiments and in our recent experiments with carbon fibres at Imperial College the early time behaviour is characterised by a radial expansion. Our optical radial streak photographs indicate an expansion velocity of 2×10^4 m/s. There are several theories for the interpretation of this expansion; the non-linear development of m=0 instabilities[20] or the triggering of lower-hybrid drift turbulence which results in an anomalous resistivity[21,22]. The former theory would lead to ion viscous heating, necessary to explain the neutron yield from deuterium experiments. The latter is dependent on the applicability of a formula for the lower hybrid drift turbulence resistivity for which there is much variation in the literature[23,24]. Recent 1-D simulations[25] suggest that the expansion and the subsequent re-compression (if any) is sensitive to the energy required for the total ionisation of the particular atomic species present; re-compression is expected to be possible for hydrogen or deuterium.

More complex loads such as multiple shell implosions[26] claim to give a reduction in the Rayleigh-Taylor instability. Similarly a plasma-on-wire (POW) configuration[27] shows an apparently stable central column, but it is uncertain as to whether very much current has transferred to this[28]. If an outer shell of a complex Z-pinch load is actually unstable there may be a benefit in that the disruption of current in this layer due to plasma erosion could lead to a more rapid transfer of current to the next stage ie. it acts as a plasma opening switch.

3. Stability

For an understanding of the Z-pinch in all the different regimes we use the I^4a-N diagram[29] shown in fig.2 for a hydrogen pinch satisfying the Bennett relation. The hatched region indicates where ideal MHD theory applies. Below this, ie. at lower values of I^4a, resistive effects might reduce the growth to a negligible value[30,31], which might explain the stable behaviour of Z-pinches developed at CERN[32] and at Darmstadt as magnetic lenses or indeed the pinch developed as an X-ray laser[13]. However before we can apply the notion of resistive stabilisation the linear theory has to overcome some apparent inconsistencies[33] and its dependence on thermal conduction[34] and the Nernst and Ettinghausen effects clarified. Indeed we know that the latter effects can have a marked effect on the equilibrium profile[35].

The important large ion Larmor radius regime to the left of fig.1 has benefitted from much progress recently. It has been shown that the growth rate of the m=0 instability is reduced by a maximum amount of 70% when a_i/a is 0.2[36]. Recently the Vlasov models have been extended to show that an 80% reduction in growth also occurs for the m=1 (kink) mode. Attention is now being focussed on the non-linear development because of the experimental evidence[12] that early saturation of the m=0 instability occurs.

Recently Bell[37] has carried out 2-D resistive MHD simulations of Z-pinches with sheared axial flow velocity for intergalactic jets and for the pinch arising in a short pulse, sharply focussed intense laser beam on a solid target. Despite the ratio of scale-lengths being 10^{25} remarkably similar enhanced stability to m=0 was found. Linear stability theory of sheared axial flow in a Z-pinch indicates that it is still unstable, though the growth rate is reduced, and it is presently conjectured that the apparent stability is a non-linear effect.

Our experiments on carbon fibres, both on a small scale Z-pinch[38] and on MAGPIE with currents up to 800kA[8,39,40] show in optical emission intense bright spots that bifurcate. The axial velocity, inferred from axial streak images would correspond to a temperature of 6kV if sonic, and therefore our present interpretation is that they are axially propagating shocks. They occur at different times at different axial location and appear to terminate on coalescing with another. At later times (>150ns), gated optical pinhole images indicate a well-developed m=0 instability, and some indication of a m=1 mode as well. X-ray images from a 4-frame camera show intense X-ray emission from the same bright optical spots. On smaller scale carbon fibre Z-pinches we have shown that this X-ray emission is correlated with electron-beam induced emission from the anode[38]. It is important to discover whether these transient bright spots are MHD in origin (instabilities in their non-linear phase), or whether the electron beam is causal, triggering a local region of anomalous resistivity which leads to the diversion of the current around it, this current then triggering a transient pinch. There is strong evidence from the

XUV emission that the electron distribution is non-Maxwellian, probably driven by the high drift parameter in our experiment.

4. Disruption

After a time varying from 100 to 200ns our carbon fibre discharges disrupt. There is a sharp negative value of dI/dt coinciding with a hard X-ray burst from the anode corresponding to an electron beam of 300 to 500keV energy. When CD_2 fibres are used there is also a burst of neutrons at this time with a total yield of about 10^9. The anisotropy suggests a non-thermal origin, probably an ion beam formed at the same time as the electron beam.

We presently conjecture that the beams form as a result of a drop in line density of plasma causing the triggering of an anomalous resistivity. This drop in line density could be due to axial loss of plasma, radial loss, or a local region of depletion of plasma due to the non-linear development of a m=0 MHD instability. The sudden occurrence of the disruption favours a linear axial loss of line density associated with singular and guiding centre motion of ions; the earlier occurrence of the elctron beam when the fibre is shortened from 3 to 2cm is consistent with this.

However we do not believe the formation of electron and ion beams to be electrostatic acceleration of non-neutral plasma ie. due to $q_v E_z$, a net charge density q_v times the axial electric field as this requires through $\nabla\left(\frac{1}{2}\varepsilon_0 E^2\right)$ too large a value of E_z. As pointed out earlier[41] ion beam formation will occur in a quasi-neutral plasma either through $J_r B_\theta$ ponderomotive force in the plasma adjacent to the anode or effective anode at a plasma neck (with an equal and opposite force in the region where J_r is reversed) or through the resonant acceleration of ions in the $\partial A/\partial t$ electric field in the neck region of an m=0 instability. In this latter case there is an energetic beam of singular ions produced, with the equal and opposite momentum in off-axis guiding centre ions. This is probably the mechanism that leads to the second and main neutron pulse in deuterium-filled plasma focus experiments. If correct the avoidance of such a disruption will be dependent therefore on preventing both non-linear development of the m=0 instability and not allowing the axial or radial loss of plasma to let the line density drop below the critical value at which the drift velocity exceeds the ion sound velocity, thus triggering anomalous resistivity.

Acknowledgements

This review and the experimental and theoretical results presented owe much to my colleagues in the Plasma Physics Group, which is in turn indebted to SERC, AWE (Aldermaston) and Sandia National Laboratory for all their assistance with the design and construction of the MAGPIE generator.

References

1. M. G. Haines, *J. Phys. D: Appl. Phys.* **11**, 1709 (1978)
2. M. G. Haines, *Physica Scripta* **T2/2**, 380 (1982)
3. R. S. Pease, *Proc. Phys. Soc. Lond.*, **B70**, 11 (1957)

4. S. I. Braginskii, *Sov. Phys.* **JETP 6**, 494 (1958)
5. M. G. Haines, *Plasma Phys. Controlled Fusion* **31**, 759 (1989)
6. A. E. Robson, *Nucl. Fusion* **28**, 2171 (1988)
7. J. P. Chittenden & M. G. Haines, *Phys. Fluids* **B2**, 1889 (1990)
8. J. M. Bayley et al., ICPP '94
9. M. G. Haines et al., in '*A Variety of Plasmas*' Indian Acad. Sc. (1991)
10. F. L. Curzon et al., *Proc. Roy. Soc. Lond.* **A257**, 386 (1960)
11. D. E. Potter, *Nucl. Fusion* **18**, 813 (1978)
12. J. M. Bayley, Ph.D. thesis, University of London (1991); see also
    M. G. Haines et al. in *Plasma Phys. & Contr. Nucl. Fus. Res. 1990*,
    Vol 2, 769 (IAEA Vienna, 1991)
13. J. J. Rocca et al., *Phys. Rev. Lett.* **73**, 2192 (1994)
14. P. Choi, M. Coppins, A. E. Dangor & M. B. Favre,
    Nucl. Fusion **28**, 1771 (1988)
15. J. Benage, M. Skowronek & P. Romeas, in '*Dense Z-pinches*'
    (AIP Conf. Proc. 299) p3 (1994)
16. B. B. Kadomtsev, in '*Reviews of Plasma Physics*' **Vol 2**,
    Consultants Bureau, New York (1966) 153
17. P. Choi, C. Deeney & C. S. Wong, *Phys. Lett.* **A128**, 80 (1988)
18. J. Hammel, in '*Dense Z-pinches*' (AIP Conf. Proc. 195) p303 (1989)
19. J. D. Sethian et al., *Phys. Rev. Lett.* **59**, 892 & 1790 (1987)
20. R. H. Lovberg, R. A. Riley & J. S. Shlachter, in '*Dense Z-pinches*'
    (AIP Conf. Proc. 299) p59 (1994)
21. J. P. Chittenden, ibid p103
22. A. E. Robson, *Phys. Fluids* **B3**, 1461 (1991)
23. J. F. Drake et al., *Phys. Fluids* **27**, 1148 (1984)
24. J. U. Brackbill et al. *Phys. Fluids* **27**, 2682 (1984)
25. J. P. Chittenden (submitted for publication)
26. R. B. Baksht et al., in '*Dense Z-pinches*' (AIP Conf. Proc. 299) p365 (1994)
27. F. J. Wessel, B. Etlicher & P. Choi, *Phys. Rev. Lett* **69**, 3181 (1992)
28. G. S. Sarkisov & B. Etlicher, private communication
29. M. G. Haines & M. Coppins, *Phys. Rev. Lett.* **66**, 1462 (1991)
30. I. D. Culverwell & M. Coppins, *Phys. Fluids* **B2**, 129 (1990)
31. F. L. Cochran & A. E. Robson, *Phys. Fluids* **B2**, 123 (1990)
32. E. Boggasch & H. Riege, CERN Report PS/85-30 (AA) (1985)
33. M. Lampe, *Phys. Fluids* **B3**, 1521 (1991)
34. N. A. Bobrova et al., in '*Dense Z-pinches*' (AIP Conf. Proc. 299) p10 (1994)
35. J. P. Chittenden & M. G. Haines, *J. Phys. D: Appl. Phys.* **26**, 1048 (1993)
36. T. D. Arber, M. Coppins & J. Scheffel, *Phys. Rev. Lett.* **72**, 2399 (1994)
37. A. R. Bell, *Phys. Plasmas* **1**, 1643 (1994)
38. S. L. Niffikeer et al., in '*Dense Z-pinches*' (AIP Conf. Proc. 299) 495 (1994)
39. M. G. Haines et al., *Int. Conf. Plasma Sc. & Tech.*, Chengdu (1994)
40. M. G. Haines et al., *15th Int. Conf. Plasma Phys. & Contr. Fusion Research*,
    IAEA-CN-60/A6/C-P-10 (1994)
41. M. G. Haines, *J. Nucl. Instruments & Methods* **207**, 179 (1983)

Captions

Fig.1   The Magpie Generator, shown schematically

Fig.2 I^4a-N diagram showing regimes in this parameter space for the validity of various models of stability

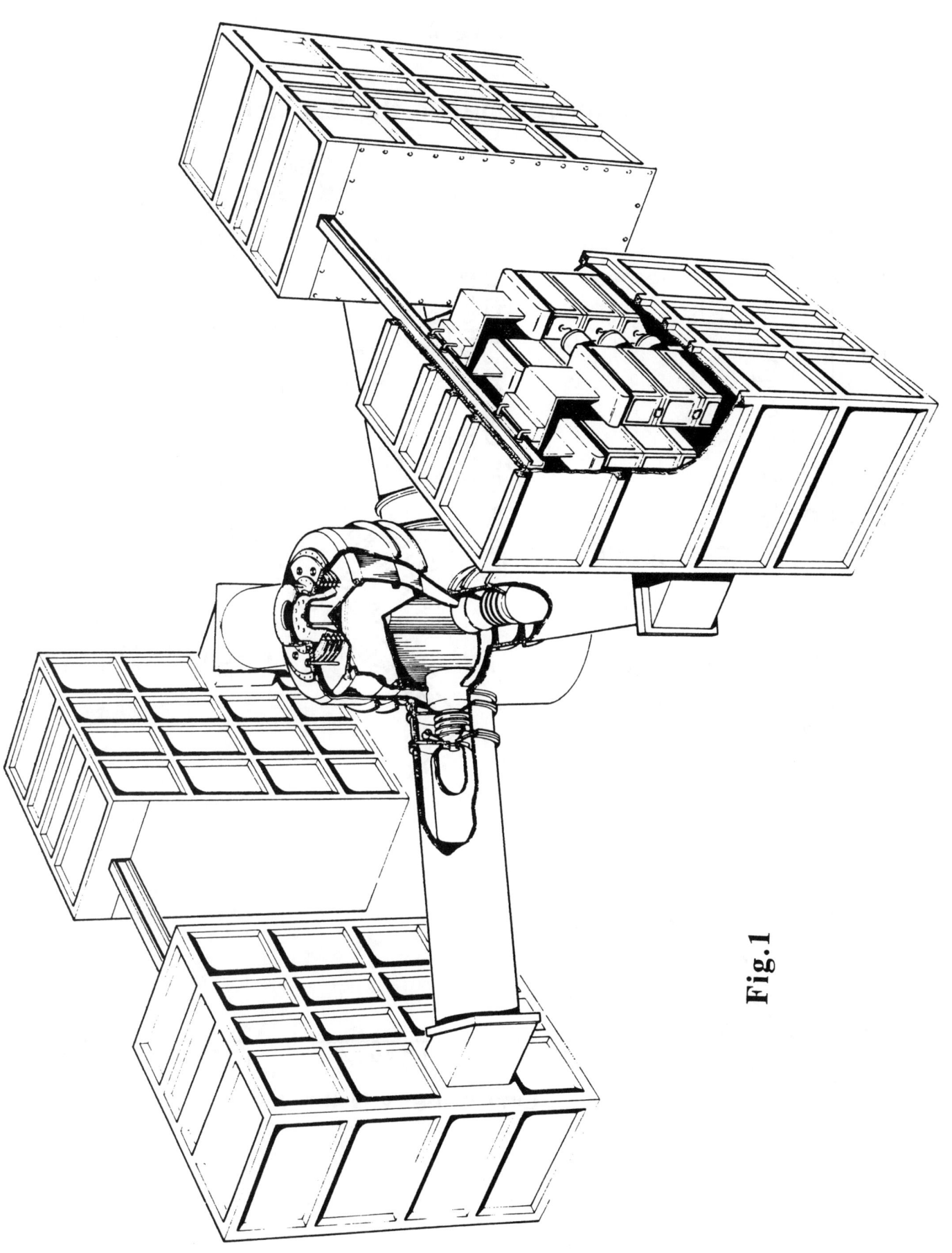
Fig.1

Fig.2

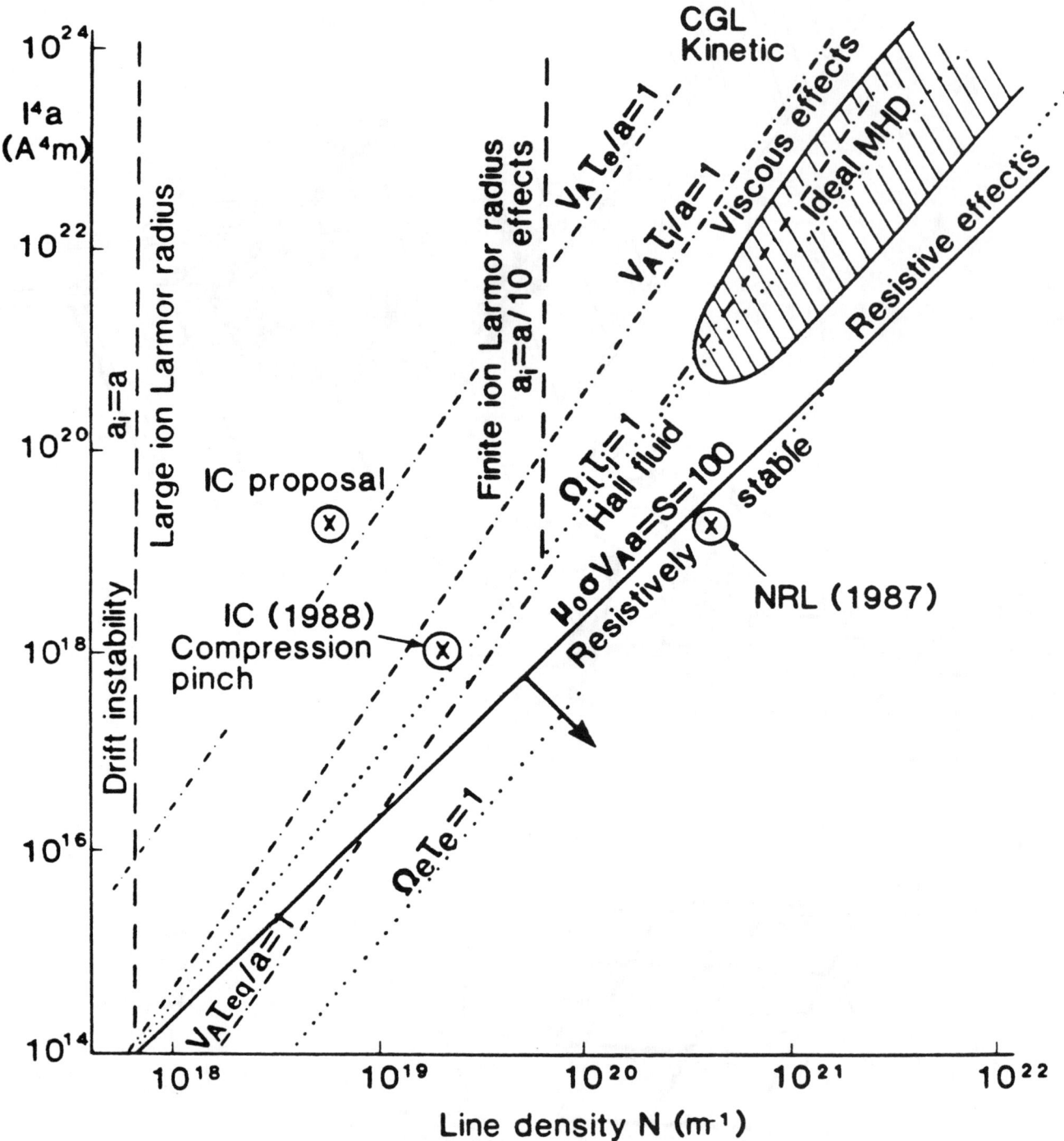

DYNAMICS OF A FIBRE Z-PINCH UNDER DIFFERENT INITIAL CONDITIONS

P. Choi[1], J. Larour[2], C. Dumitrescu-Zoita[3], P. Romeas[2] and J. Rous[2]

[1] The Blackett Laboratory, Imperial College of Science, Technology and Medicine, London SW7 2BZ, UK

[2] Laboratoire des Plasmas Denses, CNRS, Universite P. & M. Curie, case 90, F-75252 Paris Cedex 05 France

[3] Faculty of Physics, University of Bucharest, P.O.Box MG11, Magurele, Bucharest R-76900, Romania

I. INTRODUCTION

The possibility of attaining high temperature high density plasma by passing a high current through a thin wire or fibre has long been recognized. The initial current path, however, is found to flow primarily around a thin corona layer of low plasma density. This coronal plasma is heated up and expands rapidly, thus detaching from the solid fibre on axis. When the current has risen to a sufficiently high value, the hot coronal shell is recompressed onto the fibre. The recompression is unstable and leads to the well known observation of highly localized hot spots as reported in exploding wire experiments.[1,2] The availability of high voltage pulsed power generator increases the initial rate of rise of current but has been shown to be insufficient in preventing the exploding wire phenomena.

Composite pinch

The composite pinch scheme has recently been proposed to overcome the corona problem by introducing a plasma cylinder surrounding the fibre. In this scheme, the initial current flows in the plasma jacket, which is then compressed down towards the fibre as current increases. The large diameter of the jacket provides a substantially lower inductance and therefore better coupling to the driver during most of the current rise. The compression of the surrounding plasma continues until it stagnates onto the fibre. At which point, the current switches over to be shared by the fibre. By optimizing the shell mass and material, such that the shell plasma remains relatively cold at stagnation, a large proportion of the total current flowing could be transferred to the fibre in a much shorter time compared with the current delivery time. Through finite resistivity and radiation coupling, the central fibre is likely to develop a coronal plasma prior to stagnation of the outer shell. The presence of the jacket, however, serves to constrain the otherwise free expansion of the corona. The end result is that a very large current delivery rate could be established on a small diameter preformed plasma of high density. In an initial trial, an aluminium vapour jet was injected to form a concentric cylinder over a 10 μm diameter fibre. With a current of 200 kA from a small pulsed power generator, a highly uniform pinch was created with little of the localized hot spots observed.[3]

In the present experiment, another concept is proposed to produce the composite pinch configuration. Rather than the external injection of an additional plasma cylinder, an additional surface coating on the thin fibre is used. By the application of a controlled preheat current, this surface coating is then vaporised to form the additional plasma jacket over the fibre. A large variety of simple or compound materials can be used to form the coating to different thicknesses, allowing a precise control of the initial mass in the plasma jacket. By adjusting the magnitude and duration of the preheating current, a plasma cylinder of different diameter and density profile could be initiated. The scheme therefore provides a higher degree of flexibility over an injected plasma jet and inherently higher degree of uniformity. So far, the work has concentrated on the effect of an additional surface coating on a fibre pinch without separate preheat. Different coating methodologies were explored and the resulting plasma structure studied. Though preliminary in nature, the results so far demonstrate that the surface layer on a thin fibre has significant influence on the initial evolution and subsequent development of the fibre pinch.

II. EXPERIMENTAL SET UP

The experiment has been conducted on MALIX, a small pulsed power generator based on the Marx-water PFL technology, with an 8 stages Marx charging a 1.6 ohm, 100 ns water pulsed forming line in about 600 ns. A self-break switch discharges the PFL into a short transfer line section and delivers the pulse to the fibre load through a short magnetically insulated section in vacuum. The magnetically insulated section also acts as a baffle to reduce the effect of plasma radiation on the water-vacuum interface and thus preventing premature current crowbar along the perspex insulator. Most of the experiments carried out so far were done at a current of 60 - 70 kA, and a 10-90% risetime of 40 ns, into the plasma load.

The fibre load is mounted between the live cathode and an insulated brass electrode on the anode. A small separation of a couple of mm between this insulated electrode and the anode allows a separate initial current to be put on the fibre and at the same time sharpens the main voltage pulse applied to the fibre. The length of the fibre load is 13 mm. The fibre is anchored at one end with glue on a screw which is screwed into the cathode, and is threaded through a small hole on the insulated electrode before being weighed down by a small lead shot. The lead shot is sufficient to keep the fibre in tension. A variety of fibre has been tested, including simple wires of Al and Cu and coated wire of Cu with Al, PET(a plastic) with Al and PET with Agar Agar.

Apart from electrical measurements of the discharge current and voltage along different parts of the pulsed line, a number of different diagnostics were employed to monitor the emission of the plasma in the soft X-ray region. A multiple pinhole camera was used with 6 differently filtered pinholes to provide time integrated spectral information along the pinch structure. Two filtered PIN diodes were used to register the temporal development of the X-ray zones. The filters were 18 μm and 36 μm of aluminium, chosen to provide information in the 2 -20 Å region. In addition, a novel Slit-Wire camera (S-W Camera) was used to provide information on the size and number of the hot spots detected.

Instead of small apertures in the traditional pinhole camera, the S-W Camera uses a number of thin slits to provide 1-D imaging. Typically tens of micron in width and cm in length, each slit is aligned orthogonal to the axis of the plasma column to provide spatial resolution in the axial direction. Thus a single hot spot will produce a line image on the detecting film plane, being essentially the projection of the slit. This arrangement has the advantage that very small emission source will still produce an easily recognized line image. To resolve the dimension of the spot, a number of thin wires are mounted behind the slit, ranging from 10 μm to 0.5 mm. The wires provide spatial resolution, in the role of an obstacle rather than an aperture, in a direction orthogonal to the slit, i.e. along the radial dimension of the plasma. The choice of the wire diameters is such that they would cover a range from being smaller than the expected size of the hot spots to much larger than the hot spots. The size of the hot spot is then given by the amount of transmission in the shadow region of the wire for different diameter wires. By placing different filters along several similarly equipped slits, the dimension of the hot spots in different spectral region, and hence their temperatures and densities, can be assessed. Details of the S-W camera will be published elsewhere.

In the experiment, a three slit camera was used, filtered by 2.5 μm of Mylar with 400 Å Al for light blocking, 11 μm of Mylar with 800 Å Al and 7 μm of Al respectively. This arrangement allows us in particular to consider the dimension of hot spots emitting above and below the 8 Å region.

III. RESULTS AND DISCUSSION

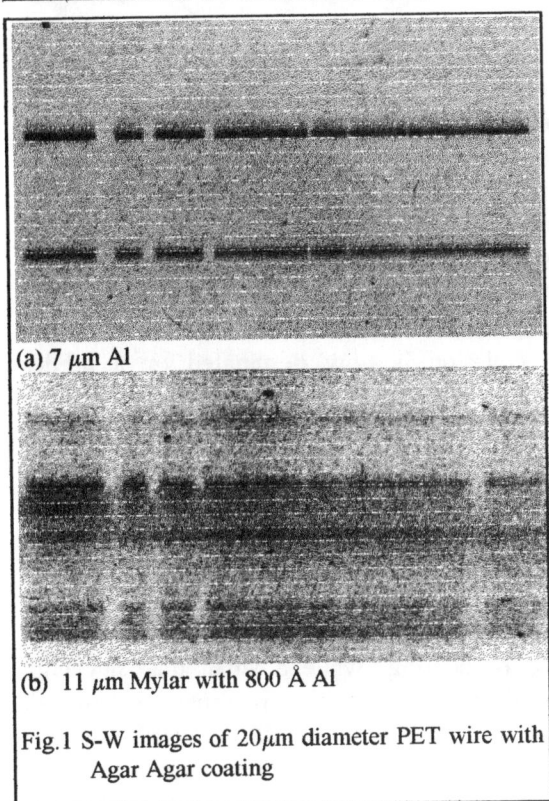

Fig.1 S-W images of 20μm diameter PET wire with Agar Agar coating

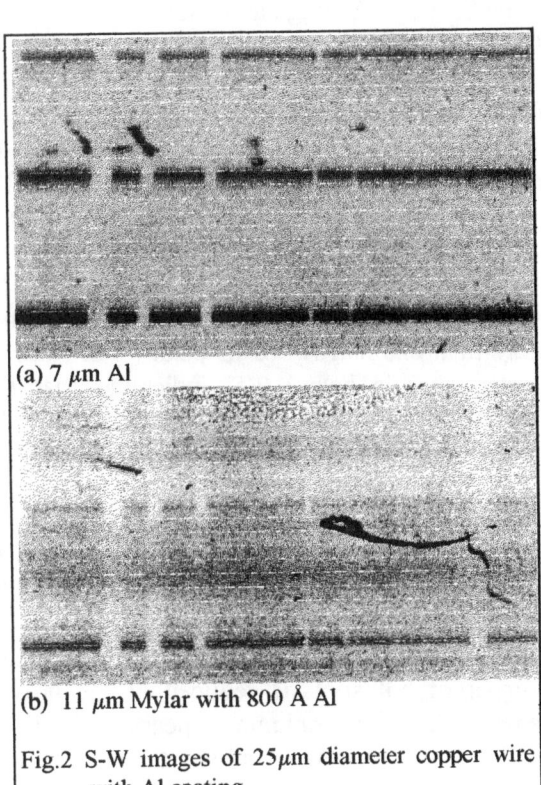

Fig.2 S-W images of 25μm diameter copper wire with Al coating

Figure 1 shows the images obtained from the S-W camera for a PET fibre of 20 μm diameter, coated with a gel like organic compound Agar Agar. The image is recorded on Kodak DEF film. The top image (a) is that obtained through 7 μm of Al which provides a higher transmission window in the 8 - 23 Å region compared with that obtained through the 11 μm Mylar filter, shown in (b). The radiation above 8 Å shows two high density structures. The dimensions of both are below 80 μm from analysing the wire shadows. The radiation below 8 Å, reveals a very different picture. Apart from the regions which correspond to the dense structures recorded in (a), there is emission coming from the bulk of the pinch region. From the level of the flux recorded, this higher temperature plasma is of lower density. From the dimension of the wire shadows, this lower density plasma is below 1 mm in diameter and is therefore not the fast electrons induced solid target X-ray traditionally observed in exploding wire experiments. This argument is supported clearly if similar observations would be obtained with higher Z fibres. Figure 2 shows the S-W images obtained through the same filter arrangement from a 25 μm diameter copper wire with a coating of about 1000 Å of aluminium. Only three prominent hot spots are recorded from (a), despite the much higher average Z value of the load. Again an extended plasma column of lower density but higher temperature can be seen from (b). Similar pictures are obtained with the 20 μm PET fibre with an aluminium coating.

These observations are in strong contrast to the conventional exploding wire results. To exclude the effect of perhaps a more efficient electrode configuration in the present experiment, fibre load from simple 25 μm diameter Al and 25 μm Cu were investigated. Substantially more hot spots were detected, with some 9 hot spots recorded in an Al discharge in the above 8 Å image and almost the same number of hot spots observed in the band below 8 Å. The low density plasma column was not found.

The temporal evolution of coated and uncoated fibre also shows strong differences. Without coating, the X-ray emission in the 2 - 20 Å region tends to start within 30 ns with an Al fibre but could be as early as 14 ns for a Cu fibre. The emission pulse tends to take some 35 ns to rise to the peak and the pulse width would be longer than 35 ns for the same Cu wire. The radiation through the lesser filtered PIN diode tends to be a factor of 10 higher than that from the stronger filtered one for the copper fibre. The emission from a similar copper fibre coated with Al, however, tends to start about 10 -15 ns later, rises to peak in about 25 ns and exhibit more than one narrow 10 ns peak. The ratio of the emission through the softer filter to the harder filter is below 2.5, indicating that the emission is now dominated by a harder component. The comparison of Al coated PET fibre and Al only discharge follows the same trend though the ratio is not so high.

SUMMARY

Despite the preliminary nature of the present experiments, the results clearly show that the surface layer on a fibre plays a dominant role in the pinch formation process. A sub-μm coating of aluminium on a 25 μm diameter copper wire is sufficient to completely modify the radiation characteristics of the resulting plasma. From the space resolved measurements, the number of hot spots is reduced in the presence of coating, while a small diameter high temperature pinch column appears. From the time resolved measurements, the overall emission time is delayed and shortened with coating. The radiation is shifted towards that of a higher temperature plasma. While no spectacular results in terms of pinch stability is observed so far, work is now in progress to extend the coating parameters and fibre diameters. Studies will also be made on the composite pinch scheme with separate pre-forming of the coating over the fibre.

Acknowledgement

The support of the British Council on the collaboration between London and Paris under an Alliance Programme is acknowledged.

References

[1] C.M. Dozier, P.G. Burkhalter, D.J. Nagel, S.J. Stephanakis and D. Mosher, J. Phys. B, **10**,(1977),L73.
[2] L.E. Aranchuk, S.L. Bogolyubskii, G.S. Volkov, V.D. Korolev et al., Sov. Phys. Plasma Phys., **12**,(1986),765.
[3] F.J. Wessel, B. Etlicher, P. Choi, Phys. Rev. Lett., **69**,(1992),3181.

BASIC PLASMA PHYSICS

Progress in Spectroscopic Plasma Diagnostics and Atomic Data

M. von Hellermann, P. Breger, W.G. Core, U. Gerstel, N.C. Hawkes, A. Howman,
R.W.T. König, C.F. Maggi, A.C. Maas, A.G. Meigs, P.D. Morgan, J. Svensson,
M.F. Stamp, H.P. Summers*, R.C. Wolf, K-D. Zastrow

JET Joint Undertaking, Abingdon, Oxon, OX14 3EA, United Kingdom.
*University of Strathclyde, Glasgow, United Kingdom.

The paper illustrates the role of quantitative spectroscopy for the diagnosis of fusion plasmas and the importance of extensive consistency checks. Four examples are given of recent spectroscopic observations which have triggered new approaches both to plasma modelling and atomic excitation processes. The examples highlight the role of passive charge exchange emission and its implication for plasma edge physics, non-thermal features in the HeII spectrum and temperature anisotropies observed in high power neutral beam heated plasmas, the importance of geometric mapping in an arbitrary magnetic configuration for the evaluation of line integrated spectra, and finally the complexity of atomic processes involved in the spectral analysis for helium transport studies.

1. INTRODUCTION

Quantitative spectroscopy plays a significant role in the deduction of plasma parameters within the overall framework of diagnosing a hot fusion plasma. The measurement of accurate ion temperatures, well resolved in space and time and the use of absolutely calibrated measurements of spectral line intensities in the subsequent deduction of local ion densities have contributed significantly to a renaissance of plasma spectroscopy and its recognition as an essential diagnostic tool, cf.[1].The practical implementation of quantitative spectroscopy can best be described as a cycle of modelling, predictions, observations, consistency checks with further refinement at each cycle (see Fig.1) cf. [2,3]. The atomic modelling of excitation processes, and its proper adaptation to the plasma environment by inclusion of collisional radiative processes, enables the prediction of local and global (line-of-sight-integrated) spectral features [4,5]. The successful deduction of plasma parameters is based on a parametric description of spectral features in terms of local plasma data (temperature, density, magnetic field strength etc.) and then the comparison of these spectral predictions with experimental observations.

High core ion temperatures (>20 keV) of present fusion devices imply that the confined plasma region (see Fig.2) contains only fully stripped ions, that is ions not radiating any line emission, with the exception of a very small fraction (<10^{-4}) of helium- or hydrogen-like high-Z impurities (Ni, Fe, etc.). Then in a narrow region close to the last closed flux surface (LCFS), or magnetic separatrix, there is evidence for line emission following passive charge exchange processes with a dense neutral particle population created by plasma ions reflected and neutralised at the plasma wall. Finally, near the vessel walls, the third type of emission process occurs, that is impact excitation of low-temperature (<0.5keV) partially ionised impurities.

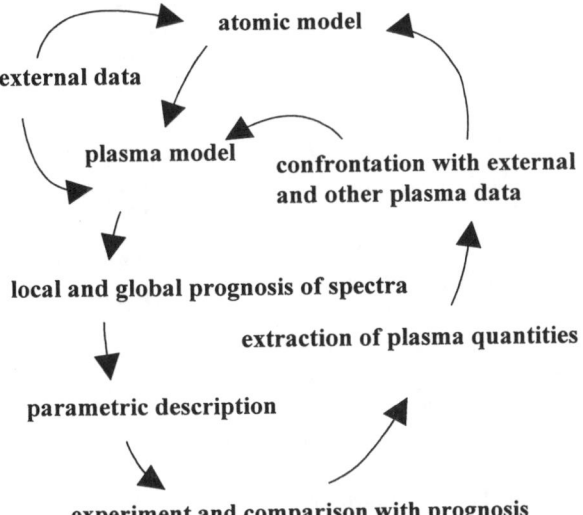

Fig.1: The diagnostic loop as basis of quantitative spectroscopy

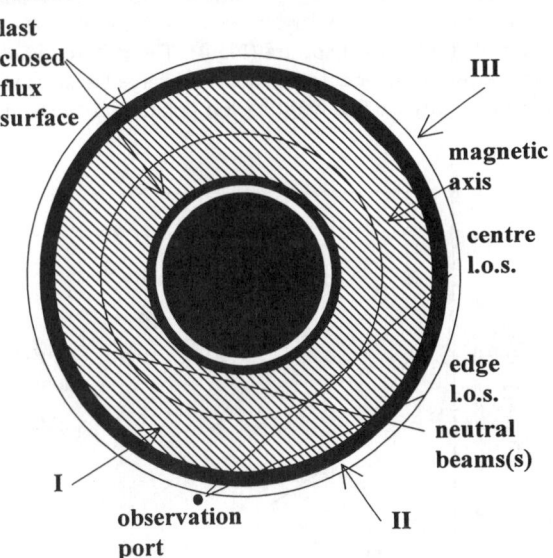

Fig.2: Schematic representation of emission layers and lines-of-sight in a hot fusion plasma (I) plasma core with fully stripped ions (T>5keV), line radiation following CX interaction with neutral beams, (II) intermediate layer close to magnetic, separatrix, T<5keV, line radiation dominantly excited by CX interactions with neutrals, (III) plasma wall, partially stripped ions, line radiation by impact excitation, T<1keV

© 1995 American Institute of Physics

2. PASSIVE CHARGE EXCHANGE FEATURES CLOSE TO THE SEPARATRIX

Low energy passive charge exchange excitation processes involving light species play an important role for the emission layers close to the outermost magnetic flux surface of a confined plasma. The present understanding of this process is that deuterium or hydrogen atoms are created as plasma ions hit the wall and then are reflected as neutrals with two or three times the thermal energy, they then interact with fully stripped light impurity ions close to the separatrix. A simple, schematic, model (Fig.3) illustrates impurity ion density profiles, the decay length of the neutral density and also the role of thermal energy in the effective charge capture coefficient. In a recent paper effective emission rates for UV and visible transitions were modelled for capture from hydrogen in its ground and excited-state [6].

A puzzling observation, which is made in most high-power JET H-mode plasmas, is a flat ion temperature profile (or even a positive ion temperature step) across the magnetic boundary, i.e. an apparent increase of ion temperature beyond the confined plasma region. Similar observations (using a poloidal viewing fan) indicating a flat ion temperature profile beyond the separatrix were reported recently [7]. In the JET case, the apparent ion temperature increase cannot be explained by a toroidal velocity shear across the CX volume. The most likely reason is that for the outermost CX l.o.s (Fig.2), which intersects the neutral beam close to the separatrix, in a region where the C^{6+} density drops rapidly, the passive CX emission overweighs the intensity of the active CX emission. The passive CX emission rate $Q(T)$ depends on the local thermal velocity, the C^{6+} ion density and on the neutral density (Fig.3). Passive charge exchange emission features can be observed from most ionisation stages of impurity ion spectra emitted close to the plasma boundary, for example, the CIII ($n=7 \rightarrow n=5$) and CVI ($n=8 \rightarrow n=7$) spectra (Fig.4). The differential effect between electron impact excitation and charge transfer in forming the highly excited emitting ion population suggests that the influence of the hydrogen donor population can be distinguished and thus a neutral hydrogen density close to the separatrix may be deduced.

3. HIGH POWER NEUTRAL BEAM HEATING AND TEMPERATURE ANISOTROPY

Many of the present fusion experiments are characterised by substantial 'auxiliary' heating. In fact the 'added' heating power may exceed the 'ohmic' heating power by more than an order of magnitude. Recent milestones of fusion yield and ion temperatures, for example, have been achieved with neutral beam powers of 20 to 30 MW. At this power level a significant proportion of the plasma ion velocity distribution function is affected within the period of the slowing-down time of the injected fast particles. A distinctive distortion of the bulk plasma velocity distribution function occurs, cf. [8]. As a result of this, apparent ion temperatures measured spectroscopically perpendicular and parallel to the magnetic field may differ considerably. This phenomenon was reported for the first time in low-target-density high-power helium-neutral-beam-heating-plasmas, cf. [9], where strongly anisotropic alpha particle thermal distribution functions were observed. That is, strong differences in temperatures deduced from essentially Maxwellian HeII spectra are observed in directions parallel and perpendicular to the magnetic field. The term 'temperature' in this context is inappropriate and should be replaced by the second moment of a generalised (not necessarily Maxwellian) anisotropic velocity distribution function:

$$T_{perp} = \tfrac{1}{2}M\langle v^2 \rangle_{perp} = \tfrac{1}{2}M\int d^3v \cdot v^2 \cdot f_{perp}(\vec{v})$$

and $T_{par} = \tfrac{1}{2}M\langle v^2 \rangle_{par} = \tfrac{1}{2}M\int d^3v \cdot v^2 \cdot f_{par}(\vec{v})$

The degree of anisotropy (i.e. $T_{perpendicular}$:$T_{parallel}$) increases with the local source rate of non-thermal particles. It is a sensitive function of the injection pitch angle (for the JET injection geometry the pitch angle is approximately 62°). The anisotropy reaches a peak value within a slowing-down-time and vanishes again in a period corresponding to one or two

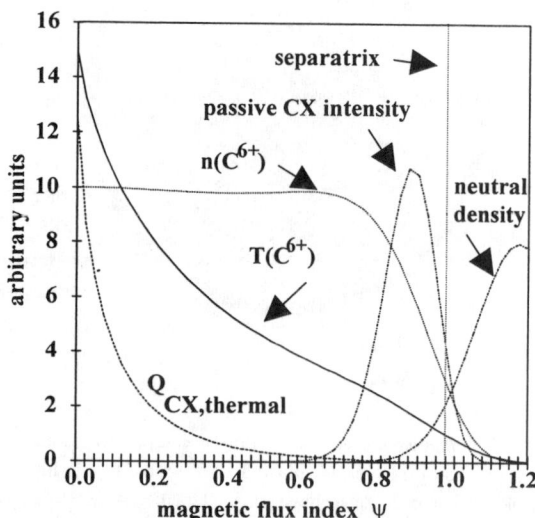

Fig.3: Schematic model for thermal CX excitation, based on a neutral density layer, a CVI density profile, and an emission rate Q which depends on the local thermal energy

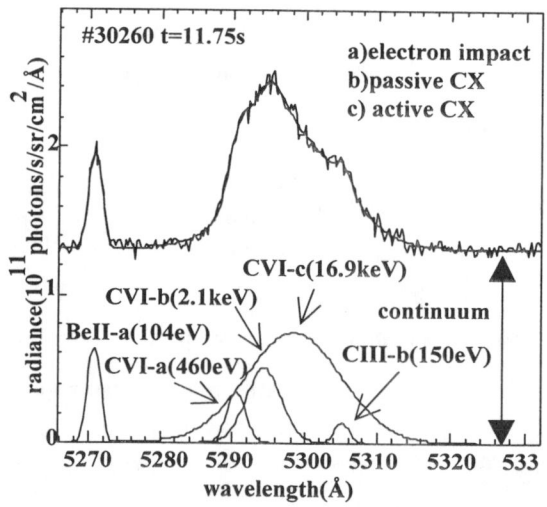

Fig.4: The CVI spectrum illustrating (a) impact excitation, (b) passive and (c) active charge exchange

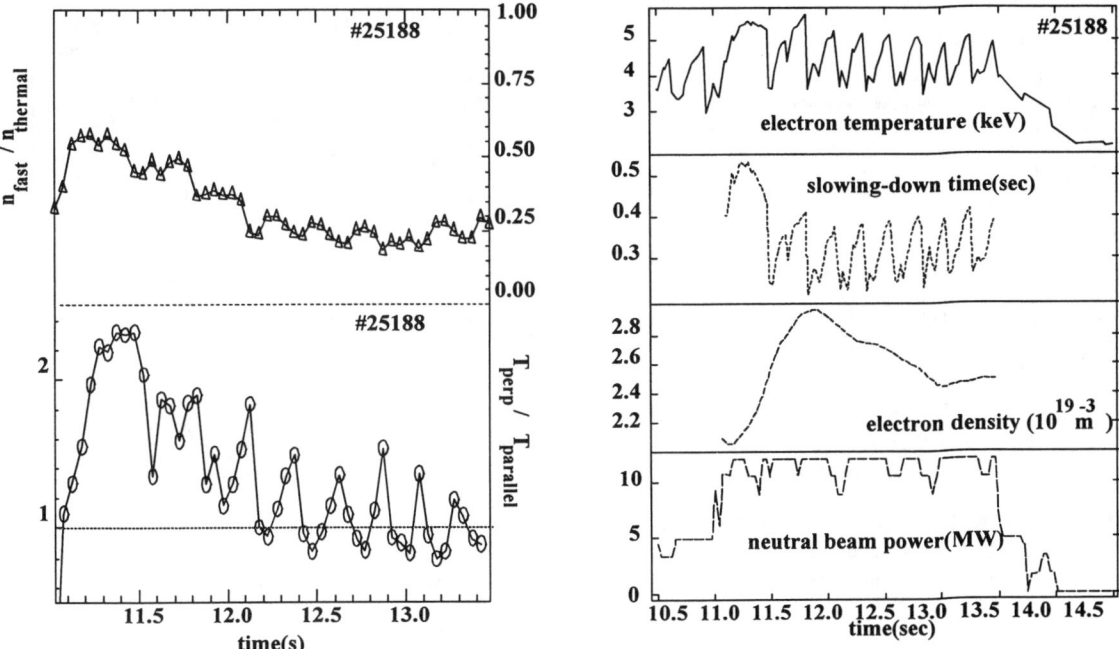

Fig.5: Temperature anisotropy observed in high power neutral helium beam heating experiment

slowing-down times. During the decay of the anisotropy the thermal population builds up continuously (the thermal particle confinement time is 1 to 3s) and the fast alpha-particle concentration is reduced to a small fraction. Since the source rate profile is strongly peaked it is expected that a noticeable anisotropy can only persist close to the magnetic axis. Sawtooth activity also strongly affects the anisotropy indicating that it is an efficient mixing mechanism removing fast particles from the plasma core. At the collapse of each sawtooth, approximately isotropic conditions are re-established, and subsequently the anisotropy builds up again to a level determined by the slow decay following the peak at about 0.5s after the high power beams were switched on (cf. Fig.5).

A conclusion which can be drawn from this observation and also from theoretical modelling of the anisotropic neutral injection Fokker-Planck equation, cf. [8], is that, for present fusion devices, where high levels of neutral beam power injected into comparatively small plasma volumes lead to a substantial fraction of non-thermal fast particles, the observed temperatures do not necessarily represent a steady state population nor an isotropic Maxwellian distribution function.

4. HELIUM TRANSPORT EXPERIMENTS

The close link between atomic data provision for spectroscopic analysis and progress in the deduction of usable plasma data is particularly striking in the case of charge exchange spectroscopy of HeII. Over the last few years new experimental data and theoretical calculations have provided an extensive database on atomic processes relevant for the interaction of neutral beams and plasma ions and electrons. The database, cf. [10,11,12] includes neutral beam attenuation cross-sections, excitation rates by neutral hydrogen or helium beams acting as a donor, collisional excitation of fast beam particles (beam emission spectroscopy, cf. [13]), the prediction of enhancement factors due to donor populations in excited or metastable states and also data required for the modelling of the neutral beam associated plumes, cf. [14]. The assessment of plume contributions requires detailed calculations including magnetic configurations, geometrical factors, and ionisation and excitation rates for He^+ ions travelling along field lines into the CX observation line. In addition, the contributions of passive charge exchange to the HeII spectrum (Fig.6), the so-called 'luke-warm' component, represents a further complication in the analysis.

It is tempting to eliminate the problems related to the complexity of passive edge emission spectra by applying subtraction or beam modulation-techniques. In the first case, spectra gained by an independent viewing system, which does not collect any active spectra, are subtracted. In the second case, spectra measured subsequently during beam-on and beam-off periods are subtracted. However, in both cases it is essential to provide confirmatory independent measurements of, for example, ion temperatures derived from X-ray spectroscopy or other CX spectra to ensure data consistency.

Experimental transport studies rely on controlled perturbations of impurity density profiles. For example, edge fuelling, cf. [15,16], or core fuelling using neutral helium beam injection, cf. [9]. In each case the transient changes of local density gradients and radial particle fluxes provide in the source-free plasma region the means for the deduction of diffusion coefficients D and convective velocity v. In order to derive gradients in time and space, usually analytical profile functions are introduced and smoothed time traces are needed. Fig.7 shows an example of helium edge fuelling, where a 100ms puff was added during a JET H-mode phase. Radial profiles $n_z(\rho)=n_z(0)\cdot(1-\rho^2)\cdot(1+p\cdot\rho^2+q\cdot\rho^4)$, where ρ is the normalised minor radius, are fitted to the experimental data. The selected time window for the evaluation of particle flows

$$\Gamma_z(\rho) = \frac{d\rho/dr}{dV/d\rho} \cdot \frac{d}{dt} \int_0^\rho dV \cdot n_z(\rho)$$

and gradient ∇n_z and the subsequent analysis of

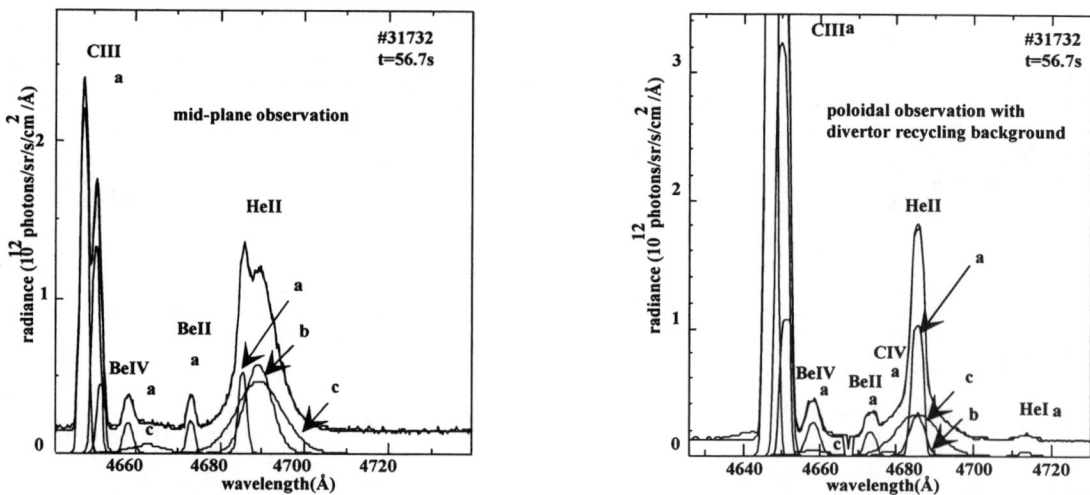

Fig.6 The HeII, CIII, BeII and BeIV spectrum with three excitation components (a) electron impact, (b) passive charge exchange (c) active charge exchange. Left: Torus mid-plane observation. Right: Vertical line-of-sight viewing parts of the JET divertor target plates. In the case of enhanced recycling weak CIV lines are observed at 4677Å(6d8f) and 4735Å(6d8p).

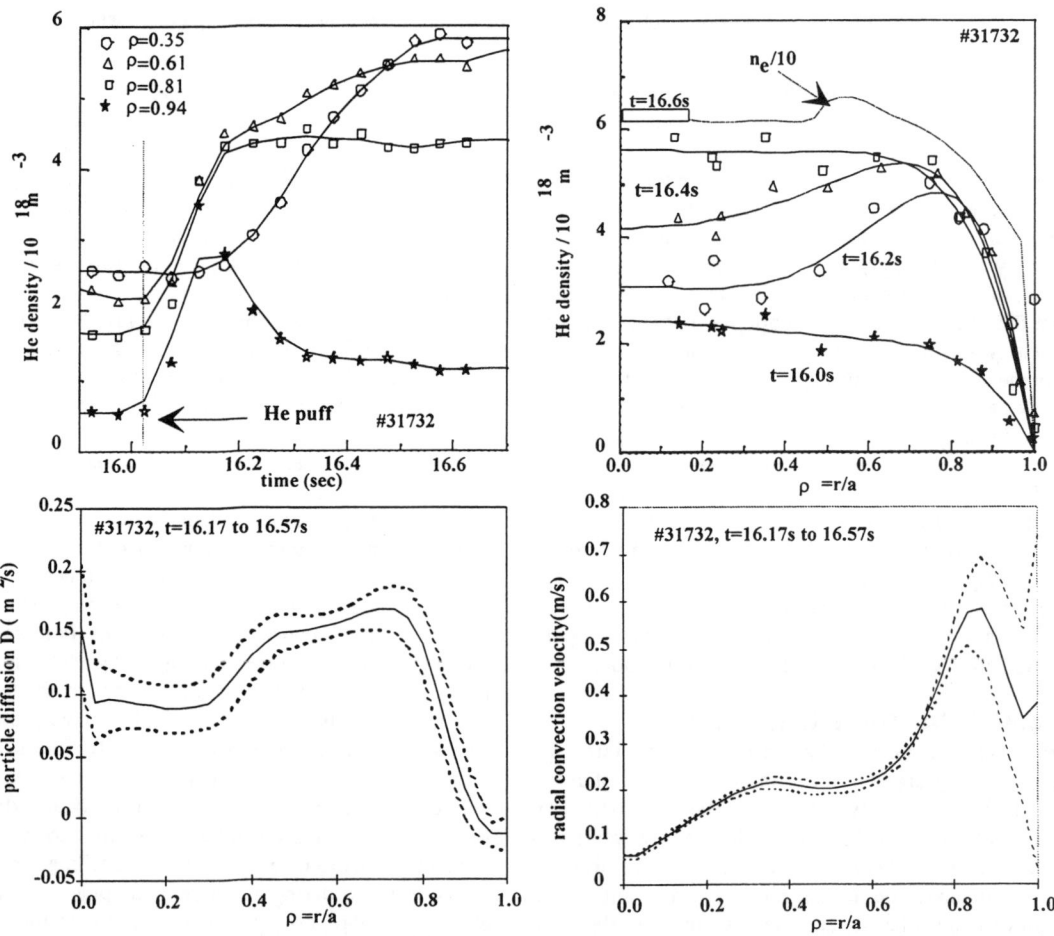

Fig.7: Helium transport studies based on plasma edge gas puffing during a H-mode phase, (a) time history at 4 different radii, (b) radial profiles following the gas puff, and for comparison, a scaled-down electron density profile ($n_e/10$). Central (H-mode) helium densities reach a steady state level in 0.5s. Particle diffusion coefficients (c) and convection velocities (b) are derived from $\Gamma(\rho)/n_z(\rho) = -D(\rho)\nabla n_z(\rho)/n_z(\rho) + v(\rho)$ for the 0.4s time interval, 16.17s to 16.57s, in which the helium density profile relaxes to an electron density-like profile.

$$\frac{\Gamma_z}{n_z} = -D\frac{\nabla n_z}{n_z} + v_z$$

may be a critical issue. Error propagation of experimental data will possibly lead to substantial uncertainties in deduced transport coefficients.

An alternative approach to the evaluation of transport experiments is the adaptation of prediction codes to experimentally observed profiles and the deduction of transport coefficients (cf. [17]).

5. MERGING OF ACTIVE AND PASSIVE DIAGNOSTIC TECHNIQUES

Local measurements are of course not the only route by which quantitative deductions can be made from spectroscopic results, although clearly preferred. The following two examples illustrate how a combination of localised data, especially electron density and electron temperature provided by Thomson scattering, atomic modelling of local emissivities and non-localised, line of sight integrated spectroscopic measurements yield useful quantitative results. The first example is the measurement of continuum radiation in the visible wavelength range and the deduction of a line averaged $<Z_{eff}>$. The second example is the deduction of a semi-localised ion temperature value from the X-ray spectrum of helium-like nickel. In both cases the mapping of the line of sight and that of electron temperature and density profiles onto a common magnetic flux surface geometry is crucial. In the first example, for each point along the line of sight, the values for electron density and temperature $n_e(\rho=r/a)$, $T_e(\rho)$ and that of the free-free Gaunt factor $g_{ff}(T(\rho),\lambda)$ need to be calculated. Assuming that each flux surface represents a constant value of Z_{eff} then any line-of-sight-integrated measurement of continuum radiation is characterised by its impact radius min, that is its smallest minor flux radius (see Fig.8). This approach enables the reconstruction of an Abel inverted $Z_{eff}(\rho)$ profile, and moreover, provides the means for cross-calibrations for any line-of-sight crossing the plasma in arbitrary geometry. The latter aspect is of great importance for most of the spectroscopic collection optics whose window transmissions may deteriorate substantially during extensive operation periods, and the access to a common radiation source, which is toroidally and poloidally symmetric, has proven to be a powerful asset for consistency checks.

A similar procedure is applied in the second example, the JET high resolution X-ray spectrometer (Figs.9 and 10). In this case, a single line of sight, and its corresponding vector of intersected flux surfaces, is used to derive an emission expectation value $<\rho(t)>$ for the location and width σ_ρ of the emission shell.

$$\langle\rho\rangle = \frac{\int \rho \cdot w(\rho) d\rho}{\int w(\rho) d\rho}$$

with: $w(\rho) = f_{NiXXVII}(T_e(\rho)) \cdot n_e^2(\rho) \cdot \varepsilon(T_e(\rho)) \cdot g(\rho)$

where f is the local fraction of helium-like nickel, ε the emissivity, and g a geometry factor. The width of the emission layer ($\sigma_\rho = \sqrt{\langle\rho^2\rangle - \langle\rho\rangle^2}$) is calculated by the second moment of the weighting function. The JET X-ray spectrometer line-of-sight passes below the magnetic mid-plane and the effective emission shell position varies in the range $0.15<\rho<0.35$ subject to changes of electron density and temperature profiles. Note, that the emissivity of NiXVVII reaches its maximum at 8keV and the peak intensity is not necessarily on-axis [5].

6. SUMMARY

Substantial progress has been achieved over the last decade in the expansion of atomic databases for fusion plasmas and in derived data used in experimental analysis. The deduction of absolute particle densities within the magnetically confined plasma based on charge exchange spectroscopy has become an advanced technique in use in many fusion laboratories. However, the need for consistency checks involving the entire

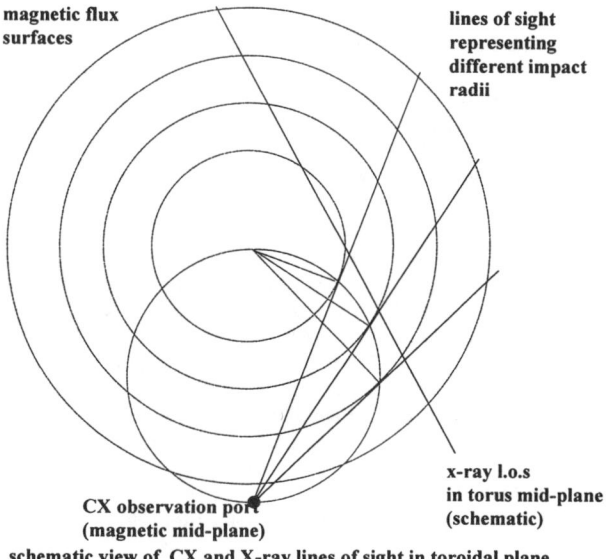
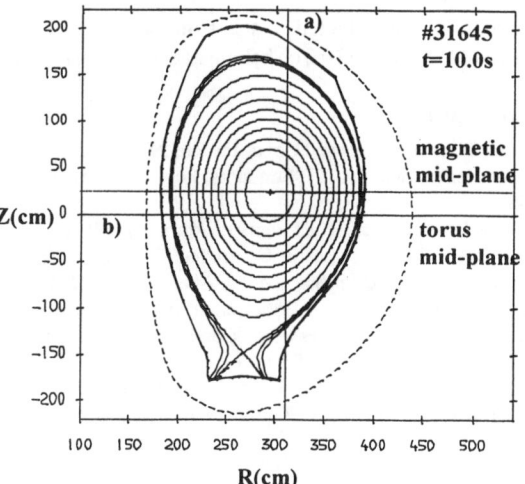

Fig.8: Schematic display of JET l.o.s. in toroidal and poloidal plane. Left: Mid-plane fan for CX and continuum, top view of X-ray l.o.s. Right: lines-of-sight in poloidal plane for (a) CX and continuum, (b) LIDAR and X-ray

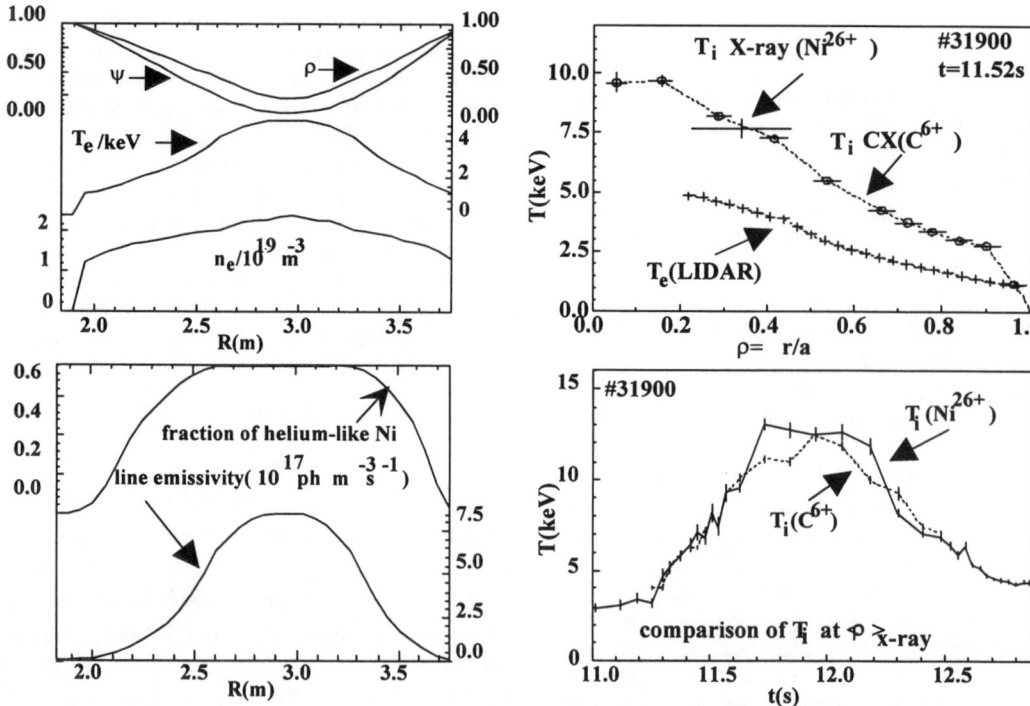

Fig.9: Top: flux index, minor radius, T_e, n_e. Bottom: line emissivities versus radius

Fig.10: Combined active and passive diagnostic techniques. The time trace shows T_i(X-ray) and T_i(CVI) at $<\rho(t)>$

plasma environment has become increasingly necessary and wide spread in their scope. This is very demanding in terms of data range, quality and integration. In particular, the plasma edge and divertor region is still difficult to assess, so that local or semi-local spectroscopic measurements leading to absolute particle densities rather than incompletely analysed photon counts, are a major challenge. In the confined plasma region, the diagnosis of high energy alpha-particles will need considerable future efforts. A combination of active and passive spectroscopic techniques is possibly the most promising way for future fusion devices with reduced access and enhanced radiation levels.

ACKNOWLEDGEMENT

The authors are indebted to the stimulating impact of D. Hillis and M. Wade (ORNL) on helium transport experiments.

REFERENCES

[1] R.Isler, Plasma Physics and Contr. Fusion, **36**, 171(1994)
[2] M von Hellermann and H.P. Summers, Rev.Sci.Instr. **63**, 5132 (1992)
[3] M. von Hellermann and H.P.Summers, Atomic and Plasma Material Interaction Processes in Controlled Thermonuclear Fusion, Ed. Janev, Elsevier Science Publishers 1993, 135-164
[4] H.P.Summers and M. von Hellermann, Atomic and Plasma Material Interaction Processes in Controlled Thermonuclear Fusion, Ed. Janev, Elsevier Science Publishers 1993, 87-117
[5] K.-D. Zastrow, H.W.Morsi, M. Danielsson, M.G. von Hellermann, E. Källne, R. König, W.Mandl, H.P.Summers, J. Appl. Phys, **70**, 6732 (1991)
[6] C.F. Maggi, L.D.Horton, R.König, M.Stamp, 21st EPS Conf. Contr. Fus. & Plasma Physics, Montpellier 1994
[7] W.Mandl, K.H.Burrell, R.J.Groebner, J.Kim, R.P.Seraydarian, M.R.Wade, J.T.Scoville, To Appear in Nuclear Fusion 1994
[8] W.G.F.Core, to be published
[9] M von Hellermann, W.G.F.Core, J.Frieling, L.D.Horton, R.W.T. König, W.Mandl, H.P.Summers, Plasma Phys. Contr. Fusion, **35**, 799, 1993
[10] H.P.Summers, 'The JET Atomic Data and Analysis Structure', JET-IR(1994)
[11] H.P.Summers, Adv. At. Mol. Opt. Physics, 33, 275 (1994)
[12] H.O.Folkerts, R. Hoekstra, L.Meng, R.E. Olson, W.Fritsch, R. Morgenstern, H.P.Summers, J.Phys.B:At.Mol.Opt.Phys.**26** (1993) 619
[13] W.Mandl, R.Wolf, M von Hellermann, Plasma Phys. Contr. Fusion, **35**, 1373, 1993
[14] R.Fonck, D.S Darrow, K.P. Jaehnig et al., Phys.Rev. (1984) 29, 3288
[15] M.R.Wade, D.L.Hillis, J.T.Hogan, M.A.Mahdavi, R.Maingi, W.P.West, N.H.Brooks, K.H.Burrell, R.J.Groebner, G.L.Jackson, C.C.Klepper, G.L.Laughon, M.M.Menon, P.K.Mioduszewski and DIII-Team, Submitted to Phys.Rev.Let., 1994, GA-A21652
[16] D.L.Hillis, M.R.Wade, J.T.Hogan, M.A.Mahdavi, R.Maingi, W.P.West, N.H.Brooks, K.H.Burrell, D.F.Finkenthal, R.J.Groebner, G.L.Jackson, C.C.Klepper, G.L.Laughon, M.M.Menon, P.K.Mioduszewski, 15th International Conf. Plasma Physics, Seville, Spain, October 1994
[17] L. Lauro-Taroni, B.Alper, R.Giannella, K.Lawson, F.Marcus, M.Mattioli, P.Smeulders, M.von Hellermann, 21st EPS Conf. Contr. Fus. & Plasma Physics, Montpellier 1994

INDUCED TRANSPARENCY AND PARAMETRIC INSTABILITIES IN THE RELATIVISTIC REGIME OF THE LASER PLASMA INTERACTION

J.C.Adam, A.Héron, S.Guérin, G.Laval, P.Mora,

Centre de Physique Théorique de l'Ecole Polytechnique,
91128 Palaiseau Cedex, France

Abstract

An electromagnetic wave is expected to propagate through an overdense plasma, provided that its amplitude is large enough to bring the electron motion into the relativistic regime. We study this induced transparency by using 1D and 2D particle simulations. We find that large chaotic longitudinal electric fields are generated and strong plasma heating takes place during the propagation. This behavior is attributed to parametric instabilities. In order to support this interpretation, the modulational and Raman instabilities of a relativistic electromagnetic wave are studied analytically as well as with the help of particle simulations. In addition it is found that a finite width beam is strongly self focused

1-INTRODUCTION

Recent progress in beam processing have increased by orders of magnitude the instantaneous power available in ultra-short laser pulses. The intensity reaches high enough levels to bring into the relativistic regime an electron interacting with the radiation. For example, the electron quiver energy reaches one Mev in a circularly polarized monochromatic electromagnetic wave whenever its intensity is $2 \times 10^{-2} \times n_c$ watts/cm^2 where n_c is the critical density in number of particles per cm^{-3}. Intensities in this range have been obtained in several experimental devices [1] by using the chirped pulse amplification process.

For such extremely high intensity levels, the electromagnetic wave propagation in a plasma becomes strongly non-linear. The electron motion is the first source of non-linearity as the electron mass is increased by the relativistic effect and the Lorentz force is not any more negligible. Moreover, this V×B force induces a longitudinal electron displacement across the plane wave. Consequently, the electron density is perturbed by the wave propagation and the electric field is no longer purely transverse. Large amplitude nonlinear longitudinal oscillations are coupled to the electromagnetic wave.

Non-linear solutions of Vlasov-Maxwell equations have been obtained by several authors. They describe periodic plane non-linear waves [2] propagating with a constant velocity in an infinite cold plasma. Other solutions provide non-linear reflection and transmission coefficients in a bounded cold plasma.[3] The most striking feature of these waves is their ability to propagate in an overdense plasma, i.e. with a frequency lower than the plasma frequency.

Among the non-linear plane periodic solutions, the most remarkable one is circularly polarized. For this wave, it is possible to write a dispersion equation and to define a refraction index N given by:

$$N^2 = 1 - \frac{\omega_p^2}{\gamma \omega^2} \quad (1)$$

with

$$\gamma^2 = 1 + \left(\frac{eE}{m\omega c}\right)^2 \qquad (2)$$

where E is the electric field amplitude and ω the wave pulsation.

For this polarization, there is no Lorentz force and consequently the electron density remains uniform. There is no longitudinal electric field. The electron velocity amplitude is constant as well as its kinetic energy. The physical meaning of Eq. (1) is easily understood by noticing that the elecron mass is increased by the transverse electron motion so that the effective plasma frequency is reduced. The plasma is transparent for $\omega_p/\sqrt{\gamma} < \omega$ and one may say that induced transparency is achieved for $\omega_p/\sqrt{\gamma} < \omega < \omega_p$

The purpose of this study is to assess the possibility of transmitting electromagnetic energy flux through an overdense plasma by using this induced transparency phenomenon. Numerical particle-in-cell simulations are used together with analytic stability computations.

2-NUMERICAL SIMULATIONS OF A 1D PLASMA SLAB.

The interaction of an ultra-high intensity electromagnetic wave with a plasma slab has been initially studied in the 1-D case of a normally incident plane wave. The electromagnetic field as well as the plasma remain invariant by translation in directions orthogonal to the direction of propagation. The particle-in-cell simulation takes into account the three components of particle velocities but the only spatial dimension in the direction of propagation. The Maxwell equations deal with a three components electric field and a magnetic field with two transverse components.

The plasma slab width is 10 times the vacuum wavelength $\lambda_0 = 2\pi c/\omega$. Its uniform electron density n is 1.5 n_c. The particle temperature is 4 keV. A small amplitude wave is totally reflected by such a layer. The temperature has been chosen on numerical grounds.

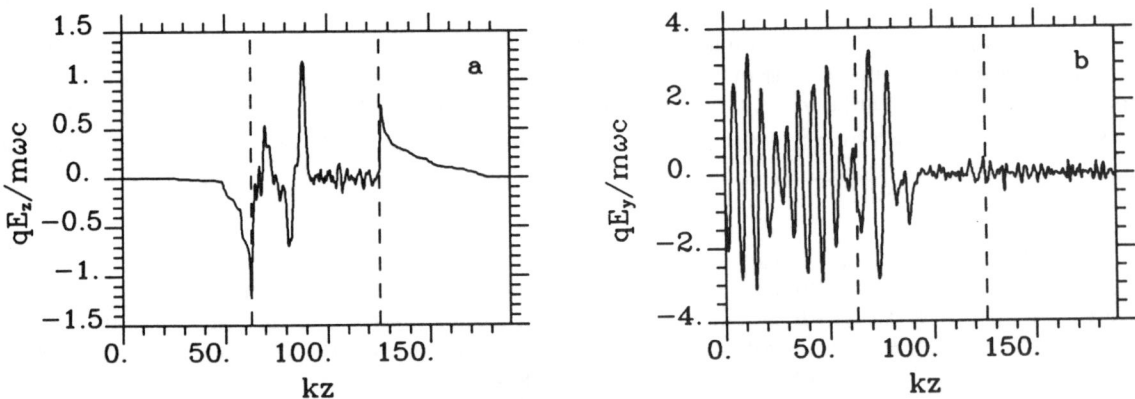

Fig.1. Electric field at time 250 ω^{-1}; the vertical dashed lines indicate the edges of the layer. -a: longitudinal electric field; -b: transverse electric field.

The launched wave reaches an intensity such that $\gamma = 2$, where γ is given by Eq.(2). Its rise time is 10 wave periods. At the left side of the simulation box, the polarization of the incoming wave is circular. It is seen on Fig.1b that the transverse electric field does propagate through the overdense plasma with intensities in the same range as the vacuum field. However several unexpected features are also displayed by the simulation. A large irregular longitudinal electric field is generated (Fig.1a).

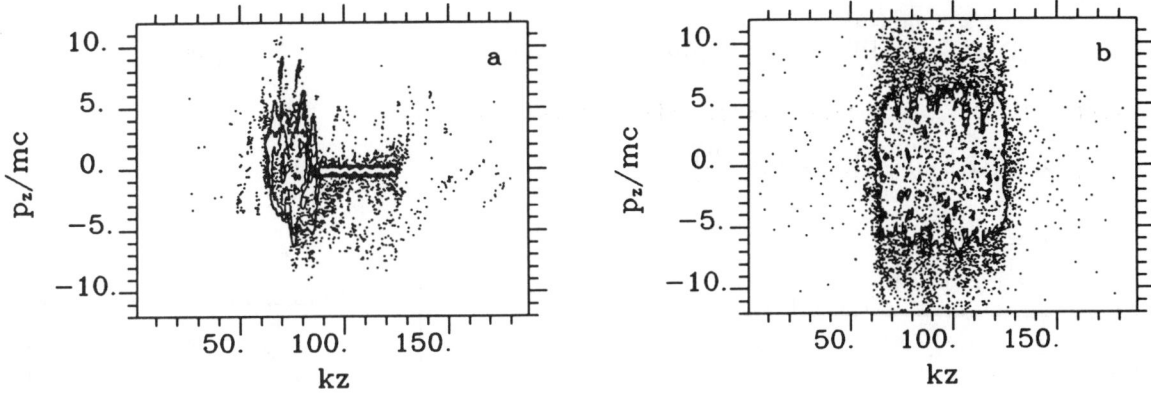

Fig.2. Electron longitudinal phase space plot. -a: $t = 250\, \omega^{-1}$; -b: $t = 1000\omega^{-1}$.

The electron phase space plot (Fig.2a) shows a strong longitudinal heating behind the wave front. As a result the absorption coefficient of the wave reaches 0.8 and the front velocity is reduced. In a cold plasma the front velocty V_f should be given by $V_f = c\, N / [\, 1 + n(\gamma-1)/2n_c\gamma(\gamma+1)]$, where N is given by Eq.(1). This expression takes into account the enegy flux which is necessary to give electrons the transverse momentum corresponding to the wave amplitude. In the numerical simulation, it yields $V_f = 0.30\, c$ while the measured one is $0.15\, c$. This last value is obtained when the longitudinal heating is added in the energy flux balance. The longitudinal heating lengthens the time which is necessary to go through the layer and therefore increases the total amount of energy which has to be contained in the incident pulse. Moreover, it lowers the effective plasma frequency so that the layer becomes almost transparent a few periods after the layer crossing.

The strong longitudinal heating could result either from the interaction of electrons with the front of the wave as it penetrates through the cold plasma or from strong parametric instabilities excited by the intense wave as soon as it penetrates through the layer. The irregular behavior of the longitudinal electric field suggests that parametric instabilities play an important role. In order to have a more rigorous evidence of their importance, we have injected a linearly polarized wave on the left side of the simulation box. The interaction is qualitatively the same as in the case of a circularly polarized wave, but the wave does not remain polarized linearly. After a short time, it looks totally depolarized, indicating obviously that the symmetry of the incident wave has been destroyed by parametric instabilities.

After the layer crossing by the front, the wave propagates almost linearly in the plasma with a relativistic temperature (Fig 2b). The wave amplitude is slowly modulated and the polarization of the incident wave is restored.

3- WAVE STABILITY IN A COLD PLASMA.

Although the parametric instabilities are more obviously identified in the numerical simulations for linear polarization, the wave stability will be studied for a circularly polarized wave. As a matter of fact, for such intensities, the usual weak coupling approximations do not apply and there is no simplifying approximations available in the general case. Fortunately, the non-linear periodic circularly polarized wave yields an exact dispersion equation [4] for pertubations propagating like the periodic wave, namely for the perturbations which are taken into account in the 1D simulations. Such perturbations are chracterized by the excitation of a longitudinal electric potential $\phi \sim \exp i(Kz - \Omega t)$, and transverse components $E \sim \exp i[(K \pm k)z - (\Omega \pm \omega)t]$. The dispersion relation is then given by:

$$D_+ D_- D_p + \frac{\omega_p^2 \left(\frac{eE}{m\omega c}\right)^2}{\gamma^3} (D_p - K^2 c^2)(D_+ + D_-) = 0, \qquad (3),$$

where

$$D_\pm = (\Omega \pm \omega)^2 - (K \pm k)^2 c^2 - \frac{\omega_p^2}{\gamma},$$

$$D_p = \Omega^2 - \frac{\omega_p^2}{\gamma}.$$

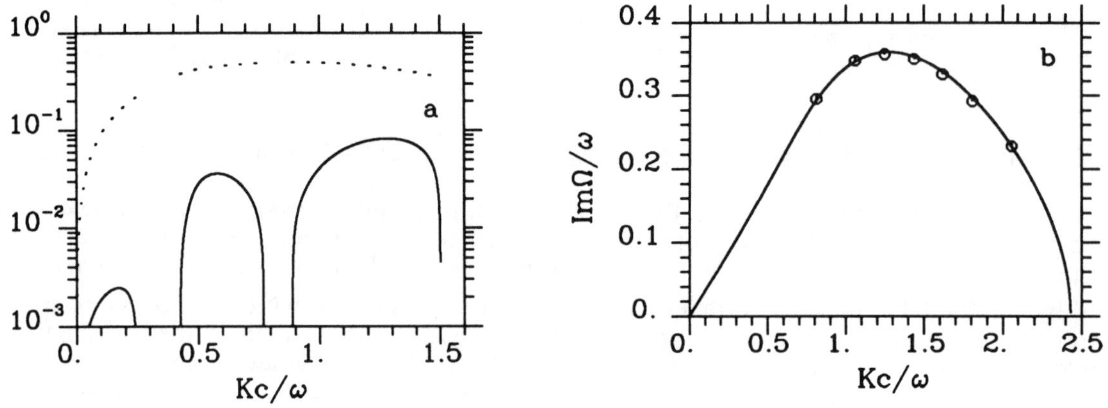

Fig.3. Growth rates (−) and frequencies (···) of parametric instabilities given by Eq. (3). -a: ω_p/ω = 0.44, for γ = 1.022; -b: ω_p/ω = 1.22 for γ = 2.

On one hand, for γ =1.022, in an underdense plasma, three parametric instabilities are found [5] (forward and backward Raman scattering, modulational instabilities). They correspond to three distinct region in K-space as shown in Fig.3a. On the other hand, Fig.3b shows the merging of these usual parametric insabilities in the overdense plasma to yield a continuous unstable domain extending from K=0 to a maximum K larger than k. In this last case, the maximum growth rate reaches 0.4 ω. The instability is found to be absolute in the wave frame with a simalar growth rates and a wave number close to k. Consequently these perturbations are generated at the wave front over a short distance. It is a possible explanation of the observed heating.

4-WAVE STABILITY IN A HOT PLASMA.

It has to be understood why the instabilities do not appear after the crossing of the layer by the wave front. In order to look for such an explanation, the stability has been studied when the wave propagates in plasma with a longitudinal electron momentum spread.

In such a plasma, a large amplitude circularly polarized electromagnetic wave propagates also sinusoidally without inducing any longitudinal electric field. Its propagation may be characterized by a refraction index N which is now $N^2 = \frac{k^2 c^2}{\omega^2} = 1 - \frac{\omega_p^2}{\omega^2} \langle \frac{1}{\gamma} \rangle$, where γ is the electron Lorentz factor and the averaging is taken over the electron momentum distribution function.

The dispersion equation for the perturbation can also be obtained in this case. Its discussion shows a reduction of the growth rates as the longitudinal momentum spread Δp increases. This stabilization is more easily seen by using a 1D particle-in-cell numerical simulation. At t=0, a large amplitude periodic circularly polarized electromagnetic wave is launched. For a cold plasma, the longitudinal field amplitude grows according to the solutions of Eq.(3) as shown on Fig.3b where the circles correspond to the measured growth rates in the simulation.

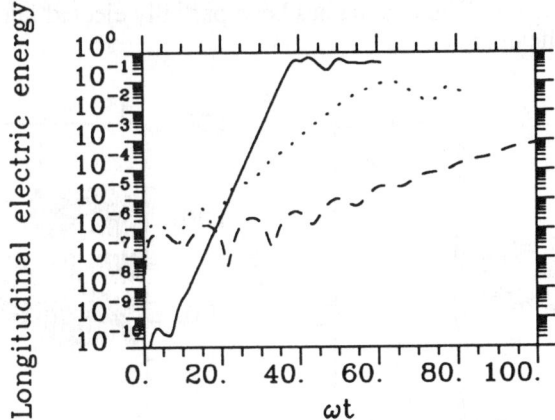

Fig.4. Longitudinal electric energy density versus time for $\omega_p/\omega=1.22$, $\gamma = 2$, $\Delta p/mc=0.005$ (—), $\Delta p/mc=1.152$ ($\cdots$), $\Delta p/mc=3.4$ (---).

The growth rates are significantly reduced provided that the longitudinal momentum spreading exceeds mc (Fig.4). This result explains the attenuation of the parametric instabilities after the layer crossing, since the observed longitudinal momentum spreading is then larger than mc.

5-NUMERICAL SIMULATION OF A 2D PLASMA SLAB.

In order to describe transverse filamentation and self-focusing of the incident beam, the particle-in-cell numerical simulation of Sec.(2) has been performed with a 2D code. The incident wave was linearly polarized and the code did not allow depolarization effects. After averaging over the transverse direction, the wave behavior and the plasma evolution did not differ markedly from the 1D case. The main new feature appears on Fig.5, showing a wave filamentation as well as a strong transverse modulation of the plasma density.

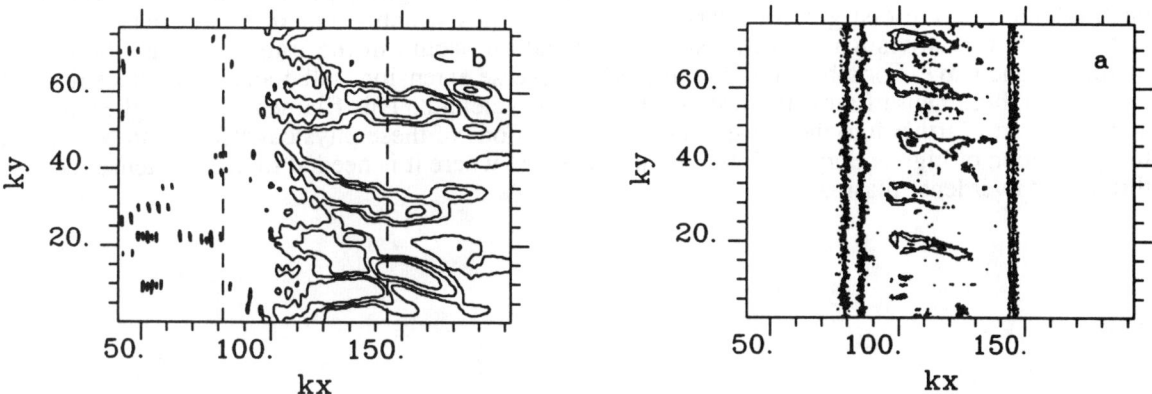

Fig.5. Plasma density (a) and transverse electric energy density (b) at time $t=400\ \omega^{-1}$ in the 2D simulation for a plane incident wave..

In order to simulate the interaction of a focused laser radiation with the layer, a finite width gaussian beam has been launched on the left side of the simulation box. The peak intensity is the same as the intensity in the previous cases. The beam is about 10 wavelength wide. A strong self-focusing is apparent on Fig.6b since the beam emerges at the right edge of the layer with a three times smaller width and a three times higher intensity. The light propagates in a hot channel in which the density is lower than the initial one (Fig.6a). The plasma has been partially ejected laterally and two dense ridges appear along the edges of the channel.

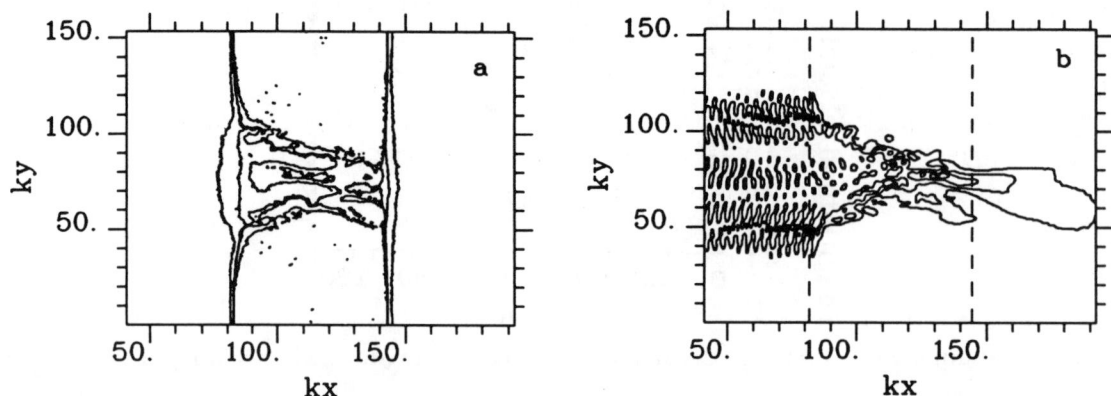

Fig.6. Plasma density (a) and transverse electric enegy density (b) at time t=400 ω^{-1} in the 2D simulation for a focused incident beam.

The plasma density profile looks somewhat similar to the hole bored in a very overdense plasma by a focused beam [6]. However the mechanisms are quite different. In the very overdense case, the light is not intense enough to induce transparency. In order to drill its hole, the light has to push the plasma electrons and ions so that the front velocity remains small compared with the velocity of light. If the intensity induces transparency of the overdense plasma, the wave front propagates much faster as the wave has to heat the electrons but is not not slowered by the ion inertia.

6-CONCLUSIONS.

A relativistic electromagnetic wave propagates through an overdense plasma. A large fraction of the incident energy flux is converted into electron thermal longitudinal motion. The origin of this heating could be the fast growing parametric instabilities which occur in a cold plasma and are stabilized in a relativistically hot plasma.Filamentation occurs in the transverse direction for an incident plane wave. For a beam with a limited transverse extension, self-focusing reduces its width to that of a thin intense filament within only a few wavelength. This filament has a higher intensity and propagates further into the dense plasma. Applications of these physical effects can be found in the generation of high energy particles and in processes where it is needed to inject intense radiation through an overdense plasma.

REFERENCES.

[1] M. Pessot, J.Squier, G. Mourou and D.J. Harter, Opt. Lett. **14,**797 (1989); M. Pessot, J.Squier, P. Bado, G. Mourou and D.J. Harter,J. Quantum Electron. **QE-25**, 61 (1989); J.P. Watteau, G. Bonand, J. Coutant, P. Dautray, A. Decoster, M. Louis-Jaquet, J. Ouvry, J. Sauteret, S. Sezuec and D. Teychenne, Phys. Fluids **B4,** 2217 (1992)

[2] A.I. Akhiezer and R.V. Polovin, Zh. Eksp. Teor. Fiz. **30,** 915 (1956) [Sov. Phys. JETP 2, 696 (1958)]; P. Kaw and J. Dawson, Phys. Fluids **13**, 472 (1970); C. Max and F. Perkins, Phys. Rev. Lett. **27**, 1342 (1971).

[3] F.S. Felber and J.H. Marburger, Phys. Rev. Lett. **36**, 1176 (1976).

[4] C.E. Max, Phys. Fluids. **16**, 1480 (1973).

[5] C.J. McKinstrie and R. Bingham, Phys. Fluids B **5**, 2626 (1992).

[6] S.C. Wilks, W.L. Kruer, M. Tabak and A.B. Langdon, Phys. Rev. Lett. **69**, 1383 (1992).

STRUCTURE OF SHORT-WAVELENGTH MODES IN A TOROIDAL PLASMA

J B Taylor and H R Wilson

UKAEA Government Division, Fusion, Culham, Abingdon, Oxon OX14 3DB, UK
(UKAEA/Euratom Fusion Association)

Abstract:

Short wavelength fluctuations may be a source of anomalous transport in toroidal plasmas. Early investigations concerned electron and ion modes that occur only at a particular radius and have a localised eigenfunction: such modes do not seem important for transport. Recently, Connor, Taylor and Wilson [1] described electron modes that occur at all radii and have extended eigenfunctions: these could lead to large transport. Similar ion modes were described by Romanelli and Zonca [2]. All these modes have discrete frequency spectra. Now, Kim and Wakatani [3] have described electron modes that occur at all radii but are localised and, it is claimed, have a continuous spectrum. We re-examine this problem of mode structure and location. We find that there are just two basic forms: 'isolated' modes that occur only at a particular radius and 'general' modes that occur at all radii. Isolated modes are always localised. In special circumstances general modes may be extended, but in typical cases they are also localised. All modes have a discrete spectrum. These results are consistent with a diffusion coefficient having the Bohm scaling and decreasing with plasma rotational velocity shear.

1. INTRODUCTION

Short wavelength fluctuations may be a source of anomalous transport in magnetically confined, toroidal, plasmas. Prominent among these fluctuations are the electron drift (ED) mode and the ion temperature gradient (ITG) mode - which have been known for more than 30 years. They were first studied in a plane slab, but it was shown by Taylor [4] that their properties may be entirely different in a torus. This led to the development of the, so-called, 'ballooning representations' which were used to investigate the ED [5] and ITG [6] modes in a torus. Early investigations naturally concentrated on the most unstable forms of the ED and ITG modes. However, these occur only at isolated radii and have a highly localised eigenfunction. Consequently they may not be the most important for transport.

Recently [1], Connor, Taylor and Wilson pointed out that there could be another, more general, class of ED modes in a torus. These have a higher stability threshold than the 'isolated' modes, but can occur at all radii. Similar ITG modes were described by Romanelli and Zonca [2]. At first sight, these general modes appear to have an extended eigenfunction, spanning a large fraction of the plasma radius. If so they might lead to large plasma transport[1]. However, we will show that this large radial extension occurs, if at all, only for a very narrow range of plasma parameters.

In the latest development, Kim and Wakatani [3] introduced an apparently different class of ED mode. Although these are not restricted to isolated radii, they resemble the isolated modes in that they have a localised eigenfunction. It is also claimed that they have a continuous spectrum. However, we will show that these modes are, in fact, a manifestation of the modes discussed by Connor, Taylor and Wilson - indeed they are the most typical form. As such they have a discrete, not continuous, spectrum.

2. THE MODEL

A standard model [7] for short wavelength plasma fluctuations with large toroidal mode number $n \gg 1$ in a large aspect ratio torus is described by

$$\left[\frac{1}{(nq')^2} \frac{\partial^2}{\partial x^2} - \frac{\sigma^2}{\Omega^2} \left(\frac{\partial}{\partial \theta} + inq'x \right)^2 - \frac{\alpha}{\Omega} \left(\cos\theta + \frac{is}{nq'} \sin\theta \frac{\partial}{\partial x} \right) - \Lambda \right] \phi(x, \theta) = 0 \qquad (1)$$

where x is the distance from some magnetic surface and θ is the poloidal angle. The first term arises from ion larmor radius, the second from ion sound and the third is the effect of toroidal coupling. The parameters of the model are: $s = rq'/q, \sigma = \varepsilon/bqs, \Omega = \omega/\omega_{*e}$ and $\alpha = 2\varepsilon/bs^2$, where $\varepsilon = L_n/R\tau, \tau = T_e/T_i$ and $b = k_\theta^2 \rho_i^2/2$. ($L_n$ is the density scale length.)

The parameter

$$\Lambda = \frac{1}{bs^2} \left[\frac{\Omega - 1}{\tau\Omega + \eta_i + 1} - b \right] \equiv \frac{\hat{\Lambda}}{bs^2} \qquad (2)$$

embodies the frequency shift $(\Omega - 1)$ (important for ED modes) and the temperature gradient $\eta_i = L_n/L_T$ (important for ITG modes).

Schlüter [8] has criticised Eq (1), suggesting that it is not based on a systematic expansion. This criticism is unfounded: Eq (1) can be derived from an asymptotic expansion with the ordering

$$b \sim \rho_i \frac{\partial^2}{\partial x^2} \sim \varepsilon \sim \hat{\Lambda} \ll 1 \qquad (3)$$

(Formally, this restricts the model to $\eta_i \gg 1$ for ITG modes.) With this ordering the parameters σ, α, and Ω are order unity and $1/n$ is the small expansion parameter. The theory below is then asymptotically correct for $1/n \to 0, n\varepsilon > 1$.

---

[1] Of course, anomalous transport is a non-linear effect and we are discussing only linear modes, but in a weakly turbulent state some influence of linear modes should persist.

3. ANALYSIS

A standard method for solving Eq (1) is by introducing a ballooning transformation,

$$\phi(x,\theta) = \sum e^{im\theta} \int e^{-im\eta} \hat{\phi}(x,\eta) d\eta \qquad (4)$$

which automatically ensures periodicity in θ, and writing

$$\hat{\phi}(x,\eta) = \xi(x,\eta) \exp[-inq'(x\eta - S(x))] \qquad (5)$$

Then for large n

$$\left[\frac{\sigma^2}{\Omega^2}\frac{d^2}{d\eta^2} + (\eta - k)^2 + \frac{\alpha}{\Omega}(\cos\eta + s(\eta - k)\sin\eta) + \Lambda\right]\xi(\eta) = \lambda\xi(\eta) \qquad (6)$$

where $k = dS/dx$. Here we have introduced an intermediate eigenvalue $\lambda(\omega, k, x)$ defined by the appropriate boundary conditions at $|\eta| \to \infty$. Note that $\lambda(k) = \lambda(k + 2\pi)$ and $\xi(k + 2\pi, \eta) = \xi(k, \eta - 2\pi)$.

The plasma eigenvalue ω satisfies

$$\lambda(\omega, k, x) = 0 \qquad (7)$$

but at this point ω is undetermined. Nevertheless, all the information needed to calculate ω and the mode structure is embodied in $\lambda(\omega, k, x)$. Within this description we can identify two classes of eigenmode: isolated (occurring only at special radii), and general (occuring at any radius).

i) **Isolated modes**

These arise when $\lambda(\omega, k, x)$ has a stationary point near which Eq (7) becomes

$$\lambda_0(\omega) + \frac{1}{2}[\lambda_{xx}(x - x_0)^2 + \lambda_{kk}(k - k_0)^2] = 0 \qquad (8)$$

This defines two branches $k_+(\omega, x)$ and $k_-(\omega, x)$ which correspond to two WKB solutions. Matching these solutions at the turning points x_1, x_2 requires

$$\frac{nq'}{2\pi}\int_{x_1}^{x_2}(k_+(\omega, x) - k_-(\omega, x))dx = \text{integer} + \frac{1}{2} \qquad (9)$$

which defines the frequency of all isolated modes. Such modes are essentially confined between the WKB turning points and have a width

$$x_1 - x_2 = \left(\frac{\pi}{nq'}\right)^{\frac{1}{2}}\left(\frac{\lambda_{kk}}{\lambda_{xx}}\right)^{\frac{1}{4}} \qquad (10)$$

(λ_0 has been eliminated using Eq (9)).

ii) **General modes**

Otherwise than at stationary points, Eq (7) becomes

$$\lambda \sim \lambda_0(\omega, k) + \lambda_x(\omega, k)(x - x_0) = 0 \qquad (11)$$

Since λ is periodic in k, this defines an infinity of branches $k_i(\omega, x)$ and the WKB solution involves all these branches. This WKB solution is discussed in [9], but it is more convenient to use an alternative method.

Instead of using the ballooning transformation we write the perturbation as

$$\phi(x, \theta) = \int \xi(p, \theta) \exp[inq'(x(p - \theta) + \hat{S}(p))] dp \qquad (12)$$

Note that this is not automatically periodic in θ. Then $\xi(p, \theta)$ satisfies Eq(6) with the substitutions $k \to p, x \to -d\hat{S}/dp$. Consequently, Eq(11) becomes

$$\lambda_0(\omega, p) - \lambda_x(\omega, p)\left(x_0 + \frac{d\hat{S}}{dp}\right) = 0 \qquad (13)$$

We now note that, because $\xi(p, \theta + 2\pi) = \xi(p - 2\pi, \theta)$, the expression (12) will be periodic in θ if $\exp i(nq'\hat{S}(p))$ is periodic in p. This requires

$$\frac{nq'}{2\pi} \oint \left(\frac{\lambda_0(x_0, p)}{\lambda_x(x_0, p)} - x_0\right) dp = \text{integer} \qquad (14)$$

which defines the frequency of all general modes. (More precisely [1,9], it defines the growth rate and relates the real frequency to the arbitrary mode centre x_0.)

The structure of the general modes can be illustrated by retaining only the first terms in the fourier expansion of $\lambda(p)$, so that Eq(13) becomes

$$\lambda_0 + \lambda_c \cos p - \lambda_x \left(x_0 + \frac{d\hat{S}}{dp}\right) = 0 \qquad (15)$$

(We will see later that this is adequate for the present discussion of ED and ITG modes.) Then, using Eq(14),

$$\hat{S}(p) = \frac{\lambda_c}{\lambda_x} \sin p \qquad (16)$$

Using a saddle-point/stationary phase approximation for the integral (12), the structure of a general mode is then

$$\phi \sim \exp\left[inq'\left(x \cos^{-1}\left(\frac{x}{g}\right) \pm g\left(1 - \frac{x^2}{g^2}\right)^{\frac{1}{2}} - x\theta\right)\right] \qquad (17)$$

with $g = -\lambda_c/\lambda_x$, and the sign is chosen so that ϕ decays at large x.

Eq (17) brings out the important point that the width of a general mode may be determined in two distinct ways. If g is real, the width is determined by the 'turning points' $X = \pm g$ and is independent of the small parameter $1/n$. Within the turning points ϕ is oscillatory, beyond them it decays - eventually as $\exp[-n\mid q'x\mid \log\mid x\mid]$.

However, if g is complex, and we expand Eq(17) for small x, we find

$$\mid \phi \mid \sim \exp[-nx^2 \mid q'g_i \mid /2 \mid g \mid^2] \tag{18}$$

where $g_i = Im(g)$. In this case, therefore, the mode decays in a distance

$$\Delta \sim \mid g \mid /(n \mid q'g_i \mid)^{\frac{1}{2}} \tag{19}$$

Note that, because n is large, this width is much less than the distance between turning points unless $Im(g)$ is close to zero. This effect of $Im(g)$ is much greater than that due to the fact that the turning points are no longer on the real axis. In fact the turning points play no part in determining the mode width, but they are important in determining the frequency, Eq(14).

4. APPLICATION TO ED AND ITG MODES

The discussion in Sec (3) applies to any high n mode. In this section we consider specifically the application to ED and ITG modes.

As we have emphasised, all the properties of modes are contained in the auxiliary quantity $\lambda(\omega, k, x)$ calculated from the ordinary differential Eq (6). Specific cases must be calculated numerically [10] but an approximate solution can be found using a quadratic approximation for the trigonometric terms. Then Eq (6) becomes a Weber equation which is readily solved to give

$$\lambda = \Lambda + \frac{\alpha \cos k}{\Omega} - \frac{\left(\frac{\alpha}{2\Omega}(1-s)\sin k\right)^2}{\left(1 - \frac{\alpha}{2\Omega}(1-2s)\cos k\right)} \pm \frac{i\sigma}{\Omega}\left(1 - \frac{\alpha}{2\Omega}(1-2s)\cos k\right)^{\frac{1}{2}} \tag{20}$$

This expression is somewhat unwieldy, but the main features can be captured if we restrict ourselves to $\alpha/\Omega \ll 1, \sigma/\Omega \ll 1$. Then

$$\lambda = \Lambda \pm \frac{i\sigma}{\Omega} + \frac{\alpha}{\Omega}\left(1 \mp \frac{i\sigma(1-2s)}{4\Omega}\right)\cos k \tag{21}$$

which is of the form of Eq(15). The last term $\sim (1-2s)$ is small in α/Ω but, as explained above, it must be retained when it is the only contribution to $Im(g)$.

i) **ED Modes**

Electron Drift modes have a frequency $\omega \sim \omega_{*e}$, so that (putting $\eta_i = 0, \tau = 1$ for brevity)

$$\lambda \simeq \frac{1}{bs^2}\left[\frac{\omega - \omega_{xe}}{2\omega_{*e}} - b \pm \frac{i\varepsilon s}{q} + 2\varepsilon\left(1 \mp \frac{i\varepsilon}{4bqs}(1-2s)\right)\cos k\right] \quad (22)$$

In this case the dominant x-dependence of λ arises from ω_{*e}.

Isolated ED modes arise where ω_{*e} has a maximum and are determined by $\lambda_{xx}/\lambda_{kk}$. From Eq(10) their width is

$$\Delta \sim \varepsilon^{\frac{1}{4}}(rL_2/nqs)^{\frac{1}{2}} \quad (23)$$

with $L_2 = (\omega_*''/\omega_*)^{1/2}$. This is typically a small fraction of the plasma radius - though it extends over many mode-rational surfaces.

Elsewhere than at the maximum of ω_* there will be general ED modes determined by the parameter

$$g = \frac{-\lambda_c}{\lambda_x} \sim 4\varepsilon\left(1 - \frac{i\varepsilon}{4qbs}(1-2s)\right)L_1 \quad (24)$$

where $L_1 = (\omega_*'/\omega_*)^{-1}$. If g were real the width of these modes would be $\sim 8\varepsilon L_1$ - which is the large width first calculated by Connor, Taylor and Wilson[1].

However, except for special values of the parameters (represented here by $(1-2s) = 0$) g is not real. Although $Im(g)$ is small, it is nevertheless important and the mode width is given by Eq(19),

$$\Delta \sim \left(\frac{brL_1}{n(1-2s)}\right)^{\frac{1}{2}} \quad (25)$$

Typically this is again a small fraction of the plasma radius. It is essentially the width of the modes described by Kim and Wakatani [3]. (But note that the frequency spectrum is not continuous as they claimed.)

ii **ITG modes**

To illustrate the nature of ITG modes in a torus we consider η_i large, formally $\sim 1/\varepsilon$. Then the frequency of the toroidal ITG mode is of order $(\varepsilon\eta_i)^{\frac{1}{2}}\omega_{*i}$ and

$$\lambda = \frac{1}{bs^2}\left[\frac{\omega}{\eta_i\omega_*} - b + \frac{i\varepsilon s\omega_*}{q\omega} + \frac{2\varepsilon\omega_*}{\omega}\left(1 - \frac{i\varepsilon\omega_*}{4qbs\omega}(1-2s)\cos k\right)\right] \quad (26)$$

Several parameters now contribute to the x-variation of λ and we assume these have a similar scale L. Isolated ITG modes arise if λ_x vanishes at some radius and their width (Eq 10) is

$$\Delta \sim (rL/nqs)^{\frac{1}{2}} \quad (27)$$

Once again this is small compared to the plasma radius (though formally $\varepsilon^{-1/4}$ larger than the width of the isolated ED modes.)

Elsewhere than at $\lambda_x = 0$ there are general ITG modes determined by the parameter $g = -\lambda_c/\lambda_x$. For ITG modes λ_x and λ_c have comparable real and imaginary parts, with

$$|\lambda_x| \sim \frac{1}{bs^2}\left(\frac{\varepsilon}{\eta_i}\right)^{\frac{1}{2}}\frac{1}{L_1} \qquad |\lambda_c| \sim \frac{1}{bs^2}\left(\frac{\varepsilon}{\eta_i}\right)^{\frac{1}{2}} \qquad (28)$$

Consequently, from Eq(19) the width of a general ITG mode is

$$\Delta \sim (rL_1/nqs)^{\frac{1}{2}} \qquad (29)$$

This is also small compared to the plasma radius (though formally $b^{-\frac{1}{2}}$ larger than the general ED modes).

If g had been real, as assumed by Romanelli and Zonca [2], the width of a general ITG mode would have been $\sim L_1$. However, bearing in mind that several plasma parameters contribute to λ_x, and that $|\lambda_x| \sim |\lambda_c|$ the conditions for $Im(g)$ to vanish seem unrealistic.

5. SUMMARY AND CONCLUSIONS

Short wavelength electron drift (ED) and ion temperature gradient (ITG) modes have two basic forms in a toroidal plasma. At particular, isolated, radii there may be ED and ITG modes whose widths are a fraction $\sim \varepsilon^{\frac{1}{4}}(1/nq)^{\frac{1}{2}}$ and $\sim (1/nq)^{\frac{1}{2}}$ of the plasma radius respectively. Because of their restriction to particular radii these modes do not seem important for transport.

At all other radii there can be ED and ITG modes that potentially have a large radial width, - comparable to the plasma radius. This would imply that they have an overwhelming effect on transport and would force the plasma to adopt a marginally stable profile, as suggested by Connor, Taylor and Wilson. However, we now see that, except for a very special circumstances, this large radial width is not realised. Instead these general ED and ITG modes have typical widths $\sim (b/n)^{\frac{1}{2}}$ and $\sim (1/nq)^{\frac{1}{2}}$ of the radius respectively. Nevertheless they remain the most important for transport because they occur at all plasma radii.

In this connection it is worth noting that, for $b = constant$ ($b = 0.1$ is a typical value for maximum instability) the width of the general ED and ITG modes can be expressed as $(\rho_i L)^{\frac{1}{2}}$. If this is taken as a step length in conjunction with the natural correlation time L/v_i, one obtains a diffusion coefficient having Bohm scaling with magnetic field and temperature.

As we mentioned in our introduction, Kim and Wakatani [3] claimed there was a third class of ED modes, also with a width $\sim (1/nq)^{\frac{1}{2}}$ but with a continuous frequency spectrum. Indeed it was this claim which stimulated the present investigation. However, we now see that this third class is included in the general modes described above and as such has a discrete spectrum. This arises from the need, overlooked by Kim and Wakatani, to ensure that the amplitude tends to zero at large distances, greater than the mode width and comparable to the distance beween turning points. As shown in Sec (3), when g is complex the amplitude decays at small distances whatever the frequency - but for it to continue to decay at large distances, beyond the turning points, the frequency must satisfy Eq (14).

Finally we would like to comment on the effect of plasma rotation due to radial electric fields. If the rotation frequency $\omega_E = kv_E$ is comparable with ω_{*e} its effect can be incorporated by replacing ω_* by $(\omega_* - \omega_E)$. Then as ω'_E increases, its first effect may be to increase or decrease the width of general modes, depending on the sign of ω'_E, but as it exceeds ω'_* the mode width is reduced by a factor ω'_E/ω'_*, - with a corresponding reduction in the estimated plasma transport. However, it must be remembered that the discussion here, and elsewhere, is based on a Ballooning approximation, or its equivalent Eq (12). This is valid only so long as the resulting mode width is much greater than the separation between mode rational surfaces. Formally, this requires $\omega'_E/\omega'_* \leq 1$, that is $v_E/v_i \leq 1/n$ (for $k\rho_i \sim 1$). When $v_E/v_i \geq 1$ the perturbations should be described in terms of Fourier modes [7,10] rather than in terms of Ballooning modes and the problem once again becomes similar to that in a plane slab, - but with the velocity shear predominant.

6. Acknowledgements

We are grateful to J W Connor for his help, guidance and support. This work was funded jointly by the UK Department of Trade and Industry and Euratom.

7. References

1 J W Connor, J B Taylor and H R Wilson, Phys Rev Lett **70**, 1803 1993
2 F Romanelli and F Zonca, Phys Fluids B **5**, 4081 1993
3 J Y Kim and M Wakatani, Phys Rev Lett **73**, 2200 1994
4 J B Taylor, *Plasma Physics and Controlled Nuclear Fusion Research 1976* (I A E A Vienna 1977) vol 2 p323
5 R J Hastie, K W Hesketh and J B Taylor, Nuclear Fusion **19**, 1223 1979
6 W Horton, Duk-In Choi and W M Tang, Phys Fluids **24**, 1077 1981
7 J W Connor and J B Taylor, Phys Fluids **30**, 3180 1987
8 M Schlüter, Plasma Phys Control Fusion **36**, 1241 1994
9 J B Taylor, J Connor and H R Wilson, Plasma Phys Control Fusion **35**, 1063 1993
10 J W Connor, J B Taylor and H R Wilson, to appear in procedings of 21st EPS Conference on Controlled Fusion and Plasma Physics, Montpellier, 1994

STRONG LANGMUIR TURBULENCE

N.L.Tsintsadze

Institute of Physics, Georgian Academy of Sciences,
3880077 Tibilisi, Republic of Georgia

The influence of relativistic effects on the nonlinear dynamics of Langmuir waves in plasmas is examined with paying particular emphasize on the modulational instability, generation of cavitons and spontaneous magnetic fields, as well as Langmuir wave collapse.

Recently, there has been a great deal of interest in studying the nonlinear interaction of strong electromagnetic waves with plasmas. The general problem of the interaction of large amplitude electromagnetic waves with plasmas is of fundamental importance, both with reference to astrophysical applications and in the physics of laser-plasma interactions. Pulsars and active galactic nuclei, as predicted by some theories, are the source of extremely strong electromagnetic waves [1,2]. On the other hand, in the field of laser fusion, the central problem is a clear understanding of the detailed mechanism by which coherent electromagnetic wave energy can be transformed into particle random energy and be used to heat the plasma [3].

As has been shown by Zakharov [4], if the energy of Langmuir oscillations which could be parametrically excited by an electromagnetic pump, satisfies the condition for the modulational instability and there appears a new dissipation mechanism associated with Langmuir wave collapse. The phenomena of the trapping of plasmons in a plasma cavity and the subsequent collapse of such a cavity have been investigated in detail, by assuming that the amplitude of the HF field. is so small that the relativistic character of the electron motion in the HF fields, namely the dependence of the electron mass on the field amplitude, can be neglected.

In this paper, we present a systematic analysis of the influence of relativistic effects on the plasma dynamics in HF fields. The cases considered cover both potential and electromagnetic HF fields. It has been found that even small relativistic effects ($v_{os}/c < 1$), where v_{os} is the electron quiver velocity and c is the speed of light, can qualitatively change the propagation characteristic of the nonlinear electromagnetic waves. For example, in the presence of the electron mass modulation, we can have new regimes of the modulational instability leading to solitary envelope pulses consisting of localized HF fields accompanied by compressional density perturbations that move with supersonic speed. It is very important also to mention that if there is no modulation of the density, then owing to the oscillation of the electron mass, self-modulation takes place even in an electron plasma. Here, one obtains a Schrödinger equation with a cubic nonlinearity for the modulated plasma wave packet. Recently, some interesting review articles as well as fundamental papers dealing with the theory of strong turbulence in plasmas have been published [5-14].

1. Let us consider the nonlinear propagation of relativistically strong circularly polarized HF electromagnetic wave along the z axis in the plasma [10]. We use the set of Maxwell and relativistic hydrodynamics equations for electrons, neglecting the thermal motion. The contribution of the latter in the equation of motion is small for strong fields as

$$\frac{T_e}{m_0 c^2}\frac{\delta n 1}{n_0 I} \ll 1,$$

where T_e, m_0, and n_0 are the temperature, the rest mass, and the equilibrium density of electrons, respectively, and $I = |eA_\perp m_0 c^2|$.

Here, e is the magnitude of the electron charge and $|A_\perp|$ is the amplitude of the electromagnetic wave. The ions are assumed to be stationary. Thus, the relevant equations for our purpose are

$$\Delta A_\perp - \frac{1}{c^2}\frac{\partial^2 A_\perp}{\partial t^2} = \frac{4\pi e^2 n_e}{m_0 c^2 \gamma}A_\perp \quad (1)$$

$$\frac{d\vec{P}_e}{dt} = -e\left(\vec{E} + \frac{1}{c}\vec{v}\times\vec{B}\right) \quad (2)$$

$$\frac{\partial n_e}{\partial t} + div\, n\vec{v}_e = 0 \quad (3)$$

$$div\,\vec{E} = 4\pi e(n_0 - n_e), \quad (4)$$

where

$$A_\perp = A_x + iA_y,\quad \vec{E} = -\frac{1}{c}\frac{\partial A}{\partial t} - \nabla\varphi,$$

$$\vec{P}_e = m_0\gamma\vec{v},\text{ and } \gamma = \sqrt{1 + \vec{P}^2/m_0^2 c^2}.$$

© 1995 American Institute of Physics

Recently, a number of papers have been published dealing with the modulational and filamentational instabilities of strong electromagnetic wave in plasmas. Such an interest is caused by the fact that in plasmas strongly nonlinear structures are formed: e.g., fine filaments, creation of cavitions and Langmuir wave collapse. However, generation mechanism for such structures are not fully understood. A detailed investigation of these phenomena is also important from the point of view of the plasma heating by means of intense laser beams.

Using the conventional procedure of the normal mode analysis and considering perturbations as plane circularly polarized electromagnetic wave $\left(\delta A_\perp \exp i\left(\vec{k}\vec{r}-\omega t\right)\right)$, we obtain from (1) to (4) the dispersion relation for arbitrarily large amplitude of the pumping field $A_{0\perp}$. The result is

$$\left(\omega^2-\omega_p^2\right)\left[\left(\omega-k v_g\right)^2 - \frac{1}{4\omega_0^2}(\omega^2-k^2c^2)\times\right.$$
$$\left.\left(\omega^2-k^2c^2+\omega_p^2\frac{\gamma_0^2-1}{\gamma_0^2}\right)\right] =$$
$$\frac{\omega_p^2}{4\omega_0^2}\left(k^2c^2-\omega^2\right)k^2c^2\frac{\gamma_0^2-1}{\gamma_0^2} \quad (5)$$

where

$$v_g = \frac{\partial \omega_0}{\partial k_0}, \quad \gamma_0 = \left(1+\left(\frac{eA_{0\perp}}{m_0 c^2}\right)^2\right)^{1/2},$$
$$\omega_p^2 = \frac{4\pi^2 n_0 e^2}{m_0 \gamma_0}; \quad \omega_0 = k_0 c + \frac{\omega_p^2}{2k_0 c}.$$

The modulation instability has been considered in [16] by assuming that $\omega \ll kc$ and $k_\parallel \ll k_\perp$. However, here we present some different interesting cases: First, we suppose that the modulated wave packet propagates in one direction ($k_\parallel \approx k$, $k_\perp = 0$). In this case, we find a new branch of the modulation instability. Assuming that $\omega = kc + \delta$ and $\omega = \omega_p + \delta$, we obtain from (5) the growth rate

$$\text{Im}\delta = \frac{1}{2}\frac{\omega_p^2}{\omega_0}\sqrt{\frac{\gamma_0^2-1}{\gamma_0^2}} \quad (6)$$

We note that in the nonrelativistic limit (viz. $\gamma_0 \approx 1$), the growth rate (6) has been found by Kruer [19]. However, for arbitrarily large amplitude pump, instability appears provided that the value of the threshold amplitude of the pumping field satisfies the inequality

$$\frac{(\gamma_0^2-1)\gamma_0^2}{(\gamma_0^2+1)} > \frac{\omega_p^2}{4\omega_0^2} \quad (7)$$

It follows from (7) that in the ultrarelativistic regime, i.e. when $\gamma_0^2 \gg 1$, the inequality, given in Eq.(7), is autometically satisfied.

In the nonrelativistic case, Eq.(7) becomes $I^2 > \omega_p^2 / \omega_0^2$.

Second, at almost perpendicular propagation of modulations, i.e. $k_\parallel \ll k_\perp$ we consider two cases: in the first case, the relativistic term dominates over the strictional effect and simple calculations show that there is an instability with the maximum growth rate

$$lm\omega = \frac{\omega_p}{4\omega_0}\frac{\gamma_0^2-1}{\gamma_0^2}\omega_p \quad (8)$$

The maximum growth rate occurs at

$$k_{0\perp} = \frac{\omega_p^2}{4c^2}\frac{\gamma_0^2-1}{\gamma_0^2} \quad (9)$$

In the second case, we suppose that the diffraction term $k_\perp^2 c^2$ is compensated by the relativistic term $\omega_p^2(\gamma_0^2-1)/\gamma_0^2$.
Here, we obtain the growth rate

$$lm\omega = \frac{\sqrt{3}}{4}\left(\frac{\omega_p}{4\omega_0}\right)^{2/3}\frac{\gamma_0^2-1}{\gamma_0^2}\omega_p \quad (10)$$

2. In order to investigate the properties of strongly turbulent plasma, the generalized set of equations have been derived in the weakly-relativistic limit ($p^2 \ll m^2c^2$), taking into account the electron nonlinearities and the spontaneously generated magnetic fields [8,14,15]. We have

$$i\vec{E}_t + \frac{3}{2}\omega_p r_p^2 \nabla \mathrm{div}\vec{E} + 2(\vec{v}_g\vec{\nabla})\frac{\partial \vec{E}}{\partial t} -$$

$$-\frac{c^2}{2\omega_0}\nabla\times\nabla\times\vec{E} - \frac{\omega_p^2 \delta n_d}{2\omega_0 n_0}\vec{E} + \frac{\omega_0^2 - \omega_p^2}{2\omega_0}\vec{E} +$$

$$+\frac{1}{2}\left\{\vec{E}\times\left(\frac{e}{m_0 c}\delta\vec{B} - \nabla\times\vec{v}_d\right) + \vec{v}_d \mathrm{div}\vec{E} +\right.$$

$$\left. + \vec{v}_2 \mathrm{div}\vec{E}^+ - \vec{\nabla}(\vec{E}^+\vec{v}_2) + \vec{\nabla}(\vec{v}_d\vec{E})\right\} +$$

$$+\frac{\omega_p^2}{2\omega_0}\frac{n_2}{n_0}\vec{E}^* + \frac{e^2}{16m^2 c^2 \omega_0}(3|E|^2\vec{E} +$$

$$+ (\vec{E}\times(\vec{E}^+\times\vec{E}))) = 0, \quad (11)$$

where r_p is the Debye radius of the electrons, $\delta\vec{B}$ is the spontaneously generated magnetic fields, $<U_e> = v_d$ and n_d are the low-frequency perturbation of the electron fluid velocity and the number density due to the action of the averaged HF potential on the low-frequency plasma motion.

As it has been shown in Ref. [15], the generated magnetic fields lead to the appearance of the term $(ie/m_0 c)\vec{E}\times\delta\vec{B}$ in Eq. (11), which is of the same order as the term connected with the relativistic effect and these two terms cancel each other in the hydrodynamic approximation. Here, the spontaneously generated magnetic fields do not directly act on the development of Langmuir plasma wave turbulence.

For determination of δn and $\vec{v}_d$, we must consider relevant processes. For the excitation of Langmuir waves by transverse electromagnetic waves with $\omega \gg \omega_p$, the slow motion is described by the following equation for δn (in this case the ions are immobile).

$$\left(\frac{\partial^2}{\partial t^2} + \omega_p^2 - v_{th}^2\Delta\right)\frac{\delta n}{n_0} = \frac{e^2}{4m_e^2\omega_0^2}\Delta|E|^2,$$

where v_{th} is the termal velocity of electrons.

On the other hand, when strong Langmuir waves excite ion-sound oscillations, then we have the equation for $\delta n_e \approx \delta n_1$ taking into account the Landau damping

$$\left(\frac{\partial^2}{\partial t^2} - v_s^2\Delta\right)\frac{\delta n}{n_0} -$$

$$-\frac{1}{(2\pi)^{2/3}}\frac{v_s^2}{v_{th}}\Delta\frac{\partial}{\partial t}\int\frac{d\vec{r}'\,\delta n(\vec{r}',t)}{|\vec{r}-\vec{r}'|^2} = \Delta\frac{|E|^2}{16\pi m_i}, (13)$$

where $v_s = (T_e/m_i)^{1/2}$ is the ion-sound speed. It follows from Eq.(13) that the nonlocal term is present only in the equation of the exited ion-sound waves. Eq. (13) represents the generalization of the well-known equation of the plasma slow motion taking into account the Landau damping of ion-sound waves on the electrons.

Regarding the equation for the self-generated magnetic field $\delta\vec{B}$, it is necessary to point out the regime (hydrodynamic or kinetic) in which the problem has been investigated. Because of improper identification of the regimes, some errors have occurred in the previous work [17,18].

In the hydrodynamic case, we obtain for $\delta\vec{B}$,

$$\left(\Delta - \frac{\omega_p^2}{c^2}\right)\delta\vec{B} = \frac{ie}{4m_e\omega_p c}\mathrm{curl}(\vec{E}\mathrm{div}\vec{E}^* - c.c), \quad (14)$$

whereas in the kinetic approximation, one easily obtains

$$\left(\Delta - \frac{\omega_{pi}^2}{c^2}\right)\delta\vec{B} - \frac{\omega_{pe}^2}{(2\pi)^{3/2}c^2 v_{th}}\frac{\partial}{\partial t}\int\frac{d\vec{r}'\,\delta B(r',t)}{|\vec{r}-\vec{r}'|^2} =$$

$$= \frac{ie}{4m_e\omega_p c}\mathrm{curl}\,\mathrm{curl}(\vec{E}\times\vec{E}). \quad (15)$$

Thus, Eqs.(11-15) form a closed system, which describes the dynamics of the HF in the presence of low-frequency plasma slow motions.

3. In his classical paper, Zakharov [4] suggested that essential nonlinear process can be developed in the plasma condensate, the so called plasma wave collapse, which should be an effective mechanism for the Langmuir wave energy dissipation in a collisionless plasma. Although at present there exist many papers dealing with the study of the physics of the collapse process, the question of the collapse dimensionality and its existence are still under discussion [20]. In Ref. [20] a criterion for of the collapse existence has been obtained in the "subsonic" regime in which the nonlinear Schrödinger equation is valid for the description of the modulated Langmuir waves. Here, we discuss the conclusion of the criterion of the collapse existence in a general

case, which is valid in both the "sub-sonic" and "super-sonic" regimes. It has been shown that in both the regimes a critical time moment exists at which the distributions of the field and the plasma density are specific.

Let us consider the wave collapse in the approximation when the potential and nonpotential high-frequency (HF) wave motion of the electrons can be separated. The collapse task in this approximation for longitudinal waves has been considered in Refs. [21,22].

For description of the nonlinear dynamics in the plasma, we shall use the set of equations consisting of Eq.(11), which neglects the electron nonlinearities, and Eqs.(13) to (15). The set of equations has following integral of motions.

$$I_1 = \int d\vec{r}|E|^2, \qquad (16)$$

$$I_2 \int d\vec{r} \left\{ \sum_{ij}(\nabla_i E_j \cdot \nabla_i E_j^+) + \varepsilon v|E|^2 + \frac{1}{2}v^2 + \frac{1}{2}u^2 - \frac{1}{6}\gamma(2|E|^4 + (\vec{E}^+)^2\vec{E}^2) - \frac{1}{6}\gamma|\vec{E}\times\vec{E}^+|^2 \right\}, \qquad (17)$$

where $\varepsilon = \dfrac{m_i}{m_e}, \gamma = \dfrac{3}{2}\dfrac{T_e}{m_e c^2}\left(\dfrac{m_i}{m_e}\right)^2, v = \dfrac{\delta n}{n_0}$, and $\vec{E} \to$

$$\to \left(\frac{m_e}{m_i}\right)^{1/2} \frac{\vec{E}}{16\pi n_0 T_e)^{1/2}}.$$

The last two terms in Eq.(17) are associated with the relativistic effect and the self-generated magnetic fields. It emerges that both these terms have a minus sign, which favours the collapse.

The electron nonlinearities may be neglected in comparison with the relativistic effect and the effect connected with δB only when $L \gg c/\omega_p$. Thus, these effects influence the collapse only at the initial stage. The electron nonlinearities arise when the characteristic dimension of the hole becomes smaller then c/ω_p.

We can determine the effective radius of perturbation localization as

$$r_{eff}^2 = \frac{1}{I_1}\int d\vec{r}(\vec{r}-<\vec{r}>)^2|E|^2. \qquad (18)$$

The calculation shows that r^2_{eff} is proportional to I_2. Thus, when $I_2 \to 0$, r_{eff} also tends to zero.

It should be noted that I_2 can go to zero only through two ways. First, for $v < 0$ the influence of relativistic and self-generated magnetic fields is weak. Second, for $v > 0$ the Langmuir collapse is due to relativistic effect and the self-generation of magnetic fields.

As has been mentioned previously in the hydrodynamic approximation for the case of Langmuir turbulence in the initial stage of collapse, we can neglect the electron nonlinearties in the supersonic regime, and Eqs. (11) and (13) have the following form

$$\Delta\left(\begin{array}{c}-\psi(t)\varphi(\vec{r},t)+\dfrac{3}{2}\omega_p\lambda_p^2\Delta\varphi(\vec{r},t)+\\ +\dfrac{\omega_p}{2}div(\beta|\nabla\varphi|^2-\dfrac{\delta n}{n_0}\end{array}\right)\vec{\nabla}\varphi=0, \quad (19)$$

$$\frac{\partial^2}{\partial t^2}\frac{\delta n}{n_0} = \frac{1}{4m_e m_i}\Delta|\nabla\varphi|^2. \qquad (20)$$

Here, we have made the substitution $iE_t \to -\psi(t)E$ in (2.1) where $\nabla\varphi = eE/\omega_p$ and $\beta = 3/8m_e^2 c^2$.

Equations (19) and (20) admit the following self-similar solutions describing the collapse

$$\psi(t) = \frac{\psi_0}{(t_0-t)^2}, \varphi = \varphi(\vec{\xi}), \frac{\delta n}{n_0} =$$

$$= \frac{1}{(t_0-t)^2}N(\vec{\xi}), \vec{\xi} = \frac{\vec{r}}{(t_0-t)}.$$

4. In conclusion, it should be noted that the theory of strong turbulence is far from being complete. There remain lots of problems including Langmuir wave collapse that are not addressed up to the date. As has been shown, the collapse theory has been developed on the assumption that the number of plasmons are conserved, i.e. when $\int d\vec{r}|E|^2$ is constant and on this basis an effective radius of the collapsed object has been found. In fact, if we take into account the third term in Eq.(11), the conserved value in this case is $\int d\vec{r}|E|^2(\omega+(\vec{v}_g\cdot\vec{\nabla})\alpha)$ (where α is the phase). Accordingly, the notion of plasmon number conservation does not probably make sense anymore.

REFERENCES

1. J.E. Gunn and J.P. Ostrike, 1969, Phys. Rev. Lett., 22, 728.
2. H.R. Miller and P. J. Witta, 1987 (eds.), Active Galactic Nuclei, Springer, Berlin.
3. M.N. Rosenbluth and R. Z. Sagdeev, 1972, Comments on Plsma Phys. and Control Fusion, 1, N4.
4. V.E. Zakharov, 1972, Sov. Phys. JETP, 35, 908.
5. R.Z. Sadeev, V.D. Shapiro and V.I. Shevchenko, Handbok of Plasma Physics, 1992, Eds. M.N. Rosenbluth and R. Z. Sagdeev, vol. 3; Physicss of Laser Plasma, edited by A. M. Rubelchik and S. Witkowski, ES Pub B.V.
6. A.M. Rubenchik and V.E. Zakharov, Handbok of Plasma Physics, 1991, EDS. M.N. Rosenbluth and R.Z. Sagdeev, vol.3; Physics of Laser Plasma, edited by A.M. Rubenchik and S. Witkowski, ES Pub. B.V.
7. P.K. Shukla, N.N. Rao, M.Y. Yu and N.L. Tsintsadze, 1986, Phys Reports, 138, pp.1-149.
8. V. I. Berezhiani, N.L. Tsintsadze and D.D. Tskahakaya, 1980, J. Plsma Phys., 24, pp. 15-23.
9. N.L. Tsintsadze, 1981, in Proc. Phenomena in Ionized Gases, July 14-18, 1981, Minsk (USSR), pp. 395-404.
10. L.N. Tsintsadze, 1990, in Proc. of 10th European School on Plasma Physics, September 2-25, Tbilisi (Georgia), edited by N.L. Tsintsadze (World Scientific, Singapore, 1991), pp.497-510.
11. L. Stenflo, 1993, J. Plasma Phys., 49, pp.237-242.
12. N.L. Tsintsadze, D.D. Tskhakaya and L. Stenflo, 1979, Phys. Lett.,72A,p.115.
13. L. Stenflo, 1991, J.Plasma Phys., 45, pp.355-359.
14. N.L. Shatashvili and N.L. Tsintsadze, 1982, Physica Scripta,2, pp.511-516.
15. V.I. Berezhiani, D.D. Tshakaya and G. Auer, 1987, J. Plasma Phys.,38, pp.139-153.
16. L.N. Tsintsadze, 1991, Sov. J. Plasma Phys., 17, 1509.
17. S.A. Bel'kov and V. N. Tsytovich, 1979, Sov. Phys., JETP, 49, 656.
18. M. Kono, M.M. Skoric and D. ter Haar, 1981, Phys. Lett. A, 78, 240.
19. W.L. Kruer, 1988, "The Physics of Laser Plasma Interaction"" (Addison-Wesley, New York).
20. M.V. Goldman, K. Rypdal and B. Hafiiizi, 1980, Phys. Fluids, 23, 945.
21 A.A. Galeev, R. Z. Sagdeev, Yu.S. Sigov, V.D. Shapiro and V. I. Schevenko, 1975, Fizika plasmi, 1, 10.
22. A.G. Litvak, G.M. Fraiman and A. D. Yuranovski, 1979, Pisma Zh. Exp. Teor. Fiz., 19,

LABORATORY EXPERIMENTS IN DUSTY PLASMAS

R. L. Merlino, A. Barkan, N. D'Angelo,
W. Xu, and B. Song

Department of Physics and Astronomy - University of Iowa
Iowa City, Iowa 52242-1479 - USA

ABSTRACT

We describe a rotating-drum dust-dispersal device, which we have used in conjunction with an existing Q machine, to produce extended, steady-state, magnetized dusty plasma columns. This device is capable of generating dusty plasmas in which as much as $\sim 90\%$ of the negative charge is attached to dust grains of $1-10\mu m$ size with dust densities up to $\sim 10^4$ cm^{-3}. Langmuir probes were used to determine how the negative charge is divided between free electrons and dust grains. This paper reports the results of three experimental investigations performed with this dusty plasma device (DPD): (1) the effect of closely packed grains, i.e., the reduction in grain charge that occurs when the intergrain spacing is on the order of or less than the plasma Debye length; (2) the effect of negatively charged dust on the electrostatic ion-cyclotron instability; and (3) the levitation of charged dust in a double layer and the observation of a 'Coulomb lattice,' i.e., strongly coupled dusty plasmas.

I. INTRODUCTION

A dusty plasma is an ionized gas containing charged dust particles. Such multicomponent plasmas are common throughout the universe since most of the solid matter in the universe exists in the form of dust grains often embedded in plasma, such as nebulas, planetary magnetospheres, and cometary environments (see, e.g., Spitzer [1]). Recent interest in dusty plasmas has been driven by the discovery of radial spokes in Saturn's B ring by the Voyager spacecraft [2], which are thought to be due to dust-plasma interactions [3]. Work on dusty plasmas, however, is motivated not only by connections with astrophysical problems, but also by problems arising in industry where plasmas are used in the manufacture of semiconductor integrated circuits. It is now known that dust particles actually grown in the plasmas [4] are a serious contamination problem in the plasma-aided manufacturing of these devices.

Much of the work on dusty plasmas has been theoretical, but recently devices have been introduced in which basic laboratory studies of dusty plasmas can be undertaken. These devices include dust shaker systems [5] and the rotating drum system [6] which we describe here.

Reviews of recent theoretical work on dusty plasmas have been published by Goertz [7], with particular emphasis on phenomena in the solar system, and also by Northrop [8]. An analysis of the dynamics of particulates in the types of discharges used for industrial processing has been given by Winske and Jones [9], in a recent paper containing 85 references to work on dusty plasmas.

In the next section (II) of this paper, we briefly review the theory of charging of dust grains in a plasma. In section III we describe our apparatus for producing a dusty plasma, and finally in section IV we present the results of three experiments we have performed in this device.

II. CHARGING OF DUST GRAINS

Charging of dust grains may be due to collection of ions and electrons from the plasma, photoelectron emission by UV radiation, ion sputtering, or emission of secondary electrons. Under typical laboratory conditions, collection of ions and electrons is the dominant charging mechanism and the situation of a dust grain is very similar to that of an electrically floating Langmuir probe in a plasma. For the case of isolated grains (intergrain spacing $d \gg$ plasma Debye length λ_D) in the plasma, the charge on the dust is easily calculated for spherical grains of radius a. Since a grain is electrically floating, it collects electrons and ions subject to the condition, $I_e = I_i$, where I_i and I_e are the magnitudes of the ion and electron currents to the grain. The balance of electron and ion currents occurs at a grain surface potential, $V_s = -\alpha(kT/e)$, where $T = T_e = T_i$ is the plasma temperature and α is a constant which depends on the ion mass. The grain charge Q is related to its surface potential by $Q = CV_s$, where C is the capacitance of a dust grain, which, for spherical grains of radius a, is given by $C = 4\pi\epsilon_0 a(1 + a/\lambda_D) \approx 4\pi\epsilon_0 a$ when the grain radius is much smaller than the Debye length, λ_D of the ambient plasma.

When there are many dust grains in the plasma, the sheath of each grain is forced closer to the grain surface thus increasing its capacitance. This effect would lead to an increase in the charge per grain, however, there is another more important consequence of the close packing of grains that reduces the grain charge, as first pointed out by Goertz and Ip [10]. In high dust density plasmas (i.e., $d \sim N^{-1/3} < \lambda_D$, where N is the dust density), the grain charge may be much less than the 'isolated grain' value, since, under these conditions, a grain does not have to be as negative (compared to the isolated grain case) to equalize the ion and electron currents to its surface. In general the charge on each grain depends on the *difference* in potential, $V_s - V_p$, between the grain surface potential and the plasma potential, i.e., $Q = 4\pi\epsilon_0 a(V_s - V_p)$. When the distance between grains is $\sim \lambda_D$, the potential between the grains is reduced and thus the grain charge is also reduced. This effect has been analyzed by a number of authors both analytically and numerically [7,10–14]. For example, Winske and Jones [9], using the theory of Whipple et al. [11], provide a typical example of the dependence of the grain charge on the ratio d/λ_D showing that when $d/\lambda_D \approx 1$, Q is reduced to about one-half the isolated grain value, Q_0, and that $Q/Q_0 < 0.05$ when $d/\lambda_D \approx 0.1$. In this paper we present results of a laboratory experiment in which this predicted reduction in the grain charge for the case of closely packed grains is observed.

III. THE DUSTY PLASMA DEVICE (DPD)

The experiment utilizes as the basic plasma source a Q machine [15] in which a fully ionized, magnetized ($B \lesssim 0.4$ T) potassium plasma column of ~ 4 cm diameter and ~ 80 cm long is produced by surface ionization of potassium atoms from an atomic beam oven on a hot (~ 2500 K) tantalum plate. The basic constituents of the ambient plasma are K$^+$ ions and electrons with approximately equal temperatures $T_i \approx T_e \approx 0.2$ eV, and densities in the range of $\sim 10^5$–10^{10} cm^{-3}. To disperse dust particles into the plasma, the plasma column is surrounded over a portion of its length by the device shown schematically in Fig. 1. This dust disperser consists of a rotating metal cylinder and a stationary screen. Dust particles, initially loaded into the bottom of the cylinder, are carried by the rotating cylinder up to the top and fall onto the screen. A series of stiff metal bristles attached to the inside of the cylinder scrape across the outer surface of the screen as the cylinder is rotated. This continuous scraping agitates the screen allowing the dust to disperse evenly throughout the plasma column. The fallen dust which collects at the bottom of the cylinder is then recycled.

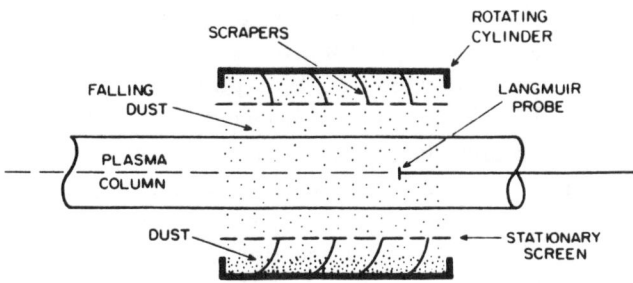

Fig. 1. Schematic diagram of the DPD.

The dust we used was hydrated aluminum silicate (kaolin) of various sizes and shapes. The screen limits the dispensed grain sizes to $< 100\,\mu$m. Samples of dust grains were collected from within the vacuum chamber, and an analysis was made of photographs taken with an electron microscope to determine their size distribution, revealing that 90% of the grains had sizes in the $1-15\,\mu$m range, with an average size $a \sim 5\,\mu$m.

The dust density, N, was inferred using two methods. First, we measured the mass of a sample of dust collected over a small time interval with a 'spoon' inserted into the device. From this measurement of the dust flux ($\mathrm{g\,s^{-1}\,cm^{-2}}$), the density was estimated using the known dust velocity (due to gravity) and the dust size distribution. For an average size $a \approx 5\,\mu$m, we obtained $N \approx 6 \times 10^4\,\mathrm{cm^{-3}}$. This measurement also showed that the dust density was uniform over the length of the dispersal system. The dust density was also estimated from the data of Fig. 2 which shows the axial variation of the plasma density within the dust cloud. The observed exponential decay with a mean-free-path for ion-dust collisions $\lambda \sim (N\pi a^2)^{-1} \approx 25$ cm yields an $N \approx 5 \times 10^4\,\mathrm{cm^{-3}}$ using $a \sim 5\,\mu$m.

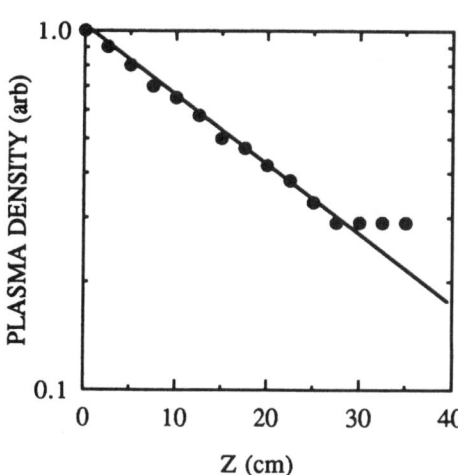

Fig. 2. Variation of the plasma density with axial position z within the DPD.

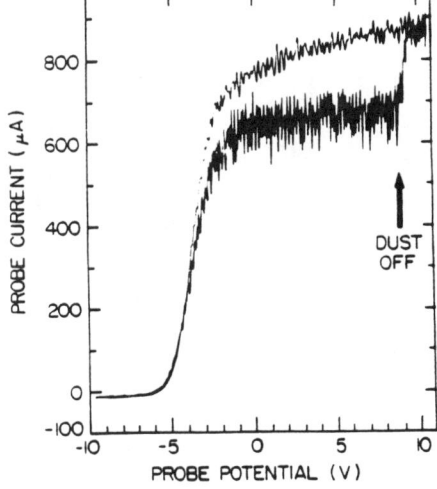

Fig. 3. Langmuir probe characteristics obtained without dust (upper curve) and with dust (lower curve).

The main diagnostic tool of the dusty plasma was a Langmuir probe movable along the axis of the plasma column (see Fig. 1) and consisting of a tantalum disk ~ 5 mm in diameter, oriented normally to the magnetic field. The Langmuir probe enables us to determine how the negative charge in the plasma is divided between free electrons and negatively charged dust grains. Figure 3 shows Langmuir characteristics obtained under identical conditions except for the absence (upper curve) or presence (lower curve) of dust, with the electron portion of the characteristic shown as positive current. When the dust is present, the electron saturation current, I_e, to a positively biased probe is smaller than the current, I_{e0}, measured without dust. This is due to the fact that electrons which attach to dust grains of extremely low mobility are not collected by the probe. The ratio $\eta \equiv (I_e/I_{e0})/(I_i/I_{i0})$ is then a measure of the fraction of negative charge present as free electrons in the dusty plasma, i.e., $\eta = n_e/n_i$, where the ratio I_i/I_{i0}, of the probe ion current with and without dust takes into account the attenuation of the plasma by the dust cloud. Careful checks were made to ensure that the probe functions properly in the dusty plasma environment, as evidenced, in Fig. 3 by the return of the electron saturation current to the 'no-dust' level when the dust is abruptly turned off.

IV. RESULTS AND DISCUSSION

A. The Effect of Closely Packed Grains

In the first experiment we investigated the effects to be expected when the dust grains are close to each other, i.e., when the average distance of a grain to its nearest neighbor, d, is comparable to or less than the plasma Debye length, λ_D. The charge on a dust grain can be determined from measurements of η. From the condition of charge neutrality, $en_i = en_e - QN$, and the definition of η we have that $-QN/e = n_i(1 - \eta)$. Thus from the measurements of the plasma density $n_i \approx n_{i0} = n_{e0}$, η, and the dust density N, the charge Q/e can be obtained. The procedure consists in measuring η as a function of plasma density while holding N, a, and kT fixed. The measured $|Q/e|$ values can then be plotted as a function of the ratio d/λ_D of the intergrain spacing to Debye length, as shown in Fig. 4. Two data sets are shown which correspond to different (but constant) dust densities. The solid and dashed curves are the theoretically predicted $|Q/e|$ values

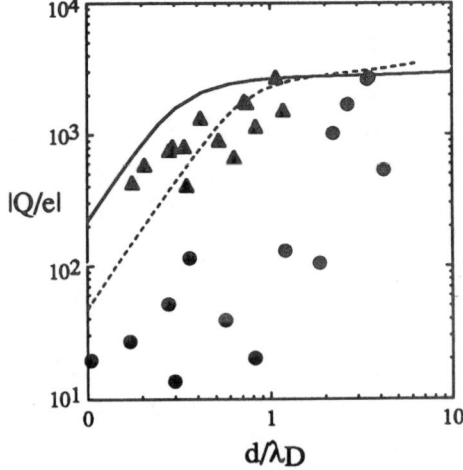

Fig. 4. $|Q/e|$ vs d/λ_D for $N = 500 \, \text{cm}^{-3}$ (triangles) and $N = 5 \times 10^4 \, \text{cm}^{-3}$ (dots). Curves are theoretical values for $N = 1000 \, \text{cm}^{-3}$ (solid) and $N = 10^5 \, \text{cm}^{-3}$ (dashed).

based on the analysis of Whipple et al. [11] as discussed by Winske and Jones [9]. For $d/\lambda_D > 1$, $|Q/e|$ approaches the value corresponding to an isolated grain in the plasma for $a = 5\,\mu$m and $kT \approx 0.2$ eV, i.e., $|Q| \approx 3000$ e. As d/λ_D decreases, $|Q|$ is significantly reduced. Note that the data points all lie below and to the right of the theoretical curves as expected. In comparing the experimental and theoretical results one should keep in mind that there are a number of simplifying assumptions in the theory that are not borne out in the actual experiment, e.g., the assumption of spherical, conducting grains all of uniform size.

B. Electrostatic Ion-Cyclotron (EIC) Waves in Dusty Plasmas

Waves in dusty plasmas have been studied theoretically for about a decade beginning with the work of Bliokh and Yarashenko [16]. A prediction of some of the theoretical work [17,18] is that the presence of a substantial amount of negatively charged dust should make the plasma more unstable to the EIC instability, lowering the critical drift of the electrons along the magnetic field lines. This prediction was tested in the DPD by measuring the EIC wave amplitude with and without dust as a function of the parameter η (the percentage of negative charge per unit volume on dust grains).

The EIC waves were excited by drawing an electron current along the axis of the plasma column to a 0.5 cm diameter disk located near the end of the column. This disk was biased at a voltage in the range from ~ 0.5 V to ~ 5 V above the space potential. The amplitude of the wave, A_0, was recorded, and without introducing any other changes in the plasma conditions, the dust dispenser was turned on producing a dusty plasma whose parameter η could be determined. The EIC wave amplitude with dust present, A_d, was then measured. If the ratio $R = A_d/A_0 \gtrsim 1$, this is taken as an indication of the enhancement of the instability produced by the dust.

The measured R vs η values are shown in Fig. 5. For $\eta \gtrsim 0.7$, the ratio R is near unity, indicating that for these conditions the dust has hardly any effect on the wave amplitude. However, as η decreases from ~ 0.7 down to ~ 0.25, the amplitude R increases by as much as a factor of ~ 2. This increase is in line with the prediction of Chow and Rosenberg [18] who show that the maximum EIC growth rate increases by about a factor of 2 as their $\delta(= \eta^{-1})$ changes from 1 to 3.

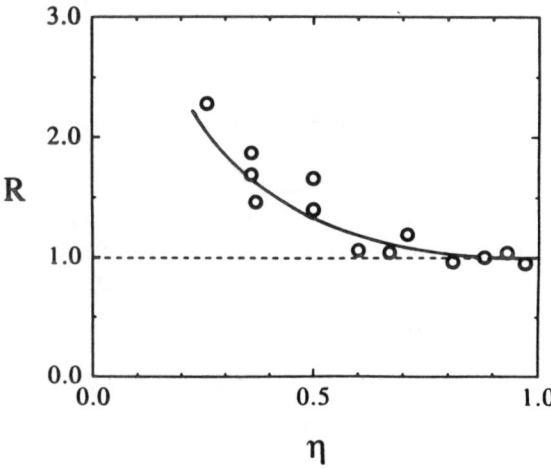

Fig. 5. Ratio $R = A_d/A_0$ of EIC wave amplitudes vs η.

C. Observations of 'Coulomb Lattices'

In the experiments described above the dominant force on the dust grains was gravity. However, since the dust grains are charged, it is possible to levitate them against gravity using electric fields. Furthermore, since the charge on a grain can be very large and the distance between grains small, the electrostatic interaction energy between grains can be comparable to their random thermal energy. In this case, the grains can form a regular lattice structure as suggested by Ikezi [19]. Coulomb crystals have recently been produced in rf plasmas [20].

In order to observe these 'Coulomb lattices' an appropriate electric field configuration (trap) must be established in the plasma to levitate and confine the dust particles. It turns out that an anode double layer produced within the plasma column is an ideal potential configuration for this purpose. A double layer can be produced within our Q machine by producing a glow discharge to a small anode disk located near the end of the plasma column. The device is filled with a neutral gas (N_2, Ar, air) to a pressure in the range of $\sim 1-25$ mTorr. When the anode disk (16 mm diameter) is biased to a positive potential, electrons from the ambient Q machine plasma are accelerated to the anode. If the potential somewhere within the anode sheath is above the ionization potential of the neutral gas, the accelerated electrons are capable of ionizing the neutral gas atoms. If sufficient ionization occurs, the sheath is transformed into an anode double layer [21]. Due to the enhanced local ionization and excitation of the gas, these double layer structures are usually visible (we refer to them as 'firerods'). A schematic diagram of this setup is shown in Fig. 6(a). The anode, which is located near the end of the dust dispersal unit produces a firerod that extends well into the device.

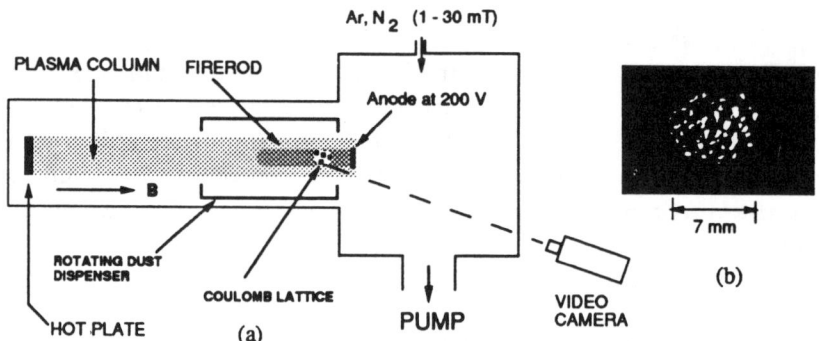

Fig. 6. (a) Schematic of setup used to observe Coulomb's lattice confined in a firerod. (b) Typical Coulomb lattice structure.

The potential contours, measured with an emissive probe, for a typical firerod are shown in Fig. 7. The potential structure of the firerod is axisymmetric and its geometry reflects the nonuniform magnetic field configuration near the end of the device. The radial electric fields associated with the firerods have strengths up to ~ 40 V/cm.

When a firerod is present and the dust dispersal unit is turned on, a cloud of negatively charged dust is trapped in the high potential region of the double layer. Typically the trapped dust cloud initially is in a turbulent state and rotates about the axis of the device in the same sense as the rotation of the dust dispersal device. If, however, the neutral gas (usually N_2) pressure is briefly increased to about 25 mTorr, the cloud relaxes to a more quiescent state due to the collisional interaction with the gas. Under these conditions,

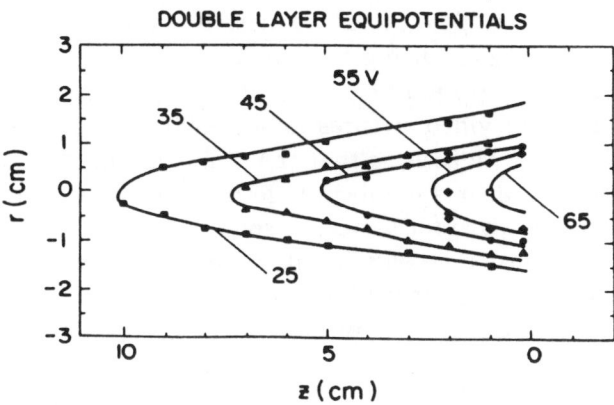

Fig. 7. Equipotential contours of a typical firerod. The anode is located at $z = 0$.

some of the dust grains undergo a transition to an ordered state arranged in a three-dimensional lattice within the firerod. This lattice is usually located ~ 5 cm in front of the disk. Using a standard video camera we are able to visually observe and record this process, since the Q machine hot plate is a sufficiently bright light source to illuminate the dust grains. A single frame snapshot of a typical lattice is shown in Fig. 6(b). These structures are typically elliptical in shape although they sometimes droop downward. Their location within the firerod and their overall shape depends somewhat on the neutral gas flow. The entire structure can be moved axially by making relatively small adjustments in the gas flow. Usually the particles within the lattice maintain a slow rotation (almost as a rigid body) about the axis, although this rotation can be arrested by careful control of the gas flow. Note that the firerod electric field provides radial confinement of the lattice and axial confinement in the direction toward the disk. We believe that the gas flow along the axis of the firerod toward the hot plate is important in preventing the lattice from collapsing into the disk. If the firerod is suddenly turned off by removing the positive voltage on the disk, the grains within the lattice usually fall, but under some conditions the lattice may actually 'explode,' presumably if the Coulomb repulsion force exceeds the gravitational force on the grains. The dust particles in the lattice appear much larger than their actual size due to problems associated with the focusing of the camera. By collecting and examining (with a microscope) some of the fallen grains, we have determined their sizes to be mainly in the 10 μm range. However, some of the grains within the lattice actually 'grow' much larger over time. These large particles are typically long branch-like structures.

It is instructive to consider the following simple numerical estimates related to these lattice structures. From the measured electric fields in the double layer, we can estimate the average size of the levitated particles from the balance of electric and gravitational forces: $QE = m_d g$, where Q is the magnitude of the grain charge, m_d is the grain mass, E is the radial electric field in the double layer, and g is the acceleration due to gravity. For Q we use the isolated grain value, $Q \approx 4\pi\epsilon_0 a V_s$, while the mass is given by $m_d = (4/3)\pi a^3 \rho$, where ρ is the mass density of the grain material. Substituting these expressions into the force balance condition and solving for a we obtain $a = [3\epsilon_0 V_s E/\rho g]^{1/2}$. The plasma inside the double layer typically has $T_e \approx 2$ eV and $T_i \approx 0.1\, T_e$, resulting in a grain surface potential of $V_s \approx -6.5$ V. Taking $E \approx 2500$ V/m and $\rho \approx 2000$ kg/m^3, we find that $a \approx 5$ μm, which is consistent with the known average dust size. Another important parameter to consider is the so-called Coulomb coupling parameter Γ, which is the ratio of the electrostatic Coulomb energy for a pair of neighboring grains to the thermal or random energy of the grains, i.e., $\Gamma = Q^2/4\pi\epsilon_0\, d\, kT_d$, where d is the intergrain spacing and kT_d is the grain thermal energy. Again, using the isolated grain value for Q with $a = 5\,\mu$m, $T_e = 2$ eV, and $T_i = 0.2$ eV, we obtain $Q/e \approx 3 \times 10^4$. The intergrain spacing, d, can be estimated from the single frame video pictures to be ~ 500–$1000\,\mu$m. For the dust temperature we take $T_d \approx 300$ K, i.e., we assume that the grains are in thermal equilibrium with the neutral gas. Putting the various numbers together yields a $\Gamma \sim 10^4$–10^5, well above the critical value discussed by Ikezi for Coulomb solid formation. Finally, in order for the dust-dust Coulomb interactions to be significant, as indicated by the large value of Γ, we expect that the intergrain spacing should be neither much larger than nor much smaller than the Debye length of the surrounding plasma in the firerod. If d were large compared to λ_D, the plasma would effectively screen the Coulomb forces and the interaction would be weak. On the other hand, if d were small compared to λ_D, then the charge on each grain would be reduced (by the effect discussed in IV.A) and again the interaction would be weak. Inside the firerod the plasma density $n \sim 10^9$ cm^{-3}, so that $\lambda_D \approx 400\,\mu$m. Since $d \approx 500$–$1000\,\mu$m, we see that the parameter $\kappa = d/\lambda_D \sim 1$, and thus we should expect a strong interaction due to the electrostatic force between the dust grains.

Work supported by the Office of Naval Research. We wish to thank Dan Winske for kindly providing the theoretical results shown in Fig. 4.

[1] L. Spitzer, Physical Processes in the Interstellar Medium (John Wiley, New York, 1978).
[2] B. A. Smith et al., Science **215**, 504 (1982).
[3] C. K. Goertz and G. Morfill, Icarus **53**, 219 (1983).
[4] K. G. Spear et al., IEEE Trans. Plasma Sci. **14**, 179 (1986).
[5] D. P. Sheehan et al., Rev. Sci. Instrum. **61**, 3871 (1990).
[6] W. Xu et al., Rev. Sci. Instrum. **63**, 5266 (1992).
[7] C. K. Goertz, Rev. Geophys. **27**, 271 (1986).
[8] T. G. Northrop, Phys. Scr. **45**, 475 (1992).
[9] D. Winske and M. E. Jones, IEEE Trans. Plasma Sci. **22**, 454 (1994).
[10] C. K. Goertz and W.-H. Ip, Geophys. Res. Lett. **11**, 349 (1984).
[11] E. C. Whipple et al., J. Geophys. Res. **90**, 7405 (1985).
[12] O. Havnes et al., J. Geophys. Res. **92**, 2281 (1987).
[13] B. Young et al., J. Geophys. Res. **99**, 2255 (1994).
[14] S. J. Choi and M. J. Kushner, J. Appl. Phys. **75**, 3351 (1994).
[15] R. W. Motley, Q-Machines (Academic Press, San Diego, CA, 1975).
[16] P. V. Bliokh and V. V. Yarashenko, Sov. Astron., Engl. Transl. **29**, 330 (1985).
[17] N. D'Angelo, Planet. Space Sci. **38**, 1143 (1990).
[18] V. W. Chow and H. Rosenberg, Planet. Space Sci., in press (1994).
[19] H. Ikezi, Phys. Fluids **29**, 1764 (1986).
[20] J. H. Chu and Lin I., Phys. Rev. Lett. **72**, 4009 (1994); H. Thomas et al., Phys. Rev. Lett. **73**, 652 (1994).
[21] S. L. Cartier and R. L. Merlino, Phys. Fluids **30**, 2549 (1987).

SHORT PULSE LASER SOLITONS

A. Sen

Institute for Plasma Research,
Bhat, Gandhinagar 382424, India.

Abstract

A class of exact nonlinear traveling wave solutions of the relativistic cold fluid model are described. They are in the form of isolated envelope solitons of electromagnetic waves coupled to electron plasma waves. Such pulses have potential applications in particle and photon acceleration schemes and may also represent nonlinear saturated states of the free electron laser instability. Some of the key physical issues pertaining to these solutions are discussed.

1 Introduction

The nonlinear propagation of intense laser pulses in plasmas is a problem of great current interest. The availability of picosecond laser pulses with energies upto tens of joules (with the near future prospect of femtosecond pulses with hundreds of joules) and their possible applications to plasma based particle accelerators [1] provide a major motivation. The basic idea in these accelerator schemes is the generation of large amplitude electrostatic plasma waves traveling close to the velocity of light which can accelerate electrons to very high energies. The plasma waves can be excited using laser waves either in the beat wave configuration or in the laser wakefield configuration [2]. A primary question of interest in either case is the nature of nonlinearly coupled electromagnetic and plasma wave excitations in a cold plasma with relativistically intense laser fields. One would also like to know the typical group velocity of such structures since that determines the phase slippage between the accelerated particle and the accelerating plasma wave fields. Other key questions pertain to stability of the wave structures, their experimental accesibility, self focusing and diffraction effects etc. To discuss these issues we consider the propagation of modulated relativistically intense light pulses coupled to nonlinear plasma waves, in a cold electron plasma. An interesting class of exact one dimensional nonlinear traveling wave solutions exist in this case [3-5], consisting of isolated envelope soliton pulses. We describe these solitons in Section II and also discuss the nonlinear relationship between their group velocity, amplitude and frequency, based on recent numerical investigations [5]. Similar soliton solutions have also been found to exist [6] in a more generalized system, consisting of a relativistic cold electron beam propagating through a helical wiggler magnetic field. They can represent nonlinear saturated states of the free-electron-laser (FEL) instability and can also form interesting candidates for particle accelerator schemes that are based on the inverse free electron laser concept [7]. The FEL solitons are discussed in section III. The concluding section briefly addresses some of the important issues related to the experimental formation of these solitons, their stability, accesibility, generalisation to higher dimensions and practical aspects pertaining to their applications.

© 1995 American Institute of Physics

2 1D Laser Pulse Solitons

The basic physical model [8-11] consists of the one dimensional relativistic set of cold fluid equations (appropriate when the intensity of the laser pulse is such that $(v_{osc}/c > 1)$), and the Maxwell's equations. For a circularly polarized wave, with the vector potential $\mathbf{A} = [\{\hat{y}a(\xi) + \hat{z}ia(\xi)\}exp(-i\lambda\tau) + c.c]$ and with a change of variables $x - \beta t = \xi$, $t = \tau$, the fluid equations can be exactly integrated to give, $n(\beta - u) = \beta$ and $\gamma(1 - \beta u) - \phi = 1$. Here $\gamma = \sqrt{(1 + A^2)/(1 - u^2)}$, u is the longitudinal velocity and the transverse velocity is eliminated by the exact integration $\mathbf{u}_\perp = \mathbf{A}/\gamma$. The normalizations are $n/n_0, u/c$, eA/mc^2, $e\phi/mc^2$, $\omega_p t$, $\omega_p x/c$ etc. The introduction of a frequency parameter λ in the phase factor allows one to identify β with the group velocity of the light pulse and also to distinguish between the group and phase speeds. The choice of circular polarisation avoids the generation of harmonics in all wave fields and also ensures that $\partial/\partial\tau = 0$ for plasma oscillations. Using these relations and further writing $a(\xi) = Rexp(i\theta)$, in the Poisson's equation and the wave equation leads to the following set of coupled nonlinear equations

$$\phi'' = \frac{u}{(\beta - u)} \tag{1}$$

$$u = \frac{\beta(1 + R^2) - (1 + \phi)[(1 + \phi)^2 - (1 - \beta^2)(1 + R^2)]^{\frac{1}{2}}}{(1 + \phi)^2 + \beta^2(1 + R^2)} \tag{2}$$

and

$$R'' + \frac{R}{1 - \beta^2}[(\lambda^2 - \frac{M^2}{R^4})\frac{1}{1 - \beta^2} - \frac{\beta}{\beta - u}\frac{1 - \beta u}{1 + \phi}] = 0 \tag{3}$$

where $M = R^2\{(1 - \beta^2)\theta' - \lambda\beta\}$, is a constant of integration (set to zero henceforth). Equations (1 - 3) describe the nonlinear propagation of the light pulse coupled to the plasma wave. They are in general difficult to analyse (except numerically) but a clue to some possible solutions can be obtained by looking at the small amplitude limit. In the limit of weak density response, the equations can be reduced to a single equation by the substitutions $\phi \approx (1 + R^2)^{\frac{1}{2}} - 1$ and $n \approx (\sqrt{1 + R^2})''$. The resulting equation is of the cubic nonlinear Schrodinger equation type and has the well known envelope soliton solution

$$A_y = R_m sech[\frac{\sqrt{1 - \beta^2 - \lambda^2}}{1 - \beta^2}(x - \beta t)]cos[\frac{\lambda\beta}{1 - \beta^2}(x - \frac{t}{\beta})] \tag{4}$$

where

$$R_m = 4\sqrt{\frac{(1 - \beta^2)(1 - \beta^2 - \lambda^2)}{4\lambda^2 - (1 - \beta^2)(3 + \beta^2)}} \tag{5}$$

Eq. (4) represents an isolated light pulse with a frequency $\omega = \lambda/(1 - \beta^2)$, wavenumber $k = \lambda\beta/(1 - \beta^2)$, phase velocity $1/\beta$, group velocity β (note $V_p V_g = c^2$), an envelope scale length (in units of c/ω_p) equal to $(1 - \beta^2)/\sqrt{1 - \beta^2 - \lambda^2}$. Note that the envelope scale length is real only when $\lambda^2 < (1 - \beta^2)$. This may be physically interpreted as follows. The wave - frequency in the frame moving with the group velocity has the Doppler shifted value $\bar{\omega} = (\omega - k\beta)/\sqrt{1 - \beta^2} = \lambda/\sqrt{1 - \beta^2}$. Thus the inequality $\lambda < \sqrt{1 - \beta^2}$ corresponds to the situation when the Doppler shifted wave frequency $\bar{\omega}/\omega_p < 1$, i.e. the electromagnetic wave finds itself in an overdense plasma and is totally trapped; this is why one gets a soliton solution and the wave does not leak out. The effective scale-length for trapping is about $\sqrt{1-\beta^2}[c/\sqrt{\omega_p^2-\bar{\omega}^2}]$. The pulse satisfies the following amplitude - group velocity relationship,

$$1-\beta^2 = \frac{-(R_m^2 - 4 + 4R_m^2\omega^2) + \sqrt{((R_m^2 - 4 + 4R_m^2\omega^2)^2 + 64R_m^2\omega^2}}{8\omega^2} \quad (6)$$

For arbitrary amplitudes one needs to analyse the full set of equations (1 - 3). Numerical investigations [4,5] seeking solutions that decay exponentially as $\xi \to \pm\infty$ show that acceptable solutions occur only at discrete values of λ i.e. for a fixed value of β finding soliton solutions is an eigenvalue problem in λ.

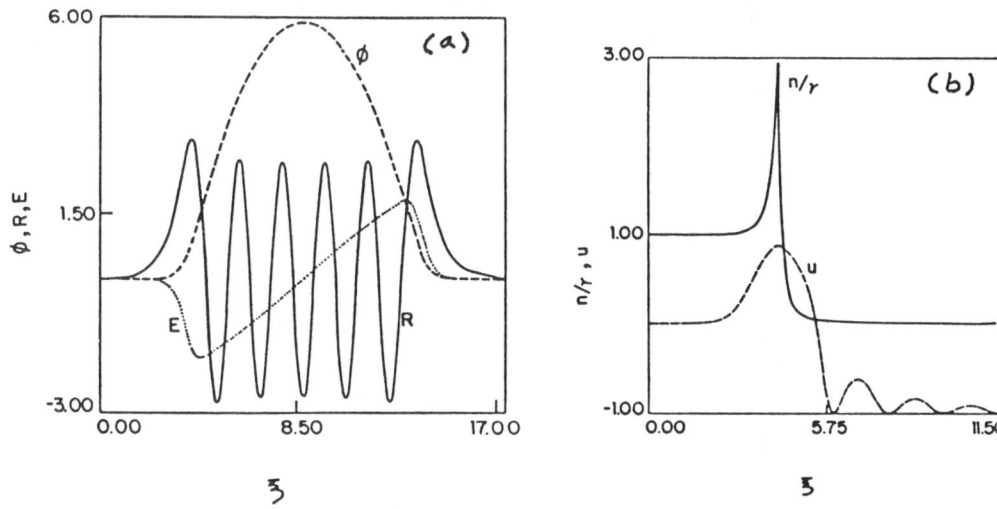

Fig. 1 (a)Electrostatic potential ϕ (dashed line), electric field E ($= -\partial\phi/\partial\xi$, dotted line) and electromagnetic wave amplitude (R) (solid line) profiles for a soliton pulse with $p = 6, \beta = 0.97$ and $\lambda = 0.224445$. (b)Variation of $\delta n/\gamma$ (solid line) and u (dashed line) for a soliton pulse with $\beta = 0.9$, $\lambda = 0.336$ and $p = 19$.

The sizes and shapes of these solitons also vary as a function of λ. Typically ϕ has a characteristic bell shape whereas R has a number of nodes. Figure 1(a) shows a typical soliton solution for $\beta = 0.97$ and $\lambda = 0.224445$. For applications such as particle acceleration or photon acceleration, the regime of interest is $\beta \to 1$, where the group velocity is close to c. The results of a detailed numerical investigation of soliton solutions in this regime [5] are summarized in Fig. 2, which is a plot of normalized group velocity V_g versus the normalized carrier frequency $\Omega(= \omega/\omega_p)$. The solid curve corresponds to soliton pulse results, the dashed line to the linear group velocity $V_{gL} = \sqrt{(1 - 1/\Omega^2)}$, the dotted line to the nonlinear group velocity for an infinite plane wave, $V_{g\infty} = \beta[1 - (\gamma_\infty - 1)/(2\Omega^2\gamma_\infty(\gamma_\infty + 1)]$, (where $\gamma_\infty = \sqrt{1 + A^2}$) and the solid triangles show the values of R_0. The nonlinear group velocities are seen to be closer to c than the linear group velocity. This is because the nonlinearity makes the electrons heavier and thereby weakens the plasma dielectric effects. Finite width soliton pulses propagate slower than the infinite plane waves. Physically this is because coupling to plasma waves acts as a drag on the electromagnetic waves and slows them down. The deviations between $V_{g\infty}$ and V_{gs} can be substantial (upto 25 percent of the difference between V_g and 1) and therefore can be significant in practical situations. Fig. 2 also indicates the complicated nonlinear dependence between the amplitude, frequency and the group velocity. An approximate analytic expression of this dependence can be obtained in the limit of large ϕ_{max}. In such a limit, the longitudinal velocity $u \to -1, n \to \beta(1+\beta)^{-1}$ and $n/\gamma \to 0$. The light wave propagates essentially in a plasma free region (because the residual electrons have become infinitely massive) and one can approximate $R = R_0 \sin(\lambda \xi/(1-\beta^2))$. Furthermore, the structure of ϕ near its maximum can be approximated by the parabolic equation ($u \to -1$ in Eq. 1) $\phi = \phi_{max} - \xi^2/2(1+\beta)$, which gives an estimate of the spatial scale size of the soliton as $\xi_{max} \simeq \sqrt{2(1+\beta)\phi_{max}}$. Using these expressions and expanding the coupled equations (1-3) to order ϕ^{-1} one can obtain an approximate analytic relationship between λ, R_0 and β.

$$R_0^2 = \frac{2(1-\beta)}{\lambda^2}[\frac{\pi^2 p^2}{2\lambda^2}(1-\beta^2)(1-\beta) - \beta] \qquad (7)$$

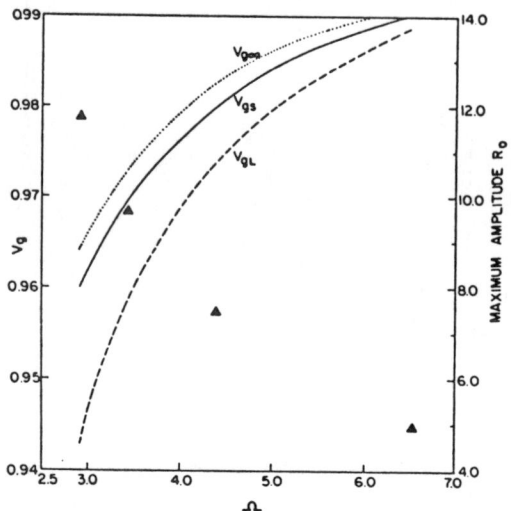

Fig. 2 Normalized group velocity vs normalized frequency curves for soliton pulses (solid line), infinite pulses (dotted line) and linear pulses (dashed line). The maximum electromagnetic wave amplitude R_0 for the soliton pulse (for $p = 18$) are indicated with solid triangles.

Physically these solutions correspond to a soliton in which standing light waves are set up in an 'effectively empty' cavity, because $n/\gamma \to 0$ in most of the region. The quantity n/γ can be made to suffer a change from 1 to a large number before it drops down to zero by making the laser pulse long. In this case, all the density swept out of the pulse piles up in layers with a width of order (c/ω_p), which is the skin depth, at either edge of the pulse (Fig. 1(b)). Such pulses would be useful for photon acceleration schemes where the large change in refractive index at the pulse edge can be effectively used for frequency multiplication [12].

3 Free Electron Laser Solitons

The relativistic cold plasma model has been extensively used in another major application area, namely, the free electron laser (FEL) device, which generates coherent electromagnetic radiation using an intense relativistic electron beam propagating through a static transverse magnetic field [13]. The model provides simple analytic estimates of the linear growth rate of the FEL instability but has surprisingly never been used to address the questions of the nonlinear development and saturation of the instability. In particular, there are no simple model descriptions of the nonlinear saturated state and an interesting question to ask is whether one can obtain soliton type solutions in the FEL situation. Ofcourse the FEL problem introduces additional complexities in the fluid model by the presence of the beam and the wiggler field which make a traveling wave analysis somewhat difficult and it is not clear that soliton solutions can exist [14]. This problem has been analysed in [6] and it has been shown that soliton solutions do indeed exist for the one dimensional cold fluid model of the FEL. These solitons will be briefly discussed in this section.

The equilibrium system now consists of a relativistic cold electron beam of uniform density propagating in the z direction through a helical wiggler magnetic field $\vec{B}^{(0)}(\vec{x}) = -\hat{B}[\cos(k_0 z)\hat{e}_x + \sin(k_0 z)\hat{e}_y]$. A key step [12] to facilitate a traveling wave analysis of the fluid equations is to transform to the so-called wiggler coordinates defined by, $\hat{e}_1 = \cos(k_0 z)\hat{e}_x + \sin(k_0 z)\hat{e}_y$, $\hat{e}_2 = -\sin(k_0 z)\hat{e}_x + \cos(k_0 z)\hat{e}_y$ and $\hat{e}_3 = \hat{e}_z$. In the new coordinate system the magnetic field assumes a constant value in the $\hat{e}_1$ direction and it is then possible to adopt the traveling wave ansatz of making all the dynamical quantities depend on z and t through the combination $\xi = z - \beta t$. Using some straightforward algebra it is then possible to reduce the set of plasma and field equations to a set of three coupled nonlinear differential equations for the variables ρ_1, ρ_2 and $Z = \beta \rho_3 - \gamma$

$$(1-\beta^2)\ddot{\rho}_1 - 2\dot{\rho}_2 - \rho_1 \left\{ 1 + \frac{\hat{\omega}_p^2 \gamma_0 |\beta - \beta_b|}{[(\beta^2-1)(1+\rho_1^2+\rho_2^2) + Z^2]^{1/2}} \right\} = -\hat{\omega}_c \gamma_0 (1 + \hat{\omega}_p^2) \quad (8)$$

$$(1-\beta^2)\ddot{\rho}_2 + 2\dot{\rho}_1 - \rho_2 \left\{ 1 + \frac{\hat{\omega}_p^2 \gamma_0 |\beta - \beta_b|}{[(\beta^2-1)(1+\rho_1^2+\rho_2^2) + Z^2]^{1/2}} \right\} = 0 \quad (9)$$

$$\ddot{Z} + \frac{\hat{\omega}_p^2 \gamma_0 |\beta - \beta_b| Z}{(\beta^2-1)[(\beta^2-1)(1+\rho_1^2+\rho_2^2)+Z^2]^{1/2}} + \frac{\hat{\omega}_p^2 \gamma_0}{(\beta^2-1)}(1-\beta\beta_b) = 0 \quad (10)$$

where the overdot denotes $d/d\zeta$ and the normalisations are $\hat{\omega}_c = \omega_c/ck_0$, $\hat{\omega}_p = \omega_p/ck_0$, $\zeta = k_0\xi$, $\rho_\alpha = p_\alpha/mc$ where $\alpha = 1,2,3$ and p_α denote the momentum components. These equations exhibit many similarities to, and important differences from, the eqautions considered in the previous section. The most striking difference is the inhomogeneous term in Eq.(8) for the ρ_1 oscillator, which results from the presence of the wiggler field. The β_b contribution arises from the beam velocity. The equilibrium solutions ($d/d\zeta = 0$) of (8-10) are given by $\rho_{01} = \hat{\omega}_c \gamma_0$, $\rho_2 = 0$ and $Z_0 = \gamma_0(\beta\beta_b - 1)$. A linear analysis around the equilibrium shows that the FEL instability corresponds to the condition $\beta < \beta_b < 1$. Numerical investigation of equations (8-10) in this regime, to obtain solutions that decay exponentially as $\zeta \to \pm\infty$, reveal that soliton solutions do indeed exist and occur for discrete values of $\hat{\omega}_p$ and β. Figure 3 shows a typical soliton solution for $\beta = 0.5$, $\hat{\omega}_c = 0.5$, $\gamma_0 = 10$. and $\hat{\omega}_p = 0.501935$. Note the similarities of the $\delta Z = (Z - Z_0)$ profile (Fig. 3a) and the $n/n_0\gamma$ profile to the corresponding profiles of ϕ and n/γ of the previous section. The $\delta\rho_1 = (\rho_1 - \rho_{01})$ (Fig. 3b) and ρ_2 (Fig. 3c) have a number of nodes and behave similarly to R (infact back transforming from the wiggler frame to the laboratory frame make their similarity more transparent). The $n/n_0\gamma$ profile (Fig. 3d) shows that the density is reduced in most of the soliton region but piles up at the edges. This suggests that physically these solitons again correspond to trapped light bubble solutions in the beam fluid. The physical explanation for the formation of these solutions can be traced back to the nonlinearities in the system - namely the striction (n/n_0) and the relativistic (γ) nonlinearities. The latter arises from the increase of mass of the electron due to its relativistic quiver motion in the high frequency field. The consequent decrease in the plasma frequency increases the local refraction coefficient leading to increased concentration of the high frequency field. The strictional nonlinearity arises from the ponderomotive force which also causes a redistribution of the plasma density. In this case, however, the plasma concentration increases in the region of high field intensity thereby decreasing the local refraction index. Thus in the cold plasma model these two nonlinearities oppose each other and their competition determines whether solutions of the soliton type are possible. For a soliton to form, the favorable relativistic nonlinearity has to overcome the unfavorable tendencies of the striction effect and push electrons out of the high field region. The pressure of this high field region is then balanced by the electrostatic restoring force arising from the local expulsion of the electrons. Since the electrostatic field plays an important role in the formation of the soliton, these solutions can only occur in the Raman regime where collective effects are significant.

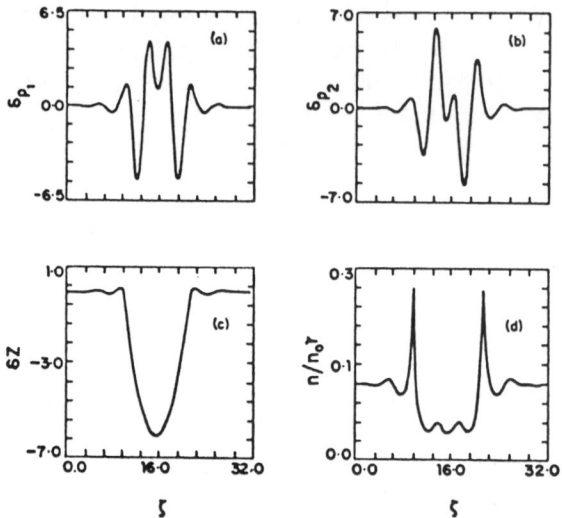

Fig. 3 Profiles of (a)$\delta\rho_1$, (b)$\delta\rho_2$, (c) δZ and (d)$n/n_0\gamma$ for a soliton pulse with $\beta = 0.5, \gamma = 10.$, $\hat{\omega}_c = 0.5$ and $\hat{\omega}_p = 0.501935$.

4 Conclusions and discussion

In conclusion, we have considered exact one dimensional solutions for modulated light pulses coupled to electron plasma waves in a cold plasma in two distinct physical situations. The exact mathematical relation between the two sets of solutions is not known but from their close similarities one suspects that they belong to a single class of generalized traveling wave solutions of the cold fluid model. These finite pulse solutions can have a large number of applications. They could provide simple model descriptions of the nonlinear saturated state of the free electron instability in the Raman regime, for example. The possibility of obtaining large electrostatic fields also makes them interesting from the point of view of particle and photon accelerators. The inverse free electron laser is one such scheme [13] where the large electric fields of the soliton solutions could in principle be used to transfer energy from a laser to a relativistic electron beam in the presence of the magnetic field of an undulator. The beat wave accelerator and the wake-field accelerator are other potential application areas. Even though these solutions are idealized, one dimensional entities, their propagation characteristics offer valuable insight about the magnitude of group velocities of finite nonlinear pulses in plasmas and the nonlinear relation between their group velocity, amplitude and frequency.

Physically, the soliton pulse may be viewed as a light wave which is trapped in a plasma wave that it generates itself. The front of the pulse generates the plasma wave as a wakefield, which is then reabsorbed by the tail of the pulse. The exchange of energy between the light wave and the plasma wave also leads to a 'chirping' of the pulse (cf Fig. 1). This suggests that to experimentally create such pulses, one must not only use characteristic values of R_0 and ω but also do some appropriate chirping of the light wave. The typical time scale for the development of these soliton solutions is related to the characteristic evolution times of the nonlinearities involved in the process. In the FEL case, for example, both the relativistic and striction nonlinearities develop on the collective electron response time scale, ω_p^{-1}, and the typical growth rate Γ for the FEL instability in the Raman regime is approximately $\omega_c\sqrt{\hat{\omega}_p/\gamma}$ whose value is a fraction of ω_p and approaches ω_p at high wiggler amplitudes. The time for an initial potential perturbation $\delta\phi_i$ to grow to an amplitude $\delta\phi_f$ can thus be approximated as $t_s \approx (1/\Gamma)ln(\delta\phi_f/\delta\phi_i)$. From Poisson's equation, $\delta\phi_i \sim 4\pi e L_s^2 \delta n_i$, where L_s is the spatial dimension of the soliton and $\delta n_i/n$ can be approximately taken as $\approx 1/(\sqrt{nL_s^3})$ for random noise level initial fluctuations. For solitons one needs $e\delta\phi_f/mc^2 \sim 1$. Using these conditions and for typical values of the soliton solutions one requires t_s to be of the order of a few plasma periods. For solitons to have enough time to form in the device, t_s should be less than $L/(c\beta_b)$, (the transit time of the electrons through the amplifier-oscillator device of length L), i.e. $Lk_0 > \beta_b\hat{\omega}_p^{-1}$. Since $\beta_b \sim 1$ and $Lk_0 >> 1$, this condition can be easily met in the Raman regime.

A number of key questions remain to be answered and are the subject of intense present investigations. Since these solitons constitute a special class of traveling wave solutions, it is pertinent to investigate their stability and accesibility. Generally this requires a study of the time dependent problem. However in some recent numerical work [15], eqns. (1-3) were solved using relaxation methods rather than the shooting techniques adopted in [5]. The initial conditions were quite arbitrary, such as a rectangular shape for R and a Gaussian shape for ϕ. Yet convergence to the soliton solutions was found to occur very rapidly. This suggests that these one dimensional solitons may be stable and accessible from fairly arbitrary initial conditions. Going beyond the present cold fluid model one has to worry whether wave-breaking or trapping of background plasma electrons can seriously jeopardize the integrity of these solutions or significantly damp them. This question is being addressed in one dimensional particle simulation studies [16]. The basic idea is to launch an initial pulse with the appropriate density and field structures corresponding to the exact fluid soliton solutions and follow their time evolution. Preliminary results indicate that the soliton structures do survive for reasonable lengths of time (several dispersive times), propagating with the appropriate group velocity and eventually disintegrate due to loss of energy to particles and creation of a series of wake fields in the trailing edge. Thus any acceleration of particles or photons has to take place before these disruptive effects take over.

Another serious limitation of these solutions is that they are one dimensional. Physically, the one-dimensional approximation is reasonable as long as transverse scale lengths are much longer than longitudinal pulse dimensions (which is a few skin depths c/ω_p). However higher dimensions introduce additional physical effects such as self-focusing and diffraction. An important theoretical advance would be to demonstrate the existence of two or three dimensional solitons in the context of the relativistic cold fluid model. This is a significantly more difficult task both analytically and numerically and remains one of the challenging problems in the area of intense laser pulse propagation in plasmas.

Acknowledgements

It is a pleasure to acknowledge and thank Drs. A. Das, G.L. Johnston, T. Katsouleas, P.K. Kaw, R. Miklazewski and N. Naumova for their advice, collaboration, suggestions and contributions.

References

1. Dawson, J.M., Sci. American **260** 54 (1989).

2. Turner, S. (Editor), "New Developments in Particle Acceleration Techniques" (Conf. Proc., Orsay 1987, CERN, Geneva, 1987), vol. II.

3. N.L. Tsintsadze and D.D. Tskhakaya,Zh. Eksp. Teor. Fiz. **72**, 480 (1977) [Sov. Phys. JETP **45**, 252 (1977)]

4. V.A. Kozlov, A.G. Litvak and E.V. Suvorov, Zh. Eksp. Teor. Fiz. **76**, 148 (1979) [Sov. Phys. JETP **49**, 75 (1979)]

5. Kaw, P.K., Sen, A. and Katsouleas, T., Phys. Rev. Letts **68**, 3172 (1992).

6. Sen, A. and Johnston, G.L., Phys. Rev. Lett. **70**, 780 (1993).

7. C. Pellegrini, Proc. ECFA-RAL *The Challenge of Ultrahigh Energies* (ed: J. Mulvey) 1982

8. P. Sprangle, E. Esarey and A. Ting, Phys. Rev. Lett. **64**, 2011 (1990) and references therein.

9. A.I. Akhiezer and R.V. Polovin, Zh. Eksp. Teor. Fiz. **30**, 915 (1956) [Sov. Phys. JETP **3**, 696 (1956)]

10. P. Kaw and J. Dawson, Phys. Fluids **13**, 472 (1970)

11. A. De Coster, Physics Reports **47**, 285 (1978)

12. S.C. Wilks, J.M. Dawson, W.B. Mori, T. Katsouleas, M.E. Jones Phys. Rev. Letts. **62**, 2600 (1989)

13. T.C. Marshall, *Free Electron Lasers* (Macmillan, New York, 1985)

14. R.C. Davidson, G.L. Johnston and A. Sen, Phys. Rev. A **34**, 392 (1986)

15. Miklazewski, R. (private communication).

16. Naumova, N. (private communication).

FLUCTUATION PHENOMENA IN DUSTY PLASMAS

A.G.Sitenko, A. G. Zagorodny

Bogolyubov Institute for Theoretical Physics
Ukrainian National Academy of Sciences, Kiev, Ukraine

V. N. Tsytovich

General Physics Institute
Russian Academy of Sciences, Moscow, Russia

1. Recently, much interest has been attracted by the effects occurring in plasmas with fine — dispersed solid component (see, e. g., the review papers [1 — 6], and references cited therein). The presence of solid objects (so — called dust particles or grains) is a rule rather than exception. They have been observed in interstellar, interplanetary, and magnetospheric plasmas, as well as under laboratory conditions. It is a matter of general knowledge that dust is a natural component of space plasmas. Metallic dust particles are usually produced in course of plasma etching. High — Z impurities in fusion plasmas may also be treated as dust particles.

Grains are describ by their sizes, charges, and number densities. The grain size usually varies within the interval $0,1 - 1\ \mu m$ for interstellar plasmas and attains tens microns for laboratory plasmas. The dust particle density depends on the nature of the system under consideration, its value is about $10^{-3} - 10^3\ cm^{-3}$. The grain charge is about $10^3 - 10^4$ telectron charges. More details of dust parameters and chemical composition are given in Refs. [1, 2, 4 — 6].

Historically, Langmuir was the first who mentioned the dust influence on the properties of a plasma column. Later, dust effects have been extensively studied in space and solar physics.

The interest to this problem was renewed at the beginning of 1980s due to the progress in space research, on one hand, and by plasma etching applications in microelectronics, on the other hand. Fairly sophisticated theories of electromagnetic processes in dusty plasmas have been worked out [7 — 16]. However, no detailed study of electromagnetic fluctuations in dusty plasmas have been reported till now.

In this paper, we develop a theory of electromagnetic fluctuations in dusty plasmas taking into account the dynamics of grain charging, namely, charging processes associated with plasma currents [16] which seem to produce the most important effects in self — consistent charging dynamics. Of course, along with plasma currents, other processes — photoemission, secondary electronic emission, etc. — also lead to grain charging. However, it is important to consider the most influential factor for the first step.

2. To involve into consideration charging dynamics, we introduce the charging cross — sections for dust particles. We assume that any grain absorbs all encountering electrons and ions (thus each accommodation coefficient is equal to one, which is completely valid for metallic grains). This means that the cross — sections are determined by the impact parameters of the particles which approach the grain to distances smaller than the particle size, i. e., by the condition $r_{min} \leq a$ (a is the grain radius, r_{min} is the minimum distance between the particle and the scattering center). The charging cross — sections are thus given by [16]

$$\sigma_e(q,v) = \pi a^2 \begin{cases} 1 - \dfrac{2qe_e}{am_e v^2}, & v^2 > \dfrac{2qe_e}{am_e} \\ 0, & v^2 < \dfrac{2qe_e}{am_e} \end{cases} \quad (1)$$

$$\sigma_i(q,v) = \pi a^2 \left(1 - \dfrac{2q e_i}{am_i v^2}\right)$$

where $q \equiv e_g$ is the grain change that is to be found self — consistently, the rest of notation is conventional.

The equilibrium value of q may be found from the condition

$$I_d(q_{eq}) = 0 \quad (2)$$

where $I_d(q)$ is the plasma current through the dust particle surface

$$I_d(q) = \sum_{\sigma=e,i} e_\sigma n_\sigma \int d\vec{v}\, v\, \sigma_\sigma(q,v) f_\sigma(\vec{v}) \quad (3)$$

n_σ and $f_\sigma(\vec{v})$ are densities and distribution functions for relevant particle species.

For equilibrium Maxwellian distributions, Eqs. (2) and (3) yield an equation for q given by [16]

$$\frac{\omega_{pe}^2}{S_e} e^{-z} = \frac{\omega_{pi}^2}{S_i}(t+z), \qquad (4)$$

where
$t = T_1/Z_i T_e$, $z = Z_g e_e^2/aT_e$, $Z_g = q/e_e$, $Z_i = |e_i/e_e|$, $S_\sigma = (T_\sigma/m_\sigma)^{1/2}$

It the distribution functions are non—Maxwellian, Eq. (2) is violated. In this case

$$I_d = \sum_\sigma e_\sigma n_\sigma \int d\vec{v}\, v\, \sigma_\sigma(q_{eq} + \delta q, v)\,[f_{0\sigma}(\vec{v}) +$$
$$+ \delta f_\sigma(\vec{r}, \vec{v}, t)] \simeq -\nu_{eq}\delta q + \qquad (5)$$
$$+ \sum_\sigma e_\sigma n_\sigma \int d\vec{v}\, v\, \sigma_\sigma(q_{eq}, v)\,\delta f_\sigma(\vec{r}, \vec{v}, t),$$

which gives rise to the equation

$$\frac{\partial \delta q}{\partial t} + \nu_{eq}\delta q = \sum_{\sigma=e,i} e_\sigma n_\sigma \int d\vec{v}\, v\, \sigma_\sigma(q_{eq}, v) \times$$
$$\times \delta f_\sigma(\vec{r}, \vec{v}, t). \qquad (6)$$

Here,

$$\nu_{eq} = \sum_\sigma e_\sigma n_\sigma \int d\vec{v}\, v\, \frac{\partial \sigma_\sigma(q_{eq}, v)}{\partial q_{eq}} f_{0\sigma}(v) =$$
$$= \frac{\omega_{pi}^2 a}{\sqrt{2}\pi S_i}(1 + t + z) \qquad (7)$$

The formal solution of Eq. (6) is given by

$$\delta q(\vec{r}, t) = \sum_\sigma e_\sigma n_\sigma \int_{-\infty}^{t} dt'\, e^{-\nu_{eq}(t-t')} \times \qquad (8)$$
$$\times \int d\vec{v}\, v\, \sigma_\sigma(q_{eq}, v)\, \delta f_\sigma(\vec{r}, \vec{v}, t').$$

3. Let us consider the equation for the microscopic phase densities of all particle species, i. e.,

$$F_\sigma(X, t) = \frac{1}{n_\sigma}\sum_{i=1}^{N_\sigma}\delta(X - X_{i\sigma}(t)),$$

where $X \equiv (\vec{r}, \vec{v})$, $X_{i\sigma}(t)$ is the phase trajectory of the i-th particle.

It is clear that in the dusty plasma case the Klimontovich equation for the electron and ion distributions must be modified in order to take into account particle adsorption by grains. To simplify the consideration, we approximate the collision term by the expression

$$\left(\frac{\partial F_\sigma}{\partial t}\right)^{coll} = -\tilde{\nu}_\sigma F_\sigma(X, t), \quad \sigma = e, i,$$

where adsorption frequencies $\tilde{\nu}_\sigma$ are determined by the cross—sections (1), i. e.,

$$\tilde{\nu}_\sigma \equiv \tilde{\nu}_\sigma(q, v) = n_g v \sigma_\sigma(q, v) = \nu_\sigma^{eq} + \delta\nu_\sigma^q + \delta\nu_\sigma^g, \quad (9)$$

where

$$\nu_\sigma^{eq} = n_g v \sigma_\sigma(q_{eq}, v), \quad \delta\nu_\sigma^q = n_g v \frac{\partial \sigma_\sigma(q_{eq}, v)}{\partial q_{eq}}\delta q$$

$$\delta\nu_\sigma^g = \delta n_g v \sigma_\sigma(q_{eq}, v)$$

Thus, the equations for $F(x, t)$ reduce to

$$\left\{\frac{\partial}{\partial t} + \vec{v}\frac{\partial}{\partial \vec{r}} - \frac{e_\sigma}{m_\sigma}\nabla\Phi(\vec{r}, t)\frac{\partial}{\partial \vec{v}}\right\}F_\sigma(X, t) =$$
$$= -\nu_\sigma F_\sigma(X, t), \quad \sigma = e, i, \qquad (10)$$

$$\left\{\frac{\partial}{\partial t} + \vec{v}\frac{\partial}{\partial \vec{r}} - \frac{q}{m_g}\nabla\Phi(\vec{r}, t)\frac{\partial}{\partial \vec{v}}\right\}F_g(X, t) = 0$$

Suppose $\qquad (11)$

$$F_\sigma(X, t) = f_{\sigma 0}(\vec{v}) + \delta f_\sigma(X, t), \quad \sigma = e, i, g$$

Then linearization of Eqs. (10) and (11) with respect to fluctuations results in the equation for the fluctuation evolution, i. e.,

$$\left\{\frac{\partial}{\partial t} + \vec{v}\frac{\partial}{\partial \vec{r}}\right\}\delta f_\sigma(X, t) -$$
$$- \frac{e_\sigma}{m_\sigma}\nabla\delta\Phi(\vec{r}, t)\frac{\partial f_{0\sigma}(\vec{v})}{\partial \vec{v}} =$$
$$= -\nu_\sigma^{eq}\delta f_\sigma(X, t) - (\delta\nu_\sigma^q + \delta\nu_\sigma^g)f_{0\sigma}(\vec{v}) \quad (12)$$

$$\left\{\frac{\partial}{\partial t} + \vec{v}\frac{\partial}{\partial \vec{r}}\right\}\delta f_g(X, t) -$$
$$- \frac{e_g}{m_g}\nabla\delta\Phi(\vec{r}, t)\frac{\partial f_{0g}(\vec{v})}{\partial \vec{v}} = 0 \qquad (13)$$

Here, $\delta\Phi(\vec{r}, t)$ is the fluctuation potential that satisfies the Poisson equation

$$\Delta\delta\Phi(\vec{r}, t) = -4\pi\sum_{\sigma=e,i,q}\delta\rho_\sigma(\vec{r}, t), \qquad (14)$$

where

$$\delta\rho_\sigma(\vec{r}, t) = e_\sigma n_\sigma \int d\vec{v}\,\delta f_\sigma(x, t), \quad \sigma = e, i,$$

$$\delta\rho_g(\vec{r}, t) = q_{eq} n_g \int d\vec{v}\,\delta f_g(x, t) +$$
$$+ \sum_{\sigma=e,i} e_\sigma n_\sigma \int_{-\infty}^{t} dt'\, e^{-\nu(t-t')} \times$$
$$\times \int d\vec{v}\, v\, \sigma_\sigma(q_{eq}\, v)\,\delta f_\sigma(x, t)$$

The formal solution of Eqs. (12), (13) may be written as

$$\delta f_\sigma(x,t) = \delta f_\sigma^{(0)}(x,t) + \sum_{\sigma'=e,i} \frac{e_{\sigma'}}{m_{\sigma'}} \times$$

$$\times \int_{-\infty}^{t} dt' \int dX' W_{\sigma\sigma'}(X, X', t-t') \times$$

$$\times \frac{\partial \delta \Phi(\vec{r}',t')}{\partial \vec{r}'} \frac{\partial f_{0\sigma}(\vec{v}')}{\partial \vec{v}'}, \quad (15)$$

where $\delta f_\sigma^{(0)}(X,t)$ is the microscopic fluctuation of the phase density in a system without interaction through the electromagnetic field, and $W_{\sigma\sigma'}(X, X', t-t')$ satisfy the set of equations

$$\left\{\frac{\partial}{\partial t} + \vec{v}\frac{\partial}{\partial \vec{r}} + \nu_\sigma^{eq}\right\} W_{\sigma\sigma'}(X,X',\tau) +$$

$$+ n_g \nu \sigma'_\sigma \sum_{\sigma''=e,i} e_{\sigma''} n\sigma'' \int_{-\infty}^{t} dt'' \, e^{-\nu(t-t'')} f_{0\sigma}(v) \times$$

$$\times \int d\vec{v} \, v\sigma_{\sigma''} W_{\sigma''\sigma'}(X,X',t''-t') +$$

$$+ \nu \sigma_\sigma f_{0\sigma}(\vec{v}) \int d\vec{v} \, W_{g\sigma'}(X,X',\tau) = 0; \; \sigma=e,i; \quad (16)$$

$$\left\{\frac{\partial}{\partial t} + \vec{v}\frac{\partial}{\partial \vec{r}}\right\} W_{g\sigma'}(X,X',\tau) = 0 \quad (17)$$

with the initial conditions

$$W_{\sigma\sigma'}(X,X',0) = \delta_{\sigma\sigma'}\delta(X-X'). \quad (18)$$

Here, $\sigma_\sigma \equiv \sigma_\sigma(q_{eq},v)$, $\sigma'_\sigma \equiv \frac{\partial \sigma_\sigma(q_{eq},v)}{\partial q_{eq}}$.

In the $\vec{k},\omega$ — representation Eq.(15) reduces to

$$\delta f_{\sigma\vec{k}\omega}(\vec{v}) = \delta f_{\sigma\vec{k}\omega}^{(0)}(\vec{v}) + \sum_{\sigma'} \frac{e_{\sigma'}}{m_{\sigma'}} \int d\vec{v}' W_{\sigma\sigma'\vec{k}\omega}(\vec{v},\vec{v}')$$

$$\vec{k} \frac{\partial f_{0\sigma}\vec{v}'}{\partial \vec{v}'} \delta\Phi_{\vec{k}\omega}, \quad (19)$$

and the solution describing the flucfuation potential is thus given by

$$\delta\Phi_{\vec{k}\omega} = \frac{4\pi}{k^2 \varepsilon(\vec{k},\omega)} \sum_{\sigma=e,i,g} \delta\rho_{\vec{k}\omega}^{\sigma(0)}, \quad (20)$$

where

$$\varepsilon(k_1\omega) = 1 + \sum_{\sigma=e,i,g} \chi_\sigma(\vec{k}_1\omega) \quad (21)$$

$$\chi_\sigma(k,\omega) = -\sum_{\sigma'} \frac{4\pi e_\sigma e_{\sigma'} n_\sigma}{k^2 m_{\sigma'}} \int dv \int dv' \times$$

$$\times W_{\sigma\sigma'\vec{k}\omega}(\vec{v},\vec{v}') \, i\vec{k} \frac{\partial f_{0\sigma}(\vec{v}')}{\partial \vec{v}'}, \quad \sigma=e,i \; (22).$$

$$\chi_g(\vec{k},\omega) =$$

$$= -\sum_{\sigma'=e,i,g} \frac{4\pi e_g e_{\sigma'} n_g}{k^2 m_{\sigma'}} \int d\vec{v} \int d\vec{v}' W_{g\sigma'\vec{k}\omega}(\vec{v},\vec{v}') \times$$

$$\times \vec{k} \frac{\partial f(v')}{\partial \vec{v}'}$$

$$- \frac{i n_g}{\omega+i\nu} \sum_{\sigma=e,i} \sum_{\sigma'=e,i,g} \frac{4\pi e_\sigma e_{\sigma'} n_\sigma}{k^2 m_{\sigma'}} \int d\vec{v} \int d\vec{v}'$$

$$W_{\sigma\sigma'\vec{k}\omega}(\vec{v},\vec{v}') \, v\sigma_\sigma \, i\vec{k} \frac{\partial f(\vec{v}')}{\partial \vec{v}'}, \quad (23)$$

$$\delta\rho_{\vec{k}\omega}^{\sigma(0)} = e_\sigma n_\sigma \int d\vec{v} \, \delta f_{\sigma\vec{k}\omega}^\omega(\vec{v}), \; \sigma=e,i \quad (24)$$

$$\delta\rho_{\vec{k}\omega}^{g(0)} = e_g n_g \int d\vec{v} \, \delta f_{g\vec{k}\omega}^{(0)}$$

$$+ \frac{i n_g}{\omega+i\nu} \sum_{\sigma=e,i} e_\sigma n_\sigma \int d\vec{v} \, v\sigma_\sigma(q_{eq},v) \delta f_\sigma^{(0)}(\chi,t);$$

$$e_g = q^{eq}.$$

As follows from Eg.(20), the quantities $\delta\rho_{\vec{k}\omega}^{\sigma(0)}$ play the role of Langevin sources for the electrostatic potential fluctuations. Since these quantities have the physical meaning of charge fluctuations in the system without interaction through self-consistent electromagnetic field, we may express the correlation function of the Langevin sources in terms of transition probabilities $W_{\sigma\sigma'}(X,X',\tau)$. In particular, the Fourier components of correlation functions for microscopic phase density fluctuations are given by

$$\langle \delta f_\sigma^{(0)}(\vec{v}) \, \delta f_{\sigma'}^{'(0)}(\vec{v}')\rangle_{\vec{k}\omega} =$$

$$= \frac{1}{n_{\sigma'}} f_{\sigma'}(\vec{v}') \, W_{\sigma\sigma'\vec{k}\omega}(\vec{v},\vec{v}') +$$

$$+ \frac{1}{n_\sigma} f_{0\sigma}(\vec{v}) W_{\sigma'\sigma\vec{k}\omega}(\vec{v}',\vec{v}). \quad (25)$$

where

$$W_{\sigma\sigma\vec{k}\omega}(\vec{v},\vec{v}') =$$

$$= \int_0^\infty d\tau \, e^{i\omega\tau} \int d(\vec{r}-\vec{r}') \, e^{-i\vec{k}(\vec{r}-\vec{r}')} W_{\sigma\sigma'}(X,X',\tau)$$

This equation, combined with Eqs. (23),(24), determine the correlation functions for the Langevin sources.

The total charge density (the source and induced contributions) is given by

$$\delta\rho^{\sigma}_{\vec{k}\omega} = \delta\rho^{\sigma(0)}_{\vec{k}\omega} - \frac{\chi_\sigma(k_1\omega)}{\varepsilon(k_1\omega)} \sum_{\sigma'=e,i,g} \delta\rho^{\sigma'(0)}_{\vec{k}\omega} \quad (26)$$

The next problem is to solve set of equations (16), (17). It is possible to show that

$$W_{\sigma\sigma'}(X,X',\tau) =$$

$$= \int \frac{d\vec{k}}{(2\pi)^3} \int \frac{d\omega}{2\pi} e^{i\vec{k}(\vec{r}-\vec{r}') - i\omega\tau} W_{\sigma\sigma'\vec{k}\omega}(\vec{v},\vec{v}') \quad (27)$$

where

$$W_{\sigma\sigma'\vec{k}\omega}(\vec{v},\vec{v}') = \frac{i\delta(\vec{v}-\vec{v}')\delta_{\sigma\sigma'}}{\omega - \vec{k}\vec{v} + i\nu^{eq}_\sigma} \quad (28)$$

$$+ \frac{vf_{0\sigma}(\vec{v})}{\omega - \vec{k}\vec{v} + i\nu^{eq}_\sigma} \left\{ \frac{in_g\sigma'_\sigma(v)}{g(k,\omega)(\omega + i\nu_{eq})} \times \right.$$

$$\times \left[\frac{e_{\sigma'}n_{\sigma'}\sigma_{\sigma'}(v')}{\omega - \vec{k}\vec{v}' + i\nu^{eq}_{\sigma'}} - \sum_{\sigma''=e,i} e_{\sigma''}n_{\sigma''} \times \right.$$

$$\times \int \frac{d\vec{v} v^2 f_{0\sigma''}(v)\sigma^2_{\sigma''}(v)\delta_{\sigma'g}}{(\omega - \vec{k}\vec{v} + i\nu^{eq}_{\sigma''})(\omega - \vec{k}\vec{v} + iO)} \right] +$$

$$+ \frac{\sigma_\sigma\delta_{\sigma'g}}{\omega - \vec{k}v + iO} \right\},$$

$$g(\vec{k},\omega) = 1 - \frac{n_g}{\omega + i\nu} \sum_{\sigma=e,1} \int \frac{d\vec{v} v^2\sigma_\sigma\sigma'_\sigma f^{(v)}_{0\sigma}}{\omega - \vec{k}\vec{v} + i\nu^{eq}_\sigma}$$

For $\sigma = g$, in Eq (28) $\nu^{eq}_g = 0$, $\sigma_g = \sigma'_g = 0$.

4. Within the context of Eq.(20), the quantity $\varepsilon(\vec{k},\omega)$ is the dielectric response function of the dusty plasma accounting for the grain charging and the random motion of the dust. After substituting $m_g \to \infty$ in Eqs. (21), (22), (28), it is easy to reproduce the earlier results for the dusty plasma with grains of infinite mass [16].

Obviously, the solutions of the equation

$$\varepsilon(\vec{k},\omega) = 0 \quad (29)$$

give the eigenfrequencies and damping rates for the waves occurring in the dusty plasma.

The analysis shows that the effect of grain charging on the dispersion properties of dusty plasmas can be significant. It produces considerable modifications of the Green's matrix $W_{\sigma\sigma'}(X,X',\tau)$ and the dielectric susceptibilities $\chi_\sigma(\vec{k},\omega)$. If we disregard the charging dynamics ($\sigma_\sigma = 0$), then the second term in Eq. (20) vanishes and expressions for $\chi_\sigma(\vec{k},\omega)$ and $\varepsilon(\vec{k},\omega)$ reduce to the traditional formulas for a multi-component collisionless plasma. In the general case, however, Eq. (20) contains additional terms with the resonance — type denominators which give rise to new collisionless damping.

For example, we consider the grain — charging influence on the propagation of ion acoustic waves. It was shown in Ref. [16] that, in the limiting case $\omega \ll \nu_{eq}$,

$$\omega \ll P\nu_{eq} \left(P = \frac{n_g}{n_e} \frac{aT_e}{e^2_e} \right) \text{ the charging}$$

processes result in the renormalization of the sound velocity

$$\omega^{ch}_k = kv^{ch}_s, \quad v^{ch}_s = \sqrt{\frac{3\pi-2}{3\pi-3}} v_s,$$

$$v_s = (T_e/m_1)^{1/2} \quad (30)$$

and there arises well pronounced new collisionless damping which can be much greater than the Landau damping.

In the opposite case, $\omega \gtrsim \nu_{eq}$ the renormalization of the ion — sound velocity is observed for the ration ω/ν_{eq} smaller than or of the order of 3. Fig.1 [16] shows this velocity as a function of the ratio ω/ν_{eq}.

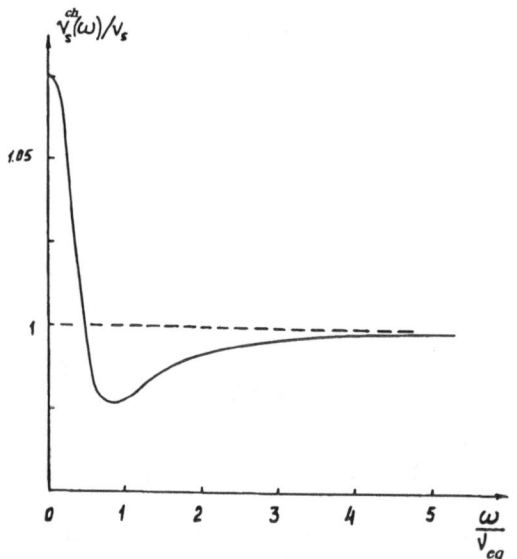

Fig.1

The damping rate for $\omega \gtrsim \nu_{eq}$ is determined mainly by particle losses associated with the absorption by grains. In this case

$$\gamma^{ch}_k = -\frac{1}{2}\Gamma_{ch}\left(1 - \frac{2}{3}\frac{(kv_s)^2}{\omega^2}\right) \quad (31)$$

where Γ_{ch} is a function of ω and z. The frequency dependence of Γ_{ch} for various values of P at $z = 25$ is shown in Fig.2.

5. With the relations between the fluctuation quantities and the Langevin sources being known, we can calculate the

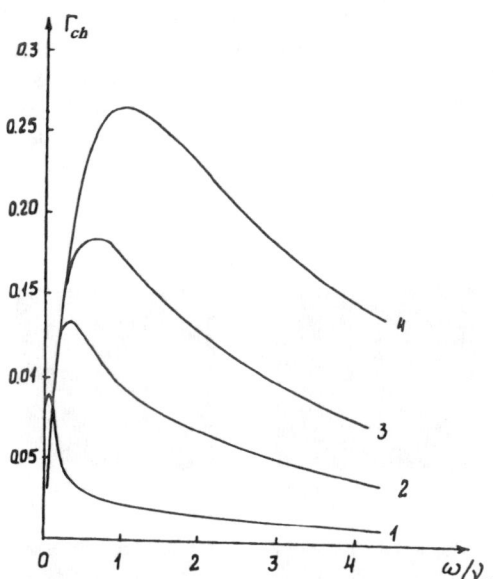

Fig.2

correlation functions for any physical quantities of interest. For example, the correlation function for the electron number density is given by

$$\langle \delta n^2 \rangle_{\vec{k}\omega}^e = \sum_{\sigma,\sigma' = e,i,g} \frac{e_\sigma e_{\sigma'} n_\sigma}{e_e^2} \int d\vec{v} \int d\vec{v}' f_{0\sigma}(\vec{v}) \times$$

$$\times \left[\left(\delta_{\sigma e} - \frac{\chi_e(k,\omega)}{\varepsilon(k,\omega)} \right) - \frac{in_g}{\omega + i\nu_{eg}} \frac{\chi_e(k,\omega)}{\varepsilon(k,\omega)} \times \right. \quad (32)$$

$$\times v\sigma_\sigma(q_{eq},v) \right] \left[\left(\delta_{\sigma' e} - \frac{\chi_e(k_1,\omega)}{\varepsilon(k_1,\omega)} \right) - \right.$$

$$\left. - \frac{in_g}{\omega - i\nu_{eq}} \frac{\chi_e(k,\omega)}{\varepsilon(k,\omega)} v'\sigma_{\sigma'}(q_{eq}v') \right] W_{\sigma\sigma'}^{(\vec{v},\vec{v}')} + c.c.$$

In the existence domain of the solutions of Eg.(29), relation (32) takes the form

$$\langle \delta n^2 \rangle_{\vec{k}\omega}^e = \sum_{\sigma,\sigma' = e,i,g} \frac{e_\delta e_{\delta'}}{\pi e_e^2} n\sigma \int d\vec{v} \int d\vec{v}' f_{0\sigma'}(\vec{v}) \cdot$$

$$\left[\left(1 + \frac{in_g}{\omega + i\nu_{eq}} v \sigma_\sigma(v) \right) \times \right.$$

$$\times \left(1 - \frac{in_g}{\omega - i\nu_{eq}} v'\sigma_\sigma(v') \right) W_{\sigma\sigma'}(\vec{v},\vec{v}') +$$

$$+ c.c. \right] \frac{|\chi_e(\vec{k},\omega)|}{\text{Im } \varepsilon(\vec{k},\omega)} \delta(\text{Re } \varepsilon(\vec{k},\omega)) \quad (33)$$

Thus, the fluctuation spectrum contains pronounced resonance maxima in the vicinity of eigenfrequencies. Resonance frequencies and widths are governed by the solutions of the renormalized dispersion equation.

If we disregard the charging dynamics, i.e., assume the dust particle charges to be fixed, then Eg.(32) yields the well known result for the multi — component plasma, i.e.,

$$\langle \delta n^2 \rangle_{\vec{k}\omega}^e = \left| \frac{1 + \chi_i(\vec{k}_1\omega) + \chi_g(\vec{k}_1\omega)}{\varepsilon(\vec{k}_1\omega)} \right|^2 \langle \delta n^{(0)2} \rangle_{\vec{k}\omega}^e$$

$$+ \left| \frac{\chi_e(\vec{k}_1\omega)}{\varepsilon(\vec{k}_1\omega)} \right|^2 \left[Z_i^2 \langle \delta n_\xi^{(0)2} \rangle_{\vec{k}\omega}^i + Z_g^2 \langle \delta n^{(0)2} \rangle_{\vec{k}\omega}^g \right]$$

where

$$\langle \delta n^{(0)2} \rangle_{\vec{k}\omega}^\sigma = 2\pi n_\sigma \int d\vec{v} f_{0\sigma}(\vec{v}) \delta(\omega - \vec{k}\vec{v}),$$

$$\sigma = e,i,g$$

$$\chi_\sigma(\vec{k},\omega) = \frac{\omega_{\rho 0}^2}{k^2} \int d\vec{v} \frac{\vec{k} \frac{\partial f_{0\sigma}(\vec{v})}{\partial v}}{\omega - \vec{k}\vec{v} + i\nu_\sigma^{eq}}$$

In the equilibrium case, the fluctuation-dissipation theorem holds, i.e.,

$$\langle \delta n^2 \rangle_{\vec{k}\omega}^e =$$

$$= \frac{Tk^2}{2\pi\omega e^2} \text{Im} \left\{ \frac{1 + \chi_i(\vec{k}_1\omega) + \chi_g(\vec{k},\omega)}{\varepsilon(\vec{k}\omega)} \chi_e(\vec{k}\omega) \right\}$$
(34)

The static form-factor is in this case given by

$$\langle \delta n^2 \rangle_{\vec{k}}^e \equiv \int \frac{d\omega}{2\pi} \langle \delta n^2 \rangle_{\vec{k}\omega}^e = n_e \frac{k^2 + k_i^2 + k_g^2}{k^2 + k_i^2 + k_e^2 + k_g^2}$$
(35)

where $k_\sigma^2 = 4\pi e_\sigma^2 n_\sigma / T$, $\sigma = e, i, g$.

Relations (32) and (35) determine the differential and integral, with respect to frequency, cross-sections of wave scattering in the plasma. As follows from (35), the static form-factor is proportional to the electron density and the wave-number-dependent coefficient varies from 1 (for large k) to $(Z_i^2 \frac{n_i}{n_e} + Z_g^2 \frac{n_g}{n_e}) / (1 + Z_i^2 \frac{n_i}{n_e} + Z_g^2 \frac{n_g}{n_e})$ (for small k). Thus, the dust does not cause any considerable modifications of the integral cross — section provided the dust particles are in equilibrium with the plasma, are described by the Maxwellian distribution, and manage to respond to the fluctuation potential during the time of high — frequency field exposure. The last assumption implies that the frequency width of the scattering field $\Delta\omega \sim 1/\tau$ (τ is the impulse durability) must be less that the quantity ks_g. The dust response is appreciable just in the case.

In the opposite case, $\tau < 1/ks_g$ the dust response may be ignored and dust particles may be treated as immovable ($m_g \to \infty$). In this approximation, formula (34) is replaced by

$$\langle \delta n^2 \rangle^e_{\vec{k}\omega} = \frac{Tk^2}{2\pi\omega e^2} \times$$

$$\times \text{Im} \left\{ \frac{1 + \chi_i(k,\omega)}{1 + \chi_i(k,\omega) + \chi_e(k\omega)} \chi_e(k\omega) \right\}$$

$$+ z_g^2 n_g 2\pi\delta(\omega) \frac{k_e^4}{(k^2 + k_e^2 + k_i^2)^2}. \quad (36)$$

Respectively, instead of (35) we obtain

$$\langle \delta n^2 \rangle^e_{\vec{k}} = n_e \frac{k^2 + k_i^2}{k^2 + k_i^2 + k_e^2} + Z_g^2 n_g \frac{k_e^4}{(k^2 + k_e^2 + k_i^2)^2} \quad (37)$$

Under the condition that $Z_g n_g / n_e \gg 1$ the second term in Eqs. (37) can become dominant, then we have

$$\langle \delta n^2 \rangle^e_{\vec{k}} \simeq \frac{Z_g n_g}{\left(1 + \frac{z_i^2 n_i}{n_e} + k^2 \lambda_{De}^2 \right)^2}, \quad \lambda_{De} = \frac{1}{k_e} \quad (38)$$

in complete agreement with the results of [17]. Actually, the same result was obtained in [18].

We note that the "enhancement" of the form factor (41) depends on the sign of the dust charge. In particular, if the dust is charged positively, in view of the quasi-neutrality requirement we have

$$\frac{\langle \delta n^2 \rangle_{\vec{k}}}{n_e} \approx \frac{Z_g n_g}{n_e} \frac{1}{1 - k^2 \lambda_{De}^2}, \quad Z_g n_g / n_e \sim 1 \quad (39)$$

This result in accordance with the conclusions of Ref. [18].

For a negatively charged dust, we have

$$\frac{\langle \delta n^2 \rangle^e_{\vec{k}}}{n_e} = \frac{Z_g^2 n_g}{n_e} \frac{1}{1 + 2Z_1 + k^2 \lambda_{De}^2}. \quad (40)$$

Thus, the enhancement codition is given by $Z_g^2 N_g / n_e \gg 1$.

It is to be noted, however, that formulas (36) — (40) correspond to uncorrelated spatial distribution of the dust. It is not difficult to show that, with the static correlation function for dust fluctuation density being given, we obtain the relation of Ref. [19], i.e.,

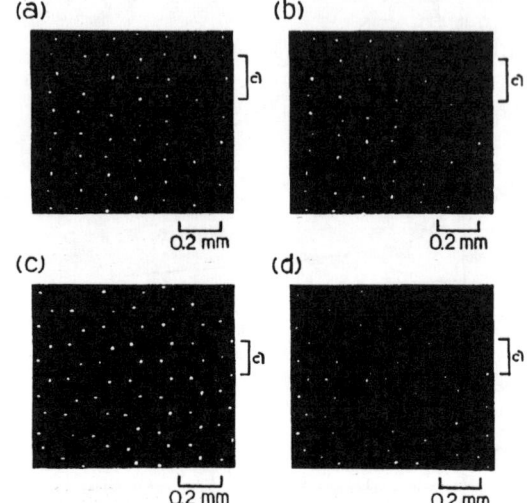

Fig.3

$$\langle \delta n^2 \rangle^e_{\vec{k}} = n_e \frac{k^2 + k_i^2}{k^2 + k_e^2 + k_i^2} +$$

$$+ \frac{k_e^4 Z_g^2}{(k^2 + k_e^2 + k_i^2)^2} \langle \delta n^2 \rangle^g_{\vec{k}} \quad (41)$$

Making use of the equilibrium values for $\langle \delta n^2 \rangle^g_{\vec{k}}$, i.e.

$$\langle \delta n^2 \rangle^g_{\vec{k}} = n_g \frac{k^2 + k_e^2 + k_i^2}{k^2 + k_i^2 + k_e^2 + k_g^2}, \quad (42)$$

we have

$$\langle \delta n^2 \rangle^e_{\vec{k}} = n_e \left\{ \frac{k^2 \lambda_{De}^2 + \frac{z_i^2 n_i}{n_e}}{k^2 \lambda_{De}^2 + 1 + \frac{z_i^2 n_i}{n_e}} + \frac{Z_g^2 N_g}{n_e} \frac{1}{\left(k^2 \lambda_{De}^2 + 1 + \frac{z_i^2 n_i}{n_e}\right)\left(k^2 \lambda_{De}^2 + 1 + \frac{z_i^2 n_i}{n_e} + \frac{Z_e^2 N_g}{n_e}\right)} \right\} \quad (43)$$

We see that in this case the enhancement does not occur.

The relations obtained may be applied to calculate the mean squared charge of the dust particle. We obtain

$$\langle \delta Z_g^2 \rangle = Z_g \frac{2Z_i(t+z)}{z(1+t+z)}, \quad (44)$$

i.e., $\langle \delta Z_g^2 \rangle \sim Z_g$.

6. Formation of dusty crystals is another problem of interest in the statistical theory of dusty plasmas. Latest experiments [20] reported the observation that, as soon as the grain density attains certain critical value, the dust particles arrange in a regular manner and thus form a crystal structure (Fig.3 [20]). The picture is similar to the crystallization of colloidal systems [21,22].

Suppose the system under consideration may be treated as a four-component mixture of charged hard spheres (dust particles), neutral point particle (molecular gas), and sizeless charged particles (electrons and ions). In the equilibrium state, calculating the static correlation functions $G_{\sigma\sigma'}(\vec{r}_1,\vec{r}_2)$ may be reduced to the problem of effective interaction potential $V_{eff}^{\sigma\sigma'}(\vec{r}_1,\vec{r}_2)$. We shall derive an equation for V_{eff} from the well-known Ornstein — Zernike equation [23], i.e.,

$$G_{\sigma\sigma'}(\vec{r}_1,\vec{r}_2) = C_2^{\sigma\sigma'}(\vec{r}_1,\vec{r}_2) + \sum_{\sigma''} n_{\sigma''} \times$$
$$\times \int d\vec{r}_3 C_2^{\sigma\sigma''}(\vec{r}_1,\vec{r}_3) G_{\sigma''\sigma'}(\vec{r}_3,\vec{r}_2) \quad (45)$$

where $C_2^{\sigma\sigma'}(\vec{r}_1,\vec{r}_2)$ is the direct correlation function. In the hyperchain approximation [24]

$$C_2^{\sigma\sigma'}(\vec{r}_1,\vec{r}_2) = G_{\sigma\sigma'}(\vec{r}_1,\vec{r}_2) -$$
$$- \ln F_2^{\sigma\sigma'}(\vec{r}_1,\vec{r}_2) - \frac{1}{T} V^{\sigma\sigma'}(\vec{r}_1,\vec{r}_2) \quad (46)$$

where $V^{\sigma\sigma'}(\vec{r}_1,\vec{r}_2)$ is the pair interaction potential, we obtain [25]

$$V_{eff}^{\sigma\sigma'}(\vec{r}_1,\vec{r}_2) = V^{\sigma\sigma'}(\vec{r}_1,\vec{r}_2) - \sum_{\sigma''} n_{\sigma''} T$$

$$\int d\vec{r}_3 \left(e^{-\frac{V_{eff}^{\sigma\sigma''}(\vec{r}_1,\vec{r}_3)}{T}} + \frac{V_{eff}^{\sigma\sigma''}(\vec{r}_1,\vec{r}_2)}{T} - \right. \quad (47)$$

$$\left. - \frac{V^{\sigma\sigma''}(\vec{r}_1,\vec{r}_3)}{T} - 1 \right) \left(e^{-\frac{V_{eff}^{\sigma''\sigma'}(\vec{r}_3,\vec{r}_2)}{T}} - 1 \right)$$

We assume that $V^{\sigma\sigma'}(\vec{R})$ and $V_{eff}^{\sigma\sigma'}(\vec{R})$ may be divided into the short-range part $V_0^{\sigma\sigma'}(R)$ and the long-range parts, $v^{\sigma\sigma'}(R)$ and $\tilde{V}^{\sigma\sigma'}(R)$ respectively:

$$V^{\sigma\sigma'}(R) = V_0^{\sigma\sigma'}(R) + v^{\sigma\sigma'}(R)$$
$$V_{eff}^{\sigma\sigma'}(R) = V_0^{\sigma\sigma'}(R) + \tilde{V}^{\sigma\sigma'}(R) \quad (48)$$

where $V_0^{\sigma\sigma'}(R) = \begin{cases} \infty, & R < a_{\sigma\sigma'} \\ 0, & R > a_{\sigma\sigma'} \end{cases}$

and $a_{\sigma\sigma'} = a_\sigma + a_{\sigma'}$, a_σ is the solid core radius for the particles species σ.

We linearize Eq. (47) with respect to the long-range addends and solve it may means of the Fourier method. Then it is not difficult to show that the Fourier component of the

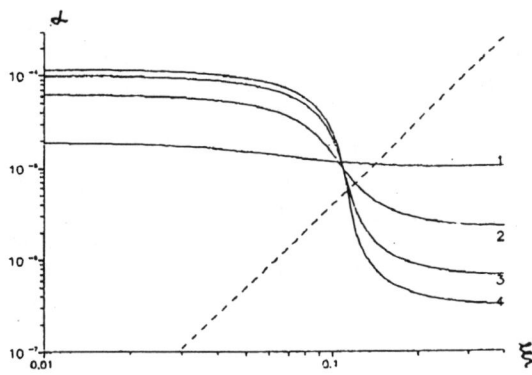

Fig.4

effective potential can be singular for real values of the wave vector components which satisfy the secular equation [26]

$$\det \Pi_{\sigma\sigma'}(\vec{k}) = 0 \quad (49)$$

This implies the possibility of ordering in the system. Here

$$\Pi_{\sigma\sigma'}(k) = \delta_{\sigma\sigma'} + n_\sigma \Lambda_{\sigma\sigma'}(k)$$

$$\Lambda_{\sigma\sigma'}(k) = \theta_k(a_{\sigma\sigma'}) + \frac{4\pi e_\sigma e_{\sigma'}}{k^2} \cos k a_{\sigma\sigma'} \quad (50)$$

$$\theta_k(a) = \frac{4\pi}{k^3}(\sin ka - ka \cos ka).$$

This equation has no real solutions for $n_g = 0$, they occur beginning from some n_g. Appearance of this solutions shows that the effective potential is oscillatory which observation may be interpreted as a tendency towards spatial ordering. If k approaches $k_0 = \frac{2\pi}{\langle R \rangle}$, $(\langle R \rangle = \left(\frac{3}{4\pi n_g}\right)^{1/3}$ is the mean distance between the dust particles), i.e., the mean interparticle distance becomes comparable to the ordering scale $R_0 = \frac{2\pi}{k_c}$ then we can assume that a crystal structure is formed.

Fig.4 shows the results of numerical computation for Eq. (52) with the dusty plasma parameters given in Ref [20]. For the sake of simplicity, calculations have been performed for $n_i \sim 2n_e$, $Z_g n_g \sim n_e$. The results are presented as $\alpha(\zeta)$ dependence, where

$$\alpha = n_g \frac{4\pi}{3} a_g^3, \quad \zeta = k a_g.$$

The dotted line shows the dependence $\bar{\alpha}(\zeta) = \left(\frac{\zeta}{2\pi}\right)^3$ which is associated with the condition $\zeta = \frac{2\pi a}{\langle R \rangle} = \frac{2\pi a}{(3/4\pi n_g)^{1/3}}$. The dotted and solid lines intersect for the density that corresponds to the mean interparticle distance equal to the ordering spatial scale. Curves 1 — 4 correspond to $Z_g = 100$, $Z_g = 350$, $Z_g = 700$, $Z_g = 1000$.

As follows from the calculations, the lattice parameter for the crystallization in the system under consideration may be of the order of $\zeta \sim 0.12 \div 0.14$. This result is in good accordance with the observed value $\zeta \simeq 0.17$ [20]

References

1. *M.E.Bailey, D.A.Williams* (Editors). Dust in the Universe. Cambr. Univ. Press. 1988
2. *C.K.Goertz.* Rev Geophys. **27**, 271 (1989)
3. *V.N.Tsytovich, G.E.Morfill, R.Bingham, U de Angelis.* Comments Plasma Phys. **13**, 153 (1990)
4. *U. de Angelis.* Phys. Scr. **45**, 465 (1992)
5. *T.G.Northrop.* Phys. Scr. **45**, 475 (1992)
6. *J.-P.Boeuf, Ph. Belenguer, T. Hbid.* In Proc.IIIICPIG XXI, p. 108. APP, Ruhr Univ. Bochum, 1993
7. *O.Havnes, G.E.Morfill, C.K.Goertz.* J. Geophys. Res. **89**, 10999 (1984)
8. *E.C.Whipple, T.G.Northrop, D.A.Mendis.* J. Geophys. Res. **90**, 7405 (1985)
9. *C.K.Goertz, W.H.Ip.* Geophys. Res. Lett. **11**, 49 (1984)
10. *O.Havnes, C.K.Goertz, G.E.Morfill, E. Griin, W.Ip.* J. Geophys. Res. **92**, 2281 (1987)
11. *U de Angelis, V.Formisano, M.Giordano.* J. Plasma Phys. **40**, 399 (1988)
12. *U de Angelis, R.Bingham, V.N.Tsytovish.* J. Plasma Phys. **42**, 429 (1989)
13. *U de Angelis, R.Bingham, V.N.Tsytovich.* J. Plasma Phys. **42**, 445 (1989)
14. *M.Salimullah, A.Sen.* Phys. Fluids B **4 (6)**, 1441 (1992)
15. *R&K.Varma, P.K.Shukla, V.Krishan.* Phys. Rev. E **47**, 3612 (1993)
16. *V.N.Tsytovich, O.Havnes.* Comments Plasma Phys. **15**, 267 (1993)
17. *V.N.Tsytovich.* Phys. Scr. **45**, 521 (1992)
18. *C LaHoz.* Phys. Scr. **45**, 529 (1992)
19. *R.Bingham, U de Angelis, V.Tsytovich, O.Havnes.* Phys. Fluids **B3**, 811 (1991)
20. *I.M.Chu, L.I.* Physica A **205**, 183 (1994)
21. *P.Pieranski.* Contemp. Phys. **24**, 85 (1983)
22. *J.Derksen, W.Vande Water.* Phys. Rev. **A45**, 5660 (1992)
23. *Yu.L.Klimontovich.* Statistical Physics. Harwood, Chur (1986)
24. *J.R.Hunsen, I.R.McDonald.* Theory of simple liquids. Academic Press, London, 1986
25. *O.I.Gerasimov, A.G.Zagorodny, Yu.L.Klimontovich.* Plasma Phys. Rep. **19**, 83 (1993)
26. *O.Gerasimov, P.Schram, A.Sitenko, A.G.Zagorodny* (to be published)

Kinetic theory of bounded plasmas

S. Kuhn
Plasma and Energy Physics Group, Institute for Theoretical Physics, University of Innsbruck, A–6020 Innsbruck, Austria

This paper deals with the "trajectory–simulation" approach to the kinetic theory of bounded plasma systems (BPS's), which naturally allows for realistic ion–electron mass ratios, the inclusion of boundary and initial conditions, and high resolution in both configuration and velocity spaces. The paper briefly reviews the basic theory elements and mathematical methods, previous results on (1d,1v) collisionless BPS's, and recent extensions to (1d,2v) collisional and (2d,3v) collisionless systems. Some perspectives for future developments are discussed.

1. INTRODUCTION

Realistic kinetic modeling of bounded plasmas — or, more appropriately, *bounded plasma systems* — is of interest in both fundamental plasma research and applied disciplines, such as plasma technology and fusion research. By definition, a "bounded plasma system" (BPS) is characterized by the self–consistent interaction of the plasma itself, its material boundaries, and the related external circuits. A full kinetic description of BPS's generally requires kinetic equations for the distribution function of each particle species, Maxwell's equations for the electromagnetic fields, boundary conditions for the particles and the fields, and external–circuit equations.

Particle simulation has by now become a routine tool for BPS studies, as is demonstrated, e.g., by the widespread and ever increasing usage of the PD ("plasma device") series of bounded–plasma particle codes[18,20,27,28] developed at U.C. Berkeley. Nevertheless, alternative methods are still desirable and needed.

This paper deals with such an alternative, more theory–oriented approach, which has been termed "trajectory simulation".[23,29-31] Along the collisionless single–particle trajectories we calculate, by means of "trajectory integration", velocity distribution functions, rather than moving single particles as in particle simulation. This approach allows quite naturally for realistic electron–to–ion mass ratios, the inclusion of boundary and initial conditions, and high resolution in both configuration and velocity spaces. It either yields *analytic solutions* or, where such are impossible, can be advantageously applied in the form of an efficient iterative scheme.

The dimensionality of a given model will be indicated in the form $(m\,d,\,n\,v)$, with m and n the numbers of the relevant space and velocity coordinates, respectively.

The basic theory elements and mathematical methods are outlined in Secs. 2, 3.A and 4, (1d,1v) collisionless BPS's are reviewed in Sec. 3, recent extensions to (1d,2v) collisional and (2d,3v) collisionless systems are presented in Secs. 5 and 6, respectively, and perspectives for future developments are briefly discussed in Sec. 7.

For more extensive reviews on BPS theory and many more references to the literature, the reader is referred to Refs. 16, 17, 23 and 29.

2. TRAJECTORY INTEGRATION OF THE KINETIC EQUATIONS

2.A. Kinetic equations

In integro–differential form, the kinetic equation for particle species s can be written

$$\frac{d_s}{dt} f^s(\mathbf{x},\mathbf{v},t) = \mathscr{C}^s(ff|\mathbf{x},\mathbf{v},t), \qquad (1)$$

where

$$\mathscr{C}^s(ff|\mathbf{x},\mathbf{v},t) = \sum_p \mathscr{C}^{sp}(f^s f^p|\mathbf{x},\mathbf{v},t) \qquad (2)$$

and

$$\frac{d_s}{dt} \equiv \frac{\partial}{\partial t} + \mathbf{v}\cdot\frac{\partial}{\partial \mathbf{x}} + \frac{e^s}{m^s}\Big[\mathbf{E}(\mathbf{x},t) + \mathbf{v}\times\mathbf{B}(\mathbf{x},t)\Big]\cdot\frac{\partial}{\partial \mathbf{v}}. \qquad (3)$$

In Eqs. (1) and (2), f^s and $\mathscr{C}^s$ are the velocity distribution function and the total collision term of species s, and $\mathscr{C}^{sp}$ is the partial collision term between species s and p. By writing the collision terms in the particular form given in (2) we indicate that they are, as is customary in plasma physics, *bilinear* functionals of the distribution functions but ultimately represent functions of position $\mathbf{x}$, velocity $\mathbf{v}$ and time t.

2.B. Collisionless trajectories

In Eq. (3), **E** and **B** are the macroscopic electric and magnetic fields, and e^s and m^s are the charge and mass of a species–s particle. The symbol d_s/dt denotes the "Lagrangian" (or "total" or "substantial") time derivative for an observer moving along a collisionless species–s particle trajectory (or "characteristic") $[\mathbf{x}(t), \mathbf{v}(t)]$, i.e., a trajectory satisfying the equations of motion

$$\frac{d_s}{dt}\mathbf{x} = \mathbf{v} \qquad (4a)$$

and

$$\frac{d_s}{dt}\mathbf{v} = \frac{e_s}{m_s}[\mathbf{E}(\mathbf{x},t) + \mathbf{v}\times\mathbf{B}(\mathbf{x},t)] . \qquad (4b)$$

In all cases considered by us, we will be interested in trajectories passing through the "current point"

$$Y \equiv [\mathbf{x},\mathbf{v},t] \equiv [P,t] , \qquad (5)$$

with

$$P \equiv [\mathbf{x},\mathbf{v}] , \qquad (6)$$

at which the distribution function is to be calculated. (More precisely, we call Y the "phase–time point" and P the "phase point".) The generic point along such a trajectory will henceforth be written as

$$\hat{Y} \equiv [\hat{\mathbf{x}},\hat{\mathbf{v}},\hat{t}] \equiv [\mathbf{x}(\hat{t}),\mathbf{v}(\hat{t}),\hat{t}] \equiv [\hat{P},\hat{t}] . \qquad (7)$$

Generally, Eqs. (1) and (4) are integrated between a suitable starting point ("origin of integration") $\hat{Y}_0$, i.e., with the "original conditions"

$$[\mathbf{x}(\hat{t}=\hat{t}_0), \mathbf{v}(\hat{t}=\hat{t}_0), \hat{t}=\hat{t}_0] = [\hat{\mathbf{x}}_0, \hat{\mathbf{v}}_0, \hat{t}_0]$$
$$\equiv [\hat{P}_0, \hat{t}_0] \equiv \hat{Y}_0 , \qquad (8)$$

and the current point Y, corresponding to the "current conditions"

$$[\mathbf{x}(\hat{t}=t), \mathbf{v}(\hat{t}=t), \hat{t}=t]$$
$$= [\mathbf{x},\mathbf{v},t] = [P,t] = Y , \qquad (9)$$

with $\hat{t}_0 \le t$, cf. Fig. 1.

2.C. Trajectory integration

By "trajectory integration of the species–s kinetic equation" we mean integrating Eq. (1) along a specific species–s trajectory, which formally leads to the "trajectory–integrated kinetic equation"

$$f^s(Y) = f^s(\hat{Y}_0) + \mathscr{J}^s(\hat{Y}_0, Y) , \qquad (10)$$

where

$$\mathscr{J}^s(\hat{Y}_0, Y) \equiv \int_{\hat{t}_0}^{t} d\hat{t}\, \mathscr{C}^s(ff|\hat{Y}) . \qquad (11)$$

In the important special case of a *collisionless* particle species, Eqs. (2), (1), (11) and (10) reduce to

$$\mathscr{C}^s(ff|\hat{Y}) \equiv 0, \qquad (12)$$

$$\frac{d_s}{dt} f^s(Y) \equiv 0 , \qquad (13)$$

$$\mathscr{J}^s(\hat{Y}_0, Y) \equiv 0 , \qquad (14)$$

and

$$f^s(Y) = f^s(\hat{Y}_0) , \qquad (15)$$

respectively, with (13) known as the "Vlasov equation".

Note that for both collisional *and* collisionless systems, we always integrate along the *collisionless* particle trajectories. Considering, in particular, an initial–value problem for $t \ge 0$, $f^s(\hat{Y}_0)$ is (i) a boundary value if $\hat{t}_0 \ge 0$ and $\hat{P}_0$ is a boundary point (Fig. 1, trajectory 1), and (ii) an initial value if $\hat{t}_0 = 0$ and $\hat{Y}_0$ is a point in volume,[14,15] (Fig. 1, trajectory 2). Hence, the integral relation (10) is equivalent to the equations of motion plus boundary and initial conditions, and thus represents a very compact and powerful formal solution of the kinetic equations.

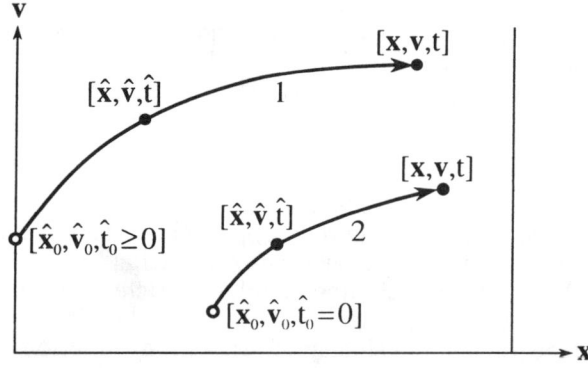

FIG. 1

2.D. Practical aspects

In practice, the basic concept of trajectory integration may be applied in several ways, depending on the problem of interest. Let us consider here the frequent situation when the fields **E** and **B** are given and f^s is to be calculated at the current point $Y = [\mathbf{x},\mathbf{v},t]$. To this end we first trace back the collisionless species–s trajectory passing through Y, trying to hit a point $\hat{Y}_0$ where f^s is known.

If we find such a point, we can proceed as follows: For a *collisionless* particle species, Eq. (15) immediately yields the distribution function *explicitly*, while for a *collisional* species ($\mathscr{C}^s$ does not vanish identically) (1) must be integrated along the trajectory from $\hat{Y}_0$ to Y, which is equivalent to evaluating (10) and (11). Except for particularly simple cases allowing for fully analytic treatment,[11] this latter step will usually be embedded in an iterative scheme in which some or all of the f's needed in the collision terms (2) are replaced with approximate ones (e.g, those obtained in the previous iteration, cf. Sec. 4.C).

However, it may happen that there exists no point $\hat{Y}_0$ where f^s is known, in which case $f^s(Y)$ must be determined from other considerations. If, e.g., in a dc situation the trajectory remains within a restricted domain of configuration space and closes on itself, we are dealing with trapped particles,[26] the distribution of which may be the result of subtle long–time evolution.[20]

2.E. Analytic solutions

1. To the present author's knowledge, *exact* analytic solutions to kinetic bounded–plasma problems have been obtained only for collisionless systems. The reason is that for these the velocity distribution functions can be written as functions of the constants of collisionless particle motion, $c_1^s(Y)$, $c_2^s(Y)$, ... :

$$f^s(Y) = \check{f}^s[c_1^s(Y), c_2^s(Y), ...] \quad (16)$$

with

$$\frac{d_s}{dt} c_m^s(Y) = 0 \quad (m = 1, 2, ...), \quad (17)$$

and if one succeeds in finding analytic expressions for c_1^s, c_2^s, ..., any function of the form (16) clearly represents an exact analytic solution to the collisionless kinetic equation (13). Again to the best of the author's knowledge, as far as bounded–plasma theory is concerned this has been achieved to date only for *(1d,1v) dc* problems, cf. Sec. 3.

2. However, there is a school of theorists from the former Soviet Union who have succeeded in treating *1d collisional* bounded–plasma systems in an approximate manner but with a remarkable degree of analyticity, including elastic collisions between charged and neutral particles,[3,4,9] charge–exchange collisions,[3] and Coulomb collisions.[2]

3. All in all, it appears that our capability of finding analytic solutions to any but the simplest kinetic bounded–plasma problems is still very limited.

3. REVIEW OF (1d,1v) COLLISIONLESS SYSTEMS

3.A. The cartesian (1d,1v) diode (collisionless plane diode, CLPD) model

In this model all quantities considered vary only in the z–direction, and only particle–velocity and electric–field components directed *along* the z–axis are considered. The "electrodes", whose surfaces are located at $z = 0$ (left–hand electrode) and $z = d$ (right–hand electrode), are slabs or planes perpendicular to the z–direction, with infinite transverse dimensions. (One often finds x instead of z, and/or L instead of d.) A very general, closed system of basic equations governing the dynamics of this cartesian (1d,1v) model has been proposed by the present author[14,16] and is presented here as an archetypal closed system for BPS kinetic theory:

Vlasov equations:

$$\left[\frac{\partial}{\partial t} + v \frac{\partial}{\partial z} + \frac{e^s}{m^s} E \frac{\partial}{\partial v}\right] f^s = 0, \quad (18)$$

Poisson's equation:
$$\frac{\partial E}{\partial z} = \sum_s \frac{e^s}{\epsilon_0} \int_{-\infty}^{+\infty} dv\, f^s, \quad (19)$$

total–current conservation:

$$\epsilon_0 \frac{\partial E}{\partial t} + \sum_s e^s \int_{-\infty}^{+\infty} dv\, v\, f^s = j_e, \quad (20)$$

particle boundary conditions at the left–hand electrode:

$$f_{l+}^s = f_{lg}^s + \sum_{s'} \int_{-\infty}^{0} dv'\, b_l^{ss'}(v,v')\, f_l^{s'}(v',t) \quad (21)$$

particle boundary condition at the right–hand electrode:

$$f_{r-}^s = f_{rg}^s + \sum_{s'} \int_{0}^{\infty} dv'\, b_r^{ss'}(v,v')\, f_r^{s'}(v',t), \quad (22)$$

and external–circuit condition:

$$-\int_0^L dz\, E = V_{es} + Z\{j_e\}, \quad (23)$$

with ϵ_0 the "permittivity of vacuum"; $j_e(t)$ the external–circuit (or "total") current density [i.e., external–circuit current per unit electrode area], $V_{es}(t)$ the potential difference produced by some external–circuit signal generator, $\hat{Z}\{j_e\}$ the external–circuit impedance operator; $f_{l+}^s(v,t) = f^s(0,v,t)$ (with $v > 0$), $f_{r-}^s(v,t) = f^s(d,v,t)$ (with $v < 0$) "emission distribution functions" [i.e., emitted–particle distribution functions at the left–hand and right–hand electrodes, respectively]; $f_{lg}^s(v,t)$, $f_{rg}^s(v,t)$ the "given" parts of the emission distribution functions [i.e., those parts of f_{l+}^s and f_{r-}^s which are determined by external mechanisms — such as thermal particle emission with half Maxwellian distribution functions (29)]; $b_l^{ss'}(v,v')$, $b_l^{ss'}(v,v')$ generalized reflection coefficients for the left–hand and right–hand electrodes, respectively [i.e., the (unnormalized) probabilities for a species–s' particle impinging with velocity v' to be "re–emitted" as a species–s particle with velocity v]. As usual, the electrostatic field E is related to the electrostatic potential Φ by

$$E(z,t) = -\partial\Phi(z,t)/\partial z , \qquad (24a)$$

with

$$\Phi(0,t) = 0 . \qquad (24b)$$

The system (18) – (24) must, of course, be complemented by *consistent* initial conditions for the distribution functions and the external circuit.[14,15] It also exhibits the complexity of BPS problems as compared to unbounded–plasma ones: In the latter case, the total system reduces to Eqs. (18) and (19) with the boundary conditions replaced by much simpler conditions, such as periodicity or the vanishing of the perturbations as $z \to \pm\infty$.

Note that the model described by the above equations is not fully electrostatic but rather represents a one–dimensional approximation to a two–dimensional, electromagnetic situation. The fact that the total current density (20) is usually nonzero implies $\nabla\times\mathbf{B} \neq 0$ and, hence, the existence of a magnetic field perpendicular to the z–direction. However, the influence of the latter on the particles is neglected in this model.

3.B. Analytic calculation of time–independent (dc) states.

Let us now demonstrate how collisionless (1d,1v) models allow for analytic treatment, at least to some extent. In a dc situation, the single–particle energy W^s is a constant of the motion (cf. Sec. 2.E), so that along the species–s particle trajectory passing through $[z,v]$ (cf. Sec. 2.B) we have

$$\frac{m^s v^2}{2} + e^s \Phi(z) = \frac{m^s \hat{v}_0^2}{2} + e^s \Phi(\hat{z}_0) . \qquad (25)$$

Then, Eq. (15) can be written in the form

$$f^s(z,v) = f^s(\hat{z}_0, \hat{v}_0)$$
$$= f^s(\hat{z}_0, \pm\sqrt{v^2 + 2e^s[\Phi(\hat{z})-\Phi(\hat{z}_0)]/m^s}) , \qquad (26)$$

where $\hat{v}_0$ has been obtained from Eq. (25) and its sign must be chosen consistently with the physical situation considered. To this end, a global qualitative assumption must be made about the shape of the potential distribution [e.g., monotonically decreasing from left to right, or with a single maximum in front of the left–hand electrode, etc.]. Assuming that $f^s(\hat{z}_0, \hat{v}_0)$ is a *known* function of $\hat{v}_0$ [e.g., the emission distributions (29)], Eq. (26) yields $f^s(z,v)$ as a *known* function of Φ and v, so that Poisson's equation (19), combined with (24), takes the form

$$d^2\Phi/dz^2 = -\rho(\Phi)/\epsilon_0 , \qquad (27)$$

with ρ a *known* function of Φ. Multiplying both sides by $d\Phi/dz$ and integrating between some suitable position z_1 and the current position z we obtain

$$\frac{d\Phi}{dz} = \pm\sqrt{E_1^2 + P(\Phi,\Phi_1)} , \qquad (28a)$$

with

$$P(\Phi,\Phi_1) \equiv -\frac{2}{\epsilon_0}\int_{\Phi_1}^{\Phi} d\Phi' \rho(\Phi') , \qquad (28b)$$

where the sign must again be chosen consistently with the situation considered.

In many situations, and especially when the emission distribution functions are half Maxwellians (29), the functions $\rho(\Phi)$ and $P(\Phi,\Phi_1)$ can be obtained analytically. The last step, however, namely integrating (28a) for the potential profile $\Phi(z)$, usually requires *numerical* evaluation.

3.C. Applications of the collisionless plane–diode model

A considerable amount of work related to the CLPD model can be found in the literature.[16,23,29]

1. The simplest and most fundamental bounded–plasma model is the **Pierce diode**, a CLPD in which a *cold* collisionless electron beam propagates between two electrodes, the interelectrode space being filled with a neutralizing immobile ion background. In this case, the cold–fluid description usually applied[19] is exactly equivalent to the kinetic one.[15] One of this model's best–known features is the "Pierce instability", an electron instability playing an important role in bounded–plasma dynamics.

2. In Ref. 14, an integral–equation method was developed for solving the *general linearized perturbation problem* for CLPD's with a uniform plasma region and thin electrode sheaths. So far, the method has been applied to the Pierce diode[15] and to the negatively biased single–ended Q–machine.[33]

3. A particularly important (1d,1v) model is the so–called *"single–emitter plasma diode"* or *"Knudsen diode with surface ionization (KDSI)"*, which has been used extensively to describe both single–ended Q–machines[12] and thermionic energy converters.[7] The emitter or "hot plate" ($z = 0$, $\Phi = 0$) whose temperature T typically exceeds 2000 K, produces electrons (via Richardson emission) and positive ions (via contact ionization of neutral alkali atoms hitting the surface), both species leaving the emitter surface at constant rates with half Maxwellian velocity distribution functions:

$$f^s(x=0, v_x>0) = 2n_{0+}^s \sqrt{\frac{m^s}{2\pi kT}} \times \exp\left[-\frac{m^s v_x^2}{2kT}\right] \quad (s = e, i), \quad (29)$$

where n_{0+}^s are the emission densities at the hot–plate surface. Particles hitting either the emitter or the collector (located at $x = d$, $\Phi = \Phi_c$) are absorbed. Typically, the particle distribution functions in this model are cut–off Maxwellians.

The dc and dynamic states of this system are conveniently classified in terms of the three dimensionless parameters $\gamma \equiv n_{0+}^i/n_{0+}^e$ ("neutralization parameter"), $\eta_c \equiv e\Phi_c/kT$ (normalized cold–plate bias), and $\delta \equiv d/l_{0+}^e$ (normalized interelectrode distance), where k is Boltzmann's constant and $l_{0+}^e \equiv \sqrt{\{\epsilon_0 kT/(n_{0+}^e e^2)\}}$ is the "emission Debye length". The possible dc potential distributions, calculated with the method of Sec. 3.B, are conveniently classified in terms of the (η_c, γ) parameter diagram,[7,12] the most important types being 1 ($\eta_c > 0$, monotonically increasing), 2 ($\eta_c > 0$, single–minimum), 1' ($\eta_c < 0$, monotonically decreasing), and 2' ($\eta_c < 0$, single–maximum).

While the negative–bias distributions 1' and 2' generally seem to be stable, the positive–bias ones 1 and 2 are usually strongly unstable, exhibiting large–amplitude "potential relaxation oscillations".[17] Very remarkably, the latter have been studied not only by particle simulations[22] but also, in great detail, by means of the "Ender–Kuznetsov method",[13] an analytic–numerical trajectory–simulation method that can handle strongly nonlinear dynamic problems involving both ion and electron timescales, with an accuracy greater than that of particle simulation.

3.D. Cylindrical and spherical (1d,1v), collisionless dc states

Basically following the procedure outlined in Sec. 3.B, *Zhykharsky* has formalized and generalized the calculation of collisionless, dc, (1d,1v) bounded–plasma states and given a systematic classification of such states in various cartesian, cylindrical and spherical geometries.[21]

3.E. Semi–bounded problems

It should be mentioned that, in addition to the fully bounded systems just discussed, the basic formalism of (1d,1v) collisionless BPS analysis has also been applied to a variety of semi–bounded problems, such as waves launched from an emitter,[1] probe problems,[25] or sheath problems.[5]

4. ITERATIVE SCHEME FOR SELF–CONSISTENT KINETIC TRAJECTORY SIMULATION OF TIME–INDEPENDENT (DC) SYSTEMS

4.A. Basic equations and boundary conditions

For illustration of the iterative method we restrict ourselves to *time–independent, electrostatic systems with given magnetic fields*. For these, the basic equations in the plasma region are (i) the kinetic equations (1) (with $\partial/\partial t = 0$) and (ii) Poisson's equation,

$$\nabla^2 \Phi(\mathbf{x},t) = -\frac{1}{\epsilon_0}\rho(\mathbf{x},t) = -\frac{1}{\epsilon_0}\sum_s e^s n^s(\mathbf{x},t), \quad (30)$$

with

$$n^s(\mathbf{x},t) = \int d^3v \, f^s(\mathbf{x},\mathbf{v},t), \quad (31)$$

where we have used $\mathbf{E} = -\nabla\Phi$, ρ is the total space–charge density, and n^s is the number density of particle species s.

To specify the boundary conditions, let us assume that we are dealing with a bounded spatial region where both the emission distribution functions (cf. (29) and (34)) and the electric potential are given at the boundaries:

$$f^s(\mathbf{x}_b,\mathbf{v},t) = f_b^s(\mathbf{x}_b,\mathbf{v},t)$$
$$[f_b^s \text{ given for } \mathbf{v}\cdot\mathbf{n}(\mathbf{x}_b) > 0] , \quad (32)$$

$$\Phi(\mathbf{x}_b,t) = \Phi_b(\mathbf{x}_b,t) \quad [\Phi_b \text{ given}] , \quad (33)$$

where $\mathbf{x}_b$ is a "boundary point", i.e., a point lying infinitesimally close to the bounding surface, and $\mathbf{n}(\mathbf{x}_b)$ is the local normal to that surface pointing into the plasma region.

4.B. The iterative scheme

The scheme to be described here has been used in various modifications both previously[8] and, more recently, in the collisional (1d,2v) and collisionless (2d,3v) studies outlined in Secs. 5 and 6, respectively. It usually requires discretization of both configuration and velocity spaces, in which case it is understood that $[\mathbf{x},\mathbf{v}]$ is a generic *grid*point in phase space (e.g., $[\mathbf{x}_{ij}, \mathbf{v}_{pqr}]$), and that the relevant equations and boundary conditions (Sec. 4.A) are actually used in their appropriate discretized forms.

The calculation is initiated by prescribing "starting" distributions $\Phi^{(0)}(\mathbf{x})$ and $f^{s(0)}(\mathbf{x},\mathbf{v})$, which should be reasonable first guesses. These are input to the "main iteration block" (Sec. 4.C), which in the first main iteration yields new, improved distributions $\Phi^{(1)}(\mathbf{x})$ and $f^{s(1)}(\mathbf{x},\mathbf{v})$. With these as input, the main iteration block is entered again (2nd iteration), to yield the next-better approximations $\Phi^{(2)}(\mathbf{x})$ and $f^{s(2)}(\mathbf{x},\mathbf{v})$, etc.. Convergence is reached if the results of iteration m coincide with those of iteration m−1 to within some prescribed accuracy.

4.C. The main iteration block

This is a package of routines calculating, on the basis of given input distributions $\Phi^{(m-1)}(\mathbf{x})$ and $f^{s(m-1)}(\mathbf{x},\mathbf{v})$ (with $m \geq 1$), new, improved output distributions $\Phi^{(m)}(\mathbf{x})$ and $f^{s(m)}(\mathbf{x},\mathbf{v})$, respectively. Its basic way of operation can be subdivided into the following five steps, of which nos. 2 and 3 are the ones requiring most of the computing time: ***Step 1:*** Get the input distributions. ***Step 2:*** Calculate $f^{s(m)}(\mathbf{x},\mathbf{v})$ via trajectory integration (Sec. 2.C), with $\mathbf{E} \rightarrow \mathbf{E}^{(m-1)}(\mathbf{x}) \equiv -\nabla\Phi^{(m-1)}(\mathbf{x})$ in (4b) and $f^s \rightarrow f^{s(m-1)}(\mathbf{x},\mathbf{v})$ in (2). ***Step 3:*** Calculate $n^{s(m)}(\mathbf{x})$ by evaluating (31), with $f^s \rightarrow f^{s(m)}(\mathbf{x},\mathbf{v})$. ***Step 4:*** Calculate $\rho^{(m)}(\mathbf{x})$ by evaluating the second equality in (30), with $n^s \rightarrow n^{s(m)}(\mathbf{x})$. ***Step 5:*** Calculate $\Phi^{(m)}(\mathbf{x})$ by solving (30), with $\rho \rightarrow \rho^{(m)}(\mathbf{x})$. For this latter purpose we use a standard relaxation scheme involving an appropriate number of sweeps across the spatial grid.

4.D. Refinements

The "basic" iteration scheme outlined in so far works in principle but is far from optimal. A refinement enhancing numerical stability consist in solving, in Step 5, Poisson's equation "not too exactly", i.e., to use a "truncated" Poisson solver involving only a relatively small number of sweeps across the spatial grid.[30,31] Another useful trick, saving a lot of computing time is the "numerical-correlation method", which rapidly leads to a very good starting distribution for initiating the "real" iterations.[31]

5. A CYLINDRICALLY SYMMETRICAL (1d,2v) Q-MACHINE MODEL INCLUDING CHARGE-EXCHANGE COLLISIONS

In a first application of the iterative scheme of Sec. 4 to *collisional* bounded plasmas, Maier and Kuhn have extended the (1d,1v), dc, *collisionless* single-emitter plasma diode model into a cylindrically symmetrical (1d,*2v*), *collisional* one.[24,30] A *finite*, uniform magnetic field is applied along the z-direction, and the ion velocity distribution function is now of the form $f^i(z,v_z,v_n)$. Instead of the *1* v emission distribution functions (29) we now have the *2* v ones

$$f^s(z=0,v_z>0,v_n) = 2n_{0+}^s \left[\frac{m^s}{2\pi kT}\right]^{\frac{3}{2}}$$
$$\times \exp\left[-\frac{m^s}{2kT}(v_z^2+v_n^2)\right] \quad (s = e, i) , \quad (34)$$

with v_n the velocity component normal to the magnetic field. The electrons are still collisionless, but the ions are now assumed to undergo symmetric charge-exchange collisions with the given neutral background gas, whose velocity distribution function $F^{bg}(z,V_z,V_n)$ is chosen to be a 2v non-drifting Maxwellian with Temperature T^{bg}, usually much lower than T. Accordingly, the ions satisfy a linearized Boltzmann equation, namely Eq. (1) with $s = i$ and a collision term (2) of the form $\mathscr{C}^i(f^i F^{bg}|z, v_z, v_n)$. The differential and total cross sections have been obtained by fitting well-known analytic approximations[6] to relevant experimental data.[10]

In Fig. 2 we present some exemplary results for a single-ended Q-machine (or thermionic diode) with a type-2 potential distribution.[30] In Fig. 2(a), the dashed curve shows the collisionless potential distribution, whereas the solid curve is the corresponding potential distribution after the potential well has filled up with ions via charge-exchange collisions. The velocity distribution function at the potential minimum, given in Fig. 2(b), clearly exhibits the cold peak corresponding to the trapped ions and the much broader distribution of the untrapped ions.

6. CARTESIAN AND CYLINDRICAL (2d,3v), COLLISIONLESS Q-MACHINE MODELS

Using the iterative scheme of Sec. 4, Pedit and Kuhn have developed a cartesian[31] and a cylindrical[32] (2d,3v) extension of the (1d,1v), dc, collisionless single-emitter plasma diode model (Sec. 3.C.3). The potential distribution again decreases monotonically in the *axial* direction, i.e., the direction of the constant,

finite applied magnetic field.

The system, shown in Fig. 3(a), is specified by prescribing the following parameters: axial length d, emitter radius R_e, collector radius R_c, conducting-wall radius R_w; emitter potential $\Phi_e \equiv 0$, collector potential Φ_c, conducting-wall potential Φ_w; emitter temperature T, electron emission density n_{0+}^e, ion emission density n_{0+}^i (or, equivalently, the neutralization parameter γ). Along the *insulating* walls between the electrodes and the conducting wall, the potential is assumed to vary linearly. The charged-particle emission distribution functions, $f^s(z=0, |r| \leq R_e, v_z \geq 0, v_n \geq 0)$, are again of the form (34). All particles reaching the collector, the emitter or the surrounding wall are absorbed. While the ion trajectories are obtained by integrating the full equations of motion, the electron ones are calculated in a *guiding-center* approach.

Figure 3(b) shows a typical 2d potential distribution for a cylindrical system, and Fig.

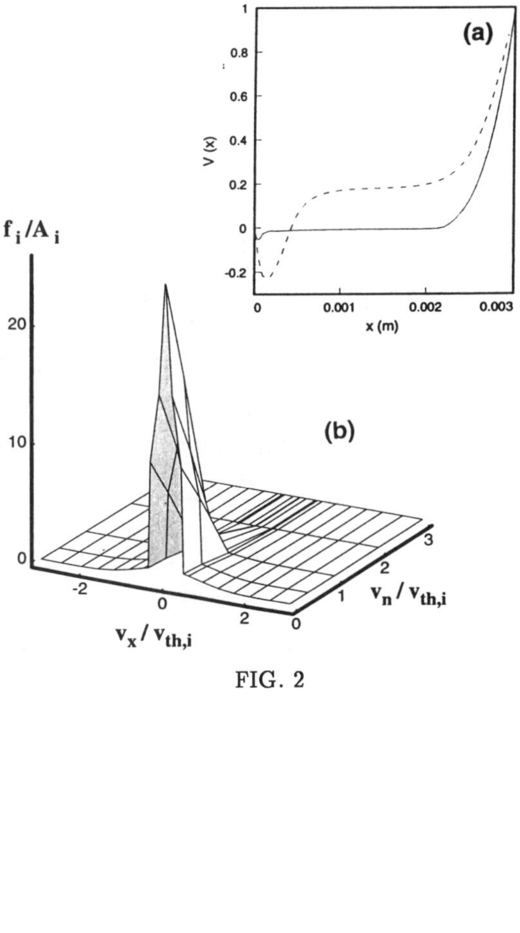

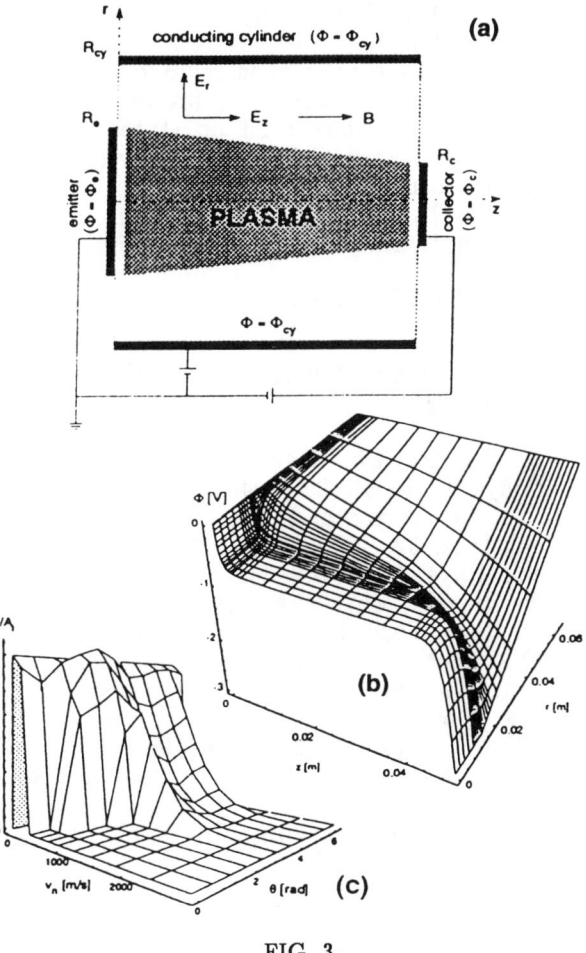

FIG. 2

FIG. 3

3(c) gives a typical ion velocity distribution function for a position lying within the plasma—vacuum transition region. These results demonstrate the high resolution the trajectory—simulation method provides in both configuration and velocity space.

7. CONCLUSIONS AND PERSPECTIVES

In this paper we have reviewed the trajectory—simulation approach to BPS kinetic theory, and have demonstrated that it permits one to model, within certain limits, even nontrivial BPS's in great detail and with high resolution in both configuration and velocity spaces. In this sense it may be viewed as being *complementary* to particle simulation.

In spite of the progress achieved recently, much more work is needed in order to develop the trajectory—simulation approach to a level of flexibility and applicability comparable to that of particle simulation. In the author's opinion, the following problems could be tackled next: (i) Further optimization of the iterative scheme of Sec. 4; (ii) inclusion of more types of collisions (up to now, only charge—exchange collisions have been included, cf. Sec. 5); (iii) linear eigenmode and stability analysis of collisional and/or 2d dc states, taking into account external—circuit properties; (iv) small—amplitude nonlinear periodic oscillations of collisional and/or 2d systems, taking into account external—circuit properties.

ACKNOWLEDGMENTS

This work was partly supported by the Austrian Science Foundation (FWF) under contract P9641.

[1] L.D. Landau, Zh. Eksp. Teor. Fiz. **16**, 574; J. Phys. USSR **10**, 25 (1946).
[2] R.Ya. Kucherov and L.E. Rikenglaz, Zh. Tekh. Fiz. **32**, 1275 (Oct 1962); Sov. Phys. Tech. Phys. **7**, 941 (1963).
[3] F.G. Baksht, Zh. Tekh. Fiz. **34**, 926 (1964); Sov. Phys. Tech. Phys. **9**, 713 (1964).
[4] R.Ya. Kucherov and Yu.A. Shuander, Zh. Tekh. Fiz. **34**, 66 (1964); Sov. Phys. Tech. Phys. **9**, 50 (1964).
[5] N. Rynn, Phys. Fluids **1**, 165 (1966).
[6] H.J. Fahr and K.G. Müller, Zeitschr. Physik **200**, 343 (1967).
[7] W. Ott, Zeitschr. Naturf. **22A**, 1057 (1967).
[8] C.W. Parker and E.C. Whipple, Jr., Ann. Phys. (USA) **44**, 126 (1967)
[9] V.I. Babanin and A.Ya. Ender, Zh. Tekh. Fiz. **44**, 102 (1974); Sov. Phys. Tech. Phys. **19**, 60 (1974).
[10] S. Sinha and J.N. Bardsley, Phys. Rev. A **14**, 104 (1976).
[11] S. Kuhn, Phys. Rev. A **22**, 2460 (1980).
[12] S. Kuhn, Plasma Phys. **23**, 881 (1981).
[13] V.I. Kuznetsov and A.Ya. Ender, Zh. Tekh. Fiz. **53**, 2329 (1983); Sov. Phys. Tech. Phys. **28**, 1431 (1983).
[14] S. Kuhn, Phys. Fluids **27**, 1821 (1984).
[15] S. Kuhn, Phys. Fluids **27**, 1834 (1984).
[16] S. Kuhn, 1987 Intl. Conf. on Plasma Phys., Kiev, 1987, Inv. Papers **2**, 954.
[17] S. Kuhn, 3rd Symp. on Plasma Double Layers and Related Topics, Bucharest, 1987. Analele Ştiinţifice ale Universităţii "Al. I. Cuza" DIN Iaşi, Vol. **34**—Supplement, p. 25.
[18] W.S. Lawson, J. Comput. Phys. **80**, 253 (1989).
[19] M. Hörhager and S. Kuhn, Phys. Fluids B **2**, 2741 (1990).
[20] T.L. Crystal, P.L. Gray, W.S. Lawson, C.K. Birdsall, and S. Kuhn, Phys. Fluids B **3**, 244 (1991).
[21] A.V. Zhykharsky, Phys. Scripta **44**, 606 (1991).
[22] F. Bauer and H. Schamel, Physica D **54**, 235 (1992).
[23] S. Kuhn, Fourth Symposium on Double Layers and Other Nonlinear Potential Structures in Plasmas, Innsbruck, Austria, 1992, p. 65.
[24] H. Maier and S. Kuhn, 1992 International Conference on Plasma Physics, Innsbruck, Austria, 1992, Contributed Papers III, 2001.
[25] A.V. Zhykharsky, Phys. Scripta **45**, 100 (1992).
[26] A.V. Zhykharsky and V.N. Shmykov, Fourth Symposium on Double Layers and Other Nonlinear Potential Structures in Plasmas, Innsbruck, Austria, 1992, p. 425.
[27] F. Greiner, T. Klinger, H. Klostermann, and A. Piel, Phys. Rev. Lett. **70**, 3071 (1993).
[28] J.P. Verboncoeur, M.V. Alves, V. Vahedi, and C.K. Birdsall, J. Comput. Phys. **104**, 321 (1993).
[29] S. Kuhn, Contrib. Plasma Phys. **34**, 495 (1994).
[30] H. Maier and S. Kuhn, 1994 Intl. Conf. on Plasma Physics, Foz do Iguaçu, Brazil, 1994, Contributed Papers **3**, p. 45.
[31] H. Pedit and S. Kuhn, Phys. Plasmas **1**, 13 (1994)
[32] H. Pedit and S. Kuhn, 1994 Intl. Conf. on Plasma Physics, Foz do Iguaçu, Brazil, 1994, Contributed Papers **3**, p. 209.
[33] N. Schupfer, S. Kuhn, A.S. de Assis, and M.A.M. Santiago, *ibidem*, **2**, p. 143.

AUTORESONANT WAVE INTERACTIONS IN NONUNIFORM PLASMAS

L. Friedland

Racah Institute of Physics, Hebrew University, 91904 Jerusalem, Israel

The theory of the autoresonance is outlined for the case of resonant two-wave interactions (mode conversion) in a nonuniform, weakly nonlinear plasma. The autoresonance is encountered when the incident wave passes the region, where it resonates with another wave supported by the medium. The new wave is then excited and the incident wave amplitude changes accordingly. The autoresonance proceeds from the characteristic trapping into the resonance in the vicinity of the linear resonance point followed by the self-preservation of the approximate nonlinear resonance relation in an extended plasma region. The broadening of the resonant region may significantly enhance the mode conversion efficiency as compared to that in the linear mode conversion case. The effect of the dissipation on the autoresonance and a multidimensional generalization are discussed.

I. INTRODUCTION

The autoresonance is a fascinating nonlinear phase locking phenomenon characteristic of resonantly perturbed nonlinear oscillating systems with adiabatically varying parameters. The simplest model problem exhibiting the autoresonance is that of the nonlinear oscillator driven by a resonant quasi-harmonic perturbation with a slowly varying frequency $\omega(t)$. The isolated resonance Hamiltonian [1] $H(I,\theta) = H_0(I) + \varepsilon g(I)\cos[\theta - \int \omega(t)dt]$ for this system expressed in terms of the action-angle variables (I,θ) of the unperturbed oscillator (ε being a small dimensionless parameter measuring the strength of the perturbation) yields the following set of the evolution equations for the action and the phase shift $\Phi = \theta - \int \omega(t)dt$:

$$dI/dt = \varepsilon g \sin\Phi, \quad d\Phi/dt = \Omega(I) - \omega(t) + \varepsilon(dg/dI)\cos\Phi. \quad (1)$$

Here $\Omega = dH_0/dI$ is the frequency of the unperturbed oscillator and system (1) differs from that describing the classical nonlinear resonance [1] only by a slow variation of ω with time. The essence of the autoresonance phenomenon is that despite the time variation of ω, under certain conditions, Eqs.(1) preserve [to $O(\sqrt{\varepsilon})$] the approximate resonance relation $\omega(t) \approx \Omega[I(t)]$, if this relation is satisfied initially.

The idea of the autoresonance was proposed by McMillan [2] and Veksler [3], and developed at early stages by Bohm and Foldy [4] in applications to relativistic particle accelerators. The term *phase stability principle* was used to describe the autoresonance in these studies. The synchrotron, synchrocyclotron [5] and later the GYRAC [6] and the SAC [7] accelerator are based on the autoresonance idea. Recently, similar ideas were applied in atomic physics [8], intense plasma wave excitation by chirped laser pulses [9], nonlinear dynamics [10-12], mode conversion [13] and resonant 3-wave interactions [14].

This work describes and further develops the concept of the autoresonance in the mode conversion representing the simplest case of wave interactions in a nonuniform plasma.

II. DYNAMIC AUTORESONANCE

The autoresonance in the aforementioned driven nonlinear oscillator problem (the term *dynamic autoresonance* or *DAR* was suggested for this case [8]) serves as the prototype for the theory of autoresonant wave interactions. Thus, we shall start by a brief review of the theory of the DAR. Consider an oscillator described by system (1), where $g = const$ and $H_0 \equiv MI^2/2$. Then, by defining $I_0(t) \equiv \omega(t)/M$ and introducing a new action variable $\delta I \equiv I - I_0$, one obtains the following evolution equations

$$d(\delta I)/dt = \varepsilon g \sin\Phi - M^{-1} d\omega/dt, \quad d\Phi/dt = M\delta I. \qquad (2)$$

This system can be viewed as generated by the effective Hamiltonian

$$H_{\mathit{eff}} = (M/2)(\delta I)^2 + \varepsilon g \cos\Phi + M^{-1}(d\omega/dt)\Phi, \qquad (3)$$

describing (in the $d\omega/dt = const$ case) the motion of a "quasi-particle" of "mass" M in a *stationary* potential of form $V_{\mathit{eff}}(\Phi) = \varepsilon g \cos\Phi + M^{-1}(d\omega/dt)\Phi$. Then, if

$$|(\varepsilon g M)^{-1}(d\omega/dt)| < 1, \qquad (4)$$

there exist regions of trapped trajectories in the $(\delta I, \Phi)$ phase space, in which the "quasi-particle" performs nonlinear oscillations around stable equilibrium points $(\delta I_e = 0, \Phi_e)$, where $\sin\Phi_e \equiv (\varepsilon g M)^{-1}(d\omega/dt)$, while $\varepsilon g \cos\Phi_e < 0$. The characteristic width of such a region is small ($\Delta I < |\varepsilon g/M|^{1/2}$). Therefore, to $O(\varepsilon^{1/2})$, the solution of the original problem is $I \approx I_0(t)$. This result can be interpreted as a continuing, automatic preservation of the nonlinear resonance relation $\omega(t) \approx \Omega[I(t)] \approx MI_0(t)$ in our system despite the time variation of the driving frequency, provided the "quasi-particle" starts in a trapped region of the phase space, i.e., satisfies the approximate resonance condition initially. This is the simplest example of the DAR and (4) can be viewed as the necessary (adiabaticity) condition for the DAR in the system. This condition must be supplemented by the *moderate nonlinearity* conditions $\Delta I/I << 1$, and $\Delta\Omega/\Omega = (d\Omega/dI)\Delta I/\Omega << 1$ of the theory of the nonlinear resonance [1], or, in our case,

$$|\varepsilon g/MI_0^2|^{1/2} << 1, \qquad (5)$$

so I_0 must be sufficiently large in the autoresonance. We have seen that the DAR is accompanied by oscillations around the exact nonlinear resonance. The frequency of these oscillations is small and of order $v_{\mathit{eff}} \sim O(|\varepsilon g M|^{1/2})$. Finally, note that the general DAR problem described by Eqs.(1) is similar to that discussed above and is characterized by the Hamiltonian of form (3), where M and g are replaced by $d\Omega/dI_0$ and $g(I_0)$ respectively and, therefore, vary (slowly) in time. Since trapped trajectories of the oscillator remain trapped under the *adiabatic* variation of the parameters, all the general features of the autoresonance are preserved in the general case, but the "quasi-particle" will perform nonlinear oscillations around the exact resonance with an *adiabatically varying* amplitude. This amplitude variation can be found from the theory of adiabatic invariants.

III. SPATIAL MODE CONVERSION AND TRAPPING INTO THE RESONANCE

In this Section we shall exploit the similarity between the DAR equations and those governing the spatial mode conversion in a nonuniform plasma. The weakly nonlinear mode conversion is a simplest two-wave resonant process described (in a one-dimensional case) by the following system of coupled, conservative, slow amplitude transport equations [13]:

$$V^a dA_a / dx + \Gamma_a A_a + i\delta D_a A_a = iH p_a A_b \exp[i\kappa x^2 / 2],$$
$$V^b dA_b / dx + \Gamma_b A_b + i\delta D_b A_b = iH^* p_b A_a \exp[-i\kappa x^2 / 2]. \quad (6)$$

Here $V^{a,b}$ are the components of the group velocities of the two waves in the direction of the nonuniformity (x-direction), $\Gamma_{a,b} = (1/2)dV^{a,b}/dx$, H is the complex linear coupling coefficient, $\delta D_{a,b} = C_{a,b}|A_{a,b}|^2$ are the lowest order nonlinear corrections to the wave dispersion functions and $p_{a,b}$ are the wave energy signs. The phases in the rhs of Eqs.(6) represent the nonuniformity of the medium and comprise the expansion of the fast phase mismatch between the waves near the resonance point (x=0), where the two modes have the same frequencies and wave vectors. Note that (6) conserves the total action flux (the Manley-Rowe relation):

$$J = p_a V^a |A_a|^2 + p_b V^b |A_b|^2 = const. \quad (7)$$

At this point, we assume $\kappa, J, V_{a,b} > 0$ and transform to dimensionless amplitudes $A'_a = (V^a/J)^{1/2} A_a$ and $A'_b = (V^b/J)^{1/2} A_b \exp[i\kappa x^2/2 + iArg(H)]$. Then (6) yields

$$dA'_a / d\xi + ic_a|A'_a|^2 A'_a = i\eta p_a A'_b,$$
$$dA'_b / d\xi - i(\xi - c_b|A'_b|^2) A'_b = i\eta p_b A'_a, \quad (8)$$

where we defined the dimensionless coordinate $\xi = \kappa^{1/2} x$ and parameters $\eta = (V^a V^b \kappa)^{-1/2}|H|$ and $c_{a,b} = [(V^{a,b})^2 \kappa^{1/2}]^{-1} C_{a,b}$. Next, we write $A'_{a,b} = B_{a,b} \exp(i\Phi_{a,b})$, where $Im B_{a,b} = 0$ and use (8) in deriving three *real* equations for $B_{a,b}$ and the phase difference $\Phi \equiv \Phi_b - \Phi_a$:

$$dB_a / d\xi = -\eta p_a B_b \sin\Phi,$$
$$dB_b / d\xi = +\eta p_b B_a \sin\Phi, \quad (9)$$
$$d\Phi / d\xi = \xi - c_b B_b^2 + c_a B_a^2 - \eta(p_a B_b / B_a - p_b B_a / B_b)\cos\Phi.$$

The Manley-Rowe relation now becomes $p_a B_a^2 + p_b B_b^2 = 1$. Note that because of this relation we have only two independent equations in (9). For example, one can express B_a in terms of B_b and view the second and the third equation in (9) as a complete set. The similarity of this set with system (1) for the driven oscillator is obvious and one can expect the *spatial* autoresonance (or SAR) in the mode conversion problem. The approximate nonlinear resonance relation

$$\xi - c_b B_b^2 + c_a B_a^2 \approx 0 \quad (10)$$

will be preserved in this case to $O(\eta^{1/2})$, if the following necessary autoresonance conditions are fulfilled [compare to Eqs.(4) and (5)]:

$$|2\eta B_a B_b (c_b p_a + c_a p_b)|^{-1} < 1, \qquad (11)$$

$$(\eta B_{b,a} / B_{a,b})^{1/2} (c_b p_a + c_a p_b)^{-1/2} \ll 1. \qquad (12)$$

We observe that the SAR yields simple approximate spatial dependencies of the wave amplitudes. Indeed, from the Manley-Rowe relation and Eq.(10) we find:

$$B_a^2 \approx (c_b - p_b \xi) / (c_a p_b + c_b p_a), B_b^2 \approx (c_a + p_a \xi) / (c_a p_b + c_b p_a). \qquad (13)$$

Thus, as functions of ξ, $B_{a,b}^2$ are straight lines and the slopes of these lines may differ only by signs (when the energy signs of the waves are different). This is a much simpler dependence than that in the *linear* mode conversion problem, where the waves are described by the parabolic cylinder functions [15]. Therefore, the autoresonance comprises a rare example, where the addition of the nonlinearity simplifies the solution significantly. Finally, we observe that, as always in the autoresonance, the wave amplitudes in the SAR regime exhibit spatial oscillations around the exact nonlinear resonance solutions (13). These oscillations are described by the effective Hamiltonian of form (3) and their characteristic frequency is of order

$$v_{\text{eff}} \sim (2\eta |c_b p_a + c_a p_b| B_a B_b)^{1/2}. \qquad (14)$$

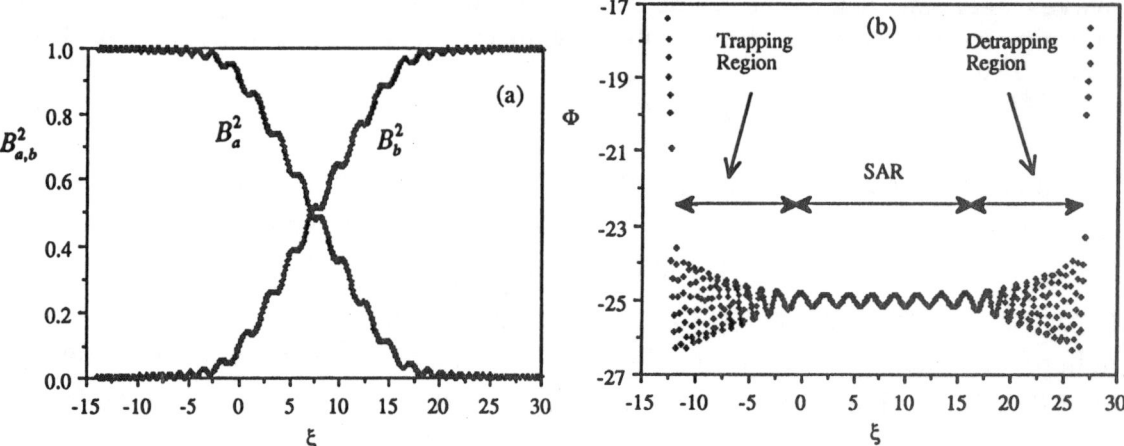

Fig.1 The spatial autoresonance (SAR) and the trapping into the resonance in mode conversion: (a) wave action fluxes vs. ξ; (b) phase difference Φ vs. ξ.

Now we proceed to the illustration of the SAR in the mode conversion. Fig.1a shows the numerical solution [13] of Eqs.(9) for the fluxes $B_{a,b}^2$ in the positive energy mode interaction case, subject to initial conditions $B_b = 0.077$ ($B_a = 0.997$), $\Phi = 0$ at $\xi = -14$ and the parameters $\eta = 0.5, c_a = 0, c_b = 15$. Note that we do not start in resonance initially. Nevertheless, one finds the autoresonant solutions (13) (the linear averaged dependence of $B_{a,b}^2$ on ξ) in the region $0 < \xi < 16$ in Fig.1a. This surprising result, seemingly contradicting our previous requirement of starting in resonance, is explained as follows. While starting off resonance, we have assigned, at the same time, a relatively *small* initial value to one of the amplitudes (mode b). This was also in violation of the autoresonance conditions (11) and (12). Nevertheless, in this case, the term with $\cos\Phi$ in the third equation in (9) becomes important despite the smallness of η and our previous ordering assumptions in (9) may be violated. It was shown in Ref.[13] that this singularity in the third equation in (9) leads to a new and important effect of the *trapping* into the resonance as one approaches the resonant region. It was shown that the smaller the initial value of one of the amplitudes, the stronger is the inequality (12) after the trapping. We illustrate this trapping effect in Fig.1b, showing the relative phase between the waves in our example. We see in the Figure that after the transition period, the relative phase is trapped at $\Phi \approx -8\pi$, while B_b becomes large enough to satisfy the autoresonance conditions (11) and (12) at $\xi \approx 0$. Beyond this point the SAR sets in and continues until the initial action flux is almost completely transferred to mode b. The amplitude of mode a then becomes small, leading to the gradual phase detrapping as the autoresonance conditions are violated.

The trapping into the resonance is a very important phenomenon. It allows to conveniently enter the autoresonant interaction stage even by starting outside the resonance. This finding [13] may be of great help in experiments (accelerators [6,7] being a good example) since (a) due the small $[O(\eta^{1/2})]$ width of the nonlinear resonance it may be difficult to tune the experimental conditions to the resonance initially and (b) the strong trapping effect removes the system from the separatrix region, where the phase detrapping may destroy the autoresonant interaction.

IV. NON-CONSERVATIVE SPATIAL AUTORESONANCE

Now, we shall consider the following non-conservative generalization of (9):

$$dB_a^2 / d\xi = -2\gamma_a B_a^2 + 2\eta p_a B_a B_b \sin\Phi,$$
$$dB_b^2 / d\xi = -2\gamma_b B_b^2 - 2\eta p_b B_b B_a \sin\Phi, \quad (15)$$
$$d\Phi / d\xi = \xi - c_b B_b^2 + c_a B_a^2,$$

where $\gamma_{a,b}$ are the linear spatial dissipation (or growth) rates and we have neglected the $O(\eta)$ term with $\cos\Phi$ in the third equation. It is also assumed that, initially, one starts in the vicinity of the resonance, i.e., $\xi - c_b B_b^2 + c_a B_a^2 \approx 0$. Now one must solve three and not just two equations as in the conservative mode conversion case. Nevertheless, for sufficiently small $\gamma_{a,b}$, we can proceed as follows. We differentiate the third equation in (15) and use the first two equations to obtain

$$d^2\Phi / d\xi^2 = 1 + 2(\gamma_b c_b B_b^2 - \gamma_a c_a B_a^2) + 2\eta(p_a c_a + p_b c_b) B_a B_b \sin\Phi. \quad (16)$$

Next, we seek quasi-equilibrium solutions $B_{a,b0}, \Phi_0$, such that

$$\xi - c_b B_{b0}^2 + c_a B_{a0}^2 \approx 0, \quad (17)$$

i.e., $d^2\Phi_0 / d\xi^2 \approx 0$ continuously. Then, from (16),

$$2\eta B_{a0} B_{b0} \sin\Phi_0 \approx -[1 + 2(\gamma_b c_b B_b^2 - \gamma_a c_a B_a^2)](p_b c_b + p_a c_a)^{-1} \quad (18)$$

and, after substituting (18) into the first two equations in (15), we obtain

$$dB_{a0}^2 / d\xi = -2c_b(c_a p_a + c_b p_b)^{-1}(\gamma_a p_b B_{a0}^2 + \gamma_b p_a B_{b0}^2) - p_a(c_a p_a + c_b p_b)^{-1},$$
$$dB_{b0}^2 / d\xi = -2c_a(c_a p_a + c_b p_b)^{-1}(\gamma_a p_b B_{a0}^2 + \gamma_b p_a B_{b0}^2) + p_b(c_a p_a + c_b p_b)^{-1}. \quad (19)$$

Finally, if one defines $Z \equiv \gamma_a p_b B_{a0}^2 + \gamma_b p_a B_{b0}^2$, Eq.(19) yields
$$dZ / d\xi = \alpha - \beta Z, \qquad (20)$$
where $\alpha = p_a p_b (\gamma_b - \gamma_a)(c_a p_a + c_b p_b)^{-1}$ and $\beta = 2(p_b c_b \gamma_a + c_a p_a \gamma_b)(c_a p_a + c_b p_b)^{-1}$. Thus,
$$Z(\xi) = (\alpha / \beta) + [Z(0) - \alpha / \beta] \exp(-\beta \xi), \qquad (21)$$
where $Z(0)$ is the value at the initial integration point (assumed to be at 0). By using this result and Eq.(17) we find that the fluxes $B_{a,b0}^2$ in the non-conservative mode conversion case comprise a nontrivial combination of linear and exponential functions of ξ. The study of the stability of these quasi-equilibrium [autoresonant, see Eq.(17)] solutions will be reported elsewhere. Here, we shall illustrate this stability by presenting a numerical example. Fig.2 shows the same case as Fig.1, but with $\gamma_{a,b} = -0.015$. Therefore, $\alpha = 0, \beta = 2\gamma_a$ and, as before, $B_{b0}^2 = \xi / c_b$. However, the competition between the linear and exponential spatial dependencies is seen in the evolution of B_{a0}^2 in Fig.2a. The trapping stage and stable oscillations around $B_{a,b0}^2$ in our example are seen in Fig.2b. Finally, we observe that the autoresonance continues indefinitely in this case.

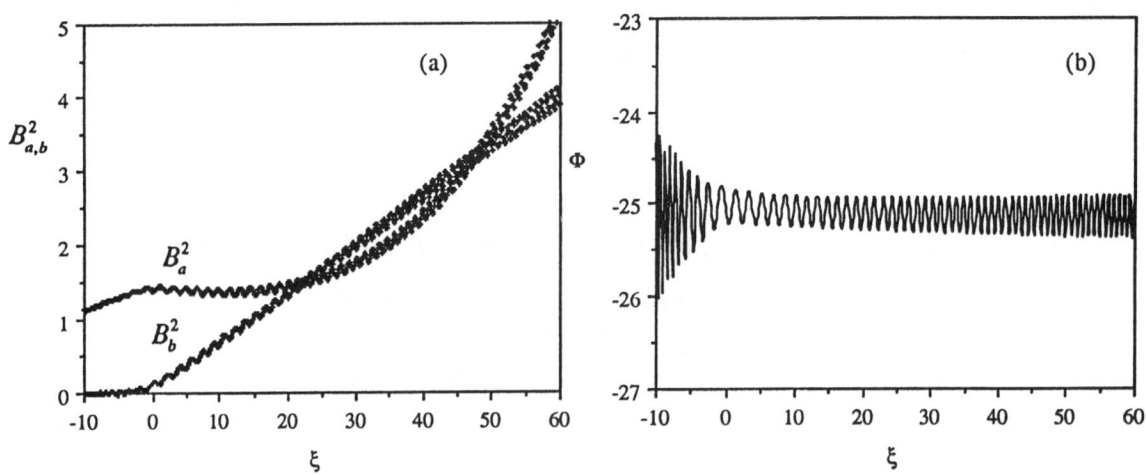

Fig.2 Non-conservative autoresonance: (a) (a) $B_{a,b}^2$ vs. ξ; (b) Φ vs. ξ.

V. MULTIDIMENSIONAL SPATIAL AUTORESONANCE

The one-dimensional theory of the autoresonant mode conversion (Sections III, IV) was based on the possibility of reducing the problem to the aforementioned driven nonlinear oscillator model. There exists, however, an important difference between the autoresonance associated with the oscillator dynamics and that in resonant wave interactions. While in the former case the system evolves in *time*, the waves may propagate in the *four-dimensional* space-time. Nevertheless, until recently, the autoresonant wave interactions were studied in media with a one-dimensional nonuniformity and, the coordinate in the direction of the nonuniformity played the role of time in treating the wave evolution similarly to the dynamics of the driven nonlinear oscillator.

Very recently [16] we have performed the first study of the multidimensional *mode conversion* problem described by the following generalization of (6):

$$V_i^a \partial_{x_i} A_a + \Gamma_a A_a + i\delta D_a A_a = iHp_a A_b \exp[i\kappa_{ij} x_i x_j / 2], \quad (22)$$
$$V_i^b \partial_{x_i} A_b + \Gamma_b A_b + i\delta D_b A_b = iH^* p_b A_a \exp[-i\kappa_{ij} x_i x_j / 2].$$

We have shown that when one of the waves is not excited externally, similarly to the one-dimensional case, the phase trapping phenomenon takes place as the second wave (the incident mode) passes the resonance region and the system enters the autoresonant stage of interaction despite the multidimensionality of the problem. An example of such a multidimensional autoresonance is given in Fig.3a, showing the fluxes $B_{a,b}^2 \equiv |A_{a,b}|^2$ in a two-dimensional, positive-negative energy ($p_b = -p_a = 1$) mode conversion problem [15], described by [compare to (8)]:

$$\partial A_a' / \partial x + ic_a |A_a'|^2 A_a' = -i\eta p_a A_b', \quad (23)$$
$$\partial A_b' / \partial y - i(q_x x + q_y y - c_b |A_b'|^2) A_b' = i\eta p_b A_a'.$$

Mode a *only* is launched from the left boundary in the Figure, the boundary conditions are uniform, $q_x = -0.45, q_y = -0.89, \eta = 0.15, c_a = 0.5$ and $c_b = -15$.

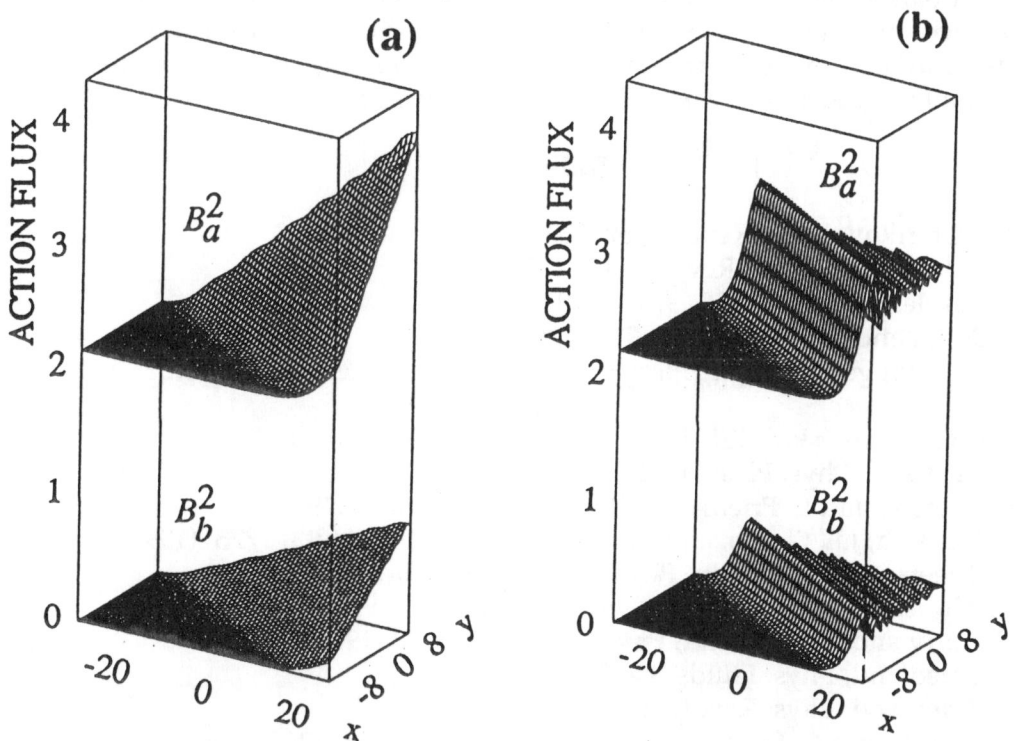

Fig.3 Two-dimensional mode conversion: (a) autoresonant interaction (b) linear mode conversion

We can see in Fig.1a that mode b is excited beyond a certain line in the x-y plane and the system enters the spatial autoresonance regime in which both $B_{a,b}^2$ are nearly linear functions of x and y, as the spatial and nonlinear dispersions are in balance (the waves are in the nonlinear resonance) in an extended plasma region. The interaction is accompanied by the characteristic of the autoresonance spatial oscillations around the exact nonlinear resonance. For comparison, Fig.1b shows the same example, but for the linear mode conversion case ($c_{a,b}=0$) where, in contrast to the autoresonance, the interaction takes place only in a narrow region in space and saturates rapidly due to the luck of the nonlinear phase shifts capable of balancing the growing spatial phase mismatch beyond the excitation line. Thus, presently, we have an evidence and a first theory [16] of the multidimensional autoresonant mode conversion.

VI. CONCLUSIONS

In conclusion: (i) We have presented and illustrated the theory of the autoresonant spatial mode conversion in a nonuniform medium. The autoresonance phenomenon is the manifestation of the self-preserved balance between the nonlinear and spatial dispersion effects. This balance increases the width of the resonant interaction region and may enhance the wave interaction efficiency. (ii) Our theory was based on the analog to the dynamic autoresonance effect in a resonantly perturbed nonlinear oscillator with adiabatically varying parameters. (iii) We have discussed the necessary conditions for the spatial autoresonance and described the important possibility of the trapping into the resonance. (iv) The effect of the dissipation or/and growth on the autoresonant mode conversion was treated in detail, and (v) recent results on the multidimensional mode conversion were discussed.

REFERENCES

1. B.V. Chirikov, Phys. Reports, **52**, 263 (1979).
2. E.M. McMillan, Phys. Rev., **68**, 143 (1945).
3. V. Veksler, J. Phys. USSR, **9**, 153 (1945).
4. D. Bohm and L. Foldy, Phys. Rev., **70**, 249 (1946).
5. M.S. Livingston, *High-Energy Accelerators* (Interscience Publ., New York, 1954).
6. K. S. Golovanivski, IEEE Trans. Plasma Sci., **PS-11**, 28 (1983).
7. L. Friedland, Phys. Plasmas, **1**, 421 (1994).
8. B. Meerson and L. Friedland, Phys. Rev. A **41**, 5233 (1990).
9. M. Deutsch, J.E. Golub, and B. Meerson, Phys. Fluids **B3**, 1773 (1991).
10. B. Meerson and S. Yariv, Phys. Rev. A **44**, 3570 (1991).
11. G. Cohen and B. Meerson, Phys. Rev. E **2**, 967 (1993).
12. S. Yariv and L. Friedland, Phys. Rev. E **48**, 3072 (1993).
13. L. Friedland, Phys. Fluids B **4**, 3199 (1992).
14. L. Friedland, Phys. Rev. Lett. **69**, 1749 (1992).
15. R. Cairns and C. Lashmore-Davies, Phys. Fluids **26**, 1268 (1983).
16. L. Friedland (to be published).

Scenario of Self–Organization

Tetsuya Sato

and

Complexity Simulation Group [†]

Theory and Computer Simulation Center

National Institute for Fusion Science

Nagoya, 464-01 Japan

Abstract

With the aim of obtaining a grand view of self-organization, we have made an extensive computer simulation study of complex plasmas. We assert that energy pumping, entropy (superfluity) expulsion, and nonlinearity are the three key processes governing self-organization. An open system exhibits characteristically different self-organization, depending on whether the energy pumping is instantaneous or continuous, or whether the produced entroy is expulsed or reserved. The nonlinearity of the system acts to bring a far-from-equilibruium state into a bifurcation point whereby a new ordered state is realized and an entropy production rate is anomalously enhanced.

[†]R. Horiuchi, K.Watanane, T.Hayashi, Y.Todo, T.H.Watanabe, A.Kageyama, H. Takamaru, H. Amo, S. P. Zhu, B. Dasgupta, and S. Bazdenkov

1. Introduction

It may not be an exaggeration to say that the most mysterious process of our universe is the creation of an ordered structure. Is there any universal law in creating the orderliness? This is the central problem of the Science of Complexity.[1]

Taylor's variational theory has casted light on the study of self-organization in plasmas.[2] The attractiveness and impact of the Taylor's theory lies in the conservation of the magnetic helicity and the selective dissipation of the magnetic energy. Because of this conceptual attractiveness plasma physisists have paid an excessive attention to the selective dissipation of the magnetic energy and the conservation of the magnetic helicity, and thought narrowly that self-organization is equivalnt to the dynamical process leading to the minimum energy state under a certain constraint. In this paper, without being enmeshed in such a peculiar feature as the helicity conservation, we wish to aim at broadering and generalizing the concept of self-organization.

We propose here that self-organization is described by three key processes. They are the energy input, the entropy expulsion and the nonlinearity in an open system. The first two processes are concerned with the information exchanges with the external world and the third one is the intrinsic internal process of the system.

2. Self-Organization by Instantaneous Energy Pumping

We start with a Taylor self-organization process which takes place in an instantaneously pumped-up system. Given initially a higher force-free energy state which is far beyond a minimum energy state, we solve its nonlinear evolution. We consider a rectangular box in which the side boundaries of the box is made of conductors and the axial boundaries are periodic. Fig. 1 shows a structural evolution of the magnetic field [3]. The final spiral structure is the self-organized one. The evolutions of the magnetic helicity and energy are given in Fig. 2. It is observed that the relaxation occurs in two steps for the magnetic energy but not for the magnetic helicity. The two-step relaxation reflects the fact that there was another force-free state between the given initial state and the final minimum energy state.

We emphasize here that each stepwise relaxation occurs in about ten of the Alfven transit time τ_A, which is quite a fast process compared with the classical diffusion time (ten thousands of the Alfven transit time). During each relaxation process the magnetic energy is rapidly converted into the thermal energy (entropy). We here assume that the produced entropy is instantaneously expulsed from the system. The main process involved in this energy conversion is the ohmic process. However, the initial magnetic Reynolds number (normalized electrical conductivity) was 10^{-4}. Thus, we can conclude that some nonlinear process, which enormously enhances the conversion rate (thousand times in this case), must have happened. This is realized by the driven magnetic reconnection which is caused by strongly excited kink flows.

In this self-organization process one notices an important fact regarding the Taylor's conjecture. A careful examination of the helicity evolution in Fig. 2 exhibits that the helicity is never conserved in a strict sense, though the conservation has been firmly believed. On looking at the helicity curve in Fig. 2, one can readily recognize that the helicity dissipation exhibits a critical slowing-down in accordance with the stepwise relaxation of the magnetic energy. We note here that both the critical slowing-down of the magnetic helicity and the stepwise relaxation of the magnetic energy are explained by the driven reconnection process.

3. Intermittency in Self-Organization by Continuous Energy Pumping

One may consider that the example given in the above is a typical self-organization process in the sense that a system relaxes in a transient fashion into a steady minimum energy state starting from an initial non-minimum energy state. A question, then, arises as to how such an initial far-from-equilibrium state can be realized without suffering any instability on the way to it. The answer will be that the state is realized if the external pumping of energy into the system is reasonably a fast action and the system cannot react upon the pumping. Namely, the pump-up time scale is shorter than the dynamic response time of the system. We call such a self-organization process as an instantaneously pumped self-organization.

In the instantaneously pumped self-organization process the system relaxes in a stepwise fashion but monotonously into a steady ordered structure. The process is a transient from the initial state to the minimum energy state. In nature, however, the energy pump in a local system from an external world is not always a fast process. It often happens that energy flows gradually and continuously from an external world. In such a situation no Taylor type variational method would be applied. Generally, any mathematical methodology would not be able to predict what will happen in the system. The evolution would be strongly dependent on the nonlinearity of the system and also would be strongly influenced by the processes of information exchange between the system and the external world.

Based on the above consideration, we present a simulation result of a new self-organization process which is stimulated by an external energy reservoir that has an ability to supply a continuous energy flow. The thermal energy generated during the evolution is assumed to be instantaneously pumped out in the same way as in the previous example.

The simulation system is the same rectangular box as the previous simulation. Uniform straight field lines are applied in the axial direction. The axial boundaries are not periodic but are the suppliers of a continuous energy flow. Specifically, at the boundaries a circular motion is steadily applied in a limited region at both ends with the opposite polarity to each other, so that the magnetic field lines are kept on twisting. This problem may find some application in the problem of solar flares.

Fig. 3 illustrates the topological evolution of the twisted field lines starting from one axial boundary. The twisted flux tubes suffer from a knot-of-tension instability and reconnection is driven to release the tension. Consequently, the twisted field lines are connected with the untwisted field lines at the twist-untwist boundary. As the twisting continues, another knot-of-tension instability occurs whereby reconnection advances from the twist-untwist boundary to an outer region in an intermittent fashion. Finally, a linked twist-untwist flux tube reconnects with a companion twist-untwist flux tube to return to an original straight topological state. Thereafter a similar process to what we have observed is repeated. Fig. 4 shows the temporal evolution of the energy conversion rate from the magnetic to kinetic energy (the work done by the J × B force) and the Ponyting flux supplied into the system by the twisting motions at the boundaries. One can clearly see that the magnetic energy is converted into the kinetic energy in bursts ($1 \sim 2\tau_A$) in an intermittent fashion and exhibits a recurrence to the original state after several intermittent bursts. The intermittency and recurrence are known to be the featuring processes for the continuous pumping.

4. Role of Entropy Expulsion on Self-Organization

On top of the way of energy pumping there is another crucial element which would change the fate of the self-organization. In the present examples, we have assumed that the thermal energy produced during the relaxation process is instantaneously pumped out. Knowing that the produced thermal energy is the entropy which is the superfluous quantity for the created ordered structure, it is rather prudent to clarify the role of the entropy on the self-organization.

We present two different simulations. In the first we attempt to examine how seriously the maintenance of the produced entropy in the system influences upon the fate of the self-organization. For this purpose we take as an example the instantaneously pumped MHD self-organization which was previously presented. Without pumping out the produced thermal energy we solve the pressure dynamics self-consistently with the plasma motion and the magnetic field evolution. The thermal conduction effect is discarded.

The simulation result has shown that the magnetic structure evolves in a very similar way to the previous case where the pressure is discarded. Nevertheless, two important new facts have been discovered. The first one is that the structure of the pressure exhibits a very similar structure to that of the magnetic field intensity. An example is shown in Fig. 5.

The other discovery is that the magnetohydrodynamic (MHD) self-organized state is not the force-free minimum energy state, but a force-balanced state, i.e., $\boldsymbol{J} \times \boldsymbol{B} = \nabla p$. the current $\boldsymbol{J}$ consists of the perpendicular and parallel components. As the relaxation proceeds, the parallel (force-free) component decreases, while the perpendicular component increases inversely. They approach nearly to the same level in the self-organized state. This result indicates that unless some pumping out of the thermal energy or a fast perpendicular thermal conduction does exist, the Taylor relaxation does not occur. Instead, the system relaxes to a force-balanced, non-Taylor, minimum energy state.

Next, we attempt to look for what will happen when the produced entropy is filtered out. We take the ion-acoustic double layer for this purpose. What we have obtained is a remarkable and striking new self-organization phenomenon caused by filtering out the produced superfluous entropy from the system. It is well known by particle simulation that weak ion acoustic double layers are generated in a one dimensional collisionless plasma where electrons have a shifted Maxwellian distribution and ions have a normal Maxwellian[4]. In the previous simulations, however, particles were regarded as periodic in one dimensional system. This periodic condition implies that the particles leaving from the downstream boundary enter into the system from the upstream boundary as they are and vice versa, no matter how seriously they are disturbed by the generation of ion acoustic double layers. In other words, the disturbed information, or disorderliness, is retained in the system.

A new particle simulation code is developed in which fresh particles continuously flows into the system from each boundary in place of dirty particles leaving the system[5]. We can thus filter out the superfluous information (entropy), when produced. The result is really striking. A giant double layer is generated. In the initial phase of the ion acoustic instability the normal ion acoustic double layers are generated. They disappear after a while, as we observed in the previous simulation. However, new ion acoustic double layers emerge at different positions. They again disappear after a while. Once again a double layer is generated. At this time, the double layer grows singly and the potential reaches to a surprisingly large amplitude (see, Fig. 6). Compared with the conventional ion acoustic double layer, the potential difference of which is of the order of one electron thermal energy, the potential amplitude of this giant double layer reaches to a surprisingly large level, say, larger than ten times of the thermal energy. This giant double layer has much longer lifetime than that of the normal weak ion acoustic double layer and exhibits a recurrent generation. In view of the fact that the drifting electron energy was only 0.36 of the electron thermal energy in this example, the obtained potential difference is really huge. This giant ion acoustic double layer, which we call "super" ion acoustic double layer, can be an efficient electron accelerator.

The implication of this simulation is of incalculable value. If the system owns a proper filtering function of the superfluous disorderliness, then a well-organized structure can be realized and sustained for a longer period. Also, this example supports a recurrent behavior for a continuous energy flow as we observed for the continuously twisting magnetic flux.

5. Scenario of Self-Organization

We are now in a position to draw a somewhat integrated picture of the grand view of self-organization when a system is pumped up to a far-from-equilibrium state. The evolution exhibits a remarkably different feature depending on the way of energy pumping. When externally pumped up faster than its dynamical response time, the system relaxes in a transient fashion to a steady self-organized state. Upon a continuous supply of an external energy flow, on the other hand, the system exhibits a recurrence in the structure and energy state while repeating intermittent bursts.

It is also seen that whether a superfluous entropy or disorderliness can be pumped out or not, the evolution and the created structure are largely changed. When the entropy or a superfluous material is filtered out, a well-organized structure can be created. The pumping way of energy and the expulsing way of entropy are both closely related to the interrelation of the system with the external world. They are thus considered to be the crucial controlling external elements of the self-organization.

On top of these controlling elements there are several important internal natures that govern the self-organization. They are the instability, bifurcation (phase transition), anomalous dissipation, etc. When a system gains an energy, gradually or instantaneously, from an external world, beyond a threshold (marginal) point, an instability arises. When the instantaneously gained energy or the energy supplied from a continuous energy budget becomes far beyond the instability threshold, the instability develops sufficiently. The nonlinearity is enormously enlarged thereby and gives rise to a large deformation of the structure. The deformation progresses in such a way that the excessively deposited energy is released in the shortest (most effective) way. In order to release it in the shortest way an anomalous dissipation, or an anomalous entropy production rate, must take place in one way or another. Simultaneously, the structure (topology) must be drastically changed, namely, a bifurcation must take place.

In the magnetohydrodynamic self-organization, the driven magnetic reconnection plays these roles at once, namely, the anomalous magnetic energy conversion into the thermal energy and the topological change of the magnetic field. In a collisionless plasma with drifting electrons, the localized anomalous resistivity plays the roles of the current redistribution and the formation of an electrostatic potential structure.

These results and others have led us to an assertion that a natural system tends to boost up itself to a bifurcation (phase transition) point and give rise to an anomalously enhanced dissipation there, or an anomalous entropy production rate, when a far-from-equilibrium state is realized by whatever process it may be.

References

1. Nicolis, G. and Prigogine, I. Exploring Complexity (W. H. Freeman and Co., New York, 1989)
2. Taylor, J. B., Phys. Rev. Lett. 33, 1139-1141 (1974)
3. Horiuchi, R. and Sato, T., Phys. Rev. Lett. 55, 211-213 (1985); Phys. Fluids 29, 1161-1168 (1986); ibid. 29, 4174-4181 (1986); ibid., 31, 1142-1152 (1988)
4. Sato, T. and Okuda, H., Phys. Rev. Lett. 44, 740-743 (1980)
5. Takamaru, H., Sato, T., Horiuchi, R., and Watanabe, K., Physics of High Energy Particles in Toroidal Systems, p 33-43 (edited by Tajima and Okamoto, American Institute of Physics, New York, 1993)

Figure Captions

Fig. 1   A typical example of a magnetohydrodynamic (MHD) self-organization when the system is instantaneously pumped up to a high energy state (far-from-equilibrium).

Fig. 2   Time evolutions of the magnetic energy (W), the magnetic helicity (K) and the ratio (W/K) for the same simulation shown in Fig. 1.

Fig. 3   An example of MHD simulation exhibiting an intermittency in self-organization when the system is in contact with a continuous energy supplier.

Fig. 4   Intermittency in the self-organization of a continuously twisted flux tube. The time evolutions of the Ponyting flux and the work done by the ($J \times B$) force.

Fig. 5   A non-Taylor MHD self-organization in which not only the magnetic field but also the pressure self-organze into a helical sturcuture.

Fig. 6   A particle simulation creating a giant potential structure (super ion acoustic double layer) by filtering out the disturbed part of the outgoing particles.

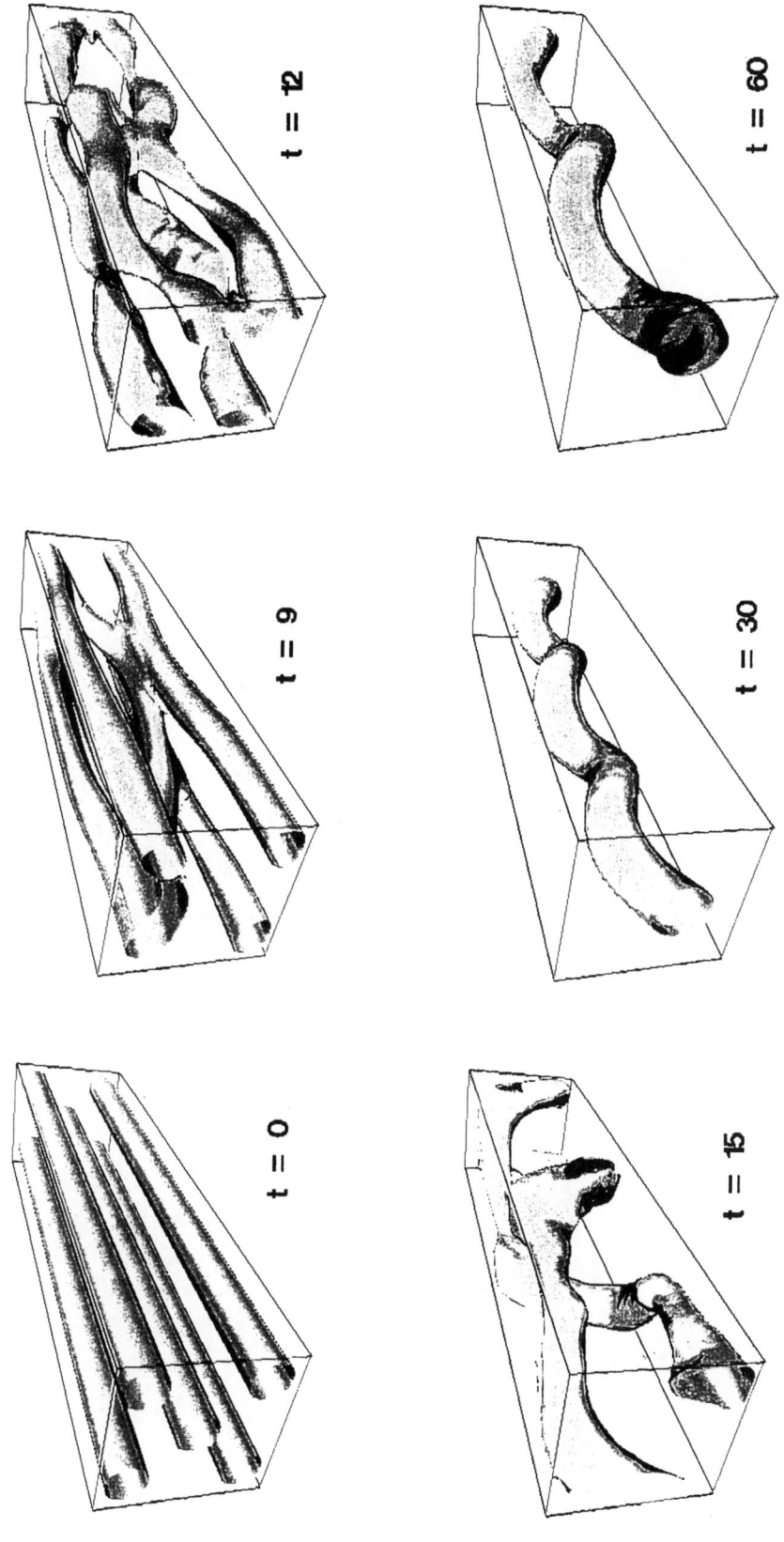

Fig. 1

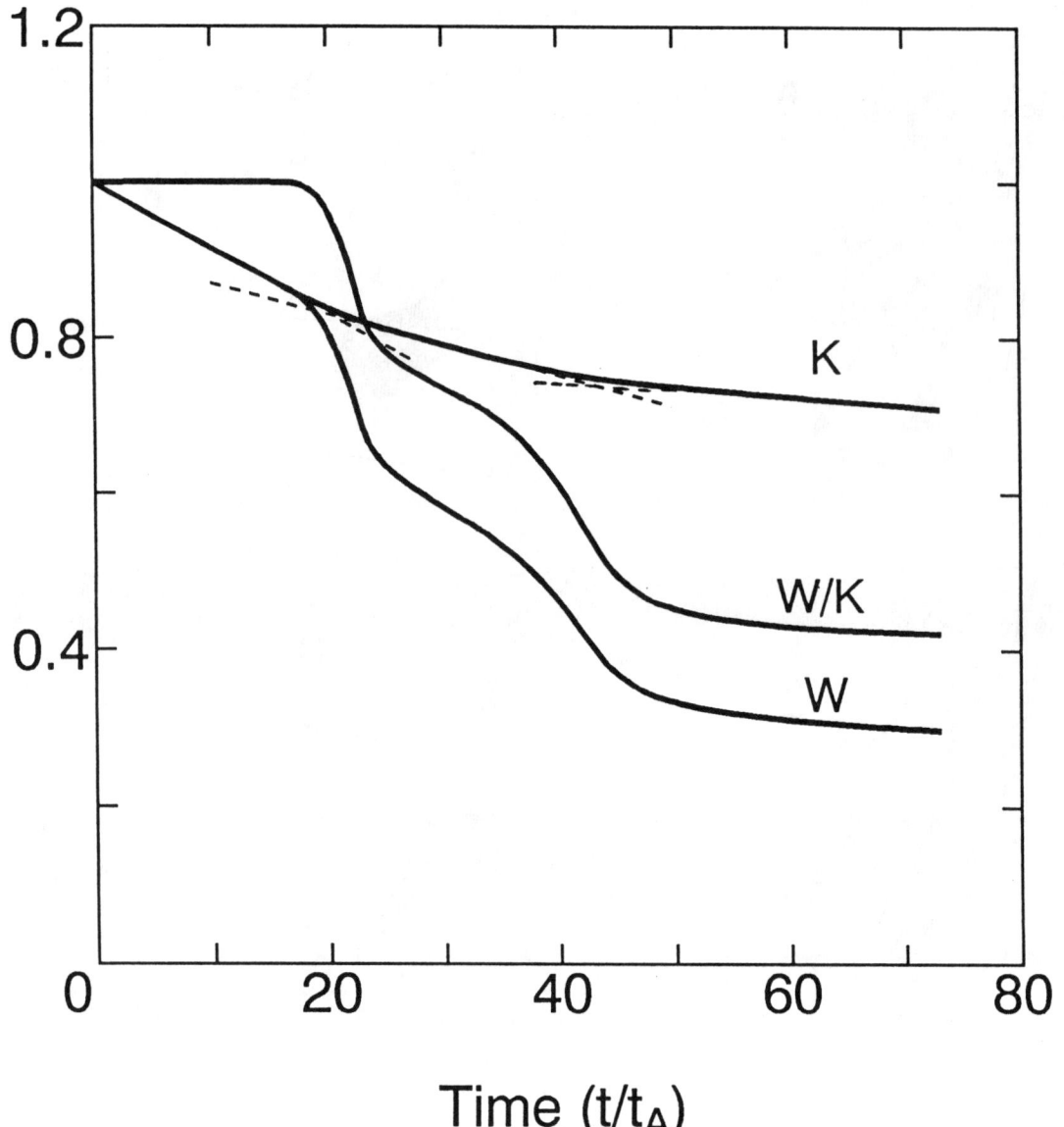

Fig. 2

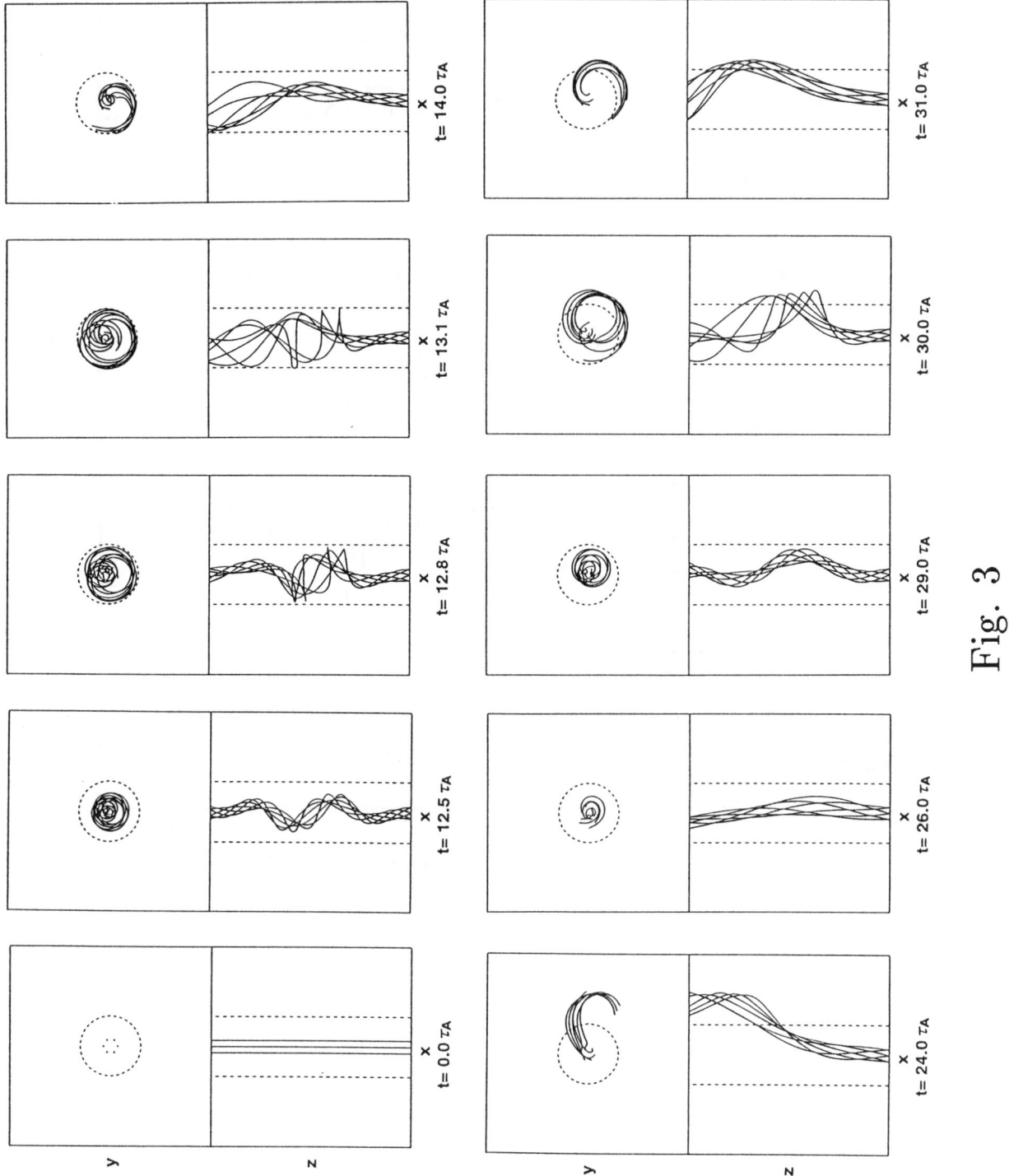

Fig. 3

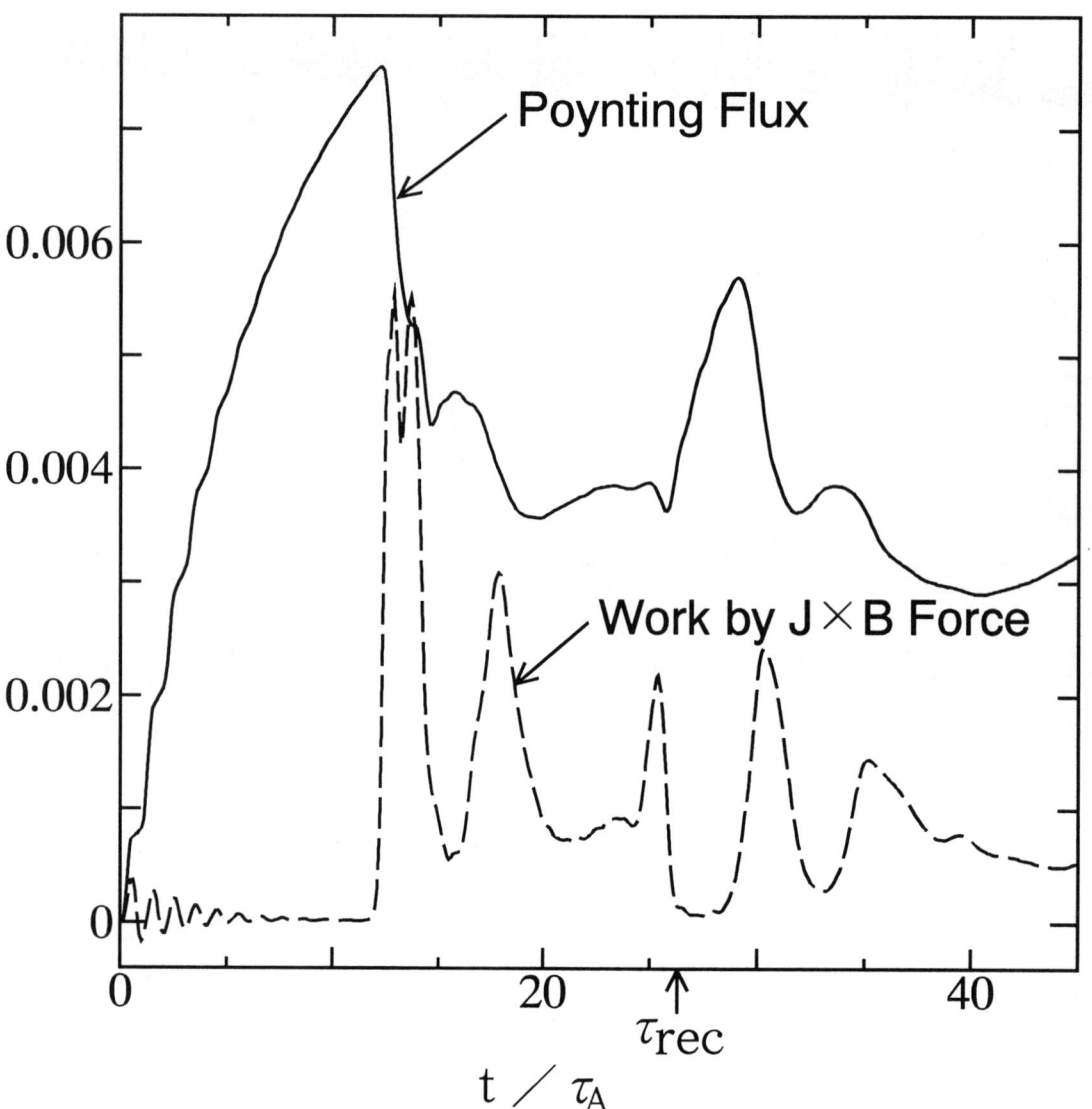

Fig. 4

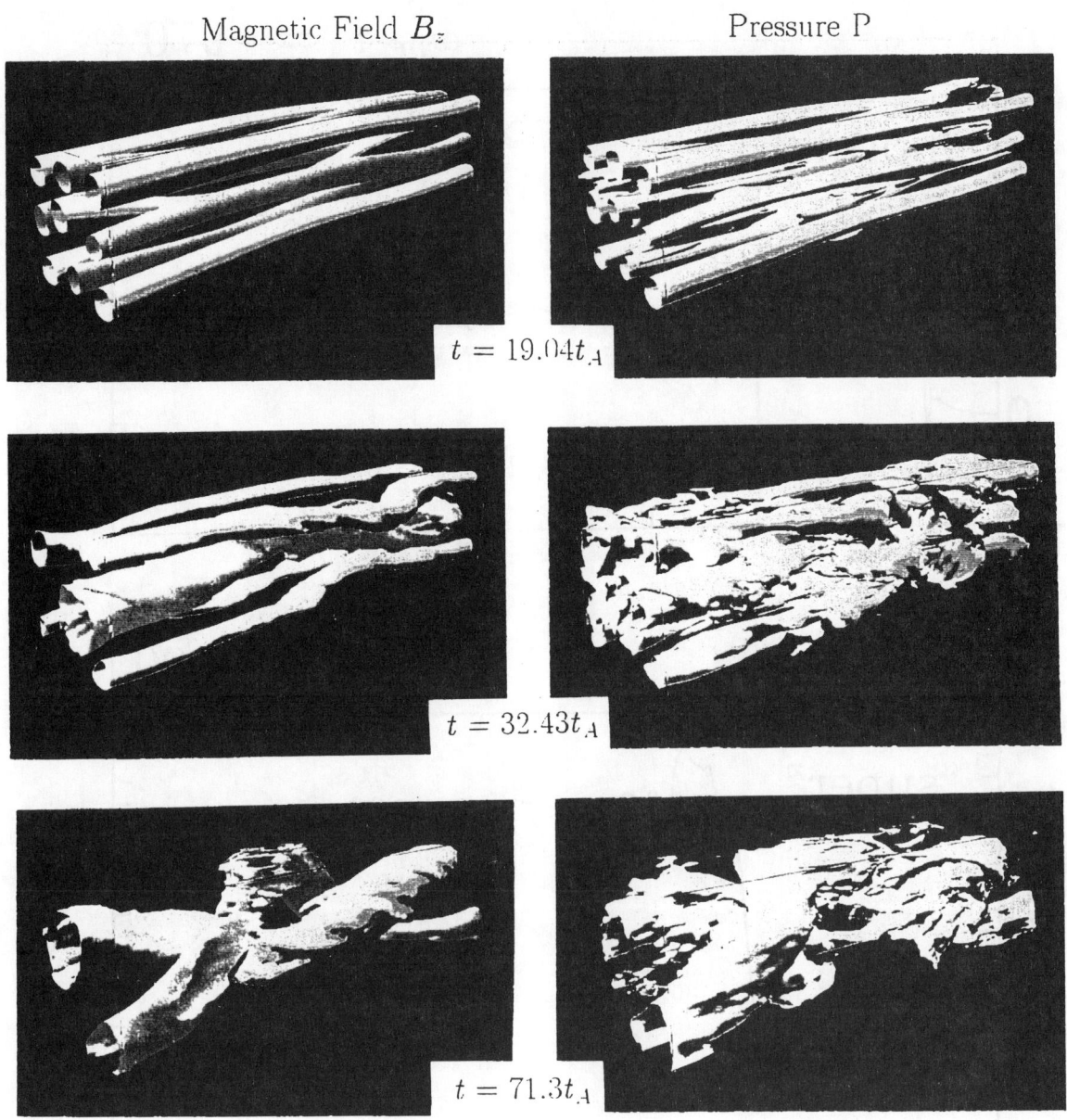

Fig. 5

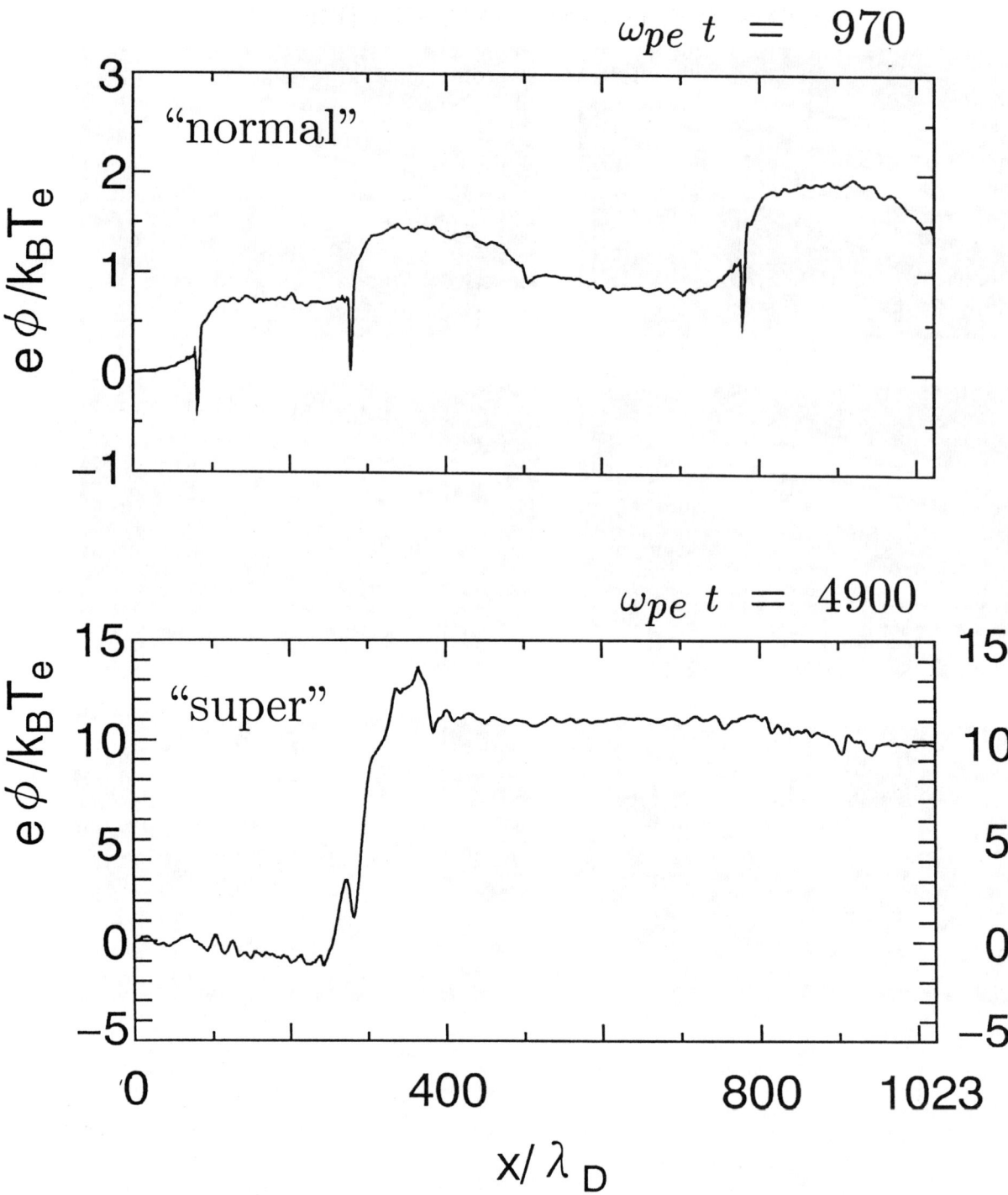

Fig. 6

OPEN FIELD LINES INSTABILITIES

Roberto Pozzoli

*Dipartimento di Fisica, Università di Milano,
via Celoria 16, 20133 Milano, Italy*

and

*Istituto di Fisica del Plasma, Consiglio Nazionale delle Ricerche,
EURATOM-ENA-CNR Association, via Bassini 15, 20133 Milano, Italy*

ABSTRACT

The results of some recent theoretical papers dealing with flute-like instabilities in the scrape-off layer of a tokamak with limiter configuration, where the magnetic field intersects conducting walls, are briefly recalled. Attention is then paid to the instability driven by the electron temperature gradient across the field in conjunction with the formation of the Debye sheath at the boundary, and to the effects due to the inclination of the end walls with respect to the magnetic field. When a divertor configuration is considered, important modifications are found owing to the strong deformations of the flux tubes passing near the x-point, which contrast the onset of flute-like perturbations, and to the stochasticity of field lines that can be excited by magnetic field perturbations.

1. INTRODUCTION

Since few years a renewed interest has arisen towards the investigation of conditions where magnetic field lines in the plasma intersect solid walls. This is related to the recognized importance of phenomena occurring in the scrape-off layer (SOL) of toroidal magnetic confinement systems with material limiter or magnetic divertor, as well as in open systems, e.g. gas dynamic trap (GDT). Space plasmas can also be found in situations analogous to those mentioned, e.g. in solar corona or in magnetosphere. Exhaustive surveys of theoretical and experimental results in this topic have been given in the review papers [1, 2]. In the following, some recent theoretical papers are shortly mentioned.

Complete line-tying at a conducting wall has a strong stabilising effect on flute perturbations in a low β plasma, as can be directly seen from the energy principle (see, e.g., Ref. 3). This seems to be in contrast to the observed strong turbulence in the tokamak SOL (see, e.g., [1,2] and references therein). However, it has been known since long time that the presence of the Debye sheath at the boundary does not allow free electron flow to the wall, so that the resulting *incomplete* line-tying considerably reduces the stabilising effect [4]. Taking this fact into account, Nedospasov has shown that in the SOL of a poloidal limiter configuration flute perturbations can be driven unstable by magnetic curvature [5]. A similar analysis including the additional stabilising effect of the polarization current has been performed by Garbet et al. [6]. The crucial dependence of stability of a GDT on the boundary conditions at the expander walls has been pointed out by Berk and Stupakov [7]. They showed that the stabilization mechanism based on favourable magnetic curvature in the expander and the large parallel pressure due to ion outflow, while effective in the case of insulating boundaries, is relaxed when conductive boundaries are considered.

A new mechanism of instability related to the boundary conditions, and effective even in the case of uniform magnetic field, has been found by Berk, et al. [8,9], extending a model originally introduced by Kadomtsev [10]. It is based on the presence, in the unperturbed system, of an electron temperature gradient (ETG) perpendicular to the magnetic field. In this case, owing to the allowed decoupling of plasma potential from the wall potential, a transverse electric field is also present in the unperturbed system: the resulting **E**×**B** flow is shown to drive the instability. The instability is quite strong, with an e-folding time much shorter than the plasma life-time.

This basic mechanism has been first investigated for electrostatic flute perturbations in the simplest model of a low β plasma in a uniform magnetic field, with field lines intersecting perpendicularly two end walls, L apart. The analysis of the instability has been then extended by Berk et al. [11], to include the electromagnetic regime and to account for finite $k_\parallel$ effects. The effect of an axial shear of the unperturbed **E**×**B** flow due to the variation of T_e along the field lines, localised near the boundaries on a length scale l « L, has been investigated by Tsidulko et al. [12]. Radial shear of the flow, due to a non uniform T_e gradient across the field, has been considered by Lotov et al. [13], and the transition between the ETG driven mode and the Kelvin-Helmhotz instability has been studied. The effect on the stability of the system of the inclination of the end plates, resulting from a non orthogonal intersection of the field lines with the wall surface, has been investigated by Farina et al. [14]. A new driver of instability is present, caused by the change of the flux tube length during its displacement in the radial direction. It can enhance or diminish the ETG driven instability, depending on the sign of **n**•∇T_e, **n** being the normal to the surface. All the above effects have been presented in a single dispersion relation by Cohen et al. [2]. A dispersion relation containing both curvature and ETG drivers has been discussed by Nedospasov and Uzdensky [15]. In the case of very small angles of incidence of the magnetic field to the end plates, ionization by electron neutral collisions within the sheath can reduce the sheath potential. The resulting stabilising effect on the ETG mode has been analysed by Xu and Cohen [16].

The stability analysis of ETG modes has been extended to include the plasma edge region adjacent to the SOL by Xu et al. [17]. A slab model is considered, where SOL and edge plasma are characterised, respectively, by boundaries a finite length apart, and periodicity conditions. Starting from fluid equations for electrostatic perturbations and solving the initial value problem it is found that the amplitude of the perturbation grows as $t^{1/2}\exp(\gamma t)$, is weakly dependent on the radial coordinate in the SOL, and decays exponentially inside the separatrix, in the edge region. Nonlinear numerical simulation of the turbulence for the above system has been performed by Xu [18], by means of a 2D fluid code in the variables $\nabla_\perp^2 \phi$ and T_e. The observed linear instability agrees with the theory, and a saturated state of turbulence is reached. Evidence of long wavelength mode penetration into the edge and inverse cascade (towards long wavelengths) of wave energy is shown. With reference to this point, a theoretical model for turbulence in SOL has been presented by Mattor and Cohen [19]. It is argued that the turbulence state is driven by a dual cascade of energy (to long wavelengths) and enstrophy (to short wavelengths) to an equilibrium state in which enstrophy is minimized for a fixed energy. Numerical simulations of the nonlinear evolution of the curvature driven (interchange) instability have been performed by Benkadda et al. [20, 21], using a 2D fluid code in the potential and density fluctuations. The process of formation and the dynamics of large scale coherent structures is observed.

The extension of the previous analysis to a divertor configuration meets some not yet solved complications, in connection with the presence of the separatrix and the x-point. Note that owing to it, a flute close to the separatrix suffers a strong squeezing, so that the existence itself of flute-like perturbations in that region becomes questionable [22].

The ETG instability in the SOL of a limiter configuration is presented in Sect. 2, together with an example of a pure MHD flute instability. Some considerations on flute perturbations in a divertor configuration and on the formation of the stochastic layer are given in Sect. 3. Conclusive remarks are made in Sect. 4.

2. THE ELECTRON TEMPERATURE GRADIENT INSTABILITY IN SOL

In the simplest model, a low β plasma in a uniform magnetic field $\mathbf{B} = B\mathbf{e}_z$, with field lines terminating on parallel end walls, normal to $\mathbf{B}$, at $z = \pm L/2$ is considered. In the unperturbed state, plasma density n and ion temperature T_i are uniform, while T_e depends on x and its gradient is characterised by the length $\Delta = |d\ln T_e/dx|^{-1}$ satisfying $\rho_i \ll \Delta \ll L$, being ρ_i the ion gyroradius. Perfectly absorbing and conducting end walls at the common potential $\phi = 0$ are assumed. In the following the derivation of the dispersion relation [9] is recalled.

The current density to the wall is established by the electron and ion flows along the magnetic field, with electrons almost totally reflected by the retarding potential of the Debye sheath. The current density is normal to the wall and reads (for the right wall) $j_n = en[v_i - \sqrt{T_e/2\pi m_e}\exp(-e\phi/T_e)]$, where v_i is the plasma flow velocity along the field lines. In the unperturbed situation $j_n = 0$, and the plasma potential ϕ is $\phi = T_e/e\Lambda$, $\Lambda \equiv \ln\sqrt{T_e/2\pi m_e v_i^2} \approx 2-4$. As ϕ depends on x, there exists an $\mathbf{E}\times\mathbf{B}$ drift, $\mathbf{v}_E = -c\nabla\phi\times\mathbf{B}/B^2$. For $T_e \approx T_i$, the drift velocity can be evaluated as $v_E \approx \Lambda v_i \rho_i/\Delta$.

Electrostatic perturbations correspond to transverse incompressible motion with $\delta\mathbf{v}_\perp = -c\nabla\delta\phi\times\mathbf{B}/B^2$ and potential perturbation $\delta\phi$ constant along the field lines. Flute perturbations of the form $\exp(-i\omega t + iky)$ are considered. The divergence of the perpendicular current density perturbation reads $\nabla\cdot\delta\mathbf{j}_\perp = icen(\Omega/\omega_{Bi})\nabla_\perp^2\delta\phi/B$, with $\Omega = \omega - \mathbf{k}\cdot\mathbf{v}_E$. From $\nabla\cdot\delta\mathbf{j} = 0$ in the plasma, the perturbation of the parallel current at the right boundary then reads $\delta j_\| = -icenL(\Omega/2\omega_{Bi})\nabla_\perp^2\delta\phi/B$. The eigenvalue equation for $\delta\phi$ is obtained equating this expression to the perturbed current density at the wall $\delta j_n = env_i\delta\phi[e/T_e + (\Lambda + 1/2)kc/(\Omega B)(d\ln T_e/dx)]$, obtained by perturbing the equation for j_n. In deriving this expression, the fact that the e-folding time of the instability is much shorter than the plasma lifetime $\tau = L/2v_i$ has been taken into account, by which plasma density and ion temperature remain unperturbed, while the electron temperature is conveyed by the flow, so that its perturbation is given by $\delta T_e = -kc\delta\phi/(\Omega B)dT_e/dx$. For local perturbations with $\Delta^{-2} \ll d^2/dx^2 \ll k^2$ (the most unstable), the following dispersion relation is obtained:

$$\tilde{\Omega}^2 + i\frac{\tilde{\Omega}}{\tilde{k}^2} + \sigma\frac{i}{\tilde{k}} = 0 \qquad (1)$$

where dimensionless variables have been introduced $\tilde{\Omega} = \Omega/\Omega_0$, $\tilde{k} = k/k_0$, with $\Omega_0 = \tau^{-1}d^{-2/3}s^{1/3}$, $k_0 = 2\rho_i^{-1}d^{1/3}s^{-2/3}$, $s = 2T_e/m_i v_i^2$, $d = (\Delta/L)/(\Lambda + 1/2)$, and $\sigma = \text{sign}(dT_e/dx)$. In this expression, the first term arises from the inertial ion drift, the second corresponds to the impedance of the Debye sheath (in the absence of the perturbation δT_e), and the last term, which comes from δT_e, is proportional to dT_e/dx and drives the instability. When T_e is uniform the last term vanishes and the mode is decaying. The real and imaginary part of Ω are of the same order of magnitude. The maximum growth rate can considerably exceed the growth rate of curvature driven flute instability. Finite Larmor radius effects, arising when the nonuninformity of the ion temperature is taken into account, reduce the growth rate. It has been shown that for realistic SOL parameters this reduction close to the maximum growth rate is very small [9].

An important modification of the above analysis is found when the inclination of the end plates is taken into account. The geometry of the system, which can represent both poloidal and toroidal limiter configurations, is shown in Fig. 1. The unperturbed field is uniform, and has both z and y components; l(x) denotes the distance along z between the plates, L the length of the field line, and $\Delta \ll l, L$ is assumed. Note that $1/2 \partial l/\partial x = -\cot\alpha$. In the case of a poloidal limiter y correspond to poloidal coordinate, z to toroidal, $l = 2\pi R$, $B_z/B_y = Aq$, where A is the aspect ratio and q the safety factor. In the case of a toroidal limiter, y corresponds to toroidal coordinate, z to poloidal, $l = 2\pi r$, $B_y/B_z = Aq$. If the angle α is taken not too small, satisfying $|\sin\alpha| > (m_e/m_i)^{1/2} B/B_z$, the current density in the flux tube is almost the same as for $\alpha = \pi/2$ [23]. Then the current density normal to the wall reads $j_n = en\sin\alpha\{v_E B_y/B + B_z/B[v_i - \sqrt{T_e/2\pi m_e} \exp(-e\phi/T_e)]\}$, where the last term represents the contribution of the electric drift. Here, it is neglected under the assumption $v_i \gg v_E B_y/B_z$. This imposes a limitation for the case of toroidal limiter $Aq < \Delta/(\Lambda \rho_i)$. In the opposite case the electric drift influences the structure of the SOL [24]. When B_y/B_z is sufficiently large, the electric drift which is mainly in the poloidal direction can cause considerable asymmetry in plasma losses [25]. Careful analyses of the role of electric and diamagnetic drifts in these situations have been given recently by Chankin and Stangeby [26], and Cohen and Ryutov [27].

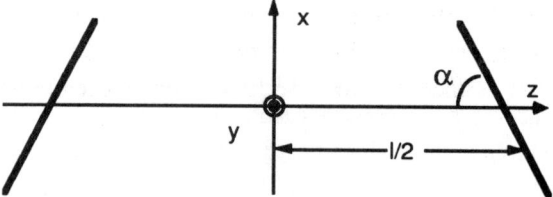

Fig. 1. The model of the "rectified" SOL.

Two modifications with respect to the previous case are introduced in the derivation of the dispersion relation: i) in addition to $\delta j_\parallel$, $\delta j_\perp$ gives a direct contribution to the current density normal to the wall, ii) the relative velocity of ion flow with respect to the wall, entering the boundary condition, becomes $v_i - (B/B_z)\delta v_x \cot\alpha$, due to the change of the flux tube length during its displacement in the radial direction. The dispersion relation for localized perturbations reads

$$\tilde{\Omega}^2 + \tilde{\Omega}\left(\frac{i}{\tilde{k}^2} + \frac{C(\Lambda+1/2)}{\tilde{k}}\right) + \sigma\left(C + \frac{i}{\tilde{k}}\right) = 0 \quad (2)$$

where **k** is perpendicular to **B** and lies in the y, z plane, and the parameter C, which depends on the geometrical characteristics of the system, reads $C = [(L/l)/(\Lambda+1/2)] d^{1/3} s^{1/3} \cot\alpha$. The term σC comes from the first modification mentioned above, and describes the rigidity of the system caused by the pressure force that acts on the flux tube through its inclined ends, the other C dependent term in the equation derives from the second modification. The new effect becomes important when $\Lambda|C|\geq 1$. This holds at small enough α's. Depending on the α value, this effect can give stabilizing or destabilizing contributions to the ETG instability.

The effect of the instability on plasma transport can be described in terms of the mixing length theory [28] which results in the following estimate for the diffusion coefficient $D = \max(\text{Im}\Omega/k^2)$. The maximum growth rate occurs at intermediate values of k, but, since the diffusion coefficient D is proportional to k^{-2}, its value is determined by k values considerably shifted to the smaller side. The plot of the diffusion coefficient normalized over the Bohm coefficient $D_{Bohm} = (1/16) cT_e/(eB)$ is shown in Fig.2 for the toroidal and poloidal limiter [29]. Note that for the chosen parameters D is much larger than D_{Bohm} for both type of limiters except in a small range of angles, close to tangential intersections. Comparing the results with the case of orthogonal intersection ($\alpha = 90°$), a considerable increase of the diffusion coefficient is found for α values in the range of 10° and a considerable decrease in the range of 170°.

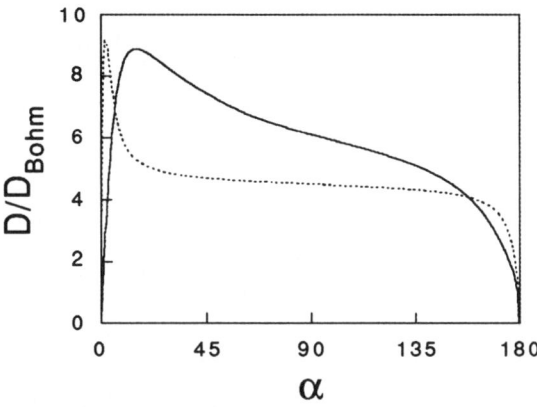

Fig 2. Diffusion coefficient D normalized over the Bohm coefficient versus the angle α for $\sigma<0$. The solid line refers to the case of a toroidal limiter, and the dotted line to a poloidal limiter. The chosen parameters are R = 1.5 m, Δ =1 cm, $Mv_i^2/2 = T_e$, A = 2.5, q = 2.5, and $\Delta/\rho_i = 30$.

The σC term in the dispersion relation, taken alone, can drive an instability analogous to the flute one. The case of flute instability in a plasma confined by perfectly reflecting end walls (i.e., $\xi_n = 0$, $\delta j_n = 0$ at the boundaries, $\xi = i\delta v/\omega$ being the displacement from the equilibrium) has been analyzed by Farina et al. [30]. In this situation the displacement of flux tubes in the x direction is accompanied by pressure perturbations arising near the wall and propagating along the flux tube. The geometry of the system corresponds to Fig. 1, with **B** along z, l = L, and pressure gradient along x. A flux tube, displaced along x by ξ_x, has a perturbed pressure $p - \gamma p(\xi_x/L)\partial L/\partial x$, while the pressure of the surrounding plasma is $p + \xi_x \partial p/\partial x$. The condition for the instability, arising when the perturbed plasma pressure is smaller than the flute pressure, can then be written in the familiar way $dp/dU + \gamma p/U < 0$, with $U = \int dz/B = L/B$. Perturbations are aperiodic. The longitudinal displacement ξ_z, representing the acoustic perturbation along z, is coupled to the transverse displacement ξ_x by the boundary conditions, so that $\xi_z = -\xi_x \cot\alpha \sin(\omega z/s)/\sin(\omega L/2s)$, where $s = (\gamma p/\rho)^{1/2}$ is the sound speed. For the localized and most unstable perturbations ($k \to \infty$), the dispersion relation reads

$$\Gamma^2 - R\cot\alpha + (\cot^2\alpha)\Gamma\coth\Gamma = 0 \quad (3)$$

where $\Gamma = -i\omega L/(2s)$ is the dimensionless growth rate, and $R = \sigma L/(2\gamma\Delta)$, with $\sigma = \text{sgn}(\partial p/\partial x)$, and $\Delta = |\partial \ln p/\partial x|^{-1}$. At a given $R \gg 1$, the maximum growth rate is $\Gamma_{max} \approx (R^2/4)^{1/3}$ at $\cot\alpha \approx (R/2)^{1/3}$. A stabilizing effect is due to the peaking of the pressure perturbation near the wall (last term in Eq. (3)), produced by the changing length of the flux tube, since $\delta p + \xi_x \partial p/\partial x = \xi_x \gamma p(\omega/s)\cot\alpha \cos(\omega z/s)/\sin(\omega L/2s)$. This phenomenon may drive non MHD effects.

3. PERTURBATIONS IN A DIVERTOR CONFIGURATION

The structure of flux tubes close to the separatrix of a divertor configuration is considerably different from that relevant to the region around the solid limiter. A flux tube, passing near the x-point suffers a very strong squeezing and acquires a ribbon-like structure [22]. This is easily seen in a rectified model with uniform toroidal field $B_z = B_0$. In a region close enough to the x-point, the poloidal field is given by $B_x = \alpha x$, $B_y = -\alpha y$, where $\alpha \approx B_p/a$, and x, y axes coincide with the separatrix asymptotes. The field lines are then simply represented by $x(\zeta) = x(\zeta')\exp(\zeta - \zeta')$, $y(\zeta) = y(\zeta')\exp(\zeta' - \zeta)$, with $\zeta = \alpha z/B_0$, which corresponds to an area preserving mapping of the plane with a uniform stretching of the coordinate x by the factor $\exp(\Delta\zeta)$, and a corresponding shrinking of y by the factor $\exp(-\Delta\zeta)$, for a positive increment $\Delta\zeta$. Thus, a cross section of a flute, circular at a given ζ, as ζ increases becomes elliptic and strongly elongated in the x direction. One of the dimensions of the flute can easily become smaller than the ion Larmor radius ρ_i. This will prevent the penetration of the flute perturbation to any significant distance from the x-point. The ribbon like structure of the flute is shown in Fig. 3, and characterizes the flux tube for almost all its length between the intersections.

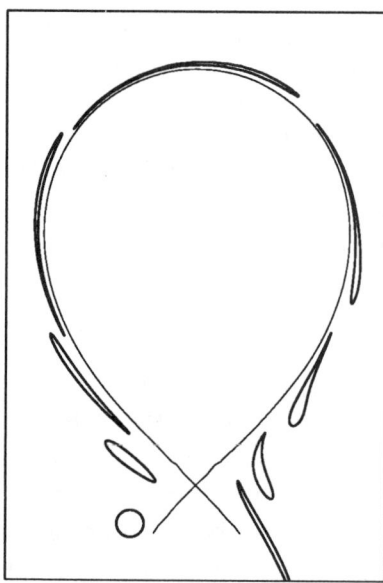

Fig. 3. Cross section of a flute tube at different toroidal positions.

Another interesting characteristic of the divertor configuration is the stochasticity of the magnetic field lines which can arise around the separatrix. The formation of a stochastic layer at the edge of a Tokamak by means of different kinds of externally induced magnetic perturbations has been investigated [31 - 34]. Here, modifications of the field structure due to magnetic perturbations which can be excited inside the plasma are presented [35].

The unperturbed field lines are described by the Hamilton equations: $\dot{x} = \partial H/\partial y$, $\dot{y} = -\partial H/\partial x$, where the dot indicates derivation with respect to z, and H is proportional to the poloidal flux ψ. In the low β plasma, magnetic perturbations represented by $\delta \mathbf{B} = \nabla \delta \psi \times \mathbf{e}_z$, being $\delta\psi$ the perturbation of the flux function, can be considered. The perturbed field lines are described by the perturbed Hamiltonian $H + \delta H$.

Having in mind the particle motion along the field lines, attention is paid to the behavior of the field lines up to their intersections with an end plate: to them the usual definition of stochasticity cannot be strictly applied. A Poincaré section at a fixed z value (mod 2π) in the region around the x-point is presented in Fig. 4. A model perturbation is used of the form $\delta\psi = \varepsilon\, q \cos nz\, (1 - \cos 2m\varphi)/m$, which vanishes at the end plates together with its gradient. In this expression φ is the unperturbed magnetic potential, $m, n \gg 1$, and the parameter ε represents a measure of $\delta B/B$. The formation of a stochastic layer, close to the separatrix, the structures characterizing the transition region, invariant curves, and chains of small islands can be distinguished. All but very few points lie inside the unperturbed separatrix. Outside the separatrix, a portion of the stochastic layer of width comparable to the part inside is not found due to the presence of the end plates, at which the integration of the field line equations is stopped. Note that all the field lines inside the stochastic layer reach a divertor plate, each with its own length of the trajectory.

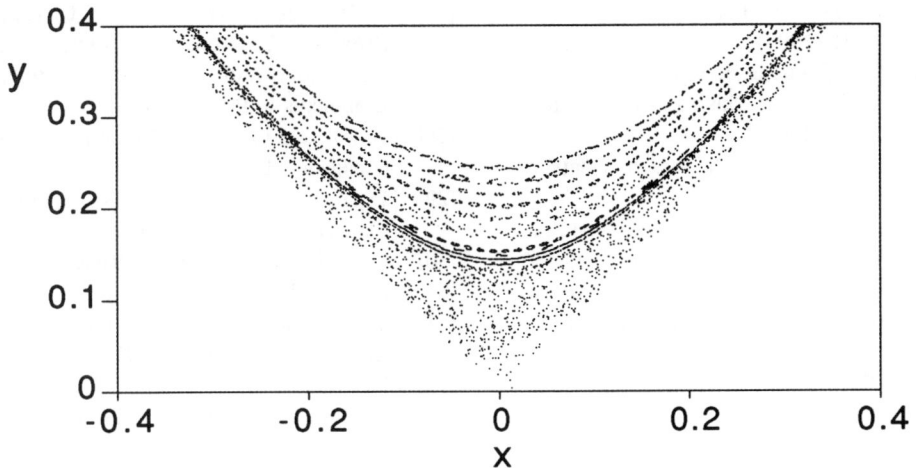

Fig. 4. Poincaré section in the x-point region.

4. CONCLUSION

In the case of the more simple limiter geometry (without the x-point) linear stability analyses have shown new driving mechanisms, or important modifications of the known ones, owing to the boundary conditions. Further linear investigation of particularly interesting cases, as that of almost tangential incidence of field lines on end walls, are required. A further development of nonlinear models, which have shown reasonable agreement with the experimental trend, is also needed.

In the case of the more complex divertor geometry, even the linear stability analysis has not yet been well developed (although some particular problems have been solved). A direct extrapolation of the above analyses made for simpler geometry seems to be inadequate.

[1] A. V. Nedospasov, J. Nucl. Mater., **196-198**, 90 (1992)

[2] R. H. Cohen, N. Mattor, X. Q. Xu, Contrib. Plasma Phys, **34**, 232 (1994)

[3] B. Kadomtsev, in *Reviews of Plasma Physics*, edited by M.A. Leontovich (Consultants Bureau, New York, 1966), vol. II, p. 153

[4] W. B. Kunkel, J. U. Guillory, *Proc. VII Intern. Conf. on Phenomena in Ionized Gases*, 1965, ed. B. Perovic and D. Tosic (Gradevinska Knijga, Beograd 1966), vol. II, p. 702

[5] A. V. Nedospasov, Sov. J. Plasma Phys., **15**, 659 (1989)

[6] X. Garbet, L. Laurent, J. P. Roubin, A. Samain, Nuclear Fusion, **31**, 967 (1991)

[7] H. L. Berk, G. V. Stupakov, Phys. Fluids B, **3**, 440 (1991)

[8] H. L. Berk, D. D. Ryutov, Yu. A. Tsidulko, JETP Lett., **52**, 23 (1990)

[9] H. L. Berk, D. D. Ryutov, Yu. A. Tsidulko, Phys. Fluids B, **3**, 1346 (1991)

[10] B. Kadomtsev, *Proc. VII Intern. Conf. on Phenomena in Ionized Gases*, 1965, ed. B. Perovic and D. Tosic (Gradevinska Knijga, Beograd 1966), vol. II, p. 610

[11] H. L. Berk, R. H. Cohen, D. D. Ryutov, Yu. A. Tsidulko, X. Q. Xu, Nucl. Fusion, **33**, 263 (1993)

[12] Yu. A. Tsidulko, H. L. Berk, R. H. Cohen, Phys. Plasmas, **1**, 1199 (1994)

[13] K. Lotov, D. Ryutov, J. Weiland, Physica Scripta, **50**, 153 (1994)

[14] D. Farina, R. Pozzoli, D. Ryutov, Plasma Phys. Control. Fusion, **35**, 1271 (1993)

[15] X. Q. Xu, M. N. Rosenbluth, P. H. Diamond, Phys. Fluids B, **5**, 2206 (1993)

[16] A. V. Nedospasov, D. A. Uzdensky, Contrib. Plasma Phys., **34**, 277 (1994)

[17] X. Q. Xu, R. H. Cohen, Plasma Phys. Control. Fusion, **35**, 1071 (1993)
[18] X. Q. Xu, Phys. Fluids B, **5**, 3641 (1993)
[19] N. Mattor, R. H. Cohen, Plasma Phys. Control. Fusion, **36**, 1115 (1994)
[20] S. Benkadda, X. Garbet, A. Verga, Contrib. Plasma Phys., **34**, 247 (1994)
[21] S. Benkadda, T. Dudok De Wit, P. Gabbai, A. D. Verga, X. Garbet, M. Endler, A. Sen, *XV Intern. Conf. on Plasma Physics and Controlled Nuclear Fusion Research*, Seville, 1994, IAEA-CN-60/D-P-I-9 (1994)
[22] D. Farina, R. Pozzoli, D. Ryutov, Nuclear Fusion, **33**, 1315 (1993)
[23] R. Chodura, Phys. Fluids, **25**, 1628 (1982)
[24] M. Tendler, V. Rozhansky, Comm. Plasma Phys. Control. Fusion, **13**, 191 (1990)
[25] R. Chodura, *Proc. 1992 International Conference on Plasma Physics*, (Geneva, EPS), vol. 16C, part II, p.871 (1992)
[26] A. V. Chankin, P. C. Stangeby, Plasma Phys. Control. Fusion, **36**, 1485 (1994)
[27] R. H. Cohen, D. D. Ryutov, UCRL-JC-117905 Preprint Lawrence Livermore National Laboratory (July 1994), to be published in Comments in Plasma Physics
[28] B. B. Kadomtsev, *Plasma Turbulence* (Academic, London, 1965).
[29] D. Farina, R. Pozzoli, D. Ryutov, Contrib. Plasma Phys., **34**, 253 (1994)
[30] D. Farina, R. Pozzoli, D. Ryutov, Phys. Fluids B, **5**, 4055 (1993)
[31] T. J. Martin, J. B. Taylor, Plasma Phys. Control. Fusion, **26**, 321 (1984)
[32] N. Reggiani, P. H. Sakanaka, Plasma Phys. Control. Fusion, **36**, 513 (1994)
[33] A. H. Boozer, A. B. Rechester, Phys. Fluids, **21**, 682 (1978)
[34] N. Pomphrey, A. Reiman, Phys. Fluids B, **4**, 938 (1992)
[35] D. Farina, S. Lasca, A. Nassigh, R. Pozzoli, *Proc. Joint Lausanne-Varenna Intern. Workshop "Theory of Fusion Plasmas"* (1994), Varenna

COHERENT STRUCTURES IN PLASMAS AND FLUIDS

J. Juul Rasmussen, J.S. Hesthaven, J.P. Lynov and A.H. Nielsen

Association EURATOM-Risø National Laboratory,
Optics and Fluid Dynamics Dept., Risø, DK-4000 Roskilde, Denmark

Abstract The formation and dynamics of coherent vortical structures in both homogeneous and inhomogeneous systems will be discussed. In particular the dynamics of monopolar and dipolar vortices in strongly magnetized plasmas and in rotating fluids is described. The role of vortical structures in connection with electrostatic plasma turbulence and the associated cross-field plasma transport will also be addressed.

1. INTRODUCTION

Coherent structures are characterized by localized regions of ordered motion. These have lifetimes that are sufficiently long to distinguish them from the surrounding random motion, i.e. the correlation time of the internal motion in the structure is much longer than the correlation time of the surrounding motion. Particular examples of coherent structures in plasmas and fluids are the so-called vortical structures, which consist of localized regions of swirling motion. Although such structures appear in general three-dimensional flows (especially in boundary layer flows)and may have significant influence on the dynamics of such flows, they are comparatively much more important in two-dimensional or quasi-two-dimensional flows, such as planetary atmospheres and oceans, laboratory experiments with rotating or stratified fluids and strongly magnetized plasmas. The vortical structures, having life times much longer than their own period of internal rotation, dominate the large scale dynamics in such situations and are naturally of great importance in connection with the transport of mass, heat and momentum. Since they can trap particles and convect them over distances much longer than their own scale size, they provide an effective transport mechanism, which in its nature is essentially different from classical diffusion dominated transport. Coherent vortical structures are likely to form after external forcing and seem to be the universal outcome of any regular as well as irregular forcing in two-dimensional systems. They are easily excited in laboratory experiments in rotating [1, 2] and stratified fluids [3]. They also appear spontaneously in two-dimensional turbulence [4] due to self-organization processes characterizing two-dimensional flows [5]. The simultaneous conservation of energy and enstrophy (mean squared vorticity) for two-dimensional flows leads to an inverse spectral cascade, by which energy is inversely cascading towards smaller and smaller wavenumbers, whereas enstrophy is cascading towards larger and larger wavenumbers. This effect is clearly demonstrated in high resolution numerical simulations of two-dimensional flows [6, 7].

Vortical structures or convective cells [8] have also been excited in magnetized plasmas [9], and the coalescence of like signed vortices, which is mediating the inverse cascade, was directly demonstrated [10]. Furthermore, vortex-like structures are observed to appear spontaneously in low-frequency electrostatic turbulence in linear [11, 12] as well as toroidal devices [13, 14] and in the scrape-off-layer of tokamaks [15]. It was directly demonstrated that they play an important role in connection with the plasma confinement by the magnetic field [11]. Finally, it should be mentioned that many features of vortex dynamics have been studied in a pure electron plasma confined by a magnetic field [16]. Thus, the understanding of the dynamics of coherent vortical structures in two-dimensional flows is not only of fundamental interest; but it has several practical implications as well.

In this contribution we describe the dynamics of individual vortices in homogeneous as well as inhomogeneous two-dimensional flows. The results apply to both magnetized plasmas and rotating fluids. The low frequency electrostatic dynamics of a strongly magnetized plasma can be described by similar dynamical models as those of a rotating fluid due to the similar effects of the two dominating forces: the Lorentz force and the Coriolis force, respectively. Vortical structures, which are strongly nonlinear entities that are not wave-like, often exist in regimes where linear waves cannot propagate, i.e. they may have propagation velocities outside the range of linear phase velocities. The vortices are then truly localized, they will not couple energy to linear waves and their life time is determined by dissipative effects. This has been invoked as a necessary requirement for the existence of steady vortex structures [17]. We should note, however, that strong quasi-steady vortices having propagation velocities coinciding with the linear phase velocity may still be sufficiently long lived to be of importance for dynamics of the flow and the associated transport. Here "long lived" designates a lifetime significantly longer than the characteristic time for dispersive spreading of a linear wave

© 1995 American Institute of Physics

packet of the same size as the vortex.

2. VORTICES IN HOMOGENEOUS MEDIA

In homogeneous situations the dynamics is governed by the two-dimensional, incompressible Navier-Stokes equations:

$$\frac{\partial \omega}{\partial t} + [\psi, \omega] = \nu \nabla^2 \omega \ , \quad \nabla^2 \psi = \omega \ , \tag{1}$$

where ψ is the streamfunction, ω is the vorticity, ν is the kinematic viscosity and $[f, g] = f_x g_y - f_y g_x$. The velocity field is related to the streamfunction as

$$\vec{v} = -\nabla \psi \times \hat{z} \ . \tag{2}$$

In the absence of viscosity Eqs. (1), which is then called the Euler equations, express the material conservation of vorticity and it poses an infinite number of conserved quantities:

$$E = \frac{1}{2} \int (\nabla \psi)^2 \, dx dy \ , \quad W = \frac{1}{2} \int \omega^2 \, dx dy \ ,$$
$$C_\omega = \int G(\omega) \, dx dy \ . \tag{3}$$

E is the energy, W is the enstrophy and C_ω (which also contain W) is often referred to as the Casimirs, here $G(\omega)$ is any continuous function. In addition to these also the linear and angular momentum are conserved in unbounded flows. The conserved quantities naturally constrain the evolution of the flow as discussed in detail in many of the references mentioned in the introduction.

For the plasma case, ψ is the electrostatic potential, the velocity $\vec{v}$ is the $\vec{E} \times \vec{B}_0$ velocity, with the background magnetic field, $\vec{B}_0$ in the z-direction, and the vorticity, ω, is proportional to the charge density.

From the Euler equations we observe that an infinite class of stationary solutions exists for which $\omega = f(\psi)$, where $f(\psi)$ is an arbitrary, integrable function of ψ. These solutions include what have appeared to be the basic vortical structures in homogeneous two-dimensional flows: The *monopolar vortex*, which is a localized, mostly circular symmetric, structure with the swirling motion of just one orientation, and is characterized by a single set of closed, concentric streamlines. The *dipolar vortex* essentially consists of two closely packed regions of opposite vorticity. The dipole has a net linear momentum, which makes it translate in contrast to the monopole. The *tripolar vortex* consists of a core vortex of elliptical shape and two satellite vortices of opposite vorticity. The tripole has an angular momentum, but no linear momentum and rotates steadily around the centre of the core vortex.

Thus, in a homogeneous flow only the dipole is propagating and able to effectively transport material. A well-known and particularly simple example of a dipolar solution to the Euler equations is the Lamb-dipole [18], where f is a linear function. Although observed dipoles often are well described by the Lamb dipole there are several examples where the function appear to be nonlinear [3, 19].

To illustrate the formation of dipolar vortices from a localized turbulent forcing we have solved Eq. (1) numerically in a double periodic domain of size $L_x \times L_y$, using a spectral scheme with a resolution $M \times N$ [19], with a turbulent patch as initial condition. The patch is characterized by its energy and enstrophy and it has a net linear momentum, but no net circulation. It may be applied as a model of the turbulent patch that appears when a jet of short duration is injected into a stratified (see e.g. [3]) or rotating fluid (see e.g. [1]). It is constructed from fluctuations with random phases and an isotropic energy spectrum $\propto \exp(-(k - k_0)^2/\Delta^2)$. The localization of the patch is provide by multiplying this wave field by $\exp(-r^4/r_0^4)$ in configuration space, after a sinusoidal in x is added to provide a finite linear momentum in the y-direction. The resulting development is shown in Fig. 1. We observe a clear self-organization of the patch into a dipolar structure.

The development in Fig. 1 is approximately described as a self-organization in a viscous fluid, noting that for this case W decays faster than E, i.e. for the coefficient $Q = W/E$ we have $\dot{Q} \leq 0$. Following the idea of Leith [20] we look for solutions minimizing the enstrophy, W, under the constraint of a given constant energy, E, and with the further constraint of a finite linear momentum, P_y. A possible solution is the Lamb dipole with a linear relation between ϕ and ω, and for that $\dot{Q} = 0$. The values of the dipole radius R and its velocity U_0 may be obtained from the given initial energy and momentum with the result $R \approx 0.24$ and $U_0 \approx 0.90$. This corresponds well with the values observed from in Fig. 1, where $R \approx 0.22$ and $U_0 \approx 0.75$. We have also monitored Q and found it to be decreasing and approaching a constant value of ≈ 154, which would correspond to a radius of $R \approx 0.22$ for a Lamb dipole. It seems that the simple self-organization approach describes the gross properties of the structure, even if the relationship between ω and ϕ (the streamfunction in the frame of reference moving with the dipole) is found to be nonlinear, f may be approximated by a cubic polynomium. The detailed structure of the emerging dipole depends on the initial condition, the more enstrophy that is contained in the random part as compared to the enstrophy of the "laminar" part (containing the linear momentum) the more nonlinear is the relation between ω and ϕ. When the initial patch does not contain linear momentum it evolves into weakly interacting monopolar vortices, which show a slow evolution.

3. VORTICES IN INHOMOGENEOUS MEDIA

Geophysical flows are anisotropic due to the variation of the Coriolis force with latitude. This introduces the so-called β-effect, which allows for Rossby wave propagation

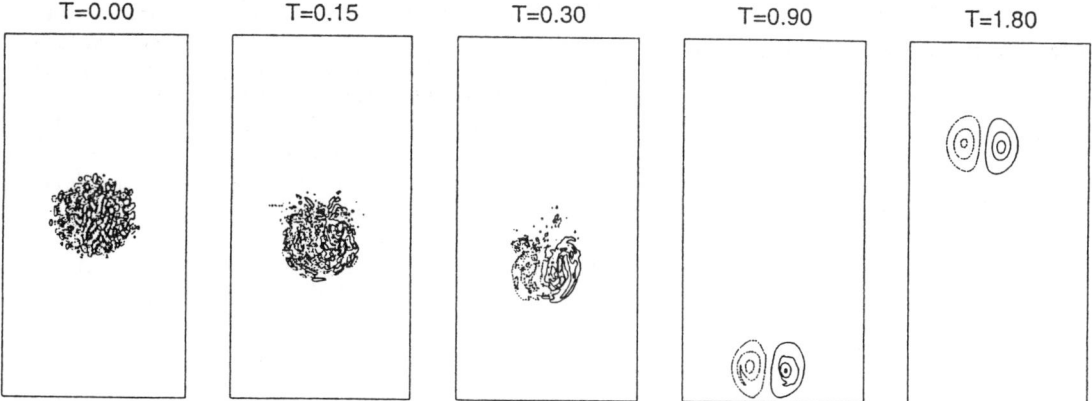

Figure 1: Evolution of a turbulent patch with initial linear momentum into a dipolar structure. The parameters defined in the text are $k_0 = 20$, $\Delta = 10$, $r_0 = 0.20$. Initially $E = 0.314$, $W = 162$, $P_y = 0.34$. The Reynolds number obtained from the developed dipole $Re = RU_0/\nu \approx 1000$. The resolution is $N_x = 128$ and $N_y = 256$.

[21], and strongly changes the dynamics of the fundamental vortical structures described in Sec. 2. In a magnetized plasma the inhomogeneity of the plasma density plays a similar role and allows for propagation of drift-waves [22, 23].

The dynamics in these cases is governed by the conservation of a generalized potential vorticity, which takes the form

$$\frac{d\Pi}{dt} = \frac{d}{dt}\frac{\omega + \omega_{ci}}{n} = 0 \quad \text{(a)} \quad \text{and}$$
$$\frac{d\Pi}{dt} = \frac{d}{dt}\frac{\omega + f}{H} = 0 \quad \text{(b)}, \qquad (4)$$

in the plasma case and the fluid case, respectively. In the plasma case, n is the density and Eq. (4a) is an exact relation for the case when the ion temperature gradient is negligible. It is often called the ion vortex equation or Ertels theorem [21], which is also used in the fluid context, where H is the depth of the fluid and $f = 2\Omega \sin \alpha$ (Ω is the rotation frequency of the planet and α is the latitude), is the Coriolis parameter. For a uniform density (or for constant f and H) Eq. (4) reduces to the Euler equation (1). Equation (4) imposes restrictions on the dynamics of vortical structures. A vortex with constant ω is bound to move along the contours of constant density in the plasma case and along constant depth and Coriolis parameter in a fluid. Thus, in inhomogeneous systems the propagation of localized coherent structures will be strongly anisotropic.

To the lowest order approximation [23] Eq. (4) may be reduced to the well-known Hasegawa-Mima equation [22] describing drift-waves in a plasma with a density gradient, or equivalently to the equation governing Rossby waves in rotating fluids with varying Coriolis force and/or depth. In this context the equation is often referred to as the equivalent barotropic vorticity equation [21]:

$$\frac{\partial}{\partial t}(\nabla^2 \phi - \phi) + [\phi, \nabla^2 \phi] + \beta \frac{\partial \phi}{\partial x} = 0 , \qquad (5)$$

For the plasma case, ϕ is the electrostatic potential (normalized by T_e/e), and β is proportional to the background density gradient, which is pointing in the negative y-direction. The background magnetic field, B_0, is in the z-direction and the linear drift waves are propagating in the negative x-direction with the maximum phase velocity given by β. Time is normalized by ω_{ci}^{-1} (ω_{ci} is the ion cyclotron frequency), and length by $\rho_s = (T_e/m_i\omega_{ci}^2)^{1/2}$. With these normalizations, β is the electron diamagnetic drift velocity normalized with the ion sound speed.

When Eq. (5) is applied to the fluid case, e.g. the large scale motion on a rotating planet, positive x is eastward and positive y is northward (the direction of the gradient in the Coriolis force). ϕ is then the geostrophic streamfunction and the geostrophic velocity is also here given by Eq. (2). In this case the time is normalized by f^{-1}, and length by the Rossby radius $\rho_R = \sqrt{gH_0/f^2}$. Here β, being proportional to the gradient of the Coriolis parameter, is the maximum phase velocity of westward propagating linear Rossby waves.

Under the assumptions used in deriving Eq. (5) the potential vorticity now reads $q = \nabla^2\phi - \phi + \beta y$. The conservation of q for each fluid element along its trajectory implies that if the fluid component, ξ, of the potential vorticity, $\xi = \nabla^2\phi - \phi$, is not changing, this fluid element is bound to stay at the same y-position and can only move along the x-direction, i.e., perpendicular to the background gradients, and no transport will take place along the gradients. Equation (5), which we shall refer to as the HM-equation, has an exact stationary solution in the form of a dipolar vortex propagating perpendicular to the gradient, i.e. along the x-direction. Monopoles are not steadily propagating since they will couple to linear waves [17]. They may, however, still be sufficiently long lived to be of importance for the flow dynamics and the associated transport. Accounting for electron temperature gradients and higher-order terms it has been shown that steadily propagating monopolar vortices may exist for drift-waves, see e.g. [23, 17] and references therein.

We have performed a detailed study of the evolution of nonstationary monopolar and dipolar vortices [24] by solving the HM-equation numerically on a two-dimensional periodic domain. In the following we summarize our main results. First we note that Eq. (5) obeys the symmetry relation $\phi(x, y, t) = -\phi(x, -y, t)$. Thus, regarding monopolar vortices, we present only the evolution of what we shall call cyclones, i.e. structures with negative amplitude of the streamfunction, i.e. positive vorticity and anticlockwise rotation (on the Northern hemisphere). In the plasma case cyclones will rotate in the same sense as the electron cyclotron rotation, and they naturally have negative potential.

We illustrate the evolution of monopolar vortices by considering an initial vortex with monotonic *Gaussian fluid potential vorticity* $\xi_{init} = \xi_m \exp(-r^2/a^2)$. A strong monopole, measured in terms of ϕ_m/β, will have closed isolines of potential vorticity, q. The monopole is not a stationary solution, but will radiate energy to linear waves. From the conservation of potential vorticity (4) we observe that for a cyclone with negative ϕ (positive ξ) the loss of energy due to radiation implies that it must move to higher y-values to conserve q, i.e., it will move slightly down the density gradient (to the north for the Rossby wave case). Similarly an anti-cyclone will move up the density gradient (to the south). More quantitatively the dynamics can be described by considering the evolution of an azimuthal $l = 1$ perturbation, which is generated due to the β-effect [24]. This perturbation sets up a secondary flow having a dipolar structure, which in turn gives rise to the propagation in the negative x-direction with a velocity, which for a strong monopole approaches the maximum linear phase velocity β, and in the positive y-direction with a comparatively much smaller velocity. The maximum excursion along the density gradient and across the magnetic field is estimated to [24]:

$$y_m = 2 \frac{\int_0^{r_c} \xi_{init} r \, dr}{\beta r_c^2} , \quad (6)$$

where r_c is the radius of the vortex core when accounting for the β-effect. For strong monopoles $\phi_m/\beta \gg 1$, $y_m \approx \phi_m/\beta$. Thus, they may provide long-distance transport across the magnetic field lines.

The numerical results for the propagation of monopoles agree with the predictions above. For a strong monopole the initially monotonic vortex core becomes elliptic and finally evolves into a tripolar structure in the potential vorticity as it moves. Its life time is many times longer than the linear wave dispersion time $T_R \approx 2\pi/\beta$ and it approach a maximum excursion in the y-direction given by Eq. (6). Thus, it may provide transport of trapped plasma over distances many times its own width. Weaker monopoles do not develop into a tripolar structure, but they still propagate a significant distance along y. The threshold amplitude for the formation of a well defined tripole is around $\phi_m = 6.5\beta$.

In Fig. 2 we show the evolution of a weak monopolar vortex $\phi_m = 2\beta$, which decays due to the dispersion at nearly linear rate (at $T = T_R \approx 60$, $\phi \approx \phi_m/3$). We observe that the vortex core in q maintains its identity significantly longer than T_R, and even longer than the core of closed streamlines exist. The appearance of closed isolines in the potential vorticity indicates the possibility of trapping and long distance transport of particles. To investigate this further we have traced particles initially released near the monopole - inside as well as outside the core of closed isolines [24]. Particles initially released deep inside the separatrix are effectively trapped and transported along with the vortex. Particles near the separatrix leave the structure after a few rotations in the vicinity of the stagnation point of the streamfunction. This confirms the leaking of particles from the core of the structure as it shrinks. For a strong monopolar vortex, where the tripolar structure develops, we have also observed trapping of particles during the propagation of the vortex. This seems to be mediated by the nonsteady oscillation of the tripolar structure. These observations clearly illustrate that the shrinking of cores of closed isolines of potential vorticity and of closed streamlines in a co-moving reference frame allows for gradual escape of transported particles induced by the non-stationary propagation of a coherent structure. Trapping and transport of particles in a monopolar vortex was also studied by Crotinger and Dupree [25]. They considered deeply trapped particles for intermediate amplitude individual monopoles and further they demonstrated that vortices, generated spontaneously in drift-wave turbulence, trap particles and transport them along.

The HM-equation (5) has an exact steadily propagating dipole solution; the so-called modon solution. This solution is only steady when propagating strictly along the x-direction, and when it propagates in the negative x-direction its velocity must be larger than β. We performed a detailed study of the influence of the β-effect on dipoles initiated as the dipole solutions for a homogeneous medium, with $\beta = 0$ in Eq. (5) [24] (f-plane dipoles). We summarize our results as follows: The influence of the β-effect is strongly dependent on the initial direction of propagation. Strong dipoles (with initial velocity $U_0 \approx \beta$) adjust to steadily propagating modon solutions either accelerating (for propagation in the negative x-direction), while its partners move closer together, decelerating (for propagation in the positive x-direction) due to slight sep-

Figure 2: The evolution of the potential vorticity, q (a) and of the streamfunction, ϕ (b) in the frame of reference moving with the vortex structure for a weak monopolar vortex with $\phi_m = 2\beta$. Dashed lines represents negative values solid lines positive. $L_x = 50$, $L_y = 25$, $M = 512$, $N = 256$.

aration of its partners or oscillating with decaying amplitude (for initial propagation in the y–direction), thereby carrying trapped particles predominantly in the positive x–direction, i.e., the direction opposite to the direction of linear wave propagation. Weak dipoles (with speeds $U_0 \ll \beta$ either decay due to drift wave radiation (negative x propagation), gradually separate (positive x propagation), or disintegrate (y–direction propagation) without significant long distance plasma transport. Tracing of test particles in a dipole showed that particles were mainly lost near the hyperbolic stagnation point at the rear end of the dipole. This is illustrated in the upper part of Fig. 3 for the case of a weak dipole propagating in the negative x-direction. The shrinking of the dipole leads to leaking of the particles. The loss rate of particles, initially released across the dipole vortex, is calculated by assuming particles to be lost, when the distance from the center of the vortex is larger than $r = 1.5$ (the initial radius of the dipole vortex is 1). To estimate the loss rate of particles we calculated the time during which the particles are trapped as a function of the initial distance, r_0, from the vortex center (Fig. 3). It is noteworthy that particles released exactly at the inner part of the separatrix propagate to the front stagnation point and stay there. Also it is observed that particles lost by the propagating dipole is peeled of the dipole near the separatrix and this loss proceeds at a near constant rate. The lower part of Fig. 3 illustrates particle trajectories for a steady propagating dipole in a homogeneous medium with $\beta = 0$. Here the trajectories are close to the streamlines and the particles are not lost.

As it has been illustrated, conditionally released particles may provide an effective diagnostic tool to follow the development of coherent vortical structures as suggested by Pécseli and Trulsen [26] for the investigations of phase space vortices. The particle trajectories may provide information on the dynamics of the structure, its influence and its amplitude.

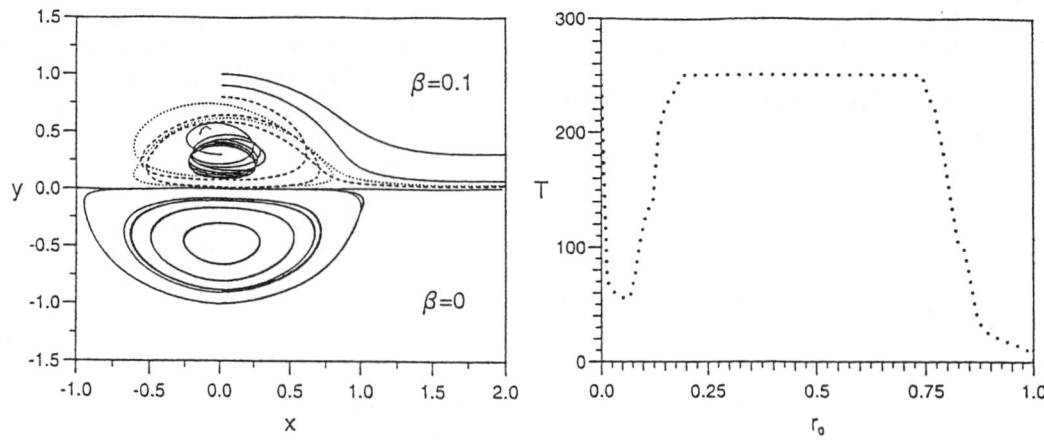

Figure 3: Selected particle trajectories initially released across a weak dipole propagating in the negative x-direction ($U_0 = 0.25\beta$) in the frame of reference following the dipole (upper half plane). Dotted and dashed lines illustrate particles that are lost and full lines particles which are effectively trapped. Lower half plane shows for comparison particle trajectories in a steadily moving dipole (left panel). Loss rate of particles initially released at a distance r_0 from the center of the dipole (right panel). T indicates the approximate time during which the particles are trapped. The total number of particles is 150.

4. DISCUSSIONS

Up to now we have considered the formation and evolution of individual vortices in relatively simple models, where the dynamics is described by one field, the electrostatic potential. For plasmas in more complex geometries, e.g., tokamaks or stellarators, these assumptions are usually not satisfied and the description of low frequency electrostatic turbulence calls for more complex models where two or more fields are coupled. The direct analogy to fluid dynamics then becomes less obvious although there are several cases where the dynamics is described by coupled fields as for instance for the evolution of baroclinic waves [21, 17]. It could be argued that large scale vortical structures would not appear for these more complex models, since usually only one quadratic conserved quantity exists and hence one should not expect the inverse cascade. Note, that coherent vortical structures appearing spontaneously in drift-wave turbulence have clearly been revealed in high resolution numerical solutions of the HM-equation including the effect of a linear instability that drives the turbulence [25, 27]. However, also for two-field models describing resistive drift waves has it been observed that vortical structures appear spontaneously in the turbulence [28]. The presence of the structures was found to significantly reduce the turbulent transport compared to what would be expected from quasilinear estimates. Recent investigations of a similar model with very high resolutions suggest, however, that the appearance of coherent structures is restricted to regimes with relatively high viscosity [29]. Also for the so-called η_i-mode turbulence, which is driven by the combined effects of a density gradient and an ion temperature gradient has the development of vortical structures been observed [30] and again the turbulent transport was markedly decreased as compared to estimates from quasilinear theories.

For models coupling two or more fields the existence of steadily propagating coherent vortical structures may be shown by a simple procedure [17]. Assuming that a localized solution does exist, its "center-of-mass" velocity is determined by using integral relations. If this velocity is outside the region of linear wave propagation, then a localized structure may exist. This method was used to show the existence of monopolar vortices for the η_i-mode [31]. The monopolar vortices were found to be steadily propagating, mainly in the direction perpendicular to the background gradients, over large distances relative to their own width for different values of η_i by direct numerical solutions of the model equations [31]. In addition it was found that the mutual interaction of vortices was markedly different from vortex interaction in one field models, i.e., no vortex merger was observed for the η_i-vortices. Also dipolar vortex solutions have been investigated for these modes [32].

In the scrape-off-layer (SOL) of Tokamaks it is found that the particle flux driven by electrostatic fluctuations is responsible for the main part of the cross-field flux [33]. The characteristics of the fluctuations in the SOL have many similarities with the low-frequency, flute-type electrostatic fluctuations observed in the edge region of linear plasma devices as for instance the Q-machine. Here the emergence of vortical structures and their strong influence on the transport was clearly demonstrated [11], although these structures only have life times on the order of their internal rotation time. One may therefore expect that vortex-like structures generated in the electrostatic turbulence would have importance also in Tokamak configurations. Up to now there exists only sparse evidence for the formation of such structures. However, the reason could be a diagnostic problem. Such structures would be difficult to detect with probes, which are only movable in a limited zone. In the ASDEX-tokamak and the W7–AS– stellarator it is found that the SOL–fluctuations are dominated by so–called "events" that appear to have a vortex-like structure [15]. They were found to have a strong influence on the transport giving rise to a burst-like turbulent flux, quite similar to the observations in the Q-machine experiments [11].

If (coherent) vortical structures indeed play a role in the SOL-turbulence it will have serious consequences for the modelling of the turbulence and transport. Linear stability considerations cannot say much about the nonlinear saturated state of the fluctuations and scaling arguments based on quasilinear transport coefficients will not be adequate. Thus, the modelling will require the full nonlinear solutions of very complex models. From such models one cannot expect to obtain "simple" power law dependent scalings for the transport coefficients.

Acknowledgements. One of the authors (JSH) was partly funded by UNI•C and partly by the National Science Academy which he gratefully acknowledges. Discussions with P. Michelsen, J. Nycander, H.L. Pécseli, G.G. Sutyrin on various aspects of the dynamics of vortices are greatfully acknowledged. This work was supported by the Danish Research Councils.

References

[1] Hopfinger, E.J. and van Heijst, G.F.J. 1993 *Ann. Rev. Fluid Mech.* **25**, 241–289.

[2] Nezlin, M.V. and Snezhkin, E.N. 1993 *Rossby vortices, spiral structures, solitons,* (Springer–Verlag, Berlin, Heidelberg).

[3] van Heijst, G.J.F. and Flor, J.B. 1989 *Nature* **340**, 212–215; Flor, J.B. 1994 *Coherent vortex structures in stratified fluids*. Thesis, The Technical University of Eindhoven, The Netherlands.

[4] Kraichnan, R.H. and Montgomery, D. 1980 *Rep. Prog. Phys.* **43**, 547–619.

[5] Hasegawa, A. 1985 *Adv. Phys.* **34**, 1–45.

[6] McWilliams, J.C. 1984 *J. Fluid Mech.* **146**, 21–42; 1990 *J. Fluid Mech.* **219**, 361–385.

[7] Matthaeus, W.H., Stribling, W.T., Martinez, D., Oughton, S. and Montgomery, D. 1991 *Physica D* **51**, 531–538; Montgomery, D., Matthaeus, W.H., Stribling, W.T., Martinez, D. and Oughton, S. 1992 *Phys. Fluids* A **4**, 3–6; Montgomery, D., Shan, X. and Matthaeus, W.H. 1993 *Phys. Fluids* A **5** 2207–2216.

[8] Shukla, P.K., Yu, M.Y., Rahman, H.U. and Spatschek, K.H. 1984 *Phys. Rep.* **105**, 227–328.

[9] Pécseli, H.L., Rasmussen, J.Juul, Sugai, H. and Thomsen, K. 1984 *Plasma Phys. Contr. Fusion* **26**, 1021–1034.

[10] Pécseli, H.L., Rasmussen, J.Juul and Thomsen, K. 1984 *Phys. Rev. Lett.* **52**, 2148–2151; 1985 *Plasma Phys. Contr. Fusion* **27**, 837–846.

[11] Huld, T., Nielsen, A.H., Pécseli, H.L. and Rasmussen, J. Juul 1991 *Phys. Fluids* B **3**, 1609–1625.

[12] Kauschke, U. 1992 *Plasma Phys. Contr. Fusion* **35**, 93–109.

[13] Singh, A.K. and Saxena, Y.C. 1994 *Phys. Plasmas* **1**, 2926–2930.

[14] Øynes, F.J., Rypdal, K., and Pécseli, H.L. 1994, In: Proceedings 1994 ICPP, Foz do Iguazu, Brazil Oct. 31 - Nov. 4, Contributed papers, Vol. 1, 245–248.

[15] Endler, M., Theimer, G., Weinlich, M., Carlson, A., Giannone, L., Neidermeyer, H., Rudyi, A. and the ASDEX-Team 1992 *Europhysics Conf. Abstracts* **16C**, part II, 787–790; 1994 In: *Transport in Fusion Plasmas, Workshop Goteborg, Sweden, June*, in press.

[16] Driscoll, C.F. and Fine, K.S. 1990 *Phys. Fluids* B , 1359–1366.

[17] Nycander, J. 1994 *Chaos* **4**, 253- -267.

[18] Meleshko, V.V. and van Heijst, G.J.F. 1994 *J. Fluid Mech.* **272**, 157–183.

[19] Hesthaven, J.S., Lynov, J.P., Nielsen, A.H., Rasmussen, J.J., Schmidt, M.R., Shapiro, E.G. and Turitsyn, S.K. 1994 "Dynamics of a nonlinear dipole vortex" submitted for publication.

[20] Leith, C.E. 1984 *Phys. Fluids* **27**, 1388–1395.

[21] Pedlosky, J. 1987 *Geophysical Fluid Dynamics* (Springer Verlag, New York).

[22] Hasegawa, A. and Mima, K. 1978 *Phys. Fluids* **21**, 87–92.

[23] Nycander, J. 1989 *Phys. Fluids* B **1**, 1788–1796.

[24] Sutyrin, G.G., Hesthaven, J.S., Lynov, J.P. and Rasmussen, J. Juul 1994 *J. Fluid Mech.* **268**, 103–131.

[25] Crotinger, J.A. and Dupree, T.H. 1992 *Phys. Fluids B* **4**, 2854–2870.

[26] Pécseli, H.L. and Trulsen, J. 1989 *Phys. Fluids B* **1**, 1616–1636.

[27] Horton, W. 1986 *Phys. Fluids* **29**, 1491–1503.

[28] Koniges A.E., Crotinger, J.A., Diamond, P.H. 1992 *Phys. Fluids B* **4**, 2785–2793.

[29] Biskamp, D., Camargo, S.J. and Scott, B.D. 1994 *Phys. Lett. A* **186**, 239–244.

[30] Ottaviani, M., Romanelli, F., Benzi, R., Briscolini, M., Santangello, P. and Succi, S. 1990 *Phys. Fluids B* **2**, 67–74.

[31] Nycander, J., Lynov, J.P. and Rasmussen, J. Juul 1993 *Europhys. Lett.* **23**, 249–255.

[32] Lynov, J.P., Michelsen, P.K. and Rasmussen, J. Juul 1994 In: Proceedings 1994 ICPP, Foz do Iguacu, Brazil, Oct. 31 - Nov. 4. Contributed papers. Vol 2, 91–94.

[33] Wootton, A.J., Tsui, H.Y.W. and Prager, S. 1992 *Plasma Phys. Contr. Fusion* **34**, 2023–2030.

MICROINSTABILITIES IN PLASMAS

A. M. EL Nadi

Electrical Engineering Department
Faculty of Engineering
Cairo University
Giza, Egypt.

Abstract

In this work we review some of the progress made in the area of plasma microinstabilities. The use of the microscopic approach allows the investigation of various features related to velocity-dependent effects such as particle trapping, wave-particle resonance and short wavelength effects. The work presented will deal mainly with those instabilities generally considered to be important in fusion devices. These include the trapped electron mode, the ion-temperature- gradient -driven mode and instabilities due to the presence of an energetic particle component. Results obtained using both slab and toroidal models will be presented.

1. INTRODUCTION

Plasma instabilities are broadly divided into two categories: macroinstabilities and microinstabilities. By macroinstabilities one means the family of instabilities which can be analyzed using one-fluid mhd equations such as those related to the presence of a current gradient or a pressure gradient in a curved magnetic field. The wavelengths considered are much longer than the size of the particle orbits, being usually of the same order as the plasma dimensions. On the other hand, plasma microinstabilities refer to the instabilities related to the non-maxwellian nature of the velocity-distribution function as is the case when density and temperature gradients are present. They require independent treatments for both electrons and ions which take into account velocity-dependent effects such as wave-particle resonance. Frequently, large growth rates are obtained when the wavelength is of the order of the ion Larmor radius, which is usually much shorter than the equilibrium scale length. It is clear that one can use the microscopic approach to study all plasma instabilities although an mhd analysis is usually carried out when velocity-dependent effects are not expected to be important and more emphasis is placed on geometrical effects. An example is the analysis of the tearing mode where one must use the microscopic approach to study the details of the inner tearing layer near the rational surface where reconnection of the magnetic field lines occurs, while an mhd analysis is adequate to study the outer layer which extends to the plasma boundary [1].

It is our objective in this work to report on some of the contributions made in the area of plasma microinstabilities with emphasis on tokamak plasmas. Many important results were produced over the years by many authors [2], and only a limited part can be mentioned here. We will begin by describing the general kinetic approach of studying linear microinstabilities in slab and toroidal geometries and then describe instabilities related to the presence of trapped electrons, an ion temperature gradient or energetic particles. Nonlinear effects, mode saturation and related anomalous transport will not be dealt with here.

2. THE SLAB LIMIT AND THE BALLOONING FORMALISM

It is clear that a three-dimensional stability analysis is desirable but very difficult to carry out, even numerically. If an axis of symmetry exists, the problem becomes two-dimensional and a Fourier decomposition of the perturbations along the ignorable coordinate can be made. It is sometimes possible to ignore the variation along a second coordinate (such as by neglecting toroidicity and considering a straight tokamak) and the stability problem is then reduced to a one-dimensional cylindrical problem in the radial variable. The slab model is the limit of the cylindrical case when curvature effects are ignored allowing the use of Cartesian coordinates. All equilibrium quantities such as the density, temperature and magnetic field intensity will thus depend on one coordinate(x) with the magnetic field being usually considered to point in the z-direction. The equilibrium velocity distribution functions of ions and electrons together with the magnetic field are chosen consistently so that Ampere's law is satisfied. This relates the magnetic shear to the plasma current along the field lines and the radial gradient of magnetic field intensity to the kinetic pressure profile. The Vlasov equation is then solved to obtain the perturbations in the velocity distribution functions if collisions can be ignored. Otherwise a suitable collision model must be used. By substituting for the source terms in Maxwell's equations, the dispersion relation may be obtained as an algebraic equation if the local approximation is made, an integral equation if the radial scale length of the perturbation is shorter than an ion Larmor radius or an ordinary differential equation in the long scale length limit.

It is interesting to note that the two-dimensional stability problem of a tokamak plasma can be reduced to a one-dimensional problem in the high toroidal mode number limit [3]. Because of toroidicity, some plasma parameters will have a strong poloidal dependence, such as magnetic drift velocity and fraction of trapped particles. This means that for a given toroidal mode number each poloidal harmonic of the perturbation will be coupled to the neighboring harmonics. If the spacing between the neighboring rational surfaces is much shorter than the radial scale length for the variation of equilibrium parameters, then to lowest order we can consider the poloidal harmonics to be identical in shape except for a constant phase factor and centered on neighboring rational flux surfaces.

When considering toroidal plasmas it is preferable to use the gyrokinetic equation instead of the Vlasov equation., where the fast gyromotion has already been accounted for [4]. The dispersion relation is obtained by solving the ion and electron gyrokinetic equations for the perturbed distribution functions and then using Maxwell's equations. Applying the ballooning mode formalism results in an ordinary differential-difference equation where the difference operator is due to the toroidal coupling terms. Following Fourier transformation one can use a one-dimensional shooting code to obtain the eigenfunctions and eigenvalues. In the limit of strong ballooning, the coupling between the neighboring harmonics is strong and this allows approximating the difference operator by a differential operator[5]. This occurs when the spacing between the rational flux surfaces is much shorter than the width of the eigenfunction in the slab limit and results in the simplification of the ordinary differential equations such that analytic solutions can be obtained. It turns out that the strongest microinstabilities have mode structures which are usually localized to the outside of the torus where the curvature is unfavorable and where the trapped particle population is localized. It should be noted that the radial structure of the mode can be obtained by superposing the different harmonics taking the equilibrium scale length into account.

3. THE TRAPPED ELECTRON INSTABILITY

Since the trapped electron instability is a drift wave which is destabilized by trapped electrons, we will begin by briefly reviewing the drift wave. It is well-known that this is a low-frequency mode which releases the available expansion free energy whenever a peaked density profile exists in a magnetized plasma. It results from the ion **E x B** motion and the electron neutralizing parallel flow along the field lines. The wave rotates in the direction of the electron diamagnetic current and feeds on the electron parallel kinetic energy through the process of inverse Landau damping. In the slab limit, stability is achieved when magnetic shear is introduced, however small the amount of shear is. The cause of stability is the outward emission of wave energy caused by the influence of shear on the radial group velocity. The picture changes when one allows for toroidal effects using the ballooning formalism. In this case two branches of eigenmodes exist, one is slab-like and the other is induced by toroidal effects. While the first branch experiences shear damping, the toroidicity- induced eigenmodes can be localized with the outward energy convection due to magnetic shear being inhibited by toroidal effects [6] . Thus drift waves can be unstable in toroidal geometry.

The trapped-electron mode arises when the effective collision frequency of trapped electrons is less than their bounce frequency . The mode frequency should also be much higher than the bounce frequency of trapped ions so that their contribution can be ignored. The presence of collisions is usually accounted for by introducing a Krook collision term into the Vlasov or gyrokinetic equations, which has been shown to be a good approximation. Magnetic shear again has a stabilizing role through the contribution of the passing ions. If grad-B drift of trapped electrons is ignored, the mode will be stabilized as the collision frequency decreases. If it is taken into account, it can be shown that resonance between the wave and electron banana precession around the torus is a more important destabilizing factor than collisions in the low collisionality limit [7]. The growth rate will then continue to rise as the trapped electron collision frequency decreases. Similar results were also obtained for a toroidal geometry using the ballooning formalism [8]. The trapped electron mode may thus be important as a source of anomalous transport in high temperature tokamaks.

4. THE ION-TEMPERATURE-GRADIENT-DRIVEN INSTABILITY

In a slab system with straight magnetic field lines, the ion-temperature-gradient-driven (ITGD) instability is related to the ion sound wave modified by the presence of the ion temperature gradient and rotates in the direction of the ion diamagnetic flow. The presence of magnetic shear merely results in the radial localization of the mode and does not produce stability [9]. The instability can be studied in the fluid limit and results from an effective negative compressibility. To obtain the stability threshold properly one must allow for the wave-ion resonance along the magnetic field lines which is usually important because of the low phase velocity of the wave near the threshold.

In a toroidal system the instability is mainly driven by the combination of the interchange and ion temperature gradient destabilizing mechanisms. In the zero-β electrostatic limit, the full kinetic analysis requires a solution of a Fredholm integral equation of the second kind which is usually carried out numerically [10] (where β is the ratio of the plasma kinetic pressure to the magnetic pressure). Various results can be obtained with different levels of sophistication with respect to obtaining the ion response. Both magnetic drift and Landau resonances can be accounted for in a radially local numerical solution. Sometimes both resonances are ignored in order to obtain an analytic result based on the fluid expansion. Other authors allowed for one type of resonance only, thus simplifying the numerical solution of the radial problem. It should also be mentioned that the use of the strong coupling approximation results in a second order ordinary differential equation which can be solved analytically [11]. The opposite limit of weak toroidicity, where the mode is stretched along the field lines with no significant ballooning, can also be treated analytically with the ion transit resonance being accounted for [12]. The effects of trapped electrons and trapped ions were also considered while some calculations specialized to the flat density H-mode plasmas. The influence of sheared toroidal and poloidal flows on the ITGD mode was investigated and it was found that the growth rate can be significantly altered by a small amount of sheared flow as shown in Fig. 1 [13].

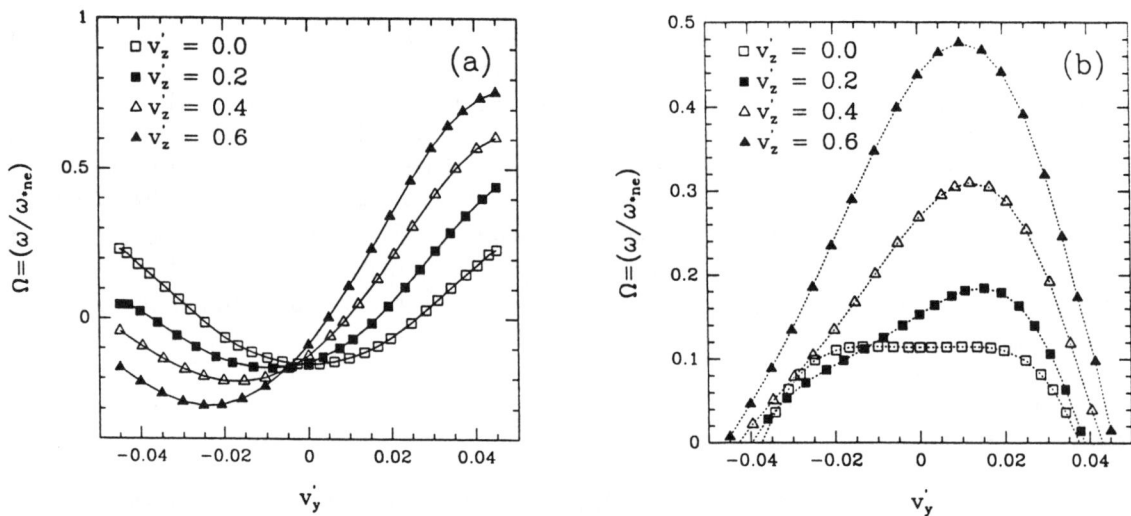

Fig. 1. (a) Normalized real frequency and (b) normalized growth rate of the ITGD mode versus normalized shear in the poloidal flow velocity [13].

The above discussion applied to the electrostatic zero-β limit. As β is increased gradually, the ITGD instability acquires first a finite radial magnetic field component driven by the perturbed current along the magnetic field lines. In the strong toroidal coupling limit one obtains the eigenfrequencies by solving a system of two coupled ordinary second order differential equations in the perturbed poloidal electric field and radial magnetic field. This coupled system describes both the finite-β modified ITGD mode and the mhd-like pressure driven ballooning mode. Numerical solutions predict a strong effect for β on the growth rate of the ITGD mode [14]. The numerical techniques must be able to distinguish between the two modes through boundary conditions. As β is increased further, one can no longer ignore both the modification in the plasma equilibrium and the compressional magnetic component of the perturbation along the magnetic field lines which results in an added coupled second order differential equation. The two effects are usually opposite in the fluid limit which results in a reduced effect.

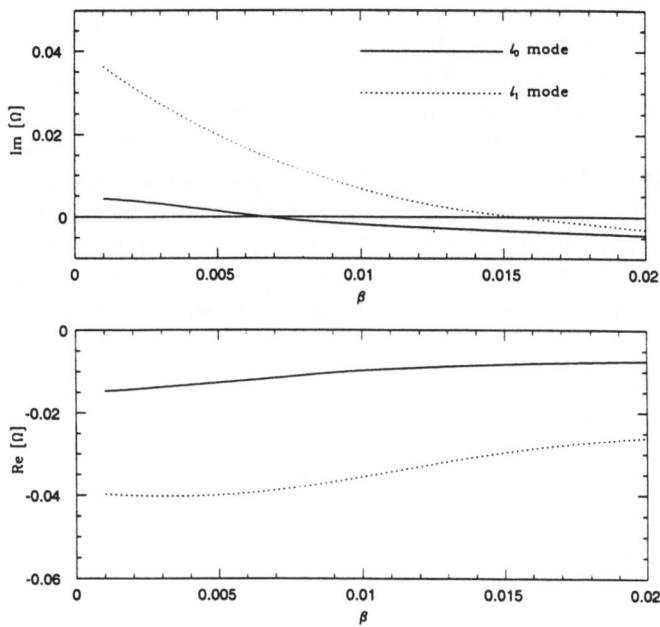

Fig. 2. Normalized real frequency and growth rate of the ITGD two lowest order radial modes as a function of β [14].

INSTABILITIES DUE TO AN ENERGETIC PLASMA COMPONENT

The interaction of an energetic particle component with a background plasma has recently received considerable attention. The energetic component can be hot electrons such as in a bumpy torus or a tandem mirror with a hot electron anchor cell, or hot ions such as in neutral-beam-injection-heated tokamaks or in a thermonuclear plasma. The hot particles will precess due to the curvature of the confining magnetic field giving rise to a negative energy precessional wave which can result in instabilities either through wave-wave coupling with a positive energy wave (such as an mhd mode) or through the presence of some positive energy absorption mechanism such as magnetic shear damping, collisions or Landau damping.

Such instabilities are obviously not desirable and should be avoided since they will be expected to result in a reduction of the density of the energetic component responsible for the instability. Only then can one expect to benefit from their presence.

An extensive amount of work was carried out trying to stabilize a main plasma confined in a magnetic field possessing a bad curvature by introducing a hot electron component [15]. It could be expected that because of their high energy and high precession frequencies, the hot electrons will not effectively take part in the low-frequency motion related to the interchange instability. This can contribute to stability in two ways. First, if the hot electron β is high enough, a rigid diamagnetic well can be created. Besides, the rigidity of the hot electrons under the influence of the low frequency wave will give rise to an asymmetry in the total electron and ion perturbed charge densities (similar to the stabilizing effect of the finite ion Larmor radius]. Obviously a high frequency analysis must be carried out to ensure the stability of the hot component. The high frequency negative energy wave related to the precession of the hot electrons can lead to instabilities by coupling to a host of positive energy waves such as the lower hybrid wave and the compressional and shear Alfvén waves. It can be shown that a high frequency interchange instability can be excited if the ratio of the energetic electron density to the background density exceeds a certain threshold as shown in Fig. 3. The validity of this threshold was confirmed experimentally through the sudden ejection of a fraction of the energetic electron population as one tries to increase their density using microwave electron resonance heating.

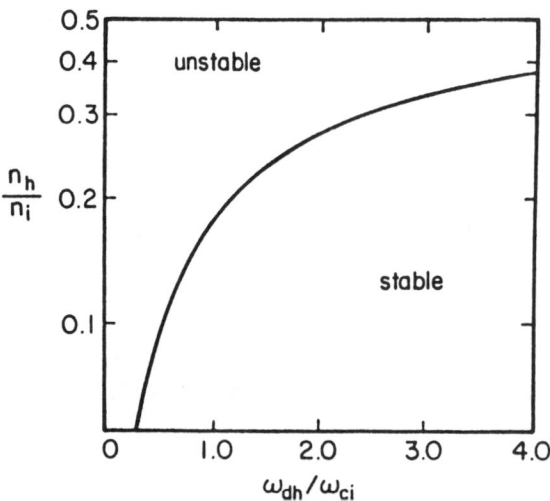

Fig. 3. Fraction of hot electrons needed to drive the high frequency interchange mode [15].

It is interesting to note that the energetic electron component is not totally rigid with respect to low frequency perturbations, and since it carries its own diamagnetism along, the energetic electron response depends on their precession in the vacuum magnetic field modified by the diamagnetism of the background plasma. This results in a magnetic perturbation which modifies the drive of the interchange mode, and which is very strong when the pressure of the background plasma is high enough to reverse the direction of the particle precession with the hot electron diamagnetism being ignored. This determines the maximum allowable pressure of the main plasma [16].

It is not difficult to study the stability of the compressional and shear Alfvén waves taking the coupling to the hot electron precessional wave into account. An increase in the plasma density lowers the Alfvén speed and hence increases the coupling to the precessional wave. Hence an upper plasma density limit can thus be obtained [17].

When high-power, nearly perpendicular beam injection is applied in tokamak experiments, bursts of large-amplitude fluctuations were observed and correlated with significant losses of energetic beam ions as shown in Fig. 4. A reasonable explanation was offered by considering the coupling of the negative energy precessional mode to the positive energy internal kink mode and a maximum limit on β was predicted for the energetic ions [18].

We finally notice that the energetic Alpha particles in a reacting deuterium-tritium tokamak plasma are born with an energy of 3.52 MeV and have a density which is centrally peaked. Unstable drift waves related to the Alpha particle distribution are then expected to be present and to be coupled to shear Alfvén waves. Transiting ions will resonate with the drift-Alfvén waves and destabilize them through inverse Landau damping [19-21]. There has been recently much interest in the study of one type of low-mode-number global (radially unlocalized) Alfvén wave called the toroidicity-induced shear Alfvén eigenmode (TAE), whose frequency lies within the gaps in the Alfvén continuum generated when toroidicity resolves the degeneracy of two Alfvén continua centered on neighboring rational surfaces as shown in Fig. 5 [22]. Because of their nonlocalized nature, they may be expected to lead to important anomalous losses for Alpha particles, but this can be confirmed only through extensive nonlinear calculations.

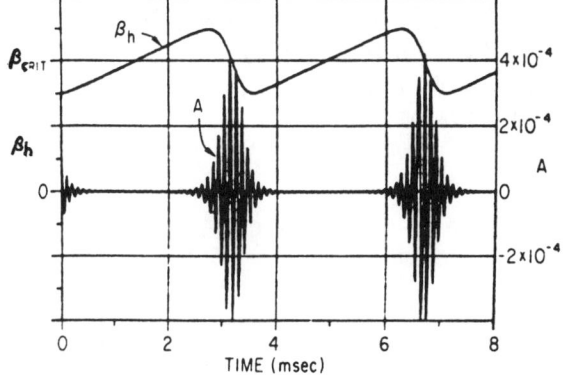

Fig. 4. Kink-mode amplitude and beam β versus time, obtained analytically [18].

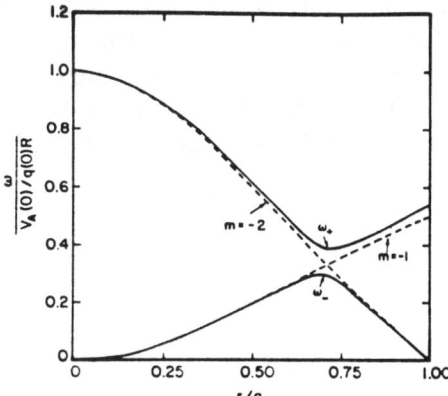

Fig. 5. Gap in the toroidal shear Alfvén continuous spectrum [22].

In conclusion we presented a summary of a part of the recent activities in area of plasma microinstabilities. This included the trapped electron instability, the ion-temperature-gradient-driven instability and a few energetic-particle-driven instabilities. The study was limited to linear analysis although only a nonlinear study can evaluate the consequences of any instability.

ACKNOWLEDGMENTS

The author wishes to acknowledge valuable discussions with A. Rogister and W. M. Tang.

REFERENCES

[1] J. Wesson, Tokamaks, Clarendon Press, Oxford (1987).
[2] A. El Nadi, Modern Plasma Physics, IAEA, Vienna, 275 (1982).
[3] J. W. Connor, R. J. Hastie and J. B. Taylor, Proc. R. Soc. 365, 1 (1979).
[4] W. M. Tang, Nucl. Fusion 18, 1089 (1978).
[5] J. B. Taylor, in *Plasma and Controlled Nuclear Fusion Research* (International Atomic Energy Agency, Vienna,1977), Vol. II, p. 323.
[6] L. Chen and C. Z. Cheng, Phys. Fluids 23, 2242 (1980).
[7] J. C. Adam, W. M. Tang and P. H. Rutherford, Phys. Fluids 19, 561 (1976).
[8] A. Rogister and G. Hasselberg, Phys. Fluids 22, 2382 (1979).
[9] B. Coppi, M. N. Rosenbluth and R. Z. Sagdeev, Phys. Fluids 10, 582 (1967).
[10] F. Romanelli, Phys. Fluids **B1**, 1018 (1989).
[11] P. N. Guzdar, L. Chen, W. M. Tang and P. H. Rutherford, Phys. Fluids 26, 673 (1983).
[12] F. Romanelli, L. Chen and S. Briguglio, Phys. Fluids **B3**, 2496 (1991).
[13} M. Artun and W. M. Tang, Phys. Fluids **B1**, 1185 (1989).
[14] J. V. W. Reynders, Phys. Plasmas 1, 1953 (1994).
[15] H. L. Berk, Phys. Fluids 19, 1255 (1976).
[16] J. W. Van Dam and Y. C. Lee, in *Proceedings of the EBT Ring Physics Workshop* (Oak Ridge National Laboratory, Oak Ridge, 1979), CONF-791228, p.471.
[17] A. El Nadi, Phys Fluids 25, 2019 (1982).
[18] L. Chen, R. B. White and M. N. Rosenbluth, Phys. Rev. Lett. 52, 1122 (1984).
[19] A. B. Mikhailovskii, Zh. Eksp. Teor. Fiz. 68,1722 (1975) [Sov. Phys. JETP 41, 890 (1975)].
[20] M. N. Rosenbluth and P. H. Rutherford, Phys. Rev. Lett. 34, 1428 (1975).
[21] Y. M. Li, S. M. Mahajan and D. W. Ross, Phys. Fluids 30, 1466 (1987).
[22] G. Y. Fu and J. W. Van Dam, Phys Fluids **B1**, 1949 (1989).

SPACE AND ASTROPHYSICAL PLASMAS

THEORY OF AURORAL ARCS

P.K. Schukla
Institut für Theoretische Physik IV, Ruhr-Universität Bochum, D-44780 Bochum, Germany

Two nonlinear models are proposed that can explain the origin of various coherent structures in the auroral zone of the earth's ionosphere. These include the sheared flow driven nonlinear drift waves, as well as nonlinear inertial Alfvén waves. It is found that the formar can account for the salient features of the black aurora, whereas the latter can successfully produce structures that resemble the fold, spiral, and curl in auroral arcs.

I. INTRODUCTION

Transitions of auroral arcs from diffuse to discrete states with subsequent formation of different types of auroral structures accompany the expansion and recovery phases of substorms. Despite extensive investigations they are, however, poorly understood. High-resolution TV observations [1-3] have shown three basic auroral forms, namely, folds, spirals, and curls, which differ from each other in their sizes, vorticities, velocities and life–times. The auroral spirals and folds occur on the dawn side as well as on the dusk side of the auroral oval [1,2] and have rotation speeds within the range of one to several tens of kilometers per second. All the observed auroral structures reported in Ref. [1] have a common vertical structure. Auroral spirals are large-scale vertical configurations in the aurora characterized by a clockwise rotation direction antiparallel to the ambient magnetic field. They have typical sizes from tens to hundred of km. Folds are characterized by a smaller scale size of about 20 km. A curl is a small-scale vortex configuration in the aurora. Its typical size is less than ten kilometers, and its rotation direction is counterclockwise.

Furthermore, optical TV observations of auroral display [4,5] also indicate the regions of dark striations to occur within discrete aurora. These regions of low auroral luminosity show east-west aligned vortex streets, as it is observed during the dynamical evolution of discrete auroral arcs. The spatial scalelengths of these structures (the black aurora) are of several kilometers. It thus emerges that the dynamic evolution of auroral arcs is characterized by the development of various types of coherent nonlinear structures.

In the past, a number of theoretical models have been proposed to explain the transition from the original state diffuse auroral arc and for describing various related phenomena. Specifically, one has to find the origin of the electron acceleration mechanism responsible for the excitation of the aurora, and to explain the characteristic spatial sizes, the geometrical forms, and the dynamics of the basic types of the auroral structures. It is widely held opinion that coherent auroral structures may be caused by some kinds of instabilities [6] such as the Kelvin-Helmholtz and the tearing instabilities. Since the instability mechanism can describe only the initial triggering of the phenomena, the formation of vortical structures is intrinsically nonlinear. Thus, nonlinear models [7-9] have been proposed for the dynamical evolution of auroral structures.

In this paper, we present descriptions [8,9] of two nonlinear models that could account for the black vortex streets and the dynamics of discrete active auroral forms. Specifically, it is shown that the salient features of the black auroral streets can be understood in terms of the sheared flow driven nonlinear drift waves, whereas the nonlinear Alfvén wave model of a cold magnetoplasma is capable of producing the structures that have been observed in the auroral regions of the earth's ionosphere.

In Section II, we present a brief derivation of the three-dimensional nonlinear mode coupling equations for electrostatic drifts waves in the presence of sheared flows. In Section III, we outline our nonlinear Alfvén wave model in which the scalar and vector potentials are nonlinearly coupled. The results of our numerical simulations are discussed in Section IV. Section V contains a summary of our investigation as well as its relevance to observational data.

II. The Model for Black Auroral Streets

Here, we discuss the nonlinear drift wave model [9] for the black aurora. Consider a nonuniform ionosphereic plasma in the presence of an equilibrium density gradient $d_x n_0$ and an equilibrium velocity gradient $(d_x v_{i0})$ the geomagnetic field-aligned ion flow velocity v_{i0} in a direction (east) transverse to the equilibrium magnetic field $B_0 \hat{z}$, where n_0 is the equilibrium plasma number density, B_0 is the strength of the geomagnetic field lines at the ionosphere heights, and **z** is the unit vector along the **z** axis, which is vertically downward. We suppose an open system with a continuous charged particle inflow from the magnetosphere into the ionosphere.

In the presence of low-frequency (in comparison with the ion gyrofrequency $\omega_{ci} = eB_0/m_i c$, where e is the magnitude of the electron charge, m_i is the ion mass, and c is the speed of light), long wavelength (in comparison with the ion gyroradius) electrostatic field **E** $(= -\nabla\phi$ where ϕ is the electrostatic potential), the perpendicular component of the electron and ion fluid velocities are given by, respectively,

$$V_{e\perp} \approx \frac{c}{B_0}\hat{z}\times\nabla\phi - \frac{c}{eB_0 n_e}\hat{z}\times\nabla(n_e T_e) \equiv v_{EB} + v_{De}, \quad (1)$$

and

$$V_{i\perp} \approx V_{EB} - \frac{c}{B_0\omega_{ci}}(\partial_t + v_i\cdot\nabla)\nabla_\perp\Phi, \quad (2)$$

where n_e (T_e) is the electron number density (constant electron temperature) and the ions are assumed to be cold.

When $|\partial_t + v_e\cdot\nabla| \ll v_{te}\partial_z, v_{D0}\cdot\nabla$, where v_{te} is the electron thermal velocity and v_{D0} is the equilibrium electron diamagnetic drift, then inertialess electrons rapidly thermalize along the **z** axis, and the corresponding electron response is given by the Boltzmann distribution

$$n_e = n_0(x)\exp(e\phi/T_e) \equiv n_0\exp(\psi). \quad (3)$$

Substituting (2) into the ion continuity equation, invoking the quasineutral approximation ($n_i = n_e$), and making use of (3) we obtain to the leading order

$$(\partial_t + v_{i0}\partial_z)(\psi - \rho_s^2 U) + v_*\partial_y\psi - \rho_s^4\omega_{ci}J(\psi,U) +$$
$$+ \partial_z v_{iz} = 0, \quad (4)$$

where

$$\rho_s = c_s/\omega_{ci}, c_s = (T_e/m_i)^{1/2}, v_* = \rho_s c_s/L_n, L_n =$$
$$= -n_0/d_x n_0 \text{ and } J(\psi, U = \nabla_\perp^2\psi) = \partial_x\psi\partial_y U - \partial_y\partial_x\psi U.$$

Here, the y direction is along the north. the parallel component of the ion fluid velocity perturbations is determined from

$$(\partial_t + v_{i0}\partial_z)v_{iz} + \rho_s^2\omega_{ci}J(\psi,v_{iz}) = -c_s^2\partial_z\psi +$$
$$+ \rho_s c_s\omega\partial_y\psi, \quad (5)$$

where $\omega_v = d_x v_{i0}$. Equations (4) and (5) govern the nonlinear dynamics of weakly interacting coupled electron drift-ion acoustic waves in nonuniform magnetoplasmas with sheared ion flow and isothermal electron distribution. However, in the presence of nonisothermal electrons, one may model the electron distribution along the **z** axis by Schamel's vortex distribution [10], which leads to the electron number density perturbation n_{e1} ($= n_e - n_0 \ll n_0$)

$$n_{e1} = n_0(x)(\psi - \frac{4}{3}b\psi^{3/2} +$$
$$+ \frac{1}{2}\psi^2 - \frac{8}{15}d\psi^{5/2} + ...), \quad (6)$$

where $b = (1-\delta)/\sqrt{\pi}, d =$
$$= (1-\delta^2)/\sqrt{\pi} \text{ and } \delta = T_e/T_{et}$$
is the ratio of free and trapped electrons. If we use (6), instead of (3), then (4) is replaced by

$$(\partial_t + v_{i0}\partial_z)(\psi - \rho_s^2 U) + v_*\partial_y\psi -$$
$$- \frac{4}{3}b\partial_t\psi^{3/2} + \partial_z v_{iz} = 0, \quad (7)$$

where $|\partial_t\psi^{3/2}|/\omega_{ci} \gg \rho_s^4 J(\psi, U)$

has been assumed. Equations (5) and (7) now constitute a pair that describe the nonlinear drift waves in the presence of the electron vortex distribution.

In the linear limit, we can neglect the nonlinear terms and obtain from (4) and (5) the local dispersion relation for

$k|L_n| \gg 1$ and $kv_{i0}/|d_x v_{i0}| \gg 1$:

$$\omega - k_z v_{i0} = -(\omega_*/2) \pm [(\omega_*/2)^2 + \omega_s^2(1 -$$
$$- k_y\omega_v/k_z\omega_{ci})]^{1/2}, \quad (8)$$

which predicts an oscillatory instability when $d_x v_{i0} > (k_z/k_y)(1 + \omega_*^2/4\omega_s^2)\omega_{ci}$.
Here
$$\omega_* = k_y v_*/(1 + k_\perp^2\rho_s^2) \text{ and } \omega_s =$$
$$= k_z c_s/(1 + k_\perp^2\rho_s^2)^{1/2}$$
are the drift wave and ion-acoustic wave frequencies, respectively. Choosing the density gradient scalelength of order 50 km and taking typical ionospheric F-layer parameters ($\omega_{ci} = 240\ s^{-1}$, $p_s = 8$ m, $c_s = 900$ m/s), we find that the maximum growth rate (with an e-folding time of roughly 130 s) is associated with electrostatic perturbations of wavelengths of about 3 km. The observed time scales of distortion of the black aurora and the observed spatial scale seem to be consistent with those involving sheared ion flow driven drift wave turbulence.

Finite amplitude drift waves nonlinearly interact among themselves and lead to the formation of black auroral streets. The dynamics of the latter can be investigated numerically by solving Eqs. (4), (5) and (7). Let us present the stationary vortex solutions of the nonlinear drift wave equations in the stationary frame $\xi = y + \alpha z - ut$, where α and u are constants. Thus, from (4) and (5), we have

$$D_\xi\left(\nabla_\perp^2 \psi - \frac{\lambda}{\rho_s^2}\psi\right) = 0, \quad (9)$$

where $D_\xi = \partial_\xi - (c/u_0 B_0)(\partial_x \phi \partial_\xi -$
$-\partial_\xi \phi \partial_x)$, $u_0 = u - v_{i0}$ and $\lambda =$
$= 1 - (v_*/u_0) - c_s^2 \alpha^2(1 - \omega_v/\omega_{ci}\alpha)/u_0^2$.
When the vortex speed is given by $u = u = v_{i0} + (v_*/2) \pm |v_*^2 + 4c_s^2\alpha^2(1-\omega_v/\omega_{ci}\alpha)|^{1/2}$, then (9) can be solved by the ansatz

$$\nabla_\perp^2 \phi = \frac{4\phi_0 K^2}{a^2}\exp\left[-\frac{2}{\phi_0}(\phi - \frac{u_0 B_0}{c}x)\right], \quad (10)$$

where ϕ_0 K and a are arbitrary constants. The solution of Eq. (10) is given by

$$\phi = \frac{u_0 B_0}{c}x + \phi_0 ln$$
$$\left[2\cosh(Kx) + 2(1-\frac{1}{a^2})\cos(Ky)\right]. \quad (11)$$

For $a^2 > 1$ the vortex profile given by (11) resembles the Kelvin-Stuart "cat's eyes" that are chains of vortices. The latter can be associated with the black auroral vortex streets.

Next, we note that the stationary Eqs. (4) and (7) can be combined to yield

$$\frac{1}{r}\frac{d}{dr}\left(r\frac{d\psi}{dr}\right) - \lambda \psi + \frac{4}{3}b\psi^{3/2} = 0, \quad (12)$$

where $r = (x^2 + \xi^2)^{1/2}/\rho_s$. For $\lambda > 0$, Eq. (12) admits localized vortex cylinder, the profile of which is similar to that presented in Ref. [10].

III. The Nonlinear Alfvén Wave model

In order to understand the formation of large-scale folds and spirals as well as small-scale curls, we here present the nonlinear Alfvén wave model. The latter involves nonlinear interaction of finite amplitude low-frequency (in comparison with ion gyrofrequency) Alfvén-like perturbations. The relevant electron and ion fluid velocities in the electromagnetic fields are

$$V_e \approx V_{EB} + V_{De} + (v_{z0} + v_{z1})B_\perp/B_0 + \hat{z}v_{z1}, \quad (13)$$

and Eq. (2), respectively. Here, v_{z1} is the electron fluid velocity perturbation in the equilibrium flow velocity v_{z0}, $B_\perp = \nabla A_z \times \hat{z}$ and A_z is the parallel (to z) component of the vector potential. The parallel component of the electron fluid velocity is determined from Ampere's law, i.e.

$$v_{c1} \approx (c/4\pi n_e e)\nabla_\perp^2 A_z, \quad (14)$$

where the contribution of the small parallel ion current has been neglected. For simplicity, the ions are assumed to be cold.

Substituting Eqs.(2) and (13) into the charge density conservation equation $\nabla \cdot j = 0$, making use of Eq. (14), and assuming $n_e = n_i$ (as $\omega_{ci} \ll \omega_{pi} = (4\pi n_0 e^2/m_i)^{1/2}$), we obtain

$$d_t \nabla_\perp^2 \phi + \frac{\omega_{ci}}{c}d_x v_{z0}\partial_y A_z + \frac{v_A^2}{c}dz\nabla_\perp^2 A_z = 0, \quad (15)$$

where a
$d_t = \partial_t + (c/B_0)\hat{z}\times \nabla\phi\cdot\nabla$, $dz = \partial_z + B_0^{-1}\nabla A_z \times \hat{z}$,
and v_A is the Alfvén velocity. Here, we have retained the most important nonlinear terms including the nonlinear ion polarization drift and the coupling of the parallel electron flow with the perturbed magnetic field. Thus, the nonlinear sources drive the ion vorticity.

On the other hand, substituting for the parallel electric field $E_z = -\partial_z - c^{-1}\partial_t A_z$ in the parallel component of the electron momentum equation, and using Eqs. (13) and (14), we obtain

$$(d_t + v_{D0}\partial_y)A_z - \lambda_e^2\nabla_\perp^2(d_t + v_{z0}\partial_z)A_z + \\ +c(\partial_z + S_v\partial_y)\phi - (cT_e/en_0)d_z n_1 = 0, \quad (16)$$

where $v_{D0} = -(c/eB_0 n_0)d_x(n_0 T_{e0})$ is the equilibrium electron diamagnetic drift, $S_v = d_x v_{z0}/\omega_{ce}$ is the electron shear parameter, $\omega_{ce} = B_0/m_e c$ is the electron, $\lambda_e = c/\omega_{pe}$ is the electron skin depth, and ω_{pe} is the electron plasma frequency.

The electron continuity equation can be written as

$$(d_t + v_{z0}\partial_z)n_1 - \frac{c}{B_0}d_x n_0 \partial_y \phi + \\ +\frac{n_0}{B_0}d_x v_{z0}\partial_y A_z + \frac{c}{4\pi e n_0}d_z \nabla_\perp^2 A_z = 0. \quad (17)$$

Equations (15) to (17) govern the dynamics of weakly interacting kinetic/inertial Alfvén waves in the presence of density and sheared electron flow. The latter can destabilize the kinetic/inertial Alfvén waves. In an uniform plasma without the sheared flow, the relevant nonlinear equations for finite amplitude inertial Alfvén waves in an extremely low $\beta(\beta \ll m_e/m_i)$ cold plasma are of the form

$$d_t \nabla_\perp^2 \phi + \frac{v_A^2}{c} d_z \nabla_\perp^2 A_z = 0, \quad (18)$$

and

$$d_t \left(1 - \lambda_e^2 \nabla_\perp^2\right) A_z + c \partial_z \phi = 0. \quad (19)$$

It is of interest to note that possible stationary solutions of (18) and (19) can be represented in the form of a dipolar vortex and vortex streets. As an illustration, when the vortex street size is much larger than λ_e, then in a stationary frame ($\xi = y + \alpha z - ut$), (18) and (19) can be combined to yield

$$\left(1 - \frac{v_A^2 \alpha^2}{uc}\right) D_\xi \nabla_\perp^2 \phi = 0, \quad (20)$$

It follows that when

$u \neq v_A^2 \alpha^2 / c$, then $D_\xi \nabla_\perp^2 \phi = 0$.

The latter is satisfied by Eq.(10), which admits the vortex street solution Eq. (11).

IV. Numerical Simulations Results

In order to model the observed raylike forms of the aurora, we consider 2D, nonlinearly interacting parallel electron current sheets which are uniform along the ambient magnetic field. Thus, in what follows, we discuss the numerical simulation results [8] of the coupled equations

$$d_t \nabla_\perp^2 \phi = J(q, \nabla_\perp^2 A_z), \quad (21)$$

and

$$d_t q = 0, \quad (22)$$

where

$d_t = \partial_t + J(\phi, ..), q = A_z - \varepsilon \nabla_\perp^2 A_z \varepsilon = m_e B_0^2 / 4\pi n_0 T_e m_i$,

which are derived from Eqs.(18) and (19) by assuming that $\partial_z \phi = 0$ and $\partial_z \nabla_\perp^2 A_z = 0$. Here, the time is normalized by ω_{ci}^{-1}, the space variables by p_s, and the potentials ϕ and A_z by T_e/e.

$\nabla_\perp^2 A_z = A_0 k^2 \ln \cosh(ky)$ and

$\phi = \phi_1 \tanh(ky) + \phi_2 \exp(-\gamma x^2)$

with

$A_0 = 0.1, k = 1.18, \phi_1 = 1, \phi_2 = 5,$ and $\gamma = 0.15$

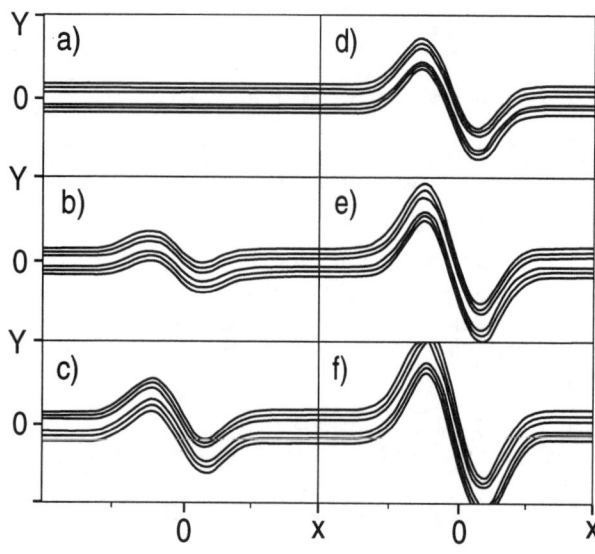

Fig.1 (after Ref.8) displays the temporal evolution of the parallel electron current sheet for the functions

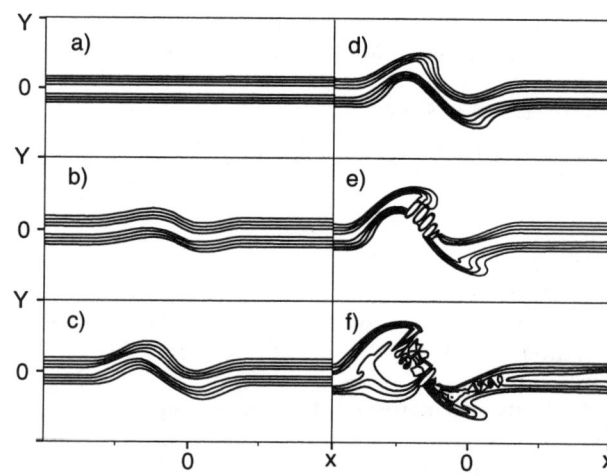

Figure 2 (after Ref.8) exhibits temporal evolution of the parallel current sheet for the same current density as in Fig.1 but assuming that $\phi=5\tanh(ky)$. Here, we see the formation of a fold and the disintegration of a fold into a vortex chain.

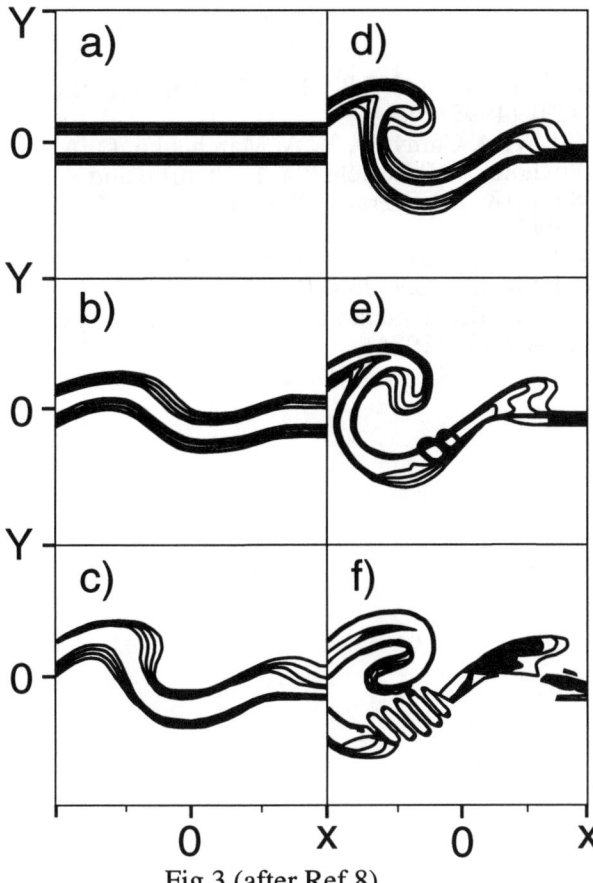

Fig.3 (after Ref.8)

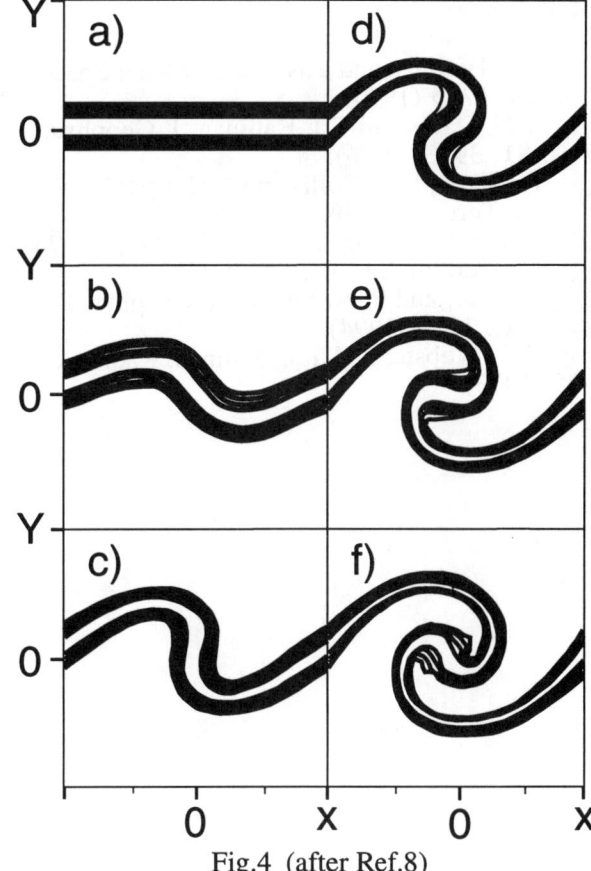

Fig.4 (after Ref.8)

Fig.3 (after Ref.8) shows the formation of a spiral (S structure) in the current sheet for the same parameters as in Fig. 1 but with an axisymmetric electric field with $\phi = \exp(-0.2r^2)$.
The formation of a spiral and the generation of a vortex chain in the spiral are depicted in Fig. 4. Here, the profile of the parallel electron current density and the other parameters are the same as in Fig.1, whereas the scalar potential has the same form as in Fig. 2. Our numerical simulations thus demonstrate that within the framework of the proposed model, one can describe all the observational stages of the development of fold, spiral, as well as the disintegration of the fold and the spiral into a vortex chain.

V. Summary

In this paper, we have presented two models that can account for various coherent nonlinear structures appearing in auroral ovals. It is found that deformation of auroral arcs from their stable configuration can be attributed to the nonlinear interaction of finite amplitude disturbances in strongly magnetized collisionless magnetoplasma. Specifically, then nonlinear drift wave model is capable of describing the basic features of the black auroral street, whereas the results of our numerical simulations show that the nonlinear mode coupling of the inertial Alfvén fluctuations can selforganize into large-scale structures, which eventually break into smaller size coherent vortex structures. The energy transfer from large vortical motion into small vortical motion is thus evident in a low-beta magnetoplasma containing inertial Alfvén wave turbulence. Our results should have relevance to the structures that are observed in the auroral region of the earth's ionosphere.

Acknowledgements-

The author is grateful to Robert Bingham, Guido Birk, Lennart Stenflo and Oleg Pokhotelov for their valuable collaboration in different parts of the work presented here. This work was partially supported by the Deutsche Forschungsgemeinschaft as well as the European Union through the network "Nonlinear Phenomena in Microphysics of collisions Plasmas. Application to Space and Laboratory Plasmas" of the Human Capital and Mobility program under the contract no. CHRX-CT.93.0356.

References

[1] T.J. Hallinan and T.N. Davis, Planet Space Sci. **18**, 1735 (1970).
[2] T.N. Davis and T.J. Hallinan, J. Geophys. Res. **81**, 3953 (1976).
[3] A. Steen, P.N. Collis and I. Häggström, J. Atmos. Terr. Phys. **50**, 301 (1988).
[4] T. Oguti, J. Geophys. Res. **79**, 3861 (1974).
[5] G. Marklund, L. Blomberg, C-G. Fälthammar, and P.-A. Lindqvist, Geophys. Res. Lett. **21**,. 1859 (1994).
[6] H.F. Webster and T.J. Hallinan, Radio Sci. **8**, 475 (1973).
[7] A. Otto and G.T. Birk. J. Geophys. Res. **97**, 8391 (1992).
[8] V.M. Chmyrev, V. A. Marchenko, O. A. Pokhotelov, P.K. Schukla, L. Stenflo, and A.V. Streltsov, IEEE Trans. Plasma Sci. **20**, 764 (1992).
[9] P. K. Shukla, G.T. Birk and R. Bingham, submitted to Geophys. Res. Lett., 1995.
[10] R. Bharutharam and P.K. Shukla, Phys Rev. A **34**, 4457 (1986).

ALFVEN WAVES IN SPACE PLASMAS

C. Uberoi

Department of Mathematics
Indian Institute of Science
Bangalore 560 012, INDIA

Abstract

A brief review of time-dependent theories of resonant mode coupling of Alfven waves in space plasmas is presented. A well behaved solution of the hydromagnetic wave equation for a general class of initial conditions is then discussed to understand the ringing, phase-mixing and variable eigenperiods of the field line oscillations in the magnetosphere. This solution is further exploited to discuss the initial transient amplification of Alfven waves in sheared magnetospheric plasma. The features of free oscillations of field lines studied here are shown to compare well with previous numerical studies and observations made for particular types of impulse excitations of the magnetosphere.

1. Introduction

Low frequency hydromagnetic waves are common feature of space plasmas. Many a phenomena in solar atmosphere or heliosphere, planetary and cometary magnetospheres can be related as a manifestation of the linear on nonlinear Alfven waves. Considering only the linear theories an important and somewhat amazing phenomena of Alfven waves is resonant mode coupling to surface or global modes of an inhomogeneous plasma. As inhomogeneity is an important characteristic feature of various regions of space plasmas the resonant mechanism of Alfven waves finds applications in modeling and understanding of phenomena like ULF waves or micropulsations, substorms, magnetosphere- ionosphere coupling, aurora formations field aligned currents and charged particle acceleration in planetary magnetospheres and in understanding of physics of outer heliosphere. Keeping in view the importance of this mechanism in the study of Alfven waves in space plasmas and my own current interest I like to give a brief review of the recent developments in the study of resonant mode coupling theories with a view to understand the impulse excitation of field line resonances and transient amplication of shear Alfven waves in the magnetospheric plasmas.

2. Steady State Theories

The theoretical understanding of the origin and characteristic properties of ULF hydromagnetic waves or micropulsations (ranging from above 1m Hz to the local proton gyrofrequency $\sim$ 1 Hz) in planetary magnetospheres has shown that most pulsations with long period (period is longer than 10 sec in the earth's magnetosphere) are generally believed to be field line oscillations excited by the external and internal source waves through the linear resonance mechanism. Due to strong factor of inhomogeneity the standing Alfven mode frequencies vary from magnetic shell to shell and purely transverse signals are highly localized to particular shell. Field line resonance is a descriptive term for these purely localized signals along the magnetic field lines of the earths dipole magnetic field.

Basically the understanding of resonant mode coupling translates into the mathematical problem of analysis of the hydromagnetic wave equation in inhomogeneous magnetic fields, and density. Here we shall consider the variation in the planes perpendicular to the magnetic field direction. There is a considerable interest in the study of resonance mechanism in two-dimensional systems and is seen that the basic concepts of resonant coupling theory holds good even in these systems. We shall not be able to consider this aspect here.

The normal mode analysis of the hydromagnetic wave equation or the steady state theories of ULF waves in the magnetosphere showed that when a monochromatic source with frequency ω_0 is present in the magnetosphere, magnetic pulsations can be excited at the frequency of the source, with the intensity maximum at a L-shell value of the resonant field line where the Alfven resonant condition $\omega_0 = k_{11}(L)V_A(L)$ is satisfied. These theories have successfully explained a majority of ground and spacecraft measurements in which the same frequency is observed at different latitudes for a given event with variations in the polarization and amplitude.

The linear resonance theory though very successful has some shortcomings which were pointed out in the literature from time to time some of these were not valid and some needed attention. The criticism pointed out by Yumoto and Saito [1] was that the resonance coupling between the surface waves excited at the magnetopause and the standing wave oscillations, never occurs in the magnetosphere. This was shown to be invalid as Uberoi [2], pointed out that the ω_0 for the surface waves was not calculated correctly. The correct value of ω_0 matched very well with the field line oscillations calculated for different values of L and propagation along θ in the work by these authors. The other major shortcoming is that the steady state theory studies the resonant mode coupling and resonant absorption only for $k_{11} \ll k_\perp$ that is, for large azimuthal numbers. Uberoi [3] has shown that if this assumption is not considered the characteristic properties of compressional surface waves and the mechanism of resonant absorption differs considerably. This needs to be looked into more seriously.

Recently, it is shown [4] that though the mathematical analysis of the Alfven wave equation does not show any variation at the non-zero or the zero singular point the role of surface waves in the physical mechanism of resonant absorption of Alfven waves is very much different at these points. This requires a certain modification in the linear resonance theory while considering the resonant mode coupling near the neutral point. The difference becomes all the more emphasized when resistivity is taken into account. At the neutral point the zero-frequency surface waves ($ka \ll 1$), which are symmetrical surface modes of the structural neutral layer couple to the tearing mode of instability of the layer. This result can be very important to the study of energy balance in tearing modes and the association of surface waves to driven magnetic reconnection.

Can these zero-frequency surface waves occur along the magnetopause? For this take $ka = 10^{-2}$, the Alfven speed 300 km/s and the neutral layer thickness 'a' varying from 100 to 300 kms. This gives the frequencies ranging from 3×10^{-2} Hz to 0.9×10^{-2} Hz. The time periods vary from 3 to 6 minutes. The surface waves with periods greater than 2 minutes have been observed along the magnetopause. Therefore it appears that the zero frequency surface waves can be an important mechanism for inducing tearing instability, which has been suggested to occur at the dayside magnetopause and consequently these waves could be important driving mechanism for magnetic reconnection [5].

The steady state linear resonance theory obviously cannot explain the features of micropulsations which are manifestation of the temporal behaviour of the hydromagnetic waves excited by time- dependent sources, although earlier a few studies were made to look at the asymptotic behaviour of the solutions of the hydromagnetic equation resulting from impulsive stimulus of the magnetospheric system the need for proper time dependent theories of this equation has been realised recently. In the next section this subject is reviewed briefly.

3. Time dependent Theories

Hasegawa et al. [7] extended the steady state approach to include a particular case of broad band source in their analysis to explain the earlier reports of ground based observations of pulsations whose period is a function of magnetic latitude for the same event. Recent spacecraft and multipoint ground based observations have increased the understanding of both external and internal sources of generation of ULF waves. It is now understood that ULF waves can be generated by impulse excitation of the magnetosphere due to solar wind-magnetosphere interaction, flux transfer events and sudden compression of the magnetosphere due to interplanetary shock. Further, the presence of such broad band impulsive sources led to a puzzle that why these sources of energy should lead to the almost monochromatic waves that are often observed. This led to proposal of the global mode oscillations of magnetospheric cavity in the presence of an external stimulus by Kivelson and Southwood [8] [see also Uberoi, [2]]. The magnetosphere reverbrates with the natural fast modes of the magnetospheric cavity formed by the magnetospause and the turning point arising due to the gradient in the Alfven velocity. The impulse excited hydromagnetic narrow band global cavity modes, stimulated by broad band solar wind disturbances, coupling to field line resonances in the magnetosphere has become a topic of great interest. Various numerical models are developed [10-13] to study the time-development of impulse excited hydromagnetic cavity and field line oscillations in a sheared magnetospheric plasma considering particular types of impulse stimulus at the magnetopause. The possibility of such excitation proved to exist but lack of substantial evidence showing the existence of cavity modes in the magnetosphere are few. Also the observed frequencies [JAU/APL Radar] or the field line resonances are much lower than those predicted by the cavity model. This gave rise to the concept of MHD waveguide model treating the magnetosphere as open waveguide [14]. The cavity here is now extended towards the magnetotail to form waveguide configuration. Recently dispersion and wave coupling in inhomogeneous waveguides is studied by Wright [15]. Based on certain observations of standing waves Harrold and Samson [16] extended the waveguide model suggesting that the bow shock rather than the magnetopause must be taken as an outer boundary of the cavity/waveguide. This model is referred to as the standing ULF waveguide model.

It is now understood that field line resonances can be driven by impulse excited surface, cavity, waveguide, standing ULF waveguide eigenmodes. In all these cases it is necessary to understand the features of impulse excited field line oscillations for a general class of initial impulse stimulus at the magnetopause. Most of the theoretical work referred above considered the particular boundary or initial perturbations while studying the solutions of hydromagnetic equation. Recently Uberoi [6] has considered the time dependent solution of the Alfven wave equation for a general class of initial conditions. Features of free oscillations of field lines are studied in detail. Some of the important results are discussed briefly in the following section.

4. Impulse excitation of field line oscillations

We start with the well known Alfven wave equation given as [for example see [17]].

$$\nabla \cdot \left[\left\{ \frac{\partial^2}{\partial t^2} - \frac{1}{4\pi\rho_0}(\vec{B}_0 \cdot \nabla)^2 \right\} \nabla v_x \right] = 0 \qquad (1)$$

where v_x is the x-component of the velocity field and the equilibrium magnetic field variation is given as

$$\vec{B}_0(x) = B_{0y}(x)\hat{y} + B_{0z}(x)\hat{z}. \qquad (2)$$

Fourier analysis of eqn(1) with space dependent as $exp[i(k_y y + k_z z)]$ gives

$$\frac{\partial}{\partial x}\left[\left(\frac{\partial^2}{\partial t^2}+\omega_A^2(x)\right)\frac{\partial v_x}{\partial x}\right] - k^2\left(\frac{\partial^2}{\partial t^2}+\omega_A^2(x)\right)v_x = 0 \tag{3}$$

where $k^2 = k_y^2 + k_z^2$ and $\omega_A^2(x) = [\vec{B}_0(x).\vec{k}]^2/4\pi\rho_0$.

Following Uberoi and Sedlacek [18] the integro-differential formulation of eqn(3) can be written as

$$\left(\frac{\partial^2}{\partial x^2} - k^2\right)\left(\frac{\partial^2}{\partial t^2}+\omega_A^2(x)\right)\frac{\partial e_k(x,t)}{\partial x} = \underbrace{k^2\frac{\partial \omega_A^2(x)}{\partial x}\int_{-\infty}^{x} e_k(s,t)ds}_{\text{coupling terms}} \tag{4}$$

where $e_k(x,t) = \frac{\partial v_x}{\partial x}$. The coupling of surface waves to the local Alfven oscillations on the left side of eqn(4) is seen to arise due to inhomogeneous Alfven frequency $\omega_A(x)$ on the right side of this eqn.

The time-dependent solution of eqn.(4) for the boundary conditions $e_k(x,t) \to 0$ as $|x| \to \infty$ and initial conditions

$$e_k(x,t)]_{t=0} = \phi(x,k) \text{ and } \left.\frac{\partial e_k}{\partial t}(x,t)\right]_{t=0} = \psi(x,k) \tag{5}$$

where $|\phi(x,k)|$ and $|\psi(x,k)| \to 0$ as $|x| \to 0$ has been obtained as

$$e_k(x,t) = \frac{\partial v_x}{\partial x} = \phi(x,k)\cos[t\omega_A(x)] + \psi(x,k)\frac{\sin[t\omega_A(x)]}{\omega_A(x)} + \int_0^t \frac{\sin[\omega_A(x)(t-\tau)]}{\omega_A(x)}K(x,\tau,k)d\tau. \tag{6}$$

where $K(x,\tau,k)$ is a convergent series of operators on $e_{k0}(x,t)$, which represent the first two terms of the solution in eqn.(5).

Let us now examine the impulse response of the incompressible, inhomogeneous magnetospheric plasma to surface wave like impulse excitations at the magnetopause. This will then give the time development of the field line resonances. As we are only looking at the free oscillation problem the first two terms in the solution (5) are taken to find the various field quantities for initial conditions (4). Calculations of v_x, v_y, vorticity component $(curl\vec{v})_z$ and current density J_z allows us to arrive at the results which are summarized as follows:

1. For a general initial perturbation of the type (4) impulse response of the magnetosphere showed the ringing of field line at Alfven frequency with amplitude decaying as $1/t$ from the initial value.

    We also note that v_y, component along the magnetic field, does not show decay with time whereas v_x decays as $1/t$. Hence, eventually for large t the wave becomes a longitudinal wave.

2. The expressions for the components of vorticity and current density were found to have secular terms with t. Thus both these physical quantities showed increase with time t reminiscent of a collisionless system driven at its resonant frequency. From the vorticity/current density expression we get the phase mixing time as:

$$t_{ph} = \frac{k\omega_{A_0}(x)}{\phi(x_0,k)} \cdot \frac{1}{d\omega_A/dx}, \qquad (7)$$

$x = x_0$ is an initial point. For a magnetosphere with a magnetic field profile

$$B_0(x) = \frac{B_0}{1 + x^2/x_0^2}, \qquad (8)$$

$x = 0$ corresponds to the equator and $x > x_0$ to the magnetopause.

In terms of Alfven transit time N_c, we get

$$N_c = \frac{\omega_{A_0}^2}{2\pi} \frac{k}{\phi(x,k)} \frac{1}{d\omega_A/dx}. \qquad (9)$$

considering $V_{A_0} = 600 km/s$ at $x = x_0$ and taking the impulse amplitude $\phi(x_0, k/k \approx 1$ we have $N_c = 138 kx_0$ for the profile (8). For the non-dimensional wavenumber $kx_0 = 3$, $N_c = 414$. In general for values $1 < kx_0 < 6$, the number of Alfven oscillations are of the order 10^2. There should be therefore good opportunity to observe ringing for time of the order of tens of Alfven transit periods near the region $x = 0.5x_0$. We note that the number N_c increases with the increase in the initial impulse amplitude.

3. The variable eigenperiods of the field line oscillations are given as $T = 2\pi/\omega_A(x)$ with

$$\omega_A(x) = V_A(x).k = V_A(x)\frac{n\pi}{\ell}$$

where ℓ is the length of the field line. Taking a particular example: $V_A = 1250 km/s$ at $x = 0.56$ and $L = 6.6$, for which field line length $\ell = 5R_E$ [19] we get the time period for the fundamental mode as

$$T = \frac{2\pi\ell}{V_A\pi} = \frac{2\ell}{V_A} = \frac{2 \times 600 \times 5}{125} = 124s.$$

The order of eigenperiods of free oscillations can be seen to vary from tens to hundreds of seconds.

Allan et al [10] considering the coupling of cavity and field line resonances have given detailed numerical results showing the spatial and temporal structures of field line resonances using a cylindrical model. The features of ringing of field lines, phase mixing and variation of oscillation frequencies with latitude as obtained from analysis of the solution (5) agree qualitatively very well with this numerical model. Finally, we like to mention that the linear growth of current density with time was pointed out as early as 1974 by Radoski [20] for the toroidal (shear) mode in his study of the asymptotic temporal behaviour of hydromagnetic waves, no attempt however, was made to discuss the phase-mixing phenomena.

5. Transient Amplification of shear Alfven waves

We shall now discuss the initial conditions for which a surface 'wavelet' when superposed on the sheared magnetospheric plasma shows the initial transient amplification by using the solutions in eqn. (5). These type of surface wavelets are caused by the passage of an interplanetary shock wave and have been observed traveling tailward, by the satellite GEOS2 [21]. They are also important for fusion plasmas [22,23].

Consider the inital disturbance (4) of the type

$$\phi(x,k) = Ae^{-k|x|}e^{iu_0 kx}, \quad \psi(x,k) = 0 \qquad (10)$$

with A as a constant.

Taking a linear Alfven speed profile

$$V_A(x) = V_A\left(\hat{z} + \frac{x}{d}\hat{y}\right) \qquad (11)$$

with (10) and (11) the eqn. (5) with zeroth order terms and $k_y >> k_z$ on integrating gives [24].

$$v_x(x,y,z,t) = \frac{1+u_0^2}{1+\left(u_0-\frac{V_A t}{d}\right)^2}e^{-k|x|}exp\left[ik_y x\left(u_0-\frac{V_A t}{d}\right)+ik_z\left(z-\frac{V_A t}{d}\right)+ik_y y\right]+D, \qquad (12)$$

Here $A = 2(1+u_0^2)k$ and D is the part of the solution which does not show any amplification.

The solution (12) shows that for $u_0 > 0$ the amplitude of v_x increases from its initial value to a large amplitude till $t = \frac{du_0}{V_A} \equiv T$. For $t > T$, the amplitude decays and tend to zero for $t \to \infty$. Considering the quantitative values for magnetospheric region close to the geomagnetic equator, $v_A = 1000km/s$ and taking $d = 1R_E$, we get $t = 6.4u_0$. Choosing $u_0 = 25$, which means the inclination $k_x/k_y >> 1$, we get $T = 160s$. The wave amplification will be maximum at time $t = 160s$, after $t = 2T = 320s$, the wave begins to turn and starts moving towards the resonant layer [25]. These quantitative results agreed very well with the observations.

Finally, I like to point out that our analysis of eqn. (5) does not give the transient amplification for the initial conditions of the type used by Lau [22]. This can perhaps be attributed to certain error in his work [24].

References

1. K. Yumoto and T. Saito, J. Geophys. Res. **87**, 5159 (1982).
2. C. Uberoi, J. Geophys. Res. **88**, 4959 (1983).
3. C. Uberoi, J. Geophys. Res. **94**, 6941 (1989).
4. C. Uberoi, J. Plasma Physics, under publication (1994).
5. C. Uberoi, L.J. Lanzerotti and A. Wolfe, to be published.
6. C. Uberoi, J. Geophys. Res. submitted for publication (1994).
7. A. Hasegawa, H.K. Tsui and A. S. Assis, Geophys. Res. Lett., **10**, 765 (1983).
8. M.G. Kivelson and D.J. Southwood, J. Geophys. Res. **91**, 4345 (1986) and Geophys. Res. Lett **12**, 49 (1985).
9. C. Uberoi, J. Geophys. Res., **94**, 2745 (1989).
10. W. Allan, S.P. White and E.M. Poulter, Planet. Space Sci., **31**, 371 (1986).
11. B. Inhester, J. Geophys. Res. **22**, 4751 (1987).
12. X. Zhu and M.G. Kivelson, J. Geophys. Res. **99**, 1479 (1989).
13. D.H. Lee and R.L. Lysak, J. Geophys. Res., **94**, 17097 (1989).
14. J.C. Samson, B.G. Harrold, J.M. Ruhoneimi, R.A. Greenwald, K.B. Baker, A.D.M. Walker, Geophys. REs. Lett., **19**, 441 (1992).
15. A.N. Wright, J. Geophys. Res. **99**, 159 (1994).
16. B.G. Harrold and J.C. Samson, Geophys. Res. Lett. **19**, 1811 (1992).
17. C. Uberoi, Phys. Fluids, **15**, 1673 (1972).
18. C. Uberoi and Z. Sedlacek, Phys. Fluids, **B4**, 6 (1992).
19. W. Allan and D.R. McDiarmid, Ann. Geophysicae, **11**, 916 (1993).
20. H.R. Radoski, J. Geophys. Res. **79**, 595 (1974).
21. W. Baumjohann, H. Junginger, G. Harendel and O.H. Bauer, J. Geophys. Res. **89**, 2765 (1984).
22. Y.Y. Lau, Phys. Rev. Lett. **42**, 779 (1979).
23. Y.Y. Lau, R.C. Davidson and B. Hui, Phys. Fluids, **23**, 1827, (1980).
24. C. Uberoi, Phys. Fluids, submitted for publication.
25. C. Uberoi, J. Geophys. Res., **93**, 7595 (1988).

REVIEW ON INTERPLANETARY-MAGNETOSPHERE INTERACTIONS

W.D. GONZALEZ, A.L. CLUA DE GONZALEZ
INSTITUTO NACIONAL DE PESQUISAS ESPACIAIS - INPE, P.O.BOX 515,
12201 SÃO JOSÉ DOS CAMPOS, SP, BRAZIL

and

B.T. TSURUTANI
JET PROPULSION LABORATORY, PASADENA, CAL., U.S.A.

ABSTRACT

This review deals with the interplanetary-magnetosphere coupling during intense geomagnetic activity, especially near solar maximum. For this purpose we use the plasma and magnetic field measurements obtained mainly by the ISEE-3 satellite during the interval of time when it was at the L_1 libration point of the sun-earth system.

During this interval 90% of the intense geomagnetic storms (Peak Dst < -100nT) were preceded by the arrival of interplanetary fast forward shocks at 1AU. For these events we show the interplanetary structures that are associated with the large-amplitude and long-duration negative B_z fields that were found in the sheath and/or driver gas regions of the shock and that are thought to be the main cause of the intense storms.

On the other hand, during this same interval, "High-Intensity, Long-Duration and Continuous Auroral Activity" events (HILDCAAs for short) were found to be caused by the southward magnetic fields of large-amplitude interplanetary Alfvén waves, presumably through magnetic reconnection with the geomagnetic field.

With respect to geomagnetic storm predictability we emphasize the need to have a satellite at the L_1 point, such as ISEE-3, continuously monitoring the solar wind, as perhaps the only reliable tool at the moment.

INTRODUCTION

Recent studies (1 - 4) indicate that the category of storms that have the largest association with interplanteray shocks, or coronal mass ejections (CMEs), are the most intense ones. This level of storm intensity can be expressed by the storm index threshold, peak Dst <-100 nT. It was shown (1-3, 5) that the main interplanetary feature associated with intense storms, accompanying the CMEs, is the presence of a large-amplitude (<-10 nT), long-duration (> 3 hours), negative Bz component of the IMF.

Figure 1 shows schematically the solar-interplanetary-magnetosphere coupling of interest. At the Sun the main ingredient is assumed to be a CME, which can also involve a low-latitude, short-lived, coronal hole (e.g. Gonzalez and Tsurutani, 6), whereas at the interplanetary medium the main responsible feature for the development of the storm is the presence of a southward IMF carried by the solar wind. At the magnetosphere this southward field reconnects with the geomagnetic field leading to an effective momentum and energy transfer via a magnetospheric dynamo. In this figure two of the most important dissipation regions within the magnetosphere are indicated, the auroral and the ring current. The former refers to the substorm process, for which the level of intensity is monitored by the auroral electrojet index AE, and the latter refers to the storm process itself with its intensity monitored by the storm index Dst.

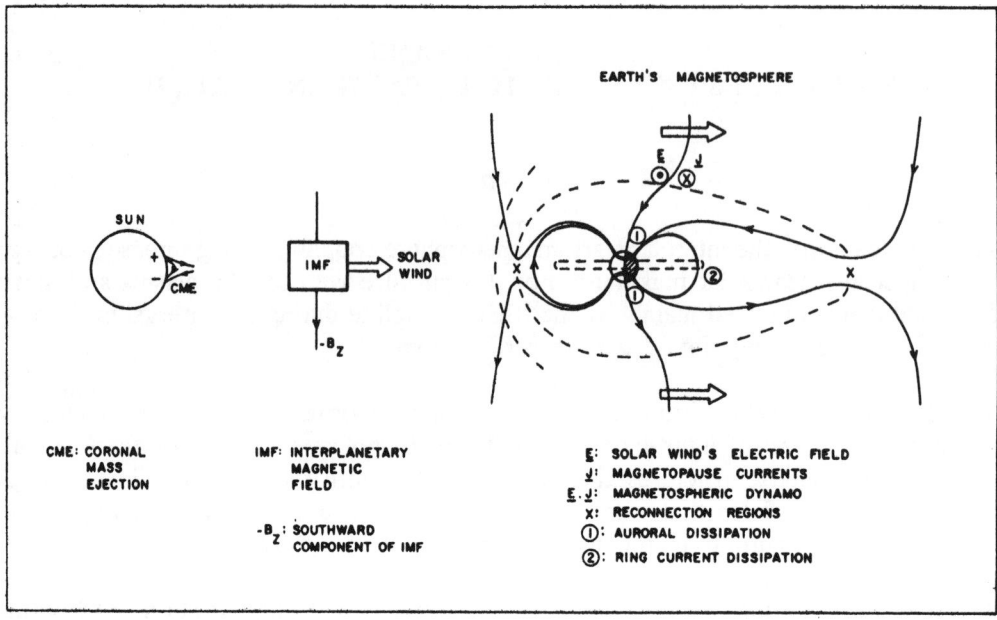

Fig. 1 - Schematic of solar-interplanetary-magnetosphere coupling during solar maximum years.

THE ISEE-3 MEASUREMENTS

The ISEE-3 satellite was situated in a halo orbit (about the L_1 libration point), at approximately 240 Earth radii in front of the Earth, and measured 56 unambiguous fast forward shocks during the interval of August 16, 1978 to December 28, 1979. The satellite provieded a full set of magnetic field (7) and plasma (8) data. From these 56 shocks Gonzalez and Tsurutani (11) reported that only nine preceded (within typical time lags) the occurrence of an intense geomagnetic storm (peak Dst <-100 nT). Thus, from the predictive point of view one can say that only about 14% of the interplanetary shocks during solar maximum are expected to lead to the development of intense storms.

On the other hand, since nine of intense storms that occurred within this interval were associated to shocks one can also say that during solar maximum 90% of the intense storms are expected to be associated with fast forward shocks within 1 AU. A similar conclusion was reached by Gosling et al. (4) for CMEs.

Figure 2 is an example of some of the solar wind parameters measured by the ISEE-3 satellite (during the interval of August 26 to September 4, 1978). From top to bottom we have: the solar wind speed, density, IMF amplitude and the By and Bz components of the IMF (in Geocentric Solar Ecliptic coordinates). The last two panels at the bottom of this figure give the auroral electrojet (AE) and the ring current (Dst) indices of geomagnetic activity.

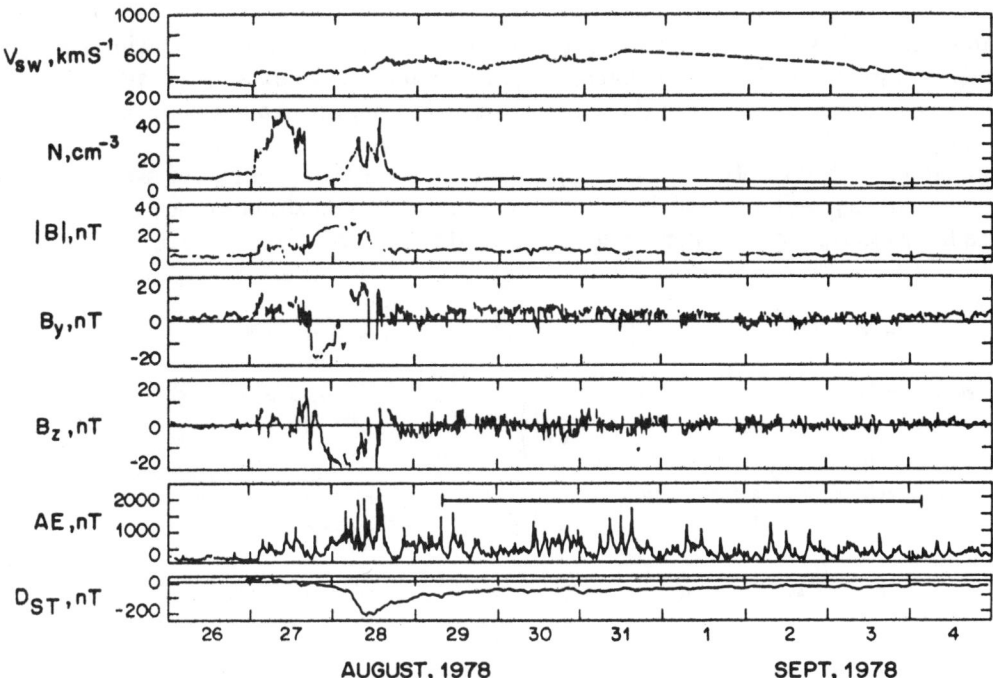

Fig. 2 - Example of the ISEE-3 measurements.

INTERPLANETARY-MAGNETOSPHERE COUPLING

From the vantage location of the ISEE-3 satellite one could study both, the "solar-interplanetary" as well as the "interplantery-magnetosphere" couplings. In this review we concentrate in the latter type.

Gonzalez and Tsurutani (1) reported that all ten intense storms (peak Dst <-100 nT) that occurred during the ISEE-3 studied interval were associated with large-amplitude (<-10 nT), long duartion (>3 hours) negative Bz fields in the interplanetary medium. Figure 2 shows one example of such an association for the August 28, 1978 storm. This figure illustrates the fast forward shock event that was observed at 02:00 UT of day 27, the compressed (and heated) sheath field region lasting approximately till 18:25 hous UT of day 27 and also a driver gas region approximately till 12:00 hours UT of day 28. The large negative Bz event is associated in this case with a driver gas-magnetic cloud structure (with a rotation in the By component).

Figure 2 also shows the occurrence of a high-intensity, long-duration and continous auroral activity (HILDCAA) event as shown by the horizontal bar in the AE panel. For the ISEE-3 studied interval Tsurutani and Gonzalez (9) reported the occurrence of 8 HILDCAA events, five of which followed intense storm events as in the case of this figure. In (9) and (10) these HILDCAA events were associated with the simultaneous occurrence of large-amplitude interplanetary Alfvenic fluctuations and it was argued that magnetic reconnection between the southward field of these fluctuations and the geomagnetic field is responsible for the magnetospheric energization.

Tsurutani et al. (2) studied the interplanetary structures that were associated with the negative Bz events responsible for the 10 intense storms of the Gonzalez and Tsurutani (1) study. Figure 3 is an updated version of those structures. They are divided in two groups: those that belong to the sheath region of the shock and those that are encountered within the driver gas region. About half of the 10 events belong to each of these two groups and can be associated with any of the suggested possibilities. Although the suggested structures are self explanatory the reader is referred to Tsurutani et al. (2) for further details.

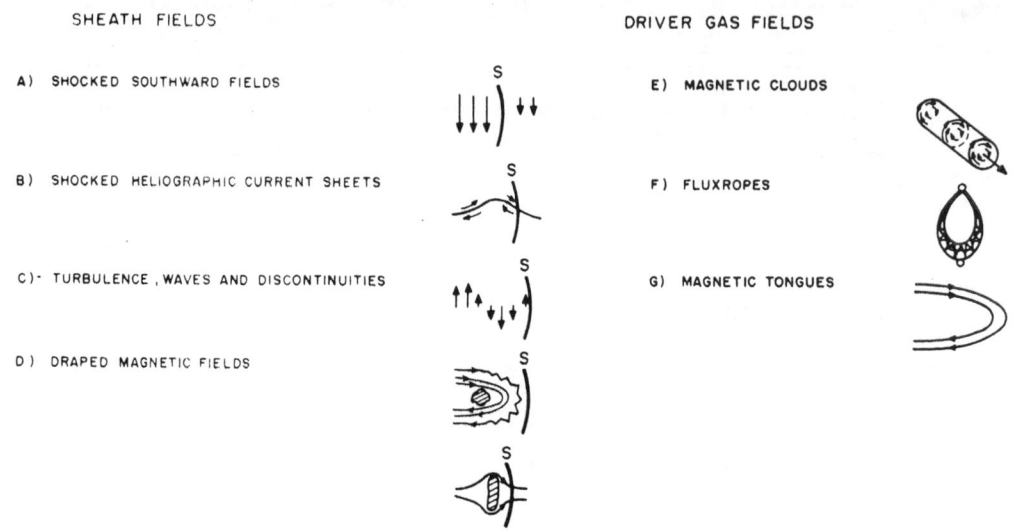

Fig. 3 - The various interplanetary features that involve large-amplitude, long-duration Bs fields during solar maximum.

For the ISEE-3 studied interval, Gonzalez and Tsurutani (1) also reported the occurrence of northward Bz events similar to the southward events but opposite in polarity (Bz $>+10$ nT, T > 3hours). These northward IMF events were similar to the southward field events in several ways: the number is about the same (11 northward events vs. 10 southward events); for both the northward and southward events, nine followed shocks within similar time lags. It is possible that these northward fields are also associated with structures similar to those shown in Figure 3 and, therefore, that the responsible physical process for generating them do so with random orientations. However, during these northward Bz events the magnetosphere is in a quiet state, namely, the level of intensity of storms or substorms, if any, is very low.

Using the ISEE-3 measurements, several authors (5, 9-11) have also studied the solar wind-magnetosphere coupling problem through coupling functions associated with the reconnection process at the magnetopause. The reader is referred to Gonzalez (12) for a review on this type of study.

In (13) and (4) the ISEE-3 measurements were also studied in association of CMEs with solar wind electron heat flux events and it was shown that this association serves as an important tool for CME identification.

For the interval of 1973-1975 (near solar minimum), it was shown (14) using IMP-7 and IMP-8 satellite measurements that intense storms were still associated to CMEs, whereas most of the moderate storms (-100 nT ≤ peak Dst <-50 nT) were found to be associated to corotating streams, thus leading to a recurrent storm behavor. The negative IMF-Bz structures accompaning these recurring events were of a different nature than those shown on Figure 3. Such Bz features were caused mainly by the interaction of high speed streams with the heliospheric current sheet. Another interesting feature found (14) for solar minimum, associated to corotating streams, was the frequent occurrence of HILDCAA events (8). This led for 1974 to an average value of the AE index higher than that for 1979, the latter being close to solar maximum.

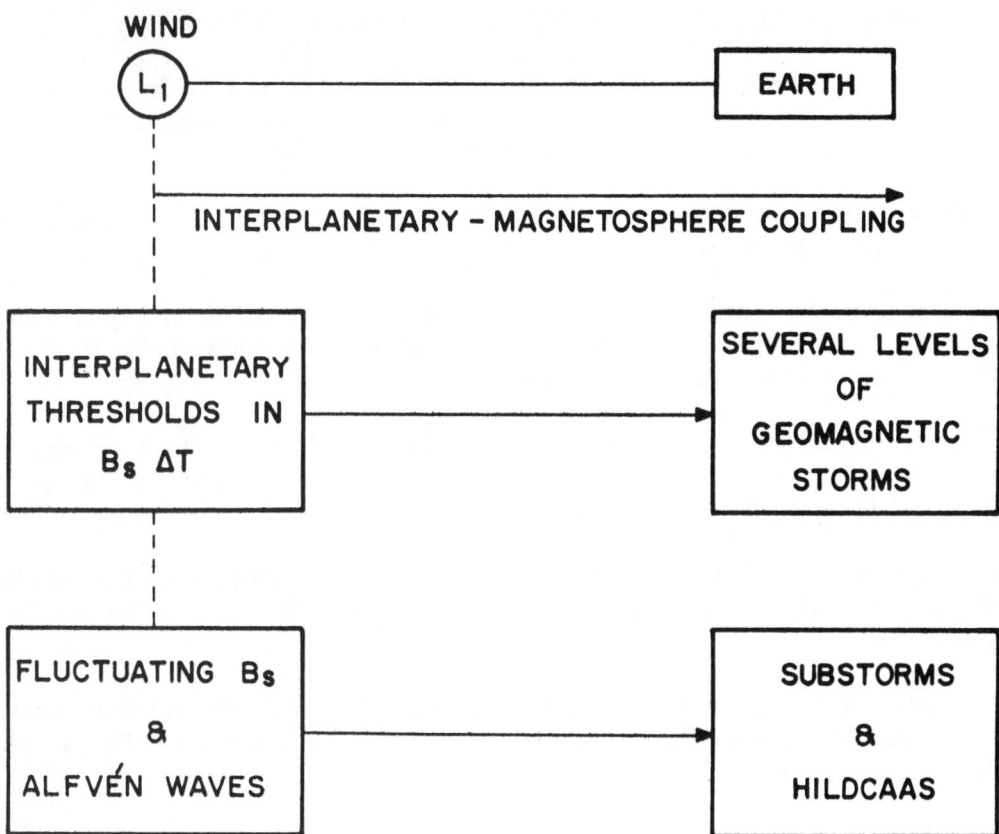

Fig. 4 - Shematic of "geomagnetic activity alert" as learned from ISEE-3.

CONCLUDING REMARKS

The ISEE-3 measurements during the interval of August 1978 - December 1979 helped to understand several interesting features of the interplanetary-magnetosphere coupling as well as of the solar-interplanetary coupling problems. Since for the former a more quantitative knowledge was advanced, there is a particular interest in extending it to a purpose of a short-term geomagnetic storm predictability or "geomagnetic activity alert". Thus, for such a purpose, Figure 4 shows a schematic with features obtained from the ISEE-3 study as applied to a coming similar space mission, such as WIND (during the ISTP program).

ACKNOWLEDGMENTS

This work was partially supported by the "Fundo Nacional de Desenvolvimento Científico e Tecnológico" of Brazil and by the Jet Propulsion Laboratory, California Institute of Technology under contract with NASA.

REFERENCES

1. Gonzalez, W.D. and B.T. Tsurutani, Criteria of interplanetary parameters causing intense magnetic storms (Dst <-100 nT), *Planet. Space Sci.*, 35, 1101, 1987.

2. Tsurutani, B.T., W.D. Gonzalez, F. Tang, S.I. Akasofu and E.J. Smith, Origin of interplanetary southward magnetic fields responsible for major magnetic storms near solar maximum (1978-1979), *J. Geophys. Res.*, 93, 8519, 1988.

3. Tsurutani, B.T., W.D. Gonzalez, F. Tang and Y. Te Lee, Great magnetic storms, *Geophys. Res. Lett.*, 19, 73, 1991.

4. Gosling, J.T., D.J. McComas, T.L. Phillips and S.J. Bame, Geomagnetic activity associated with Earth passage of interplanetary shock disturbances and coronal mass ejections, *J. Geophys. Res.*, 96, 7831, 1991.

5. Gonzalez, W.D., B.T. Tsurutani, A.L.C. Gonzalez, E.J. Smith, F. Tang and S.J. Akasofu, Solar wind-magnetosphere coupling during intense magnetic storms (1978-1979), *J. Geophys. Res.*, 94, 8835, 1989.

6. Gonzalez, W.D., B.T. Tsurutani, Interplanetary-magnetosphere coupling from ISEE-3. Proceedings of the First STEP symposium, Ed. D.N. Baker et al. pp 197-204, Pergamon Press, 1994.

7. Tsurutani, B.T. and W.D. Gonzalez, The cause of high-intensity long-duration continuous AE activity (HILDCAAs): interplanetary Alfvén wave trains, *Planet. Space Sci.*, 35, 405, 1987.

8. Tsurutani, B.T., T. Gould, B.E. Goldstein, W.D. Gonzalez and M. Sugiura, Interplanetary Alfvén waves and auroral (substorm) activity: IMP-8 *J. Geophys. Res.*, 95, 2241, 1990.

9. Baker, D.N., E.W. Hones Jr., J.B. Payne and W.C. Feldman, A high-time resolution study of interplanetary parameter correlations with AE, *Geophys. Res. Lett.*, 8, 179, 1981.

10. Tsurutani, B.T, J.A. Slavin, Y. Kamide, R.D. Zwickl, J.H. King and C.T. Russell, Coupling between the solar wind and the magnetosphere: CDAW 6, *J. Geophys. Res.*, 90, 1191, 1985.

11. Akasofu, S.I, C. Olmsted, E.J. Smith, B.T. Tsurutani, R. Okida and D.N. Baker, Solar wind variations and geomagnetic storms: A study of individual storms based on high-time resolution ISEE-3 data, *J. Geophys. Res.*, 90, 325, 1985.

12. Gonzalez, W.D., A unified view of solar wind-magnetosphere coupling functions, *Planet. Space Sci.*, 38(5), 627, 1990.

13. Gosling, J.T., D.N. Baker, S.J. Bame, W.C. Feldman, R.D. Zwickl and E.J. Smith, Bidirectional solar wind electron heat flux events, *J. Geophys. Res.*, 92, 8519, 1987.

14. Tsurutani, B.T., J.K. Arballo, W.D. Gonzalez, A.L.C. Gonzalez, F.Tang, M.Okada, Solar wind features responsible for magnetic storms and substorms during the declining phase of the solar cycle: 1974, *J. Geophys. Res.*, in press, 1994.

CHARACTERISTICS OF THE ION PRESSURE TENSOR IN THE EARTH'S MAGNETOSHEATH: AMPTE/IRM OBSERVATIONS

H. R. Lewis, X. Li, and J. LaBelle

Department of Physics and Astronomy, Dartmouth College,
Hanover, New Hampshire, USA

T.-D. Phan

Herzberg Institute of Astrophysics, National Research Council of Canada,
Ottawa, Canada

R. A. Treumann

Max-Planck-Institut für Extraterrestrische Physik, Garching, Germany

AMPTE/IRM satellite data are used to examine characteristics of the ion pressure tensor in the earth's magnetosheath. The eigenvalues and principal axes of the pressure tensor are computed, and the directions of the principal axes are compared to the direction of the independently measured magnetic field **B**. When the pressure tensor is anisotropic, as is usually the case in the magnetosheath, one of its eigenvalues is observed to be distinguishable from the other two, which are about equal to one another. Thus, the eigenvector associated with the distinguishable eigenvalue defines an axis of symmetry of the pressure tensor. The symmetry axis is generally not parallel to **B**. New features of the plasma distribution function are revealed by using the actual eigenvalues of the pressure tensor rather than the usual $p_\perp$ and $p_{||}$, where $\perp$ and $||$ denote directions perpendicular and parallel to **B**.

We report preliminary results of an analysis of ion distribution function measurements from the AMPTE/IRM satellite. The original motivation of the study was to examine the existence and nature of empirical closure relations in the earth's magnetosheath that would allow a second or third velocity moment to be expressed in terms of lower moments. Empirical closure relations for a real turbulent plasma, such as the magnetosheath plasma, could lead to a fluid description of the plasma that is based firmly on observations. As a first step in our study, we have examined the ion pressure tensor, whose elements are second velocity moments of the ion distribution function.

Various studies have been made of closure relations in space plasmas, beginning with the work of Feldman, *et al* [1] using solar wind measurements. Assuming isotropic pressure and a polytropic law in the plasma sheet, Baumjohann and Paschmann [2] and Huang *et al* [3] determined values for the polytropic index. Discrepancies among their findings were further discussed by Zhu [4] and by Goertz and Baumjohann [5]. Belmont and Mazelle [6] and Mazelle and Belmont [7], on the basis of theoretical arguments as well as data analysis, concluded that such a global index is generally not valid. There has recently been considerable interest in finding empirical relations among velocity moments in the magnetosheath region. Using AMPTE/CCE data, Anderson and Fuselier [8] identified an empirical relation between $\gamma_\perp$ and $\gamma_{||}$, which stimulated several theoretical and simulation studies (*e.g.*, Gary *et al*, [9]; Denton *et al*, [10]). Using AMPTE/IRM data, Hill *et al* [11] and Phan *et al* [12] analyzed the extent to which the marginal mirror instability criterion is satisfied. Also using AMPTE/IRM data, Hau *et al* [13] used two empirical polytropic laws to describe thermodynamic properties of the magnetosheath.

In order to study the characterization of a real plasma in terms of fluid variables, such as density, velocity, and pressure, we use a time-averaged distribution function defined with respect to a time window τ:

$$\langle f[\mathbf{r}(t), \mathbf{v}, t] \rangle_\tau = \frac{1}{\tau} \int_{-\tau/2}^{\tau/2} f[\mathbf{r}(t+t'), \mathbf{v}, t+t'] \, dt' \,, \tag{1}$$

where $\mathbf{r}(t)$ is the position vector of the satellite. This average is useful for defining fluid variables if $f[\mathbf{r}(t), \mathbf{v}, t]$ is well approximated by the form

$$f[\mathbf{r}(t), \mathbf{v}, t] = f^{(0)}[\mathbf{r}(t), \mathbf{v}, t] + f^{(1)}[\mathbf{r}(t), \mathbf{v}, t] \,, \tag{2}$$

where $f^{(0)}[\mathbf{r}(t), \mathbf{v}, t]$ varies on a timescale long compared to τ and $f^{(1)}[\mathbf{r}(t), \mathbf{v}, t]$ varies on a timescale short compared to τ. In that case, $\langle f[\mathbf{r}(t), \mathbf{v}, t] \rangle_\tau = f^{(0)}[\mathbf{r}(t), \mathbf{v}, t]$. The statistics presented in this paper of quantities derived from moments of the averaged distribution function are insensitive to the value of τ for τ in the range of 5 to 40 satellite spin periods. (The spin period is about 4.35 s; for instrumental reasons, τ is restricted to be a multiple of five times the satellite spin period.)

There are two major results of our study: 1) The ion pressure tensor is nearly always anisotropic. 2) One of the principal axes of the ion pressure tensor is approximately a symmetry axis, but that axis is generally not parallel to the magnetic field. Adiabatic theory predicts to lowest order in the adiabatic parameter ε, the ratio of the gyroradius to the characteristic distance over which fields change, that the pressure tensor should be diagonal in a frame in which **B** is along one of the axes, the diagonal components being $p_{||}$, $p_\perp$, $p_\perp$. This is a consequence of the distribution function being gyrotropic to lowest order in ε. To next order in ε, the off-diagonal components no longer vanish; there are also order-ε corrections to the diagonal components. Macmahon [14] has given expressions for these corrections. If the higher-order, not-quite-diagonal, tensor is diagonalized, the new principal axes will not be parallel and perpendicular to **B**. This should be one of the sources of the finite angle θ_1 between the symmetry axis and **B** which we observe.

The three-dimensional ion velocity distribution function was measured on the AMPTE/IRM satellite by a top-hat sensor. The energy range of the instrument is 20 eV/q - 40 keV/q. For the on-board computation of the moments (which are used in this paper), where energy has to be converted into velocity, it is assumed that all ions are protons. If another ion species is present, *e.g.*, He^{2+}, its velocities will be overestimated by a factor $\gamma = [(m_a/m_p)/z_a]^{1/2}$, (where m_a and z_a are the ion mass and the ratio of the ion charge to the proton charge), its density contribution will be underestimated by the same factor, and higher moments will also be affected. The ion density, flow velocity, pressure tensor, and heat flux vector were calculated for each spin period under the assumption that only protons were present in the ion population. The instrumental effects of not discriminating among different ion species will be discussed later in the paper. An independent flux-gate magnetometer on the satellite measured the magnetic field vector 32 times per second; for our analysis, the magnetic field measurements were averaged over a spin period. Details about the measurement apparatus on the AMPTE/IRM are given by Paschmann et al [15] and Lühr et al [16]. In this study, we used 18 magnetosheath traversals by the satellite, each of which lasted about two hours. Each crossing occurred during the local time period of 8-14 h while the satellite was within 12° of the ecliptic plane.

For each measurement of the distribution function we diagonalize the tensor

$$\langle \mathbf{T} \rangle_\tau \equiv \frac{m_p \int d^3 v \, (\mathbf{v} - \langle \mathbf{u} \rangle_\tau)(\mathbf{v} - \langle \mathbf{u} \rangle_\tau) \langle f \rangle_\tau}{\langle n \rangle_\tau}, \qquad (3)$$

where m_p = proton mass, to find its eigenvalues and principal axes. Of the three eigenvalues, we identify one distinct eigenvalue (T_1) as the one whose magnitude is most separated from the remaining two (T_2 and T_3). That is, $|T_1 - T_3| \geq |T_1 - T_2| \geq |T_2 - T_3|$. The temperatures T_2 and T_3 are usually nearly equal and quite distinct from T_1. We define a unit vector $\boldsymbol{\eta}_1$ which is parallel to the principal axis associated with T_1; the ambiguity in the direction of $\boldsymbol{\eta}_1$ is removed by arbitrarily requiring $\boldsymbol{\eta}_1 \cdot \mathbf{B} \geq 0$. We denote the angle between $\boldsymbol{\eta}_1$ and $\mathbf{B}$ by θ_1; this angle, $\theta_1 \equiv \cos^{-1}(\boldsymbol{\eta}_1 \cdot \mathbf{B}/B)$, only assumes values between 0 and $\pi/2$.

Figure 1 shows time histories for the outbound traversal of the magnetosheath on September 19, 1984, one of 18 traversals that we have analyzed. The data presented are averages over 15 spin periods. The interval shown begins just after a magnetopause crossing at UT 16:11 and ends just before a bow shock crossing at UT 18:28. The quantities displayed in the panels of the figure are: a) the magnetic field strength, B; b) ion number density, n; c) number density of ions in the energy range 8-40 keV, n_2; d) ion bulk flow speed, u; e) the eigenvalues T_1 (dashed line), T_2, and T_3 in units of $6 \times 10^6 \, °K$; f) angle between $\boldsymbol{\eta}_1$ and $\mathbf{B}$, θ_1; g) plasma beta, $\beta = 8\pi n T / B^2$, where $T = (T_1 + T_2 + T_3)/3$; h) $\delta B / B$, where δB is the rms fluctuation of B about its average over one satellite spin period; i) $T_{\|}$, $T_{\perp 1}$, and $T_{\perp 2}$ in units of $6 \times 10^6 \, °K$. $T_{\|}$, $T_{\perp 1}$, and $T_{\perp 2}$ are the diagonal elements of $\mathbf{T}$ in a coordinate system one axis of which is along $\mathbf{B}$. The orientation of the two axes perpendicular to $\mathbf{B}$ is chosen to diagonalize the 2×2 submatrix of $\mathbf{T}$ associated with those axes; $T_{\perp 1}$ and $T_{\perp 2}$ are the eigenvalues of that 2×2 submatrix. The full tensor $\mathbf{T}$ is not diagonal with respect to this coordinate system.

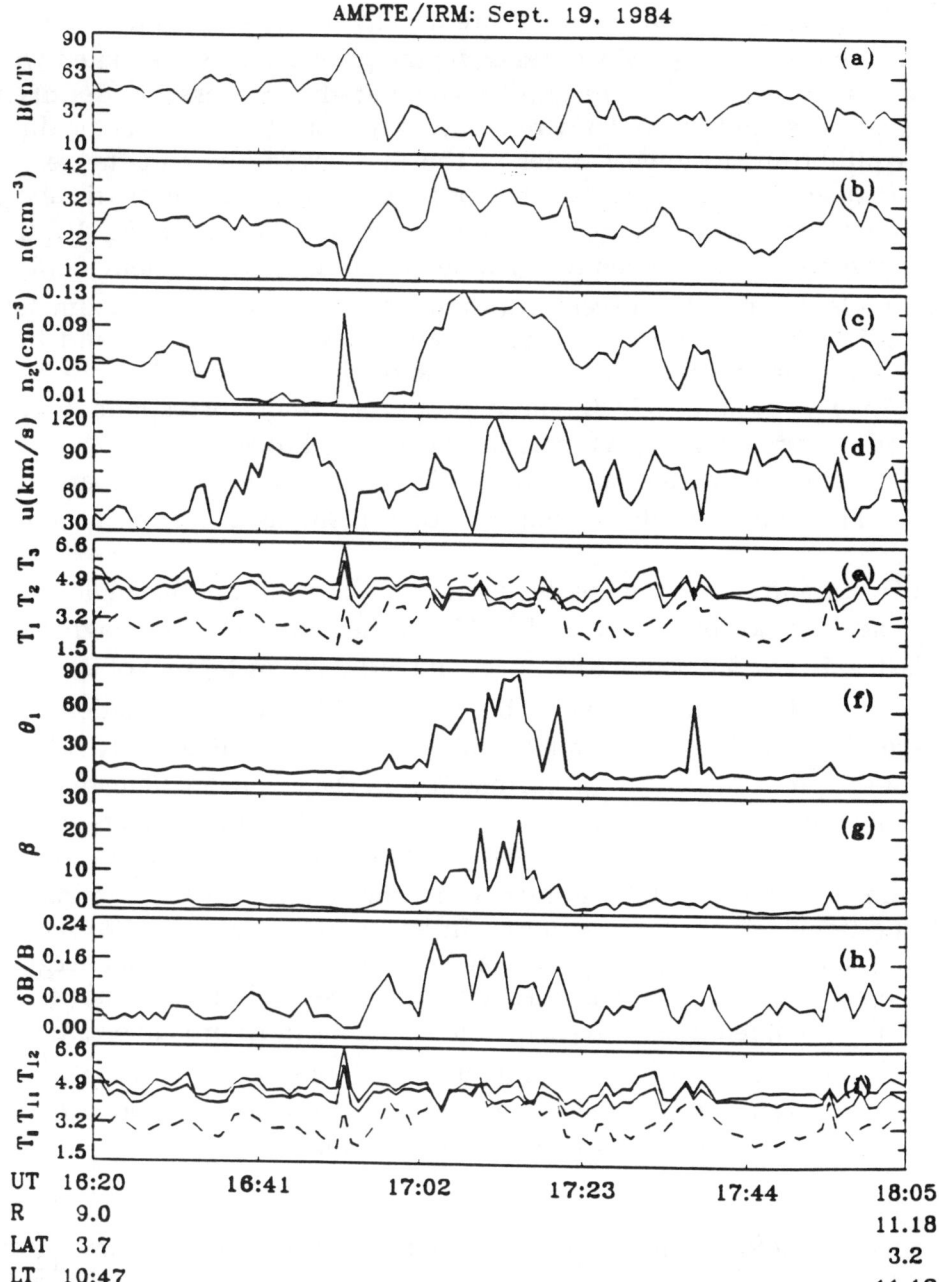

Fig. 1. Outbound traversal of magnetosheath on September 19, 1984. The interval shown begins just after the magnetopause crossing at UT 16:11 and ends just before the bow shock crossing at UT 18:28. From top to bottom, a) the magnetic field strength, B; b) ion number density, n; c) the partial density n_2 of energetic ions in the energy range 8-40 keV; d) ion bulk flow speed, u; e) the three eigenvalues of **T**: T_1 is plotted with a dashed line; f) θ_1, the angle between $\mathbf{\eta}_1$ and **B**; g) plasma beta, defined as $\beta = 8\pi n T / B^2$; h) $\delta B / B$, fractional rms fluctuation of B over one satellite spin period; i) the usual $T_{\|}$ (dashed line), $T_{\perp 1}$, and $T_{\perp 2}$.

In Fig. 1(e) we see that T_2 and T_3 are usually nearly equal and quite distinct from T_1. This means that $\boldsymbol{\eta}_1$, which lies along the principal axis associated with T_1, is nearly a symmetry axis. We emphasize that the process of identifying a symmetry axis has not involved any reference direction at all, such as the direction of **B**. In equilibrium to lowest order in the the gyroradius and the reciprocal cyclotron frequency, the ion pressure tensor is expected to be symmetric about the magnetic field direction when the plasma is anisotropic. We see in Fig. 1(f) that θ_1, the instantaneous angle between the symmetry axis and the magnetic field, is not zero. The angle is about 15°, except during the period between approximately UT 17:02 and UT 17:23. During this period, we see in figs. 1(a), 1(c), 1(g), and 1(h) that there is a significant decrease in the magnetic field strength, with corresponding increases in n_2, β, and $\delta B/B$. These properties may correspond to typical conditions downstream of a quasi-parallel shock (*e.g.*, Crooker *et al* [17], Luhmann *et al* [18]). During this period, as shown in Fig. 1(i), $T_{||}$ is not well separated from $T_{\perp 1}$ and $T_{\perp 2}$. However, even during this period, T_1 is still distinct from T_2 and T_3.

In Figs. 2 and 3 we examine the distribution of values of the angle θ_1 and of the temperature anisotropy $A \equiv (T_2 + T_3)/2T_1$ for the entire set of 18 magnetosheath traversals. Fig. 2 is a scatter plot of θ_1 vs. A and Fig. 3 is a histogram of the (number of occurrences)/$\sin\theta_1$ versus θ_1 with a bin width of 1°. The normalization by $\sin\theta_1$ accounts for the variation of solid angle with the polar angle θ_1 and provides a true measure of the probability of the occurrence of θ_1. Data from all 18 magnetosheath traversals are included, a total of 36 hours of data. Each dot represents an average over a time window τ equal to 5 satellite spin periods. Data corresponding to temperatures less than 3×10^6 °K have been omitted in order to eliminate some data of questionable angular resolution. In addition, some data (4%) have been omitted where the adequacy of the coverage of the complete range of energy of the ions by the particle detectors was questionable. These omissions do not affect the following important features: 1) As shown in Fig. 3, the most probable value of θ_1 is not zero, but approximately 5°, in contrast to the lowest-order expectation for an equilibrium configuration. 2) As shown in Fig. 2, there is a correlation of θ_1 with A; the gap around $A = 1$ indicates that the ion pressure tensor does not approach isotropy ($A = 1$).

In the magnetosheath we have observed an instantaneous symmetry axis of the ion pressure tensor which is generally not parallel to the magnetic field, and which persists despite various dynamical processes in the plasma. We have also observed that the pressure tensor tends not to be isotropic; it is either oblate ($A > 1$, likely associated with a quasi-perpendicular shock), or prolate ($A < 1$, likely associated with a quasi-parallel shock).

We have done the same analysis with the electron pressure tensor. However, we found in many cases that the symmetry axis of the electron pressure tensor is parallel to the spin axis of the spacecraft. We believe this to be an instrumental effect that invalidates such an analysis of the electron data. The ion pressure tensor does not suffer from this problem.

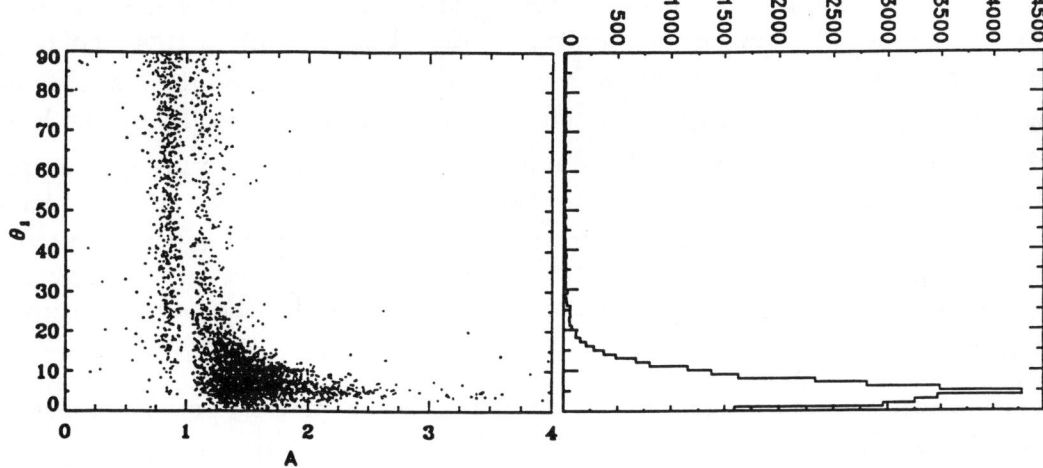

Fig. 2. Scatter plot of θ_1 vs. $A \equiv (T_2 + T_3)/2T_1$

Fig. 3. Histogram of (number of occurrences)/$\sin\theta_1$ vs. θ_1

We now address the instrumental effects of not discriminating among different ion species. In general, the relationship between flux, $\Phi(E)$, and velocity distribution function, $f(v)$, is [B. Anderson, *private communication.*, 1994]

$$\Phi(E) = (2E/m^2) f(v), \qquad (4)$$

where $E = \frac{1}{2}mv^2$. When there are both protons and He^{2+}, the total flux detected by the instrument, which is sensitive to E/q, is:

$$\Phi_{tot}(E) = (2E/m_p^2) f_p(v) + z_a (2E/m_a^2) f_a(v/\gamma), \qquad (5)$$

where $\gamma = [(m_a/m_p)/z_a]^{1/2}$. The corresponding total velocity distribution function is

$$f(v) = f_p(v) + \left[z_a / (m_a/m_p)^2 \right] f_a(v/\gamma). \qquad (6)$$

The moments n, **u** and T that are computed from the data are defined in terms of the total velocity distribution:

$$n = \int d^3\mathbf{v}\, f(v) = n_p + n_a / \left[(m_a/m_p) z_a \right]^{1/2}, \qquad (7)$$

$$\mathbf{u} = \frac{\int d^3\mathbf{v}\, v f(v)}{n} = \frac{(n_p \mathbf{u}_p + n_a \mathbf{u}_a / z_a)}{n}, \qquad (8)$$

$$\mathbf{T} = \frac{m_p \int d^3\mathbf{v}\, (\mathbf{v}-\mathbf{u})(\mathbf{v}-\mathbf{u}) f(v)}{n}. \qquad (9)$$

We have modelled the magnetosheath plasma as 96% protons and 4% He^{2+}, with both species described by bi-Maxwellian distributions with drift velocities $\mathbf{u}_p$ and $\mathbf{u}_a$. We chose a coordinate system with the z-axis parallel to the background magnetic field, and the y-axis perpendicular to both drift velocities. That is, both $\mathbf{u}_p$ and $\mathbf{u}_a$ were in the x-y plane. In the magnetosheath, the He^{2+} and proton thermal velocities are comparable, and $\mathbf{u}_p$ is about equal to $\mathbf{u}_a$ (Fuselier et al [19]). We have used this model to calculate Eqs. (7)-(9) with various sets of parameters and found that the magnitudes of the effects are far too small to account for our observations [Fig. 1(f) and Fig. (3)]. In addition, this model does not predict the gap around $A=1$ in Fig. 2.

Because of the axisymmetry of the ion pressure tensor, perhaps a useful lowest-order form for modeling ion distribution functions would be $f(\mathbf{r}, v_1, v_2^2 + v_3^2)$, where v_1 is the velocity component along the symmetry axis, and where v_2 and v_3 are two perpendicular components. This relation might provide a good basis for a more refined phenomenological description of the plasma. The corresponding reduction of the number of independent components of the heat flux tensor would simplify an investigation of empirical closure relations that involve third moments.

Acknowledgments. We would like to acknowledge some programming efforts of R. Brittain and useful discussions with M. Temerin, R. Denton, B. Sonnerup, and W. Baumjohann. We wish to thank G. Paschmann for discussion and for providing the AMPTE/IRM plasma data, and H. Lühr for providing triaxial flux gate data. We are indebted to N. Sckopke for valuable comments concerning data analysis. X. Li especially thanks B. Anderson for valuable discussions about assessing the effects of He^{2+} on the observations. This work was supported by the National Science Foundation under Grant ATM 9221669.

References

[1] W. C. Feldman, J. R. Asbridge, S. J. Bame, and H. R. Lewis, *Phys. Rev. Lett.* **30**, 271 (1973).
[2] W. Baumjohann and G. Paschmann, *Geophys. Res. Lett.* **16**, 295 (1989).
[3] C. Y. Huang, C. K. Goertz, L. A. Frank, and G. Rostoker, *Geophys. Res. Lett.* **16**, 563 (1989).
[4] X. M. Zhu, *Geophys. Res. Lett.* **17**, 2321 (1990).
[5] C. K. Goertz and W. Baumjohann, *J. Geophys. Res.* **96**, 20991 (1991).
[6] G. Belmont and C. Mazelle, *J. Geophys. Res.* **97**, 8327 (1992).
[7] C. Mazelle and G. Belmont, *Geophys. Res. Lett.* **20**, 157 (1993).
[8] B. J. Anderson and S. A. Fuselier, *J. Geophys. Res.* **98**, 1461 (1993).
[9] S. P. Gary, B. J. Anderson, R. Denton, S. Fuselier, M. E. McKean, and D. Winske, *Geophys. Res. Lett.* **20**, 1767 (1993).
[10] R. E. Denton, B. J. Brian, S. P. Gary, and S. A. Fuselier, *J. Geophys. Res.*, in press (1994).
[11] P. Hill, G. Paschmann, R. A. Treumann, W. Baumjohann, and N. Sckopke, *J. Geophys. Res.*, in press (1994).
[12] T.-D. Phan, G. Paschmann, W. Baumjohann, N. Sckopke, and H. Lühr, *J. Geophys. Res.* **99**, 121 (1994).

[13] L.-N. Hau, T.-D. Phan, B. U. Ö. Sonnerup, and G. Paschmann, *Geophys. Res. Lett.* **20**, 2255 (1993).

[14] A. Macmahon, *Phys. of Fluids* **8**, 1840 (1965).

[15] G. Paschmann, H. Loidl, P. Obermayer, M. Ertl, R. Laborenz, N. Sckopke, W. Baumjohann, C. W. Carlson, and D. W. Curtis, *IEEE Trans. Geosci. Remote Sen.*, GE-23, 262 (1985).

[16] H. Lühr, N. Klöcker, W. Oelschlägel, B. Häusler, and M. Acuña, *IEEE Trans. Geosci. Remote Sen.*, GE-23, 259 (1985).

[17] N. U. Crooker, T. E. Eastman, L. A. Frank, E. J. Smith, and C. T. Russell, *J. Geophys. Res.* **86**, 4455 (1981).

[18] J. G. Luhmann, C. T. Russell, and R. C. Elphic, *J. Geophys. Res.* **91**, 1711 (1986).

[19] S. A. Fuselier, E. G. Shelley, and D. M. Klumpar, *Geophys. Res. Lett.* **15**, 1333 (1988).

ANOMALOUS SCATTERING AND ABSORPTION OF NEUTRINOS IN DENSE PLASMAS

R Bingham, H A Bethe, J M Dawson, J J Su and V N Tsytovich

Rutherford Appleton Laboratory, Chilton, Didcot, Oxon, OX11 0QX

Abstract

A new absorption mechanism for electron neutrinos in dense plasmas is formulated and applied to the problem of reviving a stalled supernova shock. In particular we show that electron neutrinos can nonlinearly couple to electron plasma oscillations resulting in the neutrinos losing energy to the plasma oscillations which heat up the electrons. These heated electrons establish a pressure gradient within the star which is sufficient to revive the shock, pushing the material outwards and eventually ejecting the outer layers of the star producing a prodigious explosion the supernova. This new neutrino plasma interaction process is found to be important when the neutrino flux is sufficiently intense, of the order of $10^{29} W/cm^2$ for the supernova problem.

One of the fundamental problems in the field of supernovae is to understand how the gravitational energy released during the collapse or infall of the core is transferred to ejecting the outer mantle. Simply stated: how does an implosion produce an explosion? During collapse the central core gets over compressed and releases energy by bouncing back. This bounce of the core launches a shock wave which rapidly proceeds outwards and eventually ejects the outer layers of the star. In simulations of this process, however, the shock wave loses a considerable amount of energy by dissociating heavy nuclei and since the shock velocity is proportional to the energy in the shock wave the velocity decreases until the infall velocity equals the outward shock velocity. At this point, known as the sonic point, the shock stalls and fails to deliver its energy to the outer layers. It is widely believed that neutrinos are responsible for reviving the shock by delivering energy to the electrons. However, this scenario has only been shown through numerical simulations [1,2,3] to be possible by letting a few percent of the neutrino flux be absorbed by electrons. The main problem with this scenario is that the normal neutrino electron collisional scattering cross section is so small that it is impossible for this single neutrino electron scattering process to deliver enough energy to the electrons to revive the shock. However, as pointed out by Bingham et al [4] there is a possibility that a collective neutrino-electron scattering process could be operating which could result in absorbing sufficient neutrino energy by electrons.

This collective process is similar to the nonlinear photon plasma scattering process known as stimulated Raman scattering where an incident photon decays into a lower energy photon and a plasma wave. The difference in energy between the incident and scattered photon is taken up by the plasma wave. For the case of neutrinos, the incident neutrino decays into a lower energy neutrino and a plasma wave. This process does however require an extremely intense neutrino flux estimated to be of the order of or greater than $10^{29} Watts/cm^2$ which is close to the value estimated from Supernova 1987A.

During the implosion phase or collapse electron capture occurs, ie a proton and an electron coalesce to yield a neutron and a neutrino. The neutrino carries away about $10^{52} ergs sec^{-1}$ cooling the core in the process. The super-dense core has a density above $10^{12} g/cm^3$, and neutrinos become trapped ie the core is opaque to the neutrinos which scatter and diffuse slowly our in timescales of several seconds. The central core forms the neutrionsphere which radiates a Fermi distribution of neutrinos with a temperature of about $10 MeV$. Due to the rapid density fall-off as one leaves the core, the stellar material becomes transparent to neutrinos which stream freely outwards.

The neutrinosphere radiates neutrinos more or less like a black body at it's temperature with a neutrino intensity of $3 \times 10^{29} W/cm^2$ at a distance of $300 km$ from the centre. The neutrinos carry away energy and entropy allowing the collapse process to accelerate. It is widely recognized [3] that more than 1% of the total neutrino energy must be absorbed by the plasma electrons in the layers surrounding the super-dense core of the star. This produces sufficient pressure to revive the outward moving shock and eject the outer part of the star, leaving only enough mass to form a neutron star. Otherwise no supernovae would occur and only black holes would result. Previous studies[2] have concentrated on collisional energy losses between neutrinos and electrons and nucleons, however, the loss rate is marginal to produce the required heating. In this paper we describe a new neutrino interaction [4] mechanism based on the coupling between the neutrinos and collective plasma oscillations in the dense plasma surrounding the core, namely stimulated neutrino-plasmon scattering. The interaction between electrons and neutrinos takes place through the electroweak interaction, the electrons interacting via both electromagnetism and the weak interaction and the neutrinos interacting only through the weak interaction. The stimulated scattering process described in this paper has an analog with laser coupling to plasma oscillations [5], producing effects such as stimulated Raman and Brillouin scattering and parametric decay process where a photon is annihilated by creating two plasmons.

In such stimulated scattering processes the neutrino energy flux will decrease while conserving total lepton number, the interaction being determined by the energy and momentum conservation relationships,

$$\hbar\omega_{\nu o} = \hbar\omega_{\nu 1} + \hbar\omega_p \ ; \ \hbar\underline{k}_{\nu o} == \hbar\underline{k}_{\nu 1} + \hbar\underline{k}_p \tag{1}$$

where ω and $\underline{k}$ are the frequencies and wavenumbers of the wave fields (for neutrinos and Langmuir plasmons). Here the subscripts $\nu o, \nu 1$, and p refer to the incident neutrino, scattered neutrino and the plasma wave respectively.

The nonlinear coupling between the neutrino field and the Langmuir plasmon field is obtainable from the dispersion relation for neutrinos (given by equation (5) of Bethe[6]) which is

$$E^2 - p^2c^2 - m^2c^4 - 2EV = 0, \tag{2}$$

where E is the normalized energy, p is the normalized momentum, m is the normalized neutrino rest mass, and V is equivalent to a potential energy $V = G\sqrt{2}n_e$, where n_e is the electron number density and G is the Fermi constant of the weak interaction. The effect of linear dispersion and amplitude nonlinearity can be studied within the framework of a nonlinear Klein-Gordon equation obtained by making the following substitutions in equation (2)

$$p \to -i\hbar\frac{\partial}{\partial x}, \ E \to i\hbar\frac{\partial}{\partial t}.$$

This gives the following wave equation for the neutrino field ψ

$$\left(\hbar^2c^2\frac{\partial^2}{\partial x^2} - m^2c^4 - \hbar^2\frac{\partial^2}{\partial t^2}\right)\psi = i2\sqrt{2}Gn_e\hbar\frac{\partial\psi}{\partial t}, \tag{3}$$

where ψ is the neutrino amplitude wave function (not the wave function for single neutrinos) normalized so that the neutrino energy density is $W = |\psi|^2/4\pi$. Here $|\psi|^2$ is proportional to the neutrino density times their energy, $\hbar\omega$. The momentum density is $W/v_p \sim W/c$, where v_p is the phase velocity of the neutrinos passing through the plasma. The refractive index n obtained from this equation depends on the electron density and can be written as

$$n = \frac{ck_\nu}{\omega_\nu} = 1 - \frac{\sqrt{2}G}{\hbar k_\nu c} n_e \tag{4}$$

In this equation we have neglected the proton and neutron contribution to the refractive index. Equations (3) and (4) describes the neutrino field coupling into the plasma medium through the electron density: it is similar to the wave equation for electromagnetic waves coupling to a plasma medium.[1]

Writing $\psi = \psi_o + \psi_1$, where subscript $o, 1$ refers to the incident and scattered fields, equation (3) gives for ψ_1

$$\left(\hbar^2 c^2 \frac{\partial^2}{\partial x^2} - \hbar \frac{\partial^2}{\partial t^2} - \tilde{m}^2\right)\psi_1 = i2\sqrt{2}G\delta n_e \hbar \frac{\partial \psi_o}{\partial t}, \tag{5}$$

where δn_e is the perturbed electron density ($n_e = n_o + \delta n_e$), which varies as $\sim e^{i(k_p x - \omega_p t)}$. Equation (5) describes the nonlinear interaction between the incident neutrino field ψ_o, and the plasma wave δn_e, producing a scattered signal at ψ_1; with $\tilde{m}^2 = m^2 c^4 + 2\sqrt{2}G n_o \hbar \omega_{\nu 1}$.

From equation (5) and assuming the amplitudes can be written as a slowly varying part, due to the nonlinear interaction, times a linear high frequency phase, we can reduce equation 4 to a first order differential equation describing the slowly varying part of ψ_1 given by

$$\hbar \frac{\partial \psi_1}{\partial t} = -i\sqrt{2}G\delta n_e^* \psi_o, \tag{6}$$

where we have assumed $\omega_{\nu o} \simeq \omega_{\nu 1}$ and the asterisk denotes complex conjugate.

Using the fact that the plasma waves with phase velocity $\sim c$ have energy and momentum densities $n_o m_e c^2 |\delta n_e|^2/n_o^2$ and $n_o m_e c |\delta n_e|^2/n_o^2$, respectively, conservation of energy and momentum give

$$\Delta \left(n_o m_e c^2 |\delta n_e|^2/n_o^2\right) + \Delta \left(|\psi_1|^2/4\pi\right) + \Delta \left(|\psi_o|^2/4\pi\right) = 0, \tag{7a}$$

---

[1] Since neutrinos are Fermions one might question if this fact should not be taken into account here. Since the neutrinos are far from degenerate where the instability occurs, the exclusion principle plays little role.

$$\Delta\left(n_o m_e c^2 \delta n_e^2/n_o^2\right)\left(\underline{k}_p/\omega_p\right) + \Delta\left(|\psi_1|^2/4\pi\right)\left(\underline{k}_{\nu 1}/\omega_{\nu 1}\right) +$$
$$\Delta\left(|\psi_o|^2/4\pi\right)\left(\underline{k}_{\nu o}/\omega_{\nu o}\right) = 0. \tag{7b}$$

Considering only the case of forward scattering $\left(\underline{k}_p \| \underline{k}_{\nu o} \| \underline{k}_{\nu 1}\right)$, eliminating $\Delta\left(|\psi_o|^2/4\pi\right)$ between (7a) and (7b), and dividing the result by $(\omega_{\nu o} k_p - \omega_p k_{\nu o})$ (which is small but not zero) gives

$$\Delta\left(n_o m_e c^2 |\delta n_e|^2/n_o^2\right)/\omega_p = \Delta\left(|\psi_1|^2/4\pi\omega_{\nu 1}\right). \tag{8}$$

This equation simply says that the number of scattered neutrinos equals the number of plasmons generated. Assuming that δn_e and $|\psi_1|$ grow from very small amplitudes, we obtain

$$|\psi_1| = \sqrt{4\pi m_e c^2 n_o \frac{\omega_{\nu o}}{\omega_p}} \frac{|\delta n_e|}{n_o}, \tag{9}$$

where we assumed $\omega_{\nu o} \simeq \omega_{\nu 1}$.

Substituting for δn_e in equation (6) gives

$$\hbar \frac{\partial \psi_1}{\partial t} = -i\sqrt{2} G \sqrt{\frac{n_o |\psi_o|^2}{4\pi m_e c^2} \frac{\omega_p}{\omega_{\nu o}}} \psi_1. \tag{10}$$

Equation (10) gives the growth rate of stimulated scattering of neutrinos on plasma oscillations:

$$\gamma = \sqrt{2} \frac{G}{\hbar} \sqrt{\frac{n_o |\psi_o|^2}{4\pi m_e} \frac{\omega_p}{\omega_{\nu o}}}. \tag{11}$$

We can obtain an analogous equation to (6) for $\frac{\partial \delta n_e}{\partial t}$ from equations (6) and (9) which is written as

$$\frac{\partial \delta n_e}{\partial t} = -\frac{i\sqrt{2} n_o^2 G \psi_1^* \psi_o}{\hbar 4\pi m_e n_o c^2} \frac{\omega_p}{\omega_{\nu o}} \tag{12}$$

To determine the threshold of the instability we have to add the dissipation terms to equation (5) and to the corresponding equation for δn_e.

These equations then become

$$\frac{\partial \psi_1}{\partial t} + \gamma_\nu \psi_1 + \frac{i\sqrt{2} G \delta n_e^* \psi_o}{\hbar} = 0 \tag{13}$$

$$\frac{\partial \delta n_e}{\partial t} + \gamma_p \delta n_e + \frac{i\sqrt{2} n_o^2 G \psi_1^* \psi_o}{\hbar 4\pi m_e n_o c^2} \frac{\omega_p}{\omega_{\nu o}} = 0 \tag{14}$$

Solutions of equations (13) and (14) for ψ_o a constant yield the following growth rate and threshold values.

$$\gamma = \frac{1}{2}\left\{-(\gamma_\nu + \gamma_p) \pm \sqrt{(\gamma_\nu - \gamma_p)^2 + \alpha|\psi_o|^2}\right\} \quad (15)$$

where $\alpha = \frac{n_o G^2}{2\pi m_e c^2 \hbar^2}\frac{\omega_p}{\omega_{\nu o}}$. Setting $\gamma = 0$ gives the threshold for stimulated scattering of neutrinos off Langmuir plasmons

$$|\psi_o|^2 = \frac{4\gamma_p \gamma_\nu}{\alpha} \quad (16)$$

The true damping of the neutrino field is of course negligible if we only consider electron-neutrino scattering, resulting in an unrealistically low threshold and high growth rate. However, there is an effective neutrino damping rate due to dephasing of the neutrino field with the generated plasma wave. The neutrino field has a broad frequency spectrum since it is a thermal spectrum. Dephasing is very rapid except for forward scattering, where the scattered wave moves along with the incident wave (both at essentially the speed of light) and so phase coherence is maintained for long times despite the wide frequency spread. In the forward direction there is still an important phase slippage due to the angular spread ($\Delta\theta$) in the propagation direction for the neutrinos. This gives them a spread in velocities along the redial (forward) direction which is $c\Delta\theta^2$.

Here two points should be made about the above derivation. First, we have considered only three wave processes, i.e. an incident neutrino, a scattered neutrino of lower energy (analogous to Stokes scattering for EM waves), and a plasma wave. For forward scattering like this, one should include a fourth wave, forward scattered neutrinos with higher energy (analogous to anti Stokes scattering for EM waves). These come from neutrinos that are phased with respect to the plasma wave so that they gain energy rather than give up energy. Second we have considered the neutrinos to be monoenergetic, propagating in one direction and all in phase. These shortcomings of the theory can be overcome by adopting a wave kinetic equation for the neutrinos which is then coupled to the Vlasov equation for the electrons through the neutrino index of refraction. This approach can deal with the finite spectral and angular widths of the neutrino spectrum. This wave kinetic equation is given by [7]

$$\frac{\partial N_k}{\partial t} + \frac{\partial \omega_\nu}{\partial \underline{k}}\cdot\underline{\nabla} N_k - \frac{\partial \omega_\nu}{\partial \underline{x}}\cdot\underline{\nabla}_k N_k = 0, \quad (17)$$

where N_k is the wave action or neutrino number density with $\int \omega_{\nu k} N_k d^3 k = \frac{|\psi|^2}{4\pi}$. The third term in equation (17) represents the change in neutrino action due to spatial changes in the neutrino frequency: it represents a force acting on the neutrinos. In a plasma the frequency ω_ν changes with plasma density, which can be calculated from the refractive index, ε, of a neutrino in a plasma; may be obtained from Eq. 2 and is given by

$$\varepsilon = 1 - \frac{\sqrt{2}G}{k\hbar c}n_e. \tag{18}$$

The electron density n_e is a function of position such that $n_e = n_o + \delta n_e(x)$ where δn_e is the electron density perturbation due to plasmons. Using equation (18) we get

$$\frac{\partial \omega_\nu}{\partial x} = \frac{\sqrt{2}}{2}\frac{\omega_\nu^2}{ck^2}\frac{G}{\hbar c}\frac{\partial \delta n_e}{\partial x}. \tag{19}$$

The coupling of the neutrino flux to plasmons leads to a change in the neutrino number density $N_k = N_o(\underline{k}_\nu) + \delta N_k$. Linearizing equation (17) by assuming that δN varies as $\exp i\left(\underline{k}_p.\underline{x} - \Omega t\right), (k_p \ll k_\nu)$ we get

$$\delta N_k = \frac{i\frac{\sqrt{2}}{2}\frac{\omega_\nu^2}{c^2k^2}G\,\hbar\frac{\partial \delta n_e}{\partial \underline{x}}\cdot\frac{\partial N_o(\underline{k}_\nu)}{\partial \underline{k}}}{\left(\Omega - \underline{k}_p.\underline{v}_g\right)}, \tag{20}$$

where v_g is the group velocity of neutrino wave packets ($\approx c$). Substituting equation (20) into the equation for Langmuir waves, and using the cold plasma approximation leads to the following dispersion relation for plasmons coupled to neutrinos:

$$\Omega\left(\Omega + i\gamma_p\right) - \omega_p^2 = -\frac{2n_o G^2 \omega_\nu^2 \omega_p^2}{m_e c^4 \hbar^2 k_\nu^2}\int \frac{\underline{k}_p.\frac{\partial N_o(\underline{k}_\nu)}{\partial \underline{k}_\nu}}{\left(\Omega - \underline{k}_p.\underline{v}_g\right)}d\underline{k}_\nu. \tag{21}$$

We have included the effect of Langmuir wave damping by replacing Ω^2 on the left-hand side by $\Omega(\Omega + i\gamma_p)$. We note that equation (21) is exactly the dispersion relation one obtains for a small bump on the tail of the electron distribution (except for the numerical factor) and can be solved by standard plasma physics methods.

The integral has a pole at $\Omega = \underline{k}_p.\underline{v}_g$ which can give rise to an imaginary component of Ω, resulting in growth or damping, depending on sign of $\frac{\partial N_o(\underline{k}_\nu)}{\partial \underline{k}_\nu}$ of at the resonance (Landau growth or damping).

The most interesting case, however, occurs when the pole lies in a region where $\frac{\partial N_o(\underline{k}_\nu)}{\partial \underline{k}_\nu}$ is zero (analogous to the two stream instability). This is possible for the situation considered here since far from the core, all the neutrinos are moving outward near the speed of light. The spread in velocities parallel to $\underline{k}_p$ (parallel to the radius) comes from the small angular spread in the direction of neutrino propagation. We can obtain an estimate of this growth rate by assuming the neutrinos are propagating at a small angle to z with uniform angular distribution in this solid angle: we use an appropriate average of $N_o(\underline{k}_\nu)$ over $\underline{k}_\nu$ and write

$$N_o(\underline{k}_\nu) \simeq \frac{\delta(|\underline{k}_\nu| - \underline{k}_{\nu o}) \left\langle \frac{|\psi_o|^2}{4\pi\omega_{\nu o}} \right\rangle}{2\pi (1 - \cos\theta_o) k_{\nu o}^3} \qquad (22)$$

where $\theta_o = a/R$, a being the radius of the core and R the distance from the center. We will assume that the most unstable mode has a phase velocity parallel to z and just slower than $c\cos\theta_o$. Then the singularity in the denominator of equation (21) lies in the region where $\frac{\partial N_o(\underline{k}_\nu)}{\partial \underline{k}_\nu} = 0$.

Integrating equation (21) by parts over k, and using equation (22) gives

$$\Omega(\Omega + i\gamma_p) - \omega_p^2 = \frac{n_o G^2 \omega_\nu^2 \omega_p^2}{m_e c^3 \hbar^2 k_{\nu o}^3} \frac{k_p^2 \left\langle \frac{|\psi_o|^2}{4\pi\omega_{\nu o}} \right\rangle}{(1 - \cos\theta_o)} \int \frac{\sin^3\theta}{(\Omega - k_p c \cos\theta)^2} d\theta. \qquad (23)$$

The main contribution is close to θ_o because of the $\sin^3\theta$ dependence: we replace $\cos\theta$ by $\cos\theta_o$ and take θ_o small so that $\theta \simeq \sin\theta$ yields

$$\Omega(\Omega + i\gamma_p) - \omega_p^2 = \frac{n_o G^2 \omega_\nu^2 \omega_p^2 k_p^2 \left\langle \frac{|\psi_o|^2}{4\pi\omega_{\nu o}} \right\rangle \theta_o^2}{2m_e c^3 \hbar^2 \left(\Omega - k_p c + k_p \frac{c\theta_o^2}{2}\right)^2 k_{\nu o}^3} \qquad (24)$$

Letting $\Omega = \omega_p + \delta$, where δ is small, then $\Omega \simeq \pm\omega_p$ and $k_p c \left(1 - \frac{\theta_o^2}{2}\right) \omega_p$, the following growth rate of Langmuir plasmons due to the intense neutrino flux is obtained:

$$Im\delta = \left[\frac{n_o|\psi_o|^2}{8\pi m_e c^2} \frac{G^2}{\hbar^2 \omega_{\nu o}^2} \frac{\omega_p}{\gamma_p} \theta_o^2\right]^{\frac{1}{2}} \omega_p. \qquad (25)$$

For plasma densities found outside the core of $n_o = 10^{30} cm^{-3}$, and neutrino intensities of $3 \times 10^{29} W/cm^2$, at a distance of 300km we find the growth rate to be of the order of $5 \times 10^6 sec^{-1}$ when the Langmuir wave damping is $\gamma_p \simeq 10^{-3}\omega_p$ and $\theta_o \simeq 3.10^{-2}$. This corresponds to a growth distance of order $60m$ resulting in $1.2km$ for 20 e-foldings, which must be compared with a mean free path of order $10^{11} km$ for normal neutrino electron scattering in the star. From this calculation we see that the instability is quite strong and as a result we can expect to maintain a quasi-stable neutrino distribution for some distance out from the core.

We can estimate the amount of energy lost by the neutrinos to the plasma waves by using a simple diffusion model. In this model we assume that the density change produced as a result of the plasma waves scatters the neutrino beam in accordance with the refractive index which can be written as (see eq. 4) $ck/\omega = 1 - \sqrt{2}Gn_e/\hbar kc$

Using $G = 8.86 \times 10^{-47} GeV cm^3, \hbar \simeq 10^{-15} eV sec$ and $\hbar kc = 10^7 eV$, then we can write the neutrino phase velocity v_p as

$$v_p = \frac{\omega_\nu}{k_\nu} = c\left(1 + \sqrt{2}\frac{\delta n}{n}10^{-14}\right) \qquad (26)$$

As discussed earlier the phase slippage due to the angular spread ($\Delta\theta$) in the propagation direction gives rise to a velocity spread $c\Delta\theta^2$ along the beam direction. The density perturbation produces a change in the phase velocity and hence an angular rate of change given by

$$\Delta\theta = \frac{\Delta v_p}{\lambda p/2} = 2\sqrt{2} \times 10^{-14}\frac{\delta n_e}{n_o}\omega_{pe} \qquad (27)$$

for $n_o = 10^{30} cm^{-3}, \Delta\theta = 10^6 \delta n_e/n_o$. The neutrinos diffuse in angle as they scatter from the plasma waves with a diffusion coefficient given by

$$D = \Delta\theta^2 \tau_c \qquad (28)$$

Using the value for $\Delta\theta$ and the reciprocal of the growth rate eq.(25) the diffusion coefficient can be written as

$$D = 10^6 \left(\frac{\delta n_e}{n_o}\right)^2 rad^2 sec^{-1} \qquad (29)$$

for $\tau_c = 10^{-6} sec$. The change in scattering angle after a time t is given by $\Delta\theta^2 = Dt$. For the time it takes the neutrinos to reach the position of the stalled shock $R \sim 300 km$ the neutrinos scatter by an amount $\Delta\theta^2 \simeq D\frac{R}{c} = 10^{-3}D$. Using eq.29 for the diffusion coefficient the angular change can be written as

$$\Delta\theta^2 \simeq 10^3 \left(\frac{\delta n_e}{n_o}\right)^2 \qquad (30)$$

For plasma wave amplitudes with $\left(\frac{\delta n_e}{n_o}\right) \simeq 10^{-2}, \Delta\theta^2 = 10^{-1}$ and $\sqrt{\Delta\theta^2} \simeq 0.3 rad$ or $\approx 20°$. The change in parallel momentum $\Delta p_\parallel$ of the neutrinos is given by $p_o\Delta\theta^2$ where $p_o c = \epsilon_o$ and ϵ_o is the initial neutrino beam energy using $\Delta\theta^2 = 10^{-1}$ we find that the change in neutrino energy as $\Delta\epsilon = cp_o\Delta\theta^2 = 0.1\epsilon_o$, ie about 10% of the neutrino beam energy ends up in plasma waves, this is at least 5 times more than is needed but a more refined calculation should yield a smaller value.

In this paper we have shown that it is possible to nonlinearly couple an intense neutrino flux to collective plasma oscillations. The nonlinear coupling results in absorption of the neutrino energy by Langmuir waves which heat the electrons through collisional damping thus increasing the electron pressure. A simple diffusion type of calculation indicates that more than enough energy can be absorbed by the plasma and thus help to revive the shock wave to produce the supernova explosion.

Acknowledgments

This work was supported by NATO CGR 910316 and NSF PHY 91-21052.

References

1. Wilson J R., in Numerical Astrophysics, ed J M Centrella, J Le Balnc and R Bowers, Boston Jones and Bartlett, 422, (1985)

2. Wilson J, R and R W Mayle in The Nuclear Equation of State, Part A, ed W Greiner and H Stoker, New York, Plenum 731 (1989)

3. Bethe, H A., and J R Wilson, Astrophys. J., 295, 14 (1985).

4. Bingham R, J M Dawson, J J Su and H A Bethe, Phys. Lett. A., 193, 279, (1994)

5. Kaw, P K., W L Kruer, C S Liu and K Nichikawa, Advances in Plasma Physics, Vol. 6, Part 1 (1986), John Wiley and Sons.

6. Bethe, H A., Phys. Rev. Lett., 56, 1305 (1986).

7. Tappert, F D., Siam Rev. (Soc. Ind. Appl. Math), 13, 281 (1971).

ORIGIN OF ASTROPHYSICAL MAGNETIC FIELDS

George B. Field

Harvard-Smithsonian Center for Astrophysics
60 Garden Street, Cambridge, MA 02138, USA

ABSTRACT

The standard model for the origin of magnetic fields observed in stars and galaxies is the $\alpha - \Omega$ dynamo, in which a feedback loop involving differential rotation and helical turbulence leads to exponential amplification of a large-scale field. Recently this model has been criticized on the grounds that the Lorentz forces associated with the buildup of small-scale fields by the turbulence prevents the turbulent diffusion of magnetic field that is an essential part of the model.

Here we discuss the consequences for cosmology if dynamo theory is wrong, and review recent criticisms from a new perspective. We suggest new calculations that can help to decide whether the theory is right or wrong.

1. INTRODUCTION

Magnetic fields are ubiquitous in the astronomical universe. The sun, the planets, most stars, and galaxies are observed to have large-scale magnetic fields which are sustained by currents flowing in electrically conducting fluid. In our galaxy, the Milky Way, a magnetic field of several microgauss threads the interstellar medium (ISM) between the stars. Its morphology, revealed by polarization of starlight due to dust grains aligned in the magnetic field, the polarization of synchrotron radiation emitted by relativistic electrons gyrating in the field, and Faraday rotation of radio waves propagating through the field, is that of a field largely parallel to the disk of the galaxy, (whose thickness is only 1/10 of its radius) with the azimuthal component dominating radial and vertical components [1]. The sign of the azimuthal component reverses several times in the radial direction [1]. The strength is determined to be several microgauss from Zeeman splitting of the 21-cm line of neutral hydrogen, from Faraday rotation along lines of sight to pulsars for which the plasma density can be determined from dispersion of radio waves, and from the intensity of synchrotron radiation. This field strength is in rough equipartition with interstellar turbulence.

Other disk galaxies also have magnetic fields of similar morphology and strength. Some galaxies contain active galactic nuclei (AGN) which emit copious amounts of relativistic electrons. These electrons are revealed by synchrotron emission at radio wavelengths and by X- and γ-ray emission, possibly due to Compton up-scattering of low-energy photons. The sources in AGN, which are often less than a light year across, outshine by up to one hundred times the stars in the host galaxy, which is many thousand light years across. Frequently jets of relativistic electrons are observed to extend as much as a million light years from AGN's into intergalactic space. One infers magnetic fields of several microgauss in the jets, and much larger fields inside the AGN itself.

Understanding why there are large-scale magnetic fields in stars and galaxies is a fundamental problem in contemporary astrophysics. The electrical conductivity of astrophysical plasmas is high – roughly equal to that of copper [2], so magnetic flux is frozen into the plasma. The classical time for a magnetic field to diffuse a distance D is $t_D = D^2/\nu_M$, where $\nu_M = c^2\eta/4\pi$ is the classical magnetic diffusivity in terms of the resistivity η. For galaxies, t_D is 10^{20} years or more, far exceeding the age of the universe. It is expected on the basis of the flux freezing theorem that flux threading interstellar plasma today must have done so since before the origin of our galaxy. However, if this were so, the field must have originated soon after the big bang, perhaps so soon that the physical laws we recognize today did not apply at the extremely high temperatures involved. However, in spite of many attempts [3–9] to modify standard physics so that magnetic fields would be produced, none, including an attempt by me and my colleagues [8], has yet succeeded.

Most astronomers believe that the resolution of this dilemma can be found by considering not the magnetic field itself, but its ensemble average, or mean. According to mean-field magnetohydrodynamics [10], turbulence in a plasma exhibiting mean helicity can exponentially amplify a large-scale mean field by a dynamo process. This theory has been applied successfully to the stars and planets. The key insight of dynamo theory is that the mean magnetic field can violate flux-freezing even though the physical magnetic field cannot. In particular, the coefficient of turbulent diffusion, $\beta = (1/3)vL$, exceeds ν_M by 11 or more orders of magnitude [11]. This means that, on the one hand, any initial magnetic field diffuses out of a galaxy in $\sim 10^9$ years rather than $\sim 10^{20}$ years, and that a mean magnetic field can build up by a dynamo process over the same period.

To see how turbulent diffusion of the mean field works, consider the poloidal components of the mean field. Helical turbulence forms small loops of poloidal field by raising and twisting the largely toroidal mean field (the so-called "α-effect" to be discussed below). These loops are systematically of one sign, so when one averages over them, one finds that there is a large-scale mean poloidal field, even though the loops do not physically reconnect to form a large-scale field. In the words of Parker [11], "it is the lines of force of the total field

$\mathbf{B} = \langle \mathbf{B} \rangle + \delta \mathbf{B}$ that do not reconnect... The lines of $\langle \mathbf{B} \rangle$ reconnect."

Differential rotation (the "Ω-effect") stretches the large-scale poloidal field to give a large-scale toroidal field. It is essential that diffusion of the poloidal field take place so that the lines of force close outside of the disk, for if they closed within the disk, there would be no net flux for the differential rotation to work upon. The feedback between the poloidal field produced by the α-effect and the toroidal field produced by the Ω-effect results in exponential amplification of the mean field with an e-folding time of the same order as the rotation period of the galaxy, which at the position of the sun in our galaxy is about 2.5×10^8 years. This allows up to 40 powers of e amplification to occur over the lifetime of the galaxy if one uses standard parameters.

In stars, rotation periods are measured in days, not millions of years, so there is little doubt that a dynamo, if operative, has time to work. In the case of galaxies, the number of e-foldings is modest, and it becomes important to calculate growth rates accurately. We will discuss recent steps in this direction, and how further progress may be made.

2. DYNAMO THEORY OF THE GALACTIC MAGNETIC FIELD

Here we discuss the application of dynamo theory to calculating the growth of a large-scale magnetic field [12]. Unless otherwise stated, the symbol $\mathbf{B}$ denotes the mean field, which satisfies an ensemble-averaged induction equation

$$\frac{\partial \mathbf{B}}{\partial t} = \nabla \times [(\mathbf{\Omega} \times \mathbf{R}) \times \mathbf{B} + \alpha \mathbf{B} - \beta \nabla \times \mathbf{B}] , \qquad (1)$$

where

$$\alpha = -\frac{1}{3}\tau \langle \mathbf{v} \cdot \nabla \times \mathbf{v} \rangle , \qquad (2)$$

$$\beta = \frac{1}{3}\tau \langle v^2 \rangle , \qquad (3)$$

$\tau = L/v$ is the correlation time of the turbulence, and $\Omega(R)$ is the angular velocity vector of the galaxy, directed perpendicular to the disk. The term in Ω causes the poloidal field to wind up in the toroidal direction, while the term in α, which results from averaging over the turbulent motions $\mathbf{v}$ as indicated in equation (2), regenerates the poloidal field from the toroidal one; β represents turbulent diffusion.

Because the thickness of the galactic disk $2h$ is much less than its radius R, one can separate variables in equation (1), writing for the nth axisymmetric

eigenmode
$$\mathbf{B}_n = Q_n(R)\tilde{\mathbf{B}}(R, z), \tag{4}$$

where R enters $\tilde{\mathbf{B}}$ only as a parameter and $\tilde{\mathbf{B}}$ satisfies differential equations in z that apply to a disk of infinite extent but with α, β, and Ω having their local values. It is found [12] that in dimensionless variables, $Q_n(R)$ satisfies the Schrödinger equation

$$\left(\frac{h}{R}\right)^2 \frac{d}{dR}\left[\frac{1}{R}\frac{d}{dR}(RQ_n)\right] + [\gamma(R) - \Gamma_n]Q_n = 0, \tag{5}$$

where $\gamma(R)$ is the growth rate, determined from the eigenvalue problem for $\tilde{\mathbf{B}}(R, z)$, that would apply for an infinite disk, and where Γ_n is the true growth rate of the nth mode. Thus, $\gamma(R)$ acts like a potential. One finds that $Q_n(R)$ is nonvanishing only where $\gamma(R) > 0$, and that $\Gamma_n \leq \gamma(R)$ over the range of interest. Thus a particular problem of interest is to find $\gamma(R_0)$, the maximum value of the growth rate of all modes that are nonvanishing at $R_0 = 30,000$ light years, the distance of the sun from the center of the galaxy.

From the two first-order coupled differential equations in z for $\tilde{\mathbf{B}}$ one finds that $\tilde{B}_z$ is very small, that $\tilde{B}_R/\tilde{B}_\varphi \simeq \alpha/\Omega h$, a small number, that $\tilde{B}_\varphi$ has the same sign at all z, and that $\tilde{B}_R$ is negative at small z, but becomes slightly positive as $z \to h$ [12].

The details of the solution depend on α, β, and Ω. It is consistent with observation to assume that

$$\alpha = \alpha_0 \sin\left(\frac{\pi z}{h}\right), \tag{6}$$

that β is constant, and that $\Omega(R)R = constant$, so that the differential rotation that enters the local equations satisfies $Rd\Omega/dR = -\Omega(R)$. By solving the eigenvalue problem one finds [12] that γ satisfies

$$\gamma t_D = 0.8(-D)^{1/2} - 2.3, \tag{7}$$

where the dynamo number D, a dimensionless quantity, is

$$D = -\frac{\alpha_0 \Omega h^3}{\beta^2}, \tag{8}$$

and the diffusion time t_D is defined by

$$t_D = \frac{h^2}{\beta}. \tag{9}$$

Equation (7) has been verified to be valid over the range $0 < -D < 130$ [12]. From it we see that dynamo modes grow only if $-D > 8$. Is this requirement satisfied in our galaxy? Traditionally [11] α and β have been estimated from observations of chaotic motions in the ISM to be $\alpha \sim 10^4 \text{cm s}^{-1}$ and $\beta \sim 10^{26} \text{cm}^2 \text{ s}^{-1}$. Recently Ferrière [13] has studied α in more detail, averaging over the velocity fields driven by supernova explosions which are the cause of fluid helicity. Field [25] finds that $h = 1.6 \times 10^{21} \text{cm}$ gives the best fit to Ferrière's results in equation (6), in which case $\alpha_0 = 2.1 \times 10^4 \text{cm s}^{-1}$. As no detailed study of β has appeared, we may tentatively adopt $\beta = 10^{26} \text{cm}^2 \text{s}^{-1}$, in which case $-D = 6$, close to, but less than the critical value, so dynamo modes should not grow. However, as we shall see in the next section, β could well be substantially less than the classical value. For example, if β were $5 \times 10^{25} \text{cm}^2 \text{s}^{-1}$, $-D$ would be 25, larger than the critical value, and growth would take place. From equation (9), t_D would be $2 \times 10^9 \text{y}$ and γ would be 10^{-9}y^{-1}, adequate for 10 e-foldings of the mean magnetic field.

3. CRITICISMS OF DYNAMO THEORY

There is concern about the neglect of certain terms in the exact equations leading to equation (1). In particular, while it is known that the "first-order smoothing approximation," which neglects certain terms, is valid in certain limiting cases [14], turbulence in the galaxy is not one of them. Eric Blackman and I are trying to find out if the first-order smoothing approximation is also valid for the galaxy.

Other criticisms center on the fact that in deriving equation (1), the turbulent velocity field is regarded as given, rather than being consistent with the magnetic stresses that arise as the mean magnetic field grows. Presumably this is justified in the early stages of magnetic field growth, but when the mean field approaches equipartition with the turbulent energy, the turbulence may be modified so that α and/or β are drastically different. This would be true, for example, if the turbulent motions were wavelike (Alfvén waves) rather than more like the hydrodynamic turbulence that exists if the field is weak.

This question was raised in the 1970's by Piddington [15–20] soon after Parker [21] first applied dynamo theory to the galactic magnetic field. It has been discussed most recently by Kulsrud and Anderson [22], who show that if interstellar turbulence is Kolmogoroff, magnetic fields on the scale of the largest eddies (about 300 light years in the ISM) reach equipartition in a few eddy turnover times, only $\sim 10^7$ years in the ISM. Kulsrud and Anderson conclude that the assumptions underlying equation (1) are therefore not correct, so the dynamo model of the galactic magnetic field is in principle wrong.

Other authors have also criticized galactic dynamo theory on the grounds that small-scale magnetic fields will modify the turbulence, choking off turbulent diffusion and the α-effect. Cattaneo and Vainshtein [23] argue as follows. For simplicity they consider two-dimensional turbulence, and argue that their conclusions are even stronger in the three-dimensional case. If the fields are very weak, the turbulence stretches tubes of force at will, rapidly lengthening and making them thinner. The time for magnetic energy to be dissipated by ohmic losses is $t_d = \delta^2/\nu_M = (\delta^2/vL)R_M$, where δ is the width of a tube of force, ν_M is the classical magnetic viscosity, v is the turbulent velocity, L is the scale of the turbulence, and $R_M = vL/\nu_M$ is the magnetic Reynolds number of the turbulence. Flux freezing implies that $\delta = (B_L/B_S)L$, where B_L is the field on the large scale L, and B_S is the field on the small scale δ, so $t_d = (B_L/B_S)^2 L R_M/v$. In equilibrium, the rate of increase of small-scale field B_S by line-stretching, v/L, equals the rate at which it is being dissipated, so $B_S = B_L R_M^{1/2}$ ($\sim 10^{10} B_L$ in the ISM, where $R_M \sim 10^{20}$). Since $\langle B^2 \rangle^{1/2} \simeq B_S$, one would predict from this reasoning that with $B_L \sim B$, the mean field, $\langle B^2 \rangle^{1/2}/B \sim 10^{10}$ in the ISM, rather than $\sim 0(1)$ as is observed [1]. Putting aside this objection, we continue the argument of Cattaneo and Vainshtein [23].

The argument up to this point has not taken into account magnetic back-reaction, which will limit $\langle B^2 \rangle^{1/2} \sim B_S$ to $\sqrt{4\pi\rho}v = MB_L$, where $M = v/v_A(B_L) = \sqrt{4\pi\rho}v/B_L$ is the magnetic Mach number of the turbulence based on the large-scale field B_L. Hence $B_S = B_L R_M^{1/2} < B_L M$, and so $M > R_M^{1/2}$ for a field in which turbulent line stretching is in equilibrium with ohmic dissipation. (Again we note a discrepancy with the ISM, where $R_M^{1/2} = 10^{10}$, but $M = 0(1)$.)

Cattaneo and Vainshtein [23] reason that if $M > R_M^{1/2}$, the small-scale field dissipates before it reaches equipartition, and $t_d = L/v$, the classical value. If $M < R_M^{1/2}$, the reverse is true, the small-scale field reaches equipartition first, and $t_d = (B_L/B_S)^2 L R_M/v = (R_M/M^2)L/v$, a larger value. They propose an interpolation formula valid for all values of R_M and M:

$$t_d = \frac{L}{v}\left(1 + \frac{R_M}{M^2 + 1}\right). \qquad (10)$$

This reasoning yields an estimate of the time for the field to dissipate if line stretching is in balance with ohmic dissipation.

Cattaneo and Vainshtein [23] go on to define an effective diffusion coefficient

$$\beta_{\text{eff}} = \frac{L^2}{t_d} = vL\left(1 + \frac{R_M}{M^2 + 1}\right)^{-1}, \qquad (11)$$

which Vainshtein and Cattaneo [24] apply to turbulent dynamos. They point out that in the early stages of dynamo amplification, when $B_L \ll \sqrt{4\pi\rho}v (M \gg 1)$, equation (11) predicts that $\beta_{\text{eff}} \sim vL$, which agrees with the usual estimate of equation (3). Under these conditions, the galactic dynamo would presumably operate, provided α is as large as Ferrière [13] calculates for present conditions in the galaxy. When, however, B_S approaches equipartition, $M \sim R_M^{1/2}$, and β_{eff} drops below vL. The same thing will happen to α, and for both reasons, the dynamo will stop operating. Since the large scale field B_L at this point is only $R_M^{-1/2} B_S \sim R_M^{-1/2} \sqrt{4\pi\rho} v \sim 10^{-16} G$, the observed large scale field ($\sim 10^{-6} G$) could not be due to a dynamo according to Vainshtein and Cattaneo [24]. The authors therefore suggest that the galactic magnetic field may be primordial. Field [25] has questioned this reasoning (see below).

Parker [26] agrees that the growth of small-scale fields will reduce α and β and thus challenge dynamo theory, but does not believe that a primordial field is the best alternative. Rather, he argues that the agreement of dynamo theory with observation as to morphology and growth rate indicates that the parameters α and β are in fact large, and we must explain *why* they are large in spite of the growth of small-scale fields. He suggests that cosmic rays inflate loops of field out of the disk, where they reconnect, simulating turbulent diffusion. However, while this might work under present conditions, it cannot work in the early phases of dynamo amplification, when the field is still weak, because the large-scale field must have approximately the equipartition value (far greater than allowed by the arguments of Cattaneo and Vainshtein) for the instability which drives the inflation of field to operate [11].

4. DISCUSSION

We reexamine the arguments here. The first issue is this: while the dissipation time t_d and the diffusion time t_D are identical for laminar flows ($t_d = t_D = D^2/\nu_M$), the same is not true for turbulent flows, because dissipation is not required for diffusion of the mean field: recall Parker's [11] insight that in the absence of dissipation, the physical magnetic field cannot reconnect (*i.e.*, diffuse), but the mean field can. Thus, while equation (10) may be a valid estimate of the dissipation time t_d, it may not be a valid estimate of the diffusion time t_D. It would then follow that β_{eff} in equation (11), whatever else its significance, is not a valid estimate of the turbulent diffusivity, and one cannot use its small value to argue that the dynamo cannot operate.

At this point we review why a diffusivity much larger than vLM^2/R_M is essential for dynamo action. Parker [11,21] shows that a disk bounded by conducting plates is not susceptible to dynamo action, because the tops of poloidal field lines cannot then diffuse into the vacuum, leaving a net flux of poloidal field in the disk for the Ω-effect to work upon. Ruzmaikin *et. al.* [12] discuss this issue in detail.

On the other hand, when equation (7) is rewritten with the help of equations (8) and (9) in the form

$$\gamma = 0.8 \left(\frac{\alpha_0 \Omega}{h}\right)^{1/2} - \frac{2.3\beta}{h^2} \qquad (12)$$

it appears that the smaller β is, the larger is the growth rate γ. To reconcile this with the requirement of nonvanishing β, we note that equation (12) is known to be valid only for $-D < 130$, corresponding, for our values of α_0 and Ω, to $\beta > 2 \times 10^{25} \text{cm}^2\text{s}^{-1}$, larger than vLM^2/R_M for all values of $M < 3 \times 10^9$, corresponding to $B_L > 2 \times 10^{-15} G$. Hence equation (12) would cease to be valid very early in the dynamo process because β_eff from equation (11) is too small. Recall that we are not convinced that β_eff is really relevant to the discussion.

We now turn to Kulsrud and Anderson's [22] conclusion that the dynamo theory of the galactic magnetic field is wrong in principle. Indeed, it may be flawed because it ignores the terms neglected in the first-order smoothing approximation. Until this is rectified [27] we cannot be sure that its results are even approximately correct. However, this was not Kulsrud and Anderson's primary concern. Rather, they believe that because $\langle B^2 \rangle^{1/2} \gg B$ in the early stages of dynamo amplification, one cannot ignore magnetic back reaction. We agree, with the proviso that if equation (1) is correct, the only effect of back reaction is to modify α and/or β, as is claimed by Cattaneo and Vainshtein [23]. As we have shown above, a modest reduction in β would be welcome within the context of standard theory, which predicts low or even negative growth rates with the conventional value of β. We believe, therefore, that the question resolves into whether the reduction of β by back reaction is only modest, or large.

We argue that the reduction is modest, as follows. Kulsrud and Anderson followed the evolution of the spectrum of magnetic energy in response to Kolmogoroff turbulence as an initial value problem, starting with a very weak field, e.g., 10^{-17}G. They found that $\langle B^2 \rangle^{1/2}$ approached equipartition, $\sim 10^{-6}$G, in only $\sim 10^7$y. We note [11] that kinetic energy is injected by supernova explosions into the largest eddies, whose scale is ~ 300 l.y., on about the same time scale, so we should try to imagine what happens in a steady-state situation. If the magnetic energy cannot be dissipated by any means whatsoever, it will increase above the equipartition value, in which case the blast waves caused by supernova explosions will encounter a medium in which the Alfvén speed is so great that the resulting motions are sub Alfvenic, and therefore, oscillatory. It is known [14] that the α associated with such motions vanishes, and it seems likely that β would also vanish. Thus, as argued by Piddington [15-20] many years ago, an unrelenting build up of magnetic energy would defeat the dynamo, in agreement with Kulsrud and Anderson. The only way to save the dynamo is therefore to find some way to dissipate magnetic energy in a short time, say $\sim 10^7$y. We have seen [23] that if the field is dragged about by turbulent motions, the time scale for ohmic dissipation is far too long. But in a field approaching equipartition, the magnetic back reaction modifies turbulent motions, making them more wavelike; we visualize a field of Alfvén waves, waves of each length propagating a background field whose effective scale is somewhat larger. In short, we visualize Alfvén turbulence of the type described by Sridhar and Goldreich [28]. According to these authors, nonlinear wave-wave interactions result in a cascade of magnetic energy to shorter wavelengths, and if, as seems unavoidable, the cascade reaches sufficiently short wavelengths, the energy is ultimately dissipated by ohmic losses (or in the ISM, perhaps by ambipolar diffusion as waves propagate through a partly neutral gas).

The nonlinear cascade will remove magnetic energy from the largest eddies as fast as it is put into them by supernovae. This fact in itself does not allow us to calculate the steady-state magnetic energy; that requires a knowledge of the time t_M for magnetic energy to drain from the largest-scale eddies into smaller ones. If t_M is large, the magnetic energy in the largest eddies will build up, their properties will be altered, and β will be substantially modified. If, on the other hand, t_M is small, the magnetic energy in the largest eddies will be modest, and conventional estimates of β should apply.

What, then, is t_M? Here we argue that it is of the same order as L/v, the eddy turnover time in the absence of a magnetic field, for if it were much larger, the magnetic field would build up to equipartition in the largest eddies, and the motion of the eddy would become oscillatory with period L/v_A, which, because $v_A \sim v$, is $\sim L/v$. On the other hand, t_M cannot be much smaller than L/v, because an eddy cannot move far in a short time, and hence cannot induce the smaller-scale motions required to make t_M small. We conclude that t_M is of the same order as L/v, which implies that the dynamics of the largest eddies is affected moderately, but not substantially by the magnetic field. As a result, β is reduced only moderately, rather than by the huge factor posited by some authors.

5. CONCLUSIONS

Classical dynamo theory applied to differentially rotating turbulent disks can account for the morphology and strength of galactic magnetic fields provided α and β take favorable values. It is desirable to free the classical theory from having to assume the first-order smoothing approximation. Several authors have criticized the applicability of classical estimates of α and β. Ferrière [13] has attempted a realistic calculation of α. Discussions of β such as those of [29] are not yet definitive, and further calculations should be undertaken. In the spirit of Ferrière one should study in detail how the magnetic energy injected by supernova explosions is dissipated, perhaps by a nonlinear cascade of Alfvén waves. Understanding the motions involved, and in particular how much the largest-scale motions contribute to β, should then be possible.

6. ACKNOWLEDGMENTS

This work was supported by NASA under Grant NAGW 931. I am grateful for discussions with Eric Blackman, Charles Gammie, Dan Eisenstein, and Russell Kulsrud.

REFERENCES

1. Heiles, C. 1995, in Physics of the Interstellar and Intergalactic Medium, ed. A. Ferrara, PASP Conference Series, in press.

2. Spitzer, L. Jr. 1962, Physics of Fully Ionized Gases, New York: Interscience.

3. Vilenkin, A. & Leahy, D.A. 1982, Astrophys. J., 254, 77.

4. Turner, M.S. & Widrow, L.M. 1988, Phys. Rev. D, 37, 2743.

5. Quashnock, J.M., Loeb, A. & Spergel, D.N. 1988, Astrophys. J., 254, 77.

6. Vachaspati, T. 1991, Phys. Lett. B, 265, 258.

7. Tajima, T., Cable, S., Shibata, K., & Kulsrud, R.M. 1992, Astrophys. J., 390, 309.

8. Garreston, W.D., Field, G.B. & Carroll, S.M. 1992, Phys. Rev. D, 46, 5346.

9. Ratra, B. 1992a, Astrophys. J., 391, L1; 1992b, Phys. Rev. D, 45, 1913.

10. Krause, F. & Rädler, K.-H. 1980, Mean-Field Magnetohydrodynamics, Oxford: Pergamon.

11. Parker, E.N. 1979, Cosmical Magnetic Fields, Oxford: Clarendon.

12. Ruzmaikin, A.A., Shukurov, A.M., & Sokoloff, D.D. 1988, Magnetic Fields of Galaxies, Dordrecht: Kluwer.

13. Ferrière, K. 1992a, Astrophys. J., 389, 286; 1992b, *ibid*, 391, 188; 1993, *ibid*, 404, 162.

14. Moffatt, H.K. 1978, Magnetic Field Generation in Electrically Conducting Fluids, Cambridge: University Press.

15. Piddington, J.H. 1970, Aust. J. Phys., 23, 731.

16. ———. 1972a, Solar Phys., 22, 3.

17. ———. 1972b, Cosmic Electrodyn., 3, 50.

18. ———. 1972c, *ibid*, 5, 129.

19. ———. 1975a, Astrophys. & Sp. Sci., 35, 269.

20. ———. 1975b, *ibid*, 39, 157.

21. Parker, E.N. 1971, Astrophys. J., 163, 255.

22. Kulsrud, R.M. & Anderson, S.W. 1992, Astrophys. J., 396, 606.

23. Cattaneo, F. & Vainshtein, S.I. 1991, Astrophys. J., 376, L21.

24. Vainshtein, S.I. & Cattaneo, F. 1992, Astrophy. J., 393, 165.

25. Field, G.B. 1995, in Physics of the Interstellar and Intergalactic Medium, ed. A. Ferrara, PASP Conference Series, in press.

26. Parker, E.N. 1992, Astrophys. J., 401, 137.

27. Blackman, E. 1994, private communication.

28. Sridhar, S. & Goldreich, P. 1994, Astrophys. J., 432, 612.

29. Gruzinov, A.V. & Diamond, P.H. 1994, Phys. Rev. Lett., 72, 1651.

PROTOSTELLAR JETS: THE BEST LABORATORIES TO INVESTIGATE ASTROPHYSICAL JETS

Elisabete M. de Gouveia Dal Pino

University of São Paulo, Instituto Astronômico e Geofísico, Av. Miguel Stéfano 4200, São Paulo, SP 04301-904, Brazil, e-mail: dalpino@astro1.iagusp.usp.br

ABSTRACT

Highly collimated supersonic jets are observed to emerge from a wide variety of astrophysical objects, ranging from Active Nuclei of Galaxies (AGNs) to Young Stellar Objects (YSOs) within our own Galaxy. Despite their different physical scales (in size, velocity, and amount of energy transported), they have strong morphological similarities. Thanks to the proximity and relatively small timescales, which permit direct observations of evolutionary changes, YSO jets are, perhaps, the best laboratories for cosmic jet investigation. In this lecture, the formation, structure, and evolution of the YSO jets are reviewed with the help of observational information, MHD and purely hydrodynamical modeling, and numerical simulations. Possible applications of the models to AGN jets are also addressed.

1. INTRODUCTION

Supersonic collimated outflows are an ubiquitous phenomenum in the universe. They span a large range of luminosity and degree of collimation, from the most powerful examples observed to emerge from the nuclei of active galaxies (or AGNs) to the jets associated to low-mass Young Stellar Objects (YSOs) within our own Galaxy. In the intermediate scales between these two extremes one finds evidences of outflows associated to neutron stars, binary systems (with SS433 being the best example of this class), and symbiotic novae. Also, there is some evidence of a poorly collimated jet associated to our Galactic Centre (e.g., [1]).

As stressed by Königl [2], both the AGN and the YSO jets, despite their different physical scales (AGN jets have typical sizes $\geq 10^6$ pc (1 pc = 3.086 10^{18} cm), nuclear velocities ~ c (where c is the light speed), and parent source masses 10^8 $M_\odot$ ($M_\odot$ = one solar mass = 1.99 10^{33} g), while YSO jets have typical sizes < 1 pc, nuclear velocities $\leq 10^{-3}$c, and central masses ~ 1 $M_\odot$), are morphologically very similar. In both classes, the jets are: (i) highly collimated and in most cases two-sided; (ii) originate in compact objects; (iii) show a chain of more or less regularly spaced emission knots which in some cases move at high speeds away from the central source; (iv) terminate in emission lobes (with line emission in the case of the YSOs and synchrotron continuum emission in the case of the AGN jets), which are believed to be the "working surfaces" where the jets shock against the ambient medium (see below); (v) are associated with magnetic fields whose projected directions are inferred from polarization measurements (in YSOs, these vectors are generally aligned with the jet axis); and (vi) are possibly associated to accretion disks.

Extensive reviews of the observational properties of the YSO jets can be found in [1, 3-10]. Reviews of the properties of the AGN jets can be found in [e.g., 11-13]. Jets from young stars offer to investigators a number of advantages relative to their extragalactic counterparts. They are nearby and generally bright, and evolve on relatively short timescales, permitting direct observations of evolutionary changes. Furthermore, their radiation which is mainly line-emission rather than continuum, opens a vast field of observational techniques that can probe their physical conditions [e.g., 5]. Therefore, from the YSO jets we can potentially learn a great deal about the structure, evolution and driving mechanisms of the astrophysical jets as a whole. In this work, instead of discussing all the current knowledge about the various classes of astrophysical jets (which would be impossible to cover just in few pages), I focus on the YSO jets which are the closest and therefore, the best laboratories for cosmic jet investigation. I do not address here specific issues in the theory of AGN jets, which have been the subject of several extensive reviews [e.g., 10-15].

The review begins with a brief description of the observed characteristics of the YSO jets (Section 2). A discussion of the basic structural components of the jet and the analytical and numerical models which explain these features is presented in Sections 3 and 4 giving particular emphasis to the study of the jet head structure (Section 3) and the possible mechanisms for the formation of the internal knots along the jets (Section 4). The momentum transfer processes between jet and environment and the formation of molecular outflows driven by the YSO jets are addressed in Section 5. In Section 6, the effects of the propagation of the jets in non-homogeneous stratified environments are examined and in Section 7 the existing mechanisms for the origin of the jets from the central YSO object and associated accretion disk is discussed. A summary is presented in Section 8.

© 1995 American Institute of Physics

2. OBSERVATIONAL PROPERTIES OF YSO JETS

The highly collimated supersonic YSO jets have typical projected lengths between ~0.01-1 pc. Most of them show a linear chain of bright, traveling knots, frequently identified as Herbig-Haro (HH) objects (e.g., HH34, and HH111 systems), followed by an intermediate section where the emission disappears, and terminating in a bow shock-like structure identified as the working surface where the jet impacts with a slower ambient gas [e.g., 5]. In some cases (e.g., HH46/47 and HH 111 jets), there is clear evidence of two or more such bow shock structures separated by a trail of diffuse gas for many jet radii. In other cases (e.g., HH30 jet), there is no detected bow-shock like feature [8]. Some jets, instead of chains of aligned knots, show larger amplitude side-to-side "wiggles" (e.g., HH83 and HH110 jets; [15]).

All the luminous structures produce emission line spectra. Prominent features include the hydrogen Balmer lines and transitions of neutral atoms and ions. While the terminal feature may be rich in high ionization and excitation lines, the inner jet generally shows a very low excitation spectrum with [SII] and [OII] to Hα ratios substantially greater than unity [5, 15]. This implies that the temperature of the YSO jets are not much larger than ~ $1-2\times10^4$ K and the corresponding sound speeds are ~10 km/s. This yields typical Mach numbers for the emitting regions M_j=10-30 [8].

The knots move away from the sources at high speeds ~100-500 km/s, have radii $R_j \sim 3\times10^{15}$ cm, and inter-knot separation $\Delta x \sim 10^{16}$ cm ~ 3.3 Rj. They often become fainter and disappear at larger distances from the source (e.g., HH34 and HH 111 jets; [5, 8]).

The densities of the YSO jets are poorly determined. Assuming that the line emission is produced by shock-heated gas, early estimates indicated jet number densities $n_j \approx$ 20-100 cm^{-3} [3], but recent estimates seem to indicate values larger than 10^3 cm^{-3} [e.g., 17-19]. The high proper motion of the heads of the jets indicate that their density is considerably higher than the density of the surrounding medium. Although uncertain, observations suggest a jet-to-environment density ratio $\eta = n_j/n_a \approx$ 1 -20 [e.g., 3, 17]. The typical jet velocities and total lengths imply dynamical timescales L_j/v_j ~1000 yrs or larger for the jets.

The determination of the intensity of the magnetic field in stellar jets from line intensity ratios is very imprecise, but rough estimates for HH34 region [17], for example, indicate a magnetic field Å 8 µG in front of HH34. This implies an energy density in the preshock gas smaller than 10^{-3} of the ram pressure ($\rho_j v_j^2$). Thus, although of fundamental importance for the production and initial collimation of the jet (see Section 7), the magnetic field effects can be, in general, neglected in the modeling of the *large scale structure and evolution of YSO* (see Sections 4, 5, and 6).at least in the undisturbed medium.

3. THE JET STRUCTURE

The basic theoretical properties of *adiabatic and cooling* jets have been summarized by Norman [15] and Blodin et al. [20]. A supersonic jet propagating into a stationary ambient gas will develop a double shock pattern at its head also denominated *working surface*. While the impacted ambient material is accelerated by a forward bow shock, the beam outflow is decelerated in a jet shock or Mach disk. The velocity of advance of the bow shock can be estimated by balancing the momentum flux of the jet material at the jet head with the momentum flux generated by the ambient medium at the bow shock. For highly supersonic flows, the thermal pressure of the jet and ambient medium can be neglected and one obtains:

$$v_{bs} \approx v_j [1 + (\eta\alpha)^{-1/2}]^{-1} \tag{1}$$

where v_j is the jet velocity, v_{bs} is the velocity of advance of the bow shock, $\eta = n_j/n_a$ is the ratio of the jet number density to the ambient number density, and $\alpha = (R_j/R_h)^2$, where R_j is the radius of the jet beam and R_h is the radius at the jet head. Eq. 1 shows that a low-density jet ($\eta \ll 1$), which is generally believed to be the case for AGN jets, $v_{bs} \ll v_j$ and the jet material is constantly decelerated at the end of the jet (at the jet shock). The high-pressure, shocked gas drives a flow back around the jet forming a cocoon of shock-heated waste material. A dense jet, like typical YSO jets, on the other hand, for which $v_{bs} \approx v_j$ ($\eta \gg 1$), will simply plow through the ambient medium at close to the jet velocity without accumulating much waste of gas in the cocoon. The ambient material that traverses the bow shock will form a shroud of ambient shock-heated gas surrounding the beam/cocoon structure.

In contrast to AGN jets, the radiative cooling distance of the shocked gas in a YSO jet, d_{cool}, is smaller than the jet radius R_j and hence, the assumption of an *adiabatic* gas is inappropriate. Estimates of the radiative cooling time indicate $t_{cool} \sim 100$ yrs, which is, in general, smaller than the evolutionary time of the YSO jets. Defining the cooling parameter $q_s = d_{cool}/R_j$, one expects that for $q_s < 1$ the shock is fully radiative and the postshock gas loses the bulk of its internal energy in a relatively short distance. In this case, the jet is expected to develop a cold, dense shell gas formed by the cooling of the shock-heated gas in the working surface. Considering the radiative cooling function (due to collisional excitation and recombination) for a cosmic abundant gas, Blodin et al. [20] evaluated q_s for a gas cooling from $T \sim 10^6$ to 8000 K and found that for $v_{s,7} \geq 0.9$ (where $v_{s,7}$ is the shock velocity in units of 10^7 cm/s) the cooling parameter for the bow shock is

$$q_{bs} = d_{cool}/R_j \sim 4\times 10^{16} n_0^{-1} R_j^{-1} v_{j,7}^4 [1 +(\eta\alpha)^{-1/2}]^{-4} \qquad (2)$$

where n_0 is the ambient number density of nuclei, and noting that the jet shock velocity is given by $v_{js} = v_j - v_{bs}$, the cooling parameter for the jet shock is $q_{js} \sim q_{bs} \eta^{-3} \alpha^{-2}$. Assuming $\alpha \sim 1$, these relations imply that a heavy jet ($\eta >> 1$) will exhibit much stronger cooling behind the jet shock and most of the gas that accumulates in the shell in this case consists of shocked jet material. The reverse situation holds for light ($\eta << 1$) jets.

The predictions above are confirmed by multidimensional simulations [20, 22-23]. Hydrodynamical models of cooling jets are characterized by the dimensionless parameters: η (eq. 1), $M_a = v_j/c_a$ the ambient Mach number, where c_a is the ambient sound speed; and q_{bs} (eq. 2). As an example, Fig. 1 shows the results of a fully three-dimensional hydrodynamical simulation of a jet (ejected into the ambient medium with continuous velocity) from Gouveia Dal Pino and Benz [22] which was performed with the employment of the smoothed particle hydrodynamics technique (SPH).

In Fig. 1, a dense shell develops in the head of the jet formed by the cooling of the shock-heated gas in the working surface. In this case, $q_{js} \sim 0.39$ and $q_{bs} \sim 10.5$, implying that a radiative jet ($q_{js}<1$) is propagating into an adiabatic ambient medium ($q_{bs}>>1$) and the gas that accumulates in the dense shell consists essentially of shocked jet material. The little cooling of the ambient gas, results a high pressure shroud which helps to confine the beam and the cocoon. The shell is responsible for most of the emission of the jet. In the figure it becomes Rayleigh-Taylor unstable and fragments into pieces that spill out to the cocoon forming, together with the shell, an elongated plug of cold gas in the head. The fragmented shell resembles the clumpy structure observed in many HH objects at the bow shock of YSO jets (e.g., HH1, HH2, HH19, HH12).

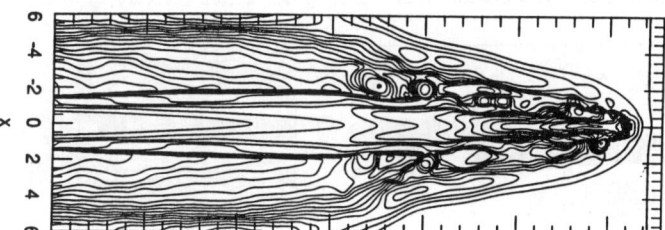

Fig. 1. The central density contours of a cooling jet with $\eta=3$, $n_j=60$ cm^{-3}, $R_j=2\times 10^{16}$ cm, $v_j=400$ km/s, initial $v_{bs}\sim 254$ km/s and $v_{js}\sim 146$ km/s, $M_a=11.55$, $M_j=v_j/c_j=20$. The contour lines are separated by a factor of 1.3 and the density scale covers the range from ~ 0.05 up to $\sim 1400/n_a$. The entire evolution corresponds to an age ~ 782 yrs (from Gouveia Dal Pino and Benz, 1993 [22]).

4. THE ORIGIN OF THE KNOTS

The precise nature of the mechanism that produces the internal knots is still controversial. Most of the available models rely on the belief that the knots represent radiative shocks excited within a jet flow.

Among the proposed models analytic and numerical calculations show that a chain of stationary oblique crossing shocks, such as those observed in laboratory jets, is produced when an initially freely expanding jet attempts to reach pressure equilibrium with its surroundings [e.g., 24-25]. However, the fact that the knots are observed to move away from the source at high speeds seems to exclude this mechanism since it necessarily produces stationary shocks.

Kelvin-Helmholtz (K-H) shear instabilities at the boundary between the jet and the surrounding medium may also excite a similar pattern of oblique internal shocks (e.g., [26-28]). This mechanism produces shocks which travel at a velocity v_p that can be substantially smaller than the fluid velocity v_j. Previous 2-D simulations of steady-state, adiabatic jets by Norman et al. [27] show that v_p/v_j increases with both increasing

Mach number and increasing jet-to-environment density ratio η, approaching the flow velocity at the limit of a highly supersonic heavy jet, as is the case of the protostellar jets. However, recent two-dimensional [20] and three-dimensional [22] hydrodynamic simulations have shown that, under the presence of *radiative cooling*, shocks driven by K-H instabilities are fainter and less numerous than in adiabatic jets. The radiative cooling reduces the thermal pressure which is deposited in the cocoon that surrounds the beam. As a result, the cocoon has less pressure to collimate, drive K-H instabilities and thus, reflect internal shocks in the beam. Although this mechanism remains viable, it probably plays a secondary role in the production of most of the internal shocks, at least in the case of the YSO jets for which the radiative cooling is important (see below, however, the formation of knots by K-H instabilities in jets propagating in stratified ambients of increasing density; Section 6). In the case of the extragalactic jets, for which the adiabatic approach is good, the simulation depicted in Fig. 2 shows that the development of the pinch and helical modes of the K-H instability can be relevant in the production of internal knots.

Time variations in the ejection mechanism that produces the jet outfow may also create traveling internal shock patterns similar to the *working surfaces* at the edges of the jets [e.g., 29-36] (For a review of the properties of models of jet flows from sources with variability in the ejection velocity and/or direction see also [8].) Strong support for this mechanism comes from recent high resolution observations of the HH34 and HH111 jets [17] which show evidence of a velocity-variable outflow with multiple ejections. The time variability of the outflow is probably associated with irruptive events in the accretion process around the protostar (see Section 7). These events could produce massive ejections along the stellar jets every 100-1000 yrs [32], suggesting that the driving sources can change on timescales shorter than the dynamical timescale of the HH objects.

Numerical simulations of jets from time-dependent sources clearly show the formation of internal shocks [32-35]. Fully 3-D numerical SPH simulations of cooling jets with time variable ejection velocity performed by Gouveia Dal Pino and Benz [34] are shown in Fig. 3. To produce the velocity variability, they assumed that the jet is periodically turned on with a full supersonic velocity and periodically turned off to a small velocity regime. Considering velocity variations with a period comparable to the transverse dynamical time scale of the jet ($\tau_{dy} = R_j/v_j$), they found the chain of regularly spaced emitting features depicted in Fig. 3. Each parcel in Fig. 3 develops a pair of shocks: a forward (downstream) shock which sweeps up the low velocity material ahead of it and propagates downstream in the jet with a velocity v_{is}, and a reverse (upstream) shock which decelerates the high velocity material behind it and propagates with a velocity $v_{rs} \approx v_j - v_{is}$.

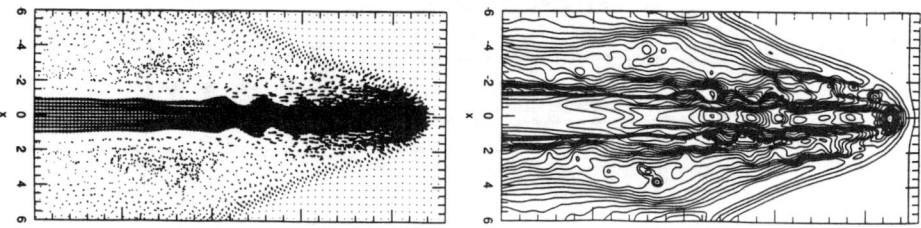

Fig. 2 Density contours and velocity distribution of an *adiabatic* jet propagating into a homogeneous ambient medium with density ratio η=1 and $M_a=v_j/c_a=20$. The evolution corresponds to a jet age ≈1115 yrs. Thanks to the high pressure cocoon of the adiabatic jet, the beam becomes K-H unstable with the appearance of reflecting pinch and helical modes which collimate the beam and drive internal traveling shocks (from Gouveia Dal Pino and Benz 1993 [22]).

The simulations show that the features (parcels) move downstream with nearly the jet velocity in the supersonic phase $v_{is} \approx v_j = 150$ km/s. This result is in agreement with the observed knots of HH34 jet. As a consequence, the propagation velocity of the reverse shock $v_{rs} \approx 0$ and the forward shock (of each parcel) is much stronger than the reverse shock and line emission between these shocks is essentially single peaked. As the peaks travel downstream, they widen and fade and eventually disappear close to the leading working surface, an effect also detected in HH 34 jet. This effect is due to the increase in the pressure of the postshock gas which causes both the separation between each pair of shocks and the expulsion of jet material laterally to the cocoon and to the rarefied portions of the beam itself. This fading explains the most frequent occurence of knots closer to the driving source.

The fact that the time dependent jet model successfully explains a number of properties of the knots seems to favor it over the K-H instability model. A way to distinguish between both could come from detailed kinematic studies [37].

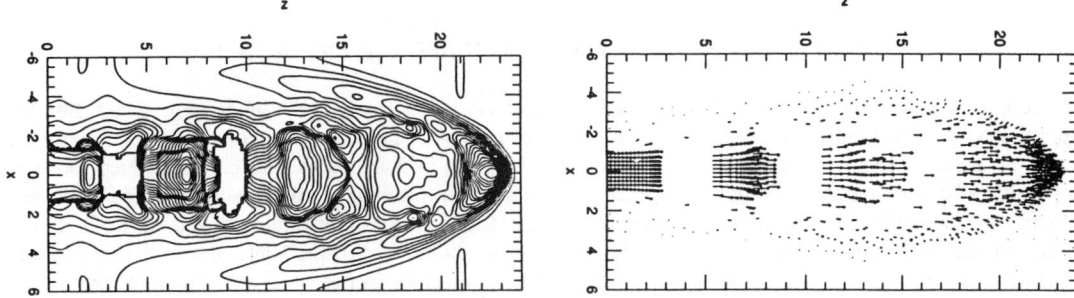

Fig. 3 Central density contours and velocity distributions of an intermittent jet periodically turned on with a supersonic velocity v_j=150 km/s and periodically turned off with a subsonic velocity 15 km/s. The *turning on* and quiescent periods are both given by $\tau_{on} = \tau_{off} = (R_j/c_a) \approx 127$ yrs = $3\ \tau_{dy}$. The initial input parameters are $\eta = 10$, n_a=1000 cm^{-3}, $R_j = 2$ X 10^{16} cm, and $M_a = 3$. The entire evolution corresponds to $t \approx 11\ R_j/c_a = 1387$ yr (from Gouveia Dal Pino and Benz 1994 [34]).

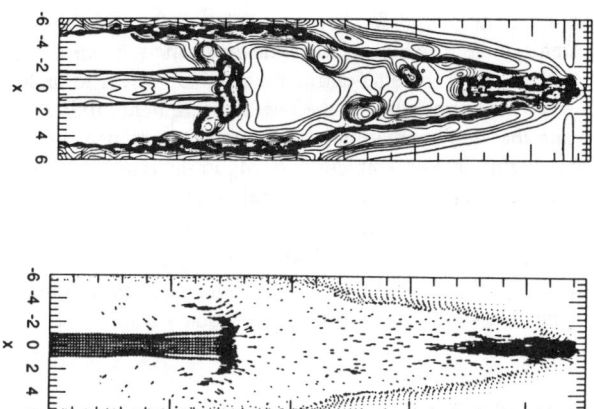

Fig. 4 Central density contours and velocity distribution of an $\eta = 3$ cooling jet with a long-period variability. The turned on period is $\tau_{on} = 22.5\ \tau_{dy} = 143$ yr with $M_a = 17.3$ and the turned off period is $\tau_{off} = 17.3\ \tau_{dy}$ with a weakly supersonic velocity. The calculation ends at $t \approx 587\ (R_j/v_j) \approx 374$ yr ($z_{max} \approx 45\ R_j$) (from Gouveia Dal Pino and Benz 1994 [34]).

Using the same scenario of time-variability one can also explain the formation of the multiple bow shock structures observed at the heads of some YSO jets. To investigate this case, Gouveia Dal Pino and Benz [34] performed a simulation with long variablity period (t>>τ_{dy}). The simulation produced a jet with a pair of bow shocks separated by a trail almost starved of gas extending for many jet radii (Fig. 4), in agreement with the observations (e.g. HH111). Also consistent with observations, the emission of the internal working surface is of lower intensity and excitation than that of the leading working surface.

All the proposed models above predict that the line emission in the knots is produced by shock-heated gas in the jet flow. Recently, however, a preliminary study by Bacciotti et al. [18], based on the diagnostics of low ionization state of the jet material (with an estimated ratio of the electron-to-hydrogen denstiy $x = n_e/n_H \approx 0.1$), suggests that the observed jet emission could be the product of soft compressions of the central portion of the flow, caused by "damped" K-H instabilities. In this picture, the emission would be connected to the heating produced not by shocks but instead by the mild compressions. Analytical and numerical calculations are still required to support or not this idea.

Finally, the possibility of the formation of the emission knots as condensations driven by thermal instabilities has been discussed by Gouveia Dal Pino & Opher [38] (see also e.g., [39-43] for a discussion on thermal instabilities in AGN jets and other astrophysical sources). However, while this mechanism may play a role in the formation of the filaments observed in AGN jets and lobes, it seems to be improbable in the case of the YSO jets as the thermal unstable perturbations in this case are overwhelmed by dynamical effects along the cooling beam, as indicated by numerical simulations [44].

5. MOMENTUM TRANSFER BETWEEN JET AND ENVIRONMENT

It is well known that many young stellar objects are associated not only to the highly collimated, fast YSO optical jets (100-500 km/s) investigated above, but also to less-collimated, slower molecular outflows (< 20 km/s) detected at radio wavelengths. The correlation between both kinds of outflows suggested since the beginning a "unified model" in which the optical jet entrains the molecular gas of the surrounding environment driving the molecular outflow [e.g., 45]. Recent determinations of the momentum carried by YSO jet have indicated that they carry, in fact, enough momentum to drive the associated molecular outflows [e.g., 18]. For example, taking a number density of ions and atoms $n \sim 1.1 \; 10^4$ cm^{-3} in HH 34 jet and $n \sim 9 \; 10^3$ cm^{-3} in HH 111 jet, one estimates momentum rates $dP/dt = \pi R_j^2 \rho_j v_j^2 \geq 6.7 \; 10^{-5}$ M$_\odot$ yr^{-1} km s^{-1} for HH 34 jet, and $dP/dt \geq 1.3 \; 10^{-4}$ M$_\odot$ yr^{-1} km s^{-1} for HH 111 jet, which are larger than the estimated momentum rates for the associated molecular and neutral outflows: $\sim 5 \; 10^{-5}$ M$_\odot$ yr^{-1} km s^{-1} and $\sim 1.7 \; 10^{-5}$ M$_\odot$ yr^{-1} km s^{-1} for the HH 111 and 34 associated outflows, respectively [18]. These determinations have motivated several authors to explore the mechanisms by which the YSO jets can transfer momentum to the molecular outflows [21, 34, 46-52].

Two basic processes of momentum transfer between a jet and a quiescent ambient medium can be distinguished: (i) the prompt entrainment, which is the transfer of momentum through the bow shock at the head of the jet; and (ii) the lateral (or turbulent) entrainment which refers to ambient gas that is entrained along the sides of the jet. This type of entrainment is the result of turbulent mixing through K-H instabilities at the interface between the jet and the ambient medium. Investigating both mechanisms through 3-D numerical simulations of steady-state jets moving into an ambient with M_a in the range $3 < M_a = v_j/c_a < 110$ and density ratios $0.3 < \eta < 100$, Chernin et al. [21] find that the lateral (turbulent) entrainment along the jet beam is important only for $M_a \leq 6$ and $\eta \leq 3$ jets. Although this mechanism can be relevant for AGN jets which fit those values of M_a and η, it is not the case for the typical YSO jets for which $10 < M_j = M_a\eta^2 < 40$ and $\eta \geq 1$. These results are illustrated in Fig. 5. So, YSO jets seem to transfer their momentum predominantly at the bow shock. These results rule out models which propose that molecular outflows are produced through turbulent mixing [46-47, 52].

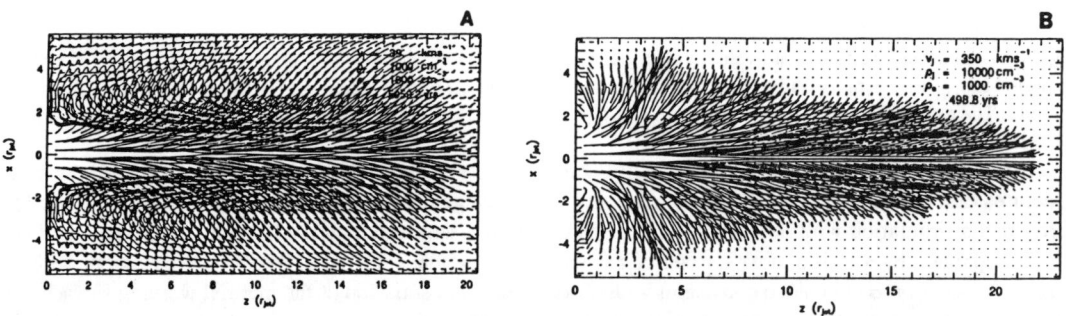

Fig. 5 The trajectories of the particles of the ambient medium. The left figure shows the case of a jet with $\eta = 1$ and M_a =1.1 and the right figure, the case of a jet with $\eta=10$ and M_a =10. The turbulent eddies in Fig. 8B are a signature of the turbulent entrainment. In Fig. 8A, the straight trajectories of the ambient particles indicate that the prompt entrainment is the dominant process (from Chernin et al. 1994 [21]).

The effect on the environment of the passage of successive bow shocks (or internal working surfaces) of a time variable jet (like those discussed in Section 4) has also been investigated. Raga and Cabrit [49] proposed that the molecular emission could be identified with the swept-up environmental gas that fills the cavity formed behind the internal working surfaces of a time-dependent ejected jet. The simulations of intermittent radiative cooling jets with small ($\tau \sim \tau_{dy} = R_j/v_j$) variability period discussed in section 4 (see Fig. 3) show no evidence that the ambient material pushed aside by the internal bow shocks have had time to fill the cavities behind them [34]. But, the jet with long variability period ($\tau > \tau_{dy}$) depicted in Fig. 4, clearly shows that the ambient gas swept up by the leading working surface eventually refills the cavity formed behind it [34], in agreement with Raga and Cabrit's model.

6. ENVIRONMENTAL EFFECTS

Jets emerging from YSOs and AGNs propagate into complex ambients [54-55]. YSO jets, in particular, are immersed in molecular clouds which are generally non-homogeneous and formed of many smaller, dense clouds. Basically all the models discussed so far have considered the propagation of the jets in initially homogeneous ambients. We now make a brief description of recent numerical results obtained for cooling steady-state jets propagating into more realistic environments [56-57]. In these models it is assumed an initial isothermal ambient medium (with a temperature $T_a=10^4$ K) with a power-law density (and pressure) distribution, $n_a(x) \propto x^\beta$, where n_a is the ambient number density, x the longitudinal jet axis, and $\beta = \pm 5/3$ for positive and negative density gradients, respectively. Such profiles are consistent with the observed density distribution of the clouds which involve protostars [58]. A negative density gradient atmosphere may represent, for example, the ambient that a jet finds when it emerges from the parent source and propagates through the cloud that involves the system. A positive density gradient atmosphere can be found when a traveling jet strikes an external cloud. The density of the cloud increases as the jet submerges into it.

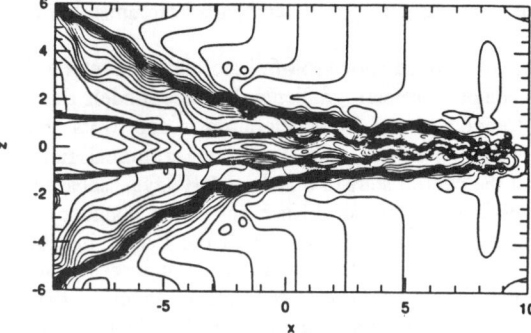

Fig. 6 Central density contours of a jet propagating into an ambient medium with positive density gradient. The initial conditions are: $\eta(x_0)=1$, $n_a(x_0)=200$ cm^{-3}, $R_j=2\times 10^{15}$ cm, $v_j=398$ km/s, $M_a=24$, $q_{bs} \approx 8.1$, $q_{js} \approx 0.3$, and $R_j/c_a=38.2$ yrs. The coordinates x and z are in units of R_j. The contour lines are separated by a factor of 1.3 and the density scales from $\approx$ 0 up to 800/n_a.(from Gouveia Dal Pino, Birkinshaw, and Benz 1994 [56]).

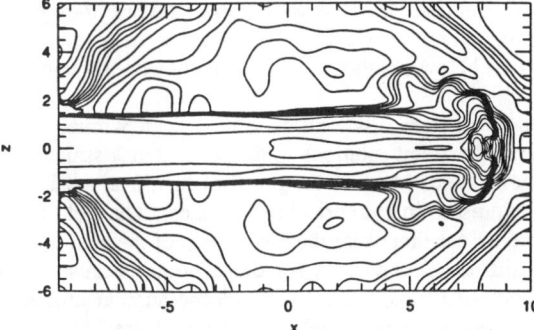

Fig. 7 Central density contours of a jet propagating into an ambient medium with negative density gradient. The initial conditions are the same of Fig. 6 (from Gouveia Dal Pino, Birkinshaw, and Benz 1994 [56]).

As an example, Fig. 6 depicts a jet propagating into an ambient medium of increasing density (pressure). In these cases, the jets have their cocoon compressed by the ambient ram pressure and the beam is highly collimated. The compressing medium induces the development of Kelvin-Helmholtz instabilities which cause the beam focusing, and wiggling and the formation of internal knots close to the head. This morphology remarkably resembles some observed YSOs jets, like Haro 6-5B [59]. (Other examples include HH30, HH24G, HH83 and HH 110.) Biro et al. [60] have performed numerical calculations of a jet ejected from a precessing source into a homogeneous ambient which also produced a wiggling beam. In their case, the mechanism producing such an effect is the "garden-hose" instability".

An example of a jet propagating in an ambient medium with negative density gradient is shown in Fig. 7. Negative density ambients produce broad and relaxed jets, with no knots and only faint radiative shocks at the head with poor emission. Therefore, they may be important in the invisible portions of the observed outflows.

Compared to a cooling jet, an adiabatic jet propagating into similar stratified ambients is slightly less affected by the stratification effects due to its higher pressure cocoon which better preserves the structure of the beam. This result may be particularly applicable to extragalactic jets [56].

7. THE ORIGIN OF THE JETS

Although considerable progress has been made toward understanding the jet structure and propagation, no consensus has been reached concerning the basic mechanism for its origin. The measured velocities and mass discharges (dM/dt $\sim 10^{-8}$-10^{-7} $M_\odot$ yr^{-1}) in the YSO jets rule out thermal or radiation pressure as the driving mechanisms [e.g., 61]. Alternative possibilities include: (i) magnetic-pressure-driven outflows; or (ii) magnetocentrifugally-driven outflows (with magnetic tapping of the rotational energy of the stellar object or of its circumstellar disk, or of both). There now exists strong evidence for the presence of nearly Keplerian accreting disks around YSOs and observational support of their correlation with the bipolar outflows [e.g., 62]. Also, polarization measurements indicate that the magnetic field of the molecular cloud cores are preferentially aligned with the symmetry axes of the embedded circumstellar disks and the associated bipolar flows [63].

In the first class of models mentioned above [64-67] the outflow material is accelerated from the surface of the circumstellar disk by magnetic pressure gradients. In these models the magnetic field at the top of the disk has a strong azimuthal component and the acceleration is due to its uncoiling like a spring above the surface. In particular, Pringle [65] proposes that the strong azimuthal field arises in the boundary layer between the accretion disk and the central star. In this boundary layer, shearing takes place and the azimuthal component of the field is amplified until it can diffuse out of the boundary layer region. Regions of strong field are expected then to decouple from the gas through the Parker instability, so that the Alfvén speed rapidly becomes larger than the escape velocity and an outflow emerges along the symmetry axis. However, these models are, in general, unsteady. In Pringle's scenario, for example, the large axial (z) gradients of the azimuthal field (B_ϕ) that are postulated to exist at the base of the outflow can be possibly established only by transient events.

Among the magnetocentrifugal models we may distinguish those in which the outflow arises from the stellar surface (pure stellar outflows), those in which it originates on the accretion disk (pure disk outflows), and those in which the outflow arises from an interacting star-disk system. These models seem to be more attractive than the magnetic pressure models discussed above because they provide a very simple and direct method for transferring angular momentum excess out of the system, thus, slowing down the protostar as required by the observations (see below), without having to invoke poorly undesrtood viscosity processes [e.g., 1].

Magnetocentrifugal models of pure disk outflows have been worked out in some detail [63, 68-71]. Wardle and Königl [63] construct explicit solutions of the vertical disk structure. They assume a disk threaded by a large scale, open magnetic field. Most of the material lies in a sub-Keplerian, quasi-hydrostatic layer close to the midplane (z=0, expressed in cylindrical coordinates). The field is coupled to the weakly ionized disk material by ion-neutral collisions that cause the inward slip of the neutral gas relative to the ions and field at a velocity v_d (ambipolar diffusion). The ions lag the neutrals because of the frictional drag. The back reaction of the ions on the neutrals causes neutral gas to loose angular momentum to the field. This in turn enables the neutral to drift toward the center and exert a radial drag of the field lines. The drag must be balanced by magnetic tension, so the field lines bend outward from the rotation axis. At higher latitudes, the field dominates the energy density. The azimuthal velocity of the field increases with the height and eventually the field lines overtake the neutral fluid and start to transfer angular momentum back to matter and to accelerate it centrifugally into a collimated wind. While attractive, the pure-disk wind model applied to YSOs has some limitations [72]: (i) the large radial drift needed in the midplane (enforced by rotation at sub-Keplerian speeds) at radius r $\sim 10^{15}$ cm (where observed disks are believed to end) leads to accretion timescales r/v_d about three orders of magnitude shorter than the values, 10^6 - 10^7 yr, inferred for YSO disks [73]; (ii) ohmic dissipation and ambipolar diffusion will possibly rid the disk interior of magnetic fields with the intensities required in the disk-outflow models for all radius r $\sim$ (0.1 - 20) 10^{13} cm. At small radius, heating by stellar photons and viscous accretion provide a good degree of ionization. Thus, effective mechanical coupling between the gas and the magnetic fields will possibly exist only in the inner regions of the disk.

In the context of magnetocentrifugal pure stellar outflow models, Hartmann and MacGregor [74] proposed that the material would stream off a contracting rotating protostar along radial magnetic field lines rooted in the star. Most of the mass loss was supposed to occur in the equatorial plane, but diverted towards the poles by magnetic pressure gradient in the azimuthal field. Shu et al. [75], modified this model to incorporate a more realistic scenario including an accretion disk. In this model the neutral molecular outflows (see Section 5) represent gas transferred from the accretion disk onto the magnetic field lines of the YSO and launched out from two narrow equatorial bands that straddle the disk, while the ionized jet originates as a normal stellar wind near the poles of the protostar. A potential difficulty with this model is that it requires the star to be rotating at high angular velocities close to breakup, which appears to be inconsistent with observations of optically revealed young stars which usually rotate at speeds an order of magnitude below breakup. In more recent work Shu et al. [72, 76-77] proposed a generalization of this previous model in order to surpass its difficulty. They assumed a viscous and imperfectly conducting disk (with larger magnetic diffusivity η_D) accreting steadily (with smaller accretion rate dM_D/dt) onto a young star with a strong magnetic field. For an aligned stellar magnetosphere, shielding currents in the surface layers of the disk will prevent stellar field lines from penetrating the disk everywhere except for a range of radii around $r = R_x$, where the Keplerian angular speed of rotation Ω_x equals the angular speed of the star Ω_s. For the low disk accretion rate and high magnetic fields associated with typical young stars, R_x exceeds the radius of the star R_s by a factor of a few, and the inner disk region is effectively truncated at a radius $R_t < R_x$. Between R_t and R_x, the closed field lines bow sufficiently inward and the accreting gas attaches to the field and is funneled dynamically down the effective potential (gravitational plus centrifugal) onto the star. The associated magnetic torques to this accreting gas may transfer angular momentum mostly to the disk. Thus, the star can spin slowly as long as R_x remains $> R_s$. For $r > R_x$ field lines threading the disk bow outward making the gas off the midplane rotate at super-Keplerian velocities. This combination drives a magnetocentrifugal wind with a mass-loss rate dM_w/dt equal to a definite fraction f of the disk accretion rate dM_D/dt. For high disk accretion rates, R_x is forced down to the stellar surface, the star is spun to breakup, and the outflow is generated in an identical manner to that proposed in the previous model of Shu et al. [75]. A schematic diagram of this model is presented in Fig. 8. They find that stellar magnetic fields of $\sim$ few kG can drive outflows with mass-loss rates of 10^{-6} $M_\odot$ yr^{-1} from rapidly accreting YSOs (rotating near breakup) and 10^{-8} $M_\odot$ yr^{-1} from slowly accreting (slowly rotating) young stars. The mechanism described can accelerate outflows from these systems to velocities $\sim$ few 100 km yr^{-1} within few stellar radii ≤ 10 R_s [76-77].

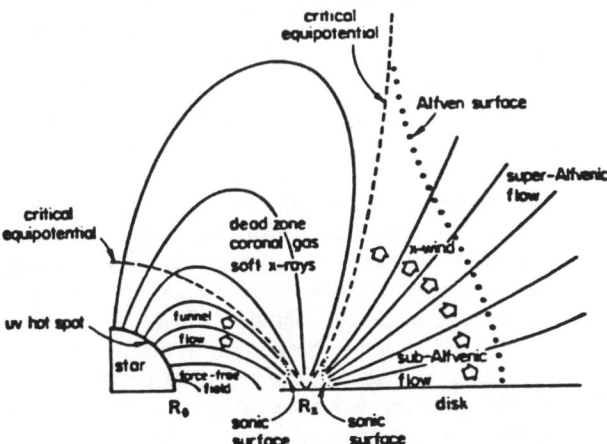

Fig. 8 Schematic diagram of the principal components of a magnetocentrifugally driven outflow from a disk-star system. Dotted curves correspond to the loci where the gas makes a transition from subsonic (or sub-Alfvénic) to supersonic (or super-Alfvénic) flow. Matter interior to the radius R_x diffuses onto field lines that bow inward and is funneled onto the star. Matter at $r > R_x$ diffuses onto field lines that bow outward and launch a magnetocentrifugally driven wind (from Shu et al. [72]).

Observational tests of the models above are very difficulty because the central sources and the associated accretion disks (which lie in a region of $\sim 10^{-3}$ pc) are not well resolved by the instruments. So, at this stage we cannot rule out the magnetic pressure models, although the magnetocentrifugal mechanisms (and, in particular, the unified star-disk model) appear to offer a more natural scenario for explaining the formation of collimated outflows. It is also possible that all these models play a role at different stages of the star-disk evolution.

More generally, these mechanisms are believed to be universal in the production of bipolar outflows from a wide range of astrophysical objetcs [e.g., 63, 67, 72]. In the case of the AGN jets, also there is now evidence for the presence of dusty molecular disks near the central nucleus [78] which gives support to disk-driven outflow models or the combined source-disk-driven outflow models.

Since the outflows are in principle capable of transporting the excess of angular momentum of the accreting matter as well as most of the gravitational energy that is liberated, one may attribute the ubiquity of bipolar outflows to the fact that the winds are a necessary ingredient in the accretion process in that they carry away the angular momentum that needs to be removed in order for the accretion to proceed [70].

8. SUMMARY

Supersonic collimated outflows are a common phenomenum in the universe and span a large range of intensities, from the powerful examples observed to emerge from the nuclei of active galaxies (or AGNs) to the jets associated to low-mass Young Stellar Objects (YSOs) within our own Galaxy. Thanks to their proximity, YSO jets constitutes the best laboratories for cosmic jet investigation. This paper tried to review the main properties of these objects and the proposed models for their origin and structure. Possible applications of the theories to the AGN jets were also briefly discussed.

Simple one-dimensional analytical approach makes possible to model the basic features of high Mach number, heavy cooling jets like the typical YSO jets (Sections 3 and 4). A supersonic jet propagating into a stationary ambient gas will develop a shock pattern at its head or working surface. The impacted ambient material is accelerated by a forward bow shock and the jet material is decelerated at a shock wave as it reaches the head, i.e., the jet shock. Numerical hydrodynamical simulations of jets propagating in initially homogeneous ambient medium confirm the analytical predictions and reveal the development of a dense shell at the head of the jet which is formed by the cooling of the shock-heated material at the working surface. The shell eventually becomes unstable and fragments into pieces which resemble the clumpy structure of observed HH objects at the heads of YSO jets (Section 3).

The chain of aligned knots observed in typical YSO jets is probably mainly formed by time-variability in the flow ejection velocity (Section 4). Supersonic velocity variations (with periods close to the transverse dynamical timescale $\tau_{dy} = R_j/v_j$) quickly evolve to form a chain of regularly spaced radiative shocks along the jet with large proper motions and low intensity spectra in agreement with the observed properties of the knots. The knots widen and fade as they propagate downstream and eventually disappear. So, this mechanism favors the formation of knots closer to the driving source in agreement with typical observations. Longer variability periods ($t \gg \tau_{dy}$) can also explain the multiple bow shocks observed in some jets. From these results we can make the conjecture that, possibly, the superluminal knots which are observed to emerge form some AGN jets in the parsec (pc) scales [e.g., 12-13] could be also the product of time-variability in the ejection mechanism from the corresponding central sources.

Shocks driven by Kelvin-Helmholtz (K-H) instability are viable candidates for knot production in jets for which the adiabatic approach is good (AGN jets). For radiative cooling jets (YSO jets), the cooling reduces the pressure in the cocoon that surrounds the beam. As a result, the cocoon has less pressure to drive K-H instabilities and thus, reflect internal shocks in the beam. However, an efficient formation of a "wiggling" chain of knots, primarily close to the jet head, by K-H instabilities is found in simulations of jets propagating into environments of increasing density (and pressure) (Section 6). In this case the cocoon is compressed by the ambient ram pressure and the beam is highly collimated. The compressing medium then induces the development of the K-H instabilities. This structure remarkably resembles some observed YSO jets (e.g., Haro 6-5B jet) which do show a wiggling structure with knot formation closer to the jet head. Ambient medium with negative density gradient, on the other hand, produces a broad and relaxed jet with no internal knots. The shocks at the head are faint in emission. Thus, this scenario may apply to the invisible portions of some observed outflows (Section 6).

YSO jets appear to carry enough momentum to drive the observed less collimated slower molecular outflows that are also associated to YSOs (Section 5). Numerical simulations indicate that for YSO jets the transfer of momentum to the molecular ambient medium occurs predominantly through the bow shocks and not by turbulent mixing through the lateral discontinuity layer. (This latter mechanism is more relevant in low Mach numbers ($M_j \leq 3$), light ($\eta \leq 3$) jets and thus, more appropriate in AGN jets.) The molecular emission could then be possibly identified with the swept-up environmental gas by the passage of the internal working surfaces (or bow shocks) of a time-dependent ejected jet.

Theoretical models for the origin of the jets involve wind acceleration from the central YSOs or from their accreting circumstellar disks (Section 7). Basically, two viable mechanisms for driving the outflows from the sources have been investigated: (i) outflow acceleration by magnetic pressure gradients; and (ii) magnetocentrifugal acceleration. Among the magnetocentrifugal models we may distinguish those in which the outflow arises from a magnetized rapid rotating protostar, or from a rotating disk with substantial poloidal magnetic fields of open geometry, or a combination of a strongly magnetized protostar with an adjoining Keplerian disk. The magnetocentrifugal models seem to be more attractive than the magnetic pressure models as they provide a very simple and direct method for transferring angular momentum excess out of the system, thus, slowing down the protostar as required, without having to invoke poorly undesrtood viscosity processes. Observational tests to check the viability of these models, however, are very difficulty because the central sources and the associated accretion disks are not observationally resolved. Also, it is not impossible, that each of these models may play a role at different stages of the star-disk evolution. These mechanisms are believed to be universal in the production of bipolar outflows from a wide range of objetcs, ranging from the YSOs to the AGNs.

While of fundamental importance in the production and initial collimation of the jets (Section 7), magnetic fields have been neglected in most of the analytical and numerical modeling of the structure of the YSO jets since the observational estimates of their intensity suggest that magnetic fields are not dynamically important along the flow. However, they can become relevant once they are amplified by compression behind the shocks. Future work should then incorporate magnetic fields in the investigation of the YSO jet structure as it has already been done for AGN jets [e.g., 15].

REFERENCES

1. Padman, R., Lasenby, A. N., and Green, D. A 1991, in: Beams and Jets in Astrophysics, P. A. Hughes (ed.), Cambridge Univ. Press, p. 484.
2. Königl, A. 1986, Can. J. Phys., 64, 362.
3. Mundt, R., Brugel, E.W., and Buhrke, T. 1987, Ap. J., 319, 275.
4. Reipurth, B. 1991, in: Physics of Star Formation and Early Stellar Evolution, NATO ASI Series, Kluwer Academic Publ., p. 497.
5. Reipurth, B. and Heathcote, S. 1993, in: STScI Symposium on Astrophysical Jets, M. Fall et al. (eds.), Cambridge Univ. Press, (in press).
6. Ray, T. P. and Mundt, R. 1993, in: STScI Symposium on Astrophysical Jets, M. Fall et al. (eds.), Cambridge Univ. Press, (in press).
7. Edwards, S., Ray, T. P. and Mundt, R. 1993, in: Protostars and Planets III, E. H. Levy and J. I. Lunine, Univ. Arizona Press (in press).
8. Raga, A. C. 1993, Astrophys. and Space. Sci., 208, 163.
9. Mundt, R. 1994, in: Disks and Outflows around Young Stars (in press).
10. Errico, L. and Vittone, A. (eds.) 1993, in: Stellar Jets and Bipolar Outflows, Dordrecht.
11. Bridle, A. H. and Perley, R. A. 1984, Annu. Rev. Astron. Astrophys., 22, 319.
12. Hughes, P. A. (ed.) 1991, Beams and Jets in Astrophysics, Cambridge Univ. Press.
13. Roland, J., Sol, H. and Pelletier, G. (eds.) 1991, Extragalactic Radio Sources from beams to Jets.
14. Begelman, M. C., Blandford, R. D. and Rees, M. J. 1984, Rev. Mod. Phys., 56, 255.
15. Norman, M.L. 1990, Ann. New York Acad. Sci., J.R. Buchler and S.T. Gottesman (eds.), 617, 217.
16. Reipurth, B. 1989, Nature, 340, 42.
17. Morse, J.A., Hartigan, P., Cecil, G., Raymond, J.C., and Heathcote, S. 1992, Astrophys. J., 339, 231.
18. Bacciotti, F., Chiuderi, C. and Oliva, E. 1994, Astron. Astrophys. (in press).
19. Hartigan, P., Morse, J. and Raymond, J. 1994, Astrophys. J. (Lett.) (in press).
20. Blondin, J.M., Fryxell, B.A., and Konigl, A. 1990, Astrophys. J., 360, 370.
21. Chernin, L., Masson, C., Gouveia Dal Pino, E.M., and Benz, W. 1994, Astrophys. J., 426, 204.
22. Gouveia Dal Pino, E.M. and Benz, W. 1993, Astrophys. J., 410, 686.
23. Stone, J. M. and Norman, M. L. 1994, Astrophys. J. (in press).

24. Blandford, R. D. and Rees, M. J. 1974, Mon. Not. Roy. ast. Soc.,169, 395.
25. Raga, A. C., Binette, L. and Cantó, J. 1990, Astrophys. J., 360, 612.
26. Ferrari, A., Massaglia, s., Trussoni, E. 1982, Mon. Not. Roy. ast. Soc., 198, 1065.
27. Norman, M. L. et al. 1984, Astron. Astrophys., 113, 285.
28. Bodo, G. et al. 1989, Astrophys. J., 341, 631.
29. Raga, A. C. et al. 1990, Astrophys. J., 364, 601.
30. Kofman, L., and Raga, A.C., 1992, Astrophys. J., 390, 359.
31. Raga, A.C., and Kofman, L., 1992, Astrophys. J., 386, 222.
32. Hartigan, P., and Raymond, J. 1993, Astrophys. J.,409, 705.
33. Stone, J. M., and Norman, M. L. 1993, Astrophys. J., 379, 676.
34. Gouveia Dal Pino, E.M. and Benz, W. 1994, Astrophys. J., 435.
35. Biro, S., and Raga, A. C. 1994, Astrophys. J (in press).
36. Falle, S.A.E.G. and Raga, A.C. 1993, Mon. Not. Roy. ast. Soc., 261, 573.
37. Eislöfel, J. and Mundt, R. 1992, Astron. Astrophys., 263, 292.
38. Gouveia Dal Pino, E.M. and Opher, R. 1990, Astron. Astrophys., 231, 571.
39. Gouveia Dal Pino, E.M. and Opher, R. 1989, Astrophys. J., 342, 686.
40. Gouveia Dal Pino, E.M. and Opher, R. 1989, Mon. Not. Roy. ast. Soc., 240, 573.
41. Gouveia Dal Pino, E.M. and Opher, R. 1991, Astron. Astrophys., 242, 319.
42. Gouveia Dal Pino, E.M. and Opher, R. 1993, Mon. Not. Roy. ast. Soc., 263, 687.
43. Bodo, G. et al. 1990, Mon. Not. Roy. ast. Soc., 244, 530.
44. Gouveia Dal Pino, E.M. 1994 (in preparation).
45. Rodríguez, L.F. 1989, Rev. Mexicana. Astron. Astrof., 18, 45.
46. Lizano, S. et al. 1988, Astrophys. J., 328, 763.
47. Stahler, S. 1993 in: Astrophysical Jets, M. Livio et al. (eds.), Cambridge U. Press,p. 183.
48. Masson, C.R., and Chernin, L.1993, Astrophys. J., 414, 230.
49. Raga, A.C., and Cabrit, S. 1993, Astron. Astrophys., 278, 267.
50. Raga, A.C. et al. J. 1993, Astron. Astrophys., 276, 539.
51. Raga, A. C., Cabrit, S. and Cantó, J. 1994 (preprint).
52. Taylor, S. D. and Raga, A. C. 1994 (preprint).
53. Bachiller, R. et al. 1990, Astron. Astrophys., 231, 174.
54. Hardee, P. E. et al. 1992, Astrophys. J., 387, 460.
55. Bally, J. and Devine, D. 1994, Astrophys. J. (Lett.) (in press).
56. Gouveia Dal Pino, E.M., Birkinshaw, M. & Benz, W. 1994 (preprint).
57. Gouveia Dal Pino, E. M., and Cerqueira, A. H. 1994, this conference.
58. Fuller, G. A., and Myers, P. C. 1992, Astrophys. J., 384, 523.
59. Mundt, R., Ray, T.P., and Raga, A.C. 1991, Astron. Astrophys., 252, 740.
60. Biro, S., Raga, A. C., and Cantó J. 1994 (preprint).
61. Königl, A., and Ruden, S. P. 1993, in: Protostars and Planets III, E. H. Levy and J. I. Lunine, Univ. Arizona Press (in press).
62. Sargent, A. I. 1994, in: Disks and Outflows around Young Stars (in press).
63. Wardle, M., and Königl, A. 1993, Astrophys. J., 410, 218.
64. Uchida, Y. and Shibata, K. 1985, PASJ, 37, 515.
65. Pringle, J. E. 1989, Mon. Not. Roy. ast. Soc., 236, 107.
66. Lovelace, R. V. E., Berk, H. L., and Contopoulos, J. 1991, Astrophys. J., 379, 696.
67. Contopoulos, J., and Lovelace, R. V. E. 1994, 429, 139.
68. Blandford, R. D. and Payne, D. G. 1982, Mon. Not. Roy. ast. Soc., 199, 883.
69. Pudritz, R. E., and Norman, C. A. 1986, Astrophys. J., 301, 571.
70. Königl, A. 1989, Astrophys. J., 342, 208.
71. Stone, J. M. and Norman, M. L. 1994, Astrophys. J. (in press).
72. Shu, F. et al. 1994, Astrophys. J., 429, 781.
73. Beckwith, S. et al. 1990, Astron. J., 99, 924.
74. Hartmann, L. and MacGregor, K. B. 1982, Astrophys. J., 259, 180.
75. Shu , F. et al. 1988, Astrophys. J., 328, L19.
76. Shu, F. et al. 1994, Astrophys. J., 429, 797.
77. Najita, J. R. and Shu, F. 1994, Astrophys. J., 429, 808.
78. Phinney, E. S. 1989, in: Theory of Accretion Disks, F. Meyer et al. (eds.), Kluwer Academic Publ., p. 457.

MHD WIND SOLUTIONS FOR ROTATING STARS

C. Ferro Fontán

Instituto de Física del Plasma, Universidad de Buenos Aires – CONICET
Ciudad Universitaria, (1428) Buenos Aires, Argentina

Abstract. An overview is presented of a known class of analytic MHD solutions for steady, rotating, axisymmetric stellar winds embedded in partially open magnetic fields [1-6]. The collimation of the outflow is achieved by assuming a magnetic configuration in which the field lines are poleward deflected. The full ideal MHD equations are reduced to a set of second order, radial, ordinary differential equations by a suitable choice of the angular dependence of the dynamic and thermodynamic variables. In the first part of the paper, spherically symmetric Mach – Alfvén surfaces are assumed, and rotating and non-rotating cases are solved. In the second part, we introduce a new [1] class of analytic MHD solutions by relaxing the hypothesis of spherical symmetry of the Mach – Alfvén surfaces. General colatitude dependent Alfvénic surfaces are developed in terms of a small ordering parameter. Approximate solutions of the resulting hierarchy of equations are obtained for a purely radial magnetic configuration and initially superalfvénic flows.

1. General formalism

The MHD equations that govern the steady dynamical interaction of a compressible, inviscid, high electrical conductivity fluid with magnetic fields in presence of a spherically symmetric gravitational field are

$$\nabla \cdot (\rho \mathbf{v}) = 0 \tag{1}$$

$$\nabla \cdot \mathbf{B} = 0 \tag{2}$$

$$\nabla \wedge (\mathbf{v} \wedge \mathbf{B}) = 0 \tag{3}$$

$$\rho(\mathbf{v} \cdot \nabla)\mathbf{v} = -\nabla \mathcal{P} + \mu_0^{-1}(\nabla \wedge \mathbf{B}) \wedge \mathbf{B} - \rho g \hat{e}_r \tag{4}$$

$$\frac{3}{2}\frac{k_B}{m}\rho(\mathbf{v} \cdot \nabla)T + \mathcal{P}\nabla \cdot \mathbf{v} = \mathcal{Q}(r,\theta) \tag{5}$$

where $\mathbf{v}$ is the velocity field, $\mathbf{B}$ the magnetic field, ρ the mean density, $\mathcal{P} = k_B \rho T/m$ the gas pressure and $g = \frac{GM_\star}{r^2}$ the gravitational acceleration. k_B, G, $M_\star$, m, T, γ denotes Boltzmann constant, gravitational constant, star mass, mean molecular weight of the fluid, temperature and polytropic index of the gas, respectively. $\mathcal{Q}(r,\theta)$ represents, as in Low and Tsinganos [3], a rate of energy interchange per unit volume of the outflow with their surroundings. This term can, in principle, represent a distribution of heating or cooling sources and it is added to consistently close the dynamic system of equations (1-4).

Some assumptions must be made on the geometry of the problem if we want to reduce it to some tractable form. Perhaps the most evident simplification is to suppose rotational symmetry, then in spherical coordinates

$$\mathbf{B} = \frac{1}{r \sin \theta} \nabla A(r, \theta) \wedge \hat{e}_\phi + B_\phi \hat{e}_\phi \tag{6}$$

$$\rho \mathbf{v} = \frac{1}{r \sin \theta} \nabla \Lambda_0(r, \theta) \wedge \hat{e}_\phi + \rho v_\phi \hat{e}_\phi \tag{7}$$

where $A(r, \theta)$ is the magnetic flux function and $\Lambda_0(r, \theta)$ is the mass flux function. It can be shown [4] that surfaces of constant A coincide with surfaces of constant Λ_0, that is, $\Lambda_0 = \Lambda_0(A)$, or, in other words, due to the fact that magnetic field lines are curves of constant A on the (r, θ) plane, $\Lambda_0(A)$ is a constant over each field line. As for the rest, $\Lambda_0(A)$ is a free function of $A(r, \theta)$.

Integration of equation (3) yields another free function of $A(r, \theta)$, namely, the electrostatic potential Ω_0

$$\mathbf{v} \wedge \mathbf{B} = \nabla \Omega_0(A) = \frac{\partial \Omega_0}{\partial A} \nabla A(r, \theta) \tag{8}$$

Let us now define the functions $\Omega = \frac{\partial \Omega_0}{\partial A}$ and $\Lambda = \frac{\partial \Lambda_0}{\partial A}$. From equation (8), by using (6) and (7) we get a general relation between the magnetic and velocity fields, say

$$\mathbf{v} = \frac{\Lambda(A)}{\rho} \mathbf{B} + \Omega(A) r \sin \theta \, \hat{e}_\phi \tag{9}$$

That is, for an equilibrium state with mass flow and with well defined magnetic surfaces we find that the flow must be parallel to $\mathbf{B}$ up to a rigid rotation of each individual flux surface.

Substituting now the poloidal and azimuthal components of the fields, given by equation (9), into the $\hat{e}_\phi$ component of equation (4) we obtain the expressions for v_ϕ and B_ϕ

$$v_\phi = \frac{L}{r \sin \theta} \left(\frac{\epsilon^2 - M}{1 - M} \right) \tag{10}$$

$$B_\phi = \frac{\mu_0 L \Lambda}{r \sin \theta} \left(\frac{\epsilon^2 - 1}{1 - M} \right) \tag{11}$$

Here

$$L(A) = \Omega(A) \, r_A^2 \sin^2 \theta_A \tag{12}$$

is another free function of $A(r,\theta)$ and the poloidal Mach-Alfvén number is

$$\frac{\mu_0 \Lambda^2}{\rho} = \frac{\rho v_p^2}{B_p^2} = M_A^2(r,\theta) \equiv M(r,\theta) \tag{13}$$

If M is less than unity at the base of the wind and further away $M \gg 1$, there exists a point $r = r_A$ where $M = 1$; we have defined

$$\epsilon^2 = \frac{\Omega(A)}{L(A)} r^2 \sin^2\theta \tag{14}$$

The suitable geometry of the magnetic field is assured with an expression of the form

$$A(r,\theta) = \mathcal{W}(r)\sin^2\theta \tag{15}$$

because it allows us to construct poleward deflected flows [5] once an appropriate function $\mathcal{W}(r)$ is selected. We will ensure that the boundary conditions are accordingly of separable variables.

2. Spherically symmetric Mach-Alfvén surfaces

A considerable simplification is obtained if we assume that M, and then ϵ, are θ-independent functions. The nondimensional functional forms chosen for $L(A)$ and $\Lambda(A)$ are ($x = r/R_\star$)

$$L(A) = \ell\left(\frac{\Psi R_\star}{\mu_0}\right)\frac{W(x)\sin^2\theta}{\Lambda(A)} \tag{16}$$

$$\Lambda(A) = \lambda(1 + jW(x)\sin^2\theta)^{1/2} \tag{17}$$

where ℓ, λ and j are three arbitrary constants. The pressure can then be expressed as

$$P(x,\theta) = \Pi_1(x) + \Pi_2(x)\sin^2\theta \tag{18}$$

where we have defined

$$\Pi_2(x) = M'\left(\frac{WW'}{x^2}\right) + M\left[\frac{WW''}{x^2} - \frac{W'^2}{2x^2}\right] + \frac{\ell^2 W^2}{x^2}\left[\frac{1}{2M}\left(\frac{\epsilon^2 - M}{1 - M}\right)^2 - \left(\frac{\epsilon^2 - 1}{1 - M}\right)^2\right] -$$
$$- \frac{W}{x^2}\left(W'' - \frac{2W}{x^2}\right) \tag{19}$$

The $\hat{e}_r$ and e_θ components of equation (4) become after some work

$$\partial_x \Pi_1(x) = -\frac{4W^2}{x^4}M' - 2M\partial_x\left(\frac{W^2}{x^4}\right) - \frac{Z^2}{Mx^2} \qquad (20)$$

$$M''\left(\frac{WW'}{x^2}\right) - M'\left[\frac{\ell^2 W^2}{2x^2 M(1-M)^2}(M^2 - 2M\epsilon^4 - \epsilon^4) + G(W)\right] -$$
$$- \frac{2\ell^2 W^2}{x^3 M(1-M)}[\epsilon^3(\epsilon - x\epsilon') - M] - \frac{\ell^2 WW'}{x^2 M(1-M)^2}[(M - \epsilon^4)(1-M) + M(\epsilon^2 - 1)^2] -$$
$$- F(W)(1-M) + \frac{Z^2}{Mx^2}jW = 0 \qquad (21)$$

where Z^2 is a constant and we have defined

$$G[W(x)] = \frac{4W^2}{x^4} + \frac{2WW'}{x^3} - \frac{2WW''}{x^2} - \frac{W'^2}{2x^2} \qquad (22)$$

$$F[W(x)] = \frac{W}{x^2}\left(\frac{8W}{x^3} - \frac{2W'}{x^2} - \frac{2W''}{x} + W'''\right) \qquad (23)$$

Equations (20) and (21) are similar to those of ref. [3] when $\ell = 0$ and $j = 0$ (no rotation and spherically symmetric density) and to those of Tsinganos and Trussoni [6] when $W(x)$ takes on a constant value.

2.1 Non-rotational solutions ($\ell = 0$)

Let us begin the discussion setting an isotropic mass density distribution, that is, $j = 0$ in equation (17). Under these assumptions let us set $W(x) = \frac{k}{\Psi R_*^2} x^q \equiv K x^q$; equations (20) and (21) will take the form

$$\partial_x \Pi_1(x) = -\frac{4K^2}{x^{4-2q}}M' - \frac{2(2q-4)K^2}{x^{5-2q}}M - \frac{Z^2}{Mx^2} \qquad (24)$$

$$M'' - \frac{\alpha(q)}{x}M' + \frac{\beta(q)}{x^2}M = \frac{\beta(q)}{x^2} \qquad (25)$$

where we have defined

$$\frac{\alpha(q)}{x} \equiv \frac{q\alpha_1(q)}{x} = \frac{x^2 G(W)}{WW'} = \frac{1}{x}\left(\frac{4}{q} + 4 - \frac{5}{2}q\right) \qquad (26)$$

$$\frac{\beta(q)}{x^2} \equiv \frac{q\beta_1(q)}{x^2} = \frac{x^2 F(W)}{WW'} = \frac{1}{qx^2}[8 - 2q^2 + q(q-1)(q-2)] \qquad (27)$$

Equation (25) can be straightforwardly solved to give

$$M(x) = C\ x^{s_1} + D\ x^{s_2} + 1 \tag{28}$$

where C and D are integration constants and s_1, s_2 are given (for $q \neq 0$) by

$$s_{1,2} = \frac{1 + \alpha_1 \mp \sqrt{(1+\alpha_1)^2 - 4\beta_1}}{2} \tag{29}$$

Note that $M(x) \geq 1 \quad \forall x$, so that within this framework the outflow is everywhere superalfvénic. Consistently, we shall take as reference surface the stellar one ($K = 1$). The only favourable situations are those with $D = 0$, which separate the breeze solutions $D < 0$ from the "explosive" solutions $D > 0$. Now we shall briefly discuss how the solutions for $M(x)$ are modified once the density becomes an anisotropic function. If we set $j \neq 0$ in expression (17) for the mass flux function and $W(x) = K\ x^q$, equation (21) becomes (for $\ell = 0$)

$$M'' - \frac{\alpha(q)}{x} M' + \frac{\beta(q)}{x^2} M = \frac{\beta(q)}{x^2} - \frac{Z^2}{K^2 q M x^{q-1}} jK \tag{30}$$

where we have introduced the mass asymmetry parameter jK. We deal now with a non-linear equation for M, so that subalfvénic flows may occur.

A more elaborated extension of the present model (for $\ell = 0, j \neq 0$) is obtained when radiation pressure effects are taken into account [2]. We adopted the Castor, Abbott and Klein model [7]. Their set of equations is easily decoupled when all the lines considered are optically thick. For simplicity, we consider a purely radial magnetic field. Mass distribution is written as $\rho(r,\theta) = \rho_o(1 + jK\sin^2(\theta))$, where ρ_o is the polar density and jK represents a mass asymmetry parameter. In Figures 1-2 we show typical solution profiles for a Wolf – Rayet star of $M_* = 30\ M_\odot$, $T_{eff} = 40,000\ K$, $\dot{M} = 5 \times 10^{-5} M_\odot yr^{-1}$ and $\log g = 3.55\ (cgs)$

2.2 Rotational solutions ($\ell \neq 0$)

The previous wind solutions are easily found owing to the non-rotational hypothesis, which makes quite simple the mathematical treatment. As it is well known, the inclusion of rotation into the wind equations leads to a complicated problem because it introduces singular points; provided the thermodynamics of the outflow is fixed through a polytropic index γ the singular "points" correspond to the surfaces where the poloidal flow matches the velocity of three MHD waves: slow, Alfvén and fast modes, respectively. In the same way, beyond the Mach-Alfvén surface, equation (21) becomes singular because of the existence of a critical point (strictly speaking, the only critical surface that can be found within this formalism). In fact, the numerical integration of the Mach-Alfvén function $M(x)$ from infinity toward the alfvénic surface can be done only up to a point at which all curves, except one, are deflected to solutions that are either unphysical or do not fulfill the wind initial conditions. The same result is achieved if numerical computation is performed from $x = x_A$ toward the external regions, since in the vicinity of that point solutions become breezes or rapidly diverge. This feature confirms the presence of a typical X-type critical point in the wind region, which filters the single solution fulfilling the boundary conditions, that is, the wind-type solution that takes on the surface values at the photosphere and reaches a plausible terminal flow speed far from the star.

3. Non-spherically symmetric Mach-Alfvén surfaces

In the previous sections we showed that the combined effect of rotation, magnetic stresses and anisotropic mass density distribution leads to a complicated problem, and some idealizations were needed in order to solve it. Axisymmetric MHD wind-type solutions were obtained by introducing an *external* mass asymmetry parameter and assuming spherically symmetric Mach-Alfvén surfaces [1,6,8]. In the present section we will consider the opposite approach, introducing the mass anisotropy through the assumption of θ-dependent Mach-Alfvén surfaces and neglecting the mass asymmetry parameter; as a result, the density profile will *intrinsically* take an anisotropic shape. An attempt to solve the wind problem in this way was done, among other authors, by Pneumann and Kopp [9] and Sakurai [10,11], but they fixed beforehand the thermodynamics of the outflow and then found numerically the geometrical structure of the magnetic field. Instead, and in order to avoid a rather arbitrary specification of the heating/cooling mechanisms along the outflow, we shall phenomenologically prescribe the magnetic structure and, *a posteriori*, we shall deduce the distribution of energy sources. In order to introduce the angular dependence of the Alfvénic Mach function, we write it in the following way

$$M(x,\theta) = \sum_{n \geq 0} \delta^n M_n(x) \sin^{2n} \theta \tag{31}$$

where δ is an ordering parameter. Integrating the pressure with respect to the variable θ we obtain

$$-P(x,\theta) = -\Pi_1(x) + [\frac{W'^2}{x^2} \sum_{n \geq 0} M_n(x)(1+2n) - \frac{2WW'}{x^3} \sum_{n \geq 0} (M_n(x) + xM_n'(x)) - \\ - \frac{2W}{x^2}(W'' - \frac{W'}{x}) \sum_{n \geq 0} M_n(x)] \frac{\sin^{2n+2} \theta}{2n+2} + \frac{W}{x^2}(W'' - \frac{2W}{x^2}) \sin^2 \theta \tag{32}$$

where $\Pi_1(x)$ depends on the radial distance and represents the isotropic part of the total pressure. Ordering the radial component of the pressure equation with respect to δ we get an infinite chain of coupled radial equations. For instance, for the case $W(x) = a$, cutting the hierarchy with the assumption $\mathcal{M}_n = 0 \ \forall n \geq 2$, one gets an ordinary, linear, first order differential equation for $M_1(x)$

$$\frac{4a^2}{x^4} M_1' - \frac{8a^2}{x^5} M_1 = 0 \qquad (33)$$

which can be inmediately integrated to get $M_1(x) = C\, x^2$ where C is a constant to be determined. For M_0 the differential equation is

$$M_0' - 2\frac{M_0}{x} + \frac{C V_e^2 M^2(1)}{2 v_{r_0}^2} \frac{x^4}{M_0^2} = -\frac{2}{x} \qquad (34)$$

where $M(1)$ refers to the value of the Alfvénic Mach function at the equator and at the base of the wind. Equation (34) only admits solutions for $C < 0$, in such a way that the Alfvénic surfaces become oblated and mass density increases toward the stellar equator. Another important point that deserves to be mentioned is the fact that $M_0(x)$ increases faster than $M_1(x)$, therefore the Alfvénic surfaces become spherical beyond a few stellar radii. It can be seen that the heating source does not extend practically beyond the stellar surface, unlike the spherical Alfvénic surfaces model, in which energy must be added to the flow along many stellar radii for the wind to take place.

Again, the previous solution has been easily found owing to the simplicity that the non-rotational hypothesis introduces in the calculations. But the $\ell \neq 0$ case implies to solve a system of highly non-linear equations and new hypotheses will be needed to keep the problem in a resolvable way. Therefore we shall integrate the system under the assumption of superalfvénic flows and a purely radial magnetic flow. The system can been written in the explicit form

$$A_1 M_0' + B_1 M_1' = C_1 \qquad (35)$$

$$A_2 M_0' + B_2 M_1' = C_2 \qquad (36)$$

where the coefficients A_k, B_k, C_k depend on x, M_0 and M_1. Therefore the integration can be performed by means of simple numerical techniques.

Summarizing, in this section we have focused the problem from a different point of view, by neglecting a free parameter (the mass asymmetry parameter) and introducing mass anisotropy only through the assumption of non-spherically symmetric Alfvénic surfaces. With this formalism, in a first step applied to purely radial magnetic fields, we have found that these surfaces could be oblated near the stellar surface and evolve to sphericity far from the star. This is a point of relevant importance, because the mass lose rate predicted within this formalism is lower than the one estimated by the spherically symmetric surfaces models (hereafter SSSM). Moreover, the dynamics of the outflow radically changes. In fact, velocity does not diverge to unbounded values as in the SSSM [6] and, furthermore, the energy injected to the wind is lower in the anisotropic model than in the SSSM. In addition, our model predicts the increment with rotation of the terminal velocity of the wind, unlike the SSSM in which the opposite phenomenon occurs.

4. Conclusions

In this paper we presented an extension [1] of a known class of analytic MHD solutions for steady, rotating, axisymmetric stellar winds. The full ideal MHD equations were reduced to a set of second order, radial, ordinary differential equations by a suitable choice of the angular dependence of the dynamic and thermodynamic variables. In this method of solution one does not assume a polytropic equation of state, but the polytropic index depends on position. It was shown that, provided the wind is properly collimated, a single wind type solution exists even in rotationless situations. In the first part of the paper, spherically symmetric Mach – Alfvén surfaces were assumed. An interesting example in this class of solutions is the case of a Wolf-Rayet star wind, obtained by combining the MHD equations with the radiatively driven outflow formalism, allowing for a mass asymmetry parameter and a radiation coupling constant. In the second part of the paper, an approximate hierarchy of equations determining a new set of solutions was obtained by relaxing the hypothesis of spherical symmetry of the Mach – Alfvén surfaces. For the first time, simple analytic solutions are presented for a purely radial magnetic configuration and initially superalfvénic flows. For this sample of solutions it was found that the terminal velocity of the wind does not diverge, and that increases with rotation. The energy needed to sustain the flow is distributed close to the photosphere, and its rate is much smaller than the black body power radiated by the star. Moreover, the temperature profile displays the typical chromospheric structure of early type stars. It must be stressed that the mass distribution that results from the angular dependence assumed for $M(x,\theta)$ is in accordance with observations from different cosmical objects, namely, Wolf-Rayet stars, Be stars, etc. Although superalfvénic flows have been assumed, so that all the "critical surfaces" remain beneath the stellar surface, it seems plausible that the prominent features of the model will remain even for curved magnetic fields and for initially subalfvénic outflows. In fact, we hope that further investigation of this solutions will take us to a deepest understanding of the behaviour of magnetohydrodynamic stellar winds.

References

1. Rotstein N., Ferro Fontán C; 1994: to be published in Ap.J.
2. Giménez de Castro G., Rotstein N, Ferro Fontán C.; 1994: Proc. of this conference
3. Low, B.C; Tsinganos, K; 1986: Ap.J, **302**, 163
4. Tsinganos, K; 1982: Ap.J, **252**, 775
5. Heyvaerts, J; Norman, C.A; 1989: Ap.J, **347**, 1055
6. Tsinganos, K; Trussoni, E; 1991: Astron.Astrophys, **249**, 156
7. Castor, J.I; Abbott, D.C; Klein, R.I; 1975: Ap. J, **195**,157
8. Tsinganos, K; Sauty, C; 1992: Astron.Astrophys, **255**, 405
9. Pneumann, G,W; Kopp, R.A; 1972: Solar Phys, **18**, 258
10. Sakurai, T; 1985: Astron.Astrophys, **152**, 121
11. Sakurai, T; 1990: Computer Physics Report, **12**, 247

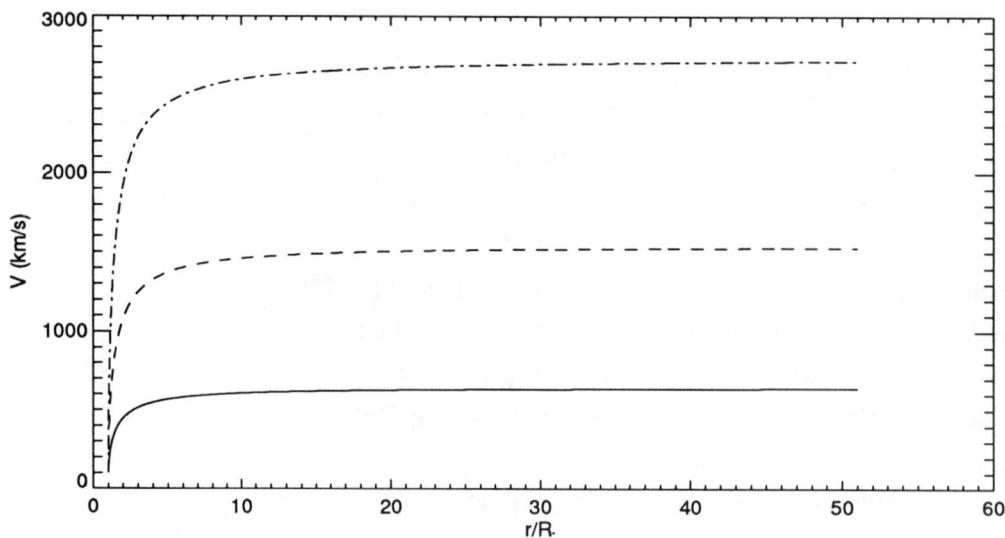

Figure 1. Radial velocity at the poles as a function of the distance from the star. The continuous line represents the case with $jK = 1$, dashed line $jK = 5$ and dot-dashed line $jK = 15$. In all cases $B_o = 500$ Gauss.

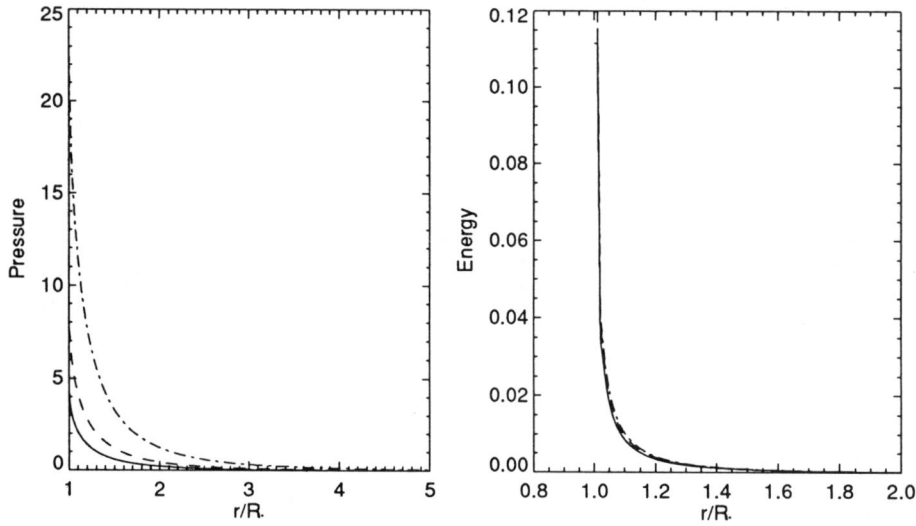

Figure 2. left: polar pressure in terms of the magnetic pressure as a function of the distance from the star. Right: energy supplied to the wind in terms of black body energy radiated by the star.

LANGMUIR TURBULENCE IN SPACE PLASMAS

Martin V. Goldman, David L. Newman and Jin-Gen Wang
Campus Box 391, University of Colorado, Boulder CO 80309
1994 International Conference on Plasma Physics, Foz do Iguazu, Brazil, 10/ 31 - 11/ 4

ABSTRACT

Recent developments in theoretical and numerical modeling of Langmuir turbulence in space plasmas are addressed. After reviewing the basic concepts of nonlinear wave-wave and wave-particle interactions, various *kinetic effects*, which have been missing from traditional Zakharov eqn.[1] models, are explored using Vlasov code simulations. Two physical examples in weakly magnetized plasmas addressed in detail are: beam-driven Langmuir waves and particle distributions in Earth's foreshock region, and radiation-driven Langmuir waves and accelerated particles in the ionosphere. The physical processes studied in the Vlasov simulations of these environments include electron-beam flattening, acceleration of particles by waves, self-consistent effects of accelerated particles on *wave damping rates*, and strong Landau damping of ion-acoustic waves.

INTRODUCTION

Naturally occurring Langmuir waves have been measured by spacecraft in the solar wind during Type III solar radio emissions,[2,3,4] in Earth's electron foreshock region,[5,6,7,8] and in the auroral ionosphere.[9,10,11,12] In these space environments the waves are driven by electron beams. Langmuir waves have also been excited artificially in the F-region of the ionosphere[13,14,15] by high powered HF electromagnetic waves launched from ground-based radar installations in the 5-10 MHz frequency range, as part of a program of *ionospheric modification*. A summary table of the measured wave energy density, driver properties, and magnetization of the plasma in these various environments is shown below:

Table 1: Langmuir wave energies and driver parameters in space plasmas.

Where?	Measured $\|E_w\|^2/8\pi n T_e$	e-beam n_b/n, v_b/v_e	Ω_e/ω_e
Solar Wind Type III (Thejappa, et al[4] -*Ulysses* data) $\lambda_e \approx 10$ m, $f_p \approx 30$ kHz	3×10^{-5} max $\tau_{res} = 1$ ms $\tau_{res} \cdot v_{sw} = 400$ m	10^{-5}, 20 (minutes)	10^{-2}
Electron foreshock $\lambda_e \approx 10$ m, $f_p \approx 30$ kHz	10^{-3} max	10^{-3}, 6	10^{-2}
Auroral region (Kintner[12], *Freja* data) $\lambda_e \approx 3$ cm, $f_p \approx 1$ MHz	10^{-2} max $\tau_{res} = 125$ ns $\tau_{res} \cdot v_{rocket} < \lambda_e$	10^{-4}, 50	O(1)
Ionosphere Modification $\lambda_e \approx 0.3$ cm, $f_p \approx 6$ MHz	No direct measurement.	EM pump, E_0. (local $E_0 \approx 1$ V/m)	≈ 0.2

The nonlinear processes underlying Langmuir wave turbulence are fairly generic; the study of this subject is relevant to a host of other wave and particle processes. The *Zakharov eqn.* description[1] of Langmuir turbulence is a two-time-scale fluid model for the evolution of unstable waves, normally consisting of two equations — an *envelope eqn.* for high-frequency waves (e.g., Langmuir waves), which may be driven and damped by passive external sources, coupled (via nonlinear refraction) to a linear *slow-timescale density eqn*, driven by the fast-time-averaged ponderomotive beat-force of high-frequency waves. Passive linear Landau damping is often added ex post facto, but nonlinear kinetic effects are generally absent. Below we distinguish those physical processes well-modeled by these eqns. from the class of kinetic effects missing from them. In this paper we examine many of these kinetic effects using Vlasov simulations.

Wave-wave processes included in Zakharov eqns.

1. Three-wave cascades, in which an unstable high-frequency wave (e.g., a Langmuir wave) undergoes multiple inelastic scattering off lower-frequency waves (e.g., ion-acoustic waves), arriving at wavenumbers where stabilization is provided by Landau damping.

2. Phase-coherent wave interactions, such as modulational instability and spatial collapse, in which nonlinear refraction and trapping of waves in density cavities can intensify and constrict a spatial wavepacket to smaller scales (higher k), where efficient dissipation (burnout) may occur. Such processes signal the onset of *strong turbulence*. In the Zakharov eqns, the dissipation is via a passive sink, in which the Landau damping is effective particles do not evolve self-consistently.

Kinetic effects missing from Zakharov eqns:

3. Flattening of bump-on-tail beams (reduction of particle free-energy) by resonant waves. Beam-flattening (or removal of other forms of particle free energy, such as velocity-anisotropy), whether due to quasilinear diffusion or due to trapping may cause unstable waves to stop growing. The time-scale for beam-flattening can be either shorter or longer than the time for wave-wave interactions to occur, depending on beam parameters.

4. Advection of beam particles, which tends to reconstitute a bump-on-tail distribution. Advection, too, competes with beam-flattening, enabling resonant waves to continue to grow.

5. Tail formation on particle distributions due to acceleration by waves and localized wavepackets. Wave-wave cascades and spatial wavepacket collapse both take wave energy out of resonance with the wave-driving particles (e.g., beam), resulting in the transfer of wave energy to other particles, especially during "burnout" of collapsing wavepackets.

6. Self-consistent influence of nonlinearly-altered distribution functions back on *waves*. Tails on the particle distribution functions self-consistently alter the dispersion, Landau damping and wave spectra.

7. Linear kinetic effects essential for an accurate treatment of slow time-scale linear *quasimode* responses, such as ion-acoustic quasimodes in the solar wind and multi-species quasimodes in the magnetosphere which interact with high-frequency waves.

Whether the Langmuir turbulence is driven by electron beams or by an electromagnetic pump, there are certain common features of the nonlinear evolution which are related to the underlying linear instabilities produced by the driver. Many of the electron beam drivers in space physics environments cause unstable Langmuir wave growth via bump-on-tail instability, whereas radiation-drivers pump unstable Langmuir wave growth via parametric decay and/or modulational instabilities. In either case, there is a wave vector at which the fastest growth occurs. The wavenumber, k, of the fastest-growing Langmuir wave, together with the strength of the driver, often control important features of the nonlinear processes which saturate the primary instability.

One mechanism for this to occur is through secondary instabilities excited by the fastest-growing Langmuir wave, which acts as a driver. Numerical solutions to the "standard" Zakharov eqns in an unmagnetized plasma reveal the following kinds of wave evolution: When $k < k^*$, where $k^*\lambda_e \equiv (2/3)\sqrt{m/M}$, and m/M is the electron/ion mass ratio, parametric decay instability of the fastest growing wave is kinematically forbidden, but modulational instability (which takes wave energy from long wavelengths to shorter wavelengths) can occur. In beam-driven instabilities, the fastest growing Langmuir wave, at $k \approx \omega_e/v_b$, drives a transient modulational instability at long wavelengths when $k < k^*$. This initiates collapse processes,[16] which are then sustained by nucleation[17,18,19] of nonlinear Langmuir wave packets in density cavities (cavitation).

For instabilities pumped by radiation at frequency ω_o, $k(z) \cdot \sqrt{3} \cdot v_e \approx \sqrt{\omega_o^2 - \omega_e(z)^2}$. When $k < k^*$, the radiation pump drives a modulationally unstable wave near ω_e, which, in turn, drives secondary instabilities. Once again, a cavitation regime sets in and collapse is sustained by nucleation.[20,21,22] For larger values of $k > k^*$, the decay instability, L -> L + I, appears, and the Langmuir daughter wave initiates a partial cascade of stimulated backscatters off ion-acoustic waves, followed by collapse. This can occur for either kind of primary driver — beam or radiation. Finally, for still larger values of k (close to the Landau damped regime), weak turbulence may result for primary drivers sufficiently close to threshold.

With the aid of the preceding brief review, we now turn to address kinetic effects missing from the Zakharov model by carrying out Vlasov simulations relevant to Langmuir wave observations in Earth's electron foreshock and in ionospheric modification.

VLASOV SIMULATIONS OF KINETIC PROCESSES

There currently exist a wealth of satellite observations of Langmuir waves in the Earth's electron foreshock.[5-8] Although less frequently measured, electron distributions that exhibit a positive slope in the 1-D reduced distribution parallel to the geomagnetic field have also been observed. One such distribution, measured[5] on ISEE-1, is reproduced below, together with results of our Vlasov simulation. The measured distribution shows both a bump-on-tail "beam" and a plateau in the backward direction. The time required to measure the electron distribution is many wave growth times, so the measured beam should also have had time to experience both flattening, due to wave-induced particle diffusion (or trapping), and beam-steepening, due to advection. The evident bump is interpreted as resulting from a balance between these two processes.

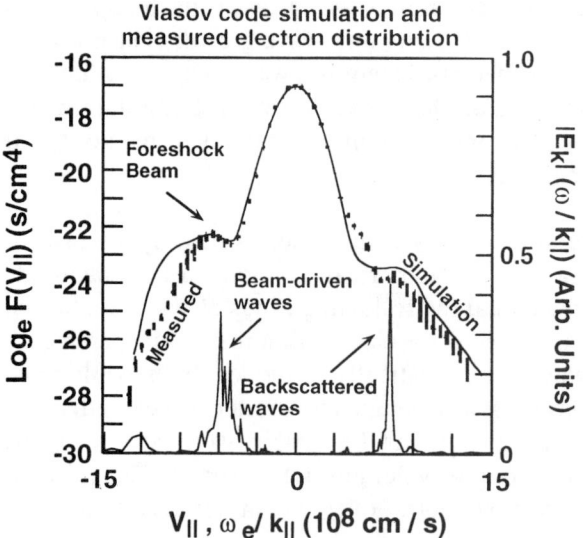

Fig. 1: Measured[5] reduced electron distribution in the electron foreshock compared to results from Vlasov simulation of wave-particle evolution. Horizontal axis is parallel velocity for particles and phase velocity, $\omega_e/k_\parallel$, for Langmuir waves. In order to expedite development of self-consistent nonlinear wave-wave scattering, a Maxwellian tail was added only after the backscattered Langmuir waves began to appear and grow into the electric field spectrum $|E_k|$ (plotted here vs. parallel wave phase velocity.)

In order to determine if the measured backward plateau is consistent with electron diffusion caused by backscattered Langmuir waves, we performed a Vlasov simulation starting with the measured beam, which was periodically replenished in order to simulate advection. (Without replenishment the simulations shows that the beam quickly flattens due to diffusion by resonant waves.) The simulation shows the growth of both beam-resonant waves and backscattered waves, and provides evidence of particle diffusion by the backscattered waves as well. The plateau in the backward direction (*opposite* to the unstable beam) can be interpreted at a consequence of electron acceleration (e.g., diffusion) by nonlinearly backscattered Langmuir waves.

In the next series of simulations[23] we address ionospheric modification by a high-power HF electromagnetic pump. Both electron and ion Vlasov equations are solved together with Poisson's eqn. The eqns are solved by a split-step method,[24] in which velocity and position operators are neglected at alternate times. The boundary conditions are semi-open — an admixture of heated and cool electrons are readmitted at each boundary to simulate the finite extent of the pump region. The pump field strength is taken to be 1.6 V/m. Two simulations are performed. Case 1 corresponds to the neighborhood of the critical altitude, where the plasma frequency equals the pump frequency, so that $k < k^*$. Case 2 corresponds to a lower altitude, where $k > k^*$. Specifically, $k \cdot \lambda_e = 0.1$ for Case 2.

Fig. 2a shows the time history of Langmuir intensity and electron temperature for Case 1. After the turbulence has fully developed, the temperature has increased by about $40 \pm 20\%$, and the average Langmuir wave intensity has settled into a statistical steady state. At the marked time of $\omega_e t = 14,800$, Fig. 2b reveals that electron phase space contains "jets" of accelerated electrons, produced by transit damping of collapsing Langmuir wave packets. The turbulence here is dominated by collapse. Fig. 2c shows the appearance of a tail on the space-averaged distribution function of electrons, which is well-fit by a v^{-2} power law tail in the range from 5 to 25 thermal velocities, v_{Te}.

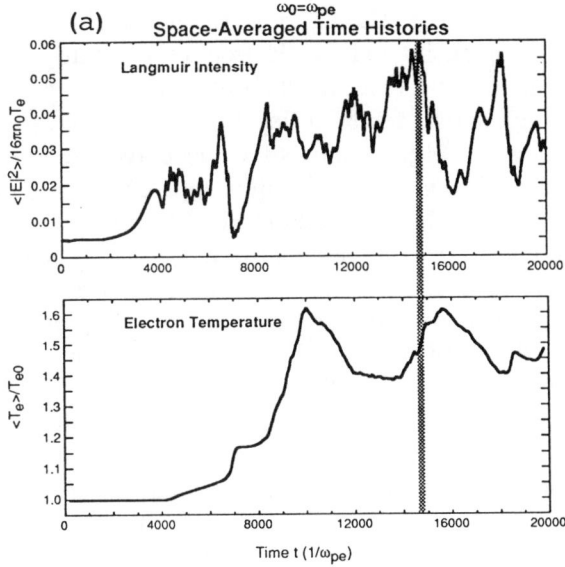
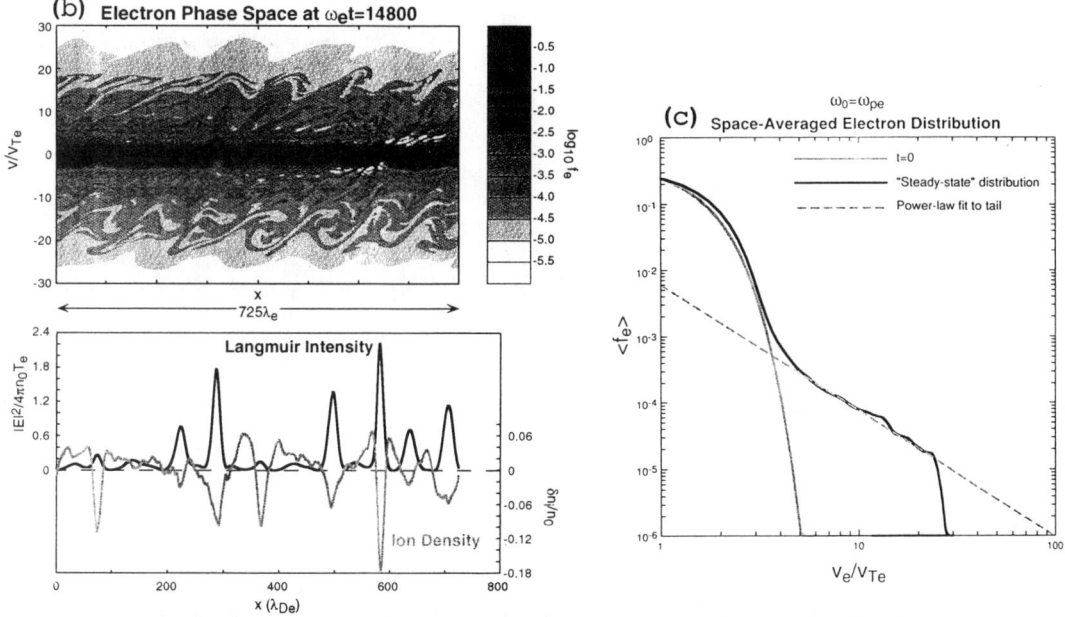

Fig. 2: Results of radiation-driven Vlasov simulation for conditions near the critical altitude (Case 1). (a) time histories of Langmuir intensity and electron temperature, (b) electron phase-space distribution (top) and spatial structure of Langmuir intensity and ion density perturbations (bottom), both at time $\omega_e t = 14800$, (c) space-averaged electron distributions at t=0 and near end of simulation, with power-law fit to developed tail.

The next set of figures corresponds to Case 2. Fig. 3a once again shows time histories of the average Langmuir wave intensity and the electron temperature. The temperature has increased by about 40% again, but exhibits shallower fluctuations. Not shown is the Langmuir turbulence, which shows fewer collapsing cavities and possible evidence of cascade to longer wavelengths. Fig. 3b shows the space-averaged velocity distribution function at a late time in Case 2. Once again, a tail is apparent from about 5 v_{Te} to 25 v_{Te}, with comparable slope but fewer electrons than in Fig. 2c. However, in this case there is increased bulk heating (at 2 v_{Te}, for example) relative to Case 1. It is important to note that the linearly unstable wave at $k \cdot \lambda_e = 0.1$ has a phase velocity at $10 \cdot v_{Te}$ — right in the middle of the tail on the electron distribution function. As a consequence, this wave can experience heavy Landau damping ($\propto k$) by the self-consistent tail. We believe it is this nonthermal Landau damping which causes the drop-out of the average Langmuir intensity at around 15,000 plasma periods, as shown in Fig. 3a. In effect, the damping reduces the growth rate (possibly even below threshold). The recovery of the Langmuir waves is enabled by the lossy boundary conditions. As heated electrons are lost from the boundaries, the damping decreases and the relative growth rate increases. Measurements[15] of oxygenic airglow during ionospheric modificaiton are thought to provide evidence for tail formation by Langmuir turbulence.

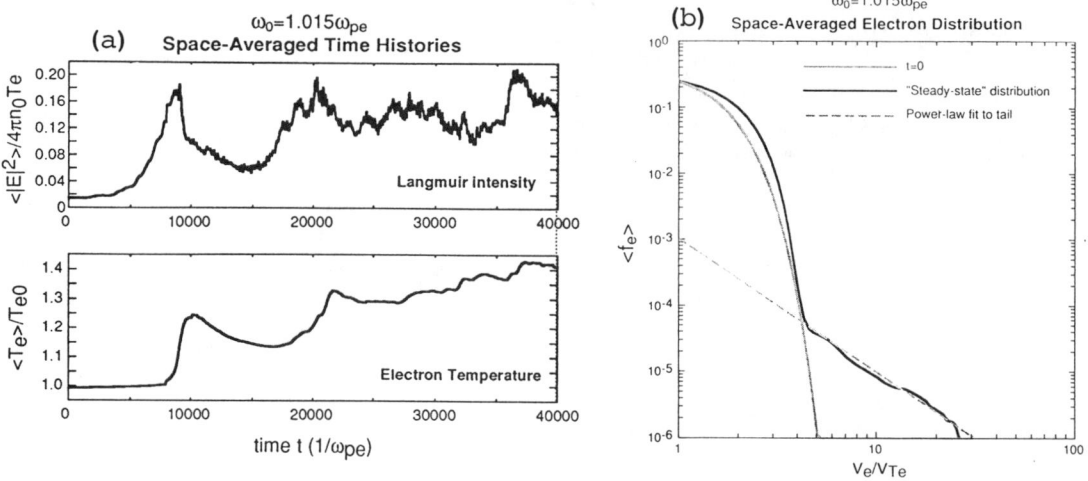

Fig. 3: Results of radiation-driven Vlasov simulation for conditions below the critical altitude (Case 2). (a) time histories of Langmuir intensity and electron temperature, (b) space-averaged electron distributions at t=0 and near end of simulation, with power-law fit to developed tail.

The kinetic effects addressed in this paper are part of the cutting edge of present research into Langmuir wave turbulence. Some progress has been made towards incorporating kinetic effects into two-time scale theories by introducing higher order moment closures[25] of fluid theories. However, space does not permit discussion of these methods in the present paper.

ACKNOWLEDGEMENT

This work was supported by grants from the Atmospheric Sciences Division of NSF.

References

[1] V.E. Zakharov, *Collapse of Langmuir waves*, Sov. Phys. JETP, Engl. Transl., **35**, 908 (1972).

[2] R.P. Lin, D.W. Potter, D.A. Gurnett and F.L. Scarf, *Energetic electrons and plasma waves associated with a solar Type III Radio Burst*, Astrophys. J. **251**, 364-73 (1981)

[3] R.P. Lin, W.K. Levedahl, W. Lotko, D.A.. Gurnett, and F.L. Scarf, *Evidence for parametric decay of Langmuir waves in solar type III radio bursts*, Astrophys. J. **308**, 954 (1986)

[4] G. Thejappa, D. Lengyel-Frey and R.G. Stone, *Evaluation of emissionmechanisms at ω_e using Ulysses observations of Type III bursts*, Astrophys. J. **416**m 831 (1993)

[5] R.J. Fitzenreiter, A.J. Klimas, and J.D. Scudder, *Detection of Bump-On-tail Reduced Electron Velocity Distributions at the Electron Foreshock Boundary*, Geophysical Research Letters, **11**, 496, (1984)

[6] G.B. Hospodarsky, D.A. Gurnett, W.S. Kurth, M.G. Kivelson, R.J. Strangeway, D.J. Williams, R.W. McEntire, *Highly Structutred Langmuir Wave Emissions: Recent Results from Galileo*, Eos supplement, American Geophysical Union Spring Meeting May 23-27, **75**, 284, (1994)

[7] J. Etcheto and M. Faucheux, *Detailed study of electron plasma waves upstream of the Earth's bow shock*, JGR **89**, 6631 (1984)

[8] P.C. Filbert and P.J. Kellogg, *Electrostatic noise at the plasma frequency beyond Earth's bos shock*, JGR, **11**, 496 (1984)

[9] D.L. Newman, M.V. Goldman, R. Ergun, and M. Boehm, "Langmuir turbulence in the auroral ionosphere: I. Linear Theory," Journal of Geophysical Research, **99** 6367-6376 (Apr., 1994).

[10] D.L. Newman, M.V. Goldman, and R. Ergun, "Langmuir turbulence in the auroral ionosphere: II. Nonlinear theory and simulations," Journal of Geophysical Research, **99** 6377-6391 (Apr., 1994).

[11] D.L. Newman, M.V. Goldman and R.E. Ergun, "Strong Langmuir turbulence in moderately magnetized Plasmas," Physics of Plasmas, **1**, 1691-1699, (May 1994).

[12] P. Kintner, *Narrow Band Langmuir Waves in the Topside Auroral Ionosphere: Observations from Freja*, invited paper, Gordon Research Conference on Collisionless Dissipation in Space Plasmas, June 23, 1994, Wolfeboro NH

[13] Cheung, Peter Y. , Alfred Y. Wong, T. Tanikawa, J. Santoru, Donald F. DuBois, Harvey A. Rose and David Russell, *Short-time-scale evidence for strong Langmuir turbulence during hf heating of the ionosphere*, Phys. Rev. Lett. **62**, 2676, (1989)

[14] Carlson, H.C., V.B. Wickwar, and G.P. Mantas, *Observations of suprathermal electrons accelerated by HF excited instabilities*, J. Atmos. Terr. Phys., **44**, 1089-1100, (1982)

[15] Bernhardt, Paul A., Lewis M. Duncan, and C.A. Tepley, *Artificial Airglow Excited by High-Power Radio Waves*, Science, Volume **242**, pp 1022-1027, 18 November (1988)

[16] P.A. Robinson, D.L. Newman, and M.V. Goldman, *Three-Dimensional Strong Langmuir Turbulence and Wave Collapse*, Physical Review Letters, **61**, 702, (1988)

[17] Doolen, Gary, D., Donald F. DuBois and Harvey A. Rose, *Nucleation of Cavitons in Strong Langmuir Turbulence*, Phys. Rev. Lett. **54**, 804, (1985)

[18] Russell, David, Donald F. DuBois, and Harvey A. Rose, *Collapsing-Caviton Turbulence in One Dimension*, Phys. Rev. Lett. **56**, 838, (1986)

[19] Russell, David, D. F. DuBois, and Harvey A. Rose, *Nucleation in Two-dimensional Langmuir Turbulence*, Phys. Rev. Lett. **60**, 581, (1988)

[20] DuBois, Donald F., Harvey A. Rose and David Russell, *Excitation of Strong Langmuire Turbulence in Plasmas Near Critical Density: Application to HF Heating of the Ionosphere*, JGR, **95**, 21221, (1990)

[21] DuBois, Donald F., David Russell, and Harvey A. Rose, *Coexistence of Parametric Decay Cascades and Caviton Collapse at Subcritical Densitites*, Phys. Rev. Lett. **66**, 1970, (1991)

[22] Newman, D.L. and M.V. Goldman, "Ionospheric Langmuir Turbulence Driven by an Electromagnetic Pump Below the Upper-Hybrid Frequency," pg 180, *Nonlinear Processes in Physics*, Springer Series in Nonlinear Dynamics, Eds. A.S. Fokas, D.J. Kaup, A.C. Newell and V.E. Zakharov, Springer-Verlag, New York, (1993)

[23] J. G. Wang, D.L. Newman and M.V. Goldman, *Vlasov simulations of electron heating by Langmuir turbulence hear the critical altitude in the radiation-modified ionosphere*, Presented at the IV Suzdal URSI Symposium on Artificial Modification of the Ionosphere, Aug. 15-20, 1994, Uppsala, Sweden — to appear in JATP

[24] J.-G. Wang, *1-D simulations of strong Langmuir turbulence in a radiation-driven plasma*, PhD Dissertation, University of Iowa (1993).

[25] M.V. Goldman, D.L. Newman & F.W. Perkins, *Fluid Closures which mimic wave-particle interactions in strong Langmuir turbulence*, Phys. Rev. Letters **70** 4075 (June, 1993)

NUMERICAL SIMULATIONS OF SUPERSONIC DIRECTED FLOWS IN ASTROPHYSICAL PLASMAS

A. Ferrari[1,2], G. Bodo[1], S. Massaglia[1], P. Rossi[1], E. Trussoni[1]

[1] *Osservatorio Astronomico di Torino, Strada dell'Osservatorio 20, I-10025 Pino Torinese, Italy*
[2] *Istituto di Fisica Matematica dell'Università, Via C. Alberto 10, I-10123 Torino, Italy*

ABSTRACT We examine the nonlinear stability properties of supersonic jets with respect to Kelvin–Helmholtz instabilities. The analysis is performed employing a finite–difference numerical code that is particularly suitable in treating shock conditions. In order to analyze the behaviour of axisymmetric and antisymmetric modes, we study the evolution of an infinite jet both in cylindrical and planar geometries, respectively. The results obtained are relevant in connection to the phenomenology of astrophysical jets, both in galactic and extragalactic environments.

1. INTRODUCTION

About 20 years after being discovered by radioastronomers, supersonic jets from active galactic nuclei and stars remain for many aspects an open problem in astrophysical plasmas. A number of basic physical questions have been understood and modelled in their general features. However, due to the presence of intrinsically highly nonlinear phenomena and to the incomplete determination of the physical parameters involved, a detailed analysis of their dynamics and emission mechanisms is far from being accomplished.

As an additional important fact, we must consider that dynamics and radiation emission in jets are combined in a peculiar way: the emission we observe is just a tiny fraction of the total energy of the flow that we estimate by assuming that the general structure is produced by the central object, either continuously or in a single outburst. This means that we have to reconstruct the dynamics observing its very partial transformation in radiation.

Nevertheless, if radiation morphologies are related to the bulk flow, we must find an explanation for its dynamics yielding a direct formation of those structures in the underlying "invisible" flow.

The common properties of jets at different scales can be summarized as follows:
a. jets are collimated over very large distances (up to some megaparsec), with aspect ratios up to 100;
b. collimation begins in the very center of the associated AGN, at distances $\lesssim 1$ pc;
c. jets remain collimated and dynamically stable even when there are signs of interaction with external environment and sharp bends;
d. strong evidence exists that in many cases jets are relativistic, especially in the region near the core;
e. correlation has been found between jets and the optical activity of the associated galactic nucleus, namely in broad and narrow emission lines, whose spatial distribution is aligned with the general direction of the jet.

More complete data on collimated outflows will most likely come from stellar jets; but not all the aspects of the two classes show a clear correspondence; this is especially true because their radiations (from which we derive our information) are well different, thermal for stellar jets and nonthermal for AGN jets. In addition the energetics are quite incommensurable, AGN jets requiring a relativistic treatment that appears to be the only way to satisfy the large demands.

To build reliable physical models with such uncertain definition of the relevant parameters may appear an impossible task. On the other hand jets represent a crucial element in our understanding of AGNs and thence galaxy evolution; at the same time, together with other types of astrophysical outflows they are an interesting challenge for plasma astrophysics.

The three basic physical questions concerning jets are:
1. origin, acceleration, collimation

2. propagation, stability, interaction with the surrounding medium, matter entrainment, energy and momentum transport, formation of radiation structures
3. termination, formation of the extended radio lobes.

Obviously along these point the origin of nonthermal and thermal emission must be considered. In this review the discussion is concentrated on some of the points listed in 2.

Many studies of jet stability have been presented in the past, most of them analytic and therefore applicable to the linear phases of evolution. Most of the calculations are based on the hydrodynamical or magneto-hydrodynamical equations, but the case of double adiabatic equations has been explored, and the validity for a fluid approximation has been tested. The main conclusion has been that supersonic flows are unstable, but the time scales of growth can be longer than the dynamical scales of propagation; a strong magnetic field can also stabilize the flow [1].

With the advent of supercomputers several authors have used hydrodynamical [2,3] and magneto-hydrodynamical numerical codes to evolve time-dependent jets. The equations used in these codes do not include any physical viscosity or resistivity, but it should be kept in mind that these effects will be in any case present because of the intrinsic (and unavoidable) numerical dissipation of the schemes adopted. This is a sharp limitation of simulation of astrophysical fluids, since their real physical viscosity must be extremely small, while the largest important spatial scales are very large, and therefore the range of spatial scales of astrophysical turbulent motions can span ten or more orders of magnitude; and it is not likely that any presently-conceived computer will allow for the required spatial resolution.

Nevertheless simulations well represent qualitatively the radiomaps, although the object of their calculations are density and velocity of the flow, while observations refer to the nonthermal electron component. One can interpret such results by saying that emission processes are strongly connected to the flow dynamics, that in its turn is dominated by fluid effects. One observes formation of conical shocks along the flow (related to knots ?) and of a large bow shock that produces in several cases a cocoon around the head of the jet (related to radio lobes ?). Shocks appear to be created by the perturbations generated at the head of the jet and hence, according to the linear theory, must be short lived transients as being backward moving modes. They seem persistent in the simulations simply because of the short scale of calculations: they are not shown to propagate backward along the whole jet, where they are observed. Instead it is quite possible that Kelvin-Helmholtz instabilities in supersonic flows do in fact evolve in a more violent way after local excitation along the whole jet; the unstable modes are basically similar to those of the quoted simulations, but are moving downstream to the flow.

In order to investigate such possibility we have performed a different type of simulation. We study a limited section of an infinite jet in pressure equilibrium with the external matter, assuming that the system is perturbed by a periodic mode: we can therefore adopt periodic boundary conditions along the flow. Consistently there are no limitations on the time for which calculations can be progressed. In these one can follow the long-term evolution of the system, which eventually attains a quasi steady state. Performing a wide coverage of the parameter space, we are able to study the physical processes leading to mixing and momentum exchange between jet and external medium and the formation of morphological structures that can be related to observations.

In this review we discuss in Section 2 the numerical scheme adopted, and in Section 3 the results obtained for the evolution of Kelvin-Helmholtz instabilities.

2. NUMERICAL SCHEME

The system of compressible, inviscid fluid (Euler) equations has been integrated with a two-dimensional version of the Piecewise Parabolic Method (PPM) of Colella & Woodward [4]. This method belongs to the class of higher-order Godunov methods [5], which are very effective in simulating the behavior of compressible fluids at transonic speeds. The scheme combines higher-order interpolants of the physical variables with an approximate Riemann solver to guarantee maximal resolution (i.e., minimal dissipation) near shock discontinuities, while preserving higher-order accuracy in smooth portions of the flow. The main advantage of using PPM lies in its special upwinding properties, which minimize the numerical dissipation required to stabilize the scheme for any given grid resolution. This means that any discontinuity in the flow, such as a shock, can be represented very accurately, without spreading it over too many grid points as it happens for other schemes, increasing the overall accuracy of the scheme [4]. As already noted in the Introduction,

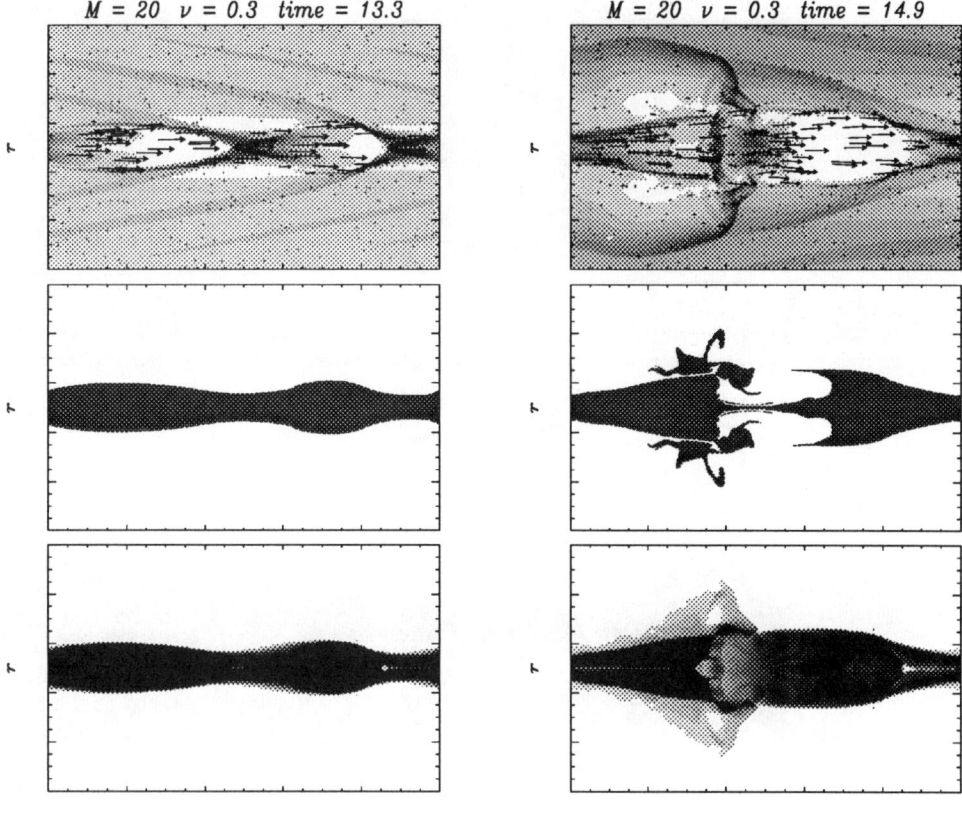

Figure 1

it is important to remember that, although the PPM scheme formally solves the inviscid, adiabatic fluid equations, both viscous and thermal disspation are introduced by the above-mentioned numerical dissipation. In practice, it is impossible to estimate the functional form of the dissipation terms by formal truncation error analysis: some estimates can be obtained by comparison with the results of other codes whose physical dissipation is known and resolved (see, e.g., [6]). It is worth noting that it has been recently conjectured that the numerical viscous and thermal dissipation introduced in monotonic finite difference schemes, such as PPM, is actually a good candidate for a subgrid scale model of turbulent dissipation; if this conjecture is correct, then schemes such as PPM are far more accurate in their depiction of fluid properties near the diffusive cutoff than is usually assumed. In any case one can regard simulations of the kind presented here as similar to laboratory experiment, whose results should be used for developing our physical intuition regarding the behaviour of astrophysical jets and understanding their complicated nonlinear fluid behaviour.

3. EVOLUTION OF KELVIN–HELMHOLTZ INSTABILITIES

In this section we present the results of numerical simulations aimed at following the long term evolution of Kelvin–Helmholtz instabilities. We have performed two kind of simulations: in the first case we investigated the evolution of the axisymmetric modes of a cylindrical jet [7], while in the second case we analyzed symmetric and antisymmetric perturbations of a 2-D slab jet in Cartesian geometry [8]. It has been often argued that the evolution of the antisymmetric modes could resemble that of the helical mode of a cylinder, so that 2-D results can give some insight into the computationally far more complex 3-D problem, although

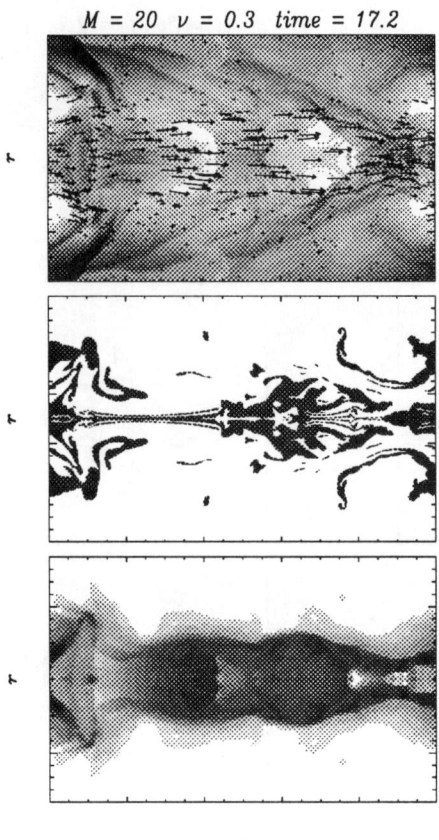

Figure 1 – Gray–scale plots of the cylindrical jet profile in the r,z plane for $M = 20$, $\nu = 0.3$ at three different times $t = 13.3, 14.9, 17.2$. Upper panel: velocity field superimposed on the density profile (dark grey: high density, light grey: low density). Medium panel: profile of the *tracer f*, marking the jet material (dark) with respect to the external medium (white). Lower panel: profile for the longitudinal velocity V_z, (dark grey: high velocity, light grey: low velocity).

the nonlinear evolution towards turbulence is conceptually different in the two cases.

The initial jet flow is separated by the external medium by a smooth velocity shear layer. The jet density can be different from that of the outside medium (the parameter ν represents the ratio between external and internal density) and its profile is identical to the velocity one.

To this initial configuration we apply a perturbation given by the superposition of longitudinally periodic transverse velocity disturbances; in this way we excite a spectrum of modes going from an upper limit given by the integration domain size to a lower limit chosen such as it can be still adequately resolved by the numerics. As stated above, the perturbations are axisymmetric in the cylindrical case, while they can be symmetric, antisymmetric or a general combination in the slab case. As for the boundary conditions, we choose periodic conditions in the longitudinal direction and outflow conditions at the outer boundaries. The choice of periodic conditions means that we are following the temporal evolution of a portion of an infinite jet. This choice, which is different from that of other authors who studied the spatial evolution of the perturbation [9], allowed us to follow the evolution of the instability for a long time until the system attains a quasi steady state.

The relevant control parameters for the jet flow are the initial jet Mach number, $M \equiv V_0/c_s$ ($V=$ jet velocity, $c_s=$ sound speed), and the density ratio ν between external medium and jet. We have performed a

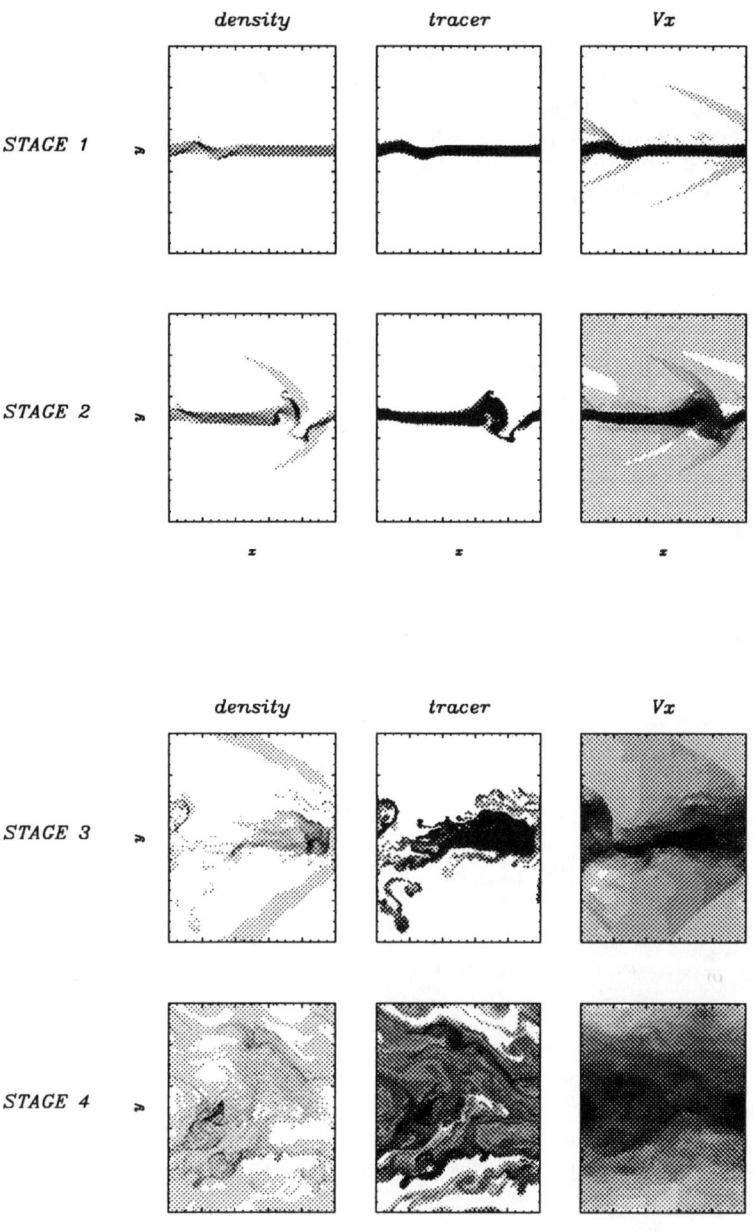

Figure 2 – Gray scale representations of the density, the "tracer," and the longitudinal velocity at four different times for the "heavy jet" case ($M = 10$, $\nu = 0.1$). The four times, chosen to be representative of the four evolution stages described in the text, are $t_1 = 6.2$, $t_2 = 8$, $t_3 = 11.6$, $t_4 = 40.3$. The gray scale is designed to represent the highest values of the variables in black, and the lowest values in white.

wide coverage of the parameter space defined by them both in the cylindrical case and in the slab case and our results show that, in both cases, we can distinguish four typical phases in the jet evolution. These phases are illustrated in the sequences of gray-scale images of the density distribution, the tracer distribution, and the longitudinal velocity (shown in Figs. 1, 2). These sequences show very distinctive qualitative changes in jet morphology, typical of the different phases, many of which can also be examined from a quantitative perspective by plotting the transverse distributions of the averaged longitudinal velocity and longitudinal momentum density for a series of time steps (Fig. 3).

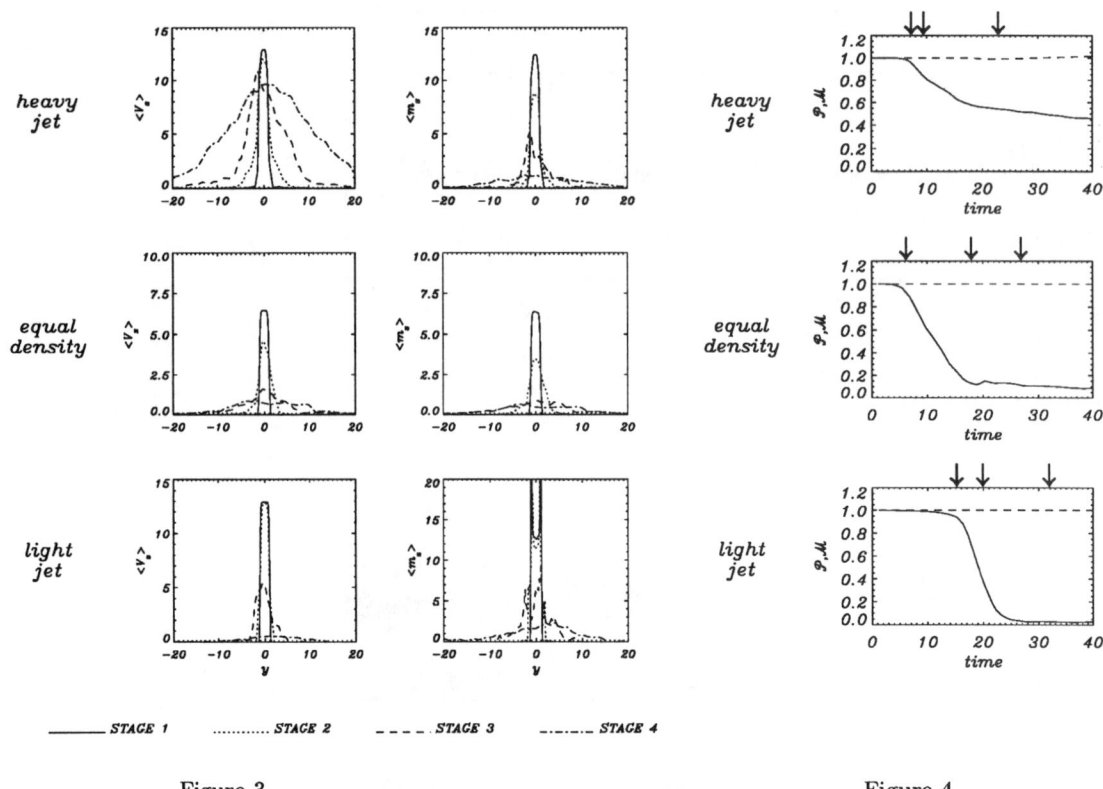

Figure 3

Figure 4

Figure 3 – Plots of the (x-)averaged longitudinal velocity $<V_x>$ (left panels) and of the (x-)averaged longitudinal momentum density $<m_x>$ (right panels) as functions of the transverse coordinate y for a slab jet for the three cases ($M = 10$, $\nu = 0.1$; $M = 5$, $\nu = 1$; $M = 10$, $\nu = 10$) chosen as representative of the heavy jet, the equal density jet, and the light jet, respectively.

Figure 4 – Plots of the total longitudinal momentum of the jet material $\mathcal{P}$ (solid lines) and of the total mass of the jet material $\mathcal{M}$ (dashed lines) as functions of time. $\mathcal{P}$ and $\mathcal{M}$ are given as fractions of their initial values. The three panels refer to the same three cases shown in the previous figures; the arrows indicate the onset of Stages 2, 3, and 4.

The first phase, which we can call linear phase, is characterized by the growth of modes which are excited by the initial perturbation, and which evolve in accord with the linear theory. In the latter portion of this stage, internal shocks form, due to the growth of the internal modes. The shocks have a biconical shape for the axisymmetric modes of the cylinder and are alternate in the case of the antisymmetric modes of the slab. In this first phase there is little exchange of energy or momentum between the jet and the external fluid. In the second phase, which we call acoustic phase, the global deformations of the jet increase in amplitude, and become sufficiently large that they drive shocks into the external medium, which in turn strongly radiate acoustic waves into this medium, thus leading to a transfer of both energy and momentum from the jet to the exterior of the jet. The momentum loss can be quantified considering the temporal evolution of the jet longitudinal momentum shown in Fig. 4. The extent to which acoustic waves manage to brake the jet depends sensitively on the density ratio: while heavy jets tend to loose very little momentum in this way, light jets can loose up to 80% of their momentum during this stage. In the third phase, which we call mixing phase, the behaviour depends sensitively on the kind of perturbation and on the density ratio. For the axisymmetric perturbations of a cylinder, we observe, for every density ratio, vigourous turbulent mixing between jet and the external medium. In the case of antisymmetric perturbations of a slab jet, we observe a break up of the jet in the light case, and again vigourous turbulent mixing in the heavy case. More precisely, in light slab jets the global oscillation seen in Stage 2 evolve into a few strong shocks which completely destroy the jet; the result is highly clumped jet material, weakly mixed with the previously external fluid. These clumps of jet material in turn interact with the external fluid via shocks, which dramatically slow them down to very low (subsonic) velocities, typically of order 10% of the original jet velocity. In the case of dense jets (small ν), these shocks merge to form one single strong shock within the periodic domain. This shock, rather than disrupting the jet, acts like a piston whose wake generates strong turbulent motions behind its head; these act to vigorously mix the jet material with the external fluid. These motions are highly effective in accelerating the external fluid, as can be seen by simply considering the gray–scale images of the longitudinal velocity (which show a clear homogenization of this velocity component over much of the computational domain during this stage); this homogenization of the longitudinal velocity component is far more effective than the mixing of either density or the tracer, whose distributions remain quite inhomogeneous. Finally, in the last stage the jet attains a statistically quasi–stationary state, whose form is sensitively dependent upon the density ratio ν. In the case of light jets, one observes a final state in which the jet is fundamentally disrupted; the turbulent flow excited in the surrounding fluid entrains the jet fluid, and dissipates virtually all of its energy and momentum. In contrast, the final state of heavy jets is simply a well-defined, much-broadened flow, in which the longitudinal velocity is quite uniform, but consists of both original jet fluid as well as entrained exterior fluid.

Although these four phases are present for any parameter choice, their relative durations are different for different values of the parameters. In particular we can note that the beginning of the mixing phase is delayed for high Mach number, light jets, while it is very quick for slow and heavy jets. This behaviour is sketched for the cylindrical case in Fig. 5, and the behaviour for the slab case is very similar. Finally we can compare in a more quantitative way the various terminal states, looking at the ratio between the peak initial and final longitudinally–averaged longitudinal velocities and momenta. In Fig. 6 these ratioes are represented as contour plots in the parameter phase plane. The figure immediately illustrates several clear trends:

1. Changes in the jet momentum depend primarily on the external fluid–to–jet density ratio, ν.

2. Changes in jet velocity have a more complex dependence. For Mach numbers roughly less than 15, the final jet velocity depends primarily on the density ratio ν. For larger Mach numbers, the jet velocity depends instead primarily on M as long as the density ratio is ≈ 1 or more, i.e., for light jets; heavy jets instead maintain their primary final velocity dependence on ν.

These trends make clear physical sense because in the case of heavy jets, entrainment of external light fluid is largely responsible for the ultimate changes in jet velocity; in contrast, this process cannot function for light jets.

4. CONCLUSIONS

In this review some examples of the study of the plasma physics of relativistic jets from AGN have been discussed. We are still at the very beginning of a complete representation of the rich variety of phenomena contributing to shape the morphology and the energetics of these objects. However the possibility of simulating nonlinear phenomena by numerical experiments is a new important resource for theorists. With the availability of fast computers many new physical processes will be introduced in the study increasing the diagnostics for interpreting the growing amount of observational data.

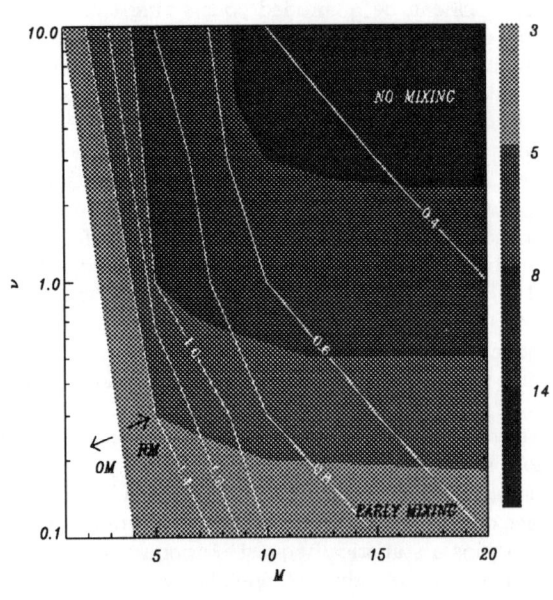
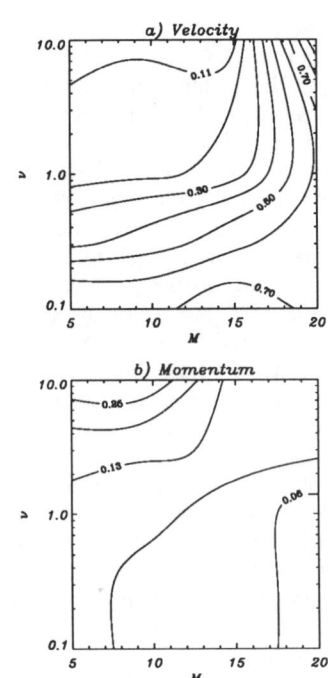

Figure 5

Figure 6

Figure 5 – Grey contour plot showing the onset of the mixing regime as function of M and ν. The labels on the contour lines mark the dominant wavenumber at the beginning of the mixing phase. The dot-dashed line separates the region where the ordinary mode (OM) and reflected modes (RM) dominate.

Figure 6 – Contour plots of the peak of the averaged longitudinal momentum and velocity (expressed as fractions of the corresponding initial values) in the parameter plane (M, ν), as attained in the terminal quasi–steady states reached during stage 4 for the anti–symmetric mode.

REFERENCES

[1] Birkinshaw M., in *Beams and Jets in Astrophysics*, Ed. P.A. Hughes, Cambridge University Press, Cambridge
[2] Norman M.L., Smarr L., Winkler K.H.A., Smith M.D., 1982, A&A 113, 285
[3] Norman M.L., Winkler K.H.A., Smarr L., 1984, in: Bridle A.H., Eilek J.A. (eds.) Physics of Energy Transport in Extragalactic Radio Sources (p. 150). NRAO Workshop 9
[4] Colella P., Woodward P.R., 1984, J. Comp. Phys. 54, 174
[5] Godunov S.K., 1959, Mat. Sb. 47, 271
[6] Cattaneo, F., Brummel, N.H., Toomre, J., Malagoli, A., Hurlburt, N.E., 1991, *Ap. J.* **370**, 282.
[7] Bodo G., Massaglia S., Ferrari A., Trussoni E., 1994a, A&A 283, 655
[8] Bodo G., Massaglia S., P. Rossi, A. Ferrari, A. Malagoli, R. Rosner, 1994, *Astron. Astrophys.*, submitted
[9] Norman M.L. and Hardee P.E., 1988, *Astrophys. J.* **334**, 80

MAGNETOHYDRODYNAMIC WAVES IN FUSION AND ASTROPHYSICAL PLASMAS

J.P. Goedbloed

FOM-Institute for Plasma Physics, Nieuwegein, The Netherlands

&

Free University Amsterdam & Utrecht University

Abstract

Macroscopic plasma dynamics in both controlled thermonuclear confinement machines and in the atmospheres of x-ray emitting stars is described by the equations of magnetohydrodynamics. This provides a vast area of overlapping research activities which is presently actively pursued.

In this lecture we wish to concentrate on some important differences in the dynamics of the two confined plasma systems related to the very different geometries that are encountered and, thus, the role of the different boundary conditions that have to be posed. As a result, the basic MHD waves in a tokamak (a periodic toroidal plasma tube) are quite different from those found in a solar magnetic flux tube (closed off by the photosphere in which the magnetic field lines are essentially anchored by the huge inertia as compared to that of the corona). The result is that, whereas the three well-known MHD waves (Alfvén, slow, and fast magnetosonic) can be traced stepwise in the curved geometry of a tokamak, their separate existence is eliminated right from the start in a line-tied coronal loop because line-tying in general conflicts with the phase relationships between the vector components of the three velocity fields. The consequences are far-reaching, viz. completely different resonant frequencies and continuous spectra (associated with the possibility of resonant Alfvén wave heating of the solar corona), absence of rational magnetic surfaces, and irrelevance of local marginal stability theory (Mercier and ballooning) for coronal magnetic loops. This qualitative difference between the two systems is responsible for the quite different behavior at the onset of instability, viz. a regime of 'enhanced MHD activity' for tokamaks and violent solar flares for coronal plasmas.

1 Introduction

1.1 Magnetic fields in astrophysics

In the standard view of nature nuclear forces are responsible for the existence of positively charged nuclei and negatively charged electrons so that beyond distances of the order of 10^{-15} m essentially only electrostatic and gravitational forces remain. In turn, the electrostatic forces are responsible for the existence of atoms and molecules, the building blocks of ordinary matter, which are electrically neutral. Hence, beyond distances of the order of 10^{-9} m only the gravitational force remains giving rise to the existence of stars ($\sim 10^8$ m), the solar system ($\sim 10^{13}$ m), galaxies ($\sim 10^{20}$ m), clusters of galaxies ($\sim 10^{23}$ m), and, finally, the universe ($\sim 10^{26}$ m) itself. This view, which might be termed the bureaucratic view of nature (a problem is solved when it is moved onto somebody else's desk), is wrong for many reasons. Not the least of them is that it makes science on the intermediate distance scales (10^{-9}–10^{8} m), where we happen to live, extremely uninteresting, degrading e.g. biophysics and solid state physics to applied sciences.

There is a more obvious reason, however, why the standard view of nature is wrong. This pertains to the fact that the universe does not consist of ordinary matter alone but, as is well known, it consists for more than 90 % of plasma. Like ordinary matter, plasma is electrically neutral, but the nuclei and electrons are not tied together in atoms or molecules but move about freely and collectively as fluids. The large-scale result is that wherever there is plasma there are electric currents and *magnetic fields*. In contrast to the electrostatic and gravitational forces, which decay like r^{-2}, the magnetic interaction of plasma and magnetic field is not governed by such simple laws. Rather, plasma and the ('frozen-in') magnetic field are literally transported together over large distances, interacting with whatever other plasmas and magnetic fields are encountered. A particularly clear example, not too far away, is the interaction of the solar wind plasma with the earth's magnetosphere. Here, at a distance of 1 AU away from the sun and 5–10 earth radii away from the earth, magnetic forces completely dominate the electrostatic and gravitational forces. This gives rise to the existence of the bow shock, the magnetopause, the peculiar stretched-out 3D structure of the magnetosphere itself (without any spatial symmetry!), and dynamical phenomena like flux transfer events (i.e. reconnection of the magnetic fields of the solar wind and of the earth) on the day-side and expulsion of plasmoids on the night-side of the magnetosphere, etc. etc. Such rich dynamics should be considered as generic for plasma astrophysics on all of the spatial scales far beyond that of the solar system, witness the many exciting new phenomena discovered during the last decades.

There is an important geometric aspect to magnetic fields which is related to the above-mentioned absence of symmetry. In fact, the spherical symmetry which pervades atomic physics (due to the Coulomb force) and gravity is not present at all on the plasma scale: the basic law $\nabla \cdot \mathbf{B} = 0$ for magnetic fields simply forbids it. This is particularly evident in the beautiful x-ray observations of the dynamics of the solar coronal plasma recently made with the Japanese satellite Yohkoh [1] showing a constant battle of the plasma (and associated magnetic field) escaping from the spherical gravitational attraction of the sun. Violent flares are amongst the most striking characteristics. A more serene illustration are the coronal streamers observed by the participants of the conference during the eclipse of November 3 1994 (the day before the oral presentation of this paper).

Here, we wish to concentrate on this geometric aspect of plasma dynamics in order to enable a comparison between astrophysical and laboratory fusion plasmas and to find out why these plasmas behave so differently with respect to macroscopic wave propagation and stability. To that end, a leading order model is formulated which maximises on the differences between the dynamics of plasma confined in tokamaks and of plasma in solar coronal magnetic flux loops. This model involves the equations of magnetohydrodynamics (Sec. 1.2) and suitable boundary conditions pertaining to the two geometries (Sec. 2.1).

1.2 Equations of magnetohydrodynamics

The equations of magnetohydrodynamics (MHD) may be considered as the confluence of the fluid or gas dynamics equations with Maxwell's equations or, rather, the pre-Maxwell equations since the displacement current can be neglected for most of plasma astrophysics. This permits one to drop Poisson's equation, to eliminate the current by means of Ampère's law, $\mathbf{j} = \mu_0^{-1} \nabla \times \mathbf{B}$, and to eliminate the electric field by means of Ohm's law in the co-moving frame, $\mathbf{E} = -\mathbf{v} \times \mathbf{B} + \eta \mathbf{j}$. The resulting equations are just the usual conservation laws of fluids (extended with magnetic field contributions) + Faraday's law (which becomes the conservation law for magnetic flux). We then have, subsequently, conservation of mass:

$$\frac{\partial \rho}{\partial t} = -\nabla \cdot (\rho \mathbf{v}), \tag{1}$$

conservation of momentum:

$$\rho \left(\frac{\partial}{\partial t} + \mathbf{v} \cdot \nabla \right) \mathbf{v} = -\nabla p + \frac{1}{\mu_0} (\nabla \times \mathbf{B}) \times \mathbf{B} + \rho \mathbf{g}, \tag{2}$$

(near) conservation of entropy:

$$\left(\frac{\partial}{\partial t} + \mathbf{v} \cdot \nabla \right) p = -\gamma p \nabla \cdot \mathbf{v} + \frac{\gamma - 1}{\mu_0^2} \eta (\nabla \times \mathbf{B})^2, \tag{3}$$

(near) conservation of magnetic flux:

$$\frac{\partial \mathbf{B}}{\partial t} = \nabla \times (\mathbf{v} \times \mathbf{B}) - \frac{1}{\mu_0} \nabla \times (\eta \nabla \times \mathbf{B}), \qquad \nabla \cdot \mathbf{B} = 0. \tag{4}$$

This set of equations determines the evolution of the macroscopic variables $\rho(\mathbf{r}, t)$, $\mathbf{v}(\mathbf{r}, t)$, $p(\mathbf{r}, t)$, and $\mathbf{B}(\mathbf{r}, t)$.

The evolution equations (1)–(4) are *nonlinear* partial differential equations, which involve *dissipation* through the Ohmic heating and magnetic diffusion terms (with η), and which acquire their full prescriptive power only when boundary conditions for a finite geometry are specified, i.e. when *inhomogeneity* is introduced. At first, neglecting these effects, we obtain the well-known three waves of linearised ideal MHD for an infinite homogeneous medium, viz. the incompressible *Alfvén waves* with respect to frequency and mode characteristics sandwiched in between the compressible *slow and fast magnetosonic waves*. This central position of the Alfvén waves is crucial for all macroscopic plasma dynamics. In particular, it guarantees that the highly localised propagation characteristics of the Alfvén waves, and the associated singular behavior of the $\mathbf{B} \cdot \nabla$ operator for inhomogeneous plasmas, favor the development of small-scale structures associated with the large-scale dynamics. This is necessary in order for the dissipative terms in Eqs. (3) and (4) to play a role at all: without such small-scale structures, the magnetic Reynolds number $R_m \equiv \mu_0 v L / \eta$ would be too large ($\sim 10^9$ in fusion machines and even larger in astrophysical plasmas because of the enormous distance scales L encountered, e.g. already 10^{13} for solar coronal

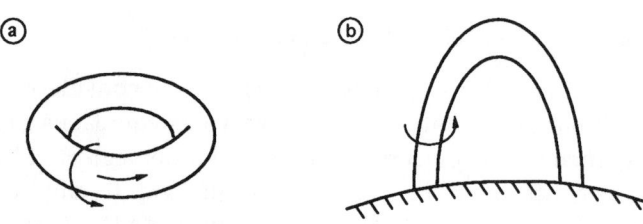

Figure 1: Generic plasma structures: a. Tokamak and b. Coronal magnetic loop (arrows indicate periodic directions).

plasmas) to permit the heating and reconnection rates observed in tokamaks and solar coronal plasmas.

Our program is to analyse the MHD waves, in particular the Alfvén waves, for the simplest configuration compatible with the tokamak and the solar coronal geometries respectively (Sec. 2) and, next, to compare their properties (Sec. 3). This will lead to an amazingly different physical picture for the two. It will also resolve a basic contradiction for Alfvén wave propagation in magnetic loops which was recently pointed out in the context of magnetospheric physics [2]. Once this most elementary part of the influence of inhomogeneity is fully understood it makes sense to further generalise to more realistic geometries and to consider the effects of dissipation and nonlinearity. This part of the program, which will only be indicated (end of Sec. 3), is presently actively pursued.

2 MHD waves in coronal magnetic loops

2.1 Influence of line-tying

Consider the two generic plasma configurations sketched in Fig. 1: a toroidal tokamak and a solar coronal loop with ends anchored in the photosphere. The latter statement reflects the enormously different physical conditions in the corona as compared to those prevailing in the photosphere: densities of the order 10^{-15}–10^{-11} kg m^{-3} in the corona and $\sim 10^{-4}$ kg m^{-3} in the photosphere. This implies that the photosphere is essentially immobile on the time scale of Alfvén wave propagation along the loop: the loop is *line-tied* at the ends. Hence, the main difference between tokamaks and coronal loops with respect to Alfvén wave propagation is due to the difference between toroidal periodicity (tokamak) and line-tying (coronal loop). This leads to a completely different dynamics, as we will see, eliminating the hope of easy success by just transferring tokamak knowledge to the field of solar physics. Let us work out the consequences of these different boundary conditions for the propagation of waves (associated with the possibility of MHD spectroscopy [3], [4]), the continuous spectra (associated with Alfvén wave heating [5]), and instabilities (MHD activity, disruptions, and flares). In particular: how do the MHD waves differ for the two configurations?

We will eliminate all effects that do not explicitly bear on this question, viz. nonlinearity, dissipation, and inhomogeneity except for that implicit in the use of periodic or line-tying boundary conditions. Neglecting all curvatures, the model then simply becomes that of a homogeneous plasma in a box ($0 \leq x \leq a$, $0 \leq y \leq b$, $0 \leq z \leq L$), confined in the x-direction, periodic in the y-direction, and either *periodic* (tokamak) or *line-tied* (coronal loop) in the z-direction. The magnetic field will be assumed to be oblique with respect to the end planes: $\mathbf{B} = (0, B_y, B_z)$. This will turn out to have quite different effects on the two model configurations. Consequently, the MHD wave equation

$$\rho \frac{\partial^2 \boldsymbol{\xi}}{\partial t^2} = \gamma p \nabla \nabla \cdot \boldsymbol{\xi} + \mathbf{B} \times (\nabla \times (\nabla \times (\mathbf{B} \times \boldsymbol{\xi}))) \tag{5}$$

is to be solved subject to the usual boundary conditions for the x- and y-directions, but with respect to the z-direction the boundary conditions are fundamentally different, viz.:

$$\boldsymbol{\xi}(0) = \boldsymbol{\xi}(L), \qquad \partial_z \boldsymbol{\xi}(0) = \partial_z \boldsymbol{\xi}(L) \qquad \text{for tokamaks}, \tag{6a}$$

and

$$\boldsymbol{\xi}(0) = \boldsymbol{\xi}(L) = 0 \qquad \text{for coronal loops}. \tag{6b}$$

The solutions may be written as

$$\boldsymbol{\xi}(x,y,z,t) = \sum_{k_x} \sum_{k_y} \tilde{\boldsymbol{\xi}}_{k_x,k_y}(z) \sin k_x x \; e^{i(k_y y - \omega t)}, \tag{7}$$

where the coefficients $\tilde{\boldsymbol{\xi}}_{k_x,k_y}(z)$ are to determined such that either the BC's (6a) or the BC's (6b) are satisfied.

For tokamaks, the BC's (6a) simply imply that $\tilde{\boldsymbol{\xi}}(z) \sim e^{ik_z z}$ with quantized longitudinal wave number: $k_z = n2\pi/L$. This leads to the usual MHD dispersion equation with the well-known *slow/Alfvén/fast*-wave triplet as solutions. The essential point is that the finite tokamak geometry to leading order just produces quantization of the wave vectors. Further geometric effects, associated with inhomogeneities, currents, field line curvatures, etc., may be introduced stepwise by perturbation of this leading order model.

For a line-tied coronal loop the procedure just outlined does not work since the BC's (6b) exclude the existence of separate slow, Alfvén, and fast waves right from the start. Instead, solutions may be obtained by means of the following procedure: Exploit the well-known dispersion diagram $(\rho a^2/B^2)\omega^2 = \lambda(k_z)$ for the three basic waves to construct six values of $\kappa_i(\lambda)$ for a given value of λ, as indicated in Fig. 2. If one now tries to fit the line-tying BC's by a superposition of the corresponding six waves (two of each kind, propagating in opposite directions),

$$\tilde{\boldsymbol{\xi}}(z) = \sum_{i=1}^{6} \hat{\xi}_i (\mathbf{e}_x - i\gamma_i \mathbf{e}_\perp - i\delta_i \mathbf{e}_\parallel) e^{i\kappa_i z}, \tag{8}$$

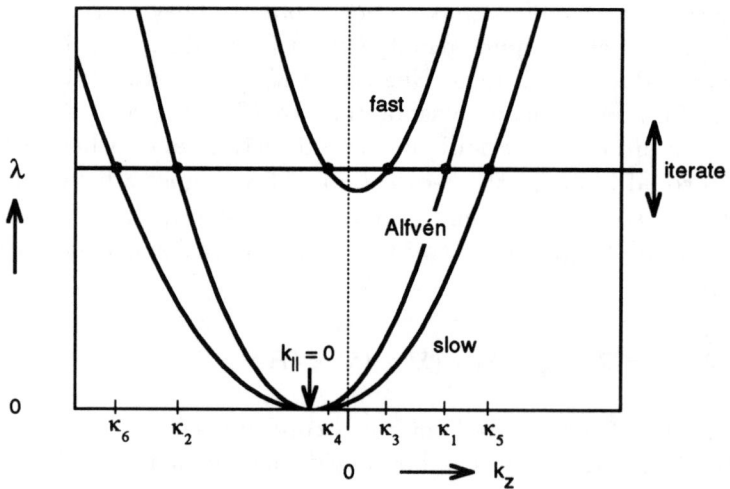

Figure 2: Construction of eigenvalues for a line-tied loop from the dispersion equation $\lambda(k_z)$ of the three MHD waves in an unconstrained plasma.

where the γ_i's and δ_i's are the known polarization factors of the three basic waves, this will lead to a 6×6 algebraic system for the six unknowns $\hat{\xi}_i$. The determinant of this system provides the required *dispersion equation*:

$$\Delta(\lambda) = 0. \tag{9}$$

This is a highly nonlinear and implicit relation in λ that has to be solved numerically by means of a root finding algorithm. This solves the general problem of MHD waves in a coronal loop.

2.2 Alfvén wave continua and resonant heating

In order to gain some understanding of the new types of waves which propagate in a line-tied coronal loop, we will simplify the problem by exploiting the fact that *the pressure is negligible in coronal loops* so that the slow wave contributions may be neglected. The original six-wave problem then transforms into a four-wave one with simple explicit expressions for the κ_i's:

$$\kappa_{1,2} = \pm\alpha - (B_y/B_z)k_y, \qquad \alpha \equiv \frac{B}{B_z}\sqrt{\lambda} \qquad \text{(Alfvén: ballooning)}, \tag{10a}$$

$$\kappa_{3,4} = \pm\varphi, \qquad \varphi = \sqrt{\lambda - k_x^2 - k_y^2} \qquad \text{(fast: no ballooning)}. \tag{10b}$$

The dispersion equation, although of a striking beauty, still remains highly transcendental:

$$\sin\alpha L \,\sin\varphi L = \tau \left[\cos\alpha L \cos\varphi L - \cos(B_y/B_z)k_y L\right], \tag{11a}$$

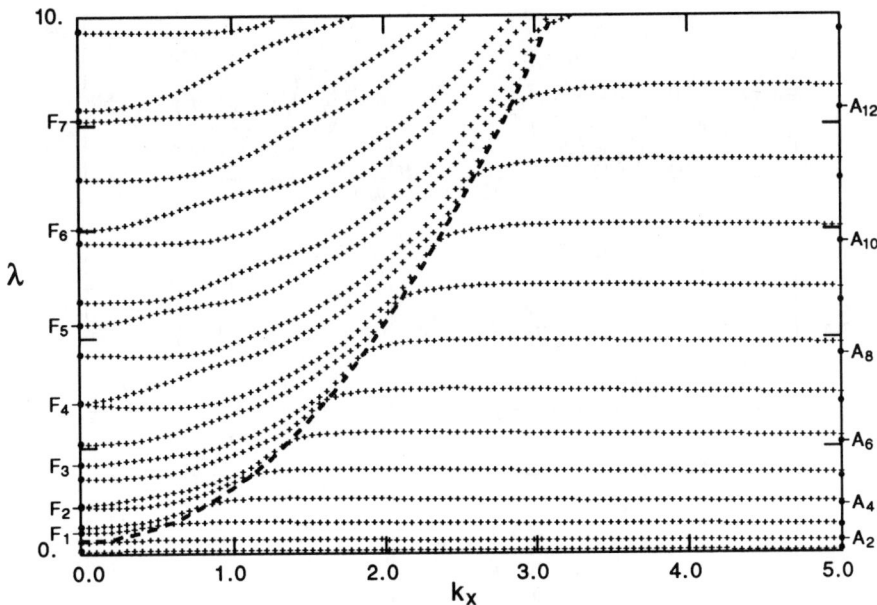

Figure 3: Dispersion diagram for a line-tied loop. The dashed line corresponds to $\varphi = 0$ where the fast wave contributions change from oscillatory to evanescent.

with a coupling parameter that also depends on the eigenvalue λ:

$$\tau \equiv \frac{2k_x^2 B_y^2 \sqrt{\lambda(\lambda - k_x^2 - k_y^2) B_z^2 B^2}}{(k_x^2 + k_y^2)^2 B^4 - [2k_y^2 \lambda + k_x^2(k_x^2 + k_y^2)] B_y^2 B^2 + \lambda(\lambda - k_x^2) B_y^4} . \tag{11b}$$

The latter expression shows why we had to introduce an oblique magnetic field: for $B_y \neq 0$ the Alfvén and fast wave components are intrinsically coupled in order to satisfy the line-tying BC's. In this process, the phase factor of the Alfvén components, which contains a constant part along the field lines (the so-called *ballooning factor*): $\kappa_{1,2} z + k_y y = \pm \alpha z + k_y [y - (B_y/B_z) z]$, is counteracted by that of the fast components, which do not have such a factor. This is why the MHD waves in a coronal flux tube become much more complicated than those in a tokamak.

Explicit numerical results for the eigenvalue $\lambda(k_x)$ are shown in Fig. 3 for a particular case (see Ref. [6] for details). For $k_x = 0$ (when $\tau = 0$), the Alfvén and fast waves are uncoupled but as soon as $k_x \neq 0$ the waves cannot be classified as either Alfvén or fast but they become a mixture. To the left of the dashed curve (where $\varphi = 0$) a web of coupled Alfvén-fast solutions is obtained. As k_x is increased, the fast wave components first become evanescent (when the curve $\varphi = 0$ is crossed) and, next, they become more and more localised in *photospheric boundary layers* at $z = 0$ and $z = L$, until finally, as $k_x \to \infty$, the pure Alfvén wave character is recovered. In the latter limit, singular Alfvén waves are obtained associated with cluster point frequencies.

The *cluster spectra* themselves are obtained by expanding the dispersion equation

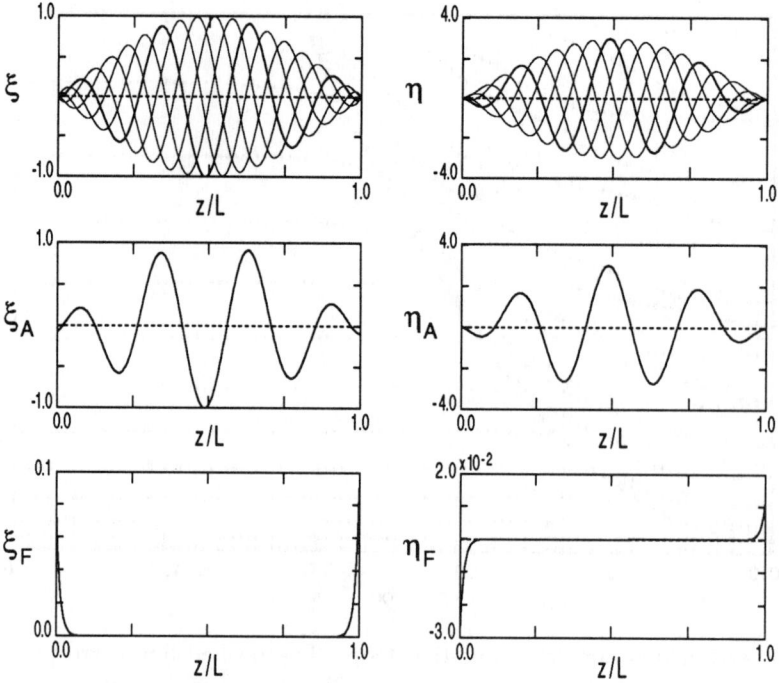

Figure 4: Eigenfunctions of cluster spectrum waves in a coronal loop ($n = 1$). Lower frames show the separate Alfvén and fast components.

for $k_x L \gg 1$ so that $\tau' \equiv -i\tau \approx [2n\pi/(|k_x|L)]B_y^2/B^2 \ll 1$:

$$\omega_n^2(k_x) \approx \left(\frac{n\pi}{L}\frac{B_z}{\sqrt{\rho}}\right)^2 \left(1 + \frac{4B_y^2/B^2}{|k_x|L}\right). \tag{12}$$

The associated *eigenfunctions* exhibit all the features that make Alfvén waves in coronal loops a species by itself, quite distinct from those in a tokamak:

$$\boldsymbol{\xi} \cdot \mathbf{e}_x \equiv \xi(x,y,z,t) \approx -\tfrac{1}{2}i\frac{B}{B_y}\tau' C \sin k_x x$$

$$\times \left[\left\{ \cos(n\pi z/L) - i\frac{B^2}{B_y B_z}\frac{k_y L}{n\pi}\sin(n\pi z/L) \right\} e^{ik_y[y-(B_y/B_z)z]} \right.$$

$$\left. - \left\{ e^{-|k_x|z} + (-1)^n e^{-i(B_y/B_z)k_y L} e^{|k_x|(z-L)} \right\} e^{ik_y y} \right], \tag{13a}$$

$$i\boldsymbol{\xi} \cdot \mathbf{e}_\perp \equiv \eta(x,y,z,t) \approx C \cos k_x x$$

$$\times \left[\left\{ (1 + \tfrac{1}{2}i\tau')\sin(n\pi z/L) - \tfrac{1}{2}\tau'(1-2z/L)\cos(n\pi z/L) \right\} e^{ik_y[y-(B_y/B_z)z]} \right.$$

$$\left. + \tfrac{1}{2}\tau'\left\{ e^{-|k_x|z} - (-1)^n e^{-i(B_y/B_z)k_y L} e^{|k_x|(z-L)} \right\} e^{ik_y y} \right]. \tag{13b}$$

Notice the rapid oscillation in the x-direction, the envelope $\sin(n\pi z/L)$ of the Alfvén component with the ballooning factor, and the exponentially decaying boundary layers of the fast component which are coupled in through the coupling parameter τ' which is small in the limit of large k_x. The illustration of Fig. 4 will help to clarify the different contributions.

The importance of the cluster spectra is that they present the simplest example of the approach to continuous spectra. For genuine inhomogeneity in the x-direction, the infinitely rapid oscillations of the homogeneous plasma are replaced by infinitely localised perturbations at the resonant layers and the cluster frequencies spread out to form *the Alfvén wave continuous spectra of line-tied coronal loops* [7], which we here present for the analogous case of a line-tied cylindrical loop:

$$\omega_n^2(r) = (n\pi/L)^2 B_z^2(r)/\rho(r). \tag{14a}$$

This expression is to be compared with that of the continua for a straight tokamak of length $L = 2\pi R$:

$$\omega_{n,m}^2(r) = \left[mB_\theta(r)/r + 2n\pi B_z(r)/L\right]^2/\rho(r). \tag{14b}$$

This enormous difference, of course, reflects that of the underlying dynamics and energetics of the two configurations. Whereas the two transverse components ξ and η of the displacement in an Alfvén wave are naturally out of phase, like in a tokamak, permitting the typical $\mathbf{k}\cdot\mathbf{B}$ dependence of Eq. (14b), the line-tying BC's have managed to force both ξ and η to vanish at the photosphere. We have seen that the fast magnetosonic boundary layer contributions are essential to do this. This explains why the differential equations describing low-β Alfvén wave continua in a coronal loop are essentially of fourth order in $\partial/\partial z$ whereas second order equations suffice for tokamaks. This resolves the problem raised by Hansen and Goertz [2], who correctly point out that the differential equations are of fourth order but incorrectly claim that this eliminates the continua. Also, it justifies the resonances found by Strauss and Lawson [8], which are of the form of Eq. (14a).

We now come to the following picture of resonant heating of line-tied coronal loops: Coronal magnetic loops are confined between granular regions of the photosphere, indicative of the underlying convective cells which continually pound at the boundaries of the loops. The kinetic energy of this convective motion is transferred to that of the compressive fast magnetosonic component which is precisely localised where it is needed, viz. in the boundary layers at the ends of the loop. The coupling to the Alfvén wave component at the resonant layer is immediate since both fast and Alfvén component are part of one and the same eigenfunction. As a result, compressive fast motion in the photosphere and below is effectively transferred to torsional Alfvén wave motion of large amplitudes in the corona (Fig. 4), where it is essentially converted into heat because of the formation of large gradients in the vortex and current sheets of the resonant layers.

One piece is missing: It has been argued [9] that the exciting convective motion (with turn-over times of minutes–hours) are too slow to couple to the resonant Alfvén waves of typical coronal loops (with periods of tens of seconds). However, this kind of energy transfer is very common for musical instruments: The violinist does not need to move his bow with the frequency of the string (fortunately), but the energy of the slow motion of the bow is transferred to the high frequencies of the string by a very effective stick-slip mechanism. This analogy suggests that we should look for similar mechanisms operating at the boundaries of the granular cells.

3 Comparison with tokamak dynamics

We have seen that the basic MHD waves in a line-tied coronal loop are quite different from those in a tokamak. This is very evident from the expressions (14a) and (14b) for the Alfvén wave continuous spectra. With respect to Alfvén wave heating, these expressions reveal a distinct advantage of coronal loops over tokamaks: the frequencies are more robust since they do not depend on the poloidal mode number m and the poloidal magnetic field B_θ. This implies that the position of the resonant layers is less dependent on details of the excitation. Moreover, whereas Alfvén wave heating in tokamaks requires an external antenna which should excite a fast compressive carrier wave over a sizeable portion of the outside volume (with concomitant generation of impurities!), Alfvén wave heating in coronal loops is intrinsically a 'cleaner' process since the compressive fast wave components are just confined to where they are needed, viz. at the photospheric boundary layers.

Another, apparently trivial, though extremely important, aspect of Alfvén waves is the fact that they become marginally stable when $\mathbf{k} \cdot \mathbf{B} \to 0$, at least if that condition can be satisfied. This leads to the usual condition of the existence of rational magnetic surfaces in tokamaks where local stability criteria are to be satisfied. The procedure is simply to expand the potential energy in powers of some small parameter, e.g. the inverse aspect ratio, which then straightforwardly leads to a huge zero order positive contribution associated with the field line bending energy of the Alfvén wave, essentially corresponding to the expression (14b). This contribution has to vanish or become small enough for pressure gradients and field line curvature contributions (which are smaller by at least one order of magnitude in the small parameter in tokamaks) to have any change of driving an instability. Hence, local stability investigations for tokamaks are concerned with regions surrounding rational magnetic surfaces where $\mathbf{k} \cdot \mathbf{B}$ (or rather $\mathbf{B} \cdot \nabla$) can be made to vanish.

From the construction of Fig. 2, where we have indicated the position $k_\parallel = 0$ where both the Alfvén and the slow wave become marginally stable ($\lambda = 0$) for a tokamak with the proper choice of the mode numbers, it is evident that this condition cannot be satisfied in a line-tied coronal loop. The presence of the stable fast wave components eliminates the possibility of constructing such marginally stable solutions. Consequently, local stability analysis (like that of interchanges and ballooning modes in tokamaks) makes no sense for coronal loops! Instead, the expression (14a) shows that the lowest energy state is that of the $n = 1$ Alfvén mode corresponding to the huge field line bending energy due to the line-tying BC's.

We come to the following qualitative picture of the behavior of tokamaks and coronal loops at the onset of instability. When local stability conditions are violated in a tokamak, in general a region of so-called enhanced MHD activity sets in which just leads to enhanced transport or local loss of confinement. Hard limits (like the Kruskal-Shafranov limit) or disruptions are carefully avoided by the choice of engineering parameters. In line-tied coronal loops, on the other hand, a current may be driven far above the Kruskal-Shafranov limit because of the presence of the stabilising field line bending energy corresponding to the lowest Alfvén wave frequency ω_1^2. However, this situation is just due to the assumption of infinite immobility of the photosphere. Of course, on a time scale determined by the finite inertia of the photosphere or dissipative processes facilitating field line slippage this assumption no longer holds and the loop discovers its lowest energy state with $\mathbf{k} \cdot \mathbf{B} = 0$: a *flare* develops.

Evidently, we have just described the leading order effects connected with the different longitudinal BC's of tokamaks and coronal loops. Further analysis is needed which takes the effects of inhomogeneity, dissipation, and nonlinearity serious. These aspects, recently extended with parallel computations [10], are extensively investigated by the scientists mentioned in the acknowledgements.

4 Conclusions

- *Line-tying completely changes the character of the basic MHD waves* occurring in a coronal loop, as compared to those in a tokamak, since it conflicts with the phase relationships between the components of $\boldsymbol{\xi}$ of the three pure waves. As a result, MHD waves of a mixed nature occur. They consist of Alfvén components with a ballooning factor, favouring minimal field line bending, and fast components without such a factor.
- The modes of the Alfvén continuous spectra consist of large amplitude Alfvén components in the corona and fast components with a small but rapidly varying amplitude in a photospheric boundary layer that develops due to the requirements of line-tying. *These modes have the right signature for effective transfer of energy from the photosphere to the corona and subsequent resonant Alfvén wave heating.*
- The usual $\mathbf{k} \cdot \mathbf{B} = 0$ condition for marginal stability at rational surfaces in tokamaks cannot be satisfied in a line-tied coronal loop. Consequently, *the concept of local stability (Suydam, Mercier, ballooning, etc.) is irrelevant for coronal loops.*
- Due to line-tying and the associated ideal fast boundary layer, solar coronal loops posses *an excess of stability* corresponding to the lowest Alfvén state

$$\omega_1^2 = (\pi/L)^2 \, B_z^2/\rho \,.$$

On a photospheric inertial-dissipation time-scale this boundary layer collapses, leading to reconnection *in the photosphere* and the release of this excess energy. Such a process might explain the very different behaviour at the onset of instability in coronal loops (*flares!*) as compared to tokamaks (just *enhanced MHD activity*).
- Similar stabilising effects due to partial line-tying are to be expected in *magnetospheric loops* and in the *scrape-off layer of divertor tokamaks*.

Acknowledgements

The work presented here is both the outflow and the starting point of extensive investigations in computational MHD of fusion and astrophysical plasmas by Stefaan Poedts, Peter Meijer, Giel Halberstadt, Hanno Holties, Sander Beliën, Ronald Nijboer (FOM-Institute for Plasma Physics, Rijnhuizen). The wider context is provided by collaborations with our colleagues Guido Huysmans, Wolfgang Kerner (JET), Marcel Goossens (KU Leuven), Tony Hearn, Max Kuperus (Astronomy Dept, Utrecht University), Henk van der Vorst (Mathematics Dept, Utrecht University), Herman te Riele, Albert Booten (CWI, Amsterdam), Ad Emmen, Jaap Hollenberg, Rik Leenders (SARA, Amsterdam), Hans Spoelder (Free University, Amsterdam), and Alexander Lifschitz (University of Illinois at Chicago). The subjects investigated are 'MHD spectroscopy in tokamaks', 'Heating of the solar corona', 'Nonlinear MHD waves in tokamaks and the solar corona', 'Parallel computations of the dynamics of thermonuclear and astrophysical plasmas', 'Visualisation of plasma dynamics', and 'Symmetries in MHD', respectively.

Stimulating discussions with Steven Lantz (Cornell Theory Center) on the coupling to convective cells are gratefully acknowledged.

This work was performed as part of the research programme of the association agreement of Euratom and the 'Stichting voor Fundamenteel Onderzoek der Materie' (FOM) with financial support from the 'Nederlandse Organisatie voor Wetenschappelijk Onderzoek' (NWO) and Euratom. A grant of the 'Koninklijke Nederlandse Akademie van Wetenschappen' is gratefully acknowledged.

References

[1] Uchida, Y. and Yohkoh Research team: 1992, 'New aspects of coronal physics derived by Yohkoh', in *G.S. Vaiana Memorial Symposium on Advances in Stellar and Solar Coronal Physics*, eds. Linsky, J. and Serio, S. (Kluwer: Dordrecht) 97.

[2] Hansen, P.J. and Goertz, C.K.: 1992, 'Validity of the field line resonance expansion', Phys. Fluids **B4**, 2713.

[3] Goedbloed, J.P.: 1991, 'MHD waves in thermonuclear and solar plasmas', in *Trends in Physics*, ed. Kaczér, J., EPS: Prague III, 827.

[4] Goedbloed, J.P., Holties, H.A., Poedts, S., Huysmans, G.T.A., Kerner, W.: 1993, 'MHD spectroscopy: Free boundary modes (ELMs) and external excitation of TAE modes', Plasma Phys. Control. Fusion **35**, B277.

[5] Poedts, S., Beliën, A.J.C., and Goedbloed, J.P.: 1994, 'On the quality of resonant heating as a coronal loop heating mechanism', Solar Phys. **151**, 271.

[6] Goedbloed, J.P. and Halberstadt, G.: 1994, 'Magnetohydrodynamic waves in coronal flux tubes', Astron. Astrophys. **286**, 275.

[7] Halberstadt, G. and Goedbloed, J.P.: 1993, 'The continuous spectrum of line-tied coronal loops', Astron. Astrophys. **280**, 647.

[8] Strauss, H.R. and Lawson, W.S.: 1989, 'Computer simulation of Alfvén resonance in a cylindrical, axially bounded flux tube', Astrophys. J. **346**, 1035.

[9] Parker, E.N.: 1992, 'The x ray corona, the coronal hole, and the heliosphere', J. Geophys. Res. **97**, 4311.

[10] Poedts, S., Meijer, P.M., Goedbloed, J.P., Van der Vorst, H.A., and Jakoby, A.: 1994, 'Parallel magnetohydrodynamics on the CM-5', in *Lecture Notes in Computer Science* **796**, 356.

PLASMA APPLICATIONS

PRODUCTION OF PLASMA WITH LARGE AREA FOR PLASMA APPLICATION

Y.Kawai, Y. Ueda, M. Tanaka and S. Shinohara

Interdisciplinary Graduate School of Engineering Sciences,
Kyushu University,
Kasuga, Fukuoka 816, Japan

ABSTRACT

An electron cyclotron resonance plasma(ECR plasma) with large area for plasma processing is produced with a multi slot antenna whose diameter is chosen independently of microwave frequency, and the uniformity is examined as a function of pressure, microwave power and magnetic field configuration. The plasma parameters are: $n_e = 2.3 \times 10^{11}$ cm^{-3}, $T_e = 5 - 10$ eV and the plasma potential $V_s = 20 - 30$ V. A circular TE$_{01}$ mode converted from the principal rectangular mode, TE$_{10}$, is also developed to produce a high density plasma with large area. The ion saturation current density of 36 mA/cm^2 is achieved for the input microwave power of 3 kW at 5×10^{-4} Torr. The uniformity of the ion saturation current density is within ± 3% over 200 mm in diameter. Furthermore, in order to avoid window contamination, a new microwave launcher with reflector plates is designed and a dense plasma of 10^{13} cm^{-3} is realized.

1. INTRODUCTION

Recently the development of plasma sources has become one of the most important subjects in plasma processing such as CVD and etching. In particular, a plasma with large diameter has been required. Above all, there has been great interest in an electron cyclotron resonance plasma(ECR plasma) because of high plasma density at low pressure.

Usually the principal mode in waveguides, TE$_{10}$ or TE$_{11}$ mode, is used to produce an ECR plasma. The wavelength in vacuum is about 120 mm at 2.45 GHz so that it is hard to produce an ECR plasma whose diameter is more than 120 mm. On the other hand, a uniform ECR plasma over 200 mm in diameter has been required as a plasma for etching from industry. Thus, the production of an ECR plasma whose diameter is more than 200 mm has become topic. Some attempts have been made and interesting results have been reported [1-7]. We have developed two types of ECR plasma sources using a multi slot antenna[5] and TE$_{01}$ mode[8] to produce a large diameter plasma. A multi slot antenna has the advantage that the diameter does not depend on the frequency of the incident microwave. Therefore, we have made the machine which consists of the chamber of 290 mm in inner diameter and a multi slot antenna of 280 mm in diameter, and a relatively uniform ECR plasma has been obtained[9].

From the point of view of high speed deposition or etching, a high plasma density such as 10^{12} cm^{-3} has been required. In order to realize a uniform and dense ECR plasma, we have developed a new type of ECR plasma source using a circular TE$_{01}$ mode microwave. The ion saturation current density of 36 mA/cm^2 has been achieved for the input microwave power of 3kW at nitrogen pressure of 5×10^{-4} Torr[10].

In practical applications, there are many problems to be solved, one of which is contamination of the microwave window. Particularly in fabrication of conductive thin films, this problem is serious, and long-term operation is impossible. To avoid window contamination, we have adopted a microwave launcher with reflector plates. It has been found that a new launcher developed here can produce a dense plasma of 10^{13} cm^{-3} plasma density[11-12].

Here we present the experimental results of the production of an ECR plasma with large area with a multi slot antenna and a TE$_{01}$ mode. Furthermore, the experiment on the production of a high density ECR plasma with a new microwave launcher with reflector plates is described.

2. EXPERIMENTS

2.1 Production of a large diameter ECR plasma with a multi slot antenna

We have produced a large diameter ECR plasma using a multi slot antenna(MSA) [5,9] to obtain large area thin films. The schematic diagram of the experimental apparatus is shown in Fig.1. The vacuum chamber is made of stainless steel with 290mm in inner diameter and 1200mm in length. An ECR plasma is produced with a multi slot antenna which is made of stainless steel of 280 mm in diameter. The length and width of slots are 70 mm and 2 mm, respectively. The magnetic coil assembly consists of six coils; four to make uniform magnetic field, while, two to form the magnetic mirror and control the current for magnetic mirror, I_m. The frequency of the microwave is 2.45GHz and the power can be varied up to 1000 W. Matching between the microwave circuit and the plasma is done with a stub tuner.

The gas used is He, and the pressures are ranged from 1×10^{-4} to 5×10^{-4} Torr. Hydrogenated amorphous silicon films are deposited on Corning #7059 glass substrates by flowing SiH$_4$ gas diluted with He gas. As well known, the uniformity of deposited films is influenced by the uniformity of the gas stream. So, we use a circular tube along the wall of the chamber as a gas feeder. The gas spouts from several holes of the tube to the center. In this experiment, the substrate which is not heated intentionally is placed at 850 mm from the multi slot antenna(z=1000mm). The diameter of the substrate holder is 270 mm, which enable a wafer of 8 inches in diameter to be setted. It has been found [9] that the radial uniformity of the ion saturation current density is within 3 % over 200 mm in diameter under the comparatively high gas pressure(2×10^{-3} Torr) and high microwave power(1000W). In the present experiment, the production of the ECR plasma and the deposition of the films at lower gas pressures($1 \sim 5 \times 10^{-4}$ Torr) are performed to look at more effect of ECR plasma on the CVD, and the comparatively high density and high electron temperature of plasma are obtained at low power because of the ECR effect.

In order to search the optimum conditions for the production of a large diameter plasma, the radial profile of the ion saturation current density is examined as a function of pressure, microwave power and current to produce the magnetic mirror, I_m. Figure 2 shows the radial profile of the ion saturation current density of He plasma for different gas pressures at the input microwave power of 200W, where (a)$I_m = 60A$ and $I_m = 90A$, respectively. As seen in Fig.2, the uniform profile over 200 mm in diameter is obtained at a certain gas pressure. Here, the ion saturation current density is large near the wall of the chamber, which is due to the fact that the electric field near the multi slot antenna is strong [5].

Figure 3 shows the radial profile of the ion saturation current density for different input microwave powers, where I_m is 90 A and the gas pressure is 2×10^{-4} Torr. When the microwave power is increased, the ion saturation current density increases, as seen in Fig.3.

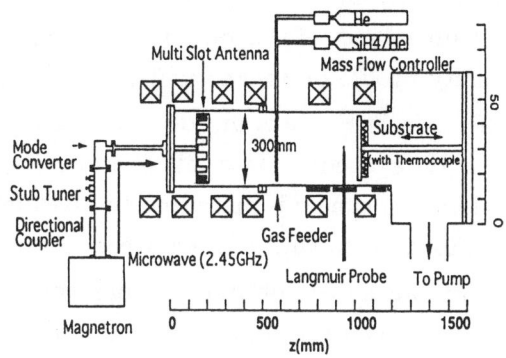

Fig.1. Experimental apparatus.

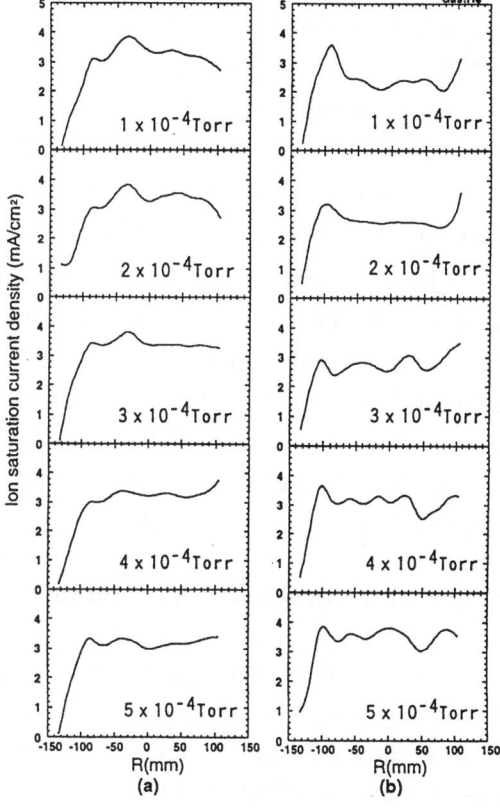

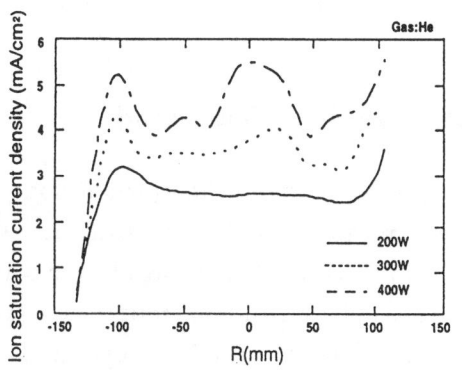

Fig.3. Dependence of the ion saturation current density profile on the input microwave power.

Fig.2. Dependence of the ion saturation current density profile on the gas pressure. ((a)Im=60A, (b)Im=90A)

Axial magnetic field configurations are illustrated in Fig.4(a). The smaller I_m is, the larger the magnetic mirror ratio is, that is, the confinement of plasma particles by the magnetic mirror field is expected. Figure 4(b) shows the radial profile of the ion saturation current density for different magnetic coil currents, I_m, where the input microwave power and the gas pressure are 200W and 2×10^{-4} Torr, respectively. When I_m is small, the ion saturation current density is high, which is due to the effect of the confinement by the magnetic mirror. When $I_m = 110$ A, the ion saturation current density at the center is low, which is due to a decrease in the mirror ratio, as seen in Fig.4(a).

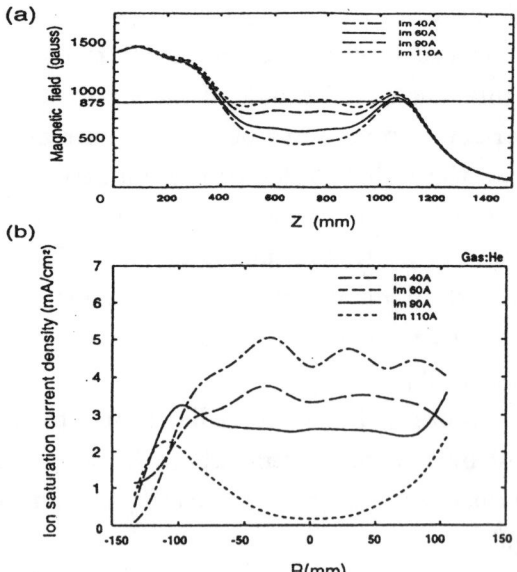

Fig.4 (a)Axial profile of magnetic fields. (b)Dependence of the ion saturation current density profile on magnetic fields.

Furthermore, hydrogenated amorphous silicon (a-Si:H) films are deposited on glass substrates by introducing SiH_4 to the He plasma. Here, the flow rates of He and SiH_4 are 18 and 2 sccm, respectively. The radial profile of film thickness is measured to investigate the conditions necessary for the deposition of uniform films with a large area.

The most uniform profile of the thickness of the film is obtained under the same condition as the ion saturation current is uniform. When the input power is increased, the thickness also increases. The radial profile of the thickness of deposited a-Si:H films is less uniform than that of the ion saturation current density. These results mean that the reaction of radicals plays an important role in the uniformity of the films, which is completely different from that of etching. The control of radicals will be necessary for more uniform deposition of the films.

2.2 Production of a large diameter ECR plasma with a TE_{01} mode

Figure 5 shows the experimental apparatus for TE_{01} mode plasma production. The microwave power of a frequency 2.45 GHz is ranged up to 5 kW. The rectangular TE_{10} mode is converted the circular TE_{01} mode, and is introduced into the ECR plasma source. The diameter and length of the ECR plasma source are 300 mm and 620 mm, respectively. The distance between the magnetic coils and the substrate is variable along the Z axis. The magnetic field profile is controlled by changing the coil current and the coil locations. In this experiment, the coil current is 260 A at maximum. The base pressure is kept less than 1×10^{-7} Torr. The ion saturation current density and plasma parameters are measured with a movable Langmuir probe.

In order to examine the uniformity of the ECR plasma, the radial profile of the ion saturation current density is measured with a movable Langmuir probe. The result is shown in Fig.6, where the abscissa stands for the radius of the ECR plasma source and the ordinate the measured ion saturation current density. In this case, the input microwave power is 3 kW and the nitrogen gas pressure is 5×10^{-4} Torr. Figure 6 shows that the ion saturation current density is uniform within $\pm 3\%$ over 200 mm in diameter and 36 mA/cm^2 is achieved with the TE_{01} mode. The measured electron temperature and the plasma density are 6 eV and 6×10^{11} cm^{-3}, respectively. The plasma potential is around 25 V and is almost uniform over 200 mm in diameter, as well. The wave propagation characteristics in the plasma are measured with an interferometer[12] to clarify the production mechanism of a uniform large diameter ECR plasma generated with a circular TE_{01} mode. It is found that the microwave introduced into the chamber with the TE_{01} mode is converted into mainly an electron cyclotron wave (whistler wave) and is almost entirely absorbed around the resonance point. Furthermore, it is experimentally proven that it is necessary to optimize the magnetic field configuration in order to produce a uniform plasma, as pointed out by Samukawa et al[3].

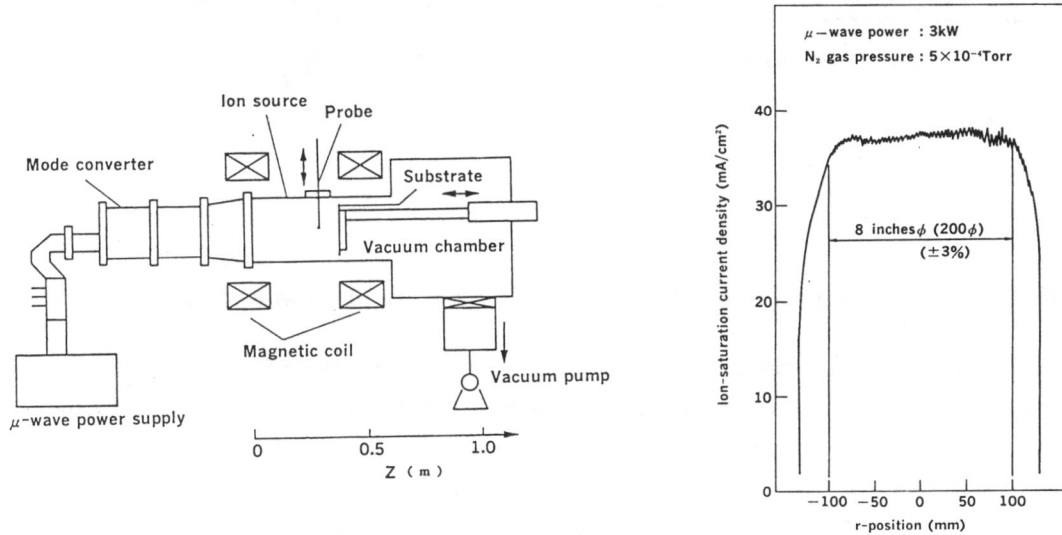

Fig.5. Experimental apparatus.

Fig.6. Radial profile of the ion saturation current density.

2.3 Multiplate Microwave Launcher for Producing High-Density Electron Cyclotron Resonance Plasmas

Figure 7 schematically shows the experimental apparatus and the magnetic field configuration. The chamber dimensions are 210 mm in diameter and 1500 mm in axial length. A magnetron oscillator, 2.45 GHz and 5 kW output, is used to excite an electron cyclotron wave (ECW). The rectangular TE_{10} mode from the oscillator is converted to the circular TE_{11} mode, and introduced into the chamber through a tapered waveguide and a quartz window. Figure 8 shows the multiplate launcher structure, in which the plates are horizontally aligned in order not to disturb the TE_{11} mode field profile (the electric field vector is vertical). We use 6 plates to cover the whole area of the window. The functions of the plates are to prevent particles from directly approaching the window surface and also to intercept the heat flux toward the window.

The magnetic field configuration satisfies high-field-side injection at the launching point ($\omega/\omega_{ce} < 1.0$), where ω is the wave frequency and ω_{ce} the electron cyclotron frequency. The ECR point is located at z = 670 mm. The wave magnetic fields, $\tilde{B}_x$ and $\tilde{B}_y$, are measured with an axially movable loop antenna introduced from the open end of the field line. The filling gas is argon, and the operation pressures are $1 \times 10^{-4} - 1 \times 10^{-2}$ Torr.

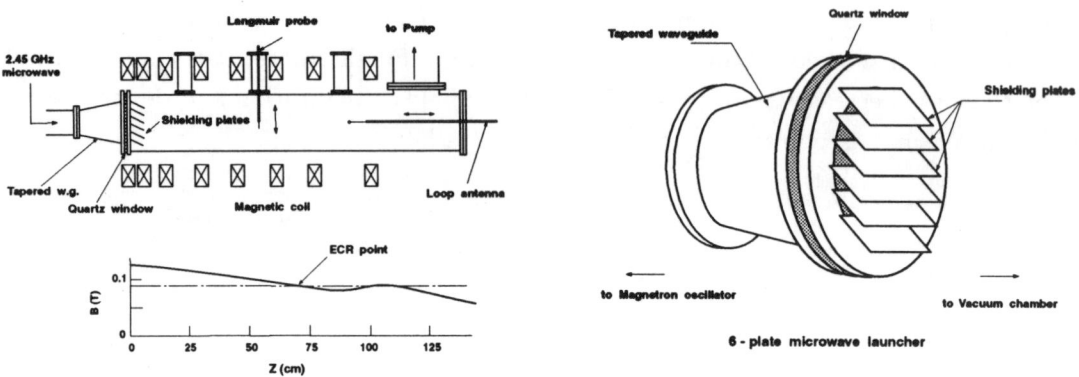

Fig.7.　Experimental apparatus.　　Fig.8.　6-plate microwave launcher.

Figure 9 shows the spatial wave pattern obtained by interferometry [13]. As seen in this figure, the wave amplitude is roughly constant in the range z < 250 mm, and then decreases along the propagation distance. This wave is fully damped before reaching the ECR point located at z = 670 mm. We have identified this wave as an ECW by comparing the experimental dispersion curve with the theoretical one [12,14,15].

The input power range is greatly extended to a high power level compared with the previous experiments with two shielding plates [11], in which the maximum input power is less than 1 kW. In the present experiments, no breakdown discharge is observed in the launcher structure even in the 5 kW input case (the maximum output power of our microwave source), and the microwave reflectivity remains low (< 10%).

Figure 10 shows the plasma density as a function of operation pressure for the 3 kW and 5 kW cases. As seen in this figure, the plasma density increases with the filling gas pressure and reaches 0.9×10^{13} cm^{-3} for the 5 kW input case. This value is comparable to the maximum density obtained with the direct microwave launching method [16] ($n_e = 1.2 \times 10^{13}$ cm^{-3}). Since the cutoff density of ordinary waves with the same frequency is 7×10^{10} cm^{-3}, the maximum density in the present experiment is two orders of magnitude higher than the ordinary-mode density scaling. The electron temperature gradually decreases with increasing plasma density, indicating that the microwave input power to the plasma is roughly constant. The radial density profile is rather flat above 3 kW input, and the flatness is less than 10% over 140 mm in diameter, in the 5 kW case.

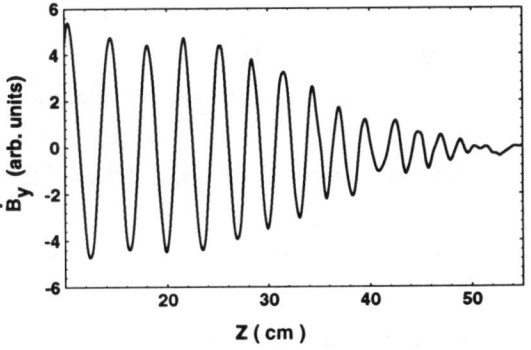 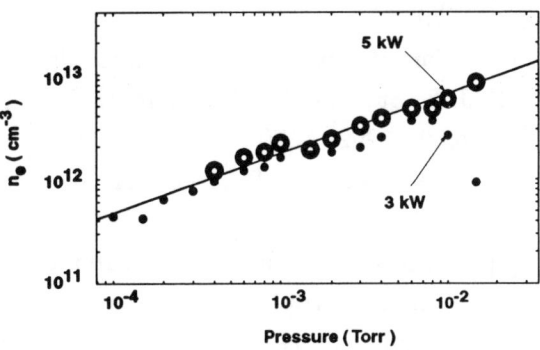

Fig.9.   Spatial waveform.   Fig.10.   Electron density.

The advantages of using the multiplate launcher are as follows: the shielding plates intercept sputtered particle flux and heat flux toward the window. The problem of window contamination and heating in realizing high-density plasmas will be reduced by using the new launcher. It is worth noting that the shielding plates can be used as an electrode. This enables dc biasing operations between the shielding plates and another electrode, adding a wide variety of operations to ECR plasma application.

3. CONCLUSION

It has been found that a uniform large diameter ECR plasma is produced with a multi slot antenna, which is useful for preparing a-Si:H films with large diameter. A circular TE_{01} mode can also produce a large diameter with high plasma density. Furthermore, it has been demonstrated that a new microwave launcher with reflector plates can produce a dense plasma of $n_e = 10^{13}$ cm^{-3}.

REFERENCES

[1] S. Matsuo and M. Kiuchi: Jpn. J. Appl. Phys., **22** (1983) L210.

[2] K. Suzuki, S. Okudaira, N. Sakudo and I. Kanomata: Jpn. J. Appl. Phys., **16** (1977) 1979.

[3] S.Samukawa and T.Nakamura: Jpn. J. Appl. Phys. **30** (1991) 3147.

[4] J. E. Stevens, Y. C. Huang, R. L. Jarecki and J. L. Cecchi: J. Vac. Sci. & Technol. **A10** (1992) 1270.

[5] A. Yonesu, Y. Takeuchi, A. Komori and Y. Kawai: Jpn. J. Appl. Phys., **27** (1988)1482.

[6] R. Itatani, Y. Yasaka, M. Kubo and A. Hatta: Proc. 8th Symp. Plasma Processing (Jpn. Sos. Appl. Phys., Nagoya 1991).

[7] S. Iizuka, P. H. Ko, N. Sato, T. Takahashi and A. Kodama: Proc. Jpn. Symp. Plasma Chem. **2** (1989) 293.

[8] Y. Kawai, M. Tanaka, R. Hidaka and N. Hirotsu: Proc. ICPIG XX (Pisa, 1991) vol.**3**, p.583.

[9] Y. Ueda, M. Tanaka, S. Shinohara and Y. Kawai, to be published in Proc. of the 7th Symposium on Plasma Science for Materials (Tokyo, 1994).

[10] R. Hidaka, T. Yamaguchi, N. Hirotsu, T. Ohshima, R. Koga, M. Tanaka and Y. Kawai: Jpn. J. Appl. Phys., **32** (1993) 174.

[11] M. Tanaka, M. Tamaoki, S. Higashi, M. Matsuoka, A. Komori and Y. Kawai: Jpn. J. Appl. Phys. **32** (1993) 1818.

[12] M. Tanaka, K. Nagao, H. Shoyama, M. Sugimoto and Y. Kawai: Jpn. J. Appl. Phys., **33** (1994) 5028.

[13] M. A. Heald and C. B. Wharton: *Plasma Diagnostics with Microwaves* (John Wiley and Sons, New York, 1965).

[14] G. Lisitano, M. Fontanesi and S. Bernabei: Phys. Rev. Lett. **26** (1971) 747.

[15] K. Ohkubo and S. Tanaka: J. Phys. Soc. Jpn. **41** (1976) 254.

[16] M. Tanaka, R. Nishimoto, S. Higashi, N. Harada, T. Ohi, A. Komori and Y. Kawai: J. Phys. Soc. Jpn. **50** (1991) 1600.

Industrial Applications of Thermal Plasmas

Roberto Nunes Szente
IPT - Instituto de Pesquisas Tecnológicas de São Paulo
Av. Prof. Almeida Prado 532 - C. Universitária
CEP 05508-901 São Paulo, S.P. , Brazil

Abstract

The main characteristics and applications of thermal plasmas are reviewed here. The industrial applications of thermal plasmas can be divided in: low power - cutting, welding, spraying; metallurgical and steelmaking; materials; environment. Some of the processes described in this article include: powder spraying, metal refining, tundish and laddle heating, production of ferroalloys and ceramic materials and treatment of residues (aluminum scrap, steel dusts, ashes, hospital wastes, electroplating mud). The use of thermal plasmas in the environment arena in particular has attracted increasingly attention as the regulations for disposal of residues become tougher. More research and development is needed particularly for decreasing the erosion of the electrodes of plasma torches and fundamental understanding of high temperature chemistry, heat transfer and electric arcs for broadening the applications of thermal plasmas.

I. Introduction

The use of thermal plasmas in industrial processes was motivated by the need of a heating source that could reach very high temperatures (around 2,000 °C) and be used inside an industrial vessel (reactor). Early attempts to establish those temperatures were based on heating elements outside the reactor or on the use of multiple gas/oil burners coupled with oxygen injection inside the reactor. External heating elements were not energy efficient; the walls of the vessels were also subjected to extreme conditions of heat flow, causing serious problems of deformations and thermal stresses. The limitations on the temperatures that the oil/gas burners could reach, as well as on the amount of energy available to the process and the presence of oxygen inside the reactor (not allowed in some processes) did not favor the use of the burners either.

On the other side, thermal plasma technology had been used since the beginning of the 50's for the simulation of the conditions of reentry for space vehicles in the atmosphere. Those tests were conducted by NASA (at Langley Air Force Base, Wright Patterson Air Force Base, Westinghouse and Avco); although the power levels used in these experiments were very high (up to 50 MW), the tests were of short duration and no attention was paid for the applicability of the devices used in these simulations for industrial processes.

At the beginning of the 60's researchers in USA and Canada started the development of plasma devices (the so called plasma torches, see below), based on the space program developments, that could stand the conditions inside industrial vessels. The first industrial processes were then developed (acetylene production - Dupont in 1963; plasma smelting reduction - Bethlehem Steel Corporation in 1967; titanium dioxide production - Linde in 1972) in North America.

It should be mentioned here that since the late 1920's Huls in Germany was producing acetylene at a total power level of 150 MW using an electric arc device, similar to some of the modern plasma torches. This process is still been presently operated. Also in the early 1900's Rjukan in Norway was using an electric arc device to produce nitrous oxide starting with air. This process was then used in several other plants in Europe and USA at the beginning of the century (some units operated until 1940). Although these two processes have many similarities with the more modern development of plasma technology nevertheless they are normally considered pre-plasma era, since the processes were based on electrical arcs operating almost with no control and limited gas flow rates. Today's development efforts in plasma technology for industrial applications (high power applications) are in the areas of production of special steels, ferroalloys and advanced materials (specially advanced ceramics), materials recovery (refractory metals), and particularly for the treatment of toxic or hazardous materials, such as hospital wastes, incineration ashes, chemical residues, metallurgical fines and others.

A brief review of the main characteristics of thermal plasmas is given below, followed by the present uses of plasma technology in industrial processes. The research needed in this field to expand the applications of thermal plasmas completes this review article.

II. Thermal Plasmas Characteristics

The main characteristics of thermal plasmas are reviewed here, followed by a description of the devices that are normally employed for the generation of the thermal plasmas, i.e., the plasma torches.

a) Thermal Plasmas

Thermal plasmas are used in industrial processes because of: a) high temperatures reached in the plasma jets; b) high energy densities of the plasmas; c) possibility of using different plasma gases depending on the application desired. Thermal plasmas are normally generated by either of three different ways: using a laser, a high frequency discharge or an electric arc. Only the last one, i.e., electric arcs, have significant importance for generating thermal plasmas for industrial applications and will be considered here.

The plasmas are created by collision processes between electrons from the electric arc (mainly) and the meloecules (or other particles) of the gas used for the process (the plasma gas). The electrons are accelerated by an electric field E (established between two electrodes) acquiring the kinetic energy that will be partially transferred in a collision with a particle of the gas. Thermal plasmas are normally generated at atmospheric pressure or close to it, thus resulting low values of E/p (p is the absolute pressure of the gas), usually in the range of 10^{-3} V/cm Pa (1).

It is convenient to consider thermal plasma in a state called LTE, i.e., a Local Thermodynamic Equilibrium. In the LTE, the plasma is considered to be thin to its own radiation and the local gradients of the plasma properties (temperature, density, heat conductivity, etc.) are sufficiently small so that a given particle diffusing from one location to another has enough time to equilibrate (2) and the collision processes (and not radiate processes) govern transitions and reactions in the plasma. Almost all the researchers in the world consider thermal plasmas in LTE for modeling work (3-6).

The electric arc and plasma surrounding it are normally divided in three regions: cathode, column, anode (2,6). The main part constituting the arc is the column. Typical current densities of 100 A/cm^2 are found in this region. In the arc column quasineutrality prevails; the temperature distribution, as well as the thermodynamics properties adjust themselves in such a way that the electric field strength required for driving a certain current becomes a minimum. Most of the useful characteristics of thermal plasmas (radiation, luminosity, heat transfer, chemical reactions) are due to the arc/plasma column. It will be shown later how the arc column can be used in an industrial process.

The electrode regions, i.e., the cathode and anode regions, contain the electrode surface, the region of the net space charge (sheath) immediately in front of the electrodes and a transition zone until the arc column. The electrode regions are characterized by much higher field strengths (up to 10^6 V/cm), temperature gradients (around 10^4 °K/cm) and current densities (up to 10^8 A/cm^2) than in the arc column. In the electrode regions one can not normally consider the plasma to be in LTE. The electrode regions are responsible for the creation and elimination of charged particles (electrons and ions). The electrode regions can also be of interest in industrial applications as it can be seen later in this article.

b) Plasma Torches

Thermal plasmas generated by electric arcs are normally produced in devices called plasma torches. These devices combine fluid flow and an electric arc to transfer heat to a gaseous or condensed phase. Plasma torches are usually divided in two categories: transferred and non-transferred. A brief description of each type is given below. More details can be found in Eschenbach et al (7).

i) Non-transferred plasma torches

In non-transferred plasma torches the electric arc strikes between two electrodes inside the torch. The electrodes can be made of different metals (or even oxides) and geometries; copper electrodes are normally used as concentric cylinders; if thoriated tungsten is chosen as the cathode material, an inserted tip or button of this material is made and an external copper cylindrical anode completes the torch. An schematic diagram of both types is given in Figure 1. The thermal plasma is obtained when passing a gas inside the torch; the gas will be heated by the electric arc, forming a plasma jet at the exit of the torch. This type of plasma torch is usually employed in powder spraying and for the treatment of liquid residues (or solid fines). The power of the torch can vary from 1 kW to 6 MW; one of the main limitations of these devices is the electrode erosion (see in Research Needed section); the amount of gas needed for the operation of the torch can be an asset or not depending on the desired application.

ii) Transferred plasma torches

In this type of plasma torches the electric arc strikes between an electrode of the torch and a metal (or other material) bath (at the beginning of operation the metal is normally solid, melting during operation) located inside the vessel as can be seen in Figure 2. The electrode can be a metal, such as thoriated tungsten or a graphite rod (which would have a central hole for the passage of gas). The gas is injected in the plasma torch and heated by the electric arc established between the inside electrode and the metal pool. The power level of these torches can be very high (more than 10 MW) and the amount of injected gas kept to a minimum if necessary. These devices are normally employed in metallurgical processes or for treatment of some solid residues (such as ashes). The transferred plasma torch has a higher energy efficiency than the non-transferred type, since almost all the heat is directed transferred to the material to be treated. These torches are however less flexible in terms of operation and need a more complex furnace (or reactor) for operation than the non-transferred systems.

III. Industrial Applications of Thermal Plasmas

Thermal plasmas are found in a diversified number of applications. The main areas of industrial applications can be divided in: a) low power - powder spraying, plasma cutting and welding, b) metallurgical / steel making, c) materials and d) environment. Low power applications (powder spraying, cutting and welding) of thermal plasmas is an established industry and only a brief review of these processes is given here. The other three areas mentioned above (high power applications) have received increased attention in the world and in particular the last one (environment); a more detailed description of the most important process in those areas is given below.

a) Low power applications - welding, cutting and powder spraying

Plasma torches operating from 1 to 100 kW are employed in welding, cutting and powder spraying. The geometries of these torches vary according to the application, but in general they all have a thoriated tungsten cathode and a cylindrical copper anode. In the case of welding and cutting the torches operate in a transferred mode (the copper anode of the plasma torch in this case is used to help the stabilization of the arc but does not carry current; the anode for the arc is the piece been cut); for powder spraying usually the plasma torches are non-transferred type. There are several manufacturers of low power plasma torches, including Metco, AVCO, Plasmadyne, Plasma Technik and others.

Thermal plasmas have been used since the beginning of the 60's for cutting metal pieces. Different plasma gases are employed for cutting: argon, argon-hydrogen, nitrogen, air, oxygen and even water (injected into the plasma column in order to decrease its diameter), depending on the material and thickness of the plate that will be cut. Plasma torches can also be used when the metal piece is submerged into a water pool in order to avoid any deformation of the plate. Of the three main techniques of cutting, i.e., plasma, hydrocarbon-oxygen and laser, plasma offers higher cutting speed (up to 20 m/min for 1 mm steel plates) and the possibility of cutting thicker pieces (up to 150 mm thick steel plates), although for precise cutting the CO_2 laser is normally used (precision of the plasma is of the order of a millimeter and of the laser is of a tenth of a milimeter). The final decision about the method that will be employed is a function of investment costs (laser is by far the most expensive, followed by plasma and hydrocarbon-oxygen), time needed for the operation and characteristics of the piece to be cut (material and thickness).

The torches used for welding are similar than for cutting, i.e., transferred arc plasma torches, although lower power levels are used (usually between 2 and 10 kW). The plasma gases used are usually argon or argon-hydrogen, in order to guarantee the integrity of the welding (avoiding the formation of oxides on the metal plate). Particularly interesting is the welding of thin sheets or pieces, where the classical TIG (tungsten inert gas) presents an erratic behavior of the electric arc due to the low currents used; for these situations plasma torches have a considerable advantage (the geometry of the torch and the use of the gas stabilize the arc column). Industries have used plasma torches for more than 30 years to weld stainless steel, copper, inconel, titanium and other metals (except for aluminum and alloys of this material). See reference (8) for more details of cutting and welding applications of thermal plasmas.

The use of plasma torches for spraying powders dates more than 35 years ago, representing today a conventional tool for such applications. The principle of operation is the injection of particulate materials (such as metals and ceramics) into the plasma jet (exiting the non-transferred plasma torch); the particles are melted or softened in the jet and accelerated towards a substrate where the particles impact, consolidating therefore and forming a coating or deposit. Plasma spraying can operate at low and atmospheric pressure, depending on the application desired. Materials such as tungsten carbide, zirconia, alumina, chromium oxide, chromium carbide, nickel-chromium, molybdenum and others are commonly deposited using plasma torches to form thermal, mechanical or chemical barriers for metal or ceramic substrates. Aircraft engines, compressors, pumps, valves, dies and others in the aerospace, automobile and chemical industries are some of the common uses of plasma spraying. Plasma Transferred Arc (PTA) have also been used for locally melting surfaces while depositing new particulate materials. For more information on plasma spraying applications see references (9-12).

b) Metallurgical / Steelmaking

The first application of high power plasma torches were in the steel making and metallurgical industries. The processes in these industries are very energy demanding; the temperatures involved in most processes are also high (from 1,000 to 2,500 ^{0}C), favoring the application of thermal plasmas. Thermal plasmas are applied in a variety of ways in those industries; rather than trying to establish a common ground for all these applications or follow a strict historical development, what is presented is a short description of the main processes presently in use in the world employing plasma torches in the metallurgical and steel making industries.

i) Melting of Iron

General Motors installed in 1989 a plasma assisted cupola at their foundry in Ohio to remelt iron turnings and scrap. Six non-transferred plasma torches of 2 MW (each) are mounted each on the end of tuyere to superheat the air blast. Melting energy is supplied partly from coke combustion and partly from the air blast. Compared to conventional cupola, the claimed advantages of the plasma system include: lower exhaust gas volume, less carryover of borings and chips, lower air pollution control costs, improved alloy yield, increased productivity, possibility of using low cost raw materials (borings) (13). Similar application has been developed at the cupola of Peugeot's foundry at Sept-Fons in the middle of 1989, with similar results (14). Both processes are still been operated.

ii) Tundish heating

Thermal plasmas are been used for heating of metals in the tundishes (and laddle) of continuous casters in steelmaking. The temperature of the molten metals are kept within 5 ^{0}C using plasma torches (transferred and non-transferred), which improves the quality of the steel to be fed to the caster, adjusting at the same time the optimal temperature for casting. Companies such as Chaparral Steel, C.F.&I. (both from USA), Nippon Steel, Kobe Steel, NKK Co (all from Japan) have plasma systems installed for this application (15, 16).

iii) Remelting of refractory and noble metals

The advantages of using thermal plasmas to remelt reactive metals, such as titanium, compared to other technologies (electron beam, vacuum arc remelting) include: independent control of power input and melt rate, operation at higher pressures avoiding volatilization of high vapor-pressure alloying elements, lower costs. Oremet (USA) since 1983 has produced more than 5,000 tons of titanium from scrap using a transferred plasma torch (800 kW). Leybold AG (Germany) developed a 2 MW, 3 transferred plasma torch furnace to remelt titanium and other reactive metals. In USA other companies such as Pratt&Whitney, Timet, Wyman Gordon, NRC are all using thermal plasma to remelt titanium, niobium and tantalum alloys, nickel (17,18).

iv) Production of ferroalloys

The production of ferroalloys is a very competitive market; processes that have proved economical feasible during some years suddenly had seen their situation changed due to forced lower prices for market domination. This has been the case of some plasma processes, such as the production of ferromanganese by La Societe du Ferromanganese de Paris-Outreau (operational between 1984 and 1991) and ferrochrome production using non-transferred torches by SKF (Plasmachrome process, operating between 1987 and 1991). Mintek (South Africa) has developed also a transferred plasma system for the production of ferroalloys (ferrochrome and ferroniobium); the processes are been in full scale operation presently. Other plants are presently under construction for the production of ferroalloys (Macalloy, USA, ferrochrome; Voest-Alpine, Austria, ferromanganese; Dow Corning, Canada, ferrosilicon) (19,20).

c) Materials Production

The production of materials using thermal plasmas combines already established processes with new developments. A short review of the main processes is given below.

i) Titanium dioxide

Tioxide (UK) has developed a non-transferred plasma system for the production of titanium dioxide, starting from titanium tetrachloride and oxygen. One plant in England (Greatham) produces 50,000 ton/year since 1982 (2 MW torches) and another plant using plasma technology is presently been considered in the USA (21).

ii) Zirconia

Zirconia and silica have been produced using a modified transferred plasma torch (see Figure 4) starting from zircon sand by Ionarc (Z-Tech Corporation presently) since 1977. The production is around 4,500 ton/year, with several 350 kW reactors. The purity of the products obtained varies from 94.5% to 99.6 depending on the operating conditions (22).

iii) Silicon

The production of silicon from silica has been demonstrated in a 200 kW pilot unit built by Voest-Alpine in collaboration with Dow Corning. Carbon and silica are fed into a reactor equipped with a transferred plasma torch (23). The claimed advantages of this process include: reduced loss of SiO from the reactor, higher silicon yield, high silicon purity, lower energy requirements. Tests in a two stages 1 MW reactor are been presently conducted in Austria.

iv) Synthesis

The synthesis of new and advanced materials using thermal plasmas is one of the most active areas of research and development in the plasma arena. Although most of these processes are still under development (and not in an industrial stage yet) it could be of some interest to mention a few examples of products presently under consideration for production using thermal plasmas. Particularly interesting is the use of thermal plasmas for the production of: AlN (24,25), Si_3N_4 (26,27), SiC (28,29), and titanium compounds -TiAl, TiC, TiN (30). The synthesis of diamond like films (and particles) using thermal plasmas have been shown to present advantages over other technologies (lower production costs, higher yields). All these processes employ non-transferred torches.

d) Environment

The number of applications of thermal plasmas in environmental related processes is rapidly expanding. A brief review of the most important industrial applications in this increasingly important field is given below.

i) Aluminum recovery

Aluminum cans, scrap and dross are normally remelted in furnaces that use oil/gas burners for recovering aluminum. In order for these burners to operate, oxygen (air) is admitted in the furnace. To avoid the oxidation of the molten aluminum, salt (NaCl, KCl) is added in large quantities (15% of the charge). The salt has to be discarded after few heats; it presents serious problems for the environment (contamination of underground water mainly). Non-transferred plasma torches have been used to replace the burners; no air is admitted inside the furnace, avoiding the use of salt. The plasma technology also gives higher aluminum recovery and lower overall production costs (31). Alcan already has a 2 MW plant (Canada) for dross treatment using plasma; Hydro-Quebec (Canada), Nupro (USA), IPT (Brazil) are marketing now similar plasma systems.

ii) Steel/metal dusts

Dusts originated from different types of furnaces used in the steelmaking industries (open hearths, electric arc furnaces) represent a significant loss of heavy metals and valuable alloying elements (around 1-2% of the charge is converted to fumes or dusts). Most of the dusts have been dumped or stockpiled since the production processes do not allow the use of the dusts as they are and any agglomeration process would have prohibitive costs. These dusts are listed as hazardous wastes in most of the countries in the world because of the leachability of its toxic constituents. As a hazardous waste, the dust must be encapsulated or transported to an industrial landfill (class I). SKF in Sweden (ScanArc presently) operates since the 1984 a 60,000 ton/year treatment of the dusts in a plasma system (3 non-transferred plasma torches, 6 MW each), recovering zinc and generating a inert ferrous slag (that is sold as scrap) (32). International Mill Service (USA) has built two plants in Florida and Tennessee (6,000 and 12,000 ton/year respectively) for treating carbon steel dust using thermal plasmas.

iii) Incineration ashes

One of the most used methods for treating domestic wastes is incineration. This method generates around 10 % of ashes (inorganic and metal materials). The ashes need to be disposed in controlled landfills because of the possibility of contamination of the soil and underground water with heavy metals. Transferred plasma torches have been used to vitrify the ashes, generating a completely inert slag, that can be used as filling material for pavements or constructions. In Japan several municipal incinerators have installed plasma furnaces for the treatment of the ashes, with capacities of around 2 MW (33). Recently the city of Bordeaux (France) signed a contract to install similar unit for treating ashes generated from incineration of domestic wastes, with capacity of 1.7 MW using a non-transferred plasma torch.

iv) Hospital wastes

The wastes from hospitals are normally incinerated in order to avoid potential risks due to pathogenic materials. Hospital wastes are made of water (40%), plastics (20%), organic materials (20%), inorganic and metals (20%). The incineration of these wastes generates ashes (which have the same problems as described before) and more importantly, favor the formation of dioxins and furans (which are some of the most carcinogenic compounds known by mankind). Non-transferred plasma torches have been used in pilot plant scale operation to treat hospital wastes (Canada, USA, France, Japan). The ashes are vitrified in the process; this fact and the inert or reducing conditions of the process completely inhibits the formation of dioxins and furans (34). The San Diego Medical Center is constructing the first commercial unit to treat hospital wastes in the USA. The unit should be operational at the end of 1995, comprising a 600 kW plasma torch (35).

v) Electroplating mud

The processes of electroplating chromium, zinc, nickel, copper and others generates a liquid effluent that requires treatment before it can be discarded into the sewage system. The treatment of the liquid effluents produces a mud which contains around 80 % water, the remaining being the metals and flux agents used during the process. These residues (electroplating mud or galvanic mud) are been stored before a suitable technology can be used for the proper disposal of these hazardous materials. The production of these residues add to more than 1,000 ton/month just in the state of São Paulo, Brazil. A pilot plant unit is under development in IPT (Brazil), employing a 200 kW transferred plasma torch for treating these residues. Laboratory tests with a 20 kW torch revealed that it is both economical and technically feasible to treat the residues with plasma (36). The process recovers some of the metals (zinc in particular) while forming an inert slag containing essentially iron. An industrial unit (1 MW) is under consideration to be installed until 1996 in Brazil.

vi) Recovery of emission catalysts

Automobiles manufactured in parts of the world have a catalytic emission control unit in order to decrease the amount of nitrous oxides emitted. The catalysts used in these units belong to the PGM group (platinum group metals). Multimetco recovers over 2 tons/year of these catalysts from about 3,500 tons of emission control units collected in the USA. The system employs a 3MW transferred plasma torch and the furnace is operated at around 1,600 ^{0}C; the materials comprising the emission unit are blended with coke and flux and smelted inside the reactor. The PGM are recovered (more than 98% recovery rate) and the slag generated in the process is sold to the steel industry as a laddle additive (37).

vii) Asbestos inertization

In France EDF and Promethee built a mobile unit for the treatment of asbestos wastes. The process, called INERTAM, has been in use since the middle of 1994 for treating asbestos from the disactivated thermal power station in Arjuzanx. The process operates at 1,600 ^{0}C; the asbestos is melted under these conditions generating a vitrified inert slag when removed from the furnace. The unit is capable of treating 1 ton/h, using a non-transferred plasma torch of 1.5 MW. The unit should treat all the asbestos wastes from the mentioned power station in two years, when it will be moved to a site near Paris for continuing the treatment of asbestos wastes from other sources in France (38). A similar process is also been developed in England, with the difference of recovering magnesium from the asbestos (smelting process) (39), besides treating the wastes.

viii) Toxic wastes destruction

A demonstration unit was developed in France (EDF, Rhone-Poulenc, Ciments Lafarge) to decompose PCB (and other organochloride compounds); the process is based on the injection of liquid products into the jet of a non-transferred plasma torch of 600 kW. The destruction rate obtained was better than 99.999%, more than an order of magnitude better than any other technology (40). The companies involved in this project are studying the construction of an industrial plant. The Plasmox process, developed by MGC Plasma AG in Switzerland should also be mentioned here. A 1 MW transferred plasma torch has been installed in a reactor for the treatment of many different residues, specially PCB. The unit can process up to 1 ton/h of hazardous wastes and should be fully operational in a near future, although it will be probably used as a demonstration plant.

Other processes using thermal plasmas for treatment of wastes include the vitrification of radioactive wastes (41) and the remediation of contaminated soils (42). Both processes are still in a development stage at this moment. The companies involved in these programs include Westinghouse, PEC, DOE and Tectronics.

IV. Research needed

Important basic plasma research areas needed for expanding the use of thermal plasma in industrial processes include: high temperature heat transfer and chemistry, plasma jet conditions, plasma heat and momentum transfer to particles, diagnostic sensors (43). These informations will be used to extend the capacity of the present models, reducing the need of some of the expensive experimental trials that are the almost only source of information presently. Applied research and development to increase the expected lifetime of electrodes in high power torches (44) and reduction of costs for electronics and power supplies will certainly add to increase the industrial uses of thermal plasmas.

Conclusions

Thermal plasmas have been used for almost 35 years in industrial processes, ranging from metallurgical to environmental applications. More than 40 industrial plants in the world use thermal plasma technology, the number increasing every year. However the use of thermal plasmas should always take into consideration an economical study in order to avoid frustration and disappointments that occurred at the beginning of the 60's, when plasma torches were seen as almost magic tools. The future should see more applications of plasma technology, specially for environmental applications, if funds are maintained for basic and applied research.

References

1) E. Pfender, Electric Arcs and Arc Gas Heaters, Gaseous Electronics, Academic Press NY, 1978.
2) A.M. Howatson, An introduction to gas discharges, Pergamon Press, Oxford, 1976.
3) J.D. Ramshaw, C.H. Chang, Plasma Chemistry and Plasma Processing, Vol. 12, 1992.
4) Y.C. Lee, Modelling work in thermal plasma processing, Ph.D. thesis, Minnesota, 1984.
5) A.H. Dilawary, J. Szekely, Mat. Res. Soc. Symp. Proc., vol. 98, 1987.
6) E. Pfender, M. Boulos, P. Fauchais, Plasma Techn. Metallurgical Processing, ed. J. Feinman, 1987.
7) R.C. Eschenbach, N.A. Barcza. K.J. Reid, Plasma Techn. Metal. Proc., ed. J. Feinman, 1987.
8) D. Eliot, Les plasmas dans l'industrie, ed. G. Laroche, EDF, Electra, pg. 242, 1990.
9) E.J. Kubel, Advanced Materials and Processes, Vol. 12, 1987.
10) P. Fauchais, J. Desmaison, J. Machet, L'industrie ceramique, n.812, Vol.1, 1987.
11) R.W. Smith, R. Novak, Powder Metallurgy International, Springer Verlag, Vols. 23-3 and 23-4, 1991.
12) S. Sampath, H. Herman, Proc. 11th International Thermal Spray Conference, June 1989.
13) S.V. Dighe, 49th EAF Conference, Canada, Nov. 1991.
14) M.J. Copsey, R.H. Johnson, First International EPRI Plasma Symposium, USA, 1990.
15) J.M. Svoboda, A. davidson, F. Kemeny, Progress Report-EPRI/CMP Research, 1991.
16) Z. Yamazaki, H. Hashimoto, First International EPRI Plasma Symposium, USA, 1990.
17) S. Stocks, Proc. First Int. EPRI Plasma Symp., 1990.
18) S.H. Reichman, Vacuum Metallurgy Conference, USA, 1989.
19) H.R. Larson, J.F. Elliot, B.R. Perkins, INFACON 6, SAIMN, 1992.
20) V.D. Dosaj, A.W. Raucholz, H.G. Muller, E. Koch, 49th Electric Arc Furnace Conf., 1991.
21) A. Hare, Arc Plasma Process, Union Internationale de l'Electrothermie, 1988.
22) G. Clarke, Materials Edge, Sept.Oct., 1987.
23) V.D. Dosaj, J.B. May, P. Matzawrakos, First EPRI Plasma Conference, USA, 1990.
24) F. Moura, Production of AlN in a transferred arc reactor, Ph.D. thesis, McGill U., Canada, 1994.
25) K. Kijima, S. Akimoto, T. Vetsuki, K. Tanaka, 9th Int. Symp. Plasma Chemistry, Italy, 1989.
26) F. Allaire, S. Dallaire, J. Material Science, Vol. 26, pg. 6736, 1991.
27) Y. Chang, R.M. Young, E. Pfender, Plasma Chemistry and Plasma Processing, Vol. 7, pg. 299, 1987.
28) J.J. Moore, J. Material Science, Vol. 24, pg. 2794, 1989.
29) K. Upadhya, J.J. Moore, K.J. Reid, Metall. Trans., vol. 17b, pg. 192, 1986.
30) P.G. Tsantrizos, Thermal Spray: International Advances in Coating Technology Conf. Proc., 1992.
31) R. N. Szente, A.C. Cruz, L. Barbosa, Metatechnies 94 Proc., France, 1994.
32) J. Thornblom, B. Gustafsson, B. Johansson, Workshop Proc., ISPC 9, Italy, 1989.
33) S.L. Camacho, PEC Technology, Metatechnies 94 Proc., France, 1994.
34) L. Laouilleau, Les plasmas dans l'industrie, ed. G. Laroche, EDF, Electra, pg. 320, 1990.
35) E.M. Krisher, Plasma Technology for a better Environment, UIE, Chapter C1, pg. 111, 1992.
36) R.N. Szente, O. Bender, L.C. Vicente, EBRATS, Brazil, Oct. 1994.
37) J. Saville, Proc. 9th International Precious Metals Conference, 1985.
38) F. Kassabji, INERTAM, Metatechnies 94 Proc., France, 1994.
39) A.M. Cameron, L.A. Lewis, J.M. Baronnet, Metatechnies 94 Proc., France, 1994.
40) C. Guillet, EDF-DER, Metatechnies 94 Proc., France, 1994.
41) J.W. Gears, R.C. Eschenbach, R.A. Hill, Waste Management, Vol. 10, 1990.
42) L. J. Circeo, PRISM, Metatechnies 94 Proc., France, 1994.
43) J. Goodwill, EPRI Plasma Symposium, USA, 1990.
44) R.N. Szente, Ph.D. thesis, McGill University, 1990.

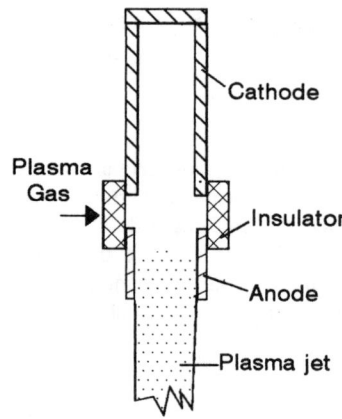

Non-Transferred Torch
Copper Electrodes

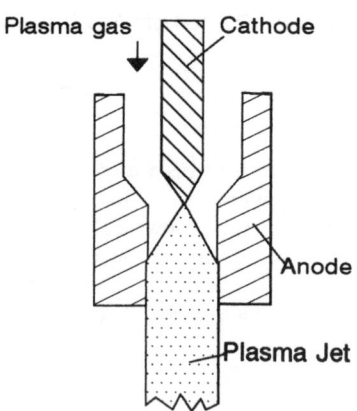

Non-Transferred Torch
Tungsten Electrode

Figure 1

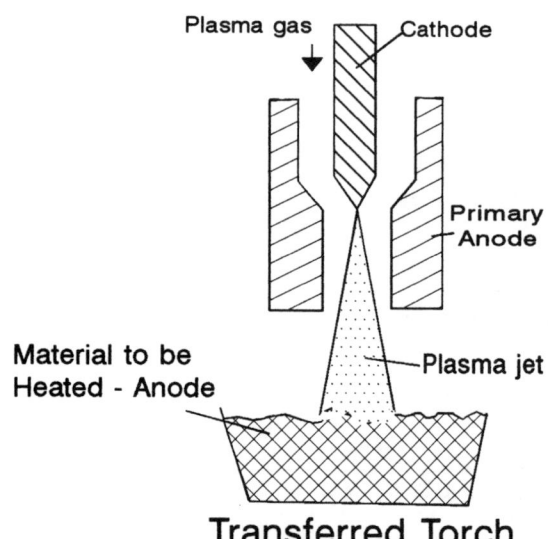

Transferred Torch

Figure 2

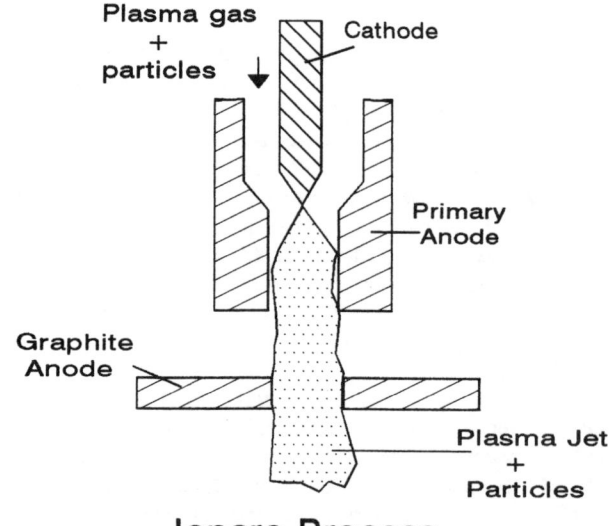

Ionarc Process

Figure 3

OBSERVATIONS OF DENSE PLASMA FORMATIONS AND BEAMS IN THE VACUUM SPARK

E.S.Wyndham

Facultad de Física, Pontificia Universidad Católica de Chile, Casilla 306, Santiago-22, Chile.

Abstract

A review mainly concentrating on the last five years of experimental work on dense plasma formations in the Vacuum Spark is presented. A comparative study is made of the micropinch and hot spot formations observed under different conditions of operation of the Vacuum Spark. The limitations of various experimental proceedures are explored. A number of simple alternatives in observational techniques suitable for low budget research are discussed.

Introduction

As early as 1897, Wood[1] observed the generation of X-rays in a vacuum electrical discharge which were, in his view, the result of electron bombardment of the Anode. Pin hole camera X-ray photographs were taken in 1937[2]. Modern research into the Vacuum Spark discharges at high currents began with work by Cohen et al.[3] with the observation of intense X-ray emission from the inter electrode volume. The observation of Fe XXV species from very small regions, in events on the ns timescale, was a complete novelty. The generation of highly ionized species in laboratory plasmas produced much interest[4]. The experimental and theoretical research into the Vacuum Spark has received attention in excellent reviews[5,6], and work has also continued into Vacuum Breakdown without preionization[7]. Although the majority of the work has been performed using generators producing in excess of 100kA, there has also been recent work on much smaller devices[8]. The principal interest, however, is concentrated in high current discharges where the phenomenon of radiative collapse gives rise to very dense and hot metallic plasmas, with dynamic processes occurring in the tens of picosecond time scale.

The physical model for radiative collapse to form plasma points or micropinches which has had most success in explaining the main energetic processes is that of Vikhrev et al.[9] This model was first applied to the contraction phase of a plasma focus[10], but was later extended to Vacuum Sparks and dense Z-pinch phenomena in general[11]. For calculations on an Iron Plasma at currents between 50 and 150kA, it is found that the onset of radiative loss by emission from line emission, which may be two orders of magnitude greater than the corresponding Bremsstrahlung component, can give rise to a very rapid collapse to micron scale plasmas with densities of up to $10^{24} cm^{-3}$. Quasiequilibrium is assumed (Bennett condition) in the radial direction but plasma outflow in the axial direction is expected. In an approximate dynamic model for a 150kA discharge, it was predicted that the micropinch would occur in two phases. The first sees the contraction of the millimetre plasma to a radius of order 0.01mm and a temperature of 50eV and a density of $10^{21} cm^{-3}$. This is followed some 30ns later by a ps collapse to a radius of 10^{-4} cm with corresponding densities and temperature of the order of $10^{24} cm^{-3}$ and 1keV respectively. Collapse is expected to cease due to anomalous Joule heating in the constriction. While this model gives a good idea of the qualitative behaviour, more physical processes could be included in a 1 or 2-D model, but so far this has not been attempted for high Z plasmas. Modelling of hydrogen plasmas has been done[12] to the extreme of radiative collapse at 10^{-4} cm. One advantage of the Vacuum Spark is that the equivalent of the Pease-Braginskii current may be approximated by the relation[13]. $I_{cr} = 10^{-4} T^{0.75} Z^{0.5}$, where I is in MA and T is in eV.

Undoubtedly one of the main attractions of the Vacuum Spark to the experimentalist is the development of diagnostic techniques able to resolve the rich variety of dynamic processes that occur in the Vacuum Spark in a sub-nanosecond and micron scale length. The conditions are quite similar to Laser-driven implosion experiments in their final phase. In practice this has meant that only the initial phase of the radiative collapse may be observed with any spatial-temporal precision and the only method available to observe the final collapse is by time integrated spectroscopic observation. Spectroscopic observations have given good data on the limit conditions of the plasma, but, as yet, are model dependant as far as detail of the intermediate phases of the discharge are concerned.

The purpose of the present work is to discuss the progress in Vacuum Spark research over the last five years or so. In fact, there has been rather less work published on the Vacuum Spark than in previous years. The emphasis of the present work will reflect the fact that the majority of the papers published have been on experimental and applied aspects of the Vacuum Spark.

Experimental Apparatus

For the main part, the experimental apparatus has not changed very significantly since early work by Cohen et al.[3] In that work, two rods of 2mm diameter separated by 3mm form the Anode and Cathode. A sliding spark discharge over a ceramic surface formed at the Cathode by a third electrode triggers the main discharge. The discharge is driven by a single 13μF Capacitor operated at up to 2.3kJ of stored energy in a circuit of 160nH giving a peak current of 176kA. An example of a more recent experiment[14] uses essentially the same apparatus, but with a lower circuit inductance of 80nH with a corresponding quarter period time of 2.2μs.

The discharge may be triggered by powerful Laser focused on to one of the electrodes, usually the Anode[15-16]. The behaviour of the current and the formation of current dips may be correlated with intense bursts of X-rays both from the Anode and the interelectrode volume and the appearance of plasma points.

Pulse power techniques have developed dramatically over the last twenty years but have been applied to the Vacuum Spark by few authors. The first published example[17] used the DON coaxial water line at the Lebedev Institute. The scheme consisted of a Marx Generator, an intermediate storage capacitor, a water switch, a 2.5 Ohm pulse forming line, a gas switch and the diode load section. The line charging voltage reached 500kV with the base pulse length of 60ns. The line current oscillated with a 150ns period and reached 100-150kA at the first maximum. A 5J Laser was focused through a conical Anode onto a flat Cathode with a separation of approximately 5mm.

Because of the rapid current switching characteristics of the Vacuum Spark a variant of the switched coaxial line has been used in which the Vacuum Spark itself switches the Line. In this "Hybrid" mode of operation[18], the applied voltage ramps up as the line is charged, whereas in the previous case, the "Normal" mode, the gas switch isolates the load from the line voltage until the gas gap breaks. The Hybrid mode shares the characteristic of the low inductance capacitor bank approach while retaining the high dI/dt of a switched line. A low power laser of about 0.3J may be focussed onto either the Anode or onto the Cathode to initiate the discharge. The two setups are compared in Figure 1. Some idea of the typical diagnostic arrangement is also shown in common with many authors. Comparative studies may be made[19] of the X-ray emission under the four possible combinations and considerable differences are found as will be referred to in following sections.

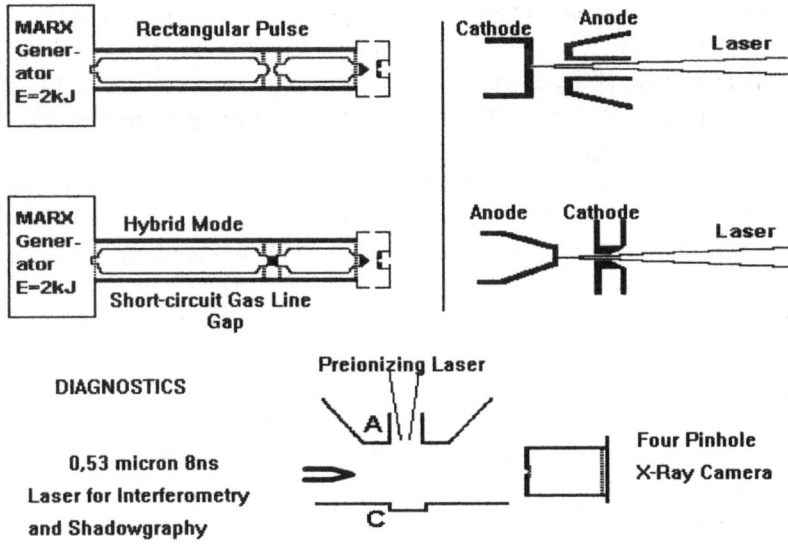

Figure 1. Comparison of the operation of the Vacuum Spark using a Coaxial Water Line. Above, in the "Normal" mode the Pulse Forming Line Gas Gap closes at the maximum of the applied charging voltage giving a rectangular pulse. Below, in the "Hybrid" mode the Gas Gap is shorted out and the discharge itself switches the line. Source impedance during the switching interval is $1.5\Omega + 30nH$.

Work has also continued on smaller Vaccuum Spark discharges[20]. In this class of device the emission is from beam target electrons, or from the plasma close to the Anode. Applications of the Vacuum Spark include its development as a multi kA laser triggered switch[21], and as soft X-ray source[22,23] for biological purposes.

There is a striking similarity between the Transient Hollow Cathode Discharge and the Vacuum Spark for the case when the Laser passes through the cathode orifice and is incident onto the Anode. The main difference is the filling pressure, 3×10^{-5} Torr instead of 40 mTorr. If, instead of illuminating the Anode, the Laser is made to hit just behind and inside the Cathode, the Vacuum Spark is found to reproduce the breakdown behaviour of the THCD[24]. That is, the hollow cathode volume physics is a determing factor for breakdown. As in the case of the Laser incident on the Anode in Vacuum Spark operation only very srong beam emission is observed. The inter-electrode plasma is not sufficiently dense to be detected with interferometry but its emission is seen with a pinhole image integrated over several shots.

Diagnostics

From the earliest results[3] on, X-ray pinhole images have revealed the existence of extremely small emitting regions. Epstein et al.[25] measured optical emission using a streak camera as well as broadband microwave emission. Optical streak photography does not show detailed features, but X-ray streak photography either direct, or by means of a fast scintillator such as Pilot-u, has been applied with success to X-pinch plasmas[26] and could be applied to the Vacuum Spark. The X-ray emission covers a very wide energy range, there is emission at energies far in excess of the applied voltage, but most is below 10keV. The limit of 10keV applies to the resonance transitions in He-like and H-like ionization states of the transition elements which are observed to form the micron scale plasma points. Combinations of PIN diodes and bare XRD's, both using filters and collimation, may be used to compare emission from different parts of the plasma. PIN diodes have little response below about 500eV, whereas the XRD response is optimum in the soft X-ray and VUV energy bands. The hot spots and plasma points emit preferentially in the 0.5-3keV range, but intense softer energy emission from the pinch column may be seen using XRD or a Pin hole photograph using, for example, a Ag filter.

Spatial and temporal resolution of the electron density has been obtained using a range of optical techniques. There are however serious limitations in these methods which, in effect, limit their use to either just before or just after the formation of the micropinch. The first limitation is the sub-nanosecond time scale of the radiative collapse phase which means that mode locked laser pulse are required if interference fringes are to be frozen. The second limitation is that size of the micropinch plasma is comparable to the wavelength of light, and, in addition, the plasma is expected to be optically dense at these times.

In spite of these limitations, there is much information that may be obtained from the pre-pinch phase. Methods include Schlieren Photography, Holography, Holographic Interferometry and Shadowgraphy. Multi-frame techniques have been applied[27], for example in shadowgraphy. Shadowgrams can give fine scale information on the density structures in dense Z-pinches[28]. X-ray backlighting might be considered as a possible means of resolving the collapsing phase of the micropinch, although the backlighting source would also have to have the same scale lengths as the micropinch. Point-Diffraction Interferometry has been applied to dense fibre pinches[29], giving quantative structural information on 10μm scale length and this method could be applied to the Vacuum Spark. The measurement of the current distribution in a Vacuum Spark has been attempted using Faraday Rotation[30] at an early phase in the conduction phase when the pinch radius was approximately 1mm.

The high cost of X-ray streak and framing cameras is usually prohibitive for small laboratories. A means of obtaining spatial information on the X-ray emission is by the use of a line of fibres in contact with a disc of scintillator which is used instead of X-ray film in a filtered Pinhole Camera[31]. A photomultiplier tube is connected to each fibre and a fast digital scope may be used to record several channels simultaneously. A system rise-time of 4ns is easily obtained. Depending on the magnification of the Pinhole Camera, axial resolution of between 0.5 and 2mm is easily obtainable. This alternative is known as the Poor Man's Streak Camera.

The measurement of the hotspot or plasma point is often made using filtered pin holes. Instead of trying to image the plasma with pinholes of 10μm or less, larger square pinholes with polished edges may be used[32], where the extent of the penumbra gives a characteristic hotspot dimension. A further variant of the pinhole camera is the Slit Wire Pin-hole Camera[31]. This allows the simultaneous measurement of hotspots and micropinches by the use of a range of calibrated thickness of wires as the obstruction.

The use X-ray emission spectroscopy has given much information on the extremal conditions of the Vacuum Spark. A comparatively recent example[33] is shown in Figure 2, where spatial and energy resolved spectra for a Vanadium plasma between 14 and 16Å are shown. The Li and Be like Vanadium lines are seen to be considerably more defined than the Ne and F like lines, a 70μm slit was used in this example. The fuzzier emission comes from a plasma column of about 0.1mm diameter and of 0.5mm height with a temperature of about 400 eV and with a density of 4×10^{20} cm^{-3}. The more defined highly ionized lines come from a smaller region whose temperature is about 800 eV and whose density is of order 10^{23} cm^{-3}.

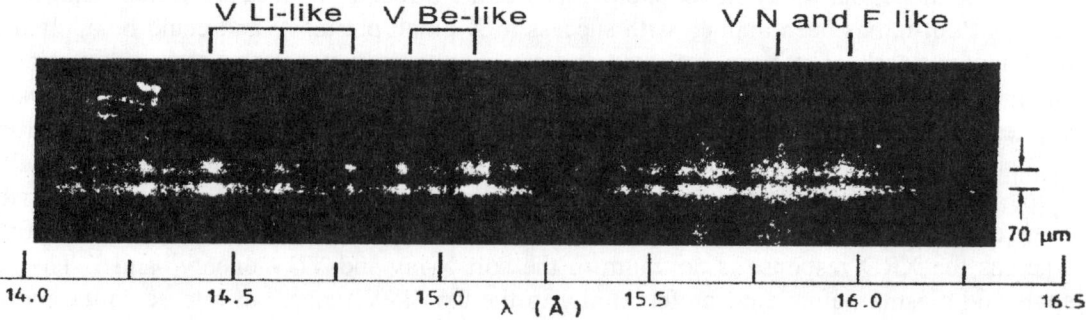

Figure 2. Space and energy resolved Vanadium spectra obtained in the 14-16 Å range observed in a low inductance Vacuum Spark. The two were recorded in a single discharge. The distance scale is indicated by the separation of the two upper traces, which is 70μm

The predicted life time using energy balance considerations of the first phase is 1 ns while that if the dense micropinch is 0.1 ns.

PLASMA POINTS AND HOT SPOTS

It has been found that there are two characteristic classes of plasma points or hot spots. Some confusion has resulted in the convention for the use of these terms. Following Koshelev and Pereira[6], we call the minute volume of ions stripped to their K-shell with a characteristic size of less than 10μm, plasma points or micropinches. Hot spots have somewhat larger volumes (radius >100μm), here radiation is insufficient to produce complete radiative collapse. The physical model of Vikhrev indicates the diifference between these two classes of phenomena in his analysis for iron plasmas. With a 50kA discharge a 1mm diameter hotspot with a 100 eV temperature is predicted whereas at 150kA, the radius is less than 1 micron and the temperature 1 keV. For a series of discharges in excess of 100kA it has been observed[3] that the plasma points diminish in size from 12μm to <3μm as Z increases from 16 to 30 and the corresponding dependence of the hotspot size is from 250 to 30μm.

In Figure 3, simultaneous shadowgrams and holographic interferograms are shown for three different instants, before during and after the pinch formation which stems from the cathode and in which the micropinches occur. The micropinches occur either at peak current or up to 20ns earlier. The pinch column that is distinguished is barely 3mm long and is less than 0.5mm in diameter. The laser is incident on the Cathode and is considerably less energetic, 0.26J, than used in previous work. The Vacuum spark is operated in the switched line mode and the Laser comes 800ns before the applied voltage. The Anode blow-off plasma is optically dense. If the electrode separation is comparable than the Anode blow-off plasma scale length, no hot spots are observed. This may be compared with the Vacuum Spark operated in the Hybrid mode as shown in Figure 4. The relative timing between the preionizing Laser and the voltage ramp and the ensuing current pulse may be appreciated from Figure 6, which is from a different shot. The Laser is applied approximately 150ns after the start of the ramp, at which point the voltage is about 80kV. The current rises to a peak of 100kA in the following 180ns. In Figure 4 the pinch channel occurs at peak current some 40ns after the appearance of a dense Anode blow off plasma. At 60 to 70ns before peak current a lower density radially expanding plasma is seen to form in which the micropinching plasma is later seen to form. In both cases the Anode material was Cu and the Cathode of a W/Cu alloy.

When the Laser is incident on the Anode with an energy of 0.3J and in both modes of use of the pulse forming line, observations of the density, using Holographic Interferometry, only show plasma formation on the Anode. No hot spot X-ray emission is seen, but intense beam-target emission is observed in long pulses. Pinhole exposures over many shots show a 2-5mm diameter emission column. The discharge is more resistive, peak currents are about 30% lower when compared to Cathode illumination.

In Figure 5a, we present a typical example of a four frame Pinhole camera image using four different filters for a Hybrid mode discharge with Titanium electrodes. The corresponding filter transmission curves are shown in Figure 5b. The cathode hot spot is visible in all four images; comparison between the Ti and the Be images show that the emission beyond 22Å is minimal. Comparison between the Al/Zn combination and the Al filters indicate that the main emission component for the three remaining micropinches is in the same wavelength region as shown in Figure 2, which is for a Vanadium plasma. The difference in the ionization energies between the two elements is approximately 10%, so this figure may be used to gain some idea of what is the line emission for a Ti Plasma.

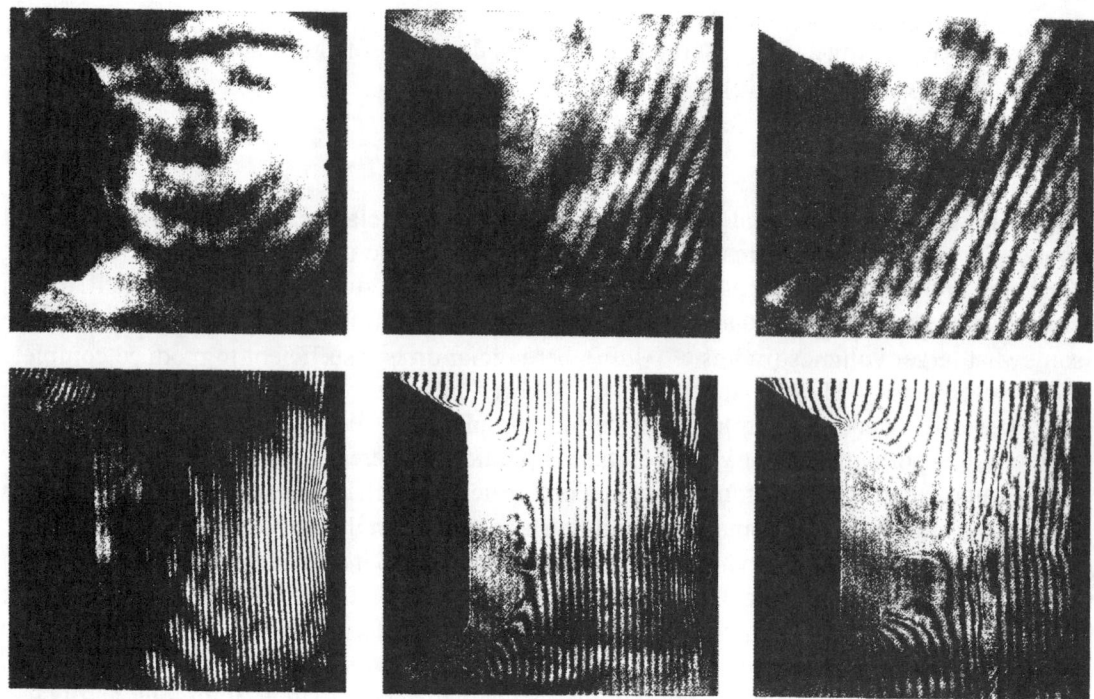

Figure 3. Simultaneous holographic interferograms and shadowgrams taken at -20, 0, 150ns with respect to the peak current in a "Normal" mode discharge. The shots were taken on different discharges. The Laser pulse was 6ns.

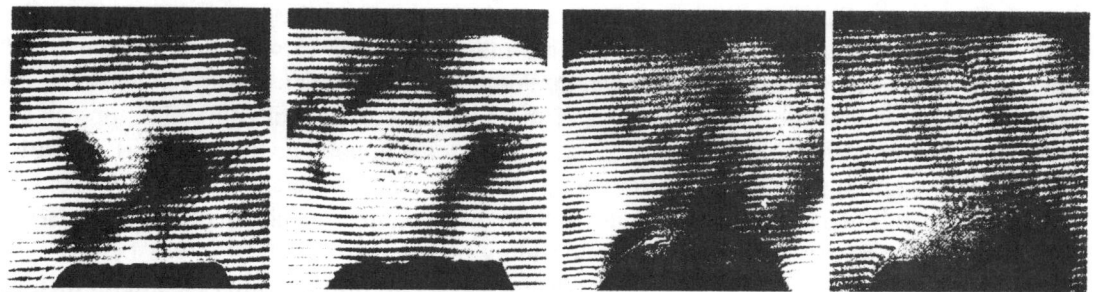

Figure 4. A series of interferometric holograms taken on different shots in the "Hybrid" mode of operation. The exposures are at -70, -60, -40 and 0ns with respect to the current pulse maximum. The diagnostic laser pulse is 20ns.

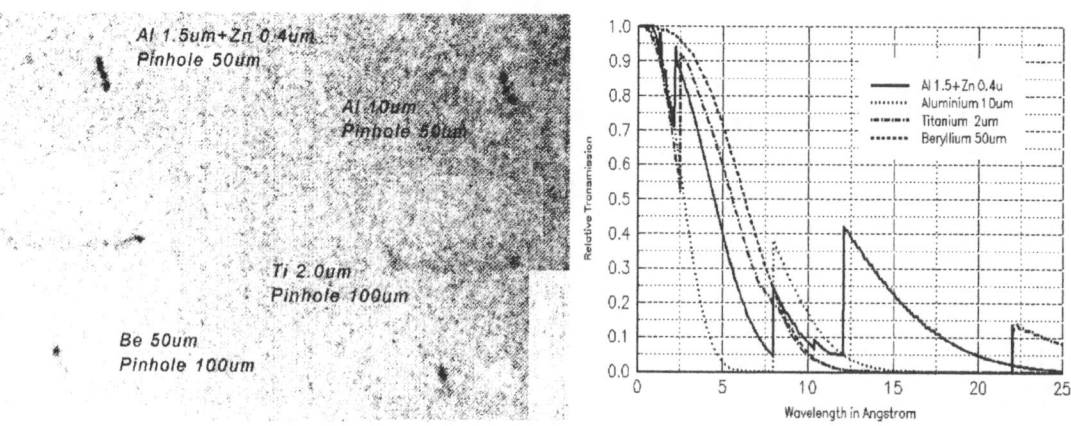

Figure 5. (a) A series of four pinhole images for a Hybrid mode discharge using Ti electrodes with a separation of 7mm. (b) The filter transmission functions for the filters.

The great complexity of the dynamic emission spectrum can be seen in Figure 6 where the emission from five zones of the discharge shown. Each trace corresponds to 1.5mm of the pinch. The Quantrad PIN diode sees the emission from the first three mm as measured from the Cathode. The PIN diode has a maximum in its response curve between 5 and 10Å, but the Al filter Scintillator combination is comparatively sensitive between 10 and 20Å. Comparison with Fig 5 indicates that the PIN diode maximum is from the a micropinch 1.5mm from the Cathode, which sustains emission over a 20ns period, whereas, apart from the micropinch in channel "F", the remaining micropinches in the plasma column emit later.

It has been found that the plasma point at 1.5mm from the cathode is very reproducible and is also always the most intense emitting region. Its spatial reproducibility is also, in essence, defined with respect to the Laser preionizing point. This is an important property if the Vaccuum Spark is to find any practical applications. The position of the hotspots has been found to be different according to the apparatus and conditions used. In some cases the hot spots may be found near the Anode[15,22], or, in the middle of the interelectrode gap[16,35], or even close to both electrodes[36]. It is not clear what are the decisive parameters as far as the position of the hot spot is concerned.

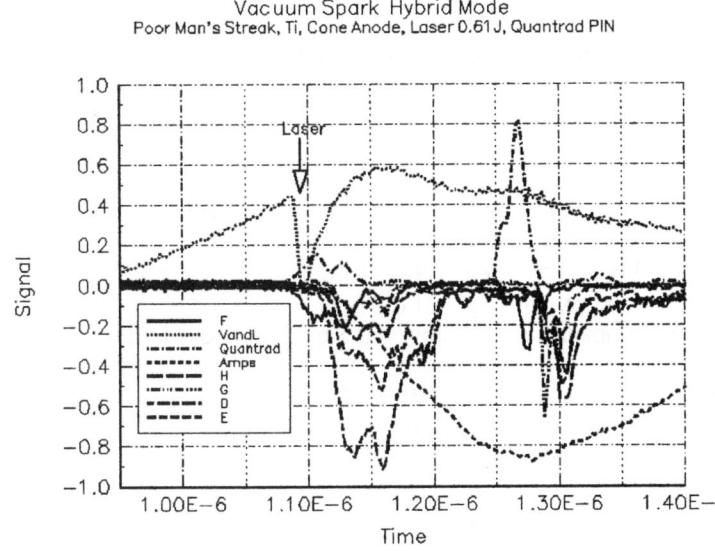

Figure 6. A Hybrid mode discharge, the Laser energy is 0.61J. Five Poor Man's Streak Camera channels resolve 1.5mm axial sections between the Cathode, "D", and the Anode, "H".
The Scintillator is Pilot-u filtered with 1.5μm Al. The Quantrad PIN diode is filtered with 5μm Al and is masked to see the 3mm closest to the Cathode. Quantrad and the Poor Man's Streak Camera temporal response is approximately 3-4ns. Peak Current is 100kA. The preionizing laser comes 22ns earlier than the arrow indicates.

CONCLUDING DISCUSSION

The Vacuum Spark is a well documented source of both hotspots and plasma points. From the experimental view, time and space resolved observations of the radiative collapse phase have yet to be achieved. The observation of features down to a scale length of a few microns appears possible and are used in other dense Z-pinch discharges. There are many similarities with Plasma Focus experiments when plasma points are seen in the second compression where the effective value of Z increases the radiation loss. The Vacuum Spark remains the easiest method to generate conditions suitable for radiative collapse as far as simplicity and the comparatively low value of current required. The micropinch dimensions of the Vacuum Spark make it a candidate as a backlighting source for other dense Z-pinch experiments. Although the energy output is quite small, its power intensity as a source is very large.

The most important characteristics of micropinch formation are in reasonable accord with theoretical models, but there are many details of the radiative collapse phase that require 2-D codes to be written. For example, it remains unclear as to why hot spots form in one place in one apparatus and in another place in a quite similar but different apparatus.

ACKNOWLEDGEMENTS

This work has been partially funded by FONDECYT grant No. 1930557. The author wishes to thank many friends and colleagues who have collaborated in the Laboratory and who have worked together in many publications on this theme.

REFERENCES

1. R.W.Wood, Phys. Rev. Ser., 1, **5**, 1 (1897)
2. J.A.Chiles, J. Appl. Phys., 8, **622** (1937)
3. L. Cohen, U.Feldman, M.Schwarz and J.H.Underwood, J. Opt. Sci. Am. 58 (6), **843**, (1968)
4. W.A.Cilliers, R.U.Datla, H Griem, Phys. Rev. A., **12** (4), 1408 (1975)
5. E.D.Korop, B.E.Meierovich, Yu.V. Sidel'nikov, S.T.Sukorukov., Sov. Phys. Usp., **22**, 727, (1979)
6. K.N.Koshelev, N.R.Pereira, J. Appl. Phys. **69** (10), R21 (1991).
7. G.A.Mesyats, D. Proskurovsky, "Pulsed Electrical Discharge in Vacuum", Ed. Springer-Verlag, Berlin, 1989.
8. C.S.Wong, S.Lee, C.X.Ong, O.H.Chin, Jap. J. Appl. Phys., **28** (7), 1264 (1989)
9. V.V. Vikhrev, V.V.Ivanov, K.N.Koshelev, Sov J. Plasma Phys. **8** (6), 688 (1982)
10. V.V.Vikhrev, JETP Lett., **27** (2), 94 (1978)
11. V.V.Vikhrev, V.V.Ivanov, V.V.Prut, Sov. J. Plasma Phys., **12** (3), 190 (1986)
12. J.P.Chittenden, M.G.Haines, Phys. Fluids B **2**, 1889, (1990)
13. K.N.Koshelev, Yu V. Sidelnikov, V.V.Vikhrev, Nucl. Instrum. Methods., B **9**, 704 (1985)
14. D.Stutman, M.Finkenthal, J.L.Schwob, Proc. 3rd Intl. Conf. Dense Pinches, London 1993, Publ. AIP Conference Proceedings No 299, New York, 1994
15. Yu A Bykovskii, G.A.Sheroziya, Sov. Phys. JETP, **56** (2), 304 (1982)
16. C.R.Negus and N.J.Peacock, J. Phys. D. Appl. Phys., **12**, 91 (1979)
17. S.M.Zakharov, G.V.Ivanenkov, A.A.Kolomenskii, S.A.Pikuz, A.I.Samokhin, Sov. J. Plasma Phys. **10**(3), 303, (1984).
18. E.Wyndham, H.Chuaqui, M.Favre, L.Soto P.Choi, J. Appl. Phys. **71**, 4164, (1992).
19. H.Chuaqui, M.Favre, L.Soto, E.Wyndham, Phys. Fluids B **5** (11) 4244 (1993)
20. M.Skowronek, P.Romeas, IEEE Trans. Plasma Science, **PS-15** (5), 589, (1987)
21. L.M.Early, G.L.Scott, IIII Trans. Plasma Sci., **18** (2), 247 (1990)
22. C.S.Wong, S.Lee, Rev. Sci. Instrum. **55** (7),1125, (1984).
23. H.Arita, K.Suzuki, Y.Kurosawa, K.Hirasawa, IEEE Trans. Plasma Sci. **18** (4), 695 (1990)
24. H.Chuaqui, M.Favre, L.Soto, E.S.Wyndham, IEEE Trans. Plasma Sci., **21** (6), 778 (1993)
25. H.M.Epstein, W.J.Gallagher, P.J.Mallozzi, T.F.Stratton, Phys. Rev. A., **2** (1), 146, 1970
26. D.H.Kalantar, D.A.Hammer, A.E.Dangor, J.M.Bayley, F.N.Beg, Proc. 3rd Intl. Conf. Dense Pinches, London 1993, Publ. AIP Conference Proceedings No 299, New York, 1994
27. V.A.Veretennikov, V.A.Gribkov, E.Ya. Kononov, O.G.Semenov, Yu.V. Sidel'nikov, Sov. J. Plasma Phys., **7** (2), 249, (1981)
28. P.Sheehey, J.E.Hammel, R.H.Lovberg et al., Phys. Fluids B **4** (11) 3698 (1992)
29. R.A.Riley, R.H.Lovberg, J.S.Schlachter D W.Scudder, Rev. Sci. Instrum. **63** (10), 5202 (1992)
30. V.A.Veretennikov et al., Sov. J. Plasma Phys., **16** (7), 475, 1991
31. See "Observations of Dense Plasma Formations in the Vacuum Spark", This conference.
32. K.N.Koshelev, et al., J.Phys. D., Appl. Phys., **21**, 1827 (1988)
33. M.Finkenthal, D.Stutman et al., J. Phys.B: At. Mol. Opt. Phys., **22**, 1133 (1989)
34. P.S.Antsiferov, K.N.Koshelev, A.E.Kramida, A.M.Panin, J.Phys. D: Appl Phys., **22**, 1073 (1989)
35. V.A.Veretennikov, A.N.Dolgov et al., Sov. J. Plasma Phys., **11**, 587 (1985)
36. S.Morita, J.Fujita, Appl. phys. Lett., **43**, 443 (1983)

1996 ICPP
FIRST ANNOUNCEMENT

1996 International Conference on Plasma Physics
Joint Conference of the 11th Kiev International Conference on
Plasma Theory and 11th International Congress on Waves and
Instabilities in Plasmas
9 - 13 September 1996
Nagoya Congress Center, Nagoya, Japan

GENERAL INFORMATION

The Eighth International Conference on Plasma Physics (1996 ICPP), Joint Conference of the 11th Kiev International Conference on Plasma Theory and 11th International Congress on Waves and Instabilities in Plamas, will be held 9-13 September 1996 in Nagoya, Japan, at the Nagoya Congress Center. Cosponsors of the conference are the Japan Society of Plasma Science and Nuclear Fusion Research and the National Institute for Fusion Science (NIFS). This conference, which began at Nagoya in 1980 as a joint conference of two previously independent conferences, brings together plasma physicists from a variety of fields including nuclear fusion, space and astrophysical plasmas, and industrial applications to discuss common problems in plasma physics which is the basis of all these fields.

SCOPE

Plasma physics dawned during the late fifties, when many branches of physics and engineering merged together to form a new stream. During the subsequent two decades this tiny stream expanded into a well-identified field of physics. During this course of development, controlled nuclear fusion research and space- and astro-plasma research were established as two separate fields of science. Recently another stream emerged in plasma physics. This is the pursuit of a truly nonlinear ("complexity") physics, based on the utilization of another tool for research, namely, computer simulation. Furthermore, there are other innovative developments in plasma physics and technologies which are concerned with advanced plasma applications. Accordingly, this conference will be organized to present the widest possible view of the vast panorama of the physics of plasmas.

TOPICS

The conference is divided into General Sessions and Plenary Sessions.
The detailed topics are as follows:

General Sessions
1. General Plasma Physics
Waves and Instabilities
Nonlinear Phenomena
Transport
Heating and Current Drive
Particle Acceleration and Beam Production
Radiation Sources and Processes
Magnetic Reconnection
Plasma Production and Confinement
Plasma-Edge Phenomena
Plasma-Material Interaction
Elementary Processes
Diagnostics
Thermodynamics of Plasma

2. New Trends in Plasmas
Self-Organizing Plasmas
Non-Neutral Plasmas
Strongly Coupled Plasmas
Non-Ideal Plasmas
Laser Plasmas and Short-Pulse Phenomena
Dust and Fine-Particle Plasmas
Reactive Plasmas for Material Processing
New Approaches

In addition to the general sessions, the following reviews and
frontier topics covering the field will be presented in the plenary sessions:

Plenary Sessions
1. Plasma Science
2. Fusion Research
3. Space and Astrophysical Plasmas
4. Advanced Plasma Applications

LANGUAGE

The official language is English.

PRELIMINARY REGISTRATION

Those who are interested in attending the conference are requested to fill out and return the enclosed preliminary registration form or e-mail to the conference secretary by the end of July, 1995.

CONFERENCE REGISTRATION AND BANQUET

The conference registration fee, which includes a banquet fee, will be 40,000 Japanese Yen per person. A considerable reduction will be available for students and participants from developing countries.

PUBLICATIONS

The proceedings of the four-page contributed papers will be distributed at the conference. Those of the eight-page invited talks will be published as a separate volume after the conference, and distributed to the participants free of charge.

OTHER INFORMATION

Information on conference registration, submission of papers, accommodations, transportation and social program will be described in the second announcement which will be circulated in September 1995.

CONFERENCE COMMITTEE

International Advisory Committee: Chairperson: Prof. Predhiman K. Kaw
Organizing Committee: Chairperson: Prof. Atsuo Iiyoshi
Steering Committee: Chairperson: Prof. Junji Fujita

CONFERENCE SECRETARY

Prof. Junji Fujita
National Institute for Fusion Science
Nagoya 464-01, Japan
Telephone: +81-52-789-4235
Facsimile: +81-52-789-4203
e-mail: icpp96@nifs.ac.jp

CALENDAR OF EVENTS

Preliminary
registration deadline: July 1995
Second announcement and call for papers: September 1995
One-page abstract deadline: March 1996

Registration and hotel reservations: June 1996
Final announcement/program: July 1996
8-page invited papers deadline: At the conference

1996 ICPP
9 - 13 September 1996
Nagoya Congress Center

PREREGISTRATION FORM

I wish to receive further information.
I wish to participate in the conference.
I wish to contribute a paper on:

Last name:
First and middle name(s):
Titles:
Affiliation:
Address:
Country:
Phone:
Facsimile:
e-mail:

Please e-mail the above items to the following address.

To: Conference Secretary:
Prof. Junji FUJITA
National Institute for Fusion Science, Nagoya 464-01, Japan
Phone: +81-52-789-4235, Facsimile: +81-52-789-4203,
e-mail: icpp96@nifs.ac.jp

AUTHOR INDEX

B

Bécoulet, A., 83
Bingham, R., 406
Boyd, T. J. M., 238
Bretz, N., 53

C

Chen, F. F., 9
Choi, P., 263
Crisanti, F., 158

D

de G. Dal Pino, E. M., C., 427

E

El-Nadi, A. M., 367
Engelmann, F., 3

F

Ferrari, A., 457
Ferro-Fontán, C., 439
Field, G. B., 416
Finken, K. H., 172
Friedland, L., 327

G

Goedbloed, J. P., 465
Golant, V. E., 143
Goldman, M. V., 449
Gonzales, W. D., 391
Groebner, R. J., 74

H

Haines, M. G., 254
Hirose, A., 127

I

Ida, K., 177
Iiyoshi, A., 206

J

Juul Rasmussen, J., 359

K

Kaw, P. K., 92
Kawai, Y., 479
Keilhacker, M., 45
Kruglyakov, E. P., 247
Kuhn, S., 319

L

Lachambre, J.-L., 185
Laval, G., 275
Lewis, H. R., 398
Livingston, W., 27
Lopes Cardozo, N. J., 135

M

Merlino, R.L., 295
Mima, K., 230

N

Nave, M. F. F., 58

O

Opher, R., 19
Ortolani, S., 222

P

Pozzoli, R., 351
Putvinski, S., 37

Q

Qian, S.-J., 150

S

Sato, T., 335
Schukla, P. K., 377
Sedlaček, Z., 119
Sen, A., 303
Sudan, R. N., 200
Sykes, A., 110
Szente, R. N., 487

T

Tamano, T., 192
Taylor, J. B., 283
Tendler, M., 185
Tsintsadze, N. L., 290

U

Uberoi, C., 383
Ushigusa, K., 66

V

Vandenplas, P. E., 166
von Hellermann, M., 269

W

Wakatani, M., 214
Wyndham, E. S., 495

Z

Zagorodny, A. G., 311

AIP Conference Proceedings

		L.C. Number	ISBN
No. 241	Scanned Probe Microscopy (Santa Barbara, CA, 1991)	91-76758	0-88318-816-3
No. 242	Strong, Weak, and Electromagnetic Interactions in Nuclei, Atoms, and Astrophysics: A Workshop in Honor of Stewart D. Bloom's Retirement (Livermore, CA, 1991)	91-76876	0-88318-943-7
No. 243	Intersections Between Particle and Nuclear Physics (Tucson, AZ, 1991)	91-77580	0-88318-950-X
No. 244	Radio Frequency Power in Plasmas (Charleston, SC, 1991)	91-77853	0-88318-937-2
No. 245	Basic Space Science (Bangalore, India, 1991)	91-78379	0-88318-951-8
No. 246	Space Nuclear Power Systems (Albuquerque, NM, 1992)	91-58793	1-56396-027-3 1-56396-026-5 (pbk.)
No. 247	Global Warming: Physics and Facts (Washington, DC, 1991)	91-78423	0-88318-932-1
No. 248	Computer-Aided Statistical Physics (Taipei, Taiwan, 1991)	91-78378	0-88318-942-9
No. 249	The Physics of Particle Accelerators (Upton, NY, 1989, 1990)	92-52843	0-88318-789-2
No. 250	Towards a Unified Picture of Nuclear Dynamics (Nikko, Japan, 1991)	92-70143	0-88318-951-8
No. 251	Superconductivity and its Applications (Buffalo, NY, 1991)	92-52726	1-56396-016-8
No. 252	Accelerator Instrumentation (Newport News, VA, 1991)	92-70356	0-88318-934-8
No. 253	High-Brightness Beams for Advanced Accelerator Applications (College Park, MD, 1991)	92-52705	0-88318-947-X
No. 254	Testing the AGN Paradigm (College Park, MD, 1991)	92-52780	1-56396-009-5
No. 255	Advanced Beam Dynamics Workshop on Effects of Errors in Accelerators, Their Diagnosis and Corrections (Corpus Christi, TX, 1991)	92-52842	1-56396-006-0
No. 256	Slow Dynamics in Condensed Matter (Fukuoka, Japan, 1991)	92-53120	0-88318-938-0
No. 257	Atomic Processes in Plasmas (Portland, ME, 1991)	91-08105	0-88318-939-9
No. 258	Synchrotron Radiation and Dynamic Phenomena (Grenoble, France, 1991)	92-53790	1-56396-008-7

No. 259	Future Directions in Nuclear Physics with 4π Gamma Detection Systems of the New Generation (Strasbourg, France, 1991)	92-53222	0-88318-952-6
No. 260	Computational Quantum Physics (Nashville, TN, 1991)	92-71777	0-88318-933-X
No. 261	Rare and Exclusive B&K Decays and Novel Flavor Factories (Santa Monica, CA, 1991)	92-71873	1-56396-055-9
No. 262	Molecular Electronics—Science and Technology (St. Thomas, Virgin Islands, 1991)	92-72210	1-56396-041-9
No. 263	Stress-Induced Phenomena in Metallization: First International Workshop (Ithaca, NY, 1991)	92-72292	1-56396-082-6
No. 264	Particle Acceleration in Cosmic Plasmas (Newark, DE, 1991)	92-73316	0-88318-948-8
No. 265	Gamma-Ray Bursts (Huntsville, AL, 1991)	92-73456	1-56396-018-4
No. 266	Group Theory in Physics (Cocoyoc, Morelos, Mexico, 1991)	92-73457	1-56396-101-6
No. 267	Electromechanical Coupling of the Solar Atmosphere (Capri, Italy, 1991)	92-82717	1-56396-110-5
No. 268	Photovoltaic Advanced Research & Development Project (Denver, CO, 1992)	92-74159	1-56396-056-7
No. 269	CEBAF 1992 Summer Workshop (Newport News, VA, 1992)	92-75403	1-56396-067-2
No. 270	Time Reversal—The Arthur Rich Memorial Symposium (Ann Arbor, MI, 1991)	92-83852	1-56396-105-9
No. 271	Tenth Symposium Space Nuclear Power and Propulsion (Vols. I–III) (Albuquerque, NM, 1993)	92-75162	1-56396-137-7 (set)
No. 272	Proceedings of the XXVI International Conference on High Energy Physics (Vols. I and II) (Dallas, TX, 1992)	93-70412	1-56396-127-X (set)
No. 273	Superconductivity and Its Applications (Buffalo, NY, 1992)	93-70502	1-56396-189-X
No. 274	VIth International Conference on the Physics of Highly Charged Ions (Manhattan, KS, 1992)	93-70577	1-56396-102-4
No. 275	Atomic Physics 13 (Munich, Germany, 1992)	93-70826	1-56396-057-5
No. 276	Very High Energy Cosmic-Ray Interactions: VIIth International Symposium (Ann Arbor, MI, 1992)	93-71342	1-56396-038-9

No. 277	The World at Risk: Natural Hazards and Climate Change (Cambridge, MA, 1992)	93-71333	1-56396-066-4
No. 278	Back to the Galaxy (College Park, MD, 1992)	93-71543	1-56396-227-6
No. 279	Advanced Accelerator Concepts (Port Jefferson, NY, 1992)	93-71773	1-56396-191-1
No. 280	Compton Gamma-Ray Observatory (St. Louis, MO, 1992)	93-71830	1-56396-104-0
No. 281	Accelerator Instrumentation Fourth Annual Workshop (Berkeley, CA, 1992)	93-072110	1-56396-190-3
No. 282	Quantum 1/f Noise & Other Low Frequency Fluctuations in Electronic Devices (St. Louis, MO, 1992)	93-072366	1-56396-252-7
No. 283	Earth and Space Science Information Systems (Pasadena, CA, 1992)	93-072360	1-56396-094-X
No. 284	US-Japan Workshop on Ion Temperature Gradient-Driven Turbulent Transport (Austin, TX, 1993)	93-72460	1-56396-221-7
No. 285	Noise in Physical Systems and 1/f Fluctuations (St. Louis, MO, 1993)	93-72575	1-56396-270-5
No. 286	Ordering Disorder: Prospect and Retrospect in Condensed Matter Physics: Proceedings of the Indo-U.S. Workshop (Hyderabad, India, 1993)	93-072549	1-56396-255-1
No. 287	Production and Neutralization of Negative Ions and Beams: Sixth International Symposium (Upton, NY, 1992)	93-72821	1-56396-103-2
No. 288	Laser Ablation: Mechanismas and Applications-II: Second International Conference (Knoxville, TN, 1993)	93-73040	1-56396-226-8
No. 289	Radio Frequency Power in Plasmas: Tenth Topical Conference (Boston, MA, 1993)	93-72964	1-56396-264-0
No. 290	Laser Spectroscopy: XIth International Conference (Hot Springs, VA, 1993)	93-73050	1-56396-262-4
No. 291	Prairie View Summer Science Academy (Prairie View, TX, 1992)	93-73081	1-56396-133-4
No. 292	Stability of Particle Motion in Storage Rings (Upton, NY, 1992)	93-73534	1-56396-225-X
No. 293	Polarized Ion Sources and Polarized Gas Targets (Madison, WI, 1993)	93-74102	1-56396-220-9
No. 294	High-Energy Solar Phenomena A New Era of Spacecraft Measurements (Waterville Valley, NH, 1993)	93-74147	1-56396-291-8

No. 295	The Physics of Electronic and Atomic Collisions: XVIII International Conference (Aarhus, Denmark, 1993)	93-74103	1-56396-290-X
No. 296	The Chaos Paradigm: Developments an Applications in Engineering and Science (Mystic, CT, 1993)	93-74146	1-56396-254-3
No. 297	Computational Accelerator Physics (Los Alamos, NM, 1993)	93-74205	1-56396-222-5
No. 298	Ultrafast Reaction Dynamics and Solvent Effects (Royaumont, France, 1993)	93-074354	1-56396-280-2
No. 299	Dense Z-Pinches: Third International Conference (London, 1993)	93-074569	1-56396-297-7
No. 300	Discovery of Weak Neutral Currents: The Weak Interaction Before and After (Santa Monica, CA, 1993)	94-70515	1-56396-306-X
No. 301	Eleventh Symposium Space Nuclear Power and Propulsion (3 Vols.) (Albuquerque, NM, 1994)	92-75162	1-56396-305-1 (Set) 156396-301-9 (pbk. set)
No. 302	Lepton and Photon Interactions/ XVI International Symposium (Ithaca, NY, 1993)	94-70079	1-56396-106-7
No. 303	Slow Positron Beam Techniques for Solids and Surfaces Fifth International Workshop (Jackson Hole, WY 1992)	94-71036	1-56396-267-5
No. 304	The Second Compton Symposium (College Park, MD, 1993)	94-70742	1-56396-261-6
No. 305	Stress-Induced Phenomena in Metallization Second International Workshop (Austin, TX, 1993)	94-70650	1-56396-251-9
No. 306	12th NREL Photovoltaic Program Review (Denver, CO, 1993)	94-70748	1-56396-315-9
No. 307	Gamma-Ray Bursts Second Workshop (Huntsville, AL 1993)	94-71317	1-56396-336-1
No. 308	The Evolution of X-Ray Binaries (College Park, MD 1993)	94-76853	1-56396-329-9
No. 309	High-Pressure Science and Technology—1993 (Colorado Springs, CO 1993)	93-72821	1-56396-219-5 (Set)
No. 310	Analysis of Interplanetary Dust (Houston, TX 1993)	94-71292	1-56396-341-8
No. 311	Physics of High Energy Particles in Toroidal Systems (Irvine, CA 1993)	94-72098	1-56396-364-7

No.	Title	LCCN	ISBN
No. 312	Molecules and Grains in Space (Mont Sainte-Odile, France 1993)	94-72615	1-56396-355-8
No. 313	The Soft X-Ray Cosmos ROSAT Science Symposium (College Park, MD 1993)	94-72499	1-56396-327-2
No. 314	Advances in Plasma Physics Thomas H. Stix Symposium (Princeton, NJ 1992)	94-72721	1-56396-372-8
No. 315	Orbit Correction and Analysis in Circular Accelerators (Upton, NY 1993)	94-72257	1-56396-373-6
No. 316	Thirteenth International Conference on Thermoelectrics (Kansas City, Missouri 1994)	95-75634	1-56396-444-9
No. 317	Fifth Mexican School of Particles and Fields (Guanajuato, Mexico 1992)	94-72720	1-56396-378-7
No. 318	Laser Interaction and Related Plasma Phenomena 11th International Workshop (Monterey, CA 1993)	94-78097	1-56396-324-8
No. 319	Beam Instrumentation Workshop (Santa Fe, NM 1993)	94-78279	1-56396-389-2
No. 320	Basic Space Science (Lagos, Nigeria 1993)	94-79350	1-56396-328-0
No. 321	The First NREL Conference on Thermophotovoltaic Generation of Electricity (Copper Mountain, CO 1994)	94-72792	1-56396-353-1
No. 322	Atomic Processes in Plasmas Ninth APS Topical Conference (San Antonio, TX)	94-72923	1-56396-411-2
No. 323	Atomic Physics 14 Fourteenth International Conference on Atomic Physics (Boulder, CO 1994)	94-73219	1-56396-348-5
No. 324	Twelfth Symposium on Space Nuclear Power and Propulsion (Albuquerque, NM 1995)	94-73603	1-56396-427-9
No. 325	Conference on NASA Centers for Commercial Development of Space (Albuquerque, NM 1995)	94-73604	1-56396-431-7
No. 326	Accelerator Physics at the Superconducting Super Collider (Dallas, TX 1992-1993)	94-73609	1-56396-354-X
No. 327	Nuclei in the Cosmos III Third International Symposium on Nuclear Astrophysics (Assergi, Italy 1994)	95-75492	1-56396-436-8
No. 328	Spectral Line Shapes, Volume 8 12th ICSLS (Toronto, Canada 1994)	94-74309	1-56396-326-4

No. 329	Resonance Ionization Spectroscopy 1994 Seventh International Symposium (Bernkastel-Kues, Germany 1994)	95-75077	1-56396-437-6
No. 330	E.C.C.C. 1 Computational Chemistry F.E.C.S. Conference (Nancy, France 1994)	95-75843	1-56396-457-0
No. 331	Non-Neutral Plasma Physics II (Berkeley, CA 1994)	95-79630	1-56396-441-4
No. 332	X-Ray Lasers 1994 Fourth International Colloquium (Williamsburg, VA 1994)	95-76067	1-56396-375-2
No. 333	Beam Instrumentation Workshop (Vancouver, B. C., Canada 1994)	95-79635	1-56396-352-3
No. 334	Few-Body Problems in Physics (Williamsburg, VA 1994)	95-76481	1-56396-325-6
No. 335	Advanced Accelerator Concepts (Fontana, WI 1994)	95-78225	1-56396-476-7 (Set) 1-56396-474-0 (Book) 1-56396-475-9 (CD-Rom)
No. 336	Dark Matter (College Park, MD 1994)	95-76538	1-56396-438-4
No. 337	Pulsed RF Sources for Linear Colliders (Montauk, NY 1994)	95-76814	1-56396-408-2
No. 338	Intersections Between Particle and Nuclear Physics 5th Conference (St. Petersburg, FL 1994)	95-77076	1-56396-335-3
No. 339	Polarization Phenomena in Nuclear Physics Eighth International Symposium (Bloomington, IN 1994)	95-77216	1-56396-482-1
No. 340	Strangeness in Hadronic Matter (Tucson, AZ 1995)	95-77477	1-56396-489-9
No. 341	Volatiles in the Earth and Solar System (Pasadena, CA 1994)	95-77911	1-56396-409-0
No. 342	CAM -94 Physics Meeting (Cacun, Mexico 1994)	95-77851	1-56396-491-0
No. 343	High Energy Spin Physics Eleventh International Symposium (Bloomington, IN 1994)	95-78431	1-56396-374-4
No. 344	Nonlinear Dynamics in Particle Accelerators: Theory and Experiments (Arcidosso, Italy 1994)	95-78135	1-56396-446-5
No. 345	International Conference on Plasma Physics ICPP 1994 (Foz do Iguaçu, Brazil 1994)	95-78438	1-56396-496-1